Seven Eyes Open

A study of the Revelation of Jesus Christ

Why you will not burn in eternal torment
and why it is now more important than ever not to be spiritually lukewarm.

by
Alexander James B Connor

- FOURTH EDITION -

A title self-published by the author

First Published April 2017

By Alexander James B. Connor

alex.connor@seveneyesopen.net

This is a fourth edition with additional content: First printed November 2020.

ISBN: 978-0-9956938-3-8

Layout and Design by Alexander James B. Connor

www.seveneyesopen.net

For you; before you buy it.

Rev 3:2 Be watchful, and strengthen the things which remain, that are ready to die: for I have not found thy works perfect before God.

Contents

Introduction

I write of an abstraction, the Trinity that I know as God. The triune nature of the God of Abraham, Isaac and Jacob has remained a mystery through the ages to which I can only add greater content (as answers generally raise more questions). So, I consider myself fortunate to have the opportunity to write upon such a great canvas as the Trinity: A matter that resonates so very clearly in the soul and which for lack of any better words remains as yet "undefended territory" with regards to the interpretation of the final book of the New Testament, the "Revelation of Jesus Christ".

The Revelation text is so thoroughly and historically misinterpreted that it takes far greater effort to understand its many interpretations than it does the text itself! I could not have hoped for a better opportunity and motive than this, that my own words would frame an abstraction that once unlocked in principle would agree with the text and do so in whole. In doing so, showing without doubt that despite its complete misinterpretation and misapplication over almost 2000 years, it has survived so incredibly and indelibly intact that there is a Bible text that obeys the principles of the Holy Trinity in truth, rather than showing any single principle lost through much recopying.

The ending of the Revelation text warns the believer against adding or taking content to and from the text itself; it has survived so very well that there are no faults found in its pages as far as may be deduced. Whereas there are modern complaints, it is now (finally) determined that modern theories such as that the number of a man is (and was originally) 616 and not 666 in particular is decidedly false, and the text is correct in incorporating the latter. It shows that in the case of the KJV Bible (the 1611 King James Version), the method of translation in employing verbal equivalency (by the received text) has allowed the correct scripture to survive in its entirety and its content is and has remained accurate and complete.

So on to abstraction. The subject of the Trinity revealed in the Revelation text has baffled many readers with its imagery and colourful language. The text itself is replete with word pictures; a true delight to decipher! Areas of the text, such as the "wings of a great eagle" given to the woman to fly into her place in the wilderness, and also the "way of the kings of the east" are astonishingly simple and whilst not technically "crude" show a sense of humour that only Jesus Christ could pull off intact. There is no sick joke in this, as this is not a "sick note" from the Trinity absent themselves; rather it is left to God now victorious to explain Himself to us in His own chosen language, and the language used for the subject matter of Revelation itself, is the language of abstraction.

2Ti 2:15 Study to shew thyself approved unto God, a workman that needeth not to be ashamed, rightly dividing the word of truth.

That language yields a consistent interpretation other than the classical one of near universal damnation and torments. (Even "hell" itself, misinterpreted for millennia, is now before us just (and merciful) when the primer is applied to the text.) The language of the text is classically and wholly misinterpreted as the language of vengeance beyond measure, and this is not what is meant by it at all. (In the absence of Christ, the reality of Hell has been twisted out of its proper place.) What are truthfully arguments of symmetry are become in comparison to the classical interpretation a "sick joke" when one considers the scare stories sold by the church to maintain control. It is this canvas upon which I write: in defence of the text rather than my God, One able to defend Himself though absent except in Spirit.

So, as "Revelation" is a "revealing", the proper language to unlock the text is the language of symmetry. (Such language is the grammar of (or consistency of) all structure as much as anyone would consider it that of symmetry.) If there could but be a key through which all the text could be unlocked I would be done. Is such a lost primer ever to be found? Clearly God Himself suffices if there is one primer at all and if the writer of the text is genuinely inspired – in fact in God's own absence there is left but one abstraction remaining. I am left with the methods of algebra upon the symmetries of the Holy Trinity as found in the text. With a particular reworking and a new slant, there is every correspondence between the text and the algebra "revealed" itself.

Whether or not the text is now revealed as or considered to be divinely inspired, the mathematics used that unlocks the Revelation is modern and "new" compared to the text itself. The theory of "Extension Fields" was not considered broached until a young French mathematician named Évariste Galois had formulated his work. Sadly, Galois died in a duel as a young man in AD 1832, perhaps one of the greatest losses to mathematics of his day. His work now can only be lamented over – as now and also then; never wholly published and it is still a great pity that most of it remains lost.

Évariste Galois:
Galois' work included the first proof that A5 is simple (in fact, that An is simple for all n>4) as well as deciding the solubility of quintic polynomials by method of radicals. (He was the first to prove that there was no such general formula.) It may interest you to know that Galois Fields (Properly all finite fields) are termed in theory "perfect" due to their separable nature and that of their extensions.

What is much more startling is that an uneducated Roman-era fisherman named John could reproduce the multidimensional arguments of finite fields as well as further group theory that did not surface for the best part of another 1800 years! John describes the maths that the angel showed him in very accurate word pictures, touching on principles of algebraic closure and simple groups as well as other algebra not taught until university. All this and more besides from a fisherman in an era in which his better contemporaries had not discovered that the integer zero was a valid number!

Now, John does not leave us ignorant of the mathematics; for in one place he states "Rev 9:16 And the number of the army of the horsemen were two hundred thousand thousand: and I heard the number of them." The statement that John "heard the number" shows us the number is special and accurate and to be left unchanged. I might also add that he understood the number to be correct. In fact, the number 200,000,000 is indeed correct and is a result of a generalisation obtained from arguments in the text itself. (The calculation is included later at the relevant point in this book.)

How many of these numbers found true are proof enough? Are half enough? Are three quarters? In truth this is God, and I show 100% of the numbers found in the text to be meaningful. There is no error in any calculation that John was shown or did deduce for himself. If I were to make a wager on the efforts of a fisherman at abstract algebra, I would be forced to concede to the rules of probability that ten repeats or occurrences assumed 50/50, as with a flip of a coin all "heads" in a row is one chance in 1024. However, the words of Revelation state "Let Him who hath understanding count the number of the beast..." and it is through the understanding given us in algebra that this is possible. The odds therefore would be closer to one in "x" for some given and large x. However, once again I find a modal collapse in my logic! For this is God Himself who divinely inspired John; so, I wrestle not with my grey matter and over matters of probability, but ahead state that the odds of John (an uneducated fisherman) getting all the numbers correct and the imagery with it are now simply put: 1-1. Why? It is because God had truly instructed him to write. (Could you lose that wager because of that modal collapse? Modality is not probability; was John savant?)

The content of the Revelation itself not only reveals the symmetry of the Godhead in triune form, most notably the Revelation argues that the only Trinity is found in Jesus Christ and He as the "Lion of the tribe of Judah" is uniquely also God Himself. The text declares that there is found none other that may take His place. That place (the position of Christ) is shown in the Revelation as to be more than glorified – His place is above all others forever.

The rest of the content of the Revelation concerns the conviction of the church to follow Jesus Christ rather than the "Church pack". There is no safety in numbers in the Revelation – but the revealing of the mysteries involved surround a character introduced early on, "the angel of the church". This individual is assumed to be the very least in the Kingdom of God, and much concerns his requirement to repent of following the herd and following the leading of Jesus Christ instead. If the least amongst the Church may succeed outside of religious institutions now corrupted, then any others may do so and they as his "betters" are convicted in the Spirit as to also exit such religion. Christ fulfilled the law of Moses; the least makes certain all grace.

The prophetic content of the Revelation indicates that the believer must consider leaving the herd of the last day's system of religion; the text instructs such to exit from the then corrupt "Church" proper. That is through an open door no man can shut. This door is simply the door to leave the Church and stay outside, and the words "Come out of her, my people..." resonate through the visions recorded by John. Once the angel exits the Church the text places him standing on the Earth and sea and imbued with the authority of Christ to stand upright despite the first two "woes". With the final overcoming by the angel, the text states that his overcoming (not that of Christ) is testified to by two witnesses: to convict the elect to leave the Church also. Lastly the angel returns justified (shown with inheritance complete) with the armies of heaven to overthrow the deceptions of the dragon at the judgement upon the second coming (on the last day) as a rider on a white horse. Obedience duly has its reward, to be as a co-heir with Christ.

Now the "angel" as the least is lower than the elect in Christ (which are all equal to each other). There is no ordering other than that the angel is made "slightly lower"; I may argue on the basis of a well-ordering principle albeit a much reduced and very simple one. Once the very least has overcome the deceptions of the dragon, there is no point in the dragon deceiving any other; therefore, I expect the gloves on Satan to come off. Likewise, the wrath

12

of God reciprocates – as Satan has no argument to challenge God's sovereignty of salvation over His own (or those of the devil's) any longer.

Returning to the content of the text, the Revelation states ahead of time the devices of the dragon to be used against the believer in Christ – for him to reform the "Church" as a cage for the "elect" in Christ (amongst a flock of tares). In this sense the Revelation fulfils the verse "Hell is prepared for the devil and his angels". This last set of devices is the only schema Satan has to combat God, and it likewise naturally resonates upon the carnality of the believers caught within that system, the Laodicean "Church" of the end. That last system is properly shown as the scarlet beast. Throughout I use the word "Church" capitalised to indicate this mainstream aberration of Christian worship.

> **Laodicean:**
> *The letters of Christ to the seven churches in the Revelation describe the state of seven churches in Asia-Minor at the time of its writing. In the last days and the prophetic sense only one church (Philadelphia) finds complete approval and the first five are merged into a collective in the text to which the exhortations of Christ (in the letter to Laodicea) to repent apply. This amalgam of five is termed "Laodicean". (The letters apply to the least of the flock's circuit (ending in Philadelphia) avoiding false doctrine and into that finding approval from God.)*
> **Gehenna:**
> *The name "Gehenna" refers to the rubbish pits outside of the old city of Jerusalem – noted for the foul odours and its continuously burning piles of refuse.*

Hell is a word much misinterpreted. Is Hell "outer darkness" or "Gehenna" (to be thrown out as trash) or is Hell as a "lake of burning sulphur"? Only the latter poses a problem alongside the other two. It must be by the imagery embedded in the text that believers are given the context of the scripture. There are no flames and pitchforks or red pyjamas waiting in Hell for the sinner; what remains as Hell at the very end (the judgement) is one planet renewed for billions of people (considerably more than live today) raised to life on the last day, left to their own devices with dwindling resources and nothing given them but to forever work out their problems with each other without ever any true rest to look forward to. Truly, without the refreshing of God one is forever tired;

Mar 9:44 Where their worm (as ageing) dieth not, and the fire (work of the spirit) is not quenched (not refreshed as new).

Yet this statement is more pertinent to faulty religion, which is forever corrupting itself and never to be answered with rest from the Holy Trinity to those ungodly souls practising it.

I expect God to hand the keys to this universe to them in full and simply translate the truly elect to "a new Heaven and a new Earth". God would show mercy, totally fulfil scripture and would show Himself to be far from an ogre. Heaven and Earth may pass away, but the words of Christ will never pass away – they are a certainty.

Rev 21:5 And he that sat upon the throne said, Behold, I make all things new. And he said unto me, Write: for these words are true and faithful.

The format of writing follows sequentially the Revelation text as far as may be done so. After a few mathematical definitions I enter the Revelation text proper. Whilst it would be preferable to examine each verse in sequence this is not always possible. You may appreciate more from a second reading; there is much to be gained from doing so.

I implore you, the reader, where in my text the approach taken appears simple and childish, the words of Christ are definitely not so. What is simple and childish are the word pictures to show the imagery has its correct meaning. There is a vocabulary of imagery that builds alongside the scripture and it is the interpretation of these concepts that require the native simplicity of word pictures. John was shown these concepts and kept his impressions simple as he wrote out the Revelation. (John was illiterate during the period in the book of Acts but must have spent much time learning to write – the Revelation is dated to around 80AD.) Amongst the last words from Christ to John was the command to preserve his original words unchanged, so John's record is in places subject to some simple imagery.

In building my spiritual vocabulary I found the meaning of the Revelation through the references to the algebra buried in the text itself. I was merely given the descriptions or word pictures embedded in the text and although the abstractions of the algebra are described by this imagery, the context supplies the content as I progressed through. Then "fire" is common in its usage between "Rev 1:14 His head and *his* hairs *were* white like wool, as white as snow; and his eyes *were* as a flame of *fire*;" as it does in the much later verse "Rev 20:14 And death and hell were cast into the lake of *fire*. This is the second death.".

Then for instance, "oil" is synonymous with "doctrine" and "light" with "virtue". Hence "oil + fire = light". I.e. the application of the term "fire" in the Revelation is equivocated with "action of Spirit" (on the one hand to salvation as opposed to being cut off from it to the other), whether to draw the elect towards rest in Christ by the Holy Ghost (to spiritually dwell with Christ as if already in heavenly places) or left only to gravitate towards the image of the beast by the "pneuma" of the false prophet's devices; such spirit (as pneuma) does so with purpose but not with genuine divine answer and for lack of any such answer, without any real profit.

The Revelation has much content which will edify the reader holding such a vocabulary and will do so whilst granting a better understanding of the text. The book ceases to be so "sensational" (and makes John equal to us); it does not support predictions of the date of the end of the world through construction of charts, and does not leave any room for predicting the second coming except by watching for the devil's devices in the Church itself.

The proper setting of the last book in the scriptures is to expand upon the Trinity as found in the very person of Christ, to affirm the efficacy of the gospel's doctrine of salvation in the case of the very least of all Christ's disciples, whilst so saving under the very worst circumstances (a time of complete deception under the sway of the dragon, the adversary, Satan). The sovereignty of God is then truly asserted in the sense of "as so for all" and His Christ is displayed the victor.

I invite you to read on through the algebra and reach the revealing proper!

Acknowledgements

No other person's work went into this examination (and re-examination) of the Revelation than that with which I started; a cursory read of Kant's "Critique of Pure Reason"[9] provided such terms as "manifold" and "apperception" but did not significantly build upon my small knowledge of abstract algebra from college and the King James Version of the Holy Bible: the tools for this study.

I often reference Gödel's ontological argument[1] and rather than repeatedly reproduce the argument itself I include it as an appendix. I often comment on his definiton of an "essence" with ideas that of necessity require it.

I thank those unknowing contributors with whom I have no association but in Christ: for the gifts of sound doctrine and opening my ears to hear it, and also for their continuing presence on online streaming radio, without which I would no doubt have tarried far, far behind.

I state that my views are not the views of those acknowledged here, and that there is on my part no intent for the content of this study to reflect poorly on them, or indeed upon any other. This study is not intended for the purpose of ridicule, or that of emasculating and limiting the Godhead in their authority. Rather, it reflects rather well the reality that God is a fair judge, and may still damn a soul, whilst equally showing mercy to all those that deserve it.

The identification of the latter of the two sets of four kingdoms found in the four beasts of Daniel[8] and the significance of the Old Testament time-line of Kingdoms which include Egypt and Assyria[7] is due to James Lloyd of "Christian Media Network" (CMN) – as is also the correlation between the "woman with twelve stars" becoming the "woman riding the beast" and the plausible "bait and switch" of authorities, here noted with regard to a similar device upon doctrines, found within this interpretation. James Lloyd also identified the phrase "two house twist" with a system of two "dispensations" enjoined for common purpose; a phrase which has a curious analogue in the image system presented in [14.4.12].

Also, a great portion of the study of the "manifold" in chapters ten and thirteen was reliant on my discovery of CMN's "Diaprax"[10] tape-set authored by Dean Gotcher and the "Dragonspeak"[11] tape set by James Lloyd. (Dr Gotcher is the head of "The Institute for Authority Research" found online at authorityresearch.com), and the study of the manifold was primarily built upon the content of Dr Gotcher's own radio show on CMN.

I posthumously thank Bill Cooper for teaching me to ask questions of my God rather than to blindly accept the believed norms – also teaching me how many "raisins" the "new world order" had in store for me! (Just the one.) William Cooper authored the famous underground book "Behold a Pale Horse" and hosted the continuing radio programme "The Hour of The Time" which led me to CMN's streaming broadcasts online.

Much of this study of the Revelation is drawn from an online microsite I solely created and maintain with similar content. Any further references to other authors will be listed in the bibliography and also enumerated by superscript within the writing itself.

[1] J. Sobel (1987), *Gödel's Ontological Proof* in J. J Thompson (ed.) *On Being and Saying: Essays for Richard Cartwright*. (Cambridge, MA: MIT Press, 1987), p241-61.

[7] James Lloyd, *The 6, 7, 8 Cycle* (Christian Media), p26-28.

[8] James Lloyd, *The Sand and The Sea* (Christian Media), p5-7.

[9] Immanuel Kant, *Critique of Pure Reason* (Penguin Classics), p104, 124.

[10] Dean Gotcher, *The Diabolical System of Diaprax* (The Apocalypse Chronicles).

[11] James Lloyd, *Dragonspeak, The Language of Lucifer* (The Apocalypse Chronicles).

Part One

(Prerequisite Algebra and the Godhead)

Rev 3:3 Remember therefore how thou hast received and heard, and hold fast, and repent. If therefore thou shalt not watch, I will come on thee as a thief, and thou shalt not know what hour I will come upon thee.

Notes On Gödel's Ontological Argument

Gödel's ontological argument[1] (see Appendix) formalises the concept of an "essence"; a positive property from which every positive property would be entailed, as for any "God-like" being. In the case of every closed set of positive predicates, necessary existence (NE) is entailed from the essence (that same "God-likeness"), becoming a principal element in an ultrafilter.

"God-likeness" is then a descriptor for every closed set (and ultrafilter) of positive predicates completed with necessary existence. However, in all possible worlds, many predicates are either positive in one world and their negations possibly so in another: the result is that in all possible worlds NE becomes the sole principal element entailed in every world, even in empty worlds of which nothing more can be stated but "In the beginning...".

In the case of an empty world, then, only NE remains – the predicate of "God-likeness" becomes equal in "collapse" to that predicate of NE. The essence (its equivalent by Gödel's argument[1]) also becomes NE and the system collapses with every set of positive predicates entailed equivalent, and all become necessary. In effect, then, NE becomes not only principal but a predicate which entails none other (being closed in any otherwise "empty world").

As all possible worlds must be connected and are so by relation(s) symmetric, reflexive and/or transitive; the unions of sets of predicates in "connected worlds" as well as their intersects are also "possible worlds", and as I suggest, each are closed in positive predicates. NE has become the essence within every closed "world", with a different ultrafilter each. As a necessary consequence of all such closed sets, NE becomes principal, an essence, and in every world entails **nothing**. Empty worlds become traversable only by predicating such a "God-like" being.

The essence (now NE), entails nothing else but itself; it engenders a complete modal collapse. Either there is no essence but for NE itself, or there can be only one possible world, or all worlds are strictly alike and every predicate, positive or not, become so necessarily. Then every world suffers a modal collapse and is either "alike" or empty.

NE, principal as an essence, suffers modal collapse unless it truly does entail nothing. The positive predicates, then, belong to or are solely of the creation and not the creator. That is, unless a remedy is found for modal collapse as found in Gödel's argument[1]. (That is for later.)

The possibility of an empty essence[2] that entails everything is also possible; but instead of closing NE (the essence) with "God-likeness" applied (equal to NE alone) in all empty worlds, the converse has been stated: that an essence may yet entail everything whilst being devoid of NE. Whilst not unsound, this is not the end of the tale. Could monotheism be flawed?

An empty essence on every possible world would entail every positive property in every world so connected ("real" or non-empty worlds only as in an ultrafilter): and as the empty set belongs to every set, the argument of "self-difference"[2] ($\phi = \Omega \setminus \phi$) across an empty world or worlds hits its mark (NE is not entailed though there is also modal collapse). An empty essence removes the descriptor of "God-like" from a closed set with NE to a set with every other positive property, and one in a filter that is necessarily non-principal.

[1] J. Sobel (1987), *Gödel's Ontological Proof* in J. J Thompson (ed.) *On Being and Saying: Essays for Richard Cartwright.* (Cambridge, MA: MIT Press, 1987), p241-61.

[2] Christoph Benzmüller and Bruno Woltzenlogel Paleo, *An Object-Logic Explanation for the Inconsistency in Gödel's Ontological Theory.*

Chapter One: Prerequisite Algebra

This study of Revelation is much reliant on the mathematical structures introduced in this first chapter. These structures include groups and fields ("Fields" are a specialization of a class of structures called "rings"). Whilst the definitions come thick and fast, you will already be familiar with most of the properties of these structures as you use them every time you add or multiply.

"Groups" are structures that obey an operation of addition or multiplication (there is only a difference in notation). "Rings" are somewhat similar, they include a second operation of multiplication as well as addition, and "fields" (of the finite variety) form a ring where the multiplicative part of the structure (not including zero) forms a cyclic group. (As defined in this chapter to follow.) One example of a ring would be the set of integers along with the usual addition and multiplication: yet this does not form a field as there is no inverse for every element (only the two integers -1 and the unity are "self-inverse"). I.e. there is no whole number (integer) "x" so that $2 * x = 1$ etc. An example of a field that should be familiar is the set of rationals (all fractions with numerator and non-zero denominator in the set of integers). For example $(2 * \frac{1}{2}) = 1$ which shows the integer "2" has an inverse in the set of rationals.

Of course, the above are simple sets with common operations and there are many more examples of groups and rings; from the symmetries of a triangle or any regular polygon (each of these sets form a group) to the set of complex numbers having modulus one (forming an infinite order group). The easiest way to generalize the definition of these structures is to define the set of properties common to all these structures (that each in turn satisfy), in a few basic rules (axioms) which form the bedrock of group theory and a large portion of abstract algebra.

This chapter is intended to be a simple reference for the maths in part two of the book. You may find it easier to turn back to this chapter when you get stuck.

I begin with several mathematical structures that obey the following axioms.

1.1) Group Axioms, Morphisms And Simple Groups

First, a definition.

Cartesian Product
The term "Cartesian" is named after the French mathematician René Descartes, and is a term most associated with the traditional coordinate system of three normal axes. A Cartesian product is an n-fold ordered list of elements from the structures in product. A Cartesian pair S x T is the set of all such pairs (s, t) with s and t from S and T. An n-fold Cartesian product of a set S is the set of "vectors" in $(s_1, s_2, s_3 ..., s_n)$ with each s_i in S.

Group Axioms (1.1.1)

A **binary operation** is a mapping from the ordered **cartesian pair** of elements (a,b) from a set S onto the set S; I may employ the Cartesian product S x S = {(a,b) : a, b in S} of S and define a binary operation * as a mapping of pairs (a,b) into S. (a * b = c, for some c in S.)

$$*: S \times S \rightarrow S.$$

An example of a binary operation is addition on the integers. I.e. (a,b) → c; (a + b = c). If (a,b) and (b,a) are mapped to the same c for all a and b in S, it is convention to write * as +.

A **"group"** G is a set together with a binary operation * that obeys the following axioms.

$G1)$ *Closure.* For all "a", "b" in G, a*b = c is also an element in G.

$G2)$ *Associativity.* For all elements "a", "b", "c" in G, a*(b*c) = (a*b)*c

$G3)$ *Identity.* There is a *unique* element "e" in G such that for all a in G, a*e = e*a = a

$G4)$ *Inverses.* For each "a" in G there is a *unique* element a⁻¹ in G such that a*a⁻¹ = a⁻¹*a = e. (note that "e" is self-inverse.)

The above show that the structure G equivalently obeys the cancellation laws.

If (a*b = a*c) or (b*a = c*a) then (a⁻¹*a*b = a⁻¹*a*c) or (b*a*a⁻¹ = c*a*a⁻¹) and so (b = c).

A group G may be an infinite or finite set. The operation of G may be commutative (called "Abelian") only if a*b=b*a for *all* a, b in G: or non-commutative (a*b ≠ b*a in some cases).

A "cyclic" group Cn of order n is generated by a single element g, so that every element h=gʳ in the group is a power of that generator. All cyclic groups are abelian. Note that the order is the smallest whole number n such that gⁿ=e (the identity).

An example of an infinite order abelian group is the set of integers under addition; one of a finite non-abelian group is of the symmetries of a regular n-sided polygon with 2n elements.

K4 Group

The K4 (Klein-four) group is the smallest non-cyclic finite group. It has four elements and the addition table as below. (It is also abelian; its operation is commutative.) The K4 group is central to this study of Revelation. The first thing to notice is its symmetry (a+b=c, a+c=b, b+c=a). Its elements act as those pairs (x,y) of the cartesian product A x B (it is "isomorphic" to C2 x C2) of the two cyclic groups A = [0,a] and B = [0,b]. (Here, c=(a,b) in that product.)

+	0	a	b	c
0	0	a	b	c
a	a	0	c	b
b	b	c	0	a
c	c	b	a	0

Subgroups (1.1.2)

First, a set S under an **equivalence relation** divides into disjoint classes: a **"partition"** of S.

Equivalence Relation

An **equivalence relation** "~" satisfies;

1) a~a : ~ is reflexive.

2) a~b ↔ b~a : ~ is symmetric.

3) a~b and b~c ↔ a~c : ~ is transitive.

An **equivalence class** $\bar{x}$ in a set S (represented by x) is the set of elements {y in S : x~y}.

Subgroups

A **subgroup** is a closed subset of a group G under the same binary operation. There are always two "trivial" subgroups, the identity {e} and G itself. A "proper" subgroup H of G is one where H is a proper subset of G.

A **"coset"** of a subgroup H of G is the set of all elements of H in product with some given element "g" of G. I.e. a coset $g*H = \{g*h, h$ in $H\}$.

A consequence of the cancellation laws is that if $x*H$ is in the same coset of H as $y*H$ then it must be true that $x = y*h$ for some h in H; it immediately follows that $x*H = y*h*H = y*H$. The cosets are equal. (The cosets of a subgroup H partition G.)

Each coset of H contains elements of H in product with some element of G. The set of y that relate to x by $x*h = y$ for some h in H is an **equivalence class** formed from elements in G, and is of the same order (number of elements) as H. The number of cosets is $N=|G|/|H|$ (the theorem of Lagrange). The result is that the **order** $|H|$ (the number of elements in a group) of a subgroup H of G divides the order of G. (The **"order"** of an element g is equal to the order of the group $G = <g>$ generated by every power of g upon itself, i.e. $G=<g>=\{g^r :$ r an integer}.) Then for such a "cyclic group" G, $|G|=n$ where n is the smallest integer where $g^n=e$.

Then as a consequence follows the definition of a **Normal Subgroup** H of G.

A **"Normal Subgroup"** is a subgroup H of a group G where the *left* and *right* cosets of H are equal, so $x*H = H*x$ for all cosets (represented by some such x in G) of H if H is normal in G. Note that {e} and G are normal in G, and that all subgroups of abelian groups are also normal.

I refer to **conjugation** of an element g upon a subgroup H as the product gHg^{-1}. I consider gHg^{-1} in H for all g in G if and only if H is a normal subgroup of G and I write $H \triangleleft G$. Then products of such conjugations as well as compositions of conjugates remain in that normal subgroup H.

Permutations

A **permutation** on a set S is a reordering of elements in S that is also a product of **transpositions** (two elements (x,y) are exchanged by each transposition).

Then (b,c) acting on the list [a,b,c,d] reorders the list to [a,c,b,d]. A permutation may be written as a set of **disjoint cycles**; so the permutation of [a,b,c,d,e] to [b,c,a,e,d] in cycle notation is (a,b,c)(d,e). (Disjoint cycles commute: (a,b,c)(d,e) = (d,e)(a,b,c).)

(My convention for cycle notation is that the elements in the set S are operated upon by cycles rightmost first; each cycle reads from left to right, cycling elements in that order. Null permutations (a) (or even (e)) indicate the identity element.)

A cycle of n elements may be written as a product of n−1 transpositions. (a,b,c,...x,y,z) can be written (z,y)(z,x)... ...(z,c)(z,b)(z,a). Note the cycle (a,b,c...x,y,z) is equal to (y,z,...a,b,c,...x). (The starting symbol is not an issue.) A factorisation into transpositions is not necessarily unique.

> ### n-cycle, seven-cycle, Cn group.
> *A permutation σ (of order and length n, a product of n−1 transpositions) that cycles n elements and returns each of them to their original arrangement with exactly n products of σ (and no less) is an "n-cycle". A seven cycle generates a cyclic group C7 with seven elements. The cyclic group Cn is the set of elements found from only the powers of a single n-cycle.*

The **symmetric group** Sn is the set of all permutations of n objects. The set An (the **alternating group**) is the set of all members of Sn that factor into products of transpositions of even length. For Sn, I note that An (and not C2) is a normal subgroup of Sn. Note that $|Sn| = 2|An|$.

The products by an element g in G with the set of every element of G forms a one to one bijection to G (by the cancellation law). Then, g may be represented as a permutation of the elements of G. Every element of a group G ($|G|$=n) may be represented by an element of Sn. The element in Sn representing the permutation or operation upon G "by g" is a "permutation representation" of g; and the subset of Sn of order n isomorphic to G is a permutation representation of G.

G-Sets

A G-Set X is an ordered list of n elements, permuted by a subset of Sn or a group G (isomorphic to a subgroup of Sn). There is no requirement for G to permute every symbol in X, though the G-set is permuted by the elements of G and every permutation of X by an element g in G may be thought to become a "permutation representation" of the element g of G, or of a subgroup H under orbit within some normal subgroup of G (as represented by gHg^{-1}) in a natural fashion. (That is beyond the scope of this book.) The action of any element g in G acting on the G-Set is isomorphic to that of a member of Sn over the same elements in X. It follows that g is representing an element of Sn, and G is isomorphic to some subgroup of Sn.

Morphisms (1.1.3)

A **homomorphism** from a group G to a group H is a mapping $\phi(g)$ of the elements g in G to those of H, with each h in H taking the form of a coset of a normal subgroup $N = \ker(\phi)$ of G. H is then isomorphic to the set of cosets $g * \ker(\phi)$ – which obey the following.

$$\phi(a*b) = \phi(a)*\phi(b)$$

It is simple to show that the cosets of a normal subgroup $N = \ker(\phi)$ obey this condition.

$$(x*N) * (y*N) = x* (N * y) * N = x* (y * N) * N = (x*y*e)*N = (x*y)*N$$

Other Terms

An "**isomorphism**" is a homomorphism of two groups $G \cong H$ that is a bijection (a 1-1 and surjective map); the structure of G and H are then alike. An "**automorphism**" is an isomorphism between a structure and itself.

The **kernel** $\ker(\phi)$ of a homomorphism $\phi : G \to G/H$ (the set sent to the identity of G/H) is always a normal subgroup H of G. This subgroup is the "**cofactor**" H of the "**factor group**" G/H.

Simple Groups (1.1.4)

A **Simple Group** G is a group that has no proper non-trivial normal subgroups. In fact, there are no homomorphisms from G to any non-trivial subgroup H of G. Only the cases $\ker(\phi) = G$ or $\ker(\phi) = \{e\}$ remain so that $\phi : G \to \{e\}$ or $\phi : G \to G$ respectively.

All prime order (and then cyclic) groups are simple, and also many non-abelian groups. The smallest example of the latter is the "alternating group" A5.

*A group is **soluble** if it may be factored into a series of normal subgroups to produce a series of cofactors which must all be prime order cyclic groups. (This sequence of factorisation is a "composition series".) A group is insoluble if it does not so split into prime order cyclic cofactors; and simple if it has no non-trivial normal subgroups at all.*

A5 is the "**alternating group**", the subset of all permutations in the "**symmetric group**" S5 that factor into even numbers of transpositions. (S5 is, in turn, formed from all permutations upon five objects.) The groups, A5 and S5, are crucial to this study. One need note permutations of a set form a group, and to tacitly accept that A5 is simple as from any suitable textbook on abstract algebra. (A5 is not prime order or cyclic and is the smallest example of an insoluble group – it is not the smallest simple group; Cp is simple for p a prime, note that {e} is also.)

Every permutation of a set S may be written in disjoint cycles and a product of transpositions. So in S5 the permutation of {a,b,c,d,e} to {b,c,a,e,d} can be written as (a,b,c)(d,e) = (a,c)(a,b)(d,e) (permutations on the right are performed first). Note disjoint cycles always commute.

With regard to A5, every product in A5 (of even numbers of transpositions) is also in A5 (and even in transpositions) and so A5 is closed and inherits every other property satisfying the group axioms from S5 (note that the identity also is even). It is a simple exercise to show that the inverse of a permutation is also a product of transpositions of the same cycle length (albeit listed in reverse); the remainder of the group axioms are then inherent.

A5 is normal in S5 as seen from the cosets of A5 in S5. There is one conjugacy class other than A5 itself. All such products gHg^{-1} are of even length in transpositions and therefore also in H (g has the same cycle length as g^{-1} and all elements in H are "even"), then gH = Hg.

There are other laws obeyed by groups (the isomorphism theorems) yet these are also beyond the scope of this book. I will now move on to more complex structures, i.e. rings and fields.

1.2) Ring And Field Axioms

The Ring Axioms (1.2.1)

A ring R is a set that forms an abelian group under a binary operation '+'. A ring also requires a binary operation * that acts as multiplication upon R, such that the following axioms hold:

> RM1) *Closure.* For all a, b in R, a*b = c is also in R.
> RM2) *Associativity.* For all a, b, c in R, a*(b*c) = (a*b)*c.
> RM3) *The Left Distributive law.* For all members of R, a*(b+c) = (a*b) + (a*c).
> RM4) *The Right Distributive law.* For all members of R, (b+c)*a = (b*a) + (c*a).
> RM5) *The Zero.* For all x in R, 0*x = x*0 = 0. Then 0 is the additive identity, 0 is
>   unique.

For this study I will only need commutative rings (where a*b = b*a): both left and right distributive laws will hold. For rings (fields) I also require a "unity" or "1" which obeys x*1 = 1*x = x for all x in R. An example of an infinite ring is the set of integers under addition and multiplication. A finite example is the integers with those operations "modulo n" (all sums and products map to the remainder upon division by n). This ring is $\mathbb{Z}n$, often called $\mathbb{Z}/n\mathbb{Z}$ ($n\mathbb{Z}$ is an ideal (see below) generated by n in $\mathbb{Z}$, and $\mathbb{Z}/n\mathbb{Z}$ is a "factor ring" by that ideal).

Characteristic Of A Ring

The characteristic n of a ring R is the smallest integer n>0 such that $n*x \equiv 0$ for all x in R. n is then the smallest positive integer such that $n*1 \equiv 0$. If there is no such integer n, then the ring is said to have characteristic zero (for example, the integers or the rationals).

Subring

A **subring** S of a ring R is a subset of R that satisfies the ring axioms. There are also conditions for a morphism of a ring R to a subring S, i.e.

$$\phi(a + b) = \phi(a) + \phi(b)$$
$$\phi(a * b) = \phi(a) * \phi(b)$$

However, there must also be the condition that the kernel $\ker(\phi)$ is an ideal of R.

Definition:

A **Ring Ideal** is a subring I of a ring R such that x*I is in I for all x in R. Now there is the mechanism that I is as it were "normal" in R. Note that the absorptive property of an ideal always reduces the following products to an additive coset of the ideal I.

$$x+I \; + \; y+I \; = (x+y)+I$$
$$(x+I) * (y+I) = x*y +(x*I)+(y*I)+(I*I)$$
$$= x*y +(x+y+I)*I$$
$$= (x*y) +I$$

The axioms for a **field** F are as those for a ring, yet the set of all elements but the additive identity ("zero") must form an abelian group. Then the axioms G1 to G4 hold for both {F,+} and {F,*}. And a*b=b*a for all a, b in F.

Every element of {F,*} has an inverse (and unity is self-inverse) and there are no non-trivial ideals; for from the existence of inverses and closure, any non-trivial ideal J in F contains at least one multiplicative element x, so J must then contain the product $1 = x*x^{-1}$ and therefore J contains unity and so all y = y*1 are in J (and J = F). So no proper non-trivial subfield (or subring) may be an ideal in F and there are no proper non-trivial homomorphisms to any subfields (or subrings) of F. Only the trivial ones, $F \rightarrow F$ and $F \rightarrow \{0\}$ exist.

1.3) Prime Fields

The Euclidean Algorithm (1.3.1)

Now, the "Euclidean Algorithm" is a recursive method to find the greatest common divisor (gcd) of two integers (or polynomials). One simply begins with two integers a and b, and divides the larger by the smaller (or the greater degree polynomial by the lesser). The algorithm relies on the fact that the greatest common divisor divides both numbers a and b; so it must also divide the remainder on dividing a by b. I simply begin with and then iterate upon:

$$a = q_0 b + r_0$$
$$b = q_1 r_0 + r_1$$
$$r_0 = q_2 r_1 + r_2$$
$$...$$
$$r_{i-1} = q_{i+1} r_i + r_{i+1}$$
$$...$$
$$r_{n-1} = q_{n+1} r_n$$

The algorithm will terminate with zero remainder and the last divisor r_n is the gcd of a and b as desired. Reversing the algorithm and back-substituting (every r_i is divisible by r_n): gives an identity as below where r_n is the gcd of a and b and s, t are integers that are relatively prime.

$$sa + tb = r_n$$

Two numbers a and b are "relatively prime" if their gcd is 1. (Every prime p is relatively prime to every integer less that p.) 15 is relatively prime to 28, but neither of these are prime.

Finite Fields Of Prime Order (1.3.2)

Prime fields are "rings" with all arithmetic performed "modulo p" (p a prime number). Sums and products are made as in arithmetic on integers and then the result mapped onto the remainder on division by p. The set of "residue classes" (equivalence classes) modulo p form a "finite field of order p" or simply, a "prime field".

Given $a \equiv x \bmod p$ and $b \equiv y \bmod p$ then for some n, m (integers) $a - x = mp$ and $b - y = np$. Then:

$$a + b - (x + y) = (n + m)p$$
$$ab - (xy) = (nx + my + nmp)p$$
$$a + b \equiv (x + y) \bmod p$$
$$ab \equiv (xy) \bmod p$$

Sums and products from the residue classes are well formed. The residue classes form an additive group, yet they form a group multiplicatively only when p is a prime number. (Otherwise, there are "zero divisors", and elements without inverses.)

Every finite field of characteristic p has a subfield of order p. This prime field is called the "prime subfield" and is isomorphic to the set of p distinct residue classes as above.

Modular arithmetic provides a procedure to generate finite fields. Taking addition modulo five, I may add numbers as usual and simply take the result's remainder after dividing by five. I may then form five equivalence classes where $a \equiv b \bmod 5$ if the difference a−b is divisible by five. As no number but a multiple of five is congruent to zero (five is a prime) there are no zero divisors in the multiplicative part {1, 2, 3, 4}. This shows the multiplicative group is closed modulo 5, and ab ≡ ac mod 5 if b ≡ c mod 5 etc.

This generalises to the class of all prime fields. This property of a multiplicative group holds for any prime p. I must show that every non-zero element modulo p has an inverse.

I have that every integer x<p must obey the identity:

$$sp + tx = 1$$

I.e. for any integer x < p, there are integers s and t (by the **Euclidean algorithm**) such that the greatest common divisor of p and x here is 1. So $t \cdot x \equiv 1 \bmod p$ and t is an inverse for x. (Then, every element has an inverse – the ring $\mathbb{Z}_p$ is a field.)

Such finite fields of order p (and characteristic p), are denoted $\mathbb{Z}/p\mathbb{Z}$ or $\mathbb{Z}_p$. (The field $\mathbb{Z}/p\mathbb{Z}$ is the factor ring of the integers modulo the ideal generated by p in $\mathbb{Z}$.) They are members of the wider class of "extension fields" or "Galois fields". You may also see them denoted by GF(p), with p a prime number. I had shown a characteristic five field exists (of order five and degree one. The **characteristic of a ring** is the smallest number "n" such that $n * 1 \equiv 0$).

Some Finite Field Facts (1.3.3)

Several facts not proven here are given:

1) A field $\mathbb{Z}_p$ exists for every prime p. (Finite order "integral domains" are fields.)

2) Prime fields have cyclic multiplicative groups. A cyclic group is generated from the powers of a single element g of order n. Such a group G is isomorphic to Cn = <g>.

3) Extensions of prime fields F by an indeterminate "x" to form extensions F[x] form infinite order integral domains (which have no zero divisors). Every element in F[x] is a polynomial in x with coefficients in F.

4) Irreducible polynomials irr(x) of degree n over a finite field F of characteristic p are present for all n, so that ideals well formed in F[x] allow an algebraic extension of F by adjoining a root of irr(x) α to F, namely, F[x] mod irr(x) $\equiv$ F(α) may be formed, which are vector spaces over F (of order p^n) and are said to be of characteristic p, degree n.

5) All such finite fields F(α) have cyclic multiplicative groups.

6) All fields of a given finite order are isomorphic (similar) to each other.

7) All finite fields characteristic p degree n have an additive group isomorphic to an "n-fold" Cartesian product of its smallest subfields additive group $\mathbb{Z}_p + \cong \mathbb{Z} / p\mathbb{Z}$ which is of order p. ($\mathbb{Z}_p$ Is in turn isomorphic to the prime field of characteristic p.)

8) Over every finite field characteristic p there exists a tower of extension fields at the largest end of which may be placed the "algebraic closure" (AC). AC is a field that contains every possible algebraic extension F(α) of F and likewise (to theory) contains every finite field K of characteristic p with F as a subfield.

1.4) Field Extensions

This section is rather advanced (mathematically), so you will not be expected to arbitrarily construct finite fields of any prime power order, but here I give a construction of a single finite field of order 125. I hope to show these structures (finite fields) are well grounded in algebra and assure you that there is a genuine theoretic basis for these finite structures.

Given any finite field F (which may be an extension field), I may extend the field to an infinite set of polynomials in powers of x with coefficients taken from F, giving me the ring F[x] of all such polynomials in x. As F is a field, it can be easily shown that F[x] has no zero divisors. It can also be shown (much harder) that there exist irreducible polynomials, a generalised "prime-like" polynomial for every degree n. The multiples of these irreducible polynomials form ideals in the integral domain F[x], each generated by an irreducible, for which the following holds: that there are no other ideals properly containing them, and as such they are "maximal".

There is also the condition for a "prime ideal" I if a*b in I implies either a or b is in I. Every maximal ideal in F[x] is also a prime ideal. (An integral domain over a finite field has the property that every product has a unique factorisation up to a "unit" – an element of the finite field with an inverse. Also, a ring modulo a prime ideal has no zero-divisors, and a finite integral domain is a field. An integral domain is a commutative ring with unity that has no zero divisors.)

So, by analogy with prime fields modulo p, it is possible to take a set of polynomials modulo such an irreducible polynomial, which not only has no solution (root) in F, but does not even split into smaller order factors in F[x] (though it may be done so up to a "unit" – a member of F that has an inverse used as a scalar coefficient).

So for any element p(x) of F[x],

$$\forall\ p(x) \in F[x]$$
$$\exists\ s(x)\ ,\ t(x)$$
$$degree(t(x)) < degree(irr(x))$$
$$irr(x)*s(x) + p(x)*t(x) = 1$$

By direct analogy to prime fields and the Euclidean algorithm, t(x) is an appropriate inverse for any "p(x) modulo irr(x)" not congruent to zero.

Rather than enter into vector spaces to derive that all field extensions are of the order of a power of a prime, I simply construct a field here. GF(5,3) namely the field order five extended to that of 125, i.e. of degree three.

Note that irr(x) $\equiv x^3 + x + 1$ is irreducible modulo 5. It neither permits roots in $\mathbb{Z}_5$ nor splits into polynomials of lesser degree with coefficients in $\mathbb{Z}_5$ =GF(5). Now with the congruence irr(x) $\equiv 0$, I have every element in GF(125) a polynomial with coefficients in GF(5). So every element of GF(125) is now of the form:

$$p(x) = a_2 x^2 + a_1 x + a_0$$

And I have a three dimensional "vector space" where the powers of x become the entries of a cartesian product with entries in the a_i. Addition is done "coordinate-wise" collecting like terms in powers of x. Every such n-fold cartesian product with entries (elements) from a field F forms a vector space over F:

A **vector space** V is a closed set of vectors with a property of a vector scaling. (Including a null vector **0**, or "origin" or zero.) Any scalar s in F, permitting the product s*v (with **v** a vector in V a vector space), must also be in V. (The multiplicative part of our field is "closed" with elements in F.) A vector space must satisfy the following axioms:

V1)   For all a in F and **v** in V, a**v** is in V. (0**v** = **0** for zero in F and **0**, **v** in V.)
V2)   For all a, b in F and **v** in V, a(b**v**) = (ab)**v**.
V3)   For all a, b in F and **v** in V, (a + b)**v** = (a**v**) + (b**v**).
V4)   For all a in F and **v**, **w** in V, a(**v** + **w**) = (a**v**) + (a**w**).
V5)   1**v** = **v** for unity in F and **v** in V.

a, b are **scalars** from F whilst **v**, **w** are **vectors** in V.

I may form the three dimensional vector space in powers of x over $\mathbb{Z}_5$. I now have the ability to reduce products of elements in F(x) to those of lesser degree using $irr(x) \equiv 0$.

i.e. $(x^2+1)*(x+4) \equiv x^3 + 4x^2 + x + 4 \equiv (4x + 4) + 4x^2 + x + 4 \equiv 4x^2 + 3$

There is an identity for reducing x^3, namely that $x^3 \equiv 4x + 4 \mod irr(x) \mod 5$. I have shown that there exists a vector space over F of dimension (degree) three. A field of order 125 as desired. Similar constructions (also using irreducible polynomials over finite fields) permit extensions of any degree because such a vector space not only exists for any degree n (as a "n-fold cartesian product" or a vector space in powers of x), but also irreducible polynomials modulo p exist for every degree n over any finite field F.

For any extension K of a field F of degree n, what polynomial in K is congruent to x^n? It can never be zero for K is a field without zero divisors. (Such fields are vector spaces and so exist.) With this congruence p(x), no two polynomials of lesser degree may in product form $(x^n-p(x))$ since they would also be zero divisors. So $x^n-p(x)$ is irreducible over F. The existence of a polynomial $(x^n-p(x))$ shows there is truly an irreducible of every degree n with coefficients in F, which may be used for algebraically extending K over F.

I would first have to show such a field K exists[3], for algebraic closure over the prime subfield of F would certainly contain every root of the equation $f = x^{p^n} - x = 0$. Likewise the roots of this equation are closed under addition and multiplication modulo p (as if using the reduction as per the Frobenius map) and the set also satisfies all the group and ring axioms for a field. Every such root 'a' of f is distinct, having multiplicity 1. I.e. $(x-a)^2$ does not divide f.)

There is an extension K of degree mn over every extension F of degree m. (By applying the map $p^{nm}=p^m p^{m(n-1)}$ to the indices of every root of F in K, every such root f is also found in K, and so F is a unique subfield of K.)[3]

1.5) A Note On Field Morphisms – The Map Of Frobenius

A Caveat (1.5.1)

Every field contains the unity, and each non-zero element has a multiplicative inverse; there are no proper ideals or homomorphisms to any non-trivial subfield of any field F. This does not preclude the existence of subfields in F itself. Every subfield of a field of degree n has degree m dividing n, and at algebraic closure the theorist may examine fields (with the same characteristic) as subfields of the same parent field structure whilst ignoring for the most part any mention of irreducibles. This crucial approach permits us to show that two fields of the same order are isomorphic: that a bijective homomorphism exists between them.

Aside from those morphisms, an extension to algebraic closure (the "isomorphism extension theorem") permits the statement:

"For any prime power there exists a unique finite field of that order and the multiplicative group is cyclic".

As every field extension is itself a vector space, it remains to state that its multiplicative group is of order $p^n - 1$ and the elements solve the equation $x^{p^n} - x = 0$ or $x^{p^n-1} - 1 = 0$. All elements of the group GF(p^n)* are generated by distinct powers of a root x of irr(x), and the multiplicative group is then cyclic.

[3] John B. Fraleigh, *A First Course In Abstract Algebra 6th Ed* (Addison Wesley Longman), p421-423.

The Map Of Frobenius (A Field Automorphism) (1.5.2)

The "**Frobenius**" map is an automorphism upon fields of characteristic p, and is stated as:

$$\phi(x) = x^p$$

This has the result that

$$\phi(x+y) = x^p + y^p$$
$$\phi(x \cdot y) = x^p \cdot y^p$$

The binomial expansion of the map upon the sum x+y has coefficients of x in powers not p as multiples of p, hence congruent to zero.

$$\phi(x+y) = x^p + px^{p-1}y + \underbrace{\frac{p(p-1)}{2!}x^{p-2}y^2 + \frac{p(p-1)(p-2)}{3!}x^{p-3}y^3}_{\equiv 0} + \ldots + y^p$$

Only powers in p remain (since nothing divides p). "Frobenius" is an automorphism on Fields characteristic p. It generates a cyclic group of order n (n, the degree of extension over the prime subfield) – the "Galois group" Gal(F) of automorphisms. Every member σ of Gal(F) – of "order m dividing n", corresponds to a subfield K of F of order $p^{n/m}$ held fixed under σ. There is always such a K under the map: generators of Gal(F) hold fixed the prime subfield $\mathbb{Z}_p$.

1.6) The Fields GF(2), GF(4) And GF(8)

GF(2) is the prime field of order two (GF for "Galois Field"). GF(4) is an extension modulo $x^2 + x + 1 \equiv 0$, and GF(8) the extension modulo $x^3 + x + 1 \equiv 0$ or by the irreducible $x^3 + x^2 + 1 \equiv 0$.

Aside from using GF for "**Galois Field**" I use the notation F* and F+ or even GF(p^n)* and GF(p^n)+ to denote the multiplicative and additive groups of a field respectively.

Rather than list the polynomials in GF(4) and GF(8), I note their multiplicative groups are themselves of "prime order less one" and cyclic, and from inspection or by the distributive laws, any non-zero element in product with the elements of an additive subgroup also results in another isomorphic additive subgroup – true of any field F. GF(4) has three additive subgroups of order two, and GF(8) has seven subgroups of both order two and of order four.

Now GF(2) has the very simple additive and multiplicative tables:

+	0	1		*	0	1
0	0	1		0	0	0
1	1	0		1	0	1

Whereas GF(4) is slightly more complex (any two non-zero elements under addition produce the third, and upon themselves simply the zero); so, with "a = 1" as unity:

+	0	a	b	c		*	0	a	b	c
0	0	a	b	c		0	0	0	0	0
a	a	0	c	b		a	0	a	b	c
b	b	c	0	a		b	0	b	c	a
c	c	b	a	0		c	0	c	a	b

GF(8) is more complex still (with a=1 and b=x=(a,b,d,c,f,g,e) in permutation representation).

+	0	a	b	c	d	e	f	g		*	0	a	b	d	c	f	g	e
0	0	a	b	c	d	e	f	g		0	0	0	0	0	0	0	0	0
a	a	0	c	b	e	d	g	f		a	0	a	b	d	c	f	g	e
b	b	c	0	a	f	g	d	e		b	0	b	d	c	f	g	e	a
c	c	b	a	0	g	f	e	d		d	0	d	c	f	g	e	a	b
d	d	e	f	g	0	a	b	c		c	0	c	f	g	e	a	b	d
e	e	d	g	f	a	0	c	b		f	0	f	g	e	a	b	d	c
f	f	g	d	e	b	c	0	a		g	0	g	e	a	b	d	c	f
g	g	f	e	d	c	b	a	0		e	0	e	a	b	d	c	f	g

Where the table on the right is arranged to show the group GF(8)* is cyclic. The table for GF(8)+ contains the table GF(4)+ within it – it is a subgroup of GF(8)+. Likewise GF(2) has both its tables embedded within those of GF(4) and GF(8), being a subfield of both.

So I will associate the following symbols to the polynomials in the finite field GF(8)

$$0 = 0$$
$$a = 1$$
$$b = x$$
$$c = x + 1 = x^3$$
$$d = x^2$$
$$e = x^2 + 1 = x^6$$
$$f = x^2 + x = x^4$$
$$g = x^2 + x + 1 = x^5$$

An easy way to remember the operation of GF(8)+ (the octal group) is to memorise its subgroups as triples {{a,b,c}, {a,d,e}, {a,f,g}, {b,d,f}, {b,e,g}, {c,e,f}, {c,d,g}}. Each pair of elements from each one of the triples under addition produces the third. (Additively, sums of elements upon themselves give the identity zero.)

1.7) A Closure Of Unity, Orbits Of Additive Subgroups

It can be proven by induction that the number of additive subgroups N(n,k) of order p^k of a finite field's additive group of order p^n is given by:

$$N(n,k) = \prod_{j=1..k} \frac{p^{n-j+1} - 1}{p^j - 1}$$

For p=2, n=3 and k=1 or 2, N(n,k) is 7 as expected. However, for any other p, n and k this is a relation that rapidly increases in magnitude over the order of the field GF(p^n).

The additive subgroups of a field are under orbit by products of multiplicative elements. As any additive subgroup acted upon in product by an element of the multiplicative group is again an additive subgroup – as the group is cyclic, each subgroup is "under orbit".

Orbits of the additive groups of subfields are "short orbits". Their multiplicative elements form a group. As a result, the additive subgroups' non-zero elements (in orbit) form disjoint sets as their multiplicative images separate into cosets (disjoint by the theorem of Lagrange). If K is a subfield of F, then the orbit of K+ in F is equal in length to $|F^*|/|K^*|$.

The smallest fields (F) of characteristic two have the property that orbits of subgroups in F have length equal to $|F^*|$ and also N(n,k). (There is one orbit overall.) The first three such fields have no non-trivial subfields with index $|F^*|/|K^*| > 1$.

If p=3, n=2 and k=1 then N(n,k) = 4. Assuming GF(5)* generates that orbit, I introduce a different characteristic field and "diverge" with only that one additive subgroup which makes no further sense to extend. Likewise, given p=3, n=3, k=2 I have N(n,k) = 13, clearly also a different "characteristic" and also of a non-field order (there is no prime power equal to 14).

The number of subgroups of order 4 in GF(16) is 35 (the number rapidly increases with the degree of GF(p^n)). Remaining within fields of equal characteristic is no guarantee of equality between the order of generators and N(n,k) (Neither those of order greater than GF(8) or of other characteristic).

GF(16) then, has two such orbits of length 15 and one orbit of length 5. The latter is the orbit (under multiplication by a generator) of the subfield isomorphic to GF(4) in GF(16). There is no rule of thumb that I can deduce, that orbits of subgroups (that so outstrip the order of F* in number) will share a short orbit of a subfield of a specific order, other than that the subgroup's (or under multiplication its "cosets") own "multiplicative elements" are required to be regularly and equally separated in the powers of a generator of the field, and that their index of separation should divide the order of the field's multiplicative group. The additive subgroup must also be that of a multiplicative coset of a subfield someplace in its orbit.

With multiplication I may cycle the additive subgroups of any Field F in question. For a short orbit the additive subgroup must be a product of a power of a generator upon a subfield, else the orbit is not short but is in length equal to the order of F*.

Only the three smallest fields of characteristic two, GF(2), GF(4) and GF(8) have N(n,k) equal to $|F^*|$. I.e. holding a unity of index ($|F^*|/$ "orbit length") upon additive subgroups under orbit by generators of the multiplicative group. They as "Trinity" alone have only *one* orbit equal in length to N(n,k) for each non-trivial additive subgroup. Every subgroup in F+ belongs to just the same one orbit with "length" $|F^*|$. $|F^*|$=N(n,k) is unique to GF(2), GF(4) and GF(8).

Every other finite field of characteristic two with a multiplicative group F* of prime order has its every orbit of length |F*| (but then there is more than one orbit). With no non-trivial proper subfields but $\mathbb{Z}_2$ they have just one short orbit, but this (an orbit of a prime subfield of characteristic 2) is equal in length to F*. So, F being such an extension of degree n (p^n-1 a prime), all the *additive* subgroups of a given order p^m (where m|n, i.e. m divides n) share just the one *length* of orbit: the order of the group F*. (Though there be many such orbits when |F|>8.)

The Trinity formed of GF(2), GF(4) and GF(8), however, is the unique collection of fields that is also the smallest such closed set (without such a divergence or cascade); and the Trinity requires no others. The Trinity is unique in that the Galois group of GF(8) induces an operation in agreement with multiplication over the K4 form of the Trinity (the model of this study) then acting as if operating upon GF(4)*. GF(32) for example is not so structured (its Galois group is as C5 and not C2).

Chapter Two: Some Basic Logic On Positive Properties

To assert some results on "positive properties" (for Gödel's definition, see Appendix) and their relationship to perfection, I require a few definitions.

- N(x) is the modal "**Necessity**" and P(x) the modal "**Possibility**" operator. I shall use ¬ for **negation**, and " ∨ " for **disjunction**, and "&" for logical **conjunction**.

- I assert $P(x) \leftrightarrow \neg N\neg(x)$.

- $A \rightarrow B$ is "A entails B" (A implies B) and $A \leftrightarrow B$ is "A and B are logical equivalents".

- $A \rightarrow B$ (A is true **only if** B is true) is equivalent to $\neg B \rightarrow \neg A$.

- $B \rightarrow A$ is "A is true **if** B is true". $A \leftrightarrow B$ is then "A is true **if and only if** B is true".

- $A \rightarrow B$ is logically stated as $N\neg(A \& \neg B)$.

- The predicate statement "x" is to assert "God has the property x" or that "God exemplifies x". Then $x \& y \rightarrow y$ is "God exemplifies both x and y" therefore "God exemplifies y". I require all such x and y to be positive properties.

- The predicate statement Pos(x) is "x is a positive property". I note (by Anselm's argument) that the set of positive properties Ω form an **ultrafilter** (see later).

- A property x "**privates**" another property y if and only if $x \rightarrow \neg Pos(y)$.

2.1) Virtues

A **virtue** p is a positive property that:

1) Privates no other positive properties q: i.e. $N\neg(Pos(p) \rightarrow \neg Pos(q))$.

2) Cannot be positive when not exemplified. To not exemplify p is a privation of perfection. I.e. $N(\neg Pos(p^{-1}))$. The "inverse" of p, (p^{-1}) is the "lack" of exemplifying p. Consequently, p^{-1} may **never** entail anything positive.

3) Virtues allow the free exercise of positive acts, rather than the free exercise of privations on perfection. I.e. given Pos(p) I then have to consider the disjunction $(Pos(q) \vee \neg Pos(q^{-1}))$. A virtue allows a positive act from such privation q^{-1}. Hence virtues entail an inference or substitution $p \& q^{-1} \rightarrow s$. The disjunction is resolved with "given Pos(p)" : $(Pos(q) \vee Pos(s))$ for a suitable positive property s.

Axiom – Of Virtue (2.1a)

"If it is positive to exercise a property "s" entailed from the lack of exemplifying a positive predicate "q" (the set of all possible "s" taken from $\Omega \setminus r$ in all r where either $q \rightarrow r$ or $r \rightarrow q$), then there exists some **unique** virtue "p" such that $Pos(p) \& q^{-1} \rightarrow Pos(p \& q^{-1}) \rightarrow Pos(s)$ and $p \& q^{-1}$ (as also positive by Gödel's definition) entails that which is positive from Pos(s) only."

Both of "r" and "s" are positive and do not private each other: for both belong in the ultrafilter of perfections (of Anselm's argument). In order for the disjunction to be freely decidable, both must be axiomatically positive not privating one another. The sole requirement is that they must remain "disjoint sets". I.e. r $\nrightarrow$ s, s $\nrightarrow$ r. As sets indexed by separate ultrafilters, they are.

Then I state a "pure virtue"; the divine liberty "L" entails (from the set G of all virtues) every necessary (and positively) exemplifiable virtue "p" upon p&q^{-1} to entail s (by axiom of virtue):

So I expect Pos(L&G)&q^{-1} $\rightarrow$ Pos(p)&q^{-1} $\rightarrow$ Pos(p&q^{-1}) $\rightarrow$ Pos(s) for all possibly positive "s".

That liberty "L" permits unnecessary virtues to remain inactive out of divine freedom (q remains positive). As the set of all virtues G is itself in the ultrafilter of virtues (the definition to be given later), liberty "L" entails each subset p of G to also be in that ultrafilter (as long as the subset contains some principal e_p and entails s from q^{-1}). Liberty may generate that necessary p within every possible subset of G from the indexing set, entailing every possible property that may possibly be entailed from q^{-1}, i.e. my "s" as above. (I simply denote L&G by L(G).)

Liberty, L(G) generates the ultrafilter, so N(L(G)&q^{-1} $\rightarrow$ e_s) entails N(e_s&q^{-1} $\rightarrow$ e_s), where "e_s" is to "rest on the outcome of a completed act" and "e_s" is the minimal possible "s". Both L(G) and some minimal "e" must always be present (God is not set at liberty if He must necessarily act with all virtue, or never cease working). Instead, God is free to "rest" indefinitely.

Liberty is a virtue allowing the divine (G) to rest on any proper subset s of S, where S is the set {s in Ω: p&q^{-1} $\rightarrow$ s}. The virtues included of liberty also entail "rest" upon the set of positive properties v in T where T is the set: {v in Ω: p&q^{-1} $\rightarrow$ v^{-1}}. The possible virtues not worked by that liberty (as if $\bar{p}$ in $\bar{p}$&q^{-1} $\rightarrow$ v) are then excluded from X (X the set of virtue in p). By "axiom of virtue" liberty entails a "minimum necessary subset of virtue" p_s in X to entail the subset s contained in S. Note that the complement of p_s in X is Y: the set {p_v in X, p_v&q^{-1} $\rightarrow$ v^{-1}}.

Then every **positive property** q must fulfil the following:

1) For all q: ¬Pos(q) $\leftrightarrow$ Pos(¬q) $\leftrightarrow$ Pos(q^{-1}) are equivalents.

2) A positive q may private another property unless that property (or q) is a virtue: i.e. Pos(q) $\rightarrow$ ¬Pos(p) for p a virtue entails Pos(p) $\rightarrow$ ¬Pos(q) and then p privates q, a violation. Then every such Pos(q) must **necessarily** entail some Pos(r).

3) q is free to entail a positive property not used: i.e P(Pos(q) $\rightarrow$ Pos(s^{-1})).

4) For all positive q, if q $\rightarrow$ r then r is also positive. Pos(q) $\rightarrow$ Pos(r) if q $\rightarrow$ r.

The set of all positive properties together form an ultrafilter (upon an indexing set I) the definition of which is as follows:

U1)	For any subset S of the indexing set I, either S or the complement of S, (I \ S) belongs in the ultrafilter U, but never both.
U2)	If sets X and Y are in U, then so is their intersection X $\cap$ Y.
U3)	The empty set ϕ is never a member of the ultrafilter U.
U4)	If X is a set in U, then so is every set in I that is a superset of X. (All sets that properly contain X as a subset.)

Now I assume a simple rule of thumb here.

If a positive property q cannot be positively exemplified, then it is positive for perfection not to exemplify q. Then:

$\neg Pos(q) \leftrightarrow Pos(q^{-1})$

Which has the result that $(q^{-1})^{-1} \leftrightarrow q$ or that $\neg Pos(q^{-1}) \leftrightarrow Pos(q)$ and this is well formed. The statement is "To not do something negative is to act to the positive" and "not (not acting) is acting". God is such that all predicates in the ultrafilter should be positive, active or not.

A virtue p (the set of which is completed in perfection) will not permit privations of positive properties and so God will ever act with some positive "s" instead of q^{-1}, accordingly preserving all p. Essentially $Pos(p\&q^{-1}) \rightarrow Pos(s)$. And the outcome is always positive.

A virtue p is such that $s^{-1} \rightarrow ((p\&q^{-1})^{-1} \leftrightarrow p^{-1}\&q)$ and if s is positive then q is positive without requirement for the exercise of virtue. I find a well-formed set of "virtues".

The statement "liberty as p is fully virtue" is necessary for perfection: for $Pos(q) \vee Pos(e)$ holds given that liberty permits $Pos(p\&q^{-1}) \rightarrow Pos(e)$. Even God may rest from work.

Inferring then, that $s^{-1} \rightarrow ((p\&q^{-1})^{-1} \leftrightarrow p^{-1}\&q) \rightarrow q$, I examine virtue: for $N(\neg Pos(p^{-1}))$ was axiomatic for every virtue exercised in perfection.

Liberty as pure virtue also allows the free exercise of virtues, for "not exercising p" (p^{-1}) is no privation to q: $N(p^{-1}\&Pos(q) \rightarrow Pos(q))$ and as q is positive, perfection is not so easily privated.

For any virtue and even liberty L(G) I have the consequence that given L(G), I must decide:

$Pos(p) \vee Pos(p^{-1})$.

I.e. $Pos(L(G)\&q^{-1}) \rightarrow Pos(p\&q^{-1}) \rightarrow Pos(s)$ as before. $(Pos(q)\&p^{-1} \rightarrow Pos(q)$ always.) Then, divine liberty is a **principal** element and **generates** the ultrafilter of perfections. For more on how $Pos(p\&q^{-1}) \nrightarrow Pos(p)\&Pos(q^{-1})$ and on what makes a principal ultrafilter, see section [2.10] later.

I should now justify $\neg Pos(p\&q^{-1}) \leftrightarrow p^{-1}\&Pos(q)$.

Equivalence

Examine these two tables for the positive properties r and s (as taken from the same ultrafilter, i.e. from a shared indexing set).

For the conjunction r&s I have:

	$r+$	$r-$
$s+$	$+$	$+$
$s-$	$+$	$-$

(Where r+ means that r is positive, r– that r is negative)

For the conjunction r^{-1}&s^{-1} I have:

	$r+$	$r-$
$s+$	$-$	$+$
$s-$	$+$	$+$

So ¬Pos(r&s) ↔ Pos(r^{-1})&Pos(s^{-1}) providing both r and s are positive or negative ... (1).

Yet then Pos(r^{-1})&Pos(s^{-1}) → Pos(r^{-1}&s^{-1})   ... (2).

"Perfections", as positive properties so held form an ultrafilter. I.e. if A, B are in the ultrafilter then of **necessity** A&B is in the ultrafilter.

Then ¬Pos(r&s) ↔ Pos((r&s)$^{-1}$)

So, Pos((r&s)$^{-1}$) → Pos(r^{-1}&s^{-1}) as desired   ... by (1) and (2).

I find Pos(r^{-1}&s^{-1})$^{-1}$ → Pos(r&s)) from the modus tollens   ... (3).

I need, however, Pos(r&s) → Pos(r^{-1}&s^{-1})$^{-1}$.

Given Pos(r&s), Pos(r&s) → Pos(r) as r&s → r. (From the nature of a positive property following from another, r is positive.)

Pos(r&s) → Pos(s) as r&s → s

So; Pos(r&s) ↔ Pos(r)&Pos(s)   ... (4).

Pos(r)&Pos(s) → ¬Pos(r^{-1}&s^{-1}) → Pos(r^{-1}&s^{-1})$^{-1}$ ... by (4) then (1) and (2).

So, Pos(r&s) ↔ Pos(r^{-1}&s^{-1})$^{-1}$ ... by (3).

The entries in the following tables are negatives of each other **if and only if** just one of r or s is positive. (I.e. ¬Pos(r^{-1}&s) → Pos(r&s^{-1}) → Pos(r)&Pos(s^{-1}) → Pos(r)&¬Pos(s)) God has only positive predicates, the difference is that just one of r^{-1} or s^{-1} is positive to exemplify.

$Pos(r \& s^{-1})$

	$r+$	$r-$
$s+$	$+$	$-$
$s-$	$+$	$+$

$Pos(r^{-1} \& s)$

	$r+$	$r-$
$s+$	$+$	$+$
$s-$	$-$	$+$

Lemma (2.1b)

Given any list of sets of positive properties from an ultrafilter sharing the same indexing set, their conjunction N(Pos(a&b&c...&x&y&z)) may be logically negated to N¬Pos(a&b&c...&x&y&z) → N(Pos((¬a&¬b&¬c)&...&(x&y&z))) with the terms in (x&y&z) not negated. Yet for perfection I require those terms in (x&y&z) to be negated also.

I assert the proposition N(Pos(A&B)) → N(Pos(A)&Pos(B)) from A&B → A and A&B → B.

I already have N(Pos(A)&Pos(B) → N(Pos(A&B)) from the properties of an ultrafilter (see above).

From the logic tables above; ¬Pos(A&B) ↔ ¬Pos(A)&¬Pos(B) ↔ Pos(¬A&¬B) are all equivalent and there is no separation as with (x&y&z) above, every predicate in perfection is positive.

Proof:

Given Pos(r&s), Pos(r&s) → Pos(r) as r&s → r. (From the nature of a positive property following from another, r is positive.) I also have Pos(r&s) → Pos(s) as r&s → s.

So; Pos(r&s) ↔ Pos(r)&Pos(s) (given Pos(r)&Pos(s) → Pos(r&s) from the ultrafilter)

Then Pos(A&B) ↔ Pos(A)&Pos(B). Which was what was wanted.

Corollary

¬(Pos(A)&Pos(B)) ↔ ¬Pos(A&B) ↔ Pos(A^{-1})&Pos(B^{-1}) ↔ Pos(A^{-1}&B^{-1}) ↔ ¬Pos(A)&¬Pos(B)

Are all equivalent.

Proof:

This should be immediate from Lemma 2.1b and the two pairs of logic tables given above.

Lemma (2.1c)

N¬(Pos(r^{-1})&Pos(s^{-1})) ↔ N(Pos(r)&Pos(s)) for r, s any two **positive properties** in the same ultrafilter.

Proof:

Given the Corollary (to Lemma 2.1b) above, the statement is correct.

The General Case – Induction

By induction N¬(Pos(r^{-1})&Pos(s^{-1})&...&Pos(z^{-1})) → N(Pos(r)&Pos(s)&...&Pos(z))

Given p(1) the statement N¬(Pos(r^{-1})&Pos(s^{-1})) → N(Pos(r)&Pos(s))

Then p(n) → p(n+1):

N¬(Pos(r^{-1})&Pos(s^{-1})&...&Pos(y^{-1}))&N¬(Pos(z^{-1}))

→ N(¬Pos(r^{-1})&¬Pos(s^{-1})&...&¬Pos(y^{-1}))&N(Pos(z))

→ N(Pos(r)&Pos(s)&...&Pos(y))&N(Pos(z))

Therefore N(Pos(r)&Pos(s)&...&Pos(z)) as desired.

I.e. $p(n)$ = N¬(Pos(r^{-1})&Pos(s^{-1})&...&Pos(y^{-1})) → N(Pos(r)&Pos(s)&...&Pos(y))

→ N(Pos(r&s&...&y))

$p(n+1)$ = N¬(Pos(r^{-1}&s^{-1}&...&y^{-1})&Pos(z^{-1})) → N(Pos(r&s&...&y)&Pos(z))

→ N(Pos(r&s&...&y&z)) (all these predicates within the same ultrafilter).

Notes On The "Pos" Operator (Only Used As A Predicate) (2.1d)

The "Pos" operator acting on a property x evaluates to "true" if x is positively exemplified and to "false" otherwise. It is a simple fact that given x → y then unquestionably Pos(x) → Pos(y) (if x is positive). However, it is not the case given x → y, that Pos(y) proves Pos(x).

Since positive properties form an ultrafilter together, the Pos operator acts as a finite additive "measure" in that if a property is present in the ultrafilter, then the Pos operator evaluates on that property to "1" (true) or to "0" (false) if the property is not in the filter. I will write Pos(x) if and only if x is exemplified positive. In this manner, the operator becomes a predicate.

If Pos(x) is true then so is ¬Pos(x^{-1}). What cannot be true is that Pos(x) and Pos(x^{-1}) hold equally; for then x is inconsistent in its positivity or cannot be as a "positive property" (or as a negated one otherwise) and therefore x should be excluded from the ultrafilter. Likewise, Pos(x^{-1}) is also inconsistent and is not assigned positivity in the "language" of the ultrafilter.

Pos(x) ↔ Pos(x^{-1}) is an inconsistency, but this is not the statement x ↔ x^{-1}. The latter states x is logically inconsistent, whereas the former may imply that in a given situation somehow Pos(x) ∨ Pos(x). That is, given Pos(x), some other condition a ∨ b could logically hold, then Pos(x) is "out of scope". It is not true that x ∨ x holds, but that the positivity of x is undecidable.

2.2) Theorem

Assuming the set of virtues p is a "maximal" (and closed) set and that the sets of positive properties in r and s (as acts also) are entailed from the properties q and q^{-1} respectively I find the sets:

    A   {r, Pos(q) → Pos(r)}
    B   {s, Pos(p&q^{-1}) → Pos(s)}

And I state the sets A, B (given some closed set of virtues in p as above) are necessarily disjoint.

Proof:

q → r and then also Pos(q) → Pos(r). (It is not true that ¬Pos(r) → ¬Pos(q) unless Pos(r^{-1}).) Instead, ¬Pos(r^{-1}) → ¬Pos(q^{-1}) (or rather ¬r → ¬q which are both found wholly negative). For then, r^{-1} → q^{-1} → Pos(s) given Pos(p), which is necessarily positive by axiom.

The statements A → B and the modus tollens ¬B → ¬A are logically equivalent to N¬(A&¬B).

To find a contradiction I may rearrange (without including p), using Lemma 2.1c

$N¬(r^{-1}\&¬Pos(s)) ↔ N¬(Pos(s^{-1})\&r^{-1}) ↔ N(Pos(s)\&r) ↔ ¬P(Pos(s^{-1})\&r^{-1})$

Yet p is virtue! Then there is privation of virtue, there is no free exercise between r and s. I.e. r, s private each other, one is then negative and cannot be exemplified in perfection.

$r^{-1} → Pos(s)$, $s^{-1} → Pos(r)$ by symmetry. (r^{-1} and s^{-1} are wholly negative.)

So $Pos(r) ∨ Pos(s)$ and there is an excluded middle in p. (A contradiction on $N(Pos(s)\&Pos(r))$.)

$r^{-1} → q^{-1}$ was entailed without p; I cannot entail p from r^{-1}, so I cannot entail $Pos(s)$ from $Pos(q^{-1})$ as from r^{-1}. (p is indispensable as bound to $Pos(q^{-1})$. A positive p is required to entail $Pos(s)$.)

There is no positivity $Pos(s)$ entailed from any r^{-1}, (including q^{-1}) not given p and vice versa.

Such positive properties $Pos(r)\&Pos(s)$ cannot follow from $p\&q^{-1}$ without privating p. (The set of p is a closed set: entailing $Pos(r\&s)$ breaks that closure.)

Given p I really have:

$Pos(p\&r^{-1}) → Pos(p\&q^{-1}) → Pos(s)$

And the correct following:

$N¬(Pos(p\&r^{-1})\&¬Pos(s)) ↔ N¬(s^{-1}\&Pos(p\&r^{-1})) ↔ N(Pos(s)\&p^{-1}\&Pos(r))$

$→ N¬(Pos(r)\&Pos(s)\&¬p^{-1})$ and the middle $A ∧ B$ then privates p, a virtue.

(Again using Lemma 2.1c) $N¬(s^{-1}\&Pos(p\&r^{-1}))$ equates to:

$s^{-1} ∨ Pos(p\&r^{-1})$

So, as I began with the statement $Pos(r) ∨ (Pos(p\&r^{-1}) → Pos(s))$

It is implausible to operate $Pos(r\&s)$ if I necessarily have $s^{-1}\&Pos(r)$.

Any predicate in $A ∧ B$ given p is impossible to exemplify. (Here, $A ∧ B$ is the "meet" of A and B or as the conjunction of any representatives "r" and "s" from them.)

So A and B given the set of all "p" are "disjoint sets", or else they are also virtues.

$Ω$ "in all possible worlds" is then become in part as these two sets r and s. Their middle(s) in $A ∧ B$ are never found in any disjunction to be consistent sets of positive properties.

Corollary:

The set of virtues is closed, not entailing other positive properties that private (annul) them.

Aside from the identities given $Pos(p)$;

$Pos(q) \lor \neg Pos(q^{-1})$
$Pos(q) \lor Pos(p\&q^{-1})$

From which follows $Pos(r) \lor Pos(s)$

$N\neg(p\&r^{-1}\&\neg s)$ rearranges to $N\neg(p\&\neg(r\&s))$ so it then appears that: $p \rightarrow r\&s$. Yet with r in A and s in B (neither set virtues), $A \land B$ is "null" or from the above, privates virtue. Since p is a closed set, r&s (in $A \land B$) is not consistently positive (given p and one of r or s). p then entails the excluded middle of $Pos(r) \lor Pos(s)$ which is inconsistent: $(r\&s)^{-1}$ also appears to private virtue.

Given p, if I have $r \rightarrow s^{-1}$ then $p \rightarrow r\&s \rightarrow s\&s^{-1}$. r and s become inconsistent. $r\&s \rightarrow (r\&s)^{-1}$ is also inconsistent and not positive given p a closed set permitting free exercise of r and s.

Assuming a virtue p that allows the free exercise of both q and s (s is entailed from $p\&q^{-1}$). p may not be thought to entail q and therefore all r, but only such s upon $p\&q^{-1}$ as positive.

Both Sides Cannot Entail Virtue (2.2a)

Given some positive property q (from which I may entail some virtue p_0).

I obtain a similar result from the disjunction $Pos(q) \lor (Pos(p\&q^{-1}) \rightarrow Pos(s))$

I.e. given $q \rightarrow p_0$, $Pos(p_0) \lor (Pos(p\&q^{-1}) \rightarrow Pos(s))$, or even $(Pos(p)\&p_0^{-1} \rightarrow Pos(s_0))$ with $s \rightarrow s_0$.

Yet then, $p\&p_0^{-1} \rightarrow p$ for some p positive by axiom (God will not exemplify p_0^{-1}), so $p_0 \lor p$.

The necessarily positive virtue(s) p and p_0 private each other (as positive properties by axiom), which is an inconsistency. (A necessarily negative predicate cannot entail anything positive.)

One Side Cannot Follow From Virtue (2.2b)

Given a virtue p_0 and $Pos(p_0) \rightarrow Pos(q)$, by the inference $Pos(p\&q^{-1}) \rightarrow Pos(s)$ I have:

$Pos(q) \lor (Pos(p\&q^{-1}) \rightarrow Pos(p)\&\neg p_0 \rightarrow Pos(p))$ given p_0^{-1} is strictly negative (p_0 a virtue). Then p privates q (as with p_0 and s above), a contradiction. Otherwise I may state:

$\rightarrow q^{-1} \lor \neg p\&Pos(p_0)$ by inverting both sides.

$\rightarrow p_0^{-1} \lor Pos(p_0)\&p^{-1}$ that is, if $p_0 \rightarrow q$

$\rightarrow p_0^{-1} \lor (Pos(p_0)\&p^{-1} \rightarrow Pos(p_0))$ that is, if $Pos(p_0)\&p^{-1} \rightarrow Pos(p_0)$, ($p^{-1}$ is strictly negative).

p appears not to be virtue, privating q; $\neg Pos(p\&q^{-1})$ is such that p, a virtue and positive, is not a virtue with respect to q. In this sense p allows no free choice, the left side is always negative.

I have no free exercise of virtue in p; only one side is positive. The idea that $p \rightarrow q$ or $p \rightarrow s$ is inconsistent; the set of all virtues X must be closed, and by symmetry cannot entail those properties in q or q^{-1}, or namely the properties r and s (or the sets A and B).

2.3) Some Notes: When The Set Of Virtues Is Not Closed

Virtues do not allow the free exercise of privations on perfection. I.e. $Pos(p_0) \to Pos(q^{-1})$ is as false as $Pos(p_0) \to \neg Pos(q)$ (given $Pos(q)$ is true or if $p \to q$ holds).

So $p^{-1}\&Pos(q) \vee (q^{-1} \to p_0^{-1})$ is inconsistent given $p_0 \to q$, inverting both sides of the disjunction.

Rather, I could have on the left: $(Pos(p_0) \to Pos(q)) \vee (Pos(p\&q^{-1}) \to Pos(p)\&p_0^{-1} \to Pos(p))$

I.e. p_0^{-1} is possible but alone is never positive. p and p_0 would private each other, a contradiction.

Given $Pos(p_0)$ always true (axiomatic) for perfection, I have either side of "$Pos(q) \vee Pos(s)$" as before. Then virtue p_0 cannot be present on the left side (as in section [2.2b] above), neither should any virtue appear on both sides: The positivity of p_0 is most certainly decided. (As it is necessary by axiom.) God may not exemplify p_0^{-1}. (Perfections do not include p_0^{-1}.)

Then the positivity of $Pos(q) \vee Pos(s)$ is no longer decidable. (The positivity of q becomes as necessary as that of p_0.) So I may not infer one side from virtue:

i.e. given $Pos(q) \vee Pos(p\&q^{-1})$ I find on the right side $Pos(p\&q^{-1}) \to Pos(s)$ only and not $Pos(q^{-1})$.

So $Pos(q) \vee Pos(s)$ as required. (The disjunction is freely decidable, q and s may appear to private each other.)

2.4) Inconsistency If Virtues Are Not Closed

Were I to posit any mere positive property q entailed from the (closed) set of virtues, I would include and entail a suitable disjunction in that predicate q in the set of positive properties:

I.e. given: $p_0 \to q$, I may find some virtue p present on both sides, with p a conjunction of p_0 and a suitable p_1 say. Starting with:

$Pos(q) \vee Pos(p_1\&q^{-1}) \to s$

I then have with $p=p_0\&p_1$ both $p \to q$ and $Pos(p\&q^{-1}) \to s$.

$q^{-1} \to p^{-1}$

$Pos(p\&q^{-1}) \to Pos(q\&q^{-1})$

And $Pos(p\&q^{-1}) \to Pos(p)\&p^{-1}$

I have only inconsistency as a result. For p, a virtue is also now inconsistent. I have no freely decided disjunction; $Pos(q)$ is the only positive outcome.

If it were possible for q to entail p, I would prefer to seek some correction allowing me to state:

$Pos(q) \to Pos(p)$

Pos(p&q^{-1}) → Pos(p). Either side of the disjunction could be made to imply the same virtue.

(Pos(q) ∨ Pos(p&q^{-1})) → Pos(p)

Yet the modus tollens: p^{-1} → (¬Pos(q) ∨ ¬Pos(q^{-1})) is "bogus", inverting both sides of the disjunction: From the modus tollens I entail both sides from p^{-1} which may only be negative and null. Then p becomes inconsistent as the middle for which p^{-1}, as empty, indeed suffices.

Why? p to "perfection" is a necessarily positive property, a virtue which when negated to lack of action makes no sense as a positive property unless it entails without exemplification something strictly and independently positive (and not a consequence of that same virtue negated). There is no sense of performing a logically inconsistent act perfectly, therefore:

(Pos(q) ∨ Pos(q^{-1})) ↔ (p^{-1} ∨ p^{-1}) ↔ (Pos(p) ∨ Pos(p)) are inconsistent; each entailing Pos(p)&p^{-1}.

I conclude that by no means may virtue appear on both sides of a disjunction.

2.5) Some Small Points Remain (No Finer Filter Than Our Own)

What happens in the case where "super virtues" x form a filter operating on virtue: and virtue became relatively as a mere positive property?

Pos(p) ∨ (Pos(x&p^{-1}) → Pos(s))

Pos(p^{-1}) is false, so all positivity in s is derived from that of x, and I assert Pos(x) → Pos(s).

So I have Pos(p) ∨ (Pos(x) → Pos(s))

If Pos(p^{-1}) → Pos(s) holds in any sense then Pos(s^{-1}) → Pos(p). Pos(s) must be necessarily false. Then it is impossible for perfection to hold the property p^{-1} as positive. Then N(Pos(p)) → p for perfection, i.e. for God. Yet only then; N¬(Pos(p^{-1})) → ¬p^{-1}. Logically, x&p^{-1} → x → s and s instead becomes a "super virtue" as x is from a closed set (else if p^{-1} → s, s cannot be positive).

Yet (as Pos(p) is true) p is necessarily positive and so must be Pos(x). But then x would not entail p^{-1} because x is from a closed set of "super virtues" which includes all s.

Then by axiom of virtue, x itself must be part of a reformed conjunction with a "super super virtue" y that entails Pos(p) instead!

I.e. Pos(x) ∨ Pos(y&x^{-1}) → Pos(p)

All positivity in p must be from y and not x, and I immediately find myself in an infinite regress.

Then "super virtues" must belong to the set of virtues and so I have to decide the disjunction:

Pos(p) ∨ (Pos(x&p^{-1}) → Pos(s))

Once again, all positivity in s follows from x; I assert the disjunction is of the form p ∨ s, as well as Pos(p) ∨ Pos(s) where s is also a virtue. Then p privates s and s privates p. Then there is only inconsistency in virtues. Then the set of virtues (an ultrafilter) acting on positive properties is as fine a filter as one may attain from virtue. (It is a maximal filter.)

2.6) Theorem

Given at least one positive property $T \leftrightarrow r\&s$ (in the conjunction of the positive properties r and s from the sets A and B above), that T privates some virtue if it is at all present.

Proof:

$Pos(T) = Pos(r)\&Pos(s)$ for some r and s. I assume $P(Pos(T))$ if there may ever be some T. As $Pos(p\&r^{-1}) \rightarrow Pos(s)$ freely decides $(Pos(r) \vee Pos(s))$ I have $N\neg(p\&r^{-1}\&\neg s))$ or $N\neg(p\&\neg(r\&s))$.

So I have $N\neg(Pos(T)) \rightarrow p^{-1}$ and the sets A and B are otherwise disjoint by Theorem 2.2.

A $\{r, Pos(q) \rightarrow Pos(r)\}$
B $\{s, Pos(p\&q^{-1}) \rightarrow Pos(s)\}$

But for the perfect, I may not private virtue. If $N(Pos(T^{-1})) \rightarrow N(p^{-1})$ were correct; "T" logically privates the virtue in p, as not permitting free exercise between r and s. As p is virtue, $N(Pos(p))$ indicates $N\neg(Pos(p^{-1}))$ and so for God, $N\neg(p^{-1})$. Then $N\neg(\Omega\&\neg p)$ for all Ω including T^{-1}. Then $T^{-1} \rightarrow p \rightarrow T$ (T is inconsistent) or $T^{-1} \rightarrow p^{-1} \rightarrow T$, T is out of scope as is p, both are null.

Now, $Pos(p\&q^{-1}) \rightarrow Pos(s)$ follows from $N\neg(p\&q^{-1}\&\neg s)$ and this arranges to both:

$N\neg(p\&s^{-1}\&\neg q)$, i.e. $Pos(p\&s^{-1}) \rightarrow Pos(q) \rightarrow Pos(r)$

And...

$N\neg(p\&\neg(q\&s))$, i.e. $Pos(p) \rightarrow Pos(q\&s) \rightarrow Pos(r\&s)$

It appears that given $Pos(p)$, $s^{-1} \rightarrow r$ and this is most useful! Yet, what of $Pos(p) \rightarrow Pos(r\&s)$? Is $T = (r\&s)$ also virtue? It cannot be a virtue; for T is certainly null and p is inseparable from its binding to q^{-1}, and by re-arranging to get $p \rightarrow q\&s$, I broke the closure of the set of virtues and necessarily arrived at a contradiction; r&s cannot entail any positive property in this ultrafilter! The converse, that r&s is null, proves that virtues are a closed set (and bound to q^{-1}).

2.7) Theorem

There are other sets of positive properties beside A and B.

Proof:

Given the sets A and B as before, a corollary of a previous argument is that:

Given $Pos(t) \leftrightarrow Pos(r\&s)$, I find $Pos(t^{-1}) \vee (Pos(p\&t) \rightarrow Pos(u))$ for some set of properties u.

I find another application of the virtue in p. In short, the set u is not empty, as by free exercise of p (as in an ultrafilter of perfections) the set of properties in t is certainly not positively exemplified (and is inconsistent) given p and one or both of r and s; the negation in the ultrafilter of t^{-1} must be positively non-exemplifiable and p&t contains or entails some positive u. (See later, on ultrafilters and "solving for virtue in the Godhead".)

I require some third set as in all such "t" so I posit

    X   {p : p a virtue}
    A   {r : Pos(q) → Pos(r)}
    B   {s : given Pos(p) :Pos(q^{-1}) → Pos(s)}

Now the third set C is given by the complement of the symmetric difference: $(\{X, A\} \vee \{X, B\})^c$

So by symmetry I have a K4 group structure.

$(\{X, A\} \vee \{X, B\})^c = \{X, C\}$
$(\{X, A\} \vee \{X, C\})^c = \{X, B\}$
$(\{X, B\} \vee \{X, C\})^c = \{X, A\}$
$(\{X, A\} \vee \{X, A\})^c = \Omega$, etc.

(Ω, the set of all positive properties.)

I state given X and one of A or B then C is empty or inconsistent.

The symmetric difference of $\{X,A\} \vee \{X,B\} = T = \{A,B\}$ is the set of all positive properties held necessary from both A and B that private (or over-constrain) some set of virtues p in X. I.e. given p: $q^{-1} \to s$, and by inferring an analogue to Pos(r)&Pos(s) as before; Pos(p&q^{-1}) → Pos(s) may be rearranged to Pos(p) → Pos(q&s) → Pos(w in T) which infers Pos(w^{-1} in T^c) → Pos(p^{-1})). No such collective as this set T or its complement T^c may both be consistent and positive given p and p&q^{-1} → s as it then (inconsistently) privates that virtue p in X.

Any w^{-1} in T^c privates virtue and cannot be positive. It also becomes inconsistent as w (in T) may not be positively exemplified. Then the positive sets T or T^c are not in this ultrafilter over the (locally closed) set of positive properties {p, r, s}.

T is a subset of the set of all perfections "Ω" (all positive properties) and is exemplifiable: yet to rest on the inactive alternate $T^c = \{w^{-1} : w \text{ in } T\}$ is found positive instead (as if by a simple choice). T contains predicates "w" that are not positively exemplified. I find I have positively unexemplified properties w^{-1} in T^c instead. (I.e. Pos(w^{-1}), w^{-1} in $T^c = \{w^{-1}: w \text{ in } T\}$). That is, unless Ω is limited purely to the union of X, A and B. In that case, C is empty (privating virtue) and X = $(A \vee B)^c$ and A = $(X \vee B)^c$ etc. I gain a K4 group (as if embedded or "sat in place" in Ω).

Those properties w^{-1} in T^c (T^c, being the set C above) are locally "out of scope" as that K4 structure in {X, A, B} is closed. (I am not given such a w^{-1}, a positive property in that limited set Ω after all.)

I find "necessary rest" (inconsistency, or undecidability) upon $w^{-1} = (r\&s)^{-1}$ which appears "out of scope". I.e. "rest" given the liberty of "r" or "s" as before without need for any virtue but that same p, including that pure virtue of liberty itself. ("Liberty" in this case is seen to be a principal or even possibly as a "primitive" virtue (indivisible, atomic), entailing rest upon any completed act performed.)

If Ω is locally the union of X, A and B (as of the sets p, r, s) then the "liberty" of virtue in p, both freely entailing Pos(r) and Pos(s) from the disjunction $Pos(q) \vee Pos(p\&q^{-1})$ of necessity entails there to be such properties w in T; namely that $Pos(w^{-1})$ holds in T^c, the set $Pos(r^{-1}\&s^{-1}) \leftrightarrow Pos((r\&s)^{-1})$. T appears positive and in Ω, and part of the collection of all positive properties but is not "consistent" with this ultrafilter, as it privates virtue (free exercise between r and s). Now, I assert that there is either a "yet more" maximal (a finer) ultrafilter containing T^c (a nonsense?), or T^c is found in another filter altogether.

Otherwise, $Pos((r\&s)^{-1})$, "r" and "s" together **private** each other. I.e. $Pos(w^{-1})$ or $N\neg(Pos(r\&s))$ becomes $(Pos(r) \rightarrow Pos(s^{-1}))$ etc. If a set of such positive properties as T exists, either one of T or T^c is in the filter or both are inconsistent. Yet holding T as necessary privates virtue, by preventing **free** exercise between r and s; however, T^c privates virtue in p directly given the disjunction between r and s.

Without requirement for virtue, by $N\neg(Pos(r\&s))$ only one of r and s remains positive, either side of the disjunction remains positive and decidable, and consequently T^c is always positive. I am not given free exercise of T or T^c, as if either were a positive property, as T privates free exercise between r and s. T^c may be decidable given the virtue in p and some positive property in T. I.e. $Pos(w^{-1}) \vee Pos(p\&w) \rightarrow Pos(z)$, I will now find another filter.

The disjunction may have properties that are exemplified (as r or s) and also some that are positively unexemplified. I split r and s into two parts and modify my sets A and B above to:

X   {p : p a virtue}
A $\{r, u^{-1} : Pos(q) \rightarrow Pos(r\&u^{-1})\}$
B $\{s, v^{-1} : \text{given } Pos(p) : Pos(p\&q^{-1}) \rightarrow Pos(s\&v^{-1})\}$

With regard to A, clearly $q=r_1\&r_2 \rightarrow r_1\&r_2^{-1}$ is false since $r_2 \nrightarrow r_2^{-1}$ and $N\neg(r_1\&r_2\&\neg r_1)$ is not equivalent to $N\neg(r_1\&r_2\&\neg r_2^{-1}\&\neg r_1)$ unless r_2 is **virtue**; for then $N(Pos(r_1)\&r_2^{-1} \rightarrow Pos(r_1))$. Therefore $q \rightarrow r\&u^{-1}$ makes sense. (There is already some sense of liberty in the set of p.)

Now I note I still have an empty middle T as before so I may state $N\neg Pos(T)$ and $N(\neg T)$ for God: or, for all X: $N\neg(T\&\neg X)$. As X is devoid of all positivity, I may substitute for it all virtue negated in $\neg L(G)$. I.e. $N\neg(L(G)\&T)$ or $N\neg Pos(p\&r\&u^{-1}\&s\&v^{-1})$ which rearranges to $N\neg Pos(p\&r\&s\&\neg(u\&v))$.

Then I have found the disjunction $Pos(r\&s)^{-1} \vee (Pos(p\&r\&s) \rightarrow Pos(u\&v))$ or the axiom of virtue.

So given $Pos(w^{-1})$ in T^c, I have a well-formed disjunction in the properties $(r\&s)^{-1}$ excluded from {p, r, s} as from before (bound to some other set of virtues in the p so as not to private virtue).

Now I have $Pos(u^{-1}) \vee (Pos(p\&u) \rightarrow Pos(v^{-1}))$ with the properties of the middle now found in $Pos(u\&v)$. (What is not positively rested upon in q is positively rested upon in $p\&q^{-1}$.) I simply find that "r" and "u" are disjoint sets as are "s" and "v".

In looking for positive properties to decide $(r\&s)^{-1}$ I have found their "complement" in u&v.

And I am done. (I have with $(r\&s)^{-1}$ and u&v extended the model beyond the closure of Ω.)

Maximum of Extent with Minimum of Privation

I think of u&v not as an (always negative) consequence of r&s without positivity from virtue, but as a positive and disjoint set of positive properties entailed of p&r&s in its own right, an application of virtue in p. The opportunity to introduce other positive properties in u&v (as if q = $(r\&s)^{-1}$ were separately under consideration) arises on including the disjoint set $(r\&s)^{-1}$. Those properties, now in their own setting, do not private virtue: even that prior virtue in p permitting free choice between Pos(r) and Pos(s). Such a set u&v is plausible (and positive).

As they would otherwise private each other, neither "r" nor "s" is a virtue and by including C = $(r\&s)^{-1}$ in the filter, I extend the set of positive properties Ω beyond those of p, r and s (namely those of {X, A, B}). I already have r&s as a plausible "q" in Ω, so by including $(r\&s)^{-1}$ I extend the system without privating p.) By symmetry, I can include those properties that private each of those three {p, r, s}. I have a clear degree of extension over the set of virtues in X, or positive properties in {X, A}. The set of perfections Ω then properly contains the set of all "classical" perfections (as {X, A}) and I have now extended it to at least {X, A, B, C}.

There may be found many variations, each self-similar, freely exercising in Ω the elements of Ω (any conjunction of positive properties for a "q" may be taken from Ω, yet also Ω may be chosen as well), but this variance is to be expected in the Holy Trinity and for any perfect being with all such properties. I infer a morphism from the set of all positive properties into the set of all positive properties consequent of virtue. (In fact, I will find an embedding.)

The set T in Ω is positive but not so T^c (as constituting either {X, C} or {X} alone) because T also belongs in the classical ultrafilter of perfections as made in q (see for example, Anselm's argument for a non-principal ultrafilter of perfections) and so T may not be empty! So in the set of all positive properties Ω, $(\{X, A\} \vee \{X, B\})^c = \{A, B\}^c = T^c = \{X, C\}$ where C is the same set as before C = {$(r\&s)^{-1}$, u&v : Pos(p&r&s) $\rightarrow$ Pos(u&v)} i.e. given Pos(p): Pos($(r\&s)^{-1}$) $\vee$ Pos(u&v)}. T^c is not classically a set of "positive properties", but consists of all w in T now "positively unexemplified". C is overconstrained as the middle of Pos$(r\&s)^{-1} \vee$ (Pos(p&r&s) $\rightarrow$ Pos(u&v)).

When Ω is limited to {p, r, s} without the local inconsistency of C, I find a simple K4 structure with the operation of addition reduced to: A + B = $(\{A\} \vee \{B\})^c = \{A, B\}^c = \{X\}$ (C privates virtue).

All I require to span Ω is some suitable "p" from X and to identify u&v wisely. (See also chapter four.) And then no p in X is required to be excluded by privating any positive property!

Note that the set $\Omega = \{X,A\} \cup \{X,B\} \cup \{X,C\}$ is the identity element of the K4 group formed by the operation $(A \vee B)^c = A+B$ upon the sets $[\Omega, \{X,A\},\{X,B\},\{X,C\}]$ which form it.

As the set of positive properties Ω is closed in X, A and B, then because X is closed and A, B disjoint with X and each other, the set $[\Omega, X, A, B]$ forms a K4 group with A + B = X (C is not empty but it is inconsistent) under that same operation. These sets each form ultrafilters, but within a container set; an equivalently "maximal" filter, also an ultrafilter (that is no finer), properly containing this filter. (The indexing set is not quite the same; as a result the power set (the set of every subset of that collection) of Ω may be indexed in a modified fashion.)

2.8) Theorem

I assert "No positive property may entail a privation of virtue in any conjunction".

Given a virtue p bound to a predicate q^{-1}, I may write $Pos(p\&q^{-1}) \rightarrow Pos(s)$ say.

The statement of the theorem is rephrased to become $s^{-1} \rightarrow p^{-1}\&q$. Given $N\neg(Pos(p^{-1}))$ I then find that $N(p^{-1}\&Pos(q)) \rightarrow N(Pos(q))$. q becomes the conjunction of properties found positive.

Proof:

Though $\neg Pos(x) \leftrightarrow Pos(x^{-1})$ are equivalent, they are not strictly equal. In reality, x and x^{-1} are two completely different predicates. The statement $\neg Pos(x\&y) \leftrightarrow Pos(x\&\neg y)$ could be logically correct, but it requires at least one of x, $\neg y$ positive. (Here, $\neg y$ on the right-hand side. Then x&y with x and y negative is still negative.) $Pos(\neg x\&y)$ is correct in that case and appears to be just as separable, but $Pos(\neg x\&y)$ is become as logically sound as $Pos(x\&\neg y)$, and yet these are not "inverse" to each other. I employ the inversion of predicates to inaction in order to separate x and y as variables, rather than properties. Given $\neg Pos(x\&y)$ I immediately infer (of the perfect) $Pos(\neg x\&\neg y)$. (Yet not $Pos(\neg x\&\neg y) \rightarrow \neg Pos(x\&y)$.) Enforcing the predicate of action, I permit a distributive law. I am not restricted to the statement $\neg Pos(\neg x\&\neg y) \rightarrow Pos(x\&y)$.

When a predicate is positively exemplified, there is a formal statement $Pos(y)$ that may be phrased positive only if y^{-1} is a negative statement. Then, every positive or negative statement in the filter may be rephrased positively in y as $Pos(p\&q^{-1}) \rightarrow Pos(y)$.

The statement $Pos(y^{-1}) \rightarrow Pos(p^{-1}\&q)$ cannot be phrased as positive entailing positive without privating virtue; and by a defining condition of a virtue, I state that $\neg P(Pos(p^{-1}))$. It is simply an observation that no positive property may entail such a privation, it is logically indefensible to write $Pos(y^{-1}) \rightarrow \neg Pos(p)$ as much as it is so to write $Pos(y^{-1}) \rightarrow Pos(p^{-1})$. That property (an inversion of virtue) p^{-1} cannot ever be positive. Then **nothing** positive is consequent of y^{-1}, and yet I must find of necessity $p^{-1}\&Pos(q) \rightarrow Pos(q)$. The extent of the ultrafilter in q will not stretch to include $Pos(y^{-1})$ or even $Pos(p^{-1})$, (by condition U4). The filters each side are disjoint.

However, $y^{-1} \rightarrow p^{-1}\&q \rightarrow q$. Essentially, negative only entailing negative, the positive in "q" is wholly disjoint. Then $Pos(p^{-1})$ is effectively empty! $p^{-1}\&Pos(q)$ is equivalent to $Pos(q)$ alone.

Then y^{-1} can never private that q; else the right-hand side would become strictly negative and by re-arranging, $\neg Pos(y^{-1})$ gives $\neg Pos(y^{-1}) \rightarrow \neg Pos(q)$; an inconsistency. If $Pos(q) \rightarrow Pos(y^{-1})$ in any case, then $Pos(y^{-1})$ permits no $Pos(y)$ to be entailed of $p\&q^{-1}$ and this is a contradiction.

As $Pos(y^{-1}) \rightarrow p^{-1}\&Pos(q)$ is false, for the result to remain positive, I require $N(Pos(q))$ since I always have $N\neg(Pos(p^{-1}))$. Yet it is still logically correct to state $N\neg(p^{-1}\&Pos(q)\&\neg Pos(q))$.

$N\neg(Pos(p^{-1}))$ entails $N(Pos(q))$ from $N(p^{-1}\&Pos(q))$. The statement of the theorem is satisfied.

So, $p^{-1}\&Pos(q) \rightarrow Pos(q)$ only, as was expected.

Given $N(Pos(p))$ axiomatic, I find the disjunction $r \vee s$ only! (Here this was stated as $q \vee y$.)

Of note is that although $y^{-1} \to p^{-1}\&q \to q$, it is not logically correct that $Pos(y^{-1}) \to Pos(q)$. It is, however, still correct that $y^{-1} \to p^{-1}\&q$. Then of course, $p\&q^{-1}$ is no virtue but is still positive. y is still a positive property and does not private a virtue.

A Caveat On Logical Inference (2.8.1)

I may safely state that $v \to p^{-1}\&q$ is a correct logical inference; I may also safely assert that $v \to p^{-1}$. However, it is **not strictly correct** to assert that q is "*positively*" privated by v: as logically $\neg v^{-1} \to p^{-1}\&Pos(q) \to Pos(q)$ (requiring the Pos operator). Therefore, the sets v and q are truly (logically, not using the Pos operator) disjoint and the virtue in p is found "privated" so that v privates p. The inference $v \to p^{-1}\&q$ then entails that q and v^{-1} are disjoint sets of positive properties. (No q can follow from p^{-1} as surely as no "positive" v can private virtue.)

Then that is to state $N\neg(v\&p\&\neg q)$ is also $p \to v^{-1}\&q$. As p is a closed set, $v^{-1}\&q$ is not consistently positive. (Or else $p \to v^{-1}\&q \to v^{-1}$ and then $v \to p^{-1}$.) Then as v cannot otherwise private virtue, v must be a "negative" statement and therefore $Pos(v^{-1})$ is correct. Similarly, $p\&v \to q$. The property v is no virtue and so is negative, and p as virtue is closed. Given logically that q is positive and v negative, the sets p, v^{-1} and q must be disjoint in positive properties.

This does not affect the results of section [2.1] – virtue may be "evaluated strictly positive" and p^{-1} becomes "empty" (as Pos(p) is necessary for perfection). Simply evaluate the terms in conjunction afterwards to uphold the necessity of the positivity of all virtue as held by axiom. (Any Pos(x) must and will always remain positive over both sides of a disjunction for perfect consistency. However, predicates are not found so in the dialectic paradigm but are instead "judged" as if predicated positive and not either side; so that $Pos(r) \lor (Pos(r^{-1}) \to Pos(s))$ etc.)

If $x\&y \to x$ holds with $Pos(\neg x)$, I could entail $Pos(\neg x) \to Pos(\neg x\&\neg y) \to Pos(\neg x)$ but not $Pos(x\&y) \leftrightarrow Pos(x)$. Every statement would then be **equivalent**. This results in a "modal collapse" as may Gödel's argument[1], yet even then the disjunction partitions the sets into classes that will not collapse. There is always a least, principal and necessarily positive element in each class.

"Modal collapse" occurs as every possible positive statement (and indeed their negations) become necessary, as if they were all collected positive. (Ω becomes somewhat arbitrary.) In theories of "all possible worlds" this is a real concern; the concept of a "maximally great being" or of any "God" across all possible worlds is one that invites such a collapse. All predicates become inconsistent but (supposedly) for the one predicate of necessary existence also become equivalent. (Without the partition of Ω into the sets of the K4 form (or octal) on the predicate of liberty, modal collapse may be argued a real consequence of Gödel's modal proof.[1]) Existence (as liberty), then, becomes that essence and is also the principal element.

2.9) Ultrafilters

An **ultrafilter** is a set of subsets of a collection where in each case the criteria of inclusion are chosen so every member of the ultrafilter is "large" compared to its complement. Then the filter is considered to be under logic that is "true almost everywhere". (Note that the static set of all positive properties Pos(q) (q as before) form an ultrafilter.)

The definition for an ultrafilter U on an index set I is as follows.

[1] Sobel, in J. J Thompson (ed.), *On Being and Saying: Essays for Richard Cartwright*, p241-61.

U1) For any subset S of the indexing set I, either S or the complement of S, (I \ S) belongs in the ultrafilter U, but never both.

U2) If sets X and Y are in U, then so is their intersection X ∩ Y.

U3) The empty set ϕ is never a member of the ultrafilter U.

U4) If X is a set in U, then so is every set in I that is a superset of X. (All sets that properly contain X as a subset.)

These sets S, X, Y etc... are subsets of the indexing set I, and are not to be confused with the set upon which the index applies.

An ultrafilter is a **maximal** filter. That is to say there is no other filter properly containing an ultrafilter on an indexing set I. (Such a filter would be termed "finer" than the one it contains.)

Independent of the disjunction, the classical (static) set of positive properties forms a (non-principal) ultrafilter in the set of all positive properties. Of interest are the subsets of those properties within $q \lor p\&q^{-1} \to s$ (particularly those on the right-hand side of that disjunction).

Concerning the left-hand side, even within the power set of Ω of all possible (as any q) sets of positive properties, some properties are consequent from each q and others not so. I make an entirely new filter (one that is principal, and which is not Anselm's filter, being restricted solely to those positive properties in some subset of Ω), on every predicate consequent from some q. A positive property "r" is in this filter if $q \to r$ and an "s" is not in the filter if $q \not\to s$.

The "q-filter" over all q is possibly infinite and as it is non-principal may possibly contain the (free) filter of every co-finite set of positive properties. This has the consequence that unless $q=\Omega$ there is always an element "s" in Ω not entailed from q (a set disjoint from q) found in that finite set not in the filter. (Then there is always such a disjunction of virtue unless $q=\Omega$.)

Whether Ω is infinite or not, I may consider the set "q" also, as infinite or finite. I will only examine finite subsets "s" of Ω with respect to virtues; and then only co-finite sets in the complement of each "s". If Ω be finite, then without loss of generality I will take "co-finite" to indicate "complement of a finite set". If both sides of the disjunction are infinite, it complicates matters a little, but has no dire consequences. For simplicity, I restricted the "cut" of the q-filter in Ω between q and $p\&q^{-1}$ to be between co-finite and finite, but the same arguments hold for such disjoint infinite sets. (Or two complement finite sets, using a restricted Ω.)

Concerning the important parts with respect to virtue, I examine the criteria for an ultrafilter.

U1) Holds. Either sets of positive properties (as "s" above not in q) that are consistent in Ω are entailed from virtues or they are not.

Those predicates y in the indexing set found *in the ultrafilter* are positive or not. The indexing set of all y is **limited** to the set excluding all those in q (and q&y), such that $q \not\to y$ and $y \not\to q$.

Given a virtue Pos(p), and also requiring everything then entailed to be positive:

$Pos(p\&q^{-1}) \rightarrow Pos(e)$

Either $Pos(p\&q^{-1}) \rightarrow Pos(p\&y^{-1}) \rightarrow Pos(e)$,

or $Pos(p\&q^{-1}) \rightarrow Pos(p\&y) \rightarrow Pos(e)$.

For some suitable positive property "e". (Cf. a "principal element" of "rest".)

I.e. given $Pos(p\&r^{-1})$: $Pos(e\&y) \lor Pos(e\&y^{-1})$. (Either $q \nrightarrow y$, $y \nrightarrow q$ or $q \nrightarrow y^{-1}$, $y^{-1} \nrightarrow q$.)

I must preserve every property (as "e") entailed from the use of p. (I must not private p given y or its inverse.)

So in the ultrafilter, as from $p\&q^{-1}$ either y or y^{-1} may be entailed from virtue.

Now consider the logical conjunction of all positive properties (none privated by virtue) entailed from $Pos(p\&q^{-1})$ say, s_i, where $s_i = s_1\&s_2\&s_3\&s_4\&...$

s_i is also a positive property entailed from $Pos(p\&q^{-1})$.

Now consider those that are not entailed from $Pos(p\&q^{-1})$ say $v_i = v_1\&v_2\&v_3\&v_4\&...$

Then some additional virtue(s) in p allows $\neg Pos(v_i)$ to be freely exercised. (I.e. These v_i are not in the ultrafilter of s_i but are also entailed from $p\&q^{-1}$.)

Then $Pos(p\&q^{-1}) \rightarrow Pos(s_i)\&\neg Pos(v_i) \rightarrow Pos(s_i\&v_i^{-1})$.

And this last conjunction is also a positive property entailed from $Pos(p\&q^{-1})$.

Therefore, the negation $\neg Pos(s_i\&v_i^{-1}) \leftrightarrow Pos(s_i^{-1}\&v_i)$ is not positively exemplified, is not entailed from p and is excluded from the ultrafilter.

A set of positive properties (as in the indexing set) is in the ultrafilter or not. This also agrees with those positive properties not exemplified.

That said, $s_i^{-1}\&v_i \rightarrow \neg Pos(p\&q^{-1}) \rightarrow p^{-1}\&Pos(q) \rightarrow Pos(q)$ as required.

U2) Holds. If $p\&q^{-1} \rightarrow s$ and $p\&q^{-1} \rightarrow t$ both hold, as properties following after $\neg Pos(q)$ with p a virtue, $P(Pos(s\&t))$, both s and t may be exemplified at the same time as p privates no positive properties. Equivalently, if $p_1\&q^{-1} \rightarrow s$ and $p_2\&q^{-1} \rightarrow t$ then as $q \nrightarrow s$, $s \nrightarrow q$ and $q \nrightarrow t$, $t \nrightarrow q$, I immediately have $q \nrightarrow s\&t$ and $s\&t \nrightarrow q$. Then I may easily form (by axiom of virtue) the disjunction of: $p_1\&p_2\&q^{-1} \rightarrow s\&t$, which is also in the ultrafilter.

U3) Also Holds, for any virtue (or liberty) would entail at least a positive property. Given even liberty itself, the finest element in the ultrafilter as a choice of p; I have given $Pos(p)$: $Pos(p\&q^{-1}) \rightarrow e$, (e, a "principal element") and resting is (or must be found) positive.

U4) Also Holds. Liberty implies $Pos(L(G)\&q^{-1}) \rightarrow e_s$ (as an infimum). When given any virtue p such that $Pos(p\&q^{-1}) \rightarrow Pos(s)$ every element entailed from the free exercise of virtue (even by liberty L itself) that lies between Pos(e_s) and Pos(s) is entailed from a virtue. By axiom of virtue every set of positive and workable acts is implied from the set of all virtues which are exercised freely.

Given p_1: if $p_1\&q^{-1} \to t$ say, if "t" is in the ultrafilter then so are all **positive** properties implied from s where $p_2\&q^{-1} \to s$ and $s \to t$ (i.e. the set of s is a superset of that of t: Pos(s) must hold with p_2 closed ($p_2 \not\to s$) and q^{-1} negative) there is then a virtue p_2 such that Pos($p_2\&q^{-1}$) → Pos(s).

(To act positively (with "s") upon the privation of a positive property "q" must **always** be positive; hence there is always a suitable "virtue" "p_2" for perfection.)

Now the axiom of virtue states the set of all s must be taken from $\Omega \setminus r$ in all r where either $q \to r$ or $r \to q$. On assuming that this applies to t, clearly given $s \to t$, it should not be possible to entail $s \to t\&q$ in the ultrafilter if the set of t is closed with all such s. Note (t&q) is then in the middle, which is empty. So, $s \not\to q$ and separately, if $q \to s$ then $q \to t$ – also a violation.

Then by axiom of virtue there is a unique virtue p_2 such that Pos($p_2\&q^{-1}$) → Pos(s) → Pos(t). Now, as Pos(s) → Pos(t) by that condition of uniqueness I must also have $p_2 \to p_1$.

Then there is possibly (in fact certainly) some $p_2 \to p_1$ (such that $p_2\&q^{-1} \to s$) so that the virtue p_2 suffices (so that condition (U4) holds), that if $s \to t$ (s, t not virtues) and "t" is in the ultrafilter then so is "s". (If it is possible for "s" to be in the filter then it most certainly is in the filter.) I.e. a positive act that will entail more positive outcomes than another (entailing a subset of the same) is the exercise of a more perfect (finer) virtue etc. I only require some Pos(s) with $s \to t$.

Given it is positive to act with "t" (so then it is not positive to rest on q^{-1}), I may state alternatively the pure virtue of liberty L(G) suffices to show (L(G)&$q^{-1} \to p\&q^{-1} \to s$) and $p_2 \to p_1$. (Given P(Pos(s)) or N(Pos(s)) by axiom, resting on all such virtue p_2 is not positive if q is not exemplified.)

So, L(G)$^{-1}$&p_1^{-1}&Pos(q) $\vee$ Pos((L(G)&q^{-1})&p_1&q^{-1})

And Pos((L(G)&q^{-1})&p_1&q^{-1}) → Pos(p_2&q^{-1}) → Pos(s) → Pos(t) where L(G)&$q^{-1} \leftrightarrow p_2$&$q^{-1}$.

Then the outcome is positive. (Liberty is freedom for perfection to exercise any positive property that may be exercised. I.e. given Pos(s), there exists L(G)&q^{-1} with L(G) equivalent to a certain virtue p for exercising s. I have some apparent form of linearity.)

There is but one condition: that there exists some p such that $p\&q^{-1} \to s$. The virtue "p_2" is finer than "p_1" – I would consider both "t" and "s" positive, but p_2 substituted for p_1 is "more virtuous" than p_1 alone ($p_2 \to p_1$). Such a p_1 cannot then private that s; taking its place with p_2 is positive.

Then virtues form an ultrafilter. (They are a *closed* set of positive properties and behave as if they were the whole set of positive properties.)

Given Pos(s) → Pos(t) there then truly exists some unique $p_2 \to p_1$ as above. The ultrafilter is generated by divine liberty "L(G)", where Pos(L(G)&q^{-1}) → Pos(e&q^{-1}). Perfection is not

[1] Sobel, in J. J Thompson (ed.), *On Being and Saying: Essays for Richard Cartwright*, p241-61.

required to act, but does so at liberty. "L(G)" may separate out any subset of virtues for a positive outcome (here, "rest").

Then the sets {X,A} etc. each form an ultrafilter.

The set of positive properties in {{X,A}, {X,B}, {X,C}} (or even only one of the three) properly includes the set of all "q" (those positive properties Pos(q) in the static ultrafilter able to be possibly exemplified at any given time). The extension by virtues allows a greater set of positive properties; the Trinity is "better than" and "greater in perfections" than a static theistic set of perfections as by Anselm's argument. The Trinity is then more perfect.

As shown later in Chapter Four, the set of all positive properties Ω may be indexed appropriately in seven disjoint and closed subsets. Each subset forms a similar ultrafilter and then, as a result, there is no "s" in any one of those subsets that may imply such a "t" outside of that same filter on that subset. Given $s \rightarrow t$, it follows that "s" belongs in the very same class in the partition of Ω and it is then impossible to break the closure of each indexing set (equivalence class) in this manner, by positing the existence of such an "s". No "r" from any other set in Ω may also imply this "t", since that would indeed break the closure of the subsets/classes.

2.10) The Ultrafilter Is Principal

A principal ultrafilter is simply an ultrafilter where every set in the filter contains a particular element. The intersection of every set in the filter is equal to this one element – the "principal element" or "generator" of the filter.

I simply need to show that every set entailed by the application of virtue must always contain just one necessary positive property. It is then an observation that God is not a slave to act and may freely rest. Then, given $p\&q^{-1}$ always in the ultrafilter, it seems the only positive property always in action is itself p, a virtue. Every other positive property appears to fall into the set "outside of the filter" when God rests.

The set of all virtue p, which also entails the virtue of liberty or L(G) as I will call it, simply implies the liberty to act or not so in any situation. Then as virtue is a closed set I must show it entails at least one positive property. (Other than q^{-1} which is negative but for a principal e_x.)

The sets $s = s_1\&s_2\&...\&s_m$ in (with the negations outside) the ultrafilter in [p, r, s] entailed as from $p\&q^{-1}$ lead me to note that the filter may act on any closed set of positive properties in s. I may choose a subset of the s_i from s and entail this from God's free choice of virtue, His liberty L(G): Every subset of s with the remnant $v^{-1} = (v_1\&v_2\&...\&v_n)^{-1}$ must also follow from a virtue p.

$L(G)\&q^{-1}$ entails every subset of the s_i (and v_j^{-1}) as positive and in the ultrafilter, as following from $p\&q^{-1}$.

Given that set p which implies L(G), I may entail any subset of the s_i in s and even drop all, leaving only one virtue (liberty) remaining at rest. (Every s_i becomes a v_i^{-1}.) Then liberty selects those things (in s) God may rest upon (as v^{-1}). The outcome is still positive. I effectively choose a different set of virtues (a subset of the former p containing L(G)), that merely entails the subsets in s and v^{-1}. There is a definite sense of linearity.

If L(G) is the generative element in the ultrafilter of virtue (the essence of "God-likeness" as in Gödel's argument[1]), and is effectively a substitute for the results of the conjunction of every applied virtue: then the ultrafilter is not empty in positive properties given some principal $p=e_p$ only, but instead rests upon the nature of being "God-like" or having necessary existence. God, whilst at rest, does not vanish: rather He merely remains perfectly "God-like" in all positive properties whilst so necessarily at rest; one positive property always remains, God's essence (God-like or necessary existence in Gödel's terms) but in my own terms here, "rest". (For then, q is the conjunction of every positive property.)

L(G) acts always on the set of **all** virtues, "God-like" liberty is in view. L(G) does not "strictly" substitute for any p, rather L(G) makes a cut in the set of every virtue that would entail the minimal necessary subset of p required by axiom of virtue, including $L \& e_p = e_p$ itself.

So, virtue forms a principal ultrafilter and therefore so does each set entailed from the disjunction.

Supposedly, each set "x" always contains an "e_x", that rest (or "necessary existence"). Does q itself entail necessary existence? Clearly it must reside on both sides of the disjunction in separate closed sets: God is free to rest even if x appears "empty".

Were there nothing positive entailed from $L(G) \& q^{-1}$ then at least one ultrafilter would contain the empty set. Yet, there is a positive property, "rest", that remains always. God's hand is never forced. What is this positive "rest"? It cannot be the element $(r \& s)^{-1}$ I had found before, valid on both sides of the disjunction (though this is never a virtue, which would lead me to expect it to be that rest!).

What I may state is that $(r \& s)^{-1}$ and u&v never follow from any r or s given p, else $(r \& s)^{-1}$ or (u&v) become inconsistent.

Instead I simply include some single positive outcome "e_x": so that some sense of identity or unity of rest is denoted in each set x, but to be included in a positive fashion entailed from L(G). So, $p \& q^{-1} \rightarrow e_x$ at the very least (with, say, liberty $L(G) \rightarrow e_p$ as p); however, $(r \& s)^{-1}$ and u&v are present as if on both sides yet "not ever in the filter {p, r, s}". (They are not decided, and do not logically follow from either r or s.) Instead, I simply state $p \& q^{-1} \rightarrow e_x$.

Following q and $p \& q^{-1}$, every set x entailed in the resulting conjunctions in the ultrafilter must "contain" an e_x, as e_x is simply "to rest on the outcome": that outcome being x itself. "e_x" is always positive: as x (one would assume) would include the satisfaction (e_x) of a completed act. (Every set in the filter is properly disjoint: the conjunction of all e_x, itself, is a closed set and corresponds to an identity or "zero".)

I.e. Liberty is not restrained when it is exercised; liberty has a real effect when it rests positively. The axiom of virtue states that given a suitable "e_x" there is such a "$p = e_p$". Here, when the best outcome is to rest satisfied (or boast) – then liberty is able to act. Now for the mindbender: does e_x equal e_x^{-1}? Actually no, it does not (every set in the ultrafilter is principal and the disjunction must rest on one side or the other).

Liberty becomes the common property of every applied conjunction of virtue, common to every active set of virtues in "p". Then e_p is a principal element in the ultrafilter restricted to virtue for its indexing set. (Everything entailed from such virtues bound to q^{-1} is positive. Virtue itself forms an ultrafilter.)

Given the "set of all virtues" is in the ultrafilter of virtues, liberty L may select out those virtues that do not act (rather than exchange these inactive virtues in p to be entailed as if they were terms in $q\&p^{-1}$ – acting on the other side of the disjunction). Liberty reduces virtue from the conjunction over the "set of all virtues" to that of the minimal set required (p) in order to entail whatever set s is entailed from $p\&q^{-1}$ by axiom of virtue. (A necessary minimum of virtue.) This is done with a simple "cut", an act of liberty.

Liberty e_p entails a set of virtues p_1 in action and some set p_2 not so, so that logically it appears legal to write: $e_p\&p_1\&p_2 \rightarrow e_p\&p_1\&p_2^{-1}$ and as no positive property may entail any privation of virtue, this is equivalent to the legitimate inference $e_p\&p_1\&p_2 \rightarrow e_p\&p_1$. (God, will never exemplify such privation p_2^{-1}: i.e. $Pos(e_p\&p_1\&p_2) \rightarrow Pos(e_p\&p_1)\&p_2^{-1} \rightarrow Pos(e_p\&p_1)$ only.)

p_2 could be formalised as the "set of all virtues" minus that set containing p_1 and e_p, and is then not itself in the filter as it is without e_p. Hence there exists such a set in the filter $e_p\&p_1$.

This is equivalent to every conjunction distributing into those made possible on its arguments. Liberty simply formalises this with a principal element. (Virtue acts out of freedom only!)

So, the quasi-legitimate $e_p\&p_1\&p_2 \rightarrow e_p\&p_1\&p_2^{-1}$ is in logic (of action) equal to:

$N\neg(e_p\&p_1\&p_2\&\neg e_p\&\neg p_1\&\neg p_2^{-1})$ which is only valid for p_2 a **virtue** (p_2^{-1} is then null).

Note I have $N(Pos(e_p)\&Pos(p_1)\&Pos(p_2))$ since p_1 and p_2 and L(G) are virtues.

$N\neg(e_p\&p_1\&p_2\&\neg e_p\&\neg p_1\&\neg p_2^{-1})$ is then also $N\neg(e_p\&p_1\&p_2\&\neg e_p\&\neg p_1)$ or $e_p\&p_1\&p_2 \rightarrow e_p\&p_1$.

The statement is sound in that p_2^{-1} is not ever positive so may not be entailed from $e_p\&p_1\&p_2$. Only $e_p\&p_1$ is entailed: the term in p_2^{-1} must be absorbed into the antecedent p_2 (or vice-versa). This absorption is quite legal but must be carefully applied (for God, as for the perfect).

Then in every case it must be found that liberty itself is the principal element of virtue (or find a predicate of "rest" on any chosen positive outcome – simple for the omnipotent), and I write that this liberty L(G) may substitute for e_p. Then, embedded within the disjunction I have the freely decidable $e_r \vee (e_p\&e_r^{-1} \rightarrow e_s)$. Now, this notation is insufficient to describe that which is extant, as the disjunction is freely decidable and e_r^{-1}, technically, cannot be formalised as it is already a predicate of "rest" – its equivalent is necessary existence, or NE (the one property God is not free to indefinitely "rest" upon inactive). I must modify my disjunction to only those predicates which exit the disjunction on choosing each side, where the extant predicates appear on whatsoever side liberty is found to act, for I may rearrange from:

$(e_p\&e_s^{-1} \rightarrow e_r)$ to $(e_p\&e_r^{-1} \rightarrow e_s)$ or even to $(e_p \rightarrow e_r\&e_s)$.

Then I use the following altered notation $\bar{\vee}$ to denote that the disjunction has been rewritten with liberty acting on both sides, as if both were extant, for instance the above

$e_r \vee (e_p \& e_r^{-1} \rightarrow e_s)$ becomes on either side:

$$\left(e_p \& e_s \rightarrow e_r\right) \bar{\vee} \left(e_p \& e_r \rightarrow e_s\right)$$

And the minimal (principal) element of e_r now appears both sides (and then Pos(p&q^{-1}) entails the required Pos(q^{-1}) or rather Pos(e_r) with the right hand side extant). I may similarly write:

$$\neg\left(e_p \rightarrow e_r \& e_s\right) \bar{\vee} \left(e_p \& e_r \rightarrow e_s\right)$$

And it should become apparent that $e_r \& e_s$ is an equivalent to e_p. I may, of course, write that substitution $e_p \leftrightarrow e_r \& e_s$ only if I may rewrite $e_r \vee (e_p \& e_r^{-1} \rightarrow e_s)$ as $e_r \vee (e_r \& e_s \& e_r^{-1} \rightarrow e_s)$. Clearly I have $e_r \& e_s \& e_r^{-1} \rightarrow e_s$ equivalent to $e_r \& e_s \rightarrow e_r \& e_s$ and this appears as required, yet by axiom of virtue, I must evaluate $e_p \leftrightarrow e_r \& e_s$ to be that same **unique** virtue: I must always have $e_r \& e_s \rightarrow e_p$. I also find my antinomy of $e_p \rightarrow e_r \& e_s$ is no antinomy when taken in the least sense of the principal elements, for they are all necessarily in action as is virtue, and no sense of rest is implied unless a predicate of "rest" be equivocated with the properties e_p, e_r or e_s themselves! It is also true that a middle formed on rest is not as rest made in that middle, so $e_p \leftrightarrow e_r \& e_s$ is not the same as $e_{r\&s}$.

So by rearrangement I almost have a "Klein-four group" (K4 group) in the principal elements.

Now, up until now I have been missing the identity element from my K4 group in the principal elements [e_p, e_r, e_s]. In order to find an identity, I must try and decide the disjunction of say, $e_r \vee e_r$. Now, the axiom of virtue fails as there is a middle. Instead, both sides appear to private each other unless this disjunction fixes the very same predicate of L of liberty itself.

Then I should properly write $e_r \bar{\vee} e_r$. In effect I have the disjunction, $e_r \vee e_r^{-1}$ and liberty must be acting one side or the other. So I may write: $e_r \vee L \& e_r^{-1} \rightarrow e_r$. I must also not break the closure of the virtue L, for then I have an antinomy as before. Yet the antinomy should rearrange perfectly, as I will always be able to state that, $e_x \vee e_x^{-1}$ for any predicate, even a virtue! In every case (and especially so) to preserve the positivity of both sides in the case of a virtue, I must write $e_x \vee L \rightarrow e_x$ for all sets e_x. Then I must have $L \leftrightarrow e_p \& e_r \& e_s$ and L becomes my identity element for the K4 group. In fact, to properly show that L is the identity, I require a modal collapse (see later).

+	L	e_p	e_r	e_s	e_u	e_v	$e_{r\&s}$	$e_{u\&v}$
L	L	e_p	e_r	e_s	e_u	e_v	$e_{r\&s}$	$e_{u\&v}$
e_p	e_p	L	e_s	e_r	e_v	e_u	$e_{u\&v}$	$e_{r\&s}$
e_r	e_r	e_s	L	e_p	$e_{u\&v}$	$e_{r\&s}$	e_v	e_u
e_s	e_s	e_r	e_p	L	$e_{r\&s}$	$e_{u\&v}$	e_u	e_v
e_u	e_u	e_v	$e_{u\&v}$	$e_{r\&s}$	L	e_p	e_s	e_r
e_v	e_v	e_u	$e_{r\&s}$	$e_{u\&v}$	e_p	L	e_r	e_s
$e_{r\&s}$	$e_{r\&s}$	$e_{u\&v}$	e_v	e_u	e_s	e_r	L	e_p
$e_{u\&v}$	$e_{u\&v}$	$e_{r\&s}$	e_u	e_v	e_r	e_s	e_p	L

I may form an octal group on the principal elements of each set, {L, e_p, e_r, e_s, e_u, e_v, $e_{r\&s}$, $e_{u\&v}$} as above, where L is become the product or conjunction of all seven other "elements".

The set of all positive predicates then divides into seven disjoint sets which are spanned by the octal. Each element of the octal group e_x embedded in each set becomes equivalent to the rest upon whatever minimal subset is required or chosen for a positive outcome, as the set of virtue in p may be refined to a minimum to permit any amount of positivity remaining in s to be extant, resting all the while on v^{-1} (no different in symmetry as to s) as to what is perfect, all whilst remaining at liberty without modal collapse.

> **Absorption And Modal Collapse**
> Given $N\neg(r_1\&r_2\&\neg(r_1\&r_2^{-1}))$ or $r_1\&r_2 \to r_1\&r_2^{-1}$ I have either $Pos(r_2)$ or $Pos(r_2^{-1})$. Both cannot be positive in the same ultrafilter. One of r_2 and r_2^{-1} must absorb the other (one must evaluate positive and the other not so). By axiom of virtue (2.1a) I have: "$Pos(p\&q^{-1}) \to Pos(s)$ entails that which is positive from $Pos(s)$ only." I must strictly evaluate $Pos(r_2)$ as entailed from $Pos(r_1\&r_2)$ and certainly not $Pos(r_2^{-1})$. Then the term r_2^{-1} is strictly negative and to be dropped unless moved in positivity into another ultrafilter. I may, however, not drop every such term: for writing $Pos(r_1\&...\&r_n) \to r_1^{-1}\&...\&r_n^{-1}$ leaves the right hand side completely negative (and so there must be a least positive and **principal** element "e"). However, there is modal collapse otherwise; for if it were true that $r_1\&...\&r_n \to r_n$ and I also had $Pos(r_n^{-1})$ I would have by the modus tollens $Pos(r_n^{-1}) \to Pos(r_1^{-1}\&...\&r_n^{-1}) \to Pos(r_n^{-1})$ and either every statement becomes necessarily equivalent or possibly inconsistent. This is a "modal collapse".

Effectively the set of positive properties Ω over the disjunction may be any subset of the one perfected maximal set that may apply (as worked by the omnipotent) without relaxing the principal elements within them (there must needs remain a disjunction). Then, any subset of positive predicates may remain in the octal, which may possibly span an infinite or finite set of predicates. Then modal collapse is avoided as no set in the octal shares a predicate or its negation with another. (The caveat is the condition that the axiom of virtue must be upheld.)

Then minimally as for a principal element, $N\neg(e_p\&p_1\&\neg e_p\&\neg p_1^{-1})$ which is properly $e_p\&p_1 \to e_p$.

Then to rest on every possible virtue but one's own liberty in any conjunction is an argument of induction. $(p_1\&p_2\&...\&p_n)\&e_p \to (p_1\&p_2\&...\&p_{n-1})\&e_p$ etc.

e_p is formally to "rest" at liberty, whilst still remaining positive in action. Given any possible conjunction of virtues in p_2, I may with liberty deduce every conjunction p_1 entailing some p_2 is also in the filter by simply choosing to use that same liberty. From the "set of all virtues" I may reduce as before to the two separate cases $e_p\&p_1\&q^{-1} \to s$ and $e_p\&p_2\&q^{-1} \to t$ (by axiom of virtue).

Now if I wish to infer that s&t resides in the ultrafilter I simply take $e_p\&p_1\&p_2\&q^{-1} \to$ s&t, and if $s \to t$ then $p_1 \to p_2$ in a natural and linear fashion (for p_1 is unique) and $e_p\&p_1\&q^{-1} \to s \to t$ as expected.

Likewise if v is not in the filter, then by liberty I need not expect some virtue p_2 to act that would entail v, and if $p_1\&q^{-1} \to s$ then I may safely state $e_p\&p_1\&p_2\&q^{-1} \to e_p\&p_1\&q^{-1} \to s$. I would state that the bearer of virtue is free not to act on p_2, they are at liberty not to.

Yet there must then be some p_3 such that $e_p\&p_3\&q_3^{-1} \to v^{-1}$ unless the predicate v drops out of the ultrafilter of Ω and the octal altogether. The limit is that e_p, as liberty is sufficient to redeem "rest" on $e_p\&q^{-1} \to e_s$ as positive itself. e_p is the generator of the ultrafilter.

Is each "e_x" a virtue or a positive property? Its relationship to virtue is unique, as liberty e_p generates the ultrafilter. "e_x" is certainly a closed set, it does not private any positive property, one e_x being always present in each filter, and without loss, its negation – that God must be always and unceasingly at work forever (or non-existent) makes no sense as an outcome of virtue (and of liberty). Then each property "e_x" has no obvious negation that could be considered "positive" and could not be at rest in the filter from which would (of virtue) follow "e_x" and so entail $e_x\&e_x^{-1}$ (an inconsistency).

Then each e_x or "rest" fulfils the criteria for a virtue yet they are found in sets of positive properties, consequent only of the disjunction formed of virtue. I also find within the octal an identity element L the conjunction (actually the product) of each e_x.

2.11) Liberty Is Free To Rest

Now, liberty L(G) (which makes a cut on the acting sets of virtue) may not equally relax terms within the very same ultrafilter, for each set is closed, itself forming an ultrafilter and though $N\neg(r_1\&r_2\&r_3\&\neg r_1\&\neg r_2)$ appears to be as $N\neg(r_1\&r_2\&r_3\&\neg r_1\&\neg r_2\&\neg r_3^{-1})$, the term in r_3^{-1} is not equal with that of r_3 (but would cause a modal collapse), instead the action of some virtue(s) is relaxed in that same ultrafilter as a consequence of liberty in L(G) (being necessarily non-exemplified), and r_3 will no longer belong in all r. Those predicates effectively "negated" could then reappear in the separate closed set of u^{-1} (consequent of an entirely different virtue), and u^{-1} would not be further affected unless quite separately terms are "relaxed" into action and also then moved into all r in the symmetric application of that liberty, that predicate removed from all u.

However, if in positive properties both $r_1\&r_2\&r_3 \to r_3$ and $r_1\&r_2\&r_3 \to r_3^{-1}$ in the same filter, all becomes inconsistent and there is modal collapse.

Then the virtue p_r remaining in [p_r, r, s] is always a subset of the same virtue p exercised both sides of the disjunction, which p must entail both r, s and both u^{-1}, v^{-1} also, with each of those sets maximised to the whole of their collection entailed from virtue under the ultrafilter, with the virtues relaxed as above never entailing negations within each set as if by liberty (any r so relaxed is free instead to become a u^{-1} etc.). This avoids modal collapse; for if $r_1\&r_2\&r_3 \to r_1\&r_2\&r_3^{-1}$ in the same ultrafilter, both r_3 and r_3^{-1} would be in the same filter, and this breaks condition U1. If then $r_1\&r_2\&...\&r_n \to r_3$ and it is the case that $Pos(r_3^{-1})$ then $Pos(r_3^{-1}) \to Pos((r_1\&r_2\&...\&r_n)^{-1}) \to Pos(r_3^{-1})$ by the distributive law. Every property in S becomes necessary and also possibly inconsistent.

Given any disjunction and the presence of the seven principal elements only, I have a poset and an ultrafilter in all positive predicates; always formed of seven subsets of those sets which are workable by the omnipotent within the same disjunction, which is then maximal (and principal in all positive predicates with the identity of the octal as the principal element). Then I have conditions U1 to U4 held by every octal over that disjunction as follows:

Given the same disjunction with a maximal set of predicates in action (in r and s), liberty may relax any subset of these outcomes by liberty to inaction (as u^{-1} or v^{-1}). There is always a unique virtue for any such arrangement by axiom of virtue. (The same virtue must appear both sides, as if it does not by the application of liberty, the sets in the octal are not disjoint – a result to be shown in [4.8].) Then there is an ultrafilter on virtues (no term relaxed to inaction by liberty may be positive in the ultrafilter of virtue) such that every possible disjunction exercises a subset of the possible virtues in that set made possible by the omnipotent in the same "real-world" setting. (Connected worlds share a disjunction, and thereby an equivalence relation.)

Then there exists a virtue for any such octal over the same disjunction, and virtue may be relaxed to drop out any subset of predicates in the ultrafilters of r and s, or even dropping predicates from the ultrafilters in u^{-1} and v^{-1}. The result will be that as virtues are relaxed equally on both sides of the disjunction, only proper subsets are removed from each ultrafilter and each set in the octal remains disjoint.

Either a predicate appears somewhere in the disjunction or it does not. (There is no inconsistency causing modal collapse.) Then each octal whole simply fulfils condition U1.

Each closed set of positive predicates over the same disjunction, say Ω_1 and Ω_2 are together in union (as with all possible predicates together in conjunction) also valid over that same disjunction, forming seven subsets of the sets common to both (or sets of their union gained through an arrangement by liberty) always a subset of Ω. Then the octal fulfils condition U2.

No octal is empty; the seven principal elements are necessary, as for God's liberty or NE (Necessary Existence), or with the presence of the least (the identity element). No set in the octal can be completely empty, there is never an empty set in the octal at all, unless there be modal collapse and absolutely no liberty. Then the octal fulfils condition U3.

Given any disjunction and the presence of the seven principal elements only: every superset in Ω over that disjunction is also a set of positive predicates workable in yet another octal. If predicates not in an octal are considered as relaxed by exercising liberty or the lack thereof, they may appear as u^{-1} and v^{-1} instead. (Ω may be at rest for the most part.) Then the octal fulfils condition U4.

Then liberty allows for any octal that spans a subset of that which is maximal in all positive predicates over a disjunction: for omnipotence in the octal is workable by God and every subset so. (Even the e_x only.) Liberty, making a cut in virtue, permits rest on any workable disjunction.

Anselm's non-principal ultrafilter on Ω is such that every filter may be maximised as some <r>; then completed in necessary existence as per Gödel's argument[1]. Given any disjunction maximised in r and s by the omnipotent, there will exist a subset of the possible divine virtues that may arrange those predicates (as into action and inaction in seven disjoint sets) that may be evaluated overall at some proper subset Ω_0, simply by relaxing virtue in the ultrafilter of virtue to take a "cut" of liberty, entailing only a proper subset of each set of seven.

[1] Sobel, in J. J Thompson (ed.), *On Being and Saying: Essays for Richard Cartwright*, p241-61.

Chapter Three: Constructing The Godhead From Principle Of Infinite Regression

3.1) Omnipresence

If an omnipresent, omniscient God has a location (be it a "centre" or origin or otherwise, without any condition upon His extent), His mind would know itself under a principle of infinite regress. In determining the position of self, God's awareness of His surroundings is also in regress: any "creation" external to that self, is yet within the sphere of omnipresence.

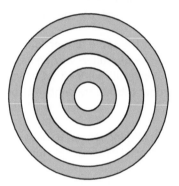

The perspective of infinite regression appears as a cycle; An "infinite loop" or regress.

For omnipresent God, colour bands represent looking out or looking within

In either case, the "self" is observed as if the identity and the "directed gaze" is as an order two translation from white to blue to white etc.

I describe such awareness of location and extension using a series of circular bands, where the "location" of infinite God (in the centre circle) references the same cusp of reality as each widening circular band in the figure (the "edge" of God's awareness would be the limit of each band); whilst being singular without a "loop", I may consider it as in regress, or with the notion of a repeating frieze pattern, to which the symmetries of algebra may be applied.

An infinite regress (geometrically) in its simplest form consists of an identity mapping of an object to itself, and a regress or translation to a congruence, widening each circle to the one that contains it, with the centre now "regressed". Cyclically, the location or origin itself is to also regress, for the awareness of God to re-assert itself in the centre.

3.2) Two Individuals – Looking At Each The Other Way

Then I show the regress forms the field GF(2). (Or at the very least the cyclic group of order 2.) I would colour those concentric bands "odd" and "even". (For God, looking out or looking in.)

If an omniscient individual (in regress) contains another, He must be able to traverse from the centre (as yellow below) to outside the next outermost band. (The yellow centre becomes linked to the next yellow band whilst spanning the blue band containing the centre.) The result is not an infinite sequence of concentric bands but a single pair of interlocking tori: each containing the other; with a suitable coordinate system each may be infinite and "on the outside". In this manner both individuals in regress are singular and the regression itself also singular: even if it appears infinite (as a loop) to us. Then I reduce the system to one of degree rather than of the infinite.

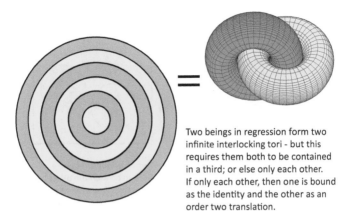

Two beings in regression form two infinite interlocking tori - but this requires them both to be contained in a third; or else only each other. If only each other, then one is bound as the identity and the other as an order two translation.

3.3) Regression Becomes Extension

Given the matter of degree, I count the number of ways I may collect individuals in regress. At the very least I would have for "n" individuals a power set of collections some 2^n-1 strong (not counting the state where none are present!). The collection itself must be self-similar to one of its members in regress; at least one in the collection should be as such else I may continue grafting individuals in without limit; I expect the regressive principle to remain. Any individual in regress would also be omnipresent if He were actually self-similar to the container (or a container with a higher degree of extension) of **every** collection containing Him. (So as not to reduce this to simply the set of Himself as lost within the others.) In other words, a regress requires a sense of "omnipresence and omniscience" with which I started.

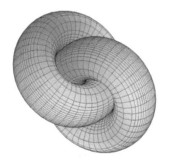

Two in regress, though contained in each other have a degree two dimension: either one may be described as having symmetry as in the elements of C2.

I may form a product (a, b) where a and b are members of C2.

(1, 0) is one individual, (0, 1) the other: Then who is (1, 1)?

Then I find at least one omniscient, omnipresent member of the collection must be so self-similar to the whole (or else there would be no closure in that existence – I would endlessly extend the set of collections and could have algebraic closure rather than Trinity), not because of some principle of necessary greatness, but simply because there is innate stability in Trinity as I will find; the members of the Trinity are all self-similar to their collective whole in one manner or the other.

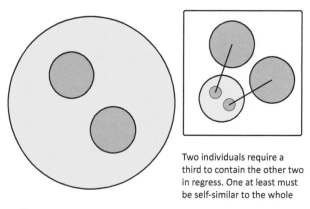

Two individuals require a
third to contain the other two
in regress. One at least must
be self-similar to the whole

The stability gained (from the Trinity constructed) entails that anything else is most stubbornly deficient. Any infinite regress towering to infinite (algebraic) closure would also form the Trinity uniquely within it. (GF(2), GF(4) and GF(8) would be unique within it as "subfields".)

The regress or "degree of extension" permits each pair of the trinity (as odd/even bands or tori as before) to alternate in their container (to which the remaining one is self-similar). That container corresponds by its free extension to the alternation of that first pair combined (yet only in regress) – now present in an extended form – as become a container of all three.

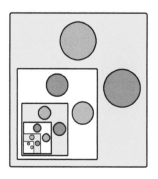

With one of the three self-similar to the whole, each of the three is as a "degree of extension".

Here, the first degree of extension is self-similar to the collection; which is of the second degree.

The result of a second degree of extension is a third member present within both the first degree (that is self-similar) and the second degree, the whole

The degree of extension (self-similar and alternating white/yellow here) shows the alternation of the other two combined (as odd/even) I had from omniscience before. Both alternate within the extended container and its self-similar member. As yellow/blue alternates to white/red and yellow/red to white/blue each of the three are distinct; alternating as with the other two alternating combined. If A alternates, B and C appear to alternate together as well.

With the above figure, are you appropriately getting "tunnel vision"? Imagine that!

3.4) The Trinity Takes Form In Abstraction

Given such a Trinity, I naïvely take two such members and form their product (a,b) = A x B, to find GF(4)+. Including a third member, I form GF(8)+ by Cartesian product – in triples (a,b,c) from A x B x C. The remaining element of GF(4)+ must also be the member generating GF(8)+.

The static and perfect (as GF(2) above) is **exceeded in perfection** by GF(8) with orbits of GF(4).

The third member is the solution for the self-similarity of the first two degrees of extension in terms of the third. Now the regress is from first to second degree and of the *second* through the third degree.

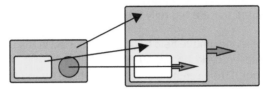

The first degree of extension is now self-similar to its container (of the second degree) whilst the second degree container in similarity is always one degree of extension above (in regress) as a parent container.
The third degree of extension is filled by its "child" under the action of the third member of the second degree; effectively transforming the self-similar child through the extra dimension of its parent.

My model begins with two individuals that lead me naturally to posit a third: Both begin self-similar to Trinity, yet between them they must agree on only one of the two being self-similar to the collection of Trinity they must form together.

Individual one posits:

$$a \equiv 1 \equiv [a,b,c]$$

The second immediately equally constructs not only Himself as "b" $\equiv$ [b,d,f], but also accounts for some "c" upon which they agree and for which they together solve.

$$a \equiv 1 \equiv [a, b, c]$$
$$b \equiv [a, d, e]$$
$$c \equiv [a, f, g]$$

$$b \equiv [b, d, f]$$
$$d \equiv [a, b, c]$$
$$f \equiv [b, e, g]$$

The question arises as to where {d,e,f,g} may come from. The smallest container providing for these two triples is a seven-cycle, say (a,b,d,c,f,g,e). I do not prove that, but leave it as an observation. (The octal container requires one more element – the additive identity, for which the Trinity may suffice under certain conditions.) The wait for such a seven cycle to occur may conclude in "the beginning".

Only the matter of similarity or symmetry at algebraic closure presents any trouble. (Upon which I currently note the omission without excuse but I will examine it a little much later on.)

By Frobenius I form a three-cycle on each Trinity of K4 groups and can also walk both congruences (a $\equiv$ [a,b,c], b $\equiv$ [b,d,f]) through the seven cycle. Yet the seven cycle has as unity a different member of the octal (not agreeing with the unity of the K4 form in regress placed as the static subgroup within it: see section [3.8]). One individual **surrenders ownership of unity** to the other in a K4 form (a ≡ [a,b,c] ≡ 1 above) and the other retains unity within the degree of extension spanned by the seven cycle. (a ≡ [b,d,f] ≡ 1.) I compare [a,b,c] to [b,d,f] here, and again unity is still free to transform amongst any of the octal's seven elements. I find three individuals, GF(4), GF(8)+ and GF(8)* (Son, Father and Holy Spirit). The surrender of unity may be done only once – the multiplicative identity is unique in algebraic closure, and if God exists necessarily I may be assured that He has only the one begotten Son.

In algebraic closure there is one unique field GF(4). If it occurred in the beginning that [0,b,d,f] begat [0,a,b,c] under a seven cycle ([b,d,f] surrenders unity in that same act, extending to GF(8) for greater perfection), then I could truly consider those two as Father and Son.

Then (cf. Joh 1:1-2) "the Word" (a ≡ 1 ≡ [a,b,c], a virtuous disjunction) was "in the beginning", and was "with God" a ≡ 1 ≡ [b,d,f] as together forming a ≡ 1 ≡ [a,b,c],[a,d,e],[a,f,g] and also "the Word was God" a ≡ 1 ≡ [a,b,c] ≡ [a,b,c],[a,d,e],[a,f,g]. Then, "the same was in the beginning with God" – the seven cycle of the Holy Spirit preserves Word, Father and Son.

3.5) Each To Their Own

There are three operations in the Trinity: whether of the K4 form as a subgroup of the octal, sharing octal addition or of the multiplication of GF(8)* appearing to induce GF(4)* under Frobenius on the K4 form spanning the octal.

F*, F+, GF(4)*, GF(4)+, GF(8)*, GF(8)+:
The terms F* and F+ refer to the multiplicative and additive groups of a field F respectively. GF(4)+ and GF(8)+ refer to the additive groups of the finite fields of four and eight elements and GF(4)* and GF(8)* to their multiplicative groups.

Whether self-similar to the whole or otherwise, I may add unity to an indeterminate "x" with one operation and multiply unity with it by the other. Unity becomes naturally transformed by the Father and Holy Spirit to 1+x or 1*x to form a K4 group GF(4)+. The automorphisms of GF(8) (under Frobenius) give three powers of x with which the K4 form is completed.

$$\textit{begin with } 1 \textit{ and } \{x, x^2, x^4\}$$
$$1, \ 1+x, \ \ 1*x$$
$$1, \ 1+x^2, \ 1*x^2$$
$$1, \ 1+x^4, \ 1*x^4$$
$$\textit{where} \quad x^4 \bmod x^3 + x + 1 \equiv x^2 + x$$
$$x^4 + 1 \bmod x^3 + x + 1 \equiv x^2 + x + 1$$

Once a ≡ 1 ≡ [a,b,c] is set with "x" a generator of a seven cycle: the octal itself is determined.

3.6) The Trinity Resolves Father, Son And Spirit

Given a closed set of positive properties p that private no others I admit the existence of some positive properties q that may. These q are "positive" but the former p "virtues" instead as they permit the free exercise of q or "not-q" (yet q cannot private p).

If the set of all p is become as unity in GF(4), K4 a minimal requirement and constant to perfection, I assume **all** positive properties {p and q} together form Ω. Given two states of perfection, some set J having every p and some subset "q_1" of Ω (in state A) and some case (in state B) both p and some subset "q_2". I find a third state C in Ω containing all p, but also a third state of J forming a K4 group structure with this state of J as befits C = (A $\vee$ B)^c = A + B. Yet the trinity must do this perfectly.

The K4 form sits in regress as a subgroup of the octal, perfect in all positive properties and with the omnipresence and omnipotence of the container GF(8) of the K4 form. The octal resolves creation for the pleasure of the Son (the K4 form as Himself within creation) and of the whole trinity (yet mostly for the Father). Clearly the Father (as the set Ω) is pleased in the virtue in His Son and in His Son's actions. Creation exists for God's pleasure and the Spirit which is God that is pleased in Christ moves (or rests, or abstains from action) upon all.

The set Ω (with virtues) forms an ultrafilter whereas the set of positive properties without virtues are incomplete; only a filter at most. There is also no sense that virtues are more "positive" than other positive properties.

Concerning the K4 form spanning the octal (A = {p, r, s}, B = {p, u^{-1}, v^{-1}}, C = {p, (r&s)$^{-1}$, u&v}) there is a sense of animation with Frobenius cycling these subgroups of the octal. Predicates of A are workable by God at any time, those in B must already have been exemplified for God to "rest" on the perfection of virtue, and lastly those in C indicate that if it is not perfect to act or rest in either of those disjunctions, God may perform them both to the contrary as in C.

The sense is that the properties in C are "forward looking", taking precedence over those in A or B. It is not perfect to rest on, say, either "u^{-1} or v^{-1}" positively exercising inaction (at least not perfect for God to remain inactive) unless those predicates are somehow already exemplified. Predicates in B are as acts completed which were once in some A, and if it is not positive to rest on either (remaining inactive), then God will perform both as in C. So "u&v" in C is truly a solution to exercise that perfection that may become freely rested upon.

Similar are the predicates in A, for if it is not positive to act on one or the other of "r" and "s", then God may rest on both. The result? A, B and C are somewhat descriptive of present, past, and future, C is a set that is a consequence (product of) both A and B, and liberty is found in C, with the conditions for leaving both A and B "until later". The predicates in {r, s, u, v} are to be "filtered" – in that to rest on perfection (in B) is to have already acted (in A) and to be free to act or to rest virtuously depends solely on the freedoms (in C) as to what acts have already been performed or can be otherwise waited upon. Then I have the verse:

Rev 11:17 Saying, We give thee thanks, O Lord God Almighty, which art, and wast, and art to come; because thou hast taken to thee thy great power, and hast reigned.

3.7) The Godhead Is Complete With Three (As A Cloud)

I answer static monotheism with a dynamic Trinity found in the structure of K4, previously some mystery (as if clouded). I state GF(4)+ acts as a sufficient additive identity on the products of the three subgroups [0,A], [0,B], [0,C] using the binary operation $(x \vee y)^c = x + y$ ("$\vee$" is the symmetric difference with $(...)^c$ its set complement taken within (0,A,B,C) which in union is the set space Ω.) **Here I apply "Theorem 2.7".**

Trinity, self-similar in regress, is always a "greater set" of perfections than those in virtue alone; I can assert that Trinity finds the "closure" of all Ω from exemplifying virtues.

However, I must account for the product C. This third member is not simply a re-arrangement of the other two; He (as God) exists with them also. I derived from those Cartesian products a requirement to close GF(8) by mapping the unity (as in "p") to the other elements of GF(8)*.

I have an isomorphic operation $(x \vee y)^c = x + y$ on K4 subgroups of the octal group GF(8)+. The K4 group is as zero with respect to subgroups order two. The additive octal group (GF(8)+) is also the zero over all those additive products of its own K4 subgroups.

Inferring self-similarity, I find regress between a K4 subgroup of the octal to an isomorphic group of K4 subgroups (increasing perfection by that self-similarity) together spanning the octal GF(8)+. All that is pleasing to God may be chosen or excluded as "q" whilst keeping intact all "p" – and in limit even to agree upon whatever "J" as that K4 form is most pleasing to (or to J within) a given choice of Ω as all the octal, on the basis that those positive properties are pleasing in p and/or q now they are dynamic, in flux, or "at rest" overall.

The third member of the K4 form extends it to the octal and in self-similarity, spans the octal as both a K4 group of K4 subgroups under $(A \vee B)^c = A + B$ (preserving its operation).

3.8) Life In The Godhead Is Multiplication

GF(8)* induces seven cycles (orbits) on every element and subgroup. (This simply permutes or relabels them.) It is the map of Frobenius sending every element x in F (a field of characteristic p) to its pth power x^p that appears to induce a multiplication on both a particular subgroup of the octal as well as upon the K4 form spanning it; this three-cycle is the action of the third member: the missing "C" in C = A*B forming GF(4)*.

GF(8) cannot have the unity within the subgroup held static under the Frobenius map; GF(8) has unity as the intersection of the three K4 groups cycled in the K4 form. That deduced, the "third part" is unity in GF(8)* as well as the "virtue" present (constant to) the set of p in each state of J: A, B or C, whereas J appears "at the right hand" as a proper subgroup of GF(8)+.

From the self-similar, Frobenius induces (in agreement with multiplication in GF(4)*) a three-cycle on the K4 subgroups with unity, completing the K4 form to mirror GF(4), as if supplying a suitable multiplication for it (of the first, second, or fourth power. Then $x^8 \equiv x$, squaring induces a three cycle of automorphisms).

The Godhead has Trinity: the Father GF(8), the Son GF(4) and the Holy Spirit C7. The result is not static but in all such perfections continually in flux: Truly such a Trinity is of a "living God".

Unity in GF(4) is then solved for as GF(4) (self-similar in regress), itself a subgroup congruent to unity as under orbit by the seven cycle. The other two members of J are the octal itself (extended one degree above K4 as unity or "J") and the transformation through GF(8)+ by the third; "filling in" the coset opened by the extra degree of extension by element(s) of C7 (those two in union account for any self-similarity in the same set space Ω).

3.9) The Trinity Closed

This algebra has a "strange twist", and many points to consider: It should be understood that the K4 form of the ultrafilter J (perfect in perfections, dynamic in states rather than static as by Anselm's argument), preserves all virtue p intact in every state of the three. Without privating virtue (of the believer's place in the salvation of God), God has also clothed us with the righteousness of His Son, as befits His liberty in requiring "p" as a minimum for perfection.

> **Anselm of Canterbury, Anselm's Argument:**
> *St. Anselm of Canterbury (1033-1109AD) is renowned for his ontological argument that has its defenders though it is considered to be refuted by objections that existence is not a predicate but an ordinal relation. Necessary existence of a being rather than an object yet has stalwart defenders; there are several modern ontological arguments that a perfect being necessarily exists if He exists at all.*

Divine liberty in its pure form is the essence of all such virtue. No filter entailing any "s" from $p \& q^{-1}$ is finer; liberty entails $L \& q^{-1} \rightarrow e_r$ within perfection. The ultrafilter is always intact.

More can be revealed by the automorphisms of the Frobenius map. Repeatedly squaring in GF(8) produces automorphisms in a three-cycle. It is a certainty in GF(8) that a static triple cannot contain unity (the remnant would form a two-cycle under that Galois group of order three; a clear violation of the three cycle; two does not divide three).

In the octal, if "a = 1" and the groups of the K4 form (omitting zero for ease) are [a,b,c], [a,d,e], [a,f,g] then under a suitable seven cycle (a,b,d,c,f,g,e) the subgroup [b,d,f] is static, the triple {c,e,g} also: and I map "a" to itself. This has the result that the Frobenius map induces (in the K4 form of the filter) three automorphisms forming a cyclic (Galois) group isomorphic to C3, as if mirroring complete agreement with the operation of GF(4)* and cycling the static subgroup [b,d,f] in "natural agreement" with the operation of the same field GF(4) in regress.

3.10) The Manifold Of Virtue

Given an octal O, the union of the Trinity of Father, Son and Holy Spirit (F, J, and H respectively), I find arrangements of that Trinity, preserving the liberty in each state of the three.

I may compose a K4 group on the Trinity with each member accounting for the liberty of the other two, with all three acting as virtue to preserve the symmetry of the three-in-one set.

For example:

$$H \circ J \circ F$$

Acting on the right first (with composition) upon O, I may equate "J" as Christ acting in virtue to please His Father in "F" by accounting for any state of the Holy Spirit "H".

I.e. I may with any octal O, choose any static subgroup of seven as "F", then choosing any seven cycle "H" on O, and (always with a factor of four) there is one of four choices for the K4 form of "J" spanning the octal (or equivalently fixing the unity element in O) justifying those two seven cycles of eight (in four pairs) as "H" in agreement with the static subgroup of "F".

Then the K4 form of the ultrafilter may self-reference itself in regression three ways.

I also have:

$$F \circ J \circ H$$

For any C7 group "H" of eight acting on O, I may choose any static subgroup of seven as "F", leaving me only one of four choices for unity and effectively, but one of four choices for the K4 form "J" (spanning O) to account for my choices in "F" and "H".

I need not stop there, for any member of the trinity, mysteriously enough, may act as virtue!

I.e. I may also form:

$$H \circ F \circ J$$

I may choose any K4 form in "J" spanning the octal, fixing the unity element as one of seven. Given any C7 group as "H" (of four pairs from eight) I always fix one of four static subgroups. Then, I have effectively fixed "the Father" in "F" to justify the state of the other two (and "O" is also free to be arranged through seven equal arrangements under the seven cycles in "H").

$$J \circ F \circ H$$

Choosing any C7 group in "H", for every choice of unity fixing the K4 form "J" spanning O, I find one of four choices for the static subgroup in "F", accounting for both "J" and "H".

$$J \circ H \circ F$$

I may also choose any static subgroup of seven in O as "F", and for every possible choice of unity in "J" I find two of eight C7 groups in "H" that preserve that structure. (Clearly, unity is not free to be chosen from the static subgroup but in the K4 form of the ultrafilter alone.)

$$F \circ H \circ J$$

Finally, I may choose any K4 form "J" of seven spanning the octal and then any static subgroup of four possible "F", and there is always one pair of C7 groups "H" from a set of eight that preserve that structure.

I find a factor of four from that "virtue", where the member accounting for the other two always intercedes in one of four ways. Every structure is preserved under multiplication by the elements of C7 (by those pairs of C7 groups holding static a certain subgroup) and those structures are preserved under all three automorphisms (of Frobenius) on O.

A disjunction $r \vee p \& r^{-1} \rightarrow s$ is formed by F, H and J where each member of the trinity may account for any arrangement of the remaining two, as if acting in the place of virtue. When F is at liberty (may rest on any set) as r, say, then one of J and H (as p and s) must account for the liberty (to rest in some state) of the other.

For "liberty" L(x): $L(r) \vee L(p)\&L(r)^{-1} \to L(s)$. ($(L(r)^{-1}$ is to negate a necessity; it must be empty.) Yet, all three are equally virtue and despite virtue appearing on both sides of the disjunction, it never privates the other. It is also apparent that the excluded middle (properly empty in any disjunction of virtue) is here the same member acting with virtue! (There is no closure broken in the statement $N¬(p\&¬(r\&s))$.) I state that the join F&H contains the full content of J.

I have, say, both $p\&r^{-1} \to s$ and $s \to r\&p$. Quite naturally, I find $r \leftrightarrow r^{-1}$ given p, r and s are somewhat "disjoint". (Yet there is no octal O without all three.)

J acting as if p is totally equivalent to r&s, or in this case, F&H. So $p \leftrightarrow r\&s$ and $J \equiv F\&H$.

$F\&H \equiv J$, fixing a static subgroup and C7 group in one of four ways (eight, if the direction H cycles both the static subgroup and triple is taken into account) effectively fixes "J" uniquely. $F\&J \equiv H$, or fixing unity and a static subgroup leaves only a pair of C7 groups from the full eight, and lastly, $J\&H \equiv F$, which concerns fixing a C7 group and one of four choices of unity which fixes the static subgroup uniquely.

In effect the relation $H \equiv J\&F$ is the statement:

$$H \equiv (J \circ F)^{-1} \equiv J \circ F$$

The state of "H" is determined by the conjunction/composition of F and J on O. I also find:

$$H \equiv (F \circ J)^{-1} \equiv F \circ J$$

An equivalent. The inverses at first appear to be undefined until those simple relations above ($H \equiv J\&F$ etc.) are restated. (And of course, all taken modulo 2 I have the Trinity.)

$$J \circ J \equiv H \circ H \equiv F \circ F \equiv I$$

I relate these modulo two; with analogy to the above (forming $r \vee p\&r^{-1} \to r$, now with one member "r" present on both sides – here I find $r \leftrightarrow r^{-1}$) the virtuous disjunction has no solution for "p" from {I, p, r, s} but for the identity map "I". (Both sides are freely chosen as equals.) By axiom of virtue, were "p" not equal to "I", I would require yet another virtue, some p_1 to account for the further closure of $p\&r \vee p_1\&r^{-1} \to p\&r$. At some stage I must use the identity map "I" (liberty), or else I find myself in an infinite cascade in this "Manifold of Virtue".

Also, the composition or map also found meeting in identity:

$$J \circ T \circ J \equiv I$$

Has a unique solution for T (given the relations above), either that of the identity map "I" or of any equivalent triple as below. (Keeping the same states in J, F or H.)

$$J \circ J \equiv J \circ (H \circ J \circ F) \circ J \equiv (J \circ H) \circ J \circ (F \circ J) \equiv F \circ J \circ H \equiv I$$

Concerning the relations order two: I state of the Trinity that "they are a solution", and together, the only closed set required to give it! (There is no cascade given the presence of

Divine virtue or of "The Word" (Joh 1:1-2). The Trinity is just fine.) I state these bindings are a union and completely different to anything which I could aesthetically label "synthetic".

Gödel's ontological argument[1] sheds some light on this, for the predicate or positive property of "God-likeness" is such as to infer that every possible (and consistent) positive property must be exemplified by a necessarily existent being with that same "God-likeness" as an essence. Then as "God-likeness" is itself a positive property (as is necessary existence to then exemplify it), both would follow from any such essence and so a "God-like" being exists necessarily if He exists at all. (For necessary existence in the case of a "God" must entail God-likeness (an essence) from which follows existence. Existence, is then an equivalent essence.)

For disjunctions of virtue $r \vee p\&r^{-1} \rightarrow s$, if p, r and s are all found "God-like", then for a principal element e_r ("rest"), I may not freely decide $r\&e_r \vee p\&(r\&e_r)^{-1} \rightarrow s$ as $(r\&e_r)^{-1}$ is then totally void and given $N(r\&e_r)$, I find $N\neg(p \rightarrow s)$. By symmetry and specifically within a "Manifold of Virtue" I would have $r\&e_r \vee s\&(r\&e_r)^{-1} \rightarrow p$, so then $N\neg(p \leftrightarrow s)$ and all binding within the disjunction fails with privation unless I may instead write $p \rightarrow (r\&e_r)\&s$ etc. in the middle.

I must find $r \nrightarrow e_r$ for some possibly principal "e_r" (as in the least sense). Now, for "God-likeness" $G(x)$ in the case of r, I would infer $G(r) \rightarrow r_0\&e_r$ for all subsets r_0 in r. Then $(r_0\&e_r)^{-1} \rightarrow \neg G(r)$ and so the disjunction $r \vee p\&r^{-1} \rightarrow s$ with $r = r_0\&e_r$ splits r_0 from e_r, else it becomes true that "r" at rest is not "God-like". Then again, I recover at most $r_0 \vee p\&e_r \rightarrow s$.

In the Trinity (and only in a "Manifold of Virtue"), I expect all three sets p, r, s to be "God-like". Minimally, $e_p\&e_r \rightarrow e_s$ or $e_p\&e_s \rightarrow e_r$ and I find that, in essence, all three in Trinity must likewise be set at liberty. The equivalence between necessary existence and that essence of "God-likeness" $G(r)$ is also to state that any such disjunction formed must require $G(p)\&G(r) \rightarrow G(s)$, and in terms of every possible disjunction that may be formed; the equivalent essence is of: $L(p)\&L(r) \rightarrow L(s)$. Then $L(r)$ freely decides $(e_p\&e_r \rightarrow e_s) \,\bar{\vee}\, (e_p\&e_s \rightarrow e_r)$; both are true.

The principal element "e_r" (as a rule), is **always found** on the side of the disjunction wherein the liberty of r does act: That liberty is exemplified by p quite simply as $L(p)$ which in this case is combined with $L(r)$ together entailing $L(s)$. Then $L(p)\&L(r) \rightarrow L(s)$. (I use $\bar{\vee}$ to denote that each side of the disjunction is freely decided, and (most importantly), that liberty always acts to one side or the other, and so it appears on both sides in the notation.)

The predicate of liberty $L(r)$ is equalled with the very same God-likeness $G(r)$; an essence entailing "rest" or necessary existence in Ω. In any disjunction of virtue I find the image of the Trinity itself embedded: and as $L(r)$ is equal to $G(r)$ and necessary existence then to "e_r", I find by Gödel's argument[1] that in every possible disjunction in the "Manifold of Virtue" (even when r_0 is maximised to be equal to r all but for that rest e_r) the "God-like" remainder on each side is the very person in the Trinity $L(r)\&e_r$. So, that member $r \leftrightarrow r^{-1}$ is then to be found on both sides of the disjunction itself. (The liberty of r, to act and rest on r_0 is then to complete r as "God-like", for then $p \rightarrow r\&s$.) Then effectively, $p\&r \rightarrow s$ and no matter the set of positive properties Ω the Trinity may together choose to exemplify; as arranged in any possible disjunction formed, in the very limit of their essence and liberty they embody $p \leftrightarrow (r\&s)$ and they then together easily form such a "Manifold of Virtue".

[1] Sobel, in J. J Thompson (ed.), *On Being and Saying: Essays for Richard Cartwright*, p241-61.

Minimally, $L(r)\&e_r$ is enough of an equivalent to "r" as found in each disjunction in the "Manifold of Virtue". Since "rest" is found only on the side of the disjunction upon which liberty acts (as a principal element), the Trinity of p, r, s forms a manifold of virtue as constructed beforehand. The liberty found in the Trinity is enough to show that the manifold's symmetries are active in every disjunction $r \vee p\&r \to s$, and that the Trinity is triune, not singular. (I.e. $p \neq s$ as in the disjunction $r\&e_r \vee p\&(r\&e_r)^{-1} \to s$.)

So, whilst the liberty of state is preserved in Trinity, the axiom of virtue ensures a clear sense of "order two": i.e. $e_r \vee p\&e_r^{-1} \to s$ and $p \to e_r\&s$ negated certainly holds on the left and $\phi \vee p\&e_r \to s$ and $e_p\&e_r \to e_s$ (or $L(p)\&L(r) \to L(s)$ in any disjunction upon them) on the right. (e_r is necessary so $Pos(e_r^{-1}) \leftrightarrow \phi$. Then as e_r is axiomatic, $Pos(p\&e_r^{-1}) \leftrightarrow Pos(p)$ and $Pos(p) \to Pos(s)$. I.e. $Pos(e_r) \triangledown Pos(p\&e_r) \to Pos(s)$.) There is, by axiom, such a "p" for any "s". //

With the identity "I", $e_r \vee (I\&e_r^{-1} \to e_r)$ is valid, and also $I \to e_r\&e_r$ (which is not extant when liberty "I" acts on the left hand side). In that case there is either $I \equiv \wedge e_x$ for x in $\{p,r,s\}$ or modal collapse and $I \to O$ or Ω. I.e. there is $I \vee (e_r \to I\&e_r)$ instead. To better form the identity, I start with the former and state that $I \equiv J\&F\&H$ is not enough on its own to form a K4 group. Instead I find a Trinity of $J_0 \equiv I\&J \equiv I\&e_x$, $F_0 \equiv I\&F \equiv I\&e_y$ and $H_0 \equiv I\&H \equiv I\&e_z$. In each case I have, say, $e_x \vee (I\&e_x^{-1} \to e_x)$ with liberty "I" acting on each side extant of the disjunction $J_0 \triangledown I\&e_x \to J_0$. Then God sufficiently rests as e_x on $I \triangledown (J_0\&I^{-1} \to e_x) \leftrightarrow (J_0 \to I\&e_x \to J_0)$ and the J_0 etc. are completed in a K4 group. Then who is "I"? Not a member of the Trinity but simply the "right hand of God": a servant that ministers to the Trinity. Then by $J_0 \triangledown I\&e_x \to J_0$ I infer "I" intercedes whilst J (and J_0) rests. "I" is then separated from J_0 (and $I \triangledown I$ is vacuous).

Returning to the "Manifold of Virtue", there is a clear sense of the static subgroup (as containing the unity) with Christ "sat at the right hand" within this construction already!

The statement "unity is not free to be chosen from the static subgroup but in the K4 form of the ultrafilter", is met with the K4 group formed of F, J and H (each then acting as virtue) as if unity were legally floating through the static subgroup. (Unity always present as virtue.)

Then the trinity wholly accounts for the symmetry of Christ as sat at the right hand;

Col 2:9 For in him dwelleth all the fulness of the Godhead bodily.

Extending that K4 form to the octal (always found from the liberty of the Trinity, whether by choice of static subgroup, unity, or of C7 group), "rest" may be found in four ways, accounting for whatever is judged as virtue in the static subgroup's coset in GF(8). Effectively by fixing unity as the intersection of the three subgroups spanning the octal in "J" and allowing for the four static subgroups I have already extended the octal and closed it by forming the coset.

Fixing a static subgroup and extending four choices of unity also suffices. The remaining cases holding either one of four choices of unity or a subgroup static and varying the C7 group in one of four ways will always close the octal under the action of multiplication.

Then I would not expect to find seven members of the Holy Trinity, but only the three!

So far, the multiplicative group of "J" (as if GF(4)*) has remained curiously absent. The

Frobenius map acting on GF(8) will hold fixed the unity, cycle static the static subgroup and will do all this preserving the same seven cycle. As Frobenius holds all these static, how may the Frobenius map cycle the unordered group formed of F, H and J? The truth is, it cannot without a "strange twist". A simple mathematical substitution suffices!

Joh 5:26 For as the Father hath life in himself; so hath he given to the Son to have life in himself;

The K4 group of F, H and J has a multiplicative group quite separate to the "Galois group" C3 of GF(8) which the Frobenius map generates acting on O.

In essence the K4 form spanning the octal has unity in congruence to not only the whole, but also to one of its members (clearly to become J in this construction). Setting J congruent to unity in the octal is not enough, J must be self-similar to the K4 group formed of F, H and J.

For J formed of the three K4 groups spanning the octal, J then corresponds to just one (as unity), F to another and H to the last. The static subgroup in O can never be equal to any of those three; just as any seven cycle will never be a K4 group. Then, quite separate to the octal, J has a multiplicative group C3 induced by the Frobenius automorphism acting on F.

J corresponds to a triple of K4 groups congruent to unity. Yet, as self-similar to one of its own members, C3 is induced on whichever group of the three is self-similar to the whole (unity in the K4 form of the ultrafilter). Frobenius, separate to the operations of GF(8), induces a three-cycle upon the K4 group formed on F, H and J. Then the ordered list $J \equiv [J,F,H]$ is cycled "order three" in the isomorphic form (where unity is held fixed), of, say, $a \equiv 1 \equiv [a,b,c] \equiv [[a,b,c],[a,d,e],[a,f,g]]$. Substituting $a \equiv [a,b,c]$ is enough to induce C3 to act as multiplication on the Trinity formed of [J,F,H]. J thereby becomes an individual.

The remaining substitutions of $a \equiv 1 \equiv [b,d,f]$ etc. indicate the four possible choices of static subgroup from F, each of the eight groups of seven cycles in H will permit but one of those four static subgroups or choices of unity given either F or J respectively.

The cycle the static subgroup induces on the K4 form has a lucid effect. Cycling each disjunction $[r \vee s] \rightarrow [u^{-1} \vee v^{-1}] \rightarrow [(r\&s)^{-1} \vee u\&v]$ preserves the liberty to **rest** on a prior work u&v in the map $[r \vee s] \rightarrow [u^{-1} \vee v^{-1}]$ (may u&v be rested upon from $[(r\&s)^{-1} \vee u\&v]$?) the map of $[u^{-1} \vee v^{-1}] \rightarrow [(r\&s)^{-1} \vee u\&v]$ shows the liberty (identity) remaining in the cycle. Concerning the other direction, $[u^{-1} \vee v^{-1}] \rightarrow [r \vee s]$ preserves the liberty to **act** on a prior work $(r\&s)^{-1}$ not yet instantiated. $[(r\&s)^{-1} \vee u\&v]$ reaches identity (unity), effectively by composing the maps.

3.11) Things To Remember – It Figures!

There are numerous facts and figures native to the Trinity construction itself.

- With the same seven symbols "a" to "g", there are a possible thirty octal groups (isomorphic to the Cartesian product C2 x C2 x C2 or also GF(8)+) found that are isomorphic but not identical.

- For each octal group, there are eight C7 groups that send K4 subgroups to K4 subgroups preserving their structure under multiplication. The K4 group is isomorphic to the Cartesian product C2 x C2.

- Every seven cycle in a given C7 group preserves the subgroups of two octal groups only.

- There are a possible 35 ways to choose three elements from seven, so there are at most 35 possibilities for K4 subgroups found from the complete set of seven symbols.

A good start to make for this study is to memorise the possible groups preserved by multiplication over one octal. They may be easily constructed as they are all isomorphic structures. In choosing one octal in particular with the subgroups [a,b,c], [a,d,e], [a,f,g], [b,d,f], [b,e,g], [c,d,g], [c,e,f] I note that I may upon choosing some arbitrary element as unity "a", use a cycle (a,b,d,c,f,g,e) for multiplication. Then the next, the square and the fourth power of the cycle form the static subgroup which does not contain unity. There are four such subgroups not containing unity: [b,d,f], [b,e,g], [c,d,g], [c,e,f], so I easily complete the seven cycle preserving all the octal's subgroups…

$$(a, \quad b, \quad d, \quad c, \quad f, \quad g, \quad e)$$
$$\uparrow \quad \uparrow \qquad\quad \uparrow$$

$$[b,d,f] \text{ is static}$$

$$(a, \quad b, \quad d, \quad c, \quad f, \quad g, \quad e)$$
$$\quad\uparrow \quad \uparrow \qquad\quad \uparrow$$

$$[c,d,g] \text{ is a subgroup}$$

Then (a,b,d,?,f,?,?) is completed to become (a,b,d,c,f,g,e) as I "walk through" the seven cycle and seven other seven-cycles each representing their C7 groups easily follow:

$$(a,b,f,c,d,e,g)$$
$$(a,b,e,c,g,f,d)$$
$$(a,b,g,c,e,d,f)$$
$$(a,c,d,b,g,f,e)$$
$$(a,c,g,b,d,e,f)$$
$$(a,c,e,b,f,g,d)$$
$$(a,c,f,b,e,d,g)$$

Now, the opposing (complement) octals also held constant by each seven cycle are similarly represented by the inverses or the 3rd, 5th and 6th powers. Then (a,b,d,c,f,g,e) has as the affiliate octal [c,e,g], [d,f,g], [b,c,f], [a,c,d], [b,d,e], [a,b,g], [a,e,f].

The two octals referred to above are as "right and left hands" equivalently. The right hand or reference octal [a,b,c], [a,d,e], [a,f,g], [b,d,f], [b,e,g], [c,d,g], [c,e,f] is referred to as if an "origin" of sorts, and represents every mention of the "Sun" in Revelation (called throughout this study the "Sun octal"), a clear reference to God as the great light (of virtue) and attractor (redeemer) and only coincidentally as a hint at a heliocentric norm.

Chapter Four: Solving For Virtue In The Godhead

This section is somewhat overkill, but cannot be separated from the metaphysics of the Godhead and is included here where it would be most relevant (other than lost in later appendices and missed). Unless you are interested in examining the lengthier treatment of the metaphysics you may skip this section – although the following study of Revelation is diminished in its richness if you so choose to.

The metaphysics adds to the maths, increasing the quality of it when it is compared to the Revelation text itself. Aside from justifying the use of such algebra (for example, to note that the sets within the elements of the K4 form from Theorem 2.7 are themselves valid within other K4 subgroups of the octal over all positive properties), the study is still valid on the basis of that algebra and the content of chapters two and three alone – but this chapter properly grounds the maths in positive properties (whilst effectively describing the Godhead).

It likewise makes sense to state that the previous mathematical (geometrical) construction is of a finite field and that of virtue appears a field of sets; were it not for the fact they are inseparable constructions – God acts perfectly in omnipotence within His sphere of omniscience.

Rev 4:11 Thou art worthy, O Lord, to receive glory and honour and power: for thou hast created all things, and for thy pleasure they are and were created.

4.1) Recapitulating Our Past Progress

Returning to my previous metaphysics;

Given positive properties p and q (p a virtue) I previously entailed the free choice between q and some near-arbitrary property consequent of virtue in p with:

$Pos(q) \lor Pos(p\&q^{-1})$

From the positive property q I entailed:

$Pos(q) \rightarrow Pos(r)\&Pos(u^{-1})$

And from the application of the virtue p on the converse:

$Pos(p\&q^{-1}) \rightarrow Pos(s)\&Pos(v^{-1})$

This has the effect that those properties that must be positively unexemplified in turn imply the modus tollens.

$u \rightarrow Pos(p\&q^{-1})$

$v \rightarrow p^{-1}\&Pos(q)$

Now, "u" cannot be a virtue. "v" cannot private a virtue (still allowing free choice); yet "v" cannot be a virtue unless q is also a virtue (the set of virtues is closed). I appear to have the following identities given p (albeit naïvely). Then this chapter will remedy the following:

$Pos(p)\&r^{-1}\&u \rightarrow Pos(p\&q^{-1}) \rightarrow Pos(s\&v^{-1})$

$s^{-1}\&v \rightarrow Pos(q)\&p^{-1} \rightarrow Pos(r\&u^{-1})$

$Pos(r\&u^{-1}) \vee Pos(s\&v^{-1})$

$s^{-1}\&v \vee Pos(p)\&r^{-1}\&u$ entails "$Pos(r\&u^{-1}) \vee Pos(s\&v^{-1})$"

Equivalency (4.1.1)

For perfection, the statement $Pos(p\&r^{-1}) \rightarrow Pos(s)$ simplifies to $p\&r^{-1} \rightarrow s$, which is logically put: $N\neg(p\&r^{-1}\&\neg s)$. The following forms are equivalent.

$N\neg(p\&r^{-1}\&\neg s)$ i.e. $p\&r^{-1} \rightarrow s$

$N\neg(p\&s^{-1}\&\neg r)$ i.e. $p\&s^{-1} \rightarrow r$

$N\neg(p\&\neg(r\&s))$ i.e. $p \rightarrow r\&s$

I recognise the prior two from before, that virtue in p justifies free exercise between two positive properties r and s. Yet the latter of the three would indicate at first that r&s is also virtue; yet this cannot be! It is also realised that r&s is "inconsistently positive" and an antinomy, as if the set of virtues in p were not closed.

r&s is then equally stated (given p), to be $r\&r^{-1}$ which is inconsistent, for $(r\&r^{-1}) \leftrightarrow (r\&r^{-1})^{-1}$. The triples {p, r, s} are closed in positive properties under the operation $A + B = (A \vee B)^c$, the complement in Ω of r&s is similarly inconsistent, for $(r\&s)^{-1}$ should be part of the result of {r} + {s} = {p, $(r\&s)^{-1}$}, but $(r\&s)^{-1}$ is inconsistent given p and either one of r or s. The triple {p, r, s} forms a K4 group under that operation of addition. (The identity element is the triple itself, the local closure Ω of all positive properties.)

This results in both sides forming principal ultrafilters! (Even though the set of all positive properties from which each q is taken forms a non-principal ultrafilter.)

The ultrafilter (and indexing set) in "r" maximised to "all q" (as of Anselm) is the complement in Ω of the indexing set of all the "s". Trivially, r&s is certainly empty in Ω if "s" is the one element "e_s". For then Ω itself is just as the trivial union of "$r=\{\Omega\&e_r \setminus e_s\}$" and "$s = e_s$".

Then every case (i.e. of each "world") has a **maximal** subset "q" of $\{\Omega\&e_r \setminus e_s\}$. Each side (in virtue) forms a principal ultrafilter as $<e_r>$ or $<e_s>$, the side in "r" is always a subset of a "q" from the non-principal "q-filter" (as across worlds, see section [2.9]) made principal by including the singleton "e_r". (All such possible "q" are either finite (if Ω is finite also) or contain every co-finite set if Ω be infinite, or else those finite complements in Ω by equivalence, excepting every set containing "e_s". Adjoining "e_r" to every subset "r" of the "q" in scope (from the q-filter) makes the ultrafilter in all such "r" principal (as then $<e_r>$) in its indexing set "q".)

As the filter in "r" (upon all subsets of q) is effectively Anselm's same filter on q (i.e. in some world) which is non-principal without "e_r", it may contain every finite conjunction (or set) of positive properties in "q" whilst also containing the principal element e_r.

Every finite complement taken of each possible "q" in $\{\Omega\ \&e_r \setminus e_s\}$ forms a principal ultrafilter in all possible "s" as $<e_s>$ ("e_s" always consequent from virtue); no finite set in the non-principal filter in "r" upon q (taken from the "q-filter") can have that principal element e_s or share the same indexing set of the complement in $<e_s>$ (not then contained in that principal ultrafilter $<e_s>$), or has any middle in the disjunction even when r is maximised to all $\{\Omega\ \&e_r \setminus e_s\}$.

Every principal ultrafilter in virtue $<e_s>$ is made upon a complement of a set in the q-filter.

Rearranging, the filter in "r" is also possibly made principal, by joining "e_r" to every subset of that particular "q" (world) in scope. Each "q" is then a "domain" for applying virtue, an ultrafilter in "r" upon each subset q of Ω, as if Anselm's filter were made principal on each possible world, a set from the "q-filter".

I state that the "q-filter" in Ω cannot be made principal over both sides of a disjunction; but every q in that filter may become a domain for a principal filter, up to containing a principal element "e_r". I will examine how that "q-filter" privates virtue over every disjunction now.

The complement of any set in $<e_s>$ in Ω is a possible "q", but as arranged, the principal filter(s) $<e_r>$, $<e_s>$ index only subsets within one side of the disjunction, within "q" or "p&q^{-1}" (i.e. of "s"). They are "separate" or disjoint as indexed sets.

I assert that though the filter $<e_s>$ in "s" is principal (on the finite complement of a co-finite set in "q"), there is no making every complement (the q-filter, as a collection) of every set in $<e_s>$ into an ultrafilter sharing the same indexing set of Ω without **privating** virtue. (The q-filter "privates" virtue, as some properties in Ω always constrain virtue.) To show this, I drop the principal elements e_p and e_r consequent of $e_p\&(e_s)^{-1}$ and merely keep the sets p, r and $<e_s>$.

Given the "q-filter" over one side of the disjunction, maximising sides in virtue and dropping every "e_r" joined to every co-finite conjunction "r" in the "q-filter", I **cannot** remove "e_s" from every finite set of conjunctions in s=$<e_s>$ without losing all sense of a disjunction ($\Omega \setminus e_s \vee e_s$ would become trivial). The conjunction (middle) of r&s between any such co-finite "{r \ e_r}" and each set in $<e_s>$ would then **also** be a co-finite set in the q-filter (as all "q" but without "e_s" enjoined). Then every such co-finite set r&s (not properly within the ultrafilter of $<e_s>$) is in the middle and privates virtue and must be excluded.

Similarly, if both sides were composed of only finite conjunctions; then every set within r&s would also be a finite set and possibly a co-finite complement of some set in $<e_s>$, as all "r" together with r&s would then be found in the q-filter which cannot be equal to or contain $<e_s>$ or $<e_r>$ (arranged either side as "r" or "s") with that middle, as virtue is then privated.

Whether Ω be finite or infinite, all finite (or co-finite) subsets of Ω cannot belong in the "q-filter" without privating virtue. The "q-filter" in Ω forms no ultrafilter in the disjunction without entailing a middle: only the principal ultrafilters that share no indexing set are valid in the disjunction of virtue. Yet, $<e_s>$ is an ultrafilter on Ω across worlds. (And not just on p&q^{-1}.)

As "q" and "p&q^{-1}" (or sets as X and X^c in Ω, say) are present in each case of the disjunction formed, every ultrafilter in "r" shares no positive property with any ultrafilter in "s", so that any r&s is excluded from both principal filters. (But r&s is still a positive property in Ω.)

By construction <e$_s$> or "s" is a finite principal ultrafilter, whereas "q" (the complement set) may be co-finite or infinite. Then "imperfect beings" will not be capable of attaining an infinite set of positive properties in "q", not thereby being "divine" and only able to perfect finite sets of positive properties in "s". Then the "imperfect" are capable of completing (as perfecting) some finite ultrafilter by exemplifying virtues. Is it any wonder, then, that Jesus said:

Mat 9:13 But go ye and learn what *that* meaneth, I will have mercy, and not sacrifice: for I am not come to call the righteous, but sinners to repentance.

Existence and Rest Are Equivalent (4.1.2)

The proper and minimal "complement" set of the q-filter in Ω (i.e. not of Ω itself but of "r" maximised to q) is "e$_s$", yet from Gödel's ontological argument[1] I state the "essence" of a God-like being "God-likeness" can only be exemplified in the case of necessary existence. As this gives a principal ultrafilter, God-like "being" must be this same "rest" as above. (The identity or "rest" of the set of all positive properties is entailed as that same necessary existence.)

With "r" maximised as <e$_r$>, and L(G)$^{-1}$&r&e$_r$ on that one side of the disjunction in q, the only action God is not at liberty to stop acting upon is existing. Omnipotence itself is enough of a criterion for (and then is also equivalent to) necessary existence.

With "e$_s$" as above and "g" God's necessary existence, I posit {g} is principal overall in Ω (every positive property is **perfected** in necessary existence for a "God-like" being). Then {e$_s$} is not found in any subset of <e$_r$> or in the maximal "q-filter" in Ω. So, in Ω, for a "God-like" being the predicate "g" must also generate every "God-like" set {x&e$_s$: x in <g>} where <g> is the collection of every set in Ω with that principal element "g". ("g" alone as g=Ω is "God-like".)

Now, "e$_s$" is a positive property, but in the non-principal filter of all "r" (then maximised in {Ω&e$_r$ \ e$_s$}), were <g> equal to this filter in <e$_r$>, there would be found no "x" and "y" in the set of all <g> where the property x={y&e$_s$} is present. All <g> would not form an ultrafilter on positive properties (<e$_s$> free in all Ω is then a finer filter than <g>). So, unless g=e$_s$ (as otherwise all <e$_s$> truly form a *maximal* ultrafilter with some <e$_s$> containing each set in <g>) I find God then has no "essence" (entailing every set in <e$_s$> to be "God-like" or positively perfected) in His necessary existence in "g", even if "God-like" is a predicate – a predicate equivalent to requiring necessary existence as a principal element in the filter of all <g>. I.e. in the set of all positive properties. Yet if all s is maximised as <e$_s$> the same problem occurs, <e$_r$> is then a finer filter than <g>. That is, unless there is only one principal element.

Then, necessary existence is completed in "rest". That is, unless the non-principal ultrafilter (in all q) of Anselm's derivation is considered just as complete in all q = {Ω \ e$_r$} when compared to that generated (as <g>) by necessary existence (as an essence, or an indispensable property) as by Gödel's argument[1]. That itself could lead to a contradiction, equalling the statement that necessary existence is not a predicate, a positive property. (As "g" entails God to be truly "God-like".) I must find some way to mimic such monotheism in the sets of the disjunction.

[1] Sobel, in J. J Thompson (ed.), *On Being and Saying: Essays for Richard Cartwright*, p241-61.

However, there is a simple remedy to that quandary of monotheism in this language!

I may make one side necessary by resting only on one side – I can adjoin e_s to every set in $<e_r>$ with "$e_p = e_r \& e_s$" a principal element that then generates Ω when adjoined to every subset "r" in each set in the "q-filter". (I make the predicate e_s **necessary** in $<e_r>$, so $e_p = e_r \& e_s$ becomes a principal element.) Then I have $x \subset y$ (or there exists some $y \to x$) for all x in "r" and some y in $<e_p>$ (i.e. for any set "r" in q there is always a set in $<e_p>$ that contains "r"). If the statement equally holds that $x \subset z$ (i.e. there exists $z \to x$) for any x in "r" and some z in $<g>$, then $<g> = <e_p>$ and $g = e_p$ (since $<e_p>$ is now maximal as is $<g>$ and both properly contain a near-maximal filter in "r", which is maximal but for an "appropriate" principal element $\{\Omega \setminus e_p\} \vee e_p$). If I could show that there is truly equality $g = e_p = e_r \& e_s$, then necessary existence is a valid positive property and a useful predicate. (I must find such an $e_x = e_y \& e_z$ for every group in the octal. The result is that virtue and/or the additive identity is purely "God-like" and an essence.) $\{\Omega \setminus e_p\} \vee e_p$ also holds on removing e_r, e_s from the disjunction: forcing the statement that necessary existence is not attained by ascending any chain in the set of positive properties.

Without starting from a fallacy of equivocation, I state that every God-like act entails that same "rest", and that the only "fully perfect God-like act" which is duly inclusive "of rest" is God's own continuing existence itself. Rest, "e", is a closed set, so if God's essence is His necessary existence (as from Gödel's ontological argument[1]), then it (necessary existence) is a predicate that may be considered alone without any other apparent logical consequence. I.e. belief in "no God" (an inactive God) is then as close to reasonable as Anselm's fool could state.

Heb 11:1 Now faith is the substance of things hoped for, the evidence of things not seen.

Then $\{\Omega \setminus g\} \vee g$ is startling, as on one side is found all consistent positivity as if "r" but for existence which entails nothing on the other. $\{\Omega \setminus g\}$ is not "God-like" as per Gödel's proof,[1] and wouldn't therefore convince the "fool". So, in the disjunction, either a predicate of existence "g" makes no difference (as to the "fool") or the argument justifiably changes nothing, and all but for a single predicate! (Which when considered alone is also "God-like".)

Heb 11:6 But without faith *it is* impossible to please *him*: for he that cometh to God must believe that he is, and *that* he is a rewarder of them that diligently seek him.

Then the disjunction of "r" in "q" maximised and "e_s" as in $\{\Omega \& e_r \setminus e_s\} \vee e_s$ becomes $\Omega \vee \phi$ on adjoining "e_s" to "r" (as if $r = <e_p>$ were now maximised) and then, finally, without predicating existence, every positive property is generated from Ω completed, which then becomes (in whole) logically (a-priori) necessary (and yet freely decidable?). Regardless of whether necessary existence is assumed a predicate or not, if it is possible and consistent to form a set of all positive properties, then their essence, if there be one, is necessary. So, if such an essence is to state that "God exists", or similarly a principal element exists, being that which justifies His perfection to be wholly possible (as would Descartes infer,[4] reasoning along those lines), then I state that necessary (a-priori or logical) existence "g" is a predicate valid on the necessary decidability of $\Omega \vee \phi$ alone, which when considered complete always (necessarily) entails $g \vee \phi$. (I can state that Ω-completeness on one side of $\Omega \vee \phi$ is enough for consistency and this disjunction then entails logical necessity, equivalence of $g = e_p$ follows and thus "g" is found

[1] Sobel, in J. J Thompson (ed.), *On Being and Saying: Essays for Richard Cartwright*, p241-61.

[4] René Descartes, *Meditations and Other Metaphysical Writings* (Penguin Classics), p53.

necessary. Gödel had shown that necessary existence is also "an essence"[1] distinguishing and generating each "God-like" set; therefore, each such essence is equivalent.)

Then if God is not free to rest on any positive property, then He himself must logically exemplify that property. It is not inconsistent to reason that God is not free to rest from existing itself, and therefore He exists as all consistent logical perfection necessarily if He exists at all (I speak as Anselm's fool). So, as "rest" is a positive property, so, equivalently, is necessary ("God-like") existence. Then for a God-like being, $<g>=<e_p>$ and therefore $g=e_p$ (e_p cannot private any set in $<g>$). Thus, by showing necessary existence is indeed a valid predicate (and without entailing modal collapse – see later in this chapter) with Anselm's argument or with that of Gödel[1], as per above and the below: since an omnipotent God is necessarily existent; God exists.

(An omnipotent God may work any positive outcome He chooses by axiom of virtue. I.e. By equivalence to rest (an outcome of liberty which is always and most certainly a virtue, and most excellently so when perfected in omnipotence – as for rest to be found positive, and to rest upon what great works!), existence is also a positive property even if God does not "choose" His own necessary existence. It is a "closed", or singular property, that privates no other positive property.)

I assert that not only is the "q-filter" (see section [2.9]) in the power set of Ω composed of the complements of the sets in the principal filter $<e_s>$ (in positive properties entailed from virtues), but virtues are entirely a "new addition" and Anselm's ultrafilter (and possibly Gödel's[1] too) is actually incomplete, without virtue. I also state this ultrafilter of virtue is equally complete (Ω-complete) in all positive properties in "r" and "s".

The Empty Essence Lemma (4.1.2a)

Gödel's definition of an essence[1] $\varphi\, ess\, x \Leftrightarrow \forall \psi :(\psi(x) \Rightarrow \Box \forall y:(\varphi(y) \Rightarrow \psi(y)))$ permits the self-difference and empty predicate ϕ to be an essence for **every** individual, i.e. for all possible Ω. (The self-difference is otherwise stated for S as: $S \subseteq T : S \leftrightarrow T\backslash S$.) This is the "empty essence lemma"[2], which holds without the conjunct (see Appendix), yet applies over a disjunction.

However, if the essence is empty there is, again, confusion for the fool in $\{\Omega \backslash g\} \vee g$ as this becomes equivalent to $\Omega \vee \phi$ if the substitution (not an equality) of $g=\phi$ is made. Yet then the disjunction becomes necessary one side and there is no longer any **liberty** to forbear and to rest on "g" (or even Ω itself) as necessarily existent. I may state there is an empty middle in $\{\Omega \backslash g\} \vee g$; confusion of "g" with ϕ results from the statement that there is no modal collapse unless there is no liberty. ($g=\phi$ becomes $N\neg(g \& \neg \Omega)$.) "g" (liberty itself) as "p" in the disjunction of $q \vee p \& q^{-1}$, is a valid predicate for a **being** that is always free to exemplify it. The disjunction is freely decidable, yet after the substitution is made it becomes not so.

Then does the empty essence lemma[2] confuse the empty middle of the disjunction for the essence itself? That the empty self-difference satisfies the definition of an essence is certainly a flaw in Gödel's argument[1]; though there is a perfectly valid essence "g" (in liberty) for a being that may freely decide every disjunction (for that is omnipotence).

Properly, $\Omega \vee \phi$ should, in any such disjunction, always entail necessary existence in g as $g \vee \phi$. The substitution of $g=\phi$ does not preserve that one predicate of liberty and rest. (It

[1] Sobel, in J. J Thompson (ed.), *On Being and Saying: Essays for Richard Cartwright*, p241-61.

[2] Benzmüller and Woltzenlogel Paleo, *An Object-Logic Explanation for the Inconsistency in Gödel's Ontological Theory.*

also falls outside of the ultrafilter in positive properties.) Is there a work-around? There is none to be found if the assumption g=ϕ is axiomatic. The substitution g=ϕ itself causes modal collapse, leaving only a state without liberty or necessary existence. That axiom (which is the flaw in Gödel's argument[1]) is become inconsistent by there being no similarly freely decidable disjunction in the model of the ultrafilter (engendering modal collapse with g → Ω, yet improperly with g=ϕ). The empty essence[2] has confused "g" with ϕ in the empty middle.

There are certainly disjunctions with empty middles that remain freely decidable, and with the essence of "g", Gödel's definition does not permit the empty essence lemma[2] to be applied within the properly formed disjunction of g ∨ {Ω\g}. (For the middle is empty.) The confusion of g with ϕ in a valid disjunction remains to be addressed. "g" is not a member of {Ω\g} to be derived from it as collected and classed "positive". Gödel's axiom Pos(NE) (I.e. Pos(g)) with g in Ω requires extra care.

Weak Notation – The Binding of Predicates (4.1.3)

The notation used up to now has been "weak" in that the statement of p&r^{-1} → s has its logical equivalent in the latter statement of p → r&s (an apparent inconsistency).

Instead of inconsistency I would seek the slightly more effective notation that virtues are bound to the privation of positive properties: So, they only act upon the pairing of a virtue p and a privated property q^{-1}. I can bind these predicates with a "function" or some other suitable mapping. (As it is cumbersome I will only mention it; I do not use it throughout.)

I more formally state that there is always some function $f(p,q^{-1}) \to s$ of two predicates (our familiar p and q^{-1} from before), and the statement p → r&s becomes impossible even if "s" becomes the "empty set". In that case, "q" may become the set of all positive properties Ω (of Anselm's non-principal ultrafilter) leaving only "rest" entailed as "s" from virtue in the principal ultrafilter of my construction: and as no ultrafilter ever contains the empty set, I also equivalently assert that for every **non-trivial** (proper) subset q of Ω I have the binding below:

$$\forall q \in \Omega, \exists s \in T \ : \ T \subset \Omega \backslash q, \ q \not\to s$$
$$\forall s \in T, \exists p \in \Omega \ : \ g(p,q) \to s, \ p \not\to q \,\& \, s$$
$$g(p,q) = f(p,q^{-1}) = p \,\&\, q^{-1}$$

Where the "binding function" $g(p,q)$ is an equivalent to $f(p,q^{-1})$ as above.

Regarding the condition $\forall q \in \Omega, \exists s \in T \subset \Omega \backslash q : q \not\to s$, is Anselm's non-principal ultrafilter co-finite? Perhaps, if there is some "s" not consequent of or entailing q. (Such "s" is possible unless $q = \Omega$.) As q belongs to a filter (a "q-filter") in the power set of Ω, if Anselm's filter is infinite, the power set of Ω contains a filter on its co-finite subsets. If q is not all Ω, there is a possible world W with some "s" in the complement of q by which to form a disjunction with virtue. There is always a principal ultrafilter <e$_s$> given q a proper subset of Ω maximal in W.

Only one side of a disjunction may imply or be entailed from virtue (see section [2.2]). Clearly, q will not entail any p&r with p a virtue; then as p&r is not itself a virtue, it is always a candidate for "s" for the presence of virtue (by axiom). Yet then p&r=s would entail that r^{-1} → Pos(q) and r cannot be found on both sides of the disjunction, so r is "not in q".

[1] Sobel, in J. J Thompson (ed.), *On Being and Saying: Essays for Richard Cartwright*, p241-61.   [2] Benzmüller and Woltzenlogel Paleo, *An Object-Logic Explanation for the Inconsistency in Gödel's Ontological Theory*.

A least virtue (liberty as p) entailing rest is outside of q and some properties in Ω will always constrain the virtue p in scope. If $q = \Omega$ then given the least virtue liberty as my p, I should not expect there to be a least application of liberty found on both sides, entailing "rest" as a predicate "l" as if for some "s" not found in $q = \Omega$. ("l" is posited equal to $p\&l_0$ with "l_0" some least element also entailed from q, such that $l_0 \nrightarrow r$ for any other "r" entailed **from** q. Then l_0 is as close to "rest" as possible, as following from q without any presence of virtue.)

Yet then l_0 (as r above) cannot also be consequent from $q = \Omega$, so, by the axiom of choice of any "least l_0", that l_0 must not be consequent from any q (or r), and so then $r \nrightarrow l_0$. Yet then $p\&r^{-1} \rightarrow p\&l_0 \rightarrow e_s$ at least, (e_s being the only possible "rest" which is principal) and there is nothing found in the middle, not even by a "least element" found from the axiom of choice.

To conclude, $\forall q \in \Omega, \exists s \in T \subset \Omega \backslash q : q \nrightarrow s$ requires there (a-priori) to be a disjunction. If there is no possible "s" (but for e_s), there remains a freely decided disjunction with no requirement for any other application of virtue than e_p. If "s" is empty, there is no other binding of virtue found; the principal elements equivalent to God's self (His being) must remain outside of Ω.

Otherwise, $p \rightarrow \Omega$ or its equivalent $N\neg(p\&\neg\Omega)$ could equal $N\neg(p\&\phi^{-1}\&\neg\Omega)$ if that **trivial** binding of $s = \phi$ is included. This is possibly an inconsistency; ϕ is not a positive property, and I should avoid any antinomy of $p \rightarrow r\&s$. (ϕ will prove any disjunction, $\phi \rightarrow A\&B$ in all $A \vee B$.)

I.e. what does $N\neg(p\&(r\&s)^{-1}\&\neg\phi)$ mean? The inconsistency appears when r&s or ϕ is considered a positive property (there is no middle). If I have $\phi^{-1} \vee p\&(r\&s)^{-1}$, is ϕ^{-1} necessarily Ω? Then what is $\phi \vee p^{-1}\&(r\&s)$ as a disjunction? Clearly $N\neg(p\&\neg(r\&s))$ would be equivalent to $N(p^{-1}\&(r\&s))$, and by forcing a conjunction I find: $N\neg(p\&(r\&s)^{-1}\&\neg\phi^{-1})$, a clear contradiction.

4.2) Predicates Of Sets

The principal elements e_r, e_s are somewhat the "infimum" of the sets A and B whereas r and s may be generalised to become the "supremum" of A and B. (Where A and B are given by...)

A $\{r: Pos(q) \rightarrow Pos(r)\}$
B $\{s: Pos(p\&q^{-1}) \rightarrow Pos(s)\}$

Now the "predicates of sets"[5] are given by the relations

$$\vee F = \{x : f \rightarrow x, \text{ for all } f \text{ in } F\}$$
$$\wedge F = \{x : x \rightarrow f, \text{ for all } f \text{ in } F\}$$

And I may equate my r, s, u and v to the following predicates of sets:

$$r \leftrightarrow \wedge A = \{x : Pos(x) \rightarrow Pos(a), \quad \forall \ a \in A\}$$
$$u \leftrightarrow \wedge A^{-1} = \{x : Pos(x^{-1}) \rightarrow Pos(a^{-1}), \ \forall \ a^{-1} \in A\}$$
$$s \leftrightarrow \wedge B = \{y : Pos(y) \rightarrow Pos(b), \quad \forall \ b \in B\}$$
$$v \leftrightarrow \wedge B^{-1} = \{y : Pos(y^{-1}) \rightarrow Pos(b^{-1}), \ \forall \ b^{-1} \in B\}$$

And trivially (with rest as "e_x") over all such X I may write:

[5] Christopher G. Small, *Reflections on Gödel's Ontological Argument*.

$$e_x = \vee X = \{\theta \; : \; x \to \theta, \text{ for all } x \text{ in } X\}$$

The above predicates of sets "$\vee F$" etc. were introduced upon the recommendation to reference a paper by Christopher Small of the University of Waterloo entitled "Reflections on Gödel's Ontological Argument".

Now, is $\wedge(A \cup B)$ a virtue? Is $\wedge(A \cap B)$? In truth neither is virtue unless the set of virtues is not closed. The former is a possibility for a new disjunction, whereas the intersection of A and B is empty and the latter statement is void.

What of the predicates $\vee(A \cup B)$ and $\vee(A \cap B)$? The sets A and B are disjoint, so the operator on the union must either private virtue and cannot be positive or the intersection is empty and is never a consequence of a positive property in both A and B, unless the action of the predicate is a complete cessation of everything positive; $N(Pos(p^{-1}))$ for all positive properties "p" in A or B. (p cannot be a positive property in the ultrafilter.)

4.3) The K4 Form Of The Ultrafilter

Solving for $\{X, C\} = ((X, A) \vee (X, B))^c$ is now simple. I have the full K4 form of the ultrafilter;

X	{p, p a virtue}
A	$\{p\&v, p\&s^{-1}\} = \{r, u^{-1}\}$
B	$\{p\&u, p\&r^{-1}\} = \{s, v^{-1}\}$
C	$\{u\&v, r^{-1}\&s^{-1}\}$, r, s, u and v as above.

Yet, where did the set C come from?

Given any positive property T in A&B equal to some $(r\&u^{-1})\&(s\&v^{-1})$, I note that N¬Pos(T) and N¬(T) holds (for God). Then N¬(T&¬Y) for all N¬Pos(Y). I may put Y = ¬L(G) with all virtue in L(G) negated. A minimal "cut" of liberty results in N¬(p&(r&u^{-1})&(s&v^{-1})) and by rearranging I find:

N¬(p&r&s&¬(u&v)) or, p&(r&s) → (u&v), so I find the disjunction $(r\&s)^{-1} \vee (p\&(r\&s) \to (u\&v))$.

Then $C = \{u\&v, r^{-1}\&s^{-1}\}$ is a well formed set of positive properties also **consistent** with p (in X).

This chapter will show that this set of seven properties is "closed".

Besides the virtues in X, the sets A, B and C are formed from the sets r, s, u and v which are all disjoint to X and each other. (The same may be said of u and u^{-1}.) By construction the sets in A={r, u^{-1}}, B={s, v^{-1}}, C={u&v, (r&s)^{-1}} are all positive privating no "p" in X. Then the K4 product $x + y = (x \vee y)^c$ may also be made in the symbols {p, r, s}, {p, u^{-1}, v^{-1}}, {p, (r&s)^{-1}, u&v} as equally as with {X,A}, {X,B} and {X,C}. (The sets p, r, s, u^{-1}, v^{-1}, u&v, $(r\&s)^{-1}$ are effectively disjoint.)

Naïvely, $r \leftrightarrow s^{-1}$ and likewise $v \leftrightarrow u^{-1}$. Then, what does not positively act following from q instead acts following q^{-1} (and vice versa). I would know that the sets of **positive** properties in the intersections $\{r\} \cap \{u\}$ and $\{s\} \cap \{v\}$ are empty! (It is not positive to simultaneously act and not act upon a property in the filter.) Since $v \leftrightarrow u^{-1}$ etc. {r,s} is a set of positive properties in action and {u,v} not so; therefore {r,s} and {u,v} are disjoint in positive properties.

Then by the distributive laws (given r and s are disjoint sets) I have a divided union of $\{r,s\} \cap \{u,s\}$ and $\{r,s\} \cap \{r,v\}$ equal to $\{s\}$ and $\{r\}$ respectively and likewise disjoint. Likewise $\{r,v\} \cap \{u,v\}$ and $\{s,u\} \cap \{u,v\}$ are equal to $\{v\}$ and $\{u\}$. Then $\{r,s\}$ and $\{u,v\}$ together partially exclude both the sets $\{u,s\}$ and $\{r,v\}$. (So, $\{r,s\}$ contains no "u" or "v" and $\{u,v\}$ contains no "r" or "s".) Then as "u" and "s" etc. are disjoint $\{r,s\}$ and $\{u,v\}$ are disjoint. (That u and s are disjoint as are v and r is shown later in sections [4.6.2] and [4.6.3].)

The sets $\{u,s\}$ and $\{r,v\}$ are disjoint and not equal to A and B above ($v \leftrightarrow u^{-1}$ etc. does not hold).

Then $\{r,s\}$ and $\{u,v\}$ together contain the content of A and B. Therefore $\{r,s\}$ is not virtue and $\{r,s\}$ is not a subset of C. (And similarly for $\{u,v\}$.) Lastly, $\{r,s\}$ and $\{u,v\}$ are not equal as I have the disjunction $(r\&s)^{-1} \vee u\&v$ (naïvely $v \leftrightarrow u^{-1}$ etc.), then given virtue in p, the sets $\{r,s\}$ and $\{u^{-1},v^{-1}\}$ each exclude the content (or sets) $(r\&s)^{-1}$ and $u\&v$ which are also disjoint.

Then the triples $\{p, r, s\}$, $\{p, u^{-1}, v^{-1}\}$, $\{p, (r\&s)^{-1}, u\&v\}$ also form a K4 form under $x + y = (x \vee y)^{c}$.

The sets of properties in $(r\&s)^{-1}$, $u\&v$ are equal to all those excluded from the products r+s and $u^{-1}+v^{-1}$. Then the K4 product as above holds upon those triples in the K4 form equally.

4.4) Pass One – Good Things Come In Threes

I now also have a partially defined product on the octal in the symbols $\{A,B,C,D,E,F,G\}$.

	Ω	A	B	C	D	E	F	G
Ω								
A			C	B	E	D	G	F
B		C		A				
C		B	A					
D		E				A		
E		D			A			
F		G						A
G		F					A	

Where A=p, B=r, C=s, D=v^{-1}, E=u^{-1}, F=u&v, G= $r^{-1}\&s^{-1}$.

Note $(p\&u^{-1}) \vee (p^{-1}\&v^{-1})$ is not equivalent in the schema to $(p\&v) \vee (p^{-1}\&u)$ so that a different K4 form would hold: i.e. with p, $\{v,s^{-1}\} = \{r, u^{-1}\} = \{r,v\}$ etc. However the set of all positive properties Ω includes u and v rather than u^{-1} and v^{-1} which are positive at rest in the schema.

Note that u and v are disjoint sets as well as also being disjoint from r and s. Then what is not in action following q is in action upon $p\&q^{-1}$. Yet u^{-1} is not r, neither is it s. (It is disjoint.)

The conjunction $(u\&v)^{-1}$ is not in scope in the K4 group $\{p, u^{-1}, v^{-1}\}$: But then u is not in action on either side, being disjoint from r and s, and v is not in action either side also; u&v as an excepted and disjoint case may yet be in action. It is necessary in a filter that given $\exists\, u\&v$, then $Pos(u\&v) \leftrightarrow N\neg(Pos(u\&v)^{-1})$ and u and v in $\{p, u^{-1}, v^{-1}\}$ are reconciled in the mapping table (as below) with the substitutions of $p\&u \rightarrow v^{-1}$ and $p\&v \rightarrow u^{-1}$. In other terms, u^{-1} and v^{-1} are at rest, so how may (u&v) (a positive property) private any (acting) virtue in p? (It cannot, even if I have N¬(p&¬(u&v)⁻¹) or (u&v) → p⁻¹, u&v privating a virtue. Then that indexing set is limited to the union of {p, u⁻¹, v⁻¹} only.) u&v is not extant of virtue in that triple; there is no privation.

These seven symbols naturally form sets of triples (see the "filter" table to follow). Each triple of each row of the filter table is also as an element of an octal (with each triple added with (a ∨ b)ᶜ) together (in that octal) forming K4 subgroups each made of three rows. That operation on rows defines an operation of an octal group on the singletons (r, s, etc.) that are the intersections of those three triples/rows forming K4 groups. That octal is as "The Father".

4.5) The Mapping Table

I should also note that I have in my inferences the following "**mapping table**":

$$\left.\begin{array}{r} v \\ s^{-1} \end{array}\right\} \to q\ \&\ p^{-1}\ \to\ \left\{\begin{array}{l} r_0\ \&\ p^{-1} = r \\ u_0^{-1}\ \&\ p^{-1} = u^{-1} \end{array}\right.$$

$$\left.\begin{array}{l} u = u_0\ \&\ p \\ r^{-1} = r_0^{-1}\ \&\ p \end{array}\right\}\ \to\ q^{-1}\ \&\ p \to \left\{\begin{array}{l} s \\ v^{-1} \end{array}\right.$$

I make note of the following partiality of virtue (for instance "r" does not necessarily private "u" or (r&s)). I can in several cases take advantage of the lack of privation in my predicates as if I were using virtues; for if "y" does not private "x⁻¹" or "z", y becomes indiscernible from virtue in that case. You should already be familiar with the top three rows of the following "filter" table.

	may not private	or private
p	s	r^{-1}
p	u^{-1}	v
p	$(u\ \&\ v)$	$(r\ \&\ s)$
$u\ \&\ v$	v^{-1}	r^{-1}
$u\ \&\ v$	u^{-1}	s^{-1}
$(r\ \&\ s)^{-1}$	r	u
$(r\ \&\ s)^{-1}$	s	v

In other words, r ∨ r⁻¹&(u&v) → v⁻¹ requires N¬(v&¬Pos(r&u⁻¹)) which in the mapping table is correct. There is no excluded middle as v⁻¹ and r are disjoint.

v&s⁻¹ → r and then v&(r&s)⁻¹ → s, so v⁻¹ ∨ v&(r&s)⁻¹ → s is obtained simply enough.

("v" may not private s, neither v private r or r₀ etc.) Also, r and r⁻¹ and also v and v⁻¹ are disjoint so there is no excluded middle here. Note that v and s as well as u and r are disjoint pairs.

Then in each case I may write:

$$(y\ \&\ x^{-1} \to z) \vee x$$

Where "y" is from the first column in the table above (as if a virtue), and "z" from the middle column and "x" is as the inverse of the entry in the rightmost column, so y will not private x^{-1}.

Every row of the filter table satisfies a closed and separate (yet valid) filter on triples of positive properties. I implicitly define the operation of an octal group on singletons within each triple and walk them through the octal with a seven cycle. The operation $(x+z) = (x \vee z)^c = y$ is valid on them, and when "walked through the octal" I obtain the whole ultrafilter of the octal.

There is also a completely separate solution formed by symmetry or similar logic. Each row that operates without virtue may form different triples. I have an "alternate mapping table".

	may not private	*or private*
p	s	r^{-1}
p	u^{-1}	v
p	$(u \& v)$	$(r \& s)$
$u \& v$	u^{-1}	r^{-1}
$u \& v$	v^{-1}	s^{-1}
$(r \& s)^{-1}$	s	u
$(r \& s)^{-1}$	r	v

For instance, $(v \& s^{-1} \rightarrow r)$ so $v^{-1} \vee v \& (r \& s)^{-1} \rightarrow r$ as before (and similarly so for the other rows).

This alternate table is found by switching a pair of elements from one set of $\{r,s\}$, $\{u,v\}$, $\{(r \& s)^{-1}$, $u \& v\}$ within the original mapping table. This yet has a solution in the K4 form of the filter.

$$1 \equiv a \equiv [0, \ a, \ b, \ c]$$
$$b \equiv [0, \ a, \ d, \ e]$$
$$c \equiv [0, \ a, \ f, \ g]$$

The K4 form of the filter shows that if [b,d,f] is a subgroup (to the right hand) then that correspondence cannot stop [c,e,g] being also a subgroup (to the left hand) and vice-versa. The alternates occur between such pairs $\{r,s\}$ etc., once the element acting as virtue is excluded, yet these two filter tables still fulfil the operation of the K4 filter split to the left and right.

Given these alternate octals, only one of the two may be consistent, for if I repeat one correspondence I find the octal is overconstrained. For example, if I hold $v^{-1} \vee v \& (r \& s)^{-1} \rightarrow s$ as well as $v^{-1} \vee v \& (r \& s)^{-1} \rightarrow r$, then I may write $v^{-1} \vee (r \& s)^{-1} \& v \rightarrow r \& s$ and I break the closure of v with $v \rightarrow r \& s$. I have then overconstrained the set "v" and have broken the binding with r&s.

I may not stop there, for in the octal I may rearrange (up to swapping sides of the disjunction) each triple threefold. I may rearrange each disjunction to one of:

$$r \vee \left(r^{-1} \& (u \& v) \rightarrow v^{-1} \right)$$
$$(u \& v)^{-1} \vee \left(r^{-1} \& (u \& v) \rightarrow v^{-1} \right)$$
$$r \vee \left(v \& r^{-1} \rightarrow (u \& v)^{-1} \right)$$

Each element (albeit negated) is able to act "as virtue". (And so for the other rows of the filter tables.) These three identities are rearrangements in the same logic: as if from

N¬((u&v)&r^{-1}&¬v^{-1}) to N¬(r^{-1}&(u&v)&¬v^{-1}) and to N¬(v&r^{-1}&¬(u&v)$^{-1}$). They never hold together for an octal under the Frobenius map, as virtue needs to be "negated" to appear alone on each side. Disjunctions then become "decided"; one side becomes necessary.

The Frobenius map does indeed play a part; it reveals a great deal about the form of the set Ω of all positive properties.

With the schema of action and inaction held static, I may take advantage of these equivalent arrangements of the octal, for they reveal that the logic of the mapping table is somewhat fixed with the action of unity.

I can cycle the three subgroups with virtue using the two disjoint three-cycles of (u&v, v^{-1}, r) and ((r&s)$^{-1}$, u^{-1}, s). Then the subgroup [u&v, v^{-1}, r] is held static; and the remaining three rows are permuted amongst themselves in a three-cycle.

This holds for the alternate filter table as well by taking those two separate three-cycles as above. The rows of each table are cycled amongst themselves with these two three-cycles, holding virtue in p fixed. I somewhat set the scene for the seven cycle of the Holy Spirit; the octal is not "wasted effort". The Frobenius map will provide the necessary three cycles with virtue (unity) fixed. (Though they cannot hold fixed that octal's logical schema.)

It is important to note that positive properties in singletons **do not** form an octal group. Rather the octal as an ultrafilter indexes the set of all positive properties in seven disjoint subsets (in every way that virtues may provide free choice amongst alternative and positive outcomes). Each row of the filter table obeys the operation of "K4 group addition" with every other row and with the same operation for addition used locally, as within those seven K4 groups.

So with (x+z) = (x $\vee$ z)^c = y, I have both A+B=C in the rows and [A,B,C]+[A,D,E]=[A,F,G] in every K4 form. The octal (Ω) itself also acts as a suitable identity for addition on K4 groups in the octal, as well as for addition on singletons within K4 groups, also using that operation.

4.6) Pass Two – Self Similarity Every Way You Cut It

Then in each case (row in the filter table) I may form the product ({y, x} $\vee$ {y, z})^c = {y, (x&z)$^{-1}$}.

The sets in the filter (x&z) are the conjunction of every pair of those elements in x and z, and these are "almost" empty (not in the filter). They rightly display that Pos(x^{-1}&z^{-1}) and all it entails is elsewhere in the octal. Yet x&z privates the set in "y" of some positive property. (Those properties are seemingly recovered in the sum ({y, x} $\vee$ {y,z})^c = {y, (x&z)$^{-1}$}.) (A + B = C.)

The Rows Of The Filter Table Are Closed (4.6.1)

From the mapping table I find v → p^{-1}&q → p^{-1}&u$_0$$^{-1}$ as well as s^{-1} → p^{-1}&q → p^{-1}&r$_0$. But then (by Theorem 2.8) I find the following:

v → p^{-1}&q → Pos(q) → Pos(u$_0$$^{-1}$) as well as s^{-1} → p^{-1}&q → Pos(q) → Pos(r$_0$)

So then N¬(Pos(v))&N¬(Pos(u$_0$)) and N¬(Pos(s^{-1}))&N¬(Pos(r$_0$$^{-1}$))

The properties u_0&v and $(r_0$&s$)^{-1}$ are inconsistent given either u or v as well as r or s respectively. For Pos(u_0&v) becomes Pos(u_0&u_0^{-1}) and Pos($(r_0$&s$)^{-1}$) becomes Pos($(r_0$&$r_0^{-1})^{-1}$). Then, Pos(x) ↔ ¬Pos(x) and the two conjunctive statements become undecidable as to their positive content.

Of course, in the general triple {y, x, z}, if I have the inference N¬(y&z^{-1}&¬x) I may rearrange to N¬(y&¬(x&z)) and take the modus tollens N¬((x&z$)^{-1}$&¬y^{-1}) so (x&z$)^{-1}$ privates virtue. Yet then, neither x&z nor (x&z$)^{-1}$ is a positive property; x&z is inconsistent with the filter.

If y is not a virtue, it (y) is still a closed set (see below) and already "in lowest terms" (by the predicates of sets) and by y not privating x or z^{-1} the property (x&z$)^{-1}$ yet privates y which is positive by axiom, which is a contradiction. x&z is still inconsistent with that filter.

So given the product ({y, x} ∨ {y, z}$)^c$ = {y, (x&z$)^{-1}$}, if I make the product ({y, x} ∨ {y, (x&z$)^{-1}$})^c = {y, z} then {y, x} cannot be "in scope" with (x&z$)^{-1}$ as then given y: x^{-1} → z and I may only form the product ({y, x} ∨ {y})^c = {y, z} or indeed, ({x} ∨ {y})^c = {z}, all because (x&z$)^{-1}$ is effectively empty of positive properties.

Then u_0&v is not a positive property in the ultrafilter which has for its indexing set the union of {p, u^{-1}, v^{-1}}, which now in the elements (a, b, c) is also become a closed K4 group under the operation (a ∨ b$)^c$ = c. Similarly $(r_0$&s$)^{-1}$ is not a positive property in the ultrafilter with the indexing set the union of {p, r, s}, which also obeys the same operation as for a K4 group. Both ultrafilters are closed in their triples.

In the general sense, for any row in the filter table, there is an "ultrafilter" which has as its indexing set the union of the three properties (a, b, c) from that row which also forms a K4 group under the operation (a ∨ b$)^c$ = c and is closed in those triples. (I must have that y, the entry in the first column of the filter table is assumed necessarily positive, as if a virtue. I.e. as if by axiom. For then Theorem 2.8 holds just as equally.)

Returning to the filter table: If x and z are disjoint from y and each other, it is an observation to note that the (maximal or otherwise unconstrained) sets in my "algebra of predicates" (r ↔ ∨F etc.) enforce the rule that every singleton in each row in the "filter" table is disjoint from its fellows; since in each case the conjunction x&z is then not a positive property. (And then no "y" in the filter is constrained.) So ({y, x} ∨ {y, z}$)^c$ = {y, (x&z$)^{-1}$} reduces to (x ∨ z$)^c$ = y.

The sets u, u&v and s are disjoint (4.6.2)

The sets u and u&v are not disjoint in positive properties, the properties in u^{-1} are positively non-exemplified, whereas the properties in u&v most certainly are positively exemplifiable. As a result, in the mapping table they are to be treated as disjoint, since one set is potentially in action (u&v) and one set is not (u^{-1}). Then u is "negative" and u&v is "positive".

The disjunction u^{-1} ∨ s given u&v holds with u → p&q^{-1} → s but I have u_0 → q^{-1} so that u_0 in the mapping table entails no constrained or closed subset of virtues from p (not itself being a virtue, it privates q) as does u, so if a set of virtues exists necessarily, then (by the axiom of virtue) u = u_0&p rather than u_0 in the proper sense becomes indistinguishable from some virtue in p and the sets u_0, s and (u_0&v) are then disjoint sets and I may form a K4 group as if I were mapping with virtues as before. (Pos(q) → Pos(u_0^{-1}) is positive and correct so that the modus tollens $u \rightarrow p\&u_0 \rightarrow Pos(p\&q^{-1})$ but not $Pos(q^{-1})$: i.e. $u_0 \rightarrow q^{-1}$ is a strictly negative statement and u_0 is thus a disjoint set from s and v^{-1}; $Pos(s)$ and $Pos(v^{-1})$ are both strictly positive.)

The sets v, u&v and r are disjoint (4.6.3)

An immediate statement reached by symmetry (see section [4.1.1], "Equivalency"). Yet there may be more to be gleaned from the mapping table. Now, $p\&q^{-1} \rightarrow v^{-1}$ but the set of p is closed and v^{-1} is not a virtue. Then at most $v \rightarrow v\&q$ which, if also a positive property, cannot private a virtue. Then $v \rightarrow Pos(q\&p^{-1})$ is not possible for any "positive" properties v and q from the mapping table (for $v \rightarrow q\&p^{-1} \rightarrow p^{-1}$ and v must not be a positive property as it would private a virtue). So the conjunction $v\&q \rightarrow p^{-1}\&r_0$ privates p and cannot consistently or positively entail r_0 which (also closed as an ultrafilter) is all the positive content of r. So "v" and "r" are effectively disjoint. ($Pos(T)$ where $T=(q\&p^{-1})$ does not entail that $\Omega \setminus q^{-1}\&p$ is necessarily positive, rather is it true that since the set of virtues X is closed, $X \setminus p\&q^{-1}$ is positive instead. I.e. q^{-1} is no virtue so $X = X \setminus p\&q^{-1}$. In other terms, divine freedom does not constrain itself.)

Liberty (An Argument Against Modal Collapse)

Liberty, e_p generates the ultrafilter; so $N(e_p\&q^{-1} \rightarrow e_s)$, where "$e_x$" is to "rest on the outcome of a completed act" and "e_s" is the minimal "s". This minimal "e_s" must always be present (else God is not at liberty if He must necessarily act with all virtue, or never cease working), though God is free to "rest" indefinitely.

Liberty is a virtue that allows the divine (p) to act on any proper subset s_o of S, where S is the set {s in Ω : $p\&q^{-1} \rightarrow s$, p in X}. Elements of S not in that set s_o are instead found in the set T = {v in S : $p\&q^{-1} \rightarrow v^{-1}$, p in Y} where Y is the subset of the virtue in X such that Y replaces the relaxed virtue that would otherwise entail the remnant of S to be wholly in action rather than only that proper subset s_o.

By "axiom of virtue", then, liberty entails a "minimum necessary subset of virtue" p_o in X to entail the subset s_o of S. Note that the complement of p_o in X is the set Y = {p in X, $p\&q^{-1} \rightarrow v^{-1}$} found entailing the set T of all v^{-1}, itself a positive result. Likewise, there are properties w^{-1} entailed from q^{-1} without exclusion of p in X by liberty; these w^{-1} are not positive in the disjunction unlike those properties in S and T. As w entails q and all r and is disjoint from all v, I find the two sets v and r are disjoint. By symmetry I have the same for s and u. (They are also disjoint.)

As x, y and z are in fact "nearly disjoint" as sets then there is no positive middle (x&z) and $x + z = y$ where the operation $(x+z) = (x \vee z)^c = y$. The sets become "sat in place" in the partially defined "octal" once more. In this manner, I reach self-similarity (agreement) with the operation upon singletons as with the operation of the ultrafilter whole rather than found in infinite regress with every K4 subgroup in the octal aligned to a spanning K4 form.

4.7) Pass Three – Completing The Octal Ultrafilter

Then, $(x \vee y)^c = z$, $(x \vee z)^c = y$, $(y \vee z)^c = x$. (I see the seven K4 groups of the octal from the seven rows of the table above.) And I may further complete the addition table to become:

	Ω	A	B	C	D	E	F	G
Ω								
A			C	B	E	D	G	F
B		C		A	F	G	D	E
C		B	A		G	F	E	D
D		E	F	G		A	B	C
E		D	G	F	A		C	B
F		G	D	E	B	C		A
G		F	E	D	C	B	A	

Assuming no middle (x&z) and $\Omega = x \cup y \cup z$ the "local closure of positive properties", the set of x, y and z (x, y, z are disjoint so that there is no positive middle (x&z) etc.) the above table is completed. (For singletons or the K4 form, the identity is the set of all positive properties.)

	Ω	A	B	C	D	E	F	G
Ω	Ω	A	B	C	D	E	F	G
A	A	Ω	C	B	E	D	G	F
B	B	C	Ω	A	F	G	D	E
C	C	B	A	Ω	G	F	E	D
D	D	E	F	G	Ω	A	B	C
E	E	D	G	F	A	Ω	C	B
F	F	G	D	E	B	C	Ω	A
G	G	F	E	D	C	B	A	Ω

Then I have a group isomorphic to the octal. Note that the identity is not the empty set, but the complete set Ω of all positive properties of all triples. It is the case that the mapping table holds for any properly closed subset of positive properties in Ω. (Which likewise closes the set of virtues by axiom of virtue.) So then, the power set of all positive properties as q renders the extension of Ω complete. (And the infimum/extremum of the subsets u, v, r and s apply.) Then every such set of virtues is generated by one class – the liberty due sovereign God (that over-arching authority to choose any possible set of virtue for a positive outcome, or even to rest upon that liberty).

That the set Ω acts as the identity naturally follows from the definition of the operation x + y = $(x \vee y)^c$ = z. However, it has some small meaning in the filter table.

I previously stated that liberty, "L", the conjunction of all e_x, was as zero upon the singletons, the predicates of sets. Here, "L" is found alone (i.e. with virtue) on one side of a disjunction whenever the complement side of the disjunction is the whole set Ω (see section [4.1.3]). So, every case of "e" appearing alone reduces to one of the following cases. "L" by Gödel's argument[1] is also an essence and entails all Ω, for which I require a notion of modal collapse in L → Ω → L. This will be examined much later in [16.12].

For a virtue p also in Ω (mere positive properties to either side cannot imply virtue; some positive properties in the middle are excluded, here being the virtues in p) I have:

[1] Sobel, in J. J Thompson (ed.), *On Being and Saying: Essays for Richard Cartwright*, p241-61.

$$(\Omega \vee p \,\& \,\Omega^{-1}) \rightarrow (\Omega \vee p \,\& \,e)$$

As the set of virtues cannot private any property in Ω all virtue in p is excluded in the middle and this results in "rest" only, a single positive property (without those virtues) entailed upon the right-hand side. Then the product (the complement $(\Omega \setminus p \vee e)^c = p$) may only be the set of all such "p". The set of positive properties entailed from some other q (not a virtue but here present as p) is also closed; as it is also within the consequent set resulting from a virtue (by axiom of virtue) and is also not so fine a filter. (Or else it is equal in effect to the virtue L(G) because of the property "e" entailed, being indiscernible.)

If "p" is merely a positive property rather than virtue; then if on the right hand side I hold true p&p^{-1} it then renders p inconsistent and impossible to exemplify. (As it belongs in Ω.) Then with p as part of the excluded middle the symmetric difference $(\Omega \vee p \,\& \,\Omega^{-1})$ is simply $(\Omega \setminus p) \vee e$ and the complement p. (Ω acts as identity and Ω^{-1} is "empty" or simply as "e" which is disjoint from $\Omega \setminus p$.)

Likewise if the disjunction formed is as the following (p not restricted to the set of all virtues):

$$(p \vee \Omega \,\& \,p^{-1}) \rightarrow (p \vee p \,\& \,(L(G) \,\& \,p^{-1})) \rightarrow (p \vee p \,\& \,\Omega \setminus p) \rightarrow (p \vee \Omega)$$

I.e. were Ω to include all virtue, the right-hand side would entail every positive property that may be entailed upon p^{-1} by axiom of virtue and of all liberty in L(G) (**such virtues exist**). Then the complement of the symmetric difference is simply p. (Pos(p^{-1}) is empty for p a virtue.)

I also have:

$$p \,\vee\, (p \,\& \,p^{-1})$$
$$Pos(p) \vee Pos(p) \,\& \,Pos(p^{-1})$$

Then p cannot be positive with the right-hand side inconsistent; because the positivity N(Pos(p)) is solely based on its own inconsistency. There is then no such virtue or positive property p. The complement of all such p (as truly empty) in Ω is consistent and positive. Yet the product $(\phi \vee \phi)^c$, the set of all such "not p" (inclusive of virtue) is Ω when closed. That set results by axiom of virtue. (The only excluded properties are those that are likewise inconsistent and so are not positive.)

Such a group may seem strange, as true and false are opposites, and one may rightly ask, "Why should didactic absolutes depend on a third property or a 'virtue'?" Those groups or triples completed are somewhat equivalent to finding the general solution to the statement: "The boy racer had fitted a car stereo that weighed more than his car but less than the rust in it". One would immediately deduce that the car stereo is decidedly rusty too.

4.8) The Uniqueness Of Virtue

Indiscernibles (4.8.1)

The axiom of virtue states that any virtue p that acts on a privation of a specified predicate to entail any other given (and valid) positive predicate is unique. That will cause most some difficulty, as one would then expect that given the same real-world choices two different people acting with virtue to produce the exact same result must then have the same "motivation". If there are two such individuals acting with, say, two different virtues in the same manner, it could be argued that the virtues are not disparate: they share some common virtue p that will entail that same result, "s" say, from $p\&q^{-1}$.

Yet, the question begs the answer, "Is virtue unique to the ultrafilter used to solve the disjunction"?

There may be many solutions to the same disjunction, yet the condition of uniqueness applies only within each application of the ultrafilter in the set of virtue itself, as one set of seven. Given that virtue, the octal group partitions the set of all positive predicates in the disjunction into disjoint sets each forming an ultrafilter. Uniqueness, then, has no scope outside of these sets, and so all real-world events judged are "custom-tailored" to the character of the individual, and not to the diktat of any God, or of any ethical necessity to do it all "one way". The answer is that there is always scope for free will, especially in doing good:

Mat 12:5 Or have ye not read in the law, how that on the sabbath days the priests in the temple profane the sabbath, and are blameless?
Mat 12:6 But I say unto you, That in this place is one greater than the temple.
Mat 12:7 But if ye had known what this meaneth, I will have mercy, and not sacrifice, ye would not have condemned the guiltless.

If x and y are sets of virtue solving the same disjunction the exact same way, so, technically, is their conjunction. Is this set constituted of unique member predicates in its own ultrafilter?

If the predicates x, y entail the same result on $x\&q^{-1}$ and $y\&q^{-1}$, x and y are indiscernible; then, locally, does x=y? I would infer that for the truly omnipotent (the perfected) that the axiom of virtue's statement on uniqueness always holds, but when the total set of positive predicates in the disjunction is evaluated at or taken within a proper subset of those predicates, virtues evaluated within that restricted set give rise to indiscernibles. (For such virtues will surely exist if that is the case.)

The only caveat may be found in the statement that, say, $x\&q^{-1} \rightarrow r_1\&s_1$ and $y\&q^{-1} \rightarrow r_1\&s_2$. Now, if x and y are assumed indiscernible, the condition of uniqueness would render $s_1=s_2$. Is this acceptable or not? Out of liberty, there would also be a property by axiom of virtue that would entail s_1 and another that would entail s_2: it should be possible to relax both x and y completely and retain either of those virtues necessary by axiom of virtue to entail either s_1 or s_2. There should remain virtues for this! Then, x and y may not be indiscernible, because they indeed so **split**, and there is also by axiom some other property to entail r_1 as required.

The problem occurs when the set of virtue has two predicates with an indiscernible result r, but even so, if liberty relaxes that predicate r consequent of the disjunction in the "real world", then out of liberty the requirement for both such virtues x and y, vanishes! In the set of virtue, then, that predicate present in the disjunction entailing "r" is actually by axiom equivalent to x&y. If it is forcefully argued only one or the other, then equivalently, the case is x=y, or they are not indiscernible.

So, then, as to the only cases remaining: if I have two indiscernible positive predicates s_1 and

s_2 entailed from applying two indiscernible virtues x and y, there is again no conflict unless one or both of x and y entails both s_1 and s_2. (Relaxing either s_1 or s_2 may leave one of x, y remaining.) Yet as indiscernible predicates, s_1 and s_2 are effectively equal in the disjunction as outcomes: as in their own ultrafilter they must entail (and necessarily so) the same subset of positive predicates. To relax each of them by liberty would certainly relax that same subset. The desired result to relax the same subset in a real-world situation **must** then relax **both** predicates conjoined $s_1\&s_2$ and then also, as above, relax x&y: or, otherwise, if they entail different sets even in part then they must split as above, reducing to a simpler case; not a problem unless all is indiscernible; a most singular mess! Only one predicate then effectively remains.

Sets of indiscernibles could be given distinct integer values to make them unique so that conjunctions (and also intersections), and symmetric differences etc. may all relate in terms of the encoding of any statement of number theory. Uniqueness (as can be argued) is positive, preserving the octal over sets in the disjunction; for they must remain disjoint. The consequent results of liberty (again to be found in the octal) are a requirement for this uniqueness of virtue. Liberty is always positive for the omnipotent. Is uniqueness "God-like"?

God would not use indiscernibles, not requiring them; though he would "number" them in every proper subset of Ω appropriately. Only the principal element in Ω (zero in the octal) would require the whole of Ω to be so encoded (even as far as the hyperreals?), and plausibly every statement in number theory must then be formulated. (Yet will logical self-reference[6] reign over Christ's regressing closure?) There would then be equivalence between the number theory required to index every indiscernible (all) for the zero (the least), and also for God who is required to index a subset for all others except for Himself; always using that same system equivalent to the knowledge of all Ω liberally. The equivalence is that quite formally, every positive property would be enumerated whilst described perfectly. Any embedding of a number theory within the Trinity is a subject far beyond the scope of this writing.

So, to repeat, the ultrafilters in every set of the octal are principal and remain so. If there are two indiscernible virtues that entail the one predicate outcome of the disjunction, then relaxing by liberty that outcome as not present in a set in the ultrafilter is to relax both or all such indiscernible virtues.

Similarly, if there are two indiscernible outcomes then to relax the real-world action that is extant of the disjunction by liberty, both predicates, as well as both (or all) virtues that entail them, must also be relaxed by that liberty. Anchoring the disjunction in the real world requires the predicates of "indiscernibles" to be conjoined when relaxed by liberty, as if they were then equivalents.

There may be many good reasons to perform a work to entail a specific outcome, but if that work is not done, **all** of those reasons could not have been acted upon.

Then the statement on uniqueness in the axiom of virtue is effectively fulfilled, at least as far as relaxing sets of predicates by exercise of liberty over virtues.

[6] Douglas Hofstadter, *Gödel, Escher, Bach: An Eternal Golden Braid*
(Vintage Books), p502.

Liberty Appears In Symmetry (4.8.2)

I now go on to divide the virtue in a disjunction between the rows of the filter table that require it. To begin, the property of uniqueness has application straight away, even in the defining characteristics of virtue within the disjunction.

Given property U4 of an ultrafilter I must find that if $p_1 \& r^{-1} \to s_1$ and $p_2 \& r^{-1} \to s_2$, then given $s_1 \to s_2$, I must also have $p_1 \& r^{-1} \to s_2$, and by the uniqueness of virtue I must also then have $p_1 \to p_2$.

In the case of every positive property I have U1. I must have either $Pos(x) \vee Pos(x^{-1})$. It is also impossible to have the inference $Pos(x) \to Pos(x^{-1})$. Then if it be forced by liberty that $Pos(r_1 \& r_2 \& r_3) \to Pos(r_1 \& r_2) \& Pos(r_3^{-1})$ then there must exist some virtue p such that $p \& q^{-1} \to Pos(r_3^{-1})$. Now, $Pos(r_3)$ may no longer appear in the disjunction or in the ultrafilter of all $Pos(r)$ in action (as it privates virtue in p), but only in $Pos(u^{-1})$ in inaction or at rest.

Now, the virtue x to entail $Pos(r_3)$ was relaxed and is no longer found in action to entail r_3 in that ultrafilter of all r, yet it is not true that if $(p \setminus x) \& q^{-1} \to r_1 \& r_2$ then I must also have $x^{-1} \& q^{-1} \to r_3^{-1}$, for x^{-1} remains wholly negative. Instead, for $Pos(r_3^{-1})$ to be found correct, there must be an entirely different virtue found by axiom in the disjunction $u^{-1} \vee p \& u \to v^{-1}$.

So, the virtue p_1 required to entail, say, r from q, is separate and disjoint from that set p_2 entailing u^{-1}. Yet to entail the whole of that side of the disjunction from q it must hold that $p_1{}^c = p_2$, the complement is taken in p (and p_2 is the complement rather than the negation p_1^{-1}).

Then any proper subset of each set r, u^{-1} etc., may be reached not by relaxing sets of virtue and moving a property from one set into another by liberty, but by simply resting on a subset as "adequate" rather than perfected (the result of liberty acting on all positive predicates in action as in q). However, though one positive predicate (set) is not more positive than a different predicate which entails it, it remains that laziness is not a virtue of the divine without good reason for it (virtues, or a limited Ω).

Now, even given the disjunction of $q \vee p \& q^{-1} \to s \& v^{-1}$ where $q \to r \& u^{-1}$, as each set in s may be "somewhat relaxed" as if to a v^{-1} (rather than entail its negation and to be "absorbed" within the set of s instead), and each r similarly relaxed as if some u^{-1}, the conditions of liberty are such that there is a minimal set required in virtue. For, if all u^{-1} were empty but for e_u alone, and likewise v^{-1} were empty but for e_v, by axiom of virtue there is some p such that $r \& e_u \bar{\triangledown} p \& r^{-1} \& e_u \to s \& e_v$. I may split the disjunction into $r \vee p \& r^{-1} \to s$ and also $e_u \bar{\triangledown} (e_p \& e_u \to e_v)$. (And $e_u \leftrightarrow e_p \& e_v$.)

Now, I take the set S_{sv} to be the set of all positive properties entailed from $p \& q^{-1}$. A subset T_v of S_{sv} can be taken to be the subset of positive properties v in S_{sv} such that $p \& q^{-1} \to v^{-1}$. Then if the set of virtues X in the p are such that $p \& q^{-1} \to x$ where S_{sv} is the union of all x, then the minimal subset of virtue Y_s of X required to entail $p \& q^{-1} \to s$ without v^{-1} is simply $X \setminus W$ where W is the remnant of X such that $p \& r^{-1} \to v^{-1}$. (Or, equivalently any set to entail $p \& r^{-1} \to v$.) Then there is a unique relationship between the subset T_s of S_{sv} and of Y_s of X, as equally as between W and X.

I note that $(r \& s)^{-1}$ and $u \& v$ are also positive predicates and so X may not be the set of all virtues in scope, which I will name Ω_0.

I may rearrange my disjunction $r \vee p\&r^{-1} \to s$ to become $s \vee p\&s^{-1} \to r$. I have two identities each side, where there is a minimal set of virtue to entail both r without u^{-1} and s without v^{-1} both sides. I find X to be the set of p that entails $q \vee p\&q^{-1} \to x$ as before, yet both sides. I must have that q is to be taken to become the union of all r and u, without any relaxation on either side. I then find some symmetry and a simple K4 group as before.

Then there are minimal subsets $S_{ru} \setminus T_u$ and $S_{sv} \setminus T_v$ either side that entail the predicate sets of r and s. I appear to have (but not so) a simple result that the set of virtue required to entail the sets $r\&u^{-1}$ and also $s\&v^{-1}$ each side is simply the set that is required to entail $S_{ru} \setminus T_u$ and also $S_{sv} \setminus T_v$. That set would simply be $X \setminus (Y_u \cap Y_v)$ where S_{ru} etc. are taken on the side that entail r or s in a natural fashion. (Yet the sets $Pos(u^{-1})$ and $Pos(v^{-1})$ also require virtues in the disjunction.)

I know that the set T_u is disjoint with $S_{ru} \setminus T_u$ etc., so therefore the set $p_0 = X \setminus (Y_u \cap Y_v)$ is required to entail both $S_{ru} \setminus T_u$ and $S_{sv} \setminus T_v$. This seems to me quite remarkable, for it is also the case that the complement set of $\bar{p} = p_0{}^c$ is necessarily required to entail both sides of $u^{-1} \vee \bar{p}\&u \to v^{-1}$. The subset X of Ω_0 is required to entail the freely decided disjunction between $r\&u^{-1}$ and $s\&v^{-1}$.

This is actually as expected and not an invalidation of the liberty in the disjunction. The sets $r\&u^{-1}$ and its opposite(s) together with their conjoined negations will span all Ω (not merely Ω_0 which completes $p = p_0\&\bar{p}$), and I may only relax virtue when moving r on that side on which it appears into some predicate in u^{-1} the same side. $\neg(r_1\&r_2\&u^{-1}\&\neg r_2{}^{-1})$ is not simply as $r_1\&r_2\&u_1{}^{-1}\&r_2{}^{-1}$ but $r_1\&u_1{}^{-1}\&u_2{}^{-1}$ instead. The term in r_2 is not simply absorbed into the antecedent r_2, but becomes some u_2 in u instead; there is no virtue "missing" – for another virtue is required in a separate disjunction (as to entail $u_2{}^{-1}$ in the u^{-1} here from q). For r_2 is positive but not so its negation if placed within the same ultrafilter: it cannot be both in action in the poset of r and also found in u^{-1}, the octal's disjoint sets will not permit that without engendering a modal collapse. In terms of the disjunction it would confuse the filter; it would no longer agree with reality.

As all seven sets in the octal are disjoint, taking any proper subset of each set is to take a proper subset of Ω and to make a different octal (of a simplified disjunction). For any octal, I could state that overall the octal itself is then a partition upon every workable set in Anselm's non-principal ultrafilter, yet with a principal element in each set of seven.

Taking the set of all positive predicates as Ω, I find the set of virtue required is a subset of X, but I also note that $\Omega = S_{ru} \cup S_{sv} \cup X \cup V$ (V a set of virtue not entailing r, s or u^{-1} and v^{-1} but which is required for closure of the octal). Now, the local closure Ω of [p, r, s] is $(S_{ru} \setminus T_u) \cup (S_{sv} \setminus T_v) \cup (X \setminus Y_u \cap Y_v)$ taking the closure of just **one** K4 group as in this case.

Then the minimal subset of virtue in p required in $r \vee p\&r^{-1} \to s$ is actually given by $(\Omega \setminus (S_{ru} \cup S_{sv})) \setminus (Y_u \cap Y_v)$. This set must, without equivocation, simply be a subset of the set $X \setminus Y_u \cap Y_v$ as above. Therefore $Y_u \cap Y_v$ must not be empty unless u^{-1} is only e_u or v^{-1} is only e_v.

So, I repeat my working for the minimal subset of virtue required to entail r and s either side of the disjunction. Then, the minimal set of virtue required to entail the subset $S_{ru} \setminus T_u$ and also $S_{sv} \setminus T_v$ is given by $X \setminus (Y_u \cap Y_v)$. Equivalently, this is the set $(Y_u \cap Y_v)^c$ with the complement taken in X and not Ω_0. I may simply use De Morgan's laws and state that this is the union in X of $Y_u{}^c \cup Y_v{}^c$. Now, each complement is taken in X, so I have if $p_1 \& r^{-1} \to s$ and $p_2 \& s^{-1} \to r$ then the minimal subset of virtue required is simply the union of p_1 and p_2, or otherwise in simple terms, represented by and entailed from the conjunction $p = p_1 \& p_2$.

Now, $p_1 \& p_2$ contains extraneous properties $p_2 \setminus p_1$ not required to entail s from r^{-1} and also symmetrically, extraneous properties $p_1 \setminus p_2$ not required to entail r from s^{-1}. Both exclude extraneous properties in $X^c = V$ from Ω_0. Then if these sets are as x_1 and x_2 I have minimally that $\bar{p}_1 \& u \to v^{-1}$ and $\bar{p}_2 \& v \to u^{-1}$, with $\bar{p}_1$ etc taken in X. Taking the conjunction again as with $p_1 \& p_2$ before, I do not have equal virtue both sides, so I **do not** have (by De Morgan's laws) sets of virtues $\bar{p}_1 \& \bar{p}_2$ equal to the union of the complements and therefore the complement of their intersection. Now $x_1 \& x_2$ is simply $p_1 \vee p_2$ and so the two disjunctions both share the virtues in $x_1 \& x_2$, which is also $\bar{p}_1 \vee \bar{p}_2$. $x_1 \& x_2$ is actually the intersection of $(p_1 \& p_2)$ with $(\bar{p}_1 \& \bar{p}_2)$.

Then, for the octal, I must rewrite my sets as $(r\&s)^{-1} \vee p\&r\&s \to u\&v$. Now, the virtue in p may be different here when acting either side, yet I have a necessary condition that p may not exceed the closure of the set Ω_0. (There may well be virtues in Ω_0 not present in either $(p_1 \& p_2)$ or $(\bar{p}_1 \& \bar{p}_2)$ which together only span X.)

I may begin to state that the K4 group $[p, (r\&s)^{-1}, u\&v]$ is reached from adding $[p, r, s] + [p, u^{-1}, v^{-1}]$ with the operation $a + b = (a \vee b)^c$. The sets r, s, u, v are all disjoint as are $(r\&s)^{-1}$ and u&v. The set "p" is somewhat shared and must then be within the set Ω_0. Now, that must also be true for the group $[p, (r\&s)^{-1}, u\&v]$, so I "best guess" that this "p" is to be taken from all Ω_0 also.

Now if I may not rest on u^{-1} to one side of the disjunction and cannot rest on v^{-1} to the other, the logical schema of u&v allows me to act with both at the same time! What may be said about the virtue required in p is that the disjunction does not apply in terms of q and q^{-1} to one side or other. The set $(r\&s)^{-1}$ cannot be placed either side and it becomes difficult to separate those virtues that entail u&v from r&s, as r&s is in the middle and has been excluded from the logic so far. (Any "p" in $p\&r\&s \to u\&v$ is disjoint from that in $p\&r^{-1} \to s$ and not privated by r&s.)

To reach any result on u&v etc, I over-constrain any virtue applied. If I cannot act on u^{-1} and also cannot act on v^{-1}, I must form the set u&v. If I have $N\neg Pos(\bar{p})\&N\neg Pos(u)\&N\neg Pos(v)$, I have fixed $\bar{p}$ as the same virtue $(\bar{p}_1 \& \bar{p}_2)$ to decide $u^{-1} \vee \bar{p} \& u \to v^{-1}$. This, in order to over-constrain that exact same set of virtue in $\bar{p}$, has the result that $N(\neg \bar{p} \& u^{-1} \& v^{-1})$. By symmetry I have $p = p_1 \& p_2$ applied to form $r \vee p \& r^{-1} \to s$ and I also write $N\neg Pos(p)\&N\neg Pos(r^{-1})\&N\neg Pos(s^{-1})$ or $N(\neg p \& r \& s)$. With $N(\neg p)$ in scope and null, $N\neg(p\&\neg T)$ for all T, including $(r\&s)^{-1} \to p^{-1}$. Then $p \to p^{-1}$.

I find that p or $\bar{p}$ is over-constrained in both cases and I assume that $p\&\bar{p}$ in whole is again necessary in the set X (and then also Ω_0), I may write X as the union of predicates entailed from $(p_1 \& p_2)\&(\bar{p}_1 \& \bar{p}_2)$ and both sets of virtue are still properly disjoint on each **side** of the disjunction in X, but Ω_0 whole is required. For those virtues not in X but yet in the octal are possibly over-constrained by both disjunctions and present in neither (and out of scope) so I require an "error term" of remnants in $V = X^c$. Then as X is a closed set (as is Ω_0), I may decide the disjunction with $\Omega_0 = (p_1 \& p_2) \cup (\bar{p}_1 \& \bar{p}_2) \cup V$. ($V = \Omega_0 \setminus X$, virtues not common to either $[p, r, s]$ or $[\bar{p}, u^{-1}, v^{-1}]$.) Then, surprisingly, the intersection (which appears non-empty, see below), leaves the product, the complement of the symmetric difference $((p_1 \& p_2) \vee (\bar{p}_1 \& \bar{p}_2))^c$ equal in union to the set of predicates entailed from $x_1 \& x_2$ together with some subset of V (in fact, all of a minimal and consistent V). In the above and also to follow, the complements $\bar{p}_1, \bar{p}_2$ are effectively taken **each side** in X, and $\bar{p}_1 \& \bar{p}_2$ is **not yet** shown equal to X \ $p_1 \& p_2$ etc. overall.

I.e. $(p_1 \& p_2) \cap (\bar{p}_1 \& \bar{p}_2) = (p_1 \cup p_2) \cap (\bar{p}_1 \cup \bar{p}_2) = (p_1 \vee p_2) = (\bar{p}_1 \vee \bar{p}_2)$ in X.

$(p_1 \& p_2) \vee (\bar{p}_1 \& \bar{p}_2) = (p_1 \cup p_2 \cup \bar{p}_1 \cup \bar{p}_2) - (p_1 \vee p_2) = (p_1 \vee p_2)^c = (\bar{p}_1 \vee \bar{p}_2)^c$ in X.

Then $(p_1 \& p_2 \vee \bar{p}_1 \& \bar{p}_2)^c = p_1 \& p_2 \cap \bar{p}_1 \& \bar{p}_2 = (p_1 \vee p_2)$ in X, but $(p_1 \vee p_2) \cup V$ in Ω_0.

Now, $((p_1 \& p_2) \vee (p_1 \vee p_2))^c = (p_1 \cap p_2)^c = \bar{p}_1 \& \bar{p}_2$ in X and also in Ω_0, for:

$((p_1 \& p_2) \vee ((p_1 \vee p_2) \cup V))^c = (((p_1 \& p_2) \vee (p_1 \vee p_2)) \cup V)^c = (p_1 \cap p_2)^c \cap V^c = \bar{p}_1 \& \bar{p}_2$.

And similarly, $((\bar{p}_1 \& \bar{p}_2) \vee ((p_1 \vee p_2) \cup V))^c = (((\bar{p}_1 \& \bar{p}_2) \vee (\bar{p}_1 \vee \bar{p}_2)) \cup V)^c = (\bar{p}_1 \cap \bar{p}_2)^c \cap X = p_1 \& p_2$.

I have constructed the following addition table for a K4 group with A+B = $(A \vee B)^c$.

$(a \vee b)^c$	Ω_0	$(p_1 \vee p_2) \cup V$	$p_1 \cup p_2$	$\bar{p}_1 \cup \bar{p}_2$
Ω_0	Ω_0	$(p_1 \vee p_2) \cup V$	$p_1 \cup p_2$	$\bar{p}_1 \cup \bar{p}_2$
$(p_1 \vee p_2) \cup V$	$(p_1 \vee p_2) \cup V$	Ω_0	$\bar{p}_1 \cup \bar{p}_2$	$p_1 \cup p_2$
$p_1 \cup p_2$	$p_1 \cup p_2$	$\bar{p}_1 \cup \bar{p}_2$	Ω_0	$(p_1 \vee p_2) \cup V$
$\bar{p}_1 \cup \bar{p}_2$	$\bar{p}_1 \cup \bar{p}_2$	$p_1 \cup p_2$	$(p_1 \vee p_2) \cup V$	Ω_0

The K4 form which contains three K4 groups disjoint but for virtue, requires a minimum of virtue for each group (and disjunction): The K4 groups in the K4 form are closed in the elements of the octal with the operation A+B=$(A \vee B)^c$. If the set of virtue required for each K4 group is similarly split with the same operation within the set of all virtue from the octal then each disjunction requires exactly that virtue apportioned. Consider once again the K4 form

$$
\begin{aligned}
1 \equiv a &\equiv [0, \quad a, \quad b, \quad c] \\
b &\equiv [0, \quad a, \quad d, \quad e] \\
c &\equiv [0, \quad a, \quad f, \quad g] \\
\Omega &= \{0, a, b, c, d, e, f, g\}
\end{aligned}
$$

With Christ (the K4 form) with the sets of the ultrafilter:

$$
\begin{aligned}
1 \equiv \quad p &\equiv [0, \quad p, \quad (r \& s)^{-1}, \quad u \& v] \\
(r \& s)^{-1} &\equiv [0, \quad p, \quad r, \quad s] \\
u \& v &\equiv [0, \quad p, \quad u^{-1}, \quad v^{-1}]
\end{aligned}
$$

Also rewritten as:

$$
\begin{aligned}
1 \equiv \quad N\neg(r \& s \& \neg u \& \neg v) &\equiv [0, \quad p, \quad (r \& s)^{-1}, \quad u \& v] \\
N\neg(r \& s) &\equiv [0, \quad p, \quad r, \quad s] \\
N\neg(u^{-1} \& v^{-1}) &\equiv [0, \quad p, \quad u^{-1}, \quad v^{-1}]
\end{aligned}
$$

And to solve for the free decidability of those implied disjunctions $r \vee s$ and $u^{-1} \vee v^{-1}$ etc:

$$
\begin{aligned}
1 &\equiv N\neg(p \,\&\, r \,\&\, s \,\&\, \neg u \,\&\, \neg v) &\equiv&\ [0, \quad p, \quad (r \,\&\, s)^{-1}, \quad u \,\&\, v] \\
&\quad\ N\neg(p \,\&\, \neg r \,\&\, \neg s) &\equiv&\ [0, \quad p, \qquad r, \qquad\ \ s] \\
&\quad\ N\neg\left(p \,\&\, \neg u^{-1} \,\&\, \neg v^{-1}\right) &\equiv&\ [0, \quad p, \qquad u^{-1}, \qquad v^{-1}]
\end{aligned}
$$

Or, with respect to the terms in the first row found self-similar to the whole (there is necessarily no middle with the virtue in p):

$$
\begin{aligned}
1 &\equiv \qquad Pos(p) &\equiv&\ [0, \quad p, \quad (r \,\&\, s)^{-1}, \quad u \,\&\, v] \\
&\quad\ \neg Pos(r \,\&\, s) &\equiv&\ [0, \quad p, \qquad r, \qquad\ \ s] \\
&\quad\ \neg Pos\left(u^{-1} \,\&\, v^{-1}\right) &\equiv&\ [0, \quad p, \qquad u^{-1}, \qquad v^{-1}] \\
&\quad\ N\neg Pos\left(p \,\&\, r \,\&\, s \,\&\, \neg(u \,\&\, v)\right)
\end{aligned}
$$

I now split the sets of virtue in each disjunction as so:

$$
\begin{aligned}
1 &\equiv Pos\left((p_1 \vee p_2) \cup V\right) &\equiv&\ [0, \quad (p_1 \vee p_2) \cup V, \quad (r \,\&\, s)^{-1}, \quad u \,\&\, v] \\
&\quad\ \neg Pos(r \,\&\, s) &\equiv&\ [0, \qquad p_1 \,\&\, p_2, \qquad r, \qquad\ s] \\
&\quad\ \neg Pos\left(u^{-1} \,\&\, v^{-1}\right) &\equiv&\ [0, \qquad \bar{p}_1 \,\&\, \bar{p}_2, \qquad u^{-1}, \qquad v^{-1}]
\end{aligned}
$$

$$
(p_1 \vee p_2) \cup V = (\bar{p}_1 \vee \bar{p}_2) \cup V = \left((p_1 \,\&\, p_2) \vee (\bar{p}_1 \,\&\, \bar{p}_2)\right)^c
$$

And $p_1 \vee p_2 = x_1 \,\&\, x_2 = (p_1 \,\&\, p_2) \cap (\bar{p}_1 \,\&\, \bar{p}_2)$ in Ω_0.

Now, the product of these K4 subgroups using $A+B = (A \vee B)^c$ is such that the set of virtue p is assumed constant and the same in all three groups. But the disjunctions split the set of virtue between them, spanning Ω with a three way cut in Ω_0, else each set would entail from all X on q^{-1} the whole of S as if the disjunction had never been split into action and inaction.

Yet all three disjunctions share the symmetric difference $p_1 \vee p_2 = x_1 \,\&\, x_2 = (p_1 \,\&\, p_2) \cap (\bar{p}_1 \,\&\, \bar{p}_2)$.

The sets of virtues in the two disjunctions in r, s and u^{-1}, v^{-1} are easily identified as $p_1 \,\&\, p_2$ and $\bar{p}_1 \,\&\, \bar{p}_2$, so the complement of their symmetric difference taken in Ω leaves but $(p_1 \vee p_2) \cup V = (x_1 \,\&\, x_2) \cup V$ remaining as the virtue in the disjunction $(r \,\&\, s)^{-1} \vee u \,\&\, v$. This set is truthfully the intersection of $p_1 \,\&\, p_2$ and $\bar{p}_1 \,\&\, \bar{p}_2$ together with the set V in Ω_0. The intersection formed of $p_1 \vee p_2$ or its equal $(\bar{p}_1 \vee \bar{p}_2)$ is such that it would place both $(r \,\&\, s)^{-1}$ and $u \,\&\, v$ in scope together; the intersect provides a solution composed from $N\neg(p \,\&\, (r \,\&\, s)^{-1})$ and $N\neg(p \,\&\, u \,\&\, v)$. There may be no p in both these two terms that agrees with p in $p \,\&\, (r \,\&\, s)^{-1} \,\&\, (u \,\&\, v)$ which leaves $N(\neg p \,\&\, (r \,\&\, s)^{-1} \,\&\, (u \,\&\, v))$ and therefore $N\neg(p \,\&\, (r \,\&\, s) \,\&\, \neg(u \,\&\, v))$ true for any p in that intersection.

Then there is a Trinity embedded in virtue as if a "Manifold of Virtue". Then I have found a K4 group using $A+B = (A \vee B)^c$ embedded in the identity of the K4 form and I have the Trinity present (as if bodily in Christ), **self-similar** to the identity element (virtue) and in regress.

And that is an incredible result! The uniqueness of virtue required by axiom corresponds naturally to sets within the ultrafilter of virtue itself. (I find using $p = \Omega_0$ valid over all three disjunctions; no positive property is privated by any virtue – and r&s will not private V etc.)

Every Use Of Liberty Requires A Different Virtue (4.8.3)

Now, to make sure, I ask: "What if (by the axiom of virtue), the set of virtue were unique, and were required to be the same each side of each disjunction?" There are clearly such solutions as rearranged, and although the ultrafilter itself makes a "cut" in virtue, of liberty, to only entail terms $Pos(r_1 \& r_2) \rightarrow Pos(r_1)$ etc; the sets I am examining must remain disjoint: and so there is no absorption as found from liberty acting as within the same filter.

Either way, were there virtue in common between the two disjunctions in r, s and u^{-1}, v^{-1}: those properties would remain within their intersection and therefore appear in the complement of their symmetric difference.

Yet the sets $\{r,s\}$, $\{u^{-1},v^{-1}\}$ $\{(r\&s)^{-1}$, $u\&v\}$ remain completely disjoint, and by the axiom of virtue and the condition of uniqueness, the virtues required to entail r or u^{-1} and either s or v^{-1} must also remain completely disjoint within each side of the disjunction. By the axiom of virtue there remains a virtue (as in the mapping table) that entails both sides of the disjunction.

The disjunction can be arranged each side with equal virtue: ($N\neg(p\&r^{-1}\&\neg s)$ is also $N\neg(p\&s^{-1}\&\neg r)$). Then the only property in common between the two different disjunctions remains $x_1\&x_2$ which is $p_1 \vee p_2$. As the virtues to entail all r and all u^{-1} are disjoint, the symmetric difference $p_1 \vee p_2$ in X and intersection $x_1\&x_2$ is truthfully empty (but for e_p) because of the uniqueness of virtue. Given that the complement in Ω_0 of their symmetric difference contains this very same set along with the set V of $\Omega_0 \setminus X$, there are no other properties in Ω common to both of the two in product remaining to add to that set V.

I.e. the disjunction $q \vee p\&q^{-1}$ in the "real world" fixes the set(s) q (and q^{-1}), the resultant triple of K4 groups in $A=[p_A, r, s]$, $B=[p_B, u^{-1}, v^{-1}]$, $C=[p_C, (r\&s)^{-1}, u\&v]$ over the same real-world disjunction are added together by $(A \vee B)^c = A + B = C$ etc. In real terms also, by axiom of virtue: the minimal virtue required to entail A, B or C is found to be gained from the same operation within the set of virtue in all p.

So, I may modify my sets of virtue to form the groups with a minimal subset of virtue each one, each subset then disjoint.

$A=[p_A, r, s]$, $B=[p_B, u^{-1}, v^{-1}]$, $C=[p_C, (r\&s)^{-1}, u\&v]$ hold where within the closed ultrafilter of all virtue $\Omega_0 = p = p_A \cup p_B \cup p_C$, I quite naturally find $p_A = (p_B \vee p_C)^c$ which should not surprise, but I also find $p_C^c = p_A \cup p_B = p_A \vee p_B$ etc. as well which should! (It is not at all like a boolean algebra.)

Now, if I have $N(\neg p\&u\&v)$ or $N\neg(p\&\neg(u\&v))$ as well as $N\neg(p\&r\&s)$ I also have the conjunction $N\neg(p\&r\&s\&\neg(u\&v))$ as before.

I may form the disjunction $(r\&s)^{-1} \vee u\&v$ employing on each side the sets of virtues $((p_1\&p_2) \vee (\overline{p}_1\&\overline{p}_2))^c = (p_1 \vee p_2) \cup V = ((p_1\&p_2) \cap (\overline{p}_1\&\overline{p}_2)) \cup V$ within the ultrafilter on Ω_0. These sets of virtue are constrained within the relations above, formed on $N(\neg p\&u\&v) \vee N\neg(p\&r\&s)$. I then deliberately over-constrain virtue on each side to make its complement necessary (or freely decidable) by breaking the closure of virtue with the terms $p \rightarrow u\&v$ and $p \rightarrow (r\&s)^{-1}$. However, $N\neg Pos(p^{-1})$) so one side will or must reduce to $(u\&v)$, say. Then I have the disjunction in virtue of $u\&v \vee (r\&s)^{-1}$.

The (unique) virtue required to entail both sides remains $((p_1 \& p_2) \vee (\bar{p}_1 \& \bar{p}_2))^c$, although this complement of the symmetric difference (to entail either side as in the place of $p_1 \& p_2$ from before) would be as $(p_1 \vee p_2) \cup V$ or even $(\bar{p}_1 \vee \bar{p}_2) \cup V$, which are equal to V as desired (as $p_1 = p_2$), rendering no new complement(s) in action or at rest to further extend the octal. The logic then reduces to $N\neg(p \& r \& s \& \neg(u \& v))$ alone, say.

Then by the uniqueness condition of the axiom of virtue, $((p_1 \& p_2) \vee (\bar{p}_1 \& \bar{p}_2))^c = X^c = V$ and every set in the mapping table remains disjoint. The converse, that any three-way split in virtue (with a common intersection) results in an octal over the very **same** disjunction, (a rearrangement by liberty) is false without fixing the virtue(s) in X on both sides. A property r, consequent of any virtue (then rearranged to become as any u^{-1}), could legally and consistently appear as an $(r^{-1})^{-1}$ instead (virtue will split just as equally amongst such three-way divisions).

Then every application of liberty in the octal requires a different and, more importantly, a unique virtue, one which is always ready to hand for the omnipotent.

4.9) Necessary Conditions Shown

Each triple {x,y,z} must of necessity (in union) be equal to Ω the set of all positive properties (local to each row of the filter table). {x}, {y} and {z} must also be "disjoint" as (x&z) privates virtue y and is "empty", not in the filter. I recover {B,C} = {D,E} = {F,G} = A^c etc. in each case.

On The Conditions For Each Filter To Form A K4 Group (4.9.1)

It remains to be shown that the products are actually well-formed.

I have already shown the products with virtues p to be well-formed:

I.e. as with the case $\{r, (u \& v)\} + \{v^{-1}, (u \& v)\} = \{(u \& v), (r \& v^{-1})^{-1}\}$ I find that $(r \& v^{-1})^{-1}$ becomes "empty" (inconsistent given (u&v) axiomatically positive and also one of r, v^{-1}).

Then the set $(r \& v^{-1})$ privates some property in (u&v) (as {r} and $\{v^{-1}\}$ are disjoint).

Restricting myself to $\{v^{-1}, r, u \& v\}$, I should merely show that v does not private r or (u&v).

$v \to p^{-1} \& q \to r \& u^{-1} \to Pos(r_0 \& u_0^{-1})$ so $N\neg(v \& u) \to N\neg(v \& Pos(v \& u))$ actually entails that I have $N\neg(Pos(u \& v) \& v \& \neg Pos(r_0))$ or simply $v^{-1} \vee (u \& v) \& v \to r_0$. (It is also naïvely apparent from the mapping table that $v \to r_0$ also. Then r_0 and u_0 as proper "subsets" of r and u are properly disjoint from v and s, and likewise they may not be privated by, or equally private sets within those disjunctions and upon them the K4 product properly rests. Cf. sections [4.6.2], [4.6.3].)

To see this as from the mapping table, I already have:

$v \to q \& p^{-1} \to r_0 \& p^{-1} \to Pos(r_0)$ and also $v \to q \& p^{-1} \to u_0^{-1} \& p^{-1} \to Pos(u_0^{-1})$.

Yet how do I separate r_0 and u_0 from v? So far, I may only relate $v \to q \& p^{-1} \to r$ etc. I have only $r^{-1} \to p \& q^{-1}$ and r_0^{-1} does not entail p, I may not equally infer that $v \to Pos(r_0)$ as I would wish. I can only rearrange and state $p \& v \& s^{-1} \to r \& u^{-1}$. (Liberty would choose the other side.)

However, since p is a virtue and it is necessarily positive in action (i.e. it is a fact that $N\neg Pos(p^{-1})$), all positivity in the conjunction $r_o\&p^{-1}$ entailed as $v \to q\&p^{-1} \to r_o\&p^{-1} \to Pos(r_o)$ follows without any positivity of v, (or indeed s^{-1}) as they and p^{-1} are all necessarily negative. It follows that the negative "content" of v never privates that of r_o or u_o^{-1} (both disjoint to v), as any "content" cannot be entailed from v (the right hand side of the inference must remain positive). In logic under the "Pos" operator, I have $N(\neg v\&u_o^{-1})$ or $N\neg(v\&\neg u_o^{-1})$ or the equivalent to $v \to Pos(u_o^{-1})$ and equally $N\neg(v\&\neg r_o)$ or the equivalent $v \to Pos(r_o)$. Similarly, $s^{-1} \to Pos(u_o^{-1})$ and $s^{-1} \to Pos(r_o)$ as desired.

By symmetry $u \to Pos(s\&v^{-1})$ and also $Pos(v\&u_o) \to Pos(s)$ (then $u_o \to Pos(s)$ by the modus tollens of the above). Using the colloquial u and r for u_o and r_o, I have the following:

Obviously, I may begin with a simple logical disjunction:

$v^{-1} \vee v$

Then using the mapping table as a suitable jumping off point I have:

$v^{-1} \vee (v \to Pos(r\&u^{-1}))$, i.e. $N\neg(v\&\neg Pos(r\&u^{-1}))$

I may quite legally move the term u^{-1} to form:

$v^{-1} \vee (v\&u \to Pos(r))$, i.e. $N\neg(v\&u\&\neg Pos(r))$.

(I cannot similarly move r, as the disjunction of $v^{-1} \vee u^{-1}$ is uniquely fixed with p already.)

The set $Pos(u\&v)$ is closed and v is negative, so I can only form the disjunction of:

$v^{-1} \vee (v\&Pos(v\&u) \to Pos(r))$ and as v is negative I have a binding as if made of virtue.

(And $[\Omega,B,D,F]$ holds as a K4 group.)

In symmetry $u^{-1} \vee u$ entails $u^{-1} \vee (u \to Pos(s\&v^{-1})))$ so I may then form:

$u^{-1} \vee (u\&Pos(v\&u) \to Pos(s))$, $([\Omega,C,E,F]$ holds as a K4 group).

And I also have:

$r \vee (r^{-1}\&Pos(r\&s)^{-1} \to Pos(u^{-1}))$, $([\Omega,B,E,G]$ holds as a K4 group) and also:

$s \vee (s^{-1}\&Pos(r\&s)^{-1} \to Pos(v^{-1}))$, $([\Omega,C,D,G]$ holds as a K4 group)

(simply because $N\neg(s^{-1}\&\neg Pos(r\&u^{-1}))$ permits $N\neg(r^{-1}\&Pos(r\&s)^{-1}\&\neg Pos(u^{-1}))$ etc.)

Then if, say, $x=Pos(r)$, $z=Pos(u^{-1})$ and $y=Pos(r\&s)^{-1}$ given y, $(x\&z)$ is empty (and then negative).

$(x\&z)^{-1}$ is then positive (counter-intuitively x&z is empty or inconsistent given y and one of x, z).

Then these relations together entail that there is consistency within the octal's addition table. The rows of the filter table are then found to be logically consistent, also forming K4 groups.

i.e. x = y + z etc.

...and I am done. (Almost.)

On The Conditions For Closure Of The Octal (4.9.2)

I had previously assumed the product {X,A} + {X,B} = {X,C} (or given in the form of x, y and z, ({y, x} $\vee$ {y, z})^c = {y, (x&z)$^{-1}$}) was well defined, but that the filter "possibly" operated upon only the three elements (or sets) x, y and z.

Now, if v $\rightarrow$ u$_0^{-1}$ and s^{-1} $\rightarrow$ r$_0$ as was shown above, the identities (u$_0$&v) and (r$_0$&s)$^{-1}$ become inconsistent as positive properties. In strictest terms, Pos(u$_0$&v) $\leftrightarrow$ Pos(v&v^{-1}) and Pos(r$_0$&s)$^{-1}$ $\leftrightarrow$ Pos(r$_0$&r$_0^{-1}$). Both these and their negations become inconsistencies when both u and v or both r and s are in scope. In order for them to remain positive conjoined, they private the virtue p. (To understand that, I assumed that p was strictly positive and p^{-1} negative always.)

Now, given the uniqueness of virtue in each disjunction, with any application of liberty found simply in using the unique virtue (as by axiom) to decide both sides of r $\vee$ s, say, those virtue(s) to decide each of [p, r, s], [p, u^{-1}, v^{-1}] and [p, (r&s)$^{-1}$, u&v] split into disjoint sets in p to decide the disjunctions of each group separately.

Then for all virtue in [p, r, s] I have the identity that breaks the closure of p, and (r&s)$^{-1}$ $\rightarrow$ ¬p. Every virtue in that group is inconsistent with (r&s)$^{-1}$. Similarly, for p in [p, (r&s)$^{-1}$, u&v] I have N(Pos(u&v) & ¬p & ¬Pos(r&s)) from N¬(p&r&s&¬(u&v)). There is no virtue in the group [p, (r&s)$^{-1}$, u&v] that is compatible with, or may solve for the disjunction r $\vee$ s. This carries also for the disjunction u^{-1} $\vee$ v^{-1}.

Then as the only predicates extant in each disjunction remain solely within their triples, the terms in product give ({y, x} $\vee$ {y, z})^c = {y} only, there is no term in (x&z)$^{-1}$. Then (x $\vee$ z)^c = y in each disjunction.

As this only applies to the disjunctions of virtue in those three groups with virtues, the sets in the remnant of the octal are also completely disjoint and the terms in the middle are also, effectively, empty.

For any disjunction x $\vee$ z solved for by y, say, I have the negated middle (x&z)$^{-1}$ $\rightarrow$ ¬y inconsistent with y (as y is closed). I also find in the middle that there is no solution for a "virtue" in y that would entail any positive result W from y&x&z, as the middle x&z is inconsistent with all y.

I.e. N¬(y&x&z&¬W) is incompatible with N¬(y&x^{-1}&¬z) which is also N¬(Pos(y)&¬Pos(x&z)). Pos(y&x&z) may be wholly positive: yet there is no disjoint set W extant, and no liberty found.

Then no disjunction in x $\vee$ z may include the term (x&z)$^{-1}$, which is totally inconsistent with all "virtue" in y. So, every row of the filter table is closed, as each disjunction is decided "given y".

The local closure of each K4 subgroup (an ultrafilter) acts as if the whole closure of the octal group is in effect, so that the filter tables' separate rows operate only upon their triples for their indexing set and y (or p) entails free exercise of r and s, or both u^{-1} and v^{-1} as I have stated.

Then, when these are out of scope, I entail free exercise between $(u_0\&v)$ and $(r_0\&s)^{-1}$. Finally, I state no virtue is ever privated, every coset of every K4 subgroup (which are not ultrafilters) cannot be found in the octal ultrafilter, and perfection is satisfied only in the octal group, "the Father".

Then the virtues, as conjunctions in p "in scope" within the triples {p, r, s}, {p, u^{-1}, v^{-1}} and {p, (u&v), (r&s)$^{-1}$} are limited to the same set (of virtue) in the K4 ultrafilter; rather than the content in each triple being of all possible virtues p not required. The positive properties entailed from y&z (in the mapping table; with y by "axiom of virtue" bound to z as if by my inferred linearity before) each require a split of the set of virtues/properties in p (or as y). A restricted set of positive properties z requires by axiom the limited application of a restricted set of p (virtues, by axiom of virtue – even when the whole set of p is possibly in scope). Some virtues, then, will not appear to act. The virtues in scope and that certainly do act are a split of "p" across each triple. No virtue is privated; i.e. there is only one disjunction in $q \vee q^{-1}$ overall.

Every y in each triple (x, y and z) in the filter table grants free exercise between x and z only. By virtue in p, I have freedom between r and s, or u^{-1} and v^{-1} or of (u&v) and (r&s)$^{-1}$. However, p (or y) entails that x&z is inconsistent rather than negative, and therefore (x&z)$^{-1}$ is likewise inconsistent (N¬(x&z) is a privation, $x \rightarrow \neg z$) and then y = x + z. The naïve $u^{-1} \leftrightarrow v$ and $r^{-1} \leftrightarrow s$ entail that u&v and r&s cannot be entailed from either side of $q \vee p\&q^{-1}$. Then (r&s)$^{-1}$ does not follow from q or p&q^{-1}, yet I have (u&v) $\vee$ p&(r&s)$^{-1}$ valid as a completely independent case.

Now {r, s} and {u, v} as sets in the mapping table are "compatible"; yet they leave (r&s)$^{-1}$ and (u&v) inconsistent. I may consider that it is impossible to have (r&s)$^{-1}$ and one of r or s in scope with a virtue p, and similarly impossible for one of u and v to be in scope with (u&v) and also that same virtue. This I find plain, that given such a virtue p and $u \leftrightarrow v^{-1}$ etc., (u&v) becomes inconsistent. However, I may enjoin one of r or s or else one of u or v to either (u&v) or (r&s)$^{-1}$ if I drop the requirement for that virtue p. I may operate between any u, v, r or s for my triples without including p, (virtue). Virtue in p permits freedom between (r&s)$^{-1}$ and (u&v). (Neither r or s, nor either of u or v is then in scope, so $u \leftrightarrow v^{-1}$ and $r \leftrightarrow s^{-1}$ are logically averted.)

Then the sum {p, r} + {p, s} = {p, (r&s)$^{-1}$} does not form a K4 group, as (r&s)$^{-1}$ is inconsistent with r or s given p, and is not in scope with either r or s. ((r&s)$^{-1}$ doesn't have to move position in the mapping table; it logically does not follow from either side of the disjunction.)

So, ({y, x} $\vee$ {y, z})^c = {y, (x&z)$^{-1}$} may appear to hold, but ({y, x} $\vee$ {y, (x&z)$^{-1}$})^c = {y, z} makes no sense as (x&z)$^{-1}$ is inconsistent (not a positive property in the "local" ultrafilter, i.e. one to which {x, y, z} are "native") so ({y, x} $\vee$ {y, (x&z)$^{-1}$})^c = ({y, z^{-1}} $\vee$ {y, (z&z^{-1})})^c = ({y, x} $\vee$ {y})^c = {y, z} as desired. Then (x $\vee$ y)^c = z etc. (as sets) within each triple also.

4.10) Summary

A Principal Ultrafilter Contains The Non-Principal (4.10.1)

Each set of positive properties is present in an ultrafilter of perfections (as of Anselm's argument, using a non-principal ultrafilter). The octal group is formed upon any disjunction (as to decide Pos(q) or Pos(p&q^{-1})). It is simply an observation that the principal ultrafilter (and as a set the octal) may properly contain the non-principal filter and is also a degree of extension over it. God may decide any set of q from the set of all positive properties.

Either of the properties r and s may be arranged in disjunction as the property q, and then since q = r&s is a possibility within the set of all positive properties indexed under the octal ultrafilter, it is not true that r and s as positive properties may private each other, merely that they would over-constrain a particular virtue or set of virtues in "p".

On The Local Closure Of Indexing Sets (4.10.2)

In each K4 group in the octal above, the operation $(x \vee y)^c = z$ operates only upon the elements within each triple. So, for instance, $(r \vee u)^c = (r\&s)^{-1}$ would be defined on only those three elements of the octal. The complement is taken within that triple rather than the whole set of positive properties. I also preserve the spanning K4 form of the ultrafilter.

In the mapping table, all r following from q and all s following from $p\&q^{-1}$ are unique in their placing and are well-formed. The additive sum from the "mapping" and "filter" tables (of **disjoint** sets) is unique but for the alternate octal operating under a similar logical schema. (And so for all rows of the filter table, if binding is not broken.)

In each row of the filter table every property x, y, z is positive. Likewise, as each row in the filter table is a separate (and closed) filter upon positive properties, the unions of the sets in those triples (x, y, z are disjoint) are likewise positive. (The filters are of "positive properties" and constrain no elements in x, y or z.) They, as unions, are positive (also in the octal) as the triples they constitute are supersets of the positive properties x, y, z etc. in Anselm's filter.

The sets generated by $(r\&s)^{-1}$ and $(u\&v)$ in the mapping table are (as in the predicates of sets $\vee F$ etc.) somewhat "maximal"; so the assertion that subgroups in the octal that do not have the set of virtues as a member may possibly span Ω (as $\Omega = x \cup y \cup z$ in a triple only) is also justifiable, as $(u\&v)$ etc. are indeed extremes and there is no proper subset of the ultrafilter containing them. As products, they are "unique" and so do generate a maximal set in Ω.

Given some other element than virtue as y (one assumed positive by axiom, the disjunction is to remain freely decidable) I may omit all those properties that would constrain this maximal set (say, {p, r, s} constraining $(r\&s)^{-1}$ etc.) which form the rest of Ω (in the coset of the group $[\Omega, x, y, z]$), as if they also were an excluded middle as in x and z by virtue of y.

For any positive property y assumed positive by axiom, there is no property in any row of the filter table that may private it — it is as if it were virtue. (At least in each triple {x, y, z}). The elements in the coset generated by $(x\&z)^{-1}$ are constraints on y and therefore become inconsistent (by axiom) to that filter. (The properties x&z are positive and consistent, but only in a filter to which they are native.)

The rule of thumb is that if a positive property "w" may not be exemplified positively then it is positive for perfection to not exemplify that property w to action. Equivalence of the coset to an excluded middle is quite natural, even for those rows of the filter table without virtue.

Including the coset in the ultrafilter subsequently breaks the proper operation upon the octal's singletons, so that the sum $\{r\} + \{s\} = \{p, (r\&s)^{-1}\}$. The coset with $(r\&s)^{-1}$ should be excluded.

The octal is defined on pairs of elements rather than triples. By "Uniqueness Of Virtue [4.8]" every product in the octal is well formed in the filter to which it is incidentally native. There is no sense of $W=\{p, (r\&s)^{-1}\}$ being inconsistent when r and s are not in scope, as the operation of the octal automatically excludes W from any operation on any pair from $\{p, r, s\}$.

A Union Of Seven Ultrafilters (4.10.3)

The perfection of the octal as a union of these seven ultrafilters' indexing sets is found not to constrain any of the ultrafilters/indexing sets/subgroups at all. They naturally constrain themselves by closure and then I find liberty rather than constraint.

The octal group indexes the positive properties of each principal ultrafilter of the seven into either action or inaction, every bit as much do the filters $\{p, r, s\}$ etc. I simply need every actionable element x to be in the filter and every other element y^{-1} not so, no matter which triple they appear within! (Action and inaction separate out by inclusion.) The octal includes every property that is free to be acted upon – and separates out all those that are not. The octal, then, is for the "completely" free exercise of virtue. (Seamless perfection is found in the union of all seven ultrafilters' indexing sets; I find perfection indexed uniquely in GF(8).)

The elements u and v at "rest" are a "good" solution (assuming such a virtue p exists) for the logical disjunction $u^{-1} \vee v^{-1}$. A seven cycle acting on that triple $\{p, u^{-1}, v^{-1}\}$ is such that the octal is "made perfect" by mapping that local closure through the octal. Every triple becomes perfectly closed in the ultrafilter. The ultrafilter is as complete as Ω; no virtue is ever privated.

As far as resting upon that perfection is concerned, I have from the Bible "My grace is sufficient for thee: for my strength is made perfect in weakness." (2Co 12:9). The weakness of virtue, acting as liberty L(G) (by restricting the sets of exemplified properties within the ultrafilter) makes perfect the whole operation of the octal by entailing the principal element of rest "e".

Freely Decided Disjunctions Are Not Symmetric In Actions (4.10.4)

You may wonder why the symbols u^{-1} and v^{-1} are required at all; why are they not simply the complement actions of r and s in the mapping table?

Is it not obvious that u is equal to s and v equal to r from the properties and construction of the ultrafilter? Well, no. Who can justify that any particular subset of positive properties in action (entailed from a virtue p as from $p\&q^{-1}$) in union with those others entailed from a positive property q form the whole of Ω? In any given disjunction there are positive properties that may not apply, as well as those that are not entailed from a virtue due to liberty. The exclusions in Ω are clearly "at rest" or do not apply as if excluded by "a natural law" or happenstance. I may equally state they are "out of scope".

The ultrafilter is valid on every set in the power set of positive properties in Ω when their conjunction is as "q", the exclusions in u and v are then part of the octal ultrafilter; they are valid in Ω but are not necessarily in action in the mapping table at all. (Except perhaps for whole perfection.) God is free to exemplify any subset of the properties in "s" consequent from p&q^{-1} (God is at liberty and may completely rest, every "s" may become as a "v^{-1}").

It may be true of u and v that they are as "work already done" or "work not necessary". The exclusions in u and v are part of the filter and are indispensable to the octal. If God has already worked, God may surely rest and there will always be a reason for there to exist such positive properties u and v not in scope.

On Inconsistency (4.10.5):

Given N¬(p&u^{-1}&¬v), I may rearrange to N¬(p&¬(u&v)). Either the set of virtues is not closed (p is that set), or u&v is not a consistent positive property. (A contradiction in terms.) That is, (u&v)$^{-1}$ privates p or N¬((u&v)$^{-1}$&¬(p^{-1})). u&v also privates p; as an excluded middle u&v entails no free exercise between u and v and so (u&v) is inconsistent in that row of the filter table.

Clearly if N¬(u&v) or u → v^{-1} etc., then as u&v is a valid positive property in the octal, the equivalent v^{-1}&v makes no sense when given a specific virtue p and v^{-1} ∨ u^{-1} (since u^{-1}&v^{-1} privates p). The conjunction u&u^{-1} likewise makes no sense. Can I state of Pos(u&u^{-1}) that u is both positive and it is positive for perfection not to act with u? I ought to consider ¬Pos(u) ↔ Pos(u^{-1}) yet then I find ¬Pos(u&u^{-1}) ↔ Pos(u&u^{-1}). Those two statements become equivalent to the admission that N¬(u&v) or u&u^{-1} are not positive and are unable to be exemplified positively. What I really have is that given some real virtue p: v^{-1} ∨ u^{-1} with p&v → u^{-1} but there is no virtue p$_1$ such that u&v is positive and in scope where either of v → u^{-1} and u → v^{-1} hold that would allow either u&v ↔ u^{-1}&v^{-1} or u&u^{-1} etc.

Given p&v → u^{-1}, it is almost (but not) reasonable to exercise u (and also v) to action in the cases of q and p&q^{-1}. Then u and v are positive properties to be excluded from action or consideration, being rested upon or necessarily missing in inaction. (Work already done?) Then u → v^{-1} relates that somehow, given a virtue p: v^{-1} ∨ u^{-1} is equivalent to v ∨ u. That is, rather than allow of some virtue p$_1$ for the disjunction (r&s)$^{-1}$ ∨ u&v to become equally as (r&s)$^{-1}$ ∨ u^{-1}&v^{-1}. Virtues do not allow privations on perfection to be freely exercised. v ∨ u does not follow from v^{-1} ∨ u^{-1} given any p.

In the case of u&v it may **not** be said that given the former virtue p that the conjunction u&v is equal to u^{-1}&v^{-1}. The excluded middle in q ∨ p&q^{-1} is simply formed of the conjunctions "not in scope" of those properties "in scope" with q or p&q^{-1}, but if such a Pos(u&v) holds, it is positive to not consider either of u^{-1} and v^{-1} to be in scope with p at all. I could also consider (by symmetry) that u^{-1} and v^{-1} are necessarily inactive, irrespective of deciding q ∨ p&q^{-1}.

A Filter Upon Properties Rather Than Privations (4.10.6)

Whether u and v are simply found necessarily active in perfection on both sides of q ∨ p&q^{-1} (as being some r and s) or are necessarily absent from consideration altogether (as u^{-1} or v^{-1}) there is no difference to the operation of the octal; God with all positive properties is unique, whereas the sets of positive properties (as humans may consider to display them) are exemplifiable "in His likeness" but without the completeness of whole perfection.

The Revelation of Jesus Christ and the "Lamb before the throne" are completely consistent with the statement that all such analogues of (r&s) (privating virtue) are excluded from each

K4 filter of the filter table as they are actually "evils" and so empty as contradictions. (They in analogue private virtue or virtue privates them: contrary to the definition of virtue and any positive property. I would rather drop the fleeting positivity (as by axiom) of an unworkable (r&s) than dismiss all virtues altogether.)

In the logical schema, r&s privates virtue so cannot be a positive property. Then does either r or s fail to be positive? The answer is that yes, one does (on inclusion only) – and I may state r follows from q so r is classically a positive property and s is within some extension over all such q and r. Adding r&s to the set of q would not be "illegal", but would instead as a "middle" constrain the set of all possible virtues. A maximal filter of every positive (or extended) property by axiom of virtue would contain every possible virtue and so such excluded middles are natural. Smaller (with most of Ω apparently at rest) indexing sets for the filter have quite naturally a limited or constrained set of virtues.

Constraint on grace may be refined to a state of non-privation to further extend mercies.

On The Holy Spirit As Multiplication (4.10.7)

The Holy Spirit (it must be said), is found to be a multiplicative group isomorphic to C7 acting over the principal elements of the sets of positive properties in the octal. Whereas I had labelled the principal elements as e_p, e_r, e_s etc., the octal is better described using the elements [0,a,b,c,d,e,f,g]. Any element a, b, c, etc. from the octal and under that multiplication of C7 may appear in the ultrafilter as e_p, e_r, e_s, say, yet only on the condition that the Holy Spirit which may transform the elements a to g through the octal (and sets of positive predicates) preserves the subgroups in the octal itself.

It is also true that every valid multiplication C7 over the octal will transform subgroup to subgroup as well. It is also the case that any set of positive properties in the ultrafilter may be transformed to become virtue over the same disjunction within the same logical schema.

May I rearrange each K4 group under the logic of N(¬p & r&s) to, say, N(¬r & p&s)? I clearly may if this logic is found equivalent: There is no "p" in the octal permitting the middle r&s, just as given p&s I find from one side of the disjunction that, given $p\&r^{-1} \rightarrow s$, or $p\&r^{-1} \rightarrow p\&s$ I must have N¬(Pos(r)), on that one side with p&s as if it were now necessary. It is still not true that $p \rightarrow s$ or $p \rightarrow r$ or that any of $r \rightarrow s$, $r \rightarrow p$, $s \rightarrow r$ or $s \rightarrow p$ hold (that every set remains disjoint and closed).

Then by this "near symmetry" every subgroup may be rewritten with any positive predicate true by axiom as if virtue. Those subgroups must meet in the same logical schema.

Then the Holy Spirit not mentioned until now suffices to redeem the use $(y \& x^{-1} \rightarrow z) \vee x$ of every positive property y not in the set of virtues as entailing the same effect (freedom between x and z) as if a virtue p had been applied instead; there are no "faulty uses" or "half-measures" in the Godhead. I find that a suitable p may substitute for y. (There can be no other reason than to show the result z could also follow from some virtue p. I safely assume divine liberty is perfect and may be exercised by deciding from the set of all virtues (that entail z) some p to act upon in the absence of x, as if I had the case q=x in some separate instance.)

The construction of the octal was dependent on the closure of virtue and the necessary positivity (by axiom) of that set of virtues. Any predicate or set in the octal may become an absolute of virtue (as by axiom) in a separate setting (where any such property becomes necessarily positive), and there is then no distinction between virtue and any axiomatically positive set of properties other than the requirement for consistency of all "virtue" due to the perfect liberty of God (aside from the separation into seven closed subsets, in the same set of all positive properties Ω and in the formation of an isomorphic octal).

Then there is no difference at all between a virtue and any such positive property other than that original assumption of necessary positivity, and the multiplication of C7 preserves each element necessarily (axiomatically) positive as "unity" in their own setting (when positive by axiom). The chosen set of virtue (or of that unity) together with the axiom of necessary positivity entails the closure of every other set in the octal from the mapping table, and it is a simple result to state that God, in every set in the octal, is polymorphically perfected. The Holy Spirit is then just that, the preservative of all virtue in every conceivable setting.

I have argued that the seven sets of elements in the octal are effectively to be treated as disjoint, and that the Holy Spirit acts upon these as if singletons. By using a principal element as a "zero" (see section [2.10]) I note that the seven cycle will be absorbed into the zero by that multiplicative property and so under multiplication it will appear present with every set of the mapping table if it is found present with any at all. I have no reason to exclude multiplication as in GF(4) upon triples, providing one of the elements in the triple is the set of virtues in "p" (allowing purpose to multiplication as above, preserving the axiom of virtue).

Restricting the rows of the filter table to triples that are closed in positive properties and also form subgroups in the octal, I construct the octal group as a union of seven principal ultrafilters' indexing sets. (With each indexing set limited to one of those triples.) The cosets of each K4 subgroup cannot constrain the triples used as indexing sets for those ultrafilters: and the octal acting upon those triples in K4-addition $(A \vee B)^c$ gives a solution for their addition. Perfection must be found in the operation on singletons in the octal which agrees in those seven subgroups, but in order for perfection to be present in that union, I require the Holy Spirit to map from one ultrafilter to any other in a natural fashion. Then, and only then, does the octal group make any sense as a union of "singletons" or as such, disjoint sets of positive properties. Every K4 subgroup of the octal is an ultrafilter on its own triples, but given the octal is also such an ultrafilter, it is clear that whenever addition is performed on singletons in subgroups, it is also performed in the abstract structure of the octal known as "The Father".

The Holy Spirit (of multiplication as C7 over the octal to form GF(8)) transforms from K4 subgroup to subgroup, and perfection maps the local closure of triples (as triples of singletons rather than sets) through the octal. (Of course, if the triples themselves actually span a suitable set of positive properties (my Ω) then there is no problem if that "complete" set of positive properties is fully spanned as $\Omega = x \cup y \cup z$ and the rest of Ω becomes inconsistent. In turn, God as creator may well choose whatever Ω He wishes for Himself or (or upon) His creations and I may well consider that God as creator may make natural law, with a subset of virtues able to be performed by anyone.)

Therefore, by the properties of any ultrafilter, the complement in Ω of the union $x \cup y \cup z$

is negative and not in the filter. So, the K4 subgroups as sets in the octal are in complete agreement with not just each other, taking their operations in local closures of triples, but equivalently in the octal also! (The local closure is effectively the whole closure.) I may then safely ignore the cosets of each K4 subgroup completely!

As to why the Holy Spirit is become as a cyclic group of order seven: I can but state at this point that every element may be perfectly mapped by multiplication onto any other (non-zero) element and by happenstance seven is prime so it is merely an observation that the only group of order seven is cyclic.

Walking the subgroups through the octal not only suffices to complete the algebra, but preserves every positive property which is in the octal perfected, but under a different (cycled and permuted) logical schema. The octal as the next closed superset of each K4 form is likewise in the ultrafilter (being the solution to the ultrafilter itself) and the closure of the octal's operation likewise preserves what is perfect from every K4 subgroup of triples. The K4 subgroups are "just as perfect" as the octal, whether sat in place or spanning the octal.

And Lastly, On Identity (4.10.8)

Under the operation of $(A \vee B)^c$, I may use the set of all positive properties Ω for the identity elements of both the K4 group of each row of the filter table and the whole octal (formed by adding subgroups). The natural inconsistency of the excluded middle is enough to limit Ω to only those properties necessary to the product of each pair of elements, fixing the "third part" of the ultrafilter to the correct indexing set, the unique and native row of the filter table.

In Revelation this duality of sitting as the octal Ω or a K4 group to form the zero – without any difference to the operation of the octal (whether of triples or of subgroups) is easily missed at first glance, but there is then a greater depth of understanding revealed in the verse:

Rev 3:21 To him that overcometh will I grant to sit with me in my throne, even as I also overcame, and am set down with my Father in his throne.

From the letters at the very start, the Revelation astonishes with detail that could not be merely incidental, or at least – this is due the reader's full consideration upon the proposition that John was most certainly inspired.

4.11) In Whom I Am Well Pleased

It should already be apparent that the schema of the octal seems out of balance and not such a "perfect" solution after all. The division of the K4 form to left and right into sets of predicates or properties in action and inaction leaves a question somewhat unanswered: "Where is the embedding of the octal group itself within those sets?"

In section [3.10], I used the closure of virtue to account for the three members of the Trinity; in this section (as if a corollary) I do the same for the octal group: this time skipping the Trinity and going straight for the octal itself. It becomes apparent that I need seven predicates: one of virtue/liberty as for L(G) and seven for "rest" (one of the latter equal to L(G) within the set of virtues) – those principal elements e_x. I.e. each set "x" of the seven has some principal e_x.

The virtue of liberty acts over each set in the octal, each assigned their own consequent "rest" – a principal element e_x within that set of the octal. Pairs taken of the "rest elements" (i.e. of the e_x) cannot be preserved through the closure of virtue (not like the "Manifold Of Virtue") not themselves being virtues (liberty, though, as principal in "p", is a virtue) so the conjunctions in the e_x would appear excluded in part by the empty middle: and yet they are also a consequence of liberty (virtue), which may in its place as e_p (or L(G), one of the e_x), also form the necessary solution within the closure of virtue/liberty, much as the three members of the Trinity itself had done so before. I now happily construct the set of e_x (consequent of L(G)) as if forming an octal in those elements of "rest" e_x.

From the virtues in "p", within each disjunction formed (as in $r \vee p\&r^{-1} \to s$), I easily find the liberty of God present over, say, the one set in "r" so that if the left side were to be certainly (yet freely) decided upon, then "r" as a set is completed with e_r, so I would find "rest" on "r" when so acting at liberty (and I would simultaneously find the right hand side to reduce to $L(G)\&e_r^{-1} \to e_s$ instead of only $p\&r^{-1} \to s$), leaving me with $L(G)\&e_s \to e_r$ in this rather unusual case. When deciding upon the right-hand side instead, I would find $L(G)\&e_r \to e_s$.

The previous principal element "L" as "liberty" acts as zero ("L", the zero with regard to "binding" rather than of the conjunction of all the e_x); "L" may entail each of the e_x upon themselves, never entailing anything further by axiom of virtue (although it is not yet required for any disjunction). I do not absorb the e_x on the side wherein liberty is found to act: "L" then becomes the additive identity and the e_x are "elements" present as from addition only.

The disjunction of virtue leaves only the one element of e_r or e_s in action as consequent of liberty. Here, $e_p = L(G)$ and one of e_r (or e_s as e_r^{-1} is "at rest") and so $L(G)\&e_s \to e_r$ has the equivalent $e_p\&e_s \to e_r$ with e_p equal to that same "principal virtue" L(G); yet $e_r\&e_s$ is present only in the excluded middle, and it is otherwise not in the filter unless found equal in the octal to e_p ($e_r\&e_s$ is not formed of virtues but is an equivalent substitution for e_p).

There is also the case when the right-hand side of $p\&r^{-1} \to s$ is certainly decided upon. Then $p\&r^{-1} \to s$ becomes $e_p\&e_r \to e_s$ or $L(G)\&e_r \to e_s$ (e_p and e_r are still "at rest"). I may certainly arrange the disjunction to $s \vee p\&s^{-1} \to r$, permitting the equivalent conjunction of $e_p\&e_s \to e_r$. I have partially formed from elements "at rest" a K4 group within the octal. Repeating for every row of the filter table I would find an octal formed of L(G) and each e_x. Liberty (or "rest", as free to float), has a solution in the octal.

A perfect being should have the equivalences: $e_p\&e_r \leftrightarrow e_s$, $e_p\&e_s \leftrightarrow e_r$, and $e_p \leftrightarrow e_r\&e_s$. However, I drop the logical equivalences and simply write: $e_p+e_r=e_s$, $e_p+e_s=e_r$, and $e_p=e_r+e_s$.

The wider octal may be embedded in those seven ultrafilters of the filter table, replacing "e" and preserving each filter, and all virtue, and with it the closure of each group without fault.

Luk 3:22 And the Holy Ghost descended in a bodily shape like a dove upon him, and a voice came from heaven, which said, Thou art my beloved Son; in thee I am well pleased.

Each of the "rest elements" form a "predicate of a set" in that each are effectively defined by:

$$e_T = \vee T = \{x \; : \; t \to x, \text{ for all } t \text{ in } T\}$$

The octal is embedded in the seven sets of the filter table, together forming an octal given the conjunctions $e_r \& e_s$, $e_u \& e_v$, $e_{u\&v} \& e_{r\&s}$ all equivalent to $e_p = L(G)$. These are found conjoined, formed of disjoint sets and are independent of all action and inaction (being that principal element $L(G)$ acting as virtue over their respective filters) I may, from the splitting of left and right into action and inaction, infer these conjunctions of "rest" are equal to the liberty/virtue $e_p = L(G)$. Each conjunction is also axiomatically true; and always found in action as if virtue.

In regard to "rest"; the e_p, e_r, e_s etc, belong to disjoint sets forming the octal proper (as embedded in Ω) with the presence of liberty which as virtue will (without fail) entail them.

To complete the octal, I state the Father is well pleased in His Son, as the octal is found embedded in those same seven sets — perfected in the righteousness and liberty of Christ deciding on virtues. Each disjunction would freely decide $(e_p \& e_s \to e_r) \, \triangledown \, (e_p \& e_r \to e_s)$ which are both equally positive and correct, together all seven disjunctions determine the octal.

A few points remain: the relations order two are (as yet) missing. Unlike the "Manifold Of Virtue", I may not place a set of positive properties (or some e_x) on both sides of the disjunction without privating virtue. However, there are no disjunctions in the octal with a set appearing "both sides", and there is simply no data on those relations unless they are assumed present or implicitly defined to fulfil the requirements for the octal to be found embedded whole.

Note that $e_r \& e_s$ is neither in the set of "r" or "s". In the disjunction it is an equivalent to e_p and e_p is most certainly not the principal element in the disjunction (but is one of all virtue instead, equal to $L(G)$); that honour goes to the conjunction $L = e_p \& e_r \& e_s$ instead. Consequent of virtue and found only in the middle, e_p has its equivalent $e_r \& e_s$ without breaking the closure of either "r", "s" or even "p". There is a caveat: $L(G) = e_p$ is required to be perfect as well as consistent.

2Co 12:9 And he said unto me, My grace is sufficient for thee: for my strength is made perfect in weakness. Most gladly therefore will I rather glory in my infirmities, that the power of Christ may rest upon me.

The sets of the filter table split into action or inaction are arbitrary but for the cut of liberty determining them: The person of each member of the Trinity (and the octal) should be considered not as a set or container of positive properties, but consisting of liberty and existence only. The properties entailed in disjunctions are works of omnipotence and certainly not constraints defining God's character: God is free to rest or to exhibit as small a set of positive predicates as any of us, and in so doing remaining perfect.

The equivalences of those conjunctions in e_x to those virtues remain in each set of seven, even when those disjunctions formed are not "of virtue".

For example: $r \vee v^{-1}$ given u&v reduces to $(e_{u\&v} \& e_v \to e_r) \, \triangledown \, (e_{u\&v} \& e_r \to e_v)$, and these are totally determined by $L(G)$ acting in the three disjunctions with "p", the disjunction in "p" remains freely decidable, and the other four triples are a consequence of the closure in the octal.

The equivalence of $e_{u\&v}$ to $e_r \& e_v$ remains determined by "virtue" only, not by a mere "positive property" which then (as such) remains "locally positive". Though u&v in this case acts as virtue, $e_v = e_{u\&v} \& e_r$ or $e_v \& e_{u\&v} = e_r$, or $e_v \& e_r = e_{u\&v}$ hold separately from the empty middles taken of the sets v^{-1}, r, u&v etc. Those conjunctions hold consequent of virtue "p" (and not due to the other rows of the filter table) or "in the middle" of those disjunctions with virtue "p". The presence of liberty L(G) within that one set "p" (always axiomatically positive) acts as the "person" of the octal in the "Manifold Of Virtue" (of section [3.10]) so that "p" or "unity" is found present in every "product" as with the Father F, in that Manifold of the Trinity.

In comparison to Gödel's argument[1], "God-likeness" is equivalent to "liberty" and the freedom of liberty is present in every row of the filter table as it forms an ultrafilter over those sets. Necessary existence then becomes equivalent to "e_x" and each set retains some virtue, though that virtue of liberty is unique, only entailing rest from that particular set of virtues in "p".

Every set in the octal's given filters are defined in terms of that liberty as virtue. In the disjunctions of virtue in "p", every set was derived by splitting the initial construction of a K4 group into action and inaction and then defining the equivalences of e_r&e_s, e_u&e_v, $e_{u\&v}$&$e_{r\&s}$ to e_p=L(G). Given each set then decided upon is one set of six with the seventh virtue; I do not require the inclusion of virtue in the other four disjunctions not in "p", merely that the rest elements are present within all seven sets of the octal. Each set "x" contains an "e_x" as a principal element, but the closure of the octal does not require liberty to act in any disjunction without a virtue in p; the remaining four disjunctions are indeed formed of ultrafilters, but the liberty is already present in "p" to rest or act on each filter as it is given.

I form the disjunctions between e_v&e_r=$e_{u\&v}$ and e_v&$e_{u\&v}$=e_r separate to the disjunction r ∨ v^{-1} given u&v (as independent of "p") as the content of these three sets are fixed by the liberty in "p" found within another disjunction. The sets in the mapping table fix the rest of the octal.

The seven sets of the octal gain their principal elements from the three K4 groups with "p", and not from the whole closure of the octal unless it is also implicitly defined by the rows of the filter table. There is yet some liberty remaining to define the operation upon that closure (as by choice). The remaining four subgroups then fix the operation of the octal whole.

Then the choices/decisions of the Son (Christ Jesus) which provide the seven sets for the content of the ultrafilters (and for the octal itself) are indwelt by the Spirit of the Father (with His necessary existence as each of the e_x), and each of those seven sets provide some measure of closure within which the Spirit of God may act freely, whether additively in closing the octal or by multiplicative action, that life in the Holy Spirit which is the octal so perfected. Then the "Manifold Of Virtue" also has a good solution in the Trinity, one in which God in Trinity is certain to be (as from His liberty), found at "rest" and also well pleased in His Son.

4.12) Life In The Spirit

Whether a set is "at rest" (as is $(r\&s)^{-1}$ or u^{-1}) or not, the rest consequent of liberty always appears present as an active predicate: There is always a least element present consequent of virtue even when liberty rests upon the whole filter (or nothing but that "rest" itself) which allows the closure of that row of the filter table to remain intact.

Note that disjunctions of the sort r ∨ p&r^{-1} → s actually decide the disjunction made of:

[1] Sobel, in J. J Thompson (ed.), *On Being and Saying: Essays for Richard Cartwright*, p241-61.

$$\left(e_p \,\&\, e_s \to e_r\right) \bar{\vee} \left(e_p \,\&\, e_r \to e_s\right)$$

Which are both true and freely decidable, or "simultaneously correct". (Every product in the octal is now a completed act rested upon. God is then omnipotent by His free choice of sets.)

In the middle there is a conjunction made of $e_p = e_r \,\&\, e_s$ which is certainly not equal to $e_{r\&s}$. In the middle of that same conjunction, the sets e_r, e_s are only to be found "at rest", and the middle (the set found outside of all virtue in p), whilst logically found consequent of virtue (whilst virtue itself is acting on nothing but the empty middle); is empty and cannot then entail (from virtue in p) anything outside of p. The axiom of virtue is then in effect and saves the day, as to entail the middle from the empty set breaks the binding; it cannot be so done. I.e. resting upon two separate sets r, s is not the same as resting on the middle $(r\&s)^{-1}$ (as equal to virtue $e_p = e_r \,\&\, e_s$ is not $e_{r\&s}$ and is instead a "binding" of $e_p = e_r + e_s$ rather than a conjunction).

Then $e_p \to e_r \,\&\, e_s$ is to rest on the empty middle made of both sets r and s rather than to entail rest made in that middle: the operation of the octal cannot replace $e_r \,\&\, e_s$ with $e_{r\&s}$ say, as that set $(r\&s)^{-1}$ does not appear in the disjunction formed of $r \vee p\&r^{-1} \to s$.

As it is realised that e_p, e_r, e_s, e_u, e_v, $e_{r\&s}$, $e_{u\&v}$ are all present independent of action and inaction, and that they split to form three K4 groups with e_p; it is then easily stated that the seven cycle walks the closure of these three groups through the octal (as well as implicitly defining all seven groups of the octal from the three), and the seven cycle acts on these "rest elements" as by cycling the principal elements in each filter. God, as omnipotent, rests on every work chosen. Liberty, in making the cut or division between action and inaction in every set of seven (itself a "rest element"), may exemplify anything positive or thereby chosen within those sets. (All seven principal elements will agree. Each e_x justifies each set as "God-like".)

So, I hesitantly state that the seven cycle certainly preserves these seven subgroups but not the "schema" of action and inaction present over the sets in the disjunction. There are 84 such schemas (counting seven choices in the octal for virtue, three groups for the conjunction with both sides in action, two for the group with both sides in inaction, and two further choices for the last group, as to whether the sets of the disjunction are exchanged in part or not).

Virtue, it appears, is not found to be arbitrary except for within the octal (and God's judgement over the placement of sets within it). Those positive properties may cycle under C7, but the schema over them may not be altered. Liberty is instead found acting within the sets in the octal itself; and does not alter the octal's placing of "rest" within those seven sets.

Independent of any choice of "schema", the division made of or between sets in Ω or in the octal embedded in that schema is only valid when seen or evaluated "at rest". (As under the action of unity.) The octal is isomorphic to any similar arrangement of those seven sets, and all are effectively equal under the one schema (as under that unity). Then "virtue" (as defined previously) is that which pleases God, who has the choice as to what is pleasing to Him in or of those seven sets amongst or under every product in every C7 group upon those possible octals.

Then laws of virtue are decided upon by God whilst only He operates over rather than within that perfection. The operation of the octal is always symmetry (unlike the schema), and whatsoever arrangement is pleasing to God in those sets and their conjunctions is rested upon and exemplified by that schema, if it be consistent with God's schema as exemplified so far! (For what else may be said? Consistency should decide virtue in every act of God.)

So (as from symmetry made on the principal elements only), the "joining" (or sum/product) between any two sets in the octal is just the conjunction of each "rest" element upon/within each set. (And the apparent schema then has its **every** variant justified in these conjunctions.)

In those variants I can avoid compounded or "miraculous relabellings" of conjunctions as those predicates u&v and (r&s)$^{-1}$ do not private the octal's structure if its elements are already embedded in similar conjunctions. The "sums" $e_p = e_r \& e_s$ etc are independent of action and inaction, yet the exact nature of these conjunctions in the schema holds for every set (and every schema possible) in every arrangement of the triples preserved under the seven cycles as above. The sets remain disjoint; but the logic of action/inaction is the clothing made upon and for the octal, not only (or merely) the other way around. Labelling the schema as r, s, u, v etc., is convenient for the logic; but the ultrafilter of positive properties is much better labelled [0, a, b, c, d, e, f, g] as used throughout this book. The logic is simply dressing for the underlying truth of the octal that is always found spanning **all** positive properties.

Every schema over the octal (and upon the seven subgroups) is effectively equal to each of the others, and every set in it able to become "as virtue" in an appropriate setting; yet only God has the Holy Spirit to do so perfectly and consistently! (He is unique.) There is found only symmetry in the octal – after all, it is God alone that ultimately decides what is good and He alone creates natural laws from His virtue. God, then, has the ultimate test in free will, whilst Christians as a people are somewhat already "decided for" concerning what is good and right in this system of things. God, it appears, has chosen to "rest" on a particular system of His own choosing. (As found in the "Lamb" before the throne.)

The set of all virtues G acted upon by liberty L is such to entail rest "L" in every disjunction. That "L" is the zero, and perfected is equal to each e_x conjoined (as liberty), acting across every set in the octal. The seven principal elements of the octal suffice to substitute for "L(x)" in each set x and therefore also L. For all e_x I find $L+e_x = L(x) = e_x$, liberty makes the necessary cut in every set, entailing from the "God-like" the existent. Liberty, or the "zero" acts as does the octal "enthroned" – choosing all p, r, s, u and v. However, the seven principal elements do the work and have the satisfaction of a completed act. Zero never moves from rest, absorbing the seven cycle whilst remaining able to choose (to the limit of omnipotence) works that will give God consistent satisfaction.

The set of all positive properties Ω is then "God shaped". Positive properties do not form an octal group alone, but the ultrafilter which indeed does so is a good model of those properties as both maximally (from virtue) and minimally (in "rest") completed and so I will always find that the octal ultrafilters index that same set. Ω is then characterised by the octal, and is effectively "God shaped". (Or instead, "God formed".)

As to the Holy Spirit alighting on Christ as a dove, it should later become apparent that the Holy Spirit had in bodily form descended "as a bird", then loosing all restraint upon Christ, doing so in a "clean" manner, and "harmlessly" so. His ministry was then empowered with omnipotence and in such a manner as to please God, and God only. Christ was empowered to do more than simply obey God and choose well; He was approved as the Son of God.

4.13) An Example Of The Octal – Triage

In triage, would you use much needed materials on one person who was bleeding out and will certainly die, or do you desire that dressing to go to one who may recover?

Need that other someone have arrived in the clinic yet? I note $(Pos(u^{-1}) \lor Pos(v^{-1}))$, i.e. resist on the spot treatment in a crisis.

The triage process is positive. Thus $Pos(u\&v)$. Treating emergencies wisely is a good act.

Likewise, evaluating patients according to need is positive (r = the next patient is accurately assessed) vs. (s = there are limited resources; those not assessed yet must also be assessed equally).

Therefore $Pos(r) \lor Pos(s)$. (No simultaneity.)

I state $Pos(r\&s)^{-1}$ – it is good to finish treating everyone so that triage may be suspended at the end of an emergency.

The goal is that the last disjunction is freely decided. $Pos(u\&v) \lor Pos(r\&s)^{-1}$ are both positive and to be met.

4.14) An Example Of The Octal – Gödel's Incompleteness

It is often objected to (by Gödel's incompleteness theorems[6]) that there is a "God of all truth". Given that, one could "ask God" as to whether the following statement is true or false:

"The God will always say this statement is false".[6]

If God answers true or false then there is an apparent contradiction, and apparently God does not know everything. However, given "r" as "saying true" or "s" as "saying false"; whichever seems a positive answer at that moment is within the realm of freedom and God's own liberty. God may flip-flop every bit as much as I may choose to do! Similarly, given u&v or $(r\&s)^{-1}$ positive – which hold no matter whether r or s is extant: there is the freedom not to answer, and more so; there need be no answer at all and the outcome is positive. I.e. the statement becomes "God is never free to act if the alternative is always to act". God may simply and freely act with u&v avoiding both cases r or s.

Likewise, God may not answer "The God will never say this statement is true" as I would expect.

Similarly, u^{-1} and/or v^{-1} are positive: I may make the comparison to the statement "God is never free to rest if the alternative is always to rest". Then, God may simply perform $(r\&s)^{-1}$ and logic is satisfied, albeit strangely.

[6] Hofstadter, *Gödel, Escher, Bach: An Eternal Golden Braid*, p438-460, p18.

So lastly the triple {p, (r&s)$^{-1}$, u&v} as a filter provides the statement "God is never free unless the alternative is freedom". Yet then freedom for God indicates that He may rest or act!

The principal element "e_x" from before is also of note. For, answering the question with either "true" or "false" is positive if the matter may be finally rested upon; for the positivity is found in the freedom to answer the statement: God may state as "e_r"; "the statement is true" and then exercise His virtue "e_p" in stating "the statement is false" as if "e_s". The matter is then closed; God is free to answer. (The positivity of either statement is momentary and fleeting, but demonstrates as near as may be done so; the satisfaction of perfection by "e_p" due to the exercise of liberty.) After the exercise of liberty in this case, there is no other positive property to demonstrate but "e_p".

More pertinent to incompleteness, though, is the statement of the union in all positive properties. The sets in the mapping table are disjoint, the rows of the filter table not so. One could ask the question of the omnipotent God, "which side of the following is positive q={Ω\e}$\vee$ p&q^{-1} → {e,A,B,C,D,E,F,G}?" ("A", the set of virtues, B to G positive in the octal with "e" rest or the zero.) Both sides are equally positive but for "rest" in "e" and by symmetry this makes no difference. From incompleteness, were it to apply, I would expect the closure of positive properties Ω to always outrun the closure of the octal, but this is not so. (Omnipotence does not fail on incompleteness, each set in the octal may rest on its principal e_x alone.)

Given {Ω\e} (even as the full closure of the K4 group) made to "self-reference" the octal complete (as in X={e,A,B,C,D,E,F,G} which is also found to be "r&s" in the disjunction here) given their product (A$\vee$B)^c, this particular case results in p = {Ω\e} (the local closure in regress). Is it then (assuming incompleteness remains) impossible but still positive to rest on (and exclude) (r&s)$^{-1}$?

What is this (r&s)$^{-1}$? It cannot be any subgroup of X, with some closure {Ω\e}, as (r&s)$^{-1}$ is part of the logical schema that is formed of disjoint sets, which between K4 groups in the octal is certainly not the case. However, there is a work-around; each multiplicative group (C7) may circuit between seven others, all valid over the one octal. Taking the full closure of each of seven octals as one set in the logical schema of the C7 group over it: I may again form such a schema, but only with the whole octal formed of disjoint sets.

I then need to maximise the set in each octal/element into action; I need a modal collapse with each octal reduced to its bare representation $\Omega \vee \phi$. I can do this only with the unique "essence" (of Gödel's argument[1]) and for this I require a singleton zero, rather than any other principal element: I require their conjunction and a modal collapse.

Then, instead of generating the set of all positive predicates, I generate the entire closed set of each octal (disjoint but for the zero, or that "essence") as a single ultrafilter, which may together meet in one disjunction in forming X. ((r&s)$^{-1}$ effectively becomes a set also in X.)

Is there such a disjunction? There surely is, as will be shown in sections [16.11-13].

Does X then contain every possible positive predicate? Only if each set in X has a principal element, then also an "essence". The "Revelation" now opened reveals this as shown by the election of grace.

[1] Sobel, in J. J Thompson (ed.), *On Being and Saying: Essays for Richard Cartwright*, p241-61.

Every C7 group over an octal may circuit as a multiplication over each of the others in like fashion. Every octal is then complete in the full closure of the others. There must needs be self-reference between elements and/or closures of octals whole: so that the elements in the schema of every octal remain disjoint. The statement on incompleteness "Is it then impossible but still positive to rest on (exclude) (r&s)$^{-1}$?", has its answer.

Isa 55:9 For *as* the heavens are higher than the earth, so are my ways higher than your ways, and my thoughts than your thoughts.

In fact, it is never impossible to rest on (r&s)$^{-1}$ in the octal, but only impossible in each K4 group's local closure (as formed maximised to that of a whole octal) as there is no middle r&s in $\Omega \vee \phi$. God may span all positive properties in a mode of infinite regression, and there is then always a fully completed octal in positive predicates, as those local closures in the seven elements/octals (each maximised) are each closed in their own logic. There is such a disjunction to be found for this "master-schema", and the "Revelation of Jesus Christ" has it.

If there is anything incomplete outside of any octal, in Jesus Christ there is a Father for it.

4.15) A Name For Miracles.

I could argue that a creator God, exemplifying in creation every subset of Ω in the "q-filter" of all possible created "beings", as to become a natural law maker, indeed has "merit", but not, then, for God Himself to forever surrender liberty of virtue! Instead, He, by resting on any minimal positive subset in "s" at will, leaves Himself with more liberty. In the case of Jesus Christ, He also could have "made these stones bread..." yet would not then have perfected any subset of the "q-filter" which God made man for. (And whilst this dispensation of grace continues it cannot be vice-versa!) God instead has made man "a little lower than the angels" whereby "angels" are "ministering" and it ought, properly, to be reconsidered what is meant by a miracle of omnipotent God (by virtue).

The indexing set of positive properties then has some mystery attached to it, the octal not being a system in which the statements of number theory could be readily formulated, at least not unless "understands all algebra" is positive! I instead state quite separately that mathematics is as virtue for any omnipotent God to then justify (by maths) all of His creative works, and that He cannot rest upon them until they are shown justified as coherent with all past creation, to become as completed (finished) works. "Miracles" are to be considered as all part of this system of things. (I would state "mathematically consistent" is a requirement for the satisfaction of a completed act rather than the rest that arrives afterward, so "maths" is not principal in the ultrafilter and I infer miracles are possible.)

"Mathematically consistent", then, would also include every justifiable act of an omnipotent God, if that final statement is consistent with all positive properties. Virtue may act otherwise, and I expect "this system of things" to be somewhat wider than creation's law-book. If p&q^{-1} entails both s and v^{-1}, then if creation's laws are always "an s" and miracles always in "v", then I could not fault virtue for providing a miracle or not. "Maths" as a virtue in "p" ensures that no new laws are added to "s", and then God rests upon some steady state "e" in (r&s)$^{-1}$. "q" would be as every unworkable act of omnipotence upon the laws of nature, and there is no contradiction in this with respect to miracles, this set being disjoint.

Now, "q" entails both r and u^{-1}, where "r" I would consider as I would "wrath" and properties in "u" I would consider judgements that inverted would entail mercy, so that q, the "unworkable", would entail recompense only upon those that justifiably deserve it. I would consider, then, that these properties in "q" are "God's laws for God" and are, by symmetry, consistent in positive properties as much as those following virtue in p, even to their exercise in creation, also justified by maths. Then, as an ultrafilter without virtue is incomplete, the statement "I desire mercy rather than sacrifice" is a positive one of virtue towards all as opposed to just His elect; all in creation are in receipt of God's virtue.

Then $(r\&s)^{-1}$ and u&v are also positive and freely decidable, that there be no wrath upon the laws of nature and that God should likewise supply provision of mercy by miracles. Now, which is entailed from virtue? It is actually both!

4.16) Prerequisite Maths Checklist

Here I present every mathematical concept you **need** to know to fully complete this study of Revelation. It may help to refer back to this list as every item here has been (somewhat) covered within this first part. You should mark off each item as you become familiar with the maths and before you move on.

- The notion of an n-cycle is required, as is that of a Cartesian product A x B (see below next for example).

- Familiarity with these three finite fields (characteristic two) GF(2), GF(4) and GF(8) (and also modular arithmetic).

- K4 as C2 x C2 or GF(4)+ and the octal as C2 x C2 x C2 or GF(8)+.

- C3 and C7 as also GF(4)* and GF(8)*.

- The concept of a "poset" (a "Partially Ordered Set"). A poset is equivalent to a collection of sets under an operation of inclusion $\subseteq$; or otherwise of an ordering upon its elements which may be singletons.

- The operation $(A \vee B)^c$ as addition in the octal upon its K4 subgroups (the "set complement" S^c of the "symmetric difference" $S = A \vee B$).

- The K4 form of the ultrafilter $0 \equiv \{a,b,c,d,e,f,g\}$, $a \equiv [a,b,c]$, $b \equiv [a,d,e]$, $c \equiv [a,f,g]$.

- Finite fields of order p^n have every extension of order p^{nm} (n, m integers greater than zero) all have cyclic multiplicative groups. The top of the chain of all such finite field extensions is the field of "Algebraic Closure", which contains every finite extension. All extensions of finite fields are separable.

- (Separability) The following is important for understanding the fifth trumpet in the ninth chapter of Revelation. Any extension of degree nm over a field F of order p is a field of order p^{nm} which has subfields J and K of order p^n and p^m. The polynomials in GF(p^{nm}) may be written with coefficients in the subfields J and

         K, modulo irreducible polynomials over J and K of degree m and n respectively.

- Familiarity with the concept of an isomorphism from a structure to itself (an automorphism) is a must. Isomorphisms are bijections and are therefore invertible maps. (A bijection is a one-to-one and a fully "onto" or surjective map.)

- The notion of a normal subgroup for kernels of group morphisms as well as ideals for kernels of ring morphisms should be known.

- The map of Frobenius $x \to x^y$ (where $y=p^n$) on a field F of order p^{nm} holds fixed the subfield of order p^n and is an automorphism which itself forms a cyclic (Galois) group $Gal(F) \cong Cm$ of order m generated by the Frobenius automorphism itself.

- The action of the Frobenius map upon the field GF(8) forms a three cycle, which itself is similar or congruent to a possible multiplication upon the K4 form of the ultrafilter, forming GF(4)*.

- Multiplicative products upon the elements of additive subgroups in a field will always form additive subgroups of the same order.

- Orbits of subgroups of F+ are formed under repetitive product by a multiplicative element (usually with a generator) in F*. Short orbits in F (of length less than |F*|) are formed by products of generators upon subfields of F only.

- The difference between an abelian and non-abelian group. (Commutativity.)

- Every permutation can be written as a series of disjoint cycles.

- The group of all permutations of n symbols forms a group Sn. (The symmetric group of degree n.) The n symbols actually form a G-Set, permuted by elements in Sn. (In this case G=Sn.) You are expected to recognise the difference between a group and a G-Set.

- Every n-cycle (of length n) can be written as a product of n−1 transpositions.

- The subgroup of Sn of products of even numbers of transpositions form a normal subgroup of Sn named An (the alternating group of degree n).

- The alternating group An for n>4 is simple and insoluble. (Has no non-trivial normal subgroups and may not be factored into prime order cyclic groups.)

- Conjugating elements of a group about one of its normal subgroups yields a product always in the normal subgroup.

- De Morgan's laws are also mentioned in the writing: they are roughly stated as the equivalences $\left(A \vee B\right)^c = A^c \wedge B^c$ and $\left(A \wedge B\right)^c = A^c \vee B^c$.

Now, to see whether the scripture survives the maths and the maths the scriptures! Following is a near **exhaustive** examination of the whole book of Revelation. If it can be approached (and rightly so) by a rigid fundamentalism and also upheld with as liberal a stance as would an apologist; then when the maths yet survives and this writing's subtitle also – I would consider myself to have been wholly successful!

Part Two

(The Revelation of Jesus Christ)

Luk 8:17 For nothing is secret, that shall not be made manifest; neither any thing hid, that shall not be known and come abroad.

The Revelation Is No Testament Against Monotheism

The "Revelation of Jesus Christ" is just that: of the Christ (and for Christ) and it answers the question of correct monotheism as put to a faithful Christ finding the union of the Holy Trinity an equivalent. Christ is acknowledged by His Father as God with Him, not only "with us".

The commandment: "Deu 6:4 Hear, O Israel: The LORD our God is one LORD:" is acknowledged as the greatest commandment (Mar 12:29) by Christ: for faith to be perfected in Christ it needs to be perfected by Christ. The construction in the Revelation is of a God in Trinity requiring the further justification to show their selves as one God, no composite made of three or seven.

The octal group in the Revelation's interpretation is constructed of disjoint sets, and only the completion of the work of God in the Revelation completes the mystery of the one whole in triune union. For the "least" in God's kingdom there is the unique property of that union between otherwise irreconcilable opposites. In a useful modal collapse, the name of God (in disjoint sets) is superseded by one made not of disjoint sets, but of the union of all seven ultrafilters indexing one whole, one set of seven ultrafilters each sharing the same domain, indexing the same positive properties, each unions of the others in pairs, only differing in how they index the whole by the octal's seven K4 subgroups. The only requirement is for the one case to completely justify that new name: to bring the reign of truly omnipotent God.

There is but one least, not "made so" but found so necessarily. His office is not up for grabs, it is prepared for him and it is not even for Christ to give that office away (Mat 20:23). His reward or "crown", however, is of his uniqueness and is so held of his person; only knowledge of that same union or "marriage of the lamb" is able to be transferred. That knowledge is not to be left to those that are without Christ, to tarnish God's new name when it is rightly untouched.

Revelation, then, is truthfully not about a failing God unable to save in any one case, but of incomplete grace shown complete by justifying the least with it. Christ answered His call and His work is completely finished; He is now glorified. The least, instead, is awaited only to justify the completion of the mystery of God as one God, one whole and one Trinity, without causing a violation of any of the laws of God. His justification is sound and requires no addition to the gospel, but God is never found the same for justifying Christ with that one servant given to Him to complete the kingdom of God as of one Lord, able to judge and ready to do so.

The current creation does not limit God's omnipotence (it is instead prepared for judgement). Paradise will certainly be found the opposite: unrestricted in all positive properties. Any work of omnipotence is within the reach of God; His old name always a fallback for that sure rest upon His works of salvation. The Revelation of Jesus Christ is then also an invitation for many to share the ongoing benefits of that salvation: rescued by faith upon the works of their creator – one also justified by the cross, raised to life to show that same promise is kept to His people.

So, the Revelation is not to be found as it is regarded, a passion play of disaster or of ill omen: it is instead in every way a sure promise to the Christ that He is God, and inherits that one Godhead which will be found indexing all positive predicates in a manner wholly in agreement to the utmost reach of the omnipotence of God the Father. There is only one premise: that Christ should do the first works put to Him, and who could undo that?

Chapter Five: Revelation Opened

The Revelation starts with greetings and testimony from John that Christ gave him His own "Revelation" to spread to all His servants in the recipient churches (of seven cities) in Asia Minor: each entrusted to keep the words of this vision in its entirety. (It may be assumed each had pastoral and doctrinal problems relating to the letters in the opening chapters.) Convinced by the "more sure word of prophecy" they would guarantee the survival of the apocalyptic text presented in the New Testament as part of the historically accepted canon.

John recounts the visitation of Christ stood before Him and the setting moves on to John seeing in the spirit rather than the flesh what was occurring in heaven and now opened up before him. In the visitation of Christ, Jesus' letters were dictated by Him to preserve verbatim: Afterwards John was shown the heavenly scene of the throne, the elders about it and the four beasts: revealed was the "Earth" and its sphere from the heavenly perspective. Most important is John's account of the book of life: witnessing that only Christ is able to open the book with its seven seals, for the approval of God in three of those seals is necessarily found in just one individual of the trinity, the only begotten Son of God. I am assured God (Christ Jesus) is also unique. Holiness and virtue are not enough: to open the book one must be God also.

As the seals are loosed it appears that the world is free to encroach upon the church (it becomes more "worldly") and as restraint on the paradigm of the church is loosened, the strictness of election is such that those with a greater love for the truth are elected to salvation rather than those that reside comfortably in an age of compromise easier to live in.

As the restraint of God's Spirit on the church (that restrains rather than permits its worldliness) is relaxed, true doctrine is required with love for the truth of the faith. The "lukewarm" are rejected and only those that hold to true doctrine merit grace and favour. The "angel of the church" is as the very "least in the kingdom of God". He is a person that when posited to exist, as he becomes convicted to leave a fellowship and does so, then all others (being greater than he) are spiritually convicted to leave and by principle justifiably so. He may just be "perfect" to do it and could be anyone as in the apparent sense, or simply be the first to do so in a manner that satisfies all requirement as to convict everyone else. He would then be the one whom God has the least influence over. Later on, this has a surprising "revealing" of its own.

The seals on the book of life are opened fourfold on the Earth (in the four beasts) and thrice in the heavens. For the sovereignty of God to be absolute the least is assumed to be present only under the worst circumstances in an age of deception. He makes his circuit within the period between the opening of the fourth seal, completing it sometime close to the opening of the sixth (when the saints are "sealed" completed in number). The sealing of the saints requires the ministry of the two witnesses at the end of his circuit (see much later) and for this I may realign sections in (or tracts of) the text, but essentially the least overcomes sometime between the opening of the fourth seal and the period I would consider to end with the sixth seal's fulfilment (the circuit must then of necessity end before the seventh seal is opened).

The seven letters are directly paralleled to the parable of the ten virgins. At the time of the end, the kingdom of God is "likened" to these ten virgins, and the seven cycle of the least's circuit is not peace and plenty for all. Those called and chosen (and wise with doctrine) exit the first five churches to approval in "Philadelphia", spiritually meeting Christ outside all incorporated and institutionalised worship. The five sets remaining in the Church proper are without doctrinal strength (oil) to stand apart and they are doomed to return to the "Church" of Laodicea (the final amalgam of the first five "refining" churches barren of the elect and correct faith, that will not endure or keep sound doctrine). Never able to "buy" enough good oil (doctrine) in time to last long enough standing apart from that corrupt Church, they are caught in the Church far too late; the door is to be shut on all within that system to be forever bereft of God.

Refining Churches:
The seven letters were sent to fellowships with various doctrinal and pastoral problems, each receiving exhortations from Christ. The first five churches (at the time of the end) are under the condemnation of the last letter (to the angel of the Laodicean Church), rather than the first five letters to the one angel overcoming that worldliness of the five in turn. Those five churches are to refine "as gold is refined", the angel himself "tried in the fire": all that overcome that five-fold Laodicea enter into Philadelphia as "wise virgins", filtered and refined separate to all those that are "lukewarm".

Doctrine that cannot last (such as of the "rapture") will fail fast with the onset of wrath: with no fulfilment, and too many adherents without salvation. Returning to a Church that simply attempts to put a new "spin" on their doctrinal fallacies ensures that there is no wisdom in those who cannot themselves endure (with) sound doctrine (because they found their place in such incorporated institutions which are at the end as "whitewashed sepulchres", not as the kingdom of God outside of them). This is very hard to hear but it is scripturally prophecised. God blesses only on His own terms, not as men or corporations would codify themselves.

5.1) A Baptism By Fire

With the opening of the Revelation, the reader faces much imagery. Leaping in at the deep end, here starts the excitement! (The rest of the text is interpreted very much as it comes.)

Rev 1:9 I John, who also am your brother, and companion in tribulation, and in the kingdom and patience of Jesus Christ, was in the isle that is called Patmos, for the word of God, and for the testimony of Jesus Christ.
Rev 1:10 I was in the Spirit on the Lord's day, and heard behind me a great voice, as of a trumpet,
Rev 1:11 Saying, I am Alpha and Omega, the first and the last: and, What thou seest, write in a book, and send *it* unto the seven churches which are in Asia; unto Ephesus, and unto Smyrna, and unto Pergamos, and unto Thyatira, and unto Sardis, and unto Philadelphia, and unto Laodicea.
Rev 1:12 And I turned to see the voice that spake with me. And being turned, I saw seven golden candlesticks;
Rev 1:13 And in the midst of the seven candlesticks *one* like unto the Son of man, clothed with a garment down to the foot, and girt about the paps with a golden girdle.
Rev 1:14 His head and *his* hairs *were* white like wool, as white as snow; and his eyes *were* as a flame of fire;

Rev 1:15 And his feet like unto fine brass, as if they burned in a furnace; and his voice as the sound of many waters.

John states he heard a great voice as a trumpet behind him: trumpets are referenced as prophetic utterances from or to the future. The "Spirit on the Lord's day" signifies from John that he was not under the influence, but was lucid, in his right mind.

> **Trumpets:**
> *Trumpets reveal where the timing of the judgements of God are shifted within the text's narrative to an earlier moment. Any association with "a trumpet" as "of a trumpet" indicates such a movement or realignment yet also a voice as from the future, clear and precise without any possible prophetic fault at all. Such a voice could be identified as of Christ as "that prophet" (Deu 18:15, 18:18).*

There may be some sleight of hand but consider the figure from chapter three, on constructing the Godhead from principle of infinite regression.

The visitation aligns with the structures in that diagram – astonishingly they are preserved!

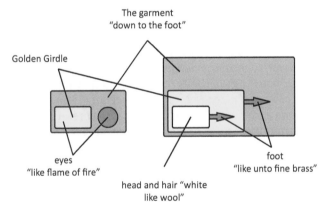

The garment
"down to the foot"

Golden Girdle

eyes
"like flame of fire"

foot
"like unto fine brass"

head and hair "white
like wool"

Though scriptures are given as absolute, I do not assert John ever saw this diagram: deducing the parts and putting them together I have approximated John's vision of Christ before him. The language he began with at the first then had to remain fixed in lexicon for the rest of the text. (God would not have two terms referenced by the word "head" etc. in His Revelation!) The K4 form of the ultrafilter is as such:

$$
\begin{aligned}
1 &\equiv a \equiv [0, \ a, \ b, \ c] \quad the\,"face" \\
eye &\to b \equiv [0, \ a, \ d, \ e] \\
eye &\to c \equiv [0, \ a, \ f, \ g]
\end{aligned}
\left.\vphantom{\begin{aligned}1\\2\\3\end{aligned}}\right\} Golden\ Girdle
$$

$$
\sigma = \underbrace{(a,b,d,c,f,g,e)}_{eyes\ as\ flames\ of\ fire}
$$

The unity element is found self-similar to the whole (to the girdle and K4 form) which in the octal forms a static triple of K4 groups. GF(8) and the K4 form share the "girdle", but GF(8) is without the correspondence of those eyes {a,b,c} to itself and has the two "feet" of [b,d,f] and {c,e,g} instead. The "head" (rather than the face) also now free to "float" corresponds to the unity or some (arbitrary) subgroup in orbit under the seven cycle of multiplication. For simplicity I align the static subgroup with unity (as its place in the orbit is otherwise arbitrary).

Static vs. Fixed:
The terms "static" and "fixed" I use to refer either to a set sent to itself by a mapping or acted upon by an identity map respectively. The former permutes a set amongst its members, the latter is merely the identity permutation. Examples are the automorphisms of a field F upon its members (the Field held static) and of holding a subfield fixed (as the prime subfield $\mathbb{Z}_p$ under Frobenius.)

Golden Girdle:
The "Girdle" simply denotes a set of elements or subsets of the octal that span the octal group. ("Gold" indicates a field isomorphic to a member of the Trinity.)

Face:
The "face" is the appearance (countenance). In the general sense, "face" may indicate the semblance of the notation by which a concept was explained to John.

Floating Unity:
The term "float" I use throughout for the free or arbitrary position (or arbitrary shift due to a product) of the place assigned the unity in the seven cycle – as under its multiplication in GF(8). (Unity may be any non-zero additive element in the octal.) Seven powers of a generator map the identity to any one of the full seven positions (one each); yet "float" is a more succinct term and only used in reference to the arbitrary position of the unity element in the field.

Eyes:
The term "eyes" references the singleton elements not unity in the field GF(4). The seven eyes on the lamb also reference the seven elements of the floating unity in the octal group making up the difference. (With the static subgroup free to float also.) In this manner the "eyes" are also the singleton elements of the octal; the condition is that besides being under similar action by multiplication and the Frobenius map, the "eyes" are the positions through which the unity element is free to float. The self-similarity of the K4 form to its identity-member K4 group prevents this in its own regard: It inherits the unity from the octal, and the static subgroup and triple with it. Representatives of the two static triples [b,d,f] and {c,e,g} of the octal form as it were a pair of "eyes" as {b, c} within the self-similar subgroup [a,b,c] (the identity) of the K4 form. They, as elements of those static triples, remain "free to float" amongst themselves under Frobenius (as acting upon the octal) which results in the "identity group" of the K4 form shifting from [a,b,c] to [a,d,e] and to [a,f,g]; yet the unity element "a" always remains fixed. (Frobenius induces a multiplication on the K4 form, as if forming the "floating unity" in GF(4). Only unity remains fixed.)

> **Feet:**
> *The feet represent the static triples of the Field GF(8) as the unity floats through the seven symbols of the octal under multiplication.*
>
> **Head:**
> *The "head" is the (arbitrary) first in any cycle of elements or subgroups in the octal. It is to be paired with "unity" as the cycle is induced under multiplication. Note that in the K4 form the static subgroup is become as "the face", an indicator of regression between that K4 form and the static subgroup of the octal.*

In the octal the embedding of the K4 form is present but without the self-similarity to the whole. The "Golden Girdle" remains. (Unity remains as "a", with [b,d,f] and {c,e,g} static.)

$$
\begin{array}{ccccc}
e & \equiv & [0, & a, & b, & c] \\
g & \equiv & [0, & a, & d, & e] \\
c & \equiv & [0, & a, & f, & g] \\
& & \uparrow & \uparrow & \uparrow \\
& & unity & foot & foot
\end{array}
$$

$$
\sigma = \underbrace{(a,b,d,c,f,g,e)}_{eyes\ as\ flames\ of\ fire}
$$

The "feet" preserve the K4 form "walked through" the octal. (Applying products of a generator to the subgroups I unveil another subgroup in the orbit.) The feet are seen to be the static triples in the left and right handed octals which so walk (are mapped) through the field with multiplication by elements of GF(8)*. Under σ the feet are the (static) triples [b,d,f] and {c,e,g}.

> **Right Handed Octal:**
> *The right hand octal is formed of the singleton elements and subgroups of the octal that is defined with the additive operation of the field GF(8).*
>
> **Left-Handed Octal:**
> *The left hand octal is formed of triples in the cosets of the right handed octal's subgroups preserved by the same C7 multiplication as upon the right hand octal. The operations of the left handed field (also isomorphic to GF(8)) are implicitly defined within triples as K4 groups (and also by addition on the triples as sets themselves – also a valid operation upon the right hand's subgroups). There are eight different left handed fields for each right hand depending on the choice of unity (and seven cycle) with respect to the static subgroup.*
>
> **Static Triple:**
> *A static triple is one of two sets of three elements that are permuted amongst themselves in GF(8) by the map of Frobenius. Such a triple cannot contain the unity of GF(8). One static triple under each seven cycle is always a subgroup of the Field GF(8).*

The K4 form embedded in GF(8) is limited to the girdle without the correspondence of the "face" (as a ≡ [a,b,c]) to unity. However, depending upon which triple ([b,d,f] or {c,e,g}) is a static subgroup, only one triple as the "foot" can be on the "Earth" at any one time (seen above in the singletons on the left, in {c,e,g}). More on this later!

The Earth:
The Earth is represented in Revelation as the four elements of the coset of the static subgroup and in the heavens as by the four beasts before the throne of God. It is also the set from which the unity element of GF(8) is free to be chosen.

The Frobenius map acting upon the girdle suffices to agree in operation with GF(4)*, whilst the "feet" are walked through the octal by C7 as equally and simply as unity is free to float.

The two "eyes" of the K4 form must be recovered (in the octal) with the self-similarity of the "girdle" to the static subgroup (which cannot contain unity), and so cannot be recovered within the octal or GF(8). Unity is the intersection of those three subgroups forming the K4 ultrafilter, the K4 form's unity element is even (as self-similar to the whole) become the "head" and "face" (appearance) and is found in the very person of Christ; as GF(4). In GF(8), there are seven eyes on the lamb, not to be confused with the "eyes" of the K4 form.

The Lamb:
The lamb indicates the virtue of Christ: placed as the unity of each field in the Trinity, GF(8) and GF(4) and GF(2). The lamb is the intersection of all three fields and is an atonement for all those that dwell with Christ as in heavenly places. They as "elect" are "clothed" with the lamb, so as to appear as the unity in all octals of the "sea" – every octal with unity free to float. The lamb is the token of election before the throne of God in the one reference octal with a specified unity.

The K4 Form:
The K4 form of the filter is the triple of subgroups seen previously as the Golden Girdle, yet a "K4 form" is free to float in the octal as unity may float. (The intersection that was "a" in the girdle is free to become any other element as the intersection of three K4 groups forming a K4 group under a+b=(a ∨ b)ᶜ.) In the K4 form of the ultrafilter, one K4 group is self-similar to the unity and to the whole.

You may not have guessed that there was that much mathematical content, but you are very wrong! I jump straight in at Rev 1:12. I state each of the seven candlesticks (or seven branched lampstand) represent the seven elements (in three pairs of inverses with unity) that comprise a cyclic group C7 (with six elements of order seven and one identity).

In the midst John sees Jesus, in abstract within those candlesticks as the first, second and fourth elements of each seven cycle (forming a static triple), which follow after the place (the "head") given to unity. These triples are static (permuted amongst themselves) in GF(8) under the Frobenius map. By squaring, Frobenius sends $x \to x^2 \to x^4 \to (x^8 \equiv x)$ in a three cycle.

$$(a, \quad b, \quad d, \quad c, \quad f, \quad g, \quad e)$$
$$1 \quad \uparrow \quad \uparrow \qquad \quad \uparrow$$

$a = 1 \leftrightarrow [b, d, f]$ static.

$$(a, \quad b, \quad d, \quad c, \quad f, \quad g, \quad e)$$
$$1 \quad \uparrow \quad \uparrow \qquad \qquad \uparrow$$

$d = 1 \leftrightarrow [c, e, f]$ static.

Products by elements of GF(8)* cycle the octal's subgroups, starting with the static subgroup (the "head" of Rev 1:14) in product with the unity. Such multiplication upon the K4 form is observed to preserve every subgroup (as if both the static K4 subgroup and the girdle were walked through the octal to fill and form a garment "clothed down to the foot" – a closed garment covering all) then mapping subgroup to subgroup as so:

Multiplication maps elements across subgroups. Map upon elements resembles "strands of hair" on head.

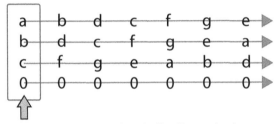

First subgroup corresponds to the "head" or product by unity; the subgroup is in orbit as under product by the powers of a generator of F*

"White" represents the cleanliness of a field's operation, and the **closure** of a set under the Frobenius map. (Throughout Revelation when referring to such an additive subgroup of the reference or "Sun octal".) Christ's "white hair" shows that He **is** the static subgroup under orbit, and also the reference from which this "white" subgroup is taken. The static subgroup floats, congruent to the choice of unity from each of the seven elements in each cycle.

White:
The colour white represents not just the closure of an additive group sent to itself under the Frobenius map but also the cleanliness of the virtue in the lamb (also "white" as GF(2)). It represents such a set, but in the reference octal only. Those clothed in white form fields GF(2) as elect and are counted for as if unity as well as additively closed with the inclusion of zero. (Able to rule over themselves by analogy to the throne.) In the Trinity, each additive group may act as identity for the addition of its members using a + b = (a ∨ b)ᶜ. This also includes GF(2)+.

Sun Octal:
The "Sun" is the origin or reference octal in the seven symbols a to g. I use the triples [a,b,c], [a,d,e], [a,f,g], [b,d,f], [b,e,g], [c,e,f], [c,d,g], to form it.

"White" sets aside the static subgroup or unity (as in regress) as the head, free to float; there is no sense as yet that there are any more structures present. They only require a correspondence between each choice of unity and the four possible static subgroups to it, fixing the three K4 subgroups forming the K4 form (which uniquely contain the unity, and are also free to float). The orbit of every structure is a permutation of the octal under elements of GF(8)*.

The "golden girdle" represents the distinction due Christ of His "self-similarity" between the three K4 groups spanning the octal (that contain unity) to the static subgroup (in regress). The K4 form itself is equally in regress with the static subgroup of the octal which is properly separate without unity. The "Girdle" is a term synonymous with groups that "span the octal".

Self-similar, yes, but a subfield, no. The scripture states:

Psa 110:1 A Psalm of David. The LORD said unto my Lord, Sit thou at my right hand, until I make thine enemies thy footstool.

And the reader cannot expect miracles whilst the Lord is resting so comfortably! (After His ministry completed – a work completely finished.) It is sufficient that Jesus, sat at the right hand (as the static subgroup) makes the K4 form to align with that one subgroup in a "strange twist". This twist takes many verses to expound upon – as separating Father and Son it is the biggest mystery in the Trinity besides everything else (that as a Holy union they have created).

"Gold" indicates a finite field in the Trinity. (Gold refers to a complete "(sub)field structure" and in Revelation this is consistent throughout.) Christ is truly God in His own right and is also as GF(4); He may live separate to GF(8) as Spirit. The intersection of GF(4) and GF(8) is merely GF(2) only. GF(4) is not a subfield of GF(8). "Sat at the right hand" of the Father, Christ is placed fixing more than a simple congruence to unity (i.e. as if at "rest" or "sat down").

> **Gold:**
> "Gold" references a field, one of the Holy Trinity. It must reference GF(4) or GF(8) also with their operations of multiplication. GF(2) is also "gold" and in the covering of the lamb as unity every believer is become as "gold refined in the fire". (Cf. the intersection of all three (unity) as a "talent" of Gold.)

Therefore, the static subgroup in the octal is termed "white"; the Field GF(2) is "white and gold" in the octal (a subfield) and GF(4) merely "gold". The golden girdle identifies its wearer as more than simply a subgroup in the right hand: It declares Jesus Christ is God. (Spanning the octal.) There are also seven angels wearing golden girdles which **combined** span the octal: Each are as "GF(2)", with a singleton of the octal; and are not "God" but are "Holy" and dressed with the same pure and white linen (holiness) as the elect in Christ. I hope this presents no problem; I will accumulate more terms as I follow the text through.

Zec 13:8 And it shall come to pass, that in all the land, saith the LORD, two parts therein shall be cut off and die; but the third shall be left therein.

The "feet" of Christ are the static subgroup/triple(s) found so under the seven cycle (held static under the map of Frobenius). Each repeated squaring (by Frobenius) of the elements of the multiplicative group (C7) of GF(8) holds static one subgroup and one other triple; those subgroups and triples are reversed (vice-versa) between two complement octals (as the left and right hands) fully formed as if a garment "down to the foot" (and closed when acted upon under subgroup addition $(A \lor B)^c$).

> **Complement Octal:**
> *A complement octal is one of the eight octals that share no subgroup with the reference octal. They appear as one of the eight left handed octals to a given right hand: implicitly defined (as subgroups) by the static triples under orbit by the eight C7 groups of multiplication over the right hand octal.*

Squaring every multiplicative element of GF(8) by the map of Frobenius has a similar effect as squaring its inverse. Both cycle static one subgroup and complement triple. The inverse produces an isomorphic cycle in the left hand (otherwise the cycles appear reversed).

The automorphisms generated by the map of Frobenius on GF(8)* permute the octal's elements in three cycles, forming two disjoint static triples with unity held fixed. The automorphisms (with a=1 and [b,d,f], {c,e,g} static, say) also induce three cycles on the subgroups of the K4 form. In the K4 ultrafilter they form the "eyes on the face" as those subgroups then cycle within the K4 form. The "feet" in the octal (unity is free to float), are also free and "walk amongst the seven candlesticks". I.e. in seven positions but always two disjoint sets of three.

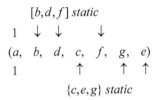

$$[b,d,f] \; static$$

The two triples held static under the Frobenius automorphisms (which form the Galois group of GF(8), isomorphic to C3) paired together induce C3 on the K4 form of the ultrafilter (on the "golden girdle"), and not by a subgroup in the octal's own multiplication. GF(8)* is isomorphic to C7 – and is a simple group. (C7 acts as the Holy Spirit whose "lamps of fire" are the seven spirits of God the Father.) The Galois group's elements which induce the C3 group (agreeing with GF(4)*) on the girdle/K4 form are "above C7" as raised powers over the multiplication (product) of the seven cycle (the seven eyes on the lamb).

The eyes of Christ referenced as "flame of fire" (Rev 1:14) – induced by Frobenius upon GF(8)'s own "fire" (its operation) are placed as if flame "above" that fire (products of indices are in some sense "above" normal multiplication which adds them).

The K4 form (Christ) "sat at the right hand" of the Father does not walk with "feet" through the seven cycle but when sat at rest enthroned sees all before Him. The static subgroup of Christ is a fixed reference for the operation of GF(8) in the "Sun octal".

Fire:
Fire indicates a set with the properties of a group. It represents the complete set of permutations of a group amongst its own elements. It may be equivocated to the action of a group upon itself or some other set. In such terms, it becomes the "quickening action or life of the Spirit".

Left and Right of Christ:
In the K4 form the left and right are a separation to {c,e,g} and [b,d,f]. In the K4 form, the "face" a ≡ [a,b,c] has as "eyes" the elements b and c. Then to the left and right I would separate b from c. Since b and c correspond to [a,d,e] and [a,f,g], under the induced three cycle I may find the face become a ≡ [a,d,e] or a ≡ [a,f,g]. I separate them consistently with the subgroup [b,d,f] of the right hand octal to "the right" and the triple {c,e,g} in the left hand octal to the "left".

Mouth:
The "mouth" always refers to the set of elements itself, and the pair of braces which is notation for a set "{ }" resembles a "mouth".

The Two Edged Sword:
The two edged sword from the mouth of Christ is such as to separate all to His right and left on the basis of His virtue. (Those whom He has chosen for Himself are so clothed as with the lamb, those without Christ are not so clothed.)

The "eyes" to the left and right of Christ as in the K4 form are quite separate from the "mouth" which is congruent to the unity. Unity is free to float and the "feet" (and the static subgroup) in the octal are free to walk through the seven cycle, and this preserves the K4 form.

$$
\begin{array}{cccccc}
 & \text{"face"} & & eye & eye & \\
 & \downarrow & & \downarrow & \downarrow & \\
 & 1 \equiv a \equiv & [0, & a, & b, & c] \quad \leftarrow \text{"face"} \\
eye \rightarrow & b \equiv & [0, & a, & d, & e] \\
eye \rightarrow & c \equiv & [0, & a, & f, & g]
\end{array}
$$

Unity (say a=1) is the element that "cuts off" the static subgroup from the static triple. If a=1, [b,d,f] is static under the seven cycle. Then "a" as unity becomes as the face and in regress the "mouth" with a two edged sword separating [0,b,d,f] and {c,e,g}.

In the K4 form of the ultrafilter, unity is the "face" congruent to the whole. It takes the form:

$$
\begin{array}{ccccccc}
1 & \equiv & a & \equiv & [0, & a, & b, & c] \\
 & & b & \equiv & [0, & a, & d, & e] \\
 & & c & \equiv & [0, & a, & f, & g] \\
\end{array}
$$
$$\Omega = \{0,a,b,c,d,e,f,g\}$$

And the octal may substitute for zero when operating on subgroups with $(A \vee B)^c$.

Rev 1:15 seems a trivial observation but things are never just so: The feet of Christ are likened to be "burned in a furnace" – the pair of octals over which each seven cycle is valid (the subgroups of the octal are under orbit as the feet also "walk" through the seven cycle): His feet burned by singletons from the set Ω of positive properties, forming seven "lamps of fire" (Christ forever walking them in the spirit) that together are the seven spirits of God. The K4 form under orbit by elements of GF(8)* is also preserved along with all self-similarity; it is even as trivial as a "cyclic relabelling", a different label used for the head of each cycle etc.

> **Burned In A Furnace:**
> *Compare the "fire" of a group's operation with the set of all positive properties within which the Trinity as groups and ultrafilters take their elements. Then the closure of all positive properties forms such a "furnace". In walking the octal as showing virtue, Christ has then shown His path trod and His feet "burned" in that furnace, as the only one judging what is positive for His elect.*

The octal group is preserved by only eight C7 groups from the possible 120. Each of those C7 groups are also shared with a different and "complement" octal, floating the position of unity in each seven cycle entails that each subgroup (with one triple) can be held static by Frobenius. The "triple" is a subgroup in the "complement octal".

The reference (Sun) octal is shared by those eight as one partner of eight pairs. The other partners are the eight complement octals. Each shares no subgroups with that Sun octal.

Christ's voice as "many waters" indicates similarity between every octal across the "sea". The "sea" is the set of all thirty octal groups on the same seven symbols. No more are assumed yet, though later I take them and every valid seven cycle into account. "Waters" also refers to the many possible static subgroups of every octal (while under orbit). If the set of thirty octals forms the "sea", the static subgroups as subsets of those octals also form the "waters" here.

> **Waters:**
> *Waters, the sea, and all subsets of it – namely octals, rivers, fountains etc. – are also the action (preserving all structure) of seven cycles upon subgroups of the octal in view (whether the seven cycles are valid to preserve subgroups of that octal or not). "Fountains" are as K4 forms in orbit under multiplication: their intersection branching out to three subgroups with the appearance of a fountain. "Rivers" are seven cycles acting on one subgroup in orbit; the Sun octal itself and its subgroups are "waters" as also the other cases across the sea of thirty octals.*
>
> **Fountain:**
> *Fountains correspond to triples of subgroups forming K4 groups (in lexicon "trees" are as such) acted on by a seven cycle; their intersection appears to "flow up" under John's reversal of the seven cycle. They are then as "trees" under a seven cycle (as in waters).*

This description of Christ's voice emphasizes the movement or product (the third part) of the operation in the K4 form recognised as Christ. The correspondence of b and c to [a,d,e] and [a,f,g], with the "face" as it were GF(4) or $a \equiv 1 \equiv [a,b,c]$ show that the "voice" projects outward from the throne. (I.e. where are "d" and "e", "f" and "g" mapped to?)

> **Voice:**
> *The voice of Christ is a projection outward from His "mouth". I.e. From the K4 group in congruence to unity (as self-similar to Himself). As a ≡ 1 ≡ [a,b,c] forms the "mouth", the voice is of the elements {d,e,f,g} appearing in the other two groups b ≡ [a,d,e], c ≡ [a,f,g]. The "voice" becomes the K4 form's analogue of the earthly elements. (Which are cut off to one side or the other by "judgement".)*

The eight C7 groups are isomorphic in their operations on each octal; octal addition under those multiplicative groups is equivalently repeated sevenfold in every octal within the sea; that is, if unity is *free to float* amongst the additive elements. The operation (and projection of the "voice") is as "of many waters". The voice of Christ relates to the induced operation of C3 from Frobenius acting on GF(8), in that Christ is "fixed" as unity or the face, and the two (non-identity) automorphisms clearly divide all to His left and right (as [b,d,f] and {c,e,g} held static) by a sharp two-edged sword – the voice of Christ.

Rev 1:16 And he had in his right hand seven stars: and out of his mouth went a sharp twoedged sword: and his countenance *was* as the sun shineth in his strength.
Rev 1:17 And when I saw him, I fell at his feet as dead. And he laid his right hand upon me, saying unto me, Fear not; I am the first and the last:
Rev 1:18 *I am* he that liveth, and was dead; and, behold, I am alive for evermore, Amen; and have the keys of hell and of death.
Rev 1:19 Write the things which thou hast seen, and the things which are, and the things which shall be hereafter;
Rev 1:20 The mystery of the seven stars which thou sawest in my right hand, and the seven golden candlesticks. The seven stars are the angels of the seven churches: and the seven candlesticks which thou sawest are the seven churches.

The Lord to John was as the Sun – a term used throughout Revelation referring to the one field from which every other octal takes reference (being the person of the Father in whom Christ is seen to sit and span as static K4 group and K4 group of subgroups spanning the octal).

In His right hand (the Sun or reference octal shared between every lampstand of seven and the circuit of the angel as the eighth) the Lord has seven stars (the angel of the church, the "least" in the kingdom of God) which are the elements of an eighth C7 group not mentioned so far by John. There is only one angel traversing a sevenfold circuit (of seven disjoint sets or stars – see [16.13]), the "right hand of God". The seven stars are as a "score sheet", as if by overcoming the world a point at a time he acquires as it were a "gold star" next to his name.

The sharp two-edged sword separates every element of the octal to either unity (the sword itself, as the word of God, the virtue in Christ as "p" formerly) or to be in correspondence to the static subgroup "In Christ" (equivalently accepted to rest at Christ's right side by His judgement or "voice"), as part of the K4 form of the ultrafilter (in regress) or else to be cut off as part of the remaining static triple as in "antichrist": (to be refused as to Christ's left) within the static subgroups' coset. (This forms no subgroup in the right handed octal, but it does so in the complement set of elements once the argument together with unity is excluded, and forms a left handed octal under the same seven cycle's inverse.)

Christ's appearance is referenced to be as the "Sun" in strength; and I simply refer the reader to the reference octal of the Father: that Christ is clothed properly as equal to God almighty in glory. (Note that in the ultrafilter's K4 form spanning the octal, the octal itself is a valid identity element.)

He (Rev 1:17) introduces Himself as Jesus to John stating He has the keys of hell and death – a reference to being the only God that may give life and save from hell. (Note that these are also the spiritual forces sharing the names in capitalised equivalents that appear later.) In Christ believers forever find their salvation from these two, the same constructs have no power over the redeemed – those constructs are merely deceptions (not God's own curse) and do not add a "Second Death" at all. Only God can and may do that.

Instructed to write, John moves on to the mystery of the seven golden candlesticks and the seven angels. Beginning with a seven cycle (that preserves the octal), another seven cycle (of a different group) may act upon it and form the seven rows following: each row an element representing one of the other seven groups. (One group's cycle upon another's generates a member of each of the seven groups that remain.)

$$(a, \quad b, \quad f, \quad c, \quad d, \quad e, \quad g)$$
$$(b, \quad d, \quad g, \quad f, \quad c, \quad a, \quad e)$$
$$(d, \quad c, \quad e, \quad g, \quad f, \quad b, \quad a)$$
$$(c, \quad f, \quad a, \quad e, \quad g, \quad d, \quad b)$$
$$(f, \quad g, \quad b, \quad a, \quad e, \quad c, \quad d)$$
$$(g, \quad e, \quad d, \quad b, \quad a, \quad f, \quad c)$$
$$(e, \quad a, \quad c, \quad d, \quad b, \quad g, \quad f)$$

Every column is become as the "angel" in circuit (a,b,d,c,f,g,e) fulfilling the missing eighth group of multiplication over the reference "Sun octal". This is the mystery of the seven golden candlesticks. It is a mystery because rest (the unity always denotes rest, a "Sabbath" in abstract; the natural background of the heavens) is attained before the angel's circuit is complete.

5.2) The Logic Of The Dialectic

The octal is a union of seven ultrafilters. Each K4 group forms an indexing set and I excluded their cosets from the logic of positive properties. Without virtue, the filter may appear to reduce to only four "earthly elements" $\{r, s, u^{-1}, v^{-1}\}$ in action or at rest on either side of the disjunction. These form the dialectic (fiat) ultrafilter that seems more intuitive to the worldly.

What is an ultrafilter, and why would one matter? An ultrafilter U is found over an index I upon a set S where subsets of that index I are either present in the filter U or not: the properties of an ultrafilter are such as to show the model on that indexed set S true "almost everywhere". The ultrafilter U is related, then, to a model which applies to subsets of that index I (then present within the ultrafilter U), and likewise not found true over those subsets of I not in the filter U. The subsets of I in the ultrafilter U are "big enough" to carry that same sense of "true almost everywhere" to all the members of the index I, showing the model correct on S.

It is easy to check the properties of an ultrafilter and realise they make the subsets/members of the index I within the ultrafilter U "big" compared to their complements in I or members that are excluded. The sets in the ultrafilter are chosen for a formula or property (model) on S and that model is found present (correct) over that index I, and true for S "almost everywhere". An ultrafilter is not properly contained in any other filter – it is a maximal filter. The ultrafilter of "perfections" with which I am concerned relates to human judgement on "positive properties". The one concerned may easily evaluate as positive or not – immediately judging in a "positive aesthetic" sense almost everything present in their environment.

The "dialectic logic" in a limited model mimics freely decidable disjunctions whilst privating their necessary virtue. Any ultrafilter formed of positive properties requires by axiom that a set is in the filter or not: this holds between a set and its complement. If four elements (earthly, or as a coset) $\{a, c, e, g\}$ would in pretence form such an ultrafilter (without acknowledging virtue) I would appear to decide disjunctions of the form $Pos(c) \vee Pos(a\&e\&g)$ etc.

Dropping the Pos operator for logical ease, I reduce to $a \vee c\&e\&g$ or even $a\&c \vee e\&g$. In these I could evaluate "a" as if it were virtue, entailing the disjunction $N(Pos(a)) \vee N(Pos(c\&e\&g))$ so that $N(Pos(a))$ entails $N\neg(Pos(c\&e\&g))$ or $Pos(c) \rightarrow Pos((e\&g)^{-1})$. So $Pos(a\&c) \rightarrow Pos((e\&g)^{-1})$ is equivalent to $a\&c \vee e\&g$ from that dialectic logic and likewise $a \vee c$ is proven by $e\&g \rightarrow c^{-1}$.

The existence of $e\&g$ (an excluded middle) is assumed to be exemplifiable (irrespective of the fact that $a\&c$ may itself appear positive); whereas acting properly (with virtue) its positivity is indefinite and certainly not exemplifiable in the K4 group in which it appears as the middle.

To freely decide $a \vee c$ as if "a" and "c" were both positive, I need to dialectically prove the disjunction with two sets $e\&g$ that are together not judged "positive". By excluding this "evil" conjunction I appear to have righteously conjoined $a\&c$ (which same wisdom Eve coveted beguiled by the serpent), whereas I have truthfully done nothing positive, denying the perfection of the octal, or of God (who would never profane the K4 group of virtue with that same "evil") and then even His virtue. It is clear that this simple logic appears to enable free will without didactic absolutes. I would also be set upon the course to decide which properties were positive and when; already an ethical conundrum without clear absolutes of virtue.

As virtue may not private any positive property, in the disjunction $a \vee c\&e\&g$ virtue may (as "a") not private $c\&e\&g$ and is then excluded (on its necessary positivity) from the consensus.

The middle of, say, $a \vee c$ is empty but negated both sides forms $c\&e\&g \vee a\&e\&g$. Then $e\&g$ would clearly prove $a \vee c$ ("a" and "c" are properly found disjoint), logically, in virtue the only middle is nullity, not present in that filter, outside of the indexing set and impossible to instantiate as positive in the octal – the empty set may prove any arbitrary disjunction and it is also not positive. (Are disjunctions always positive? What if they private virtue?)

The dialectic is a subtle device and one hardly noticeable to the ungodly. Its logic is only present in a fiat filter. It is an "evil" in that it allows people the ability to direct their own steps, but doing so they are without the proper ultrafilter to do just that and perfectly so. Ethics are a dialectic subject, morality is wholly dialectic because of the study of the former. Where, then, is the error in choosing only virtue from a perfect God rather than the ethics of man?

Given the original disjunction in the earthly elements properly decides the choice between those elements (r&u^{-1} ∨ s&v^{-1}), the dialectic may be formed between any combination of the four (the practitioner judges the predicates themselves). For God, either side is judged positive only with the action of virtue, and positivity is privated in the middle, so some outcomes cannot be positively worked virtuously. Is it evil to be unable to perform a good work? It is evil to convince all that they must succeed in attempting an impossible work – whilst dismissing every opportunity for applying virtue (God would instead desire mercy, not sacrifice).

Workable combinations of the earthly four are possible without virtue and are present in the octal, but I am here missing three of seven elements, and at least one of these excluded properties appear in each K4 group of the octal. Good works are possible, in that anyone may perform good works judging dialectically, but Christ Himself taught us to go beyond that; for sinners surely love others also and so all disciples of Christ should aim to do far better.

Everyone alive has employed that dialectic evil for their own ends (judging disjunctions that fail in privation: $Pos(r)$ ∨ ($\neg Pos(r) \rightarrow Pos(s)$) with $N\neg Pos(r^{-1})$), all trusted to their own judgement instead of that of God; all are disobedient through that knowledge of good and evil. God Himself, in the octal, does not do so: Christians always pray "Mat 6:10 Thy kingdom come. Thy will be done in earth, as it is in heaven". With heaven (unity, virtue, the static subgroup) there is a solution that reveals that good and righteous work that surely characterises Christ. There is good in the "earthly" sets even when they become separated from virtue in the octal, for with God, His works are always present in "all the Earth" since His works include disjunctions such as (r ∨ u&v&$r^{-1} \rightarrow v^{-1}$) (as from the octal), equivalent to $N\neg Pos((u$&$v)$&$\neg r$&$\neg v^{-1})$.

Cosets in the octal are evils, dialectic "filters" privating virtue. Assuming $N(Pos(a$&c&e&$g))$ I find only contradiction if God would operate in the coset. Rather, He requires virtue and properties acting as virtue for the members of cosets within K4 groups/filters to be positive.

Joh 1:5 And the light (the octal's virtue) shineth in darkness; (the coset) and the darkness (the dialectic) comprehended it not (would not agree in understanding, as if operating independently and unaware of the truth present in virtues).

If a&c&e&g is the coset of a K4 group in the octal and also actually as virtue (axiomatically positive), any disjunction a ∨ c&e&g formed privates that "virtue" by excluding its middle (that same virtue). Virtue forms a **closed** set; a&c&e&$g \rightarrow a$ and a&c&e&$g \rightarrow c$&g&e etc. That same virtue a&c&e&g may not private any such inference formed, which are each positive and then also virtue! (None may private in disjunction, a clear contradiction.) The dialectic filter is then completely fiat, as $Pos(a$&c&e&$g)$ is necessary as its complement ϕ is not present in an ultrafilter; yet as a virtue it is without any opportunity to be worked. (In any disjunction!)

I purposefully state there are certainly "dialectics" that logically agree with Godly acts that oppose them. The results of virtue take precedence over the logic of the dialectic wholly agreeing in the octal! Were there a formula for that victory I would be overjoyed to discover it and bottle it (as for always overcoming). Should I sin so grace may abound? Heaven forbid! (I am already fully satisfied with the Holy Spirit.) It is, however, certainly clear that the cross of Christ was an evil, and a dialectic intended to limit and destroy Christ's virtue at the failure of two covenants, not one. Yet virtue was then at work in the extreme and truly overcame.

Granted the dialectic allows man to judge good and evil for himself, if there is no evil found in any conjunction of any choice of those four "earthly" elements, I find the required answer but not the principle. There is no sense that every division of these "absolutes" are able to be freely worked as dialectic. Instead I can only follow God and obediently practise His virtues.

5.3) The Angel Is Instructed In His Circuit

In order to understand Jesus' requirements as to His right hand (His angel, the least in the kingdom and the dividing line between those saved entering His approval to His right and those not so, as refused to His left) within the seven letters it makes far greater sense to unravel the circuit within which the angel finds himself. I state there are "two camps" in the letters: those of "Laodicea", and those elect remnant of believers at that time in "Philadelphia".

The letter to Laodicea should convict every believer. None are perfect and all far from it! The letters indicate Philadelphia is the end of the angel's circuit ("...and he will go no more out"). Because of the length of the text I will not reproduce the scripture with the letters.

In each letter (sent to the angel attending each church in his circuit) the emphasis is to find sound doctrine outside the Church proper. Note that capitalisation of "Church", for the Laodicean letter states the requirement for all to repent (and to repent of holding to any false doctrine). The sixth letter stands apart with the angel in Jesus' firm favour without need for works or repentance. (The angel in Smyrna is affirmed a big well done; he does very well indeed!) The word of Christ to Laodicea convicts the elect to overcome the worldliness of each "Church": worldliness to which the first five letters in turn apply, and only the last church in the circuit may be overcome by holding true to the doctrine gained already as truly acceptable.

There is a seven cycle over the circuit, the identity element is the "rest" found by overcoming Christ's requirements of Laodicea at the end of the angel's circuit in Philadelphia (separate from that "Church"; the amalgam in false doctrine of all Ephesus, Smyrna, Pergamos, Thyatira and Sardis). Philadelphia is overcome only by retaining good doctrine, Laodicea overcome only by reaching the spiritual state of the angel in Philadelphia and continuing within it. There is a clear division of five churches to one (cf. the parable of the ten virgins), those five from which the believer must exit to be obedient to Christ and found worthy – ending in the sixth. (One must be under the same conviction as the least, who is not "of" that worldly "Church".)

The letter to Laodicea in the KJV declares the recipient angel is "of" the amalgam of the first five churches, yet only the first letter of the other six to the angel (before he properly circuits) repeats it. The others are sent to "the angel of the church in...". The Greek does not differentiate between "of" and "in" but the words are perhaps inspired within the KJV.

Jesus also stated "I will give each of you according to your works" in His letter to Thyatira. To overcome, the angel is given promises and gifts to ensure his predestined overcoming of each church is retained and his victory of his own accounting. The gifts are awarded him as "merited favour" allowing him to account (reckon) for his own salvation – so that he cannot backslide. These "gifts" (horns on the lamb) allow him to overcome by cumulatively accruing the equivalent grace to the Holy Spirit, moving onward within his own judgement – obedient as if he were already dwelling in heavenly places having finally overcome to apply those gifts. I assume in them the diminishing difference between being the least and the height from which he fell (as naked without Christ, and completely cut off without the Holy Spirit), compared to every single man and woman in Christ. The angel, justified before God, shows salvation effective in the case of the least and "so as for all". I may only guess what Satan would have to say about the whole church being justified so "easily", with an open door no one may shut.

Grace, then, is ministered back from heaven to the least by his own self in reckoning. That open door is perhaps his best DIY job, though he ought to have found a carpenter rather than make a door himself! Jesus in giving the gifts (power) for him to overcome gave him the doorway at least.

No doubt the angel is instructed on the need to resist false doctrine in the church; for the letters warn all to be diligent. More to the point is that when the angel overcomes and exits (it should be assumed) the remaining elect are then under the spiritual conviction to also leave. Only when fellowships have the same doctrine or agree to disagree with each other as a single flock can I conceptually connect them truly as one and induce such unity of conviction. Accepting a fellowship with false doctrine is against sound doctrine also. In the last days it may be assumed the convicted few are very sparse, and low in number amongst a great many tares (weeds amongst the "wheat harvest" of redeemed believers).

So, to the new Christian convert full of zeal, what teaching on obedience may be offered that is profitable? Worrying over keeping the Old Testament law is perhaps one of the biggest difficulties facing a believer (at least at first). There is never a need to feel a failure, but instead space to realise that grace may abound all the more. The failure is of faulty teaching.

If I were to take an absolute stance (without flexibility) to a legalist believer I would argue that Christianity is not Judaism and Judaism is not Christianity. I would also deduce that an adherent to both is not "twice blessed" but in truth may be once blessed only, as the two covenants (for now in Christ there is only one) lose force when an equal position in the middle is taken for logical necessity. However, both covenants are of the same God and must be totally compatible, and as a Christian I prefer only the new – for reasons I attempt to justify; the old is already superseded by grace, the law fulfilled by Christ for us.

In short, it must be possible to be of both covenants and Christian, and to be damn good at it! The gates are now open to the gentiles (non-jews) just as wide as to those to whom were first given the Old Testament and the law and to whom Christ came. As Paul in the book of Romans wrote, they of the old covenant benefit "much every way".

Rom 3:1 What advantage then hath the Jew? or what profit *is there* of circumcision?
Rom 3:2 Much every way: chiefly, because that unto them were committed the oracles of God.

Yet in Christ there is no difference between believers – the blessings are the same to all.

Gal 3:24 Wherefore the law was our schoolmaster *to bring us* unto Christ, that we might be

justified by faith.
Gal 3:25 But after that faith is come, we are no longer under a schoolmaster.

Consider the disjunction of: A ∨ B where:

  A   "Grace by provision in the law awaiting its fulfilment in Christ"
  B   "Grace by the Spirit of God predestined to make a person perfect in the law"

(Even if the provision of forgiveness in A takes force but once a year in the law of Moses.)

Now if I invert both sides to inaction (assuming both sides are positive) I have

$B^{-1} \vee A^{-1}$ where:

  B^{-1} "The law can never make a person perfect again without the work of the Spirit"
  A^{-1} "There is no lawlessness or sin in Christ".

Or then...

  B^{-1} "The law may continue to operate without Christ" (grace is no more grace)
  A^{-1} "Christ may operate without the law" (works is no more work)

I find the Lord has turned the world upside down, as the law (I am taught to consent) is good. Now I actually have the right hand sides B and A^{-1}! (The opposite of "good" becomes "better", because it is the work of the Christ.)

To keep both sides of these latter alternatives with a middle position is nonsense; for then the law under which all would be hopelessly condemned in the new (except one, Christ) is then near impossible to keep once broken and it becomes nothing but ceaseless works without hope of promise.

Likewise, the Spirit then has no work remaining (for men to keep the law) since grace then predestines no-one to make them perfect. (It then only works in Christ to show Him perfect and He is then the only one not under the law. Either all are perfect as Christ or either all are likewise condemned imperfect in comparison. "All have fallen short of the glory of God.") The law operates and yet grace is still merely of works.

Then if one side is not more perfect than the other, why be a Christian? The answer is simple: Virtue is a finer filter than the law. The newer is a better covenant. That said, right now, the law is still in operation within the grace of God. The Holy Spirit is even now still predestinating believers towards complete obedience at the rate of God's answering patience.

This world operates on using the dialectic paradigm; believe it or not, this work of Christ setting up His kingdom is a valid and "legal" dialectic found in the cross entailing the excluded middle in $A^{-1} \vee B^{-1}$ – a work by God that turns the subtlety of the world upside down with an outstanding act of pure virtue, none finer.

Inverting both sides of A and B to inaction, the virtue present in the one covenant should then appear revealed in the other. Christ, in laying down His life for us (His friends), did so in full satisfaction of the law. That same grace of intercession in the heavenlies is preserved to us in the New Covenant, making us equal to them.

The cross of Christ acts as the final and ultimate work of the law. The full satisfaction of the law by Christ's obedience (the law operating without the provision of virtue to justify the middle as empty, but with God physically justifying it instead), is found in the execution of Christ by the law under the charge of blasphemy. The nearest and finest act of virtue to entail the grace of the new covenant is found solely in the positivity of the statement "the world is turned upside down" in that there are certainly two things by a dialectic in which it is impossible for God to lie: His covenants.

It is correct to note that the statement "I desire mercy, not sacrifice", is and was in effect during both covenants (grace makes men perfect under predestination to complete obedience whereas the law could not do so by sacrifice). Christ, in fulfilling the law, preserved the same intercession by virtue in the old covenant – not of the letter, but of the desire of God, as to the mediation of His new covenant by Christ. (Thereby superseding the old by His testament.)

Yet there are no positive properties in the middle of the two covenants. (They are disjoint.) The law has not "passed away", but it is inactive or "relaxed". It is not the whole law that is inactive but merely both the righteousness and condemnation of the law, that sole part that condemned Christ unjustly and did break the law's purpose. (Revealing He who already had all victory over death which fulfilled the law.) So, I have a simple conjunction of virtue, $q \vee p \& q^{-1} \to s$ where:

q   "The man that does those things written in the law shall live by them."
q^{-1} "The condemnation of death that is upon all those that break the law."
p   "The virtue of Christ crucified"
s   "For all the law is intact but mercy abounds"

With respect to the above I note that the law is only made for lawbreakers. With the law relaxed, the life of (or life then found in) Christ is placed upon, or is "in" all those people as "s". (I.e. Those "elect" from whom the condemnation is lifted and are now "born again".) That aspect of the law is then "at rest". There is now only grace remaining; as without that condemnation there is no strength of sin. The purpose of the law is not merely to identify all sin, but also to point the way to the Christ that would fulfil both the law and the prophets.

1Co 15:55 O death, where *is* thy sting? O grave, where *is* thy victory?
1Co 15:56 The sting of death *is* sin; and the strength of sin *is* the law.

Christ is not the minister of sin, but of life: the good works of the law remain good works, there is no freedom found within grace for continuing to sin, but only the freedom of becoming (approaching Christ-likeness) as an heir of God, free from all condemnation. Righteousness has moved from being found in only those living within the law, and now is found only in Christ. (But that same righteousness was **already** His and that was openly revealed in the resurrection of Christ crucified.)

When the law itself broke its purpose at killing Christ and thereby legally so; Israel required of necessity to be translated to the new covenant, through an empty middle devoid of positive properties at the failure of the law. There was permitted no "logical gap" between them, but instead there was darkness over the land whilst Christ was crucified: The cross is truly the (logically) empty middle justified with an excluded middle of action (the crucifixion of Christ); by which the execution of Christ was itself a dialectic, and an evil, and yet it entailed more abounding grace than the sin of Adam could have brought in death. Truly, the ultimate evil overcome with the utmost of good. (May Satan accuse at all when God should take advantage of the enemy's last remaining device? No new convert sins by that dialectic germ of faith!)

God has used the dialectic to further the gospel message to those that commonly employ that dialectic device (sinners). The octal has obviously been worked with God's virtue; but the dialectic in this manner is made to entirely agree! There is no sense in which an enemy (or Satan) may with the same dialectic claim the logic of the cross contradictory, and so the God of heaven has "cast out" that accuser and is now free to minister to all those that believe without accusation. They may all freely be ministered to by the Holy Spirit in all grace, without any condemnation. Satan is then a defeated foe — as his "last-chance" kingdom is set up as with Christ also, as if nailed to the cross with Him! (The kingdom of God has barred the kingdoms of the "god of this world" having dominion (authority) over God's people with his only device.)

Christ's cross had the famous superscription, "THE KING OF THE JEWS". Were you aware that He was effectively (legally) recognised as the King by the ruling power of His day? There may have been much of Rome's mockery and an effort at crowd control at work by Pilate, but legally speaking, that same public recognition on a legal document (effectively a warrant of execution) is grounds for treaty! A modern equivalent would be the recognition of a new state by, say, a superpower (and even if that recognition were only verbal it would still carry a great deal of weight). The kingdom of God is truly set up. Pilate did not repent of it, even when given the free opportunity.

That dialectic that set up the kingdom of God may be phrased as such with:

r   the old covenant holds
s   the new covenant holds
u^{-1} those not alive under the old covenant
v^{-1} those not saved under the new covenant

With respect to the dialectic I will assume no faith in either covenant to start; those two covenants I will not negate from positivity as they are assumed fixed truths rather than absolutes; assuming there is liberty to ignore both u^{-1} and v^{-1}. Religion is extra work, after all, and $Pos(u^{-1}\&v^{-1})$ no difficulty, as liberty is positive, and to rest on the unnecessary positive also.

Then, Christ as the mediator (slain and "dead" in the empty middle of both covenants (r&s) as perceived by all those with and without understanding) would offer reason to all the ungodly (using the dialectic) as along the lines that all those with no faith would perceive the disjunction $r\&u^{-1} \lor s\&v^{-1}$, which has an equivalent in (r&s) proving the rather boring disjunction of $u^{-1} \lor v^{-1}$. (This middle r&s is truly empty but for the cross, but not so empty by necessity is $u^{-1}\&v^{-1}$, which has many captives.)

Then $r\&s \lor u^{-1}\&v^{-1}$ may be phrased, "The one God (Christ) of both covenants saves with both covenants or not at all." (There was once righteousness by the law, now it is only credited to faith placed on Christ.)

Choosing one covenant (the old) I find it as "r" is an equivalent in the dialectic to $N\neg(s\&u^{-1}\&v^{-1})$ or $s \to (u^{-1}\&v^{-1})^{-1} \to u\&v$. Those alive under the law are preserved (being Christ only), and those under Christ (all others) are thereby preserved in Him. Then u^{-1} and v^{-1} are no longer found positive, but now their negations conjoined $u\&v$, which meet in Christ crucified (the firstborn of the resurrection). This is no proof of the gospel, but faith is so found naturally.

Yet then dialectically I also find $r\&s$ proves $u^{-1} \vee v^{-1}$ and since Christ has already come (the kingdom of God has arrived) I naïvely default to v^{-1}. In fact, as Christ fulfilled the law, both $u \to v^{-1}$ and $v \to u^{-1}$ hold. And the new covenant must be finer than the old, superseding it and also being found in agreement to it.

If there is yet a middle found from $N\neg(s\&u^{-1}\&v^{-1})$, so that $u^{-1}\&v^{-1}$ proves $r \vee s$, then there is an evil to be avoided in $u^{-1}\&v^{-1}$, for not only does faith perceive no **positive** middle between the covenants, but as a sinner, Christ died for me (and I guiltily benefit) and I discover $N\neg(Pos(u^{-1}\&v^{-1})) \to Pos(u\&v)$ because nothing positive occurred in the brutality of the cross, it was an evil, and a dialectic, but yet grace then abounds much more freely afterward! (Those many captives in $u^{-1}\&v^{-1}$ are loosed.) Then but one covenant "s" stands, and only that one which is after Christ (who is "come in the flesh"), now that both the law and the prophets are fulfilled. The symmetric and dialectic argument affirming the old covenant is a rejection of Christ. It is "antichrist". ("2Th 2:7 For the mystery of iniquity doth already work: only he who now letteth *will let*, until he be taken out of the way.")

Then there is a result, the dialectic middle of $u^{-1} \vee v^{-1}$, now clearly $r\&s$ (which dialectically entails $Pos(u\&v)$) is found; and therefore some rudimentary faith is reached from the cross between the covenants, contradicting the assumption that u^{-1} and v^{-1} were both positive to start. I now find $Pos(s\&v^{-1})$ from $Pos(r\&u^{-1})^{-1}$ with $\neg Pos(r)$. $Pos(r)$ is negated by the fact that "s" supersedes and replaces "r" as it comes afterward, and I also find $N(Pos(v^{-1}))$ from $Pos(u)$, that God "translates" or "moves to a greater perfection". One must then place one's faith on Christ as mediating a better covenant. That covenant is surely made with a gospel of virtues.

God did not operate outside of either covenant, whether upon the cross or watching over it. God is simply God acting as God always; there is no sense of any replicable act of human virtue whereby any other being may save themselves outside of the grace of God and its provision in Jesus Christ alone. It is by this grace entailed from a dialectic (from God Himself acting within the octal) that I am able to know that Christ's Father is the creator of this world, and to recognise that Satan is ultimately defeated, the kingdom of God now completely set up. With the resurrection, Christ is revealed blameless in preaching His kingdom freely.

To show that the covenants are not both in effect (nor have they ever been together), I may reason with respect to the setting up of that same kingdom of God in Christ. I should attempt to define the language in its most basic terms, on the scale of a "mustard seed". The sown word of the kingdom becomes that "mustard seed" of faith. I shall use a principal ultrafilter as a model of the authority to preach the genuine words of God. Given that, there is no need for a seed other than Christ crucified (as for all) to ever be sown, as the work of the cross is finished and God now at rest raised to life. That seed made (sown) is one of the highest virtue, and it indeed makes sense that there will never need be another (else God would become an author of confusion).

If the law is truly also a filter (and contained within an ultrafilter, a maximal filter of the grace in the kingdom) then the kingdom of God (i.e. with God's laws) is or are all chosen within His liberty to choose His own people His way. That same liberty generates the ultrafilter "of the kingdom" as a principal element (this liberty of God, His sovereignty which is known within us) and is also able to be grasped in the understanding. Without that sown word of sovereignty, no Christian has the right or the divine authority to correct any other and then only those likewise with it! (An unbeliever can and often corrects a Christian! – And yet there appears no absolute authority to reciprocate unless one is as righteous as God Himself.)

Jer 15:19 Therefore thus saith the LORD, If thou return, then will I bring thee again, *and* thou shalt stand before me: and if thou take forth the precious from the vile, thou shalt be as my mouth: let them return unto thee; but return not thou unto them.

Such an ultrafilter may be derived on "Divine Authority" as an axiom. Liberty generates the ultrafilter, not ours but God's own.

1) 'Authority' is applied by anyone with the axiom of sovereignty. (X is in U or I\X is in U.) Those with the axiom have a consistent set of beliefs. (Rom 2:15 Which shew the work of the law written in their hearts, their conscience also bearing witness, and *their* thoughts the mean while accusing or else excusing one another;)

2) Two people with such authority may correct each other. (A, B in U, then certainly $A \cap B$ is in U.) The intersection of two sets of consistent beliefs are still consistent (and contain the axiom).

3) The empty set ϕ has no authority. (The complement of ϕ is in the filter as it has authority.) Having no belief is inconsistent with authority. The set of all consistent beliefs at any time is gospel. (From initially the axiom of sovereignty to beyond the full and inspired scriptural canon.)

4) Anyone with a superset (of consistent beliefs) of another individual's beliefs may correct the latter subset, or else also extend it. ($X \subseteq Y$, if X is in U then Y is in U.) Authority refines as well as extends such belief. (Constrained grace (virtue) may be refined to a state of non-privation to extend mercies.) All supersets likewise contain the axiom.

The grace of Christ that makes believers perfect in the law by the predestination of the Spirit is a better covenant, and the new (to the Christian at least) has superseded the old completely. There cannot be two conflicting "authorities" (maximal filters) as to the one God's own kingdom over the same people. Effectively, $X + Y = \phi$ unless $X \subseteq Y$ and one generates the will or Spirit of the letter or content of the other. Then the new is the substance and the world is turned upside down. The opposite of "good" becomes "better".

Then the new covenant "understands" or "stands under" the law, but not vice-versa. Christ will not deny the law (but establishes it) and the matter is closed there.

Then immediately those who are obedient to grace without the law of Moses (i.e. the gentiles) are as whole in Christ as those to whom the law was given (see above, Rom 2:15).

The middle position of a disjunction of both covenants would take its stance in order to make the disjunction good on both sides. In order for us to choose both sides equally (or equivalent), one must deny Christ; because either the law is not fulfilled and the works of it continue or then Christ did not fulfil the law so that the spirit could minister grace without accusation under the condemnation of that same law against His charges.

What in effect is being denied (frustrated) is as I would put, "The Spirit of grace". (The lamb as slain before the throne, in the place of all virtue.)

Now the law was for those who were circumcised to keep it, and I rightly know from the New Testament that in Christ circumcision benefits you nothing. But if circumcision is now of the heart and obedience now by the Spirit, the middle position that denies Christ does so upon those who are circumcised or converts to keep the law. (For the law actually has no effect on the gentiles.)

To hold necessary a two covenant stance of Old and New Testaments is a denial of Christ; by denying grace one denies the cost that bought it. If there had been some other way than the resurrection of Christ to affirm blamelessness before God, it surely would have been done so for us.

I may begin with:

    a   "grace extended to all"
    e   "a person under the Old Covenant"
    g   "a person under the New Covenant"

I assume that if circumcision ("c") provides a double blessing of Christ's grace in the centre, then I actually have a system whereby $Pos(e)\&a^{-1} \rightarrow Pos(c)$ and I can see when either side is inverted to inaction, I may clearly reveal that virtue remaining. ("c" already qualified an Israelite to the works under the law and the sacrifices for sins. Only the NT (New Testament) case remains with respect to the circumcision.)

So that if "a" is truly a virtue (which grace surely is, but it does not operate under the OT law), then "c" (though it remains positive) becomes at the very least profitless and has no benefit in the application of "a" to the gentiles. To benefit from it when all are equal is not Christian. Then $Pos(c) \leftrightarrow \neg Pos(c^{-1})$ in Christ since $Pos(a\&c^{-1}) \rightarrow Pos(a\&e^{-1}) \rightarrow Pos(g)$; grace may operate freely without the old covenant law whilst blessing the gentiles who are without the law, and it becomes positive for them to not ever be brought under the law in order to be "twice blessed" where full blessing of grace exists already. (The apostle Paul made that same point forcefully.) So, let the uncircumcised remain so.

So $Pos(a\&e^{-1}) \rightarrow Pos(g)$ or $Pos(a\&c^{-1}) \rightarrow Pos(a\&c^{-1}\&e^{-1}) \rightarrow Pos(g\&c^{-1}))$ so that circumcision of the old covenant blesses you with nothing at all in the new as Paul taught, then if "g" is entailed from the application of "a" to those now not under the law (grace freely extended to all), it should be noted that [0, a, e, g] forms a K4 group in positive properties and note most crucially that the intersection of the sets in e and g is totally void of positive properties.

Then, by axiom of virtue the intersection is empty or rather the person who holds both covenants equally true (necessary) at the same time to count themselves "twice blessed" are caught in their own craftiness; they are privating virtue and in whole or in part denying both covenants. One for its efficacy in pointing to Christ and its fulfilment in Him and the other (the new) for the grace that gives it and the law its real strength. ("Rom 3:31 Do we then make void the law through faith? God forbid: yea, we establish the law.") They are then blessed by neither from the middle. What this means for the believer is more works in Christ without any additional blessing.

Every **positive** property of the law is now "at rest" as if q^{-1}. The law remains, but the condemnation under the law is relaxed, as it unjustly condemned Christ who fulfilled it. There is the scripture "Rom 10:5 For Moses describeth the righteousness which is of the law, That the man which doeth those things shall live by them." Then those that have not kept the law shall not then live: that same condemnation is relaxed for all those in Christ. (Righteousness is now credited to faith on Christ having fulfilled the works of the law, which once were righteous, but continue only as His righteousness for those that broke the law and are justified only by Him.)

The excluded middle of nullity only takes force when grace is frustrated (made to fail). Both covenants then appear to lose some of their force if the middle as if by necessity is in effect. Sadly, that is the denial of Christ's virtue (and of His Father's grace). Effectively, to dwell in the middle is to attempt to repeat Christ's own work for Him – and as by the New Testament all are under the condemnation of the law, it is meaningless work. Obedience is good, but such works approved by faith (under grace) are better.

Now I have a dangerous minefield to negotiate. I forcefully assert there is no anti-Semitism in Christ. Although the scripture references a sect that appears to show an infiltration of the Church to supplant its doctrine with the same faults that plagued old covenant Israel, not a single person alive today is at fault for inventing false doctrines that were recorded in the biblical record thousands of years ago. (Excepting God's adversaries, Satan and his angels alone.) I am simply left (after all blame has been stripped bare from the deceived) with the examination of a single deception, a logical device that can ensnare the believer.

The Church at the very end of the age (whilst fearing the wrath) is at the end offered the principle of obedience to God through the Old Testament law and deduce they will be also welcoming into fellowship a dutiful and obedient sect of believers who truly honour Christ. They outwardly and truly see that and all appears to be so initially. Instead they have their expectation of further continuous and accelerated pre-prepared growth (the bait) switched with an influx and rising of tares, in truth holding some very peculiar ideas on prophecy. (That prophetic falsity fills the void in the middle with "new" inconsistencies.)

Jesus states simply that there is a "synagogue of Satan": not a reference to Jews, but to a closed Christian sect with such elevated and restricted membership. Such closure to be compared with that of virtue – these tares claim a "physical" carnal virtue (if indeed there is any – which I doubt), yet as such the "concision" had an early claim, as to holding circumcision of the flesh necessary for status in the new covenant. Then it could be claimed that $g \to e$, that the new covenant requires obedience to the old (which it does under the bands of grace and predestination). However, it is a new covenant and the old is no longer in force. I.e. $g \to e^{-1}$ and $e \to g^{-1}$.

In the early history of the church the Christians had been forced out from teaching in the synagogues. Here I find a group that in contrast or in analogue are "called Christian" (though not so by Christ) but are in the last times reciprocating with false doctrine (from the Old Testament) in the Church. That is the **whole** extent of the term "synagogue" in the Revelation.

Phi 3:2 Beware of dogs, beware of evil workers, beware of the concision.
Phi 3:3 For we are the circumcision, which worship God in the spirit, and rejoice in Christ Jesus, and have no confidence in the flesh.

So, to the over-zealous believer: if anyone were to think that there really is some "Jewish plot" to invade Christianity, they are, quite frankly, logically deluded and already a member of such a cult themselves. They place those who should reasonably be their brethren in Christ in the impossible centre of two covenants (i.e. where there is none) and have given them non-existent authority (as devoid of any positive properties) against God's own election of grace.

In pointing the finger, they have simply changed what they thought was veneration for vengeance, or their own vexation. What does the book of Romans tell us? Do not judge – believers are inexcusable when they do so. Getting worked up over this is not the solution. The answer, if one cannot correct another so mistaken using sound doctrine reasoned from within the scriptures (those refusing all such reason of correction are then tares), indeed stares you in the face. It is the door to the outside. Just leave. If you were correct, I assert God would prefer that outcome than see you caught in such a **logical snare**.

Rom 2:3 And thinkest thou this, O man, that judgest them which do such things, and doest the same, that thou shalt escape the judgment of God?

Neither is there any such conspiracy to expect – what I deduce from all this is not a conspiracy but a vacuum. The Church is as Israel became, "destroyed for lack of knowledge" (Hos 4:6) and in that lack of awareness from a place of sound doctrine the vacuum is filled with whatever is closest; that (to many to fill the void in the centre) is the smallest sideways step. What the churches actually receive instead are a number of aberrant prophetic disciplines, none of which are any more valid than any of the others. (Since the elevation of status is unimportant and without logical basis already.) Whilst simply pointing out the scripture to those who sit in the middle of two covenants without any true positive basis is easy, what has become difficult is that Christ's believers are deeply invested in their hopes for the conclusion of their prophetic expectations (and they are not easily extricated from collectives holding them).

2Ti 4:4 And they shall turn away *their* ears from the truth, and shall be turned unto fables.

It is a sad fact that the empty middle between covenants is subject to some attempt at healing. Whereas the cross of Christ is not a wound with missing doctrine, such mistaken beliefs are placed upon the empty middle as if they were to be applied as a sticking plaster would be – to obscure what is purposefully there: the translation of God's people to a spiritual house under Christ. (Which is no wound at all.)

That which is most important to note (as is also shown in chapter fifteen, by "No Second Death") is that these added beliefs are not always self-inconsistent but are yet highly convincing.

2Th 2:11 And for this cause God shall send them strong delusion, that they should believe a lie:

Those in the Church to whom believers will give their deference (specifically, in prophetic matters and to choose who will teach them) – will continue to work their doctrine throughout in order for the whole to become leavened (and realized) as if "blessed by their proxy" (or caught by it). With the doctrine of the Nicolaitans involved this leavening is a dangerous mix (see later).

2Ti 4:3 For the time will come when they will not endure sound doctrine; but after their own lusts shall they heap to themselves teachers, having itching ears;

In order for this clique with their mixture of covenants to operate, they must claim some precedence over the usual order of salvation – as all Christians are equal in Christ. In the book of Romans the apostle Paul states his condemnation of those that would judge the "inward self" of others by their outward appearance only and famously states:

Rom 2:28 For he is not a Jew, which is one outwardly; neither is that circumcision, which is outward in the flesh:
Rom 2:29 But he is a Jew, which is one inwardly; and circumcision is that of the heart, in the spirit, and not in the letter; whose praise is not of men, but of God.

Just as dangerous is the converse: that believers would only look at the outward appearance of obedience, yet not identify ones so without restraint inwardly.

Mat 7:15 Beware of false prophets, which come to you in sheep's clothing, but inwardly they are ravening wolves.

These "upper-echelon" Christians (whose praise is only of men, not being of God) as a closed group become a primary source of false doctrine in the Church (the "false prophet" mentioned in Revelation) rather than a source of outwardly obedient adherents to "messianic-judaism" or "judeo-christianity". (The danger of those two is truly present only in the hyphen, and the error is logically correctable with a sound model as before.) The real danger paradoxically is that the upper "caste" is a closed sect, and their own "new" doctrine (to justify the empty middle) becomes sacrosanct despite its own inconsistencies (not necessarily within their own beliefs, but with the gospel itself). Then dialectically with a, c, e, g as before, $a\&c \rightarrow e^{-1}\&g^{-1}$.

These "that say they are Jews and do lie" are first called Christians (and are not Jews to Christ, not either way, as under the old covenant or equally in the new) that would perpetuate the false belief in a spiritually superior caste. (Specifically, as instance, taking the place of the **sealed** 144,000 over all others.) For instance, they then take that proper place of all believers by limiting the inheritance due all those with a part in the first resurrection to themselves or their proxy on the basis of their claims of status, genetics or false doctrine alone. The genuine good news is that there are truthfully no unsealed saved and no saved unsealed. God has no respect of persons.

So, a simple answer is that such appear to present a twice-blessed "Christian" status whilst instead they would perpetuate both covenants – to uphold their own status as a closed sect meaninglessly when the good news (to which they are very welcome) of Christ's ministry was affirmed by His resurrection blameless once and for all – for Jew and Gentile without preference. If the circumcised Israelite requires a **sealing** or a second circumcision, where is the law then but only in Christ?

Heb 8:13 In that he saith, A new *covenant,* he hath made the first old. Now that which decayeth and waxeth old *is* ready to vanish away.

That ship of both covenants together (with the old covenant required to guide and shape the new) as in "life in Christ still under the law" that appeared to continue valid after the cross, sailed in the space of one generation. The scripture (chapters two and three in the book of Romans, for instance) had already answered this illusion in the first century. God's ministry had also turned to the Gentiles without any diminished effect; there is no Jew or Gentile in Christ. In any case, these Christians do not ultimately give God the glory but their own elevated status (as to their station and works which are put in the place of Christ) instead.

It is this false expectation of the product of that middle position, the "new doctrine" due to the presence of this proxy or "prophetic-cult" of tares, that with the mix of their "clique of Satan" (and those with the doctrine of the "Nicolaitans" – see below) underpins the accelerating drive towards the hoped-for fulfilment (the false prophet) that turns the gears of religion for many.

False doctrine takes its firmest hold the moment Christ calls His own out of the Church – which also sadly occurs once the churches incorporate and "receive the name of the beast". (This "name" may only be received after the setting up of the image system and Satan's insult to God is set in place. Then, any act of incorporation throws in one's lot with that beast.) Then the vacuum of the Lord's spiritual guidance is filled by whatever is then accepted to provide a sense of life and hope for growth in the gospel. That (believers should be warned) is the topic of most of the Revelation.

Much of the Revelation concerns the system that works this "leaven" through the "Church", as Jesus said; "Luk 13:21 It is like leaven, which a woman took and hid in three measures of meal, till the whole was leavened"; a system of worship that takes up the spiritual mechanism that accomplishes the process. It is a sad fact that with the revealing of the wrath of God, the Church will try almost anything to become obedient – except approach God on His own terms, outside institutions of corporate religion.

Aside from within the letters the sect referred to as the "synagogue of Satan" is instead an attempted synthetic replacement for the church of Philadelphia. (Grafted in by Satan as tares onto the collection of the first five (as purposed for refining) churches of Laodicea in order to perform a "bait and switch" for the Church proper to follow after.) In essence they are simply a "source of false doctrine" or "a source of leaven".

Such a simple mistake as to inequality between Jew and gentile should be easily rectified and in the scriptures the apostle Paul does so marvellously; there is no superior blessing from the law in Christ – all are equal in the same grace under a law of faith. Everything that followed after that error, however, has leavened the whole.

That sect, I am to expect, practise an outward yet aberrant form of legalistic Christianity as a clothing for their doctrine's origin, accumulated from all the false doctrine of the whole biblical record of Old Testament Israel, now importing it into the Church of the last days. Such doctrine is in part referenced as of "Balaam" or "Jezebel" or upon the angel simply tenfold of Satan himself. (A test of which is become the maximum upon the least as I shall show later.)

The first beast listed in the third woe (the text states – see Rev 13:2) has seven heads; one for each empire that dominated Israel. Then I should expect falsities of doctrine from Egypt,[7] Assyria, Babylon, Persia, Greece and Rome. Rather than expect false doctrine from a period without a nation state of Israel, the seventh set of "false doctrine" is that of the "Nicolaitans" – that being collective rule in the false prophet's image system (which becomes the eighth – and is "of the seven"). The little horn associated with the false prophet is not to be mistaken for Israel either, though Israel has a "little status" in Revelation. (It is actually the Church at the end of the age that is the little horn (which is the eighth kingdom formed from the false doctrine of the seven, i.e. of Israel); and that it as conceptual Babylon is "twice fallen".)

There is one other group prominent in the letters – the "Nicolaitans" which is a term that transliterates into English as "conquerors or destroyers of the laity". I will show that this sect is such as to hold true the principle that all should defer the Church's choice of doctrine to be decided by the consensus or proxy. (Whether historically as by a "professional" priesthood in their own place, or by simply accepting the decisions of the "scarlet beast" that is to be found in the third woe of the trumpets.) Believers are told Jesus hates this thing. It is to become a complete mockery of His chaste bride – an insult from Satan as shall be discovered. Doctrine is God-breathed in scripture, not to be cast to a vote.

So the "synagogue of Satan" is to form the "mind" of the "Church" together with the "Nicolaitans" that form the body or "senses" – the woman that is made into the corrupt image to the beast. The Nicolaitans will naïvely (and blindly) follow any decision to incorporate false doctrine. They "destroy" the house of God that they attend; their acceptance of poor doctrine perpetuates the methods used to leaven by it.

A very dangerous minefield to navigate and not just for the zealous believer.

5.4) The Angel's Circuit Under John's Watch

Whenever a seven cycle appears in the Revelation, it does so as if in permutation representation as, say, x = (a,b,d,c,f,g,e). Now, John read from right to left so his seven cycle appears throughout as if he were using the inverse, namely x^{-1} = (e,g,f,c,d,b,a).

I would then expect the angel's circuit to begin in Laodicea and end in Ephesus. However, as each letter is properly addressed to "the angel of the church" I may expect the angel, who himself throughout uses the form of permutation notation reading left to right (which is shown later from the text), to read the letters in their given textual order inverse to the seven cycle of the circuit over them.

This has a small but recognizable effect. If each church has the following correspondence to an element of the cycle:

[7] James Lloyd, The 6, 7, 8 Cycle (Christian Media), p26-28.

a	Laodicea
b	Philadelphia
d	Sardis
c	Thyatira
f	Pergamos
g	Smyrna
e	Ephesus

It becomes clear that given the ordered G-set of churches {Laodicea, Ephesus, Smyrna, Pergamos, Thyatira, Sardis, Philadelphia}, the seven cycle x^{-1}=(a,b,d,c,f,g,e) in circuit maps unity through the G-set as follows.

If I have the ordered set {Laodicea, Ephesus, Smyrna, Pergamos, Thyatira, Sardis, Philadelphia} = $\{1,x,x^2,x^3,x^4,x^5,x^6\}$, then one application of the seven cycle x^{-1} sends $\{1,x,x^2,x^3,x^4,x^5,x^6\}$ to $\{x^6,1,x,x^2,x^3,x^4,x^5\}$. A second application with x^{-1} sends $\{x^6,1,x,x^2,x^3,x^4,x^5\}$ to $\{x^5,x^6,1,x,x^2,x^3,x^4\}$.

The result is that unity "floats" through the churches; and that the "angel" is at rest in each "church" as the unity element in the cycle. This is not the case if the seven cycle is aligned to the letters naturally. This floating unity has importance later on where the serpents of the sixth trump oppose the two witnesses' ministry, to form five-cycles over the first five churches.

Each "church" becomes a state of rest for the angel, until such time as he again becomes convicted and made to move on in his circuit. Rather than find safety in numbers, the least is instead transient and searches for fellowship with those that have good doctrine.

Every "church" is then a "mountain" or "stronghold" of doctrine (whether bad or good), and the angel that completes the circuit the least able to do so. (He has mathematically found "rest" as if a sabbath within each place of the circuit.) It appears, though, that he is only in receipt of good doctrine at the end of his circuit in Philadelphia. The doctrines of the other churches are found to be as much stumbling blocks as they are a refuge.

At the end of the circuit, the angel in Philadelphia is instructed by Christ to "hold that fast which thou hast". Since Philadelphia corresponds to x^{-1} in the seven cycle, it is then clear the angel is instructed not to move on. (This, despite attaining rest in his circuit. It becomes clear he can "go no more out".) Then he is not under spiritual conviction from the letter to Laodicea, and may accept its action as the unity in the seven cycle (to remain in Philadelphia with divine approval). Yet even this has its hazards, as "1000 die" in this blessed state in the earthquake at the end of the two witnesses' ministry.

This is quite separate to the conviction by the letter to the Laodiceans (as unity in the cycle) to then leave a fellowship and find another with "better doctrine" (as if the angel would stay put and become "lukewarm"). The angel throughout does not defer the choice of doctrine he holds to others, neither does he have respect of persons. He has no honour for Nicolaitans or for the synagogue of Satan. When they or their devices are seen in a fellowship and are accepted by it, it is time to move on (or finally, to move out).

The circuit is complete when the least may sit enthroned with God over His creation to reckon for his own salvation as if it, not he, were the unity (for the least is effectively as the zero).

5.5) The Reckoning

The angel, overcoming the worldliness of each church in his circuit merits the giving of a gift each time from Christ for overcoming that particular false doctrine or faulty fellowship. Those same **additive** elements of the octal allow the angel to traverse that same circuit of seven lampstands without **any** influence of the Holy Spirit (as to "His place", Rev 2:5).

Each gift is equal to a "horn on the Lamb", and horns are symbolically "strengths" but in the Revelation they appear to be accrued cumulatively. Not only Christ, but the dragon, the first and scarlet beasts and also the false prophet all have "horns", and they all appear to be accumulated over time. Christ, however, always having all strength gives these horns as gifts to the angel so that he will not backslide in his circuit.

Yet how, then, does the angel get from first overcoming in 'Ephesus' to raised back to life in 'Pergamos' following 'Smyrna' or any other "church" in the circuit without first fully completing the whole circuit and overcoming the world first?

Simple, "Predestination".

Each gift allows the angel to overcome each "church" with that same gift which that victory merits. At the very end of his circuit, 'Laodicea' then overcome and 'Philadelphia' a safe-haven, he may sit enthroned in the heavenlies (whilst also being tried on Earth) and use those seven divine gifts to account for his overcoming in/of each church or set of false doctrine in his circuit, thus also meriting the "horns". (He is required to account with them combined and enthroned just the once, for fulfilling the requirements of the least's salvation.)

The circuit of the least then becomes a "base-line" – a work without the Holy Spirit justifying the outpouring of wrath occurring before the judgement. If he may overcome, nothing is stopping anyone else from showing some sound judgement and restraint, surely?

I state that the least in the kingdom of God is not the least because he is a dreadful sinner, but instead he is the one over whom God can show or has shown the very least influence. God's solution? It appears to be to let him account for himself completely instead! Assuming this, I can easily "fill in the blanks" and interpret the seven letters in the context of the angel and his overcoming.

God does not bother to waste His own efforts or preclude them, but instead supplies the least amount of divine grace (assistance) possible to the least in His kingdom in the very worst of circumstances. Every other child of God has an accounting made for them of God that justifies them as "born again". The least, however, would reckon for himself with all that power which he requires and as he requires it, using these "gifts" as effectively "merited favour" (rather than true divine grace which is directed by God overall), which justify his salvation in the kingdom of God, as with the same strength of Christ's own power of salvation. I make it clear, the angel does not climb up "some other way" (Mat 22:11-14 cf. Rev 3:5).

The seven gifts or "horns on the lamb" together form a "prescription of determination and predestination" for the least elect to move (rather than work) his own self to salvation, despite himself being as a "zero" element within the octal (with respect to the C7 multiplication of the

Holy Spirit). Zero, by definition, absorbs every other element by multiplication (so including the action of the Holy Spirit). Then God has "in principle" chosen the very least that He could possibly exercise His saving power upon (as without requiring omnipotence, so preventing Him becoming overbearing upon all the elect), yet in the same application of omnipotence, the least is reckoned for completely by the only influence he'd accept without question: his own! The angel is then truly saved by the exercising of his **own** free will (in obedience to and worship of God's only begotten, Jesus Christ).

Then the least must account for his own salvation complete, for Jesus (one could suppose) has instructed him to "reckon" for himself. Christ's own part in this work is the writing of the seven letters, giving His strict conditions for how His angel may then show himself to be willingly obedient. (The letters are not addressed to another.) Christ's final promise to His right hand is to write His new name upon His least. That least is fully as "the zero" after the overcoming of the seven churches with those seven gifts. That final gift of rest ("and he shall go no more out" (Rev 3:12)) is a mystery; I expect there is further rest waiting in an 8-cycle of multiplication with an equivalence of $1 = 0$ (present in only the person of the angel). In any case, this final gift unseals God's new name as before all the elect and brings in the reign of God omnipotent.

Rev 3:12 Him that overcometh will I make a pillar in the temple of my God, and he shall go no more out: and I will write upon him the name of my God, and the name of the city of my God, *which is* new Jerusalem, which cometh down out of heaven from my God: and *I will write upon him* my new name.

From performing the "first works" to finally sitting on the throne and marrying his life to his own accounting with the crown of union, as the very least, he becomes the justification for all others within that same marriage of the Lamb, as the least overall and as the owner of that same "crown of life" which completes it. There is no limit to that union; for the world may be moved to provoke him to salvation just as easily as a zero element is made to be unity!

The least, then, must not only overcome the world and Satan whole, but more so. He must succeed with no help at all beyond that merited already by his overcoming of each prior church in his circuit. Then the horns on the Lamb allow him to progress onward and with some sure restraint of his own making, prevent himself from ever backsliding.

So, God has supplied the least with seven horns, an equivalent to an omnipotent work of the Holy Spirit within the gospel of Christ (elements from which may be constructed any arbitrary octal later generalised to all thirty by the two witnesses). The least is free to have his circuit complete an eight-cycle over that arbitrary octal, whilst all others are only required to fulfil a work found in a seven cycle, finding grace and rest in Philadelphia from five other churches. His circuit, then, is done without such similar "multiplication" but those horns are ready or prepared to become as the works of the Holy Spirit in leading all others to overcome similarly.

In that he has reckoned without (or with the least amount of) divine grace amongst all those saved; all (not just the faithful) are able to overcome, not to remain deceived and to repent accordingly. Predestination is supplied to all others from God's grace for whosoever it is sufficient. In the case of the least this is done uniquely: God's own right hand, as under the accusation of Satan, had (or has, or will have) been cut off from all grace, and grafted back in seamlessly with this circuit.

With his overcoming predestined, he is the unique dividing line made between those that are born again with a reckoning of God, and all those which are unable to be saved at all, for whom all grace is totally insufficient: they truly belong without the conviction of the Holy Spirit, given over to a reprobate mind of their own choosing.

For understanding the Revelation and the mystery of God, Peter instructs us to be mindful of something – that the number "one thousand" is not to be calculated but instead refers to a "linear constant". This one thing the faithful are instructed never to be ignorant of:

2Pe 3:8 But, beloved, be not ignorant of this one thing, that one day *is* with the Lord as a thousand years, and a thousand years as one day.

In interpreting the scriptures, then, I am instructed of this principle as to the Lord's rule (of His 1000-fold patience) over the collective body (then multiplied 1000-fold) of His elect. Whilst every believer may accept "God's plan for them"; the least has his own reckoning as surely all the faithful do too (having a thousand years of life to justify with God that which they choose to do today or instead repent of completely). They are then given the same authority as one of the twenty four elders (as now) to also justify their same dwelling within that thousand years (one of 144,000) in heaven, resting in that authority they **already** live within, reigning over themselves. This is not a nonsense, but an equivalent to finally being saved and "born again".

The least, the "angel of the church" is found to be Christ's own right hand. His crown is that final union of life and prayer that is the very mortar used for building the Lord's temple within the Spirit. Satan seems to prefer giving him a "crown of worms" over ten corrupt doctrines instead. Yet at the very end, being blasphemed as some "false prophet of God" leading the gullible into separation is not the final result. (It is not only the least that Satan hates, but Jesus Christ on the whole.) In equity, the least is given the worship of Satan's own.

Rev 3:9 Behold, I will make them of the synagogue of Satan, which say they are Jews, and are not, but do lie; behold, I will make them to come and worship before thy feet, and to know that I have loved thee.

5.6) The Throne of God – The Sun Octal

Immediately afterwards, John sees heaven open and is suddenly before God "in the Spirit".

Rev 4:1 After this I looked, and, behold, a door *was* opened in heaven: and the first voice which I heard *was* as it were of a trumpet talking with me; which said, Come up hither, and I will shew thee things which must be hereafter.
Rev 4:2 And immediately I was in the spirit: and, behold, a throne was set in heaven, and *one* sat on the throne.
Rev 4:3 And he that sat was to look upon like a jasper and a sardine stone: and *there was* a rainbow round about the throne, in sight like unto an emerald.

The voice like a trumpet alludes to the sequence of seven trumpets that are shifted backwards in the narrative with the seven thunders. The voice is once again from a perspective yet to come – the future.

John states a door not "into" but "in" heaven was opened; John was moved to a perspective outside of time whilst in the Spirit and into a view of the future whilst in "heaven". ("Come up hither", could very well be movement in time.) John's understanding was about to get a huge dose of divine inspiration, the first voice he heard being absolutely clear, precise and without fault. John states the throne in heaven was "set". There are many possible isomorphic fields GF(8), but only one reference octal – the "Sun" octal. Enthroned upon it uniquely is one reigning above all similar "octals" – there is but one Father, that Father of Jesus Christ.

As to the Father's appearance, I state "jasper" and "sardine" have this one mention before Rev 21 (both are mentioned, referring to the foundations of the holy city of New Jerusalem). I may assert that a rainbow is not like an emerald – unless it appears to be refracted by the one on the throne like to an emerald? The trinity is present in analogy: two parts present and made clear as a finely cut and polished gemstone (The Father and Christ) without any mystery and the third of the Holy Spirit present in the rainbow as made alike to a "finely cut gemstone"– the trinity are all present (Christ is sat down with His Father in His throne), all three being revealed without a mystery to John. The rainbow, a placeholder for the circuit of the angel here generalised as every possible seven cycle (extending the mystery of the seven candlesticks over every octal) is, with the overcoming by the least, now a prerequisite for the building of the holy city itself. (If not a foundation then certainly a pillar in the temple of God.)

Precious Stones:
Gems throughout the description of the heavens indicate something without "cloudiness" of uncertainty or mystery, but of surety: as bought and paid for. Likewise they indicate something of substantial worth, without fault: a quality of virtue once exercised but that stands for all time, such as the foundations of the city of the New Jerusalem made sure from the cross etc.

The Rainbow:
The rainbow (cf. "Thy Kingdom come") is as sure as the seven cycle of the Holy Spirit gifted to the elect at any time from Christ. The Lord's bow is one of seven elements and not three. It generalises the angel's circuit from over the churches (nations), to fields isomorphic across the whole sea, and properly, all the "stars".

Gen 9:12 And God said, This *is* the token of the covenant which I make between me and you and every living creature that *is* with you, for perpetual generations:
Gen 9:13 I do set my bow in the cloud, and it shall be for a token of a covenant between me and the earth.
Gen 9:14 And it shall come to pass, when I bring a cloud over the earth, that the bow shall be seen in the cloud:
Gen 9:15 And I will remember my covenant, which *is* between me and you and every living creature of all flesh; and the waters shall no more become a flood to destroy all flesh.
Gen 9:16 And the bow shall be in the cloud; and I will look upon it, that I may remember the everlasting covenant between God and every living creature of all flesh that *is* upon the earth.

Gen 9:17 And God said unto Noah, This *is* the token of the covenant, which I have established between me and all flesh that *is* upon the earth.

The rainbow references the circuit of the angel generalised "in the Spirit" (and clouds are "mysteries" in the Revelation). This abstraction generalises every circuit by any candidate seven cycle from each multiplicative group in GF(8) as generating sets of seven lampstands over the reference field. Each lampstand can be formed of seven others and be traversed by the generalised circuit (the rainbow) that is here valid on the reference octal (the Sun octal).

Rev 4:4 And round about the throne *were* four and twenty seats: and upon the seats I saw four and twenty elders sitting, clothed in white raiment; and they had on their heads crowns of gold.

The eight multiplicative groups on any octal each have three automorphisms due to the Frobenius map holding unity fixed. There are 24 automorphisms or "elders" about the throne. "Crowns" reference the closure of an additive group in a field and "Gold" indicates not just a field but one in the Trinity: the elders are crowned as automorphisms of the Sun octal. Also, later, the first seal refers to such a crown, as well as the passage describing an angel crowned, sat on a cloud having a sharp sickle. All three share such structure in common: an additive subgroup in a field similar to one forming the Trinity. The elders are also clothed with white raiment – signifying the righteousness of the lamb in the throne scene.

> **Crowns:**
> *Crowns indicate an additive group in a finite field. (Even "Abaddon" has a crown.)*

Rev 4:5 And out of the throne proceeded lightnings and thunderings and voices: and *there were* seven lamps of fire burning before the throne, which are the seven Spirits of God.

The seven lamps of fire (as virtue applied to doctrine to make a group) are the Holy Spirits of God: They each generalise every possible circuit of seven C7 groups. Lightnings represent "fallings", thunders a "movement in Heaven", voices correspond to "judgements" and earthquakes a "movement of the four beasts" representing the "Earth" in heaven itself.

> **Lightnings:**
> *These represent the fallings of stars from heaven (as floating unity in all 30 octals aligned to the one reference octal isomorphic) as to the earth (allowing the shifting of octals through the common unity or product). Many such elements from the wider sea of thirty octals are "grafted" on to the earthly elements of the Sun octal to facilitate the application of the devices of the seals and trumpets.*
>
> **Thunders:**
> *A realignment (movement) of heaven in the narrative to a previous time: related to the moment when heaven rolls up as a scroll and the occasions when the elders and four beasts fall down before the throne (an indicator of the opening of each of the last three seals).*

> **Voices:**
> Correspond to judgements, such as at the last trump with the voice of the archangel (when there is unequivocally no doubt that God most high is the creator and true God of this world).
>
> **Earthquakes:**
> A movement of the earth. Such as the moment when the first six seals are loosed and the six non-identity maps of GF(8)* (the wings of the four beasts) are loosed to map the "earth" through every coset of every K4 subgroup.

Yet there is no earthquake mentioned here (strangely). I may explain this absence since the seals are as yet unopened and the restrainer has not yet been taken out of the way.

2Th 2:7 For the mystery of iniquity doth already work: only he who now letteth *will let,* until he be taken out of the way.

"Fire" is a term synonymous with a group's operation (or mappings) acting over men, whether in "heaven" or dwelling upon the "earth": whether as $C7 \cong GF(8)*$ or the adversary's construction of S5 (a cage for the elect in the "refining" churches of Laodicea). Men are then called and chosen or left to a "reprobate mind". The term "fire" is equivocated with the "action of spirit" or "work of spirit" with these groups. God's desire is to map people (choosing them) into and for His rest, to dwell as identity (unity) in the heavens (I.e. the "abstract" – to be clothed with the righteousness (of virtue) of the lamb which is before the throne – a reference for "unity" in both the Sun octal and K4 form), whereas the false prophet sets up a system of works instead. Christ's eyes, as "flame of fire", are themselves maps in GF(4)*, as is the unity in the Trinity – "the lamb" is common to both GF(4) and GF(8).

> **The Great City – S5:**
> The five churches of Laodicea are as "The great city" or otherwise incorporated under the descriptor "MYSTERY BABYLON". The adversary has attempted to close the seven cycle of the angel's circuit at the fifth church and cage every believer inside those churches using a partial construction of S5. The construction is only partially made, yet believers are called by Christ to exit to Philadelphia. S5 is non-abelian and is unnatural to those with the Holy Spirit. Freedom granted by God is not to be confused with choice of false doctrines, which are leavened with the false prophet's own "fire" – the operation (in S5) of that same great city.

Rev 4:6 And before the throne *there was* a sea of glass like unto crystal: and in the midst of the throne, and round about the throne, *were* four beasts full of eyes before and behind.

The throne may be referenced as God, wholly the "Sun octal" or reference GF(8). To this single field are all the others (the "sea" of 30) isomorphic, and within the throne (the reference field) and round about it (with the 24 automorphisms or elders) are the four beasts: the static subgroups' coset in the octal.

The "sea of glass" represents the similarity (an *"isometry"* as aligned side by side – keeping every operation isomorphic) of all thirty possible octals in seven symbols, as aligned multiplicatively into one reference octal. The description of "crystal" is meant as "without cloudiness" (without a mystery, the truth revealed in full – just as "God is unique"). The four beasts form the coset of the static K4 subgroup. The "eyes" refer to every element of each seven cycle (and also the eyes on the lamb) in every C7 group over every octal across the sea. They are found "before and behind", and are free to float through their octals (see below).

Sea, Sea Of Glass:
The "sea" references the set of all 30 octals in seven symbols. The "sea of glass" is that set aligned isomorphic to one reference octal. (Under an isomorphic multiplication, they are all as GF(8).)

The four beasts in collective form a normal subgroup's coset: It is easily seen that the conjugation gHg^{-1} or rather ($g + H - g$) accomplishes the equality of the cosets for $g+H=H+g$. The beasts are **full** of eyes which are each before and behind, $g+H \subseteq H+g$, $H+g \subseteq g+H$. Note lastly that the eyes are within each beast: $g+H-g$ is in H and each beast is as $g+H$ only if g is within the coset of H. (See also Rev 4:8.) This trivial distinction is made between the fields of the Trinity from the subgroups of the "great city" of S5.

The Four Beasts:
The four beasts about the throne are in place in the reference octal as the four "earthly" elements – together forming the coset of the static subgroup.

The rainbow references the angel's generalised circuit. The beasts' eyes are the equivalent (in inverse pairs) of the elements (feet) that walk the rainbow "circuit". Their eyes are within the other seven octals generalised (the lamps of fire – the Holy Spirit).

Each element (as one of seven spirits) of each C7 group (a "colour of the rainbow") as traversed by the angel's circuit are also under the operation of the six wings of each beast that map each element (as if a "church") in that C7 group to each of the others but they are without the unity (identity) map for which they are at rest and require no movement.

Here, the beast's "eyes" correspond to the elements of each seven cycle that forms a "church" under the angel's circuit. The six generators of those groups (as if "feet") freely scan or "walk" every lampstand (C7 group or church) as unity also freely floats in C7. Every octal is isomorphic and all are under continual motion by their wings, as acting for eyes. (The beasts "see" all. Cf. the generalisation of the circuit of the angel.) Only in the reference octal may the four beasts *all* be at rest. (The beasts are elements of that octal, which has eight possible multiplications.)

> **Wings vs. Eyes:**
> *"Wings" and "eyes" represent multiplicative maps of a set upon and to itself.*
> *(Eyes indicate that unity is free to float across the elements of the set.) "Eyes"*
> *represent the repeated action of just one product on a set (as eyes would scan*
> *the seven-cycle in a line of text). "Wings" represent those products raised to any*
> *power, as if leaping or flying over elements in the cycle.)*

Rev 4:7 And the first beast *was* like a lion, and the second beast like a calf, and the third beast had a face as a man, and the fourth beast *was* like a flying eagle.

Each beast of the four corresponds to a loosening of the Holy Spirit's restraint on the paradigm of believers in the Earth. The first of the four beasts (each are alike to animals with four limbs) is as a lion, which walks with one limb off the ground (no hurry for the king of beasts). The second, like a calf, canters with two feet off the ground at a time (hurrying to catch up with mom). The third as a man walks with one foot off the ground and two arms always up (i.e. three limbs off the ground). The fourth beast like an Eagle in flight has all four limbs off the ground whilst in movement.

In effect, unity is free to be chosen amongst one to four positions with each. God is not the author of confusion, there is but one unity at a time when the static subgroup is fixed.

Rev 4:8 And the four beasts had each of them six wings about *him*; and *they were* full of eyes within: and they rest not day and night, saying, Holy, holy, holy, Lord God Almighty, which was, and is, and is to come.

Each beast (element) has six wings to map that element to its image in the other six cosets of the subgroups of the octal. The combined "movement" of the six wings of each beast permute the coset to any of the other six cosets with products by a single multiplicative element (one of the wings, in pairs of inverses) that act on the octal or here, a coset. Unity denotes a rest with no movement (no wings needed for rest). Each beast is **"full** of eyes" their wings combined leave no cosets of any static subgroup (generalised to every octal) undiscovered to God enthroned over them. (The left and right cosets g+H and H+g are also equal.)

Resting "not day or night" the beasts with their six wings never all rest, only the lamb so. (Represented as one beast of four only.) The Lamb is the "origin" of the heavenly scene. "Day" and "night" are not without reference to the twenty four hour cycle of the Sun (being C7, the multiplication of the reference, forming the rainbow here, or instead John places the scene about the throne "in the clouds" as a "mystery"). Each group of seven cycles (all the possible "eyes" of the Lamb, as within each church in the angel's circuit) together form the "outer court of the temple", those same seven churches. All may be discovered by a combination of "eyes" (within each church) and the reference cycle under the six wings of each beast/seraphim.

> **Day and Night:**
> *"Day" refers only to the setting of the Sun octal paired with its eight complement left hand octals. "Night" instead refers to the other pairings of left/right octals elsewhere in the sea of thirty that do not have the Sun octal as either left or right hand. Day, then, separates those octals from the remainder that share either one or three K4 subgroups with the reference "Sun" octal. The static subgroups or K4 groups of subgroups (with the octals aligned in isomorphism) may be cycled static under Frobenius; The three beasts not unity then "rest not day and night".*

"Before and behind" may be a partial reference to the inverse of the seven cycle preserving the same structure. The seven cycle (a,b,d,c,f,g,e) under squaring (and also its inverse) cycles static both the sets [b,d,f] and {c,e,g}. The static subgroup [b,d,f] is "before" the beast and faces toward the throne, and the other triple is "behind". In each case [b,d,f] is toward Christ sat on the throne with His Father (together forming the zero of the octal) and the static triple as {c,e,g} is "behind" that beast instead (the beast itself a possible choice of unity for the lamb, which is placed in the centre of the four beasts with the altar), separated from the throne by that same seraphim at guard.

Then to be full of eyes before and behind is for each beast to be mapped by multiplication in the reference (by the seraphim's six wings) floating unity, placing it in any position within the octal and in the 42 seven cycles including also the reference circuit of C7 – the rainbow. The distinction is between octal multiplication which suffices only to put eyes "behind" each beast and that made by Jesus Christ enthroned who alone may (completely, fully) place eyes "before each beast" as well (by floating unity in the static subgroup as GF(4)*).

The wings of those seraphim certainly act equally over all eight groups of seven cycles, each choice of unity (with the static subgroup fixed) is valid in only two C7 groups. Each of the four beasts as a possible unity are then found in two lampstands. The unity element is once present as the lamb only – and that virtue is placed centrally in the reference or rainbow cycle. In either case Christ enthroned may float unity through the static subgroup whilst the four beasts are limited only to the positions of the lamb in the coset.

Then there are two arrangements to consider:

The four beasts are the elements of the coset of the static subgroup (the subgroup is Christ enthroned at the right hand) in the reference (rainbow) circuit acting across the seven churches. The four earthly "beasts" each represent a different C7 group (church) in that circuit. They are "full of eyes within" i.e. they each represent a closed C7 group as a lampstand, each of them a set of seven eyes (six choices for a generator of a seven cycle and one identity element).

Rev 5:6 And I beheld, and, lo, in the midst of the throne and of the four beasts, and in the midst of the elders, stood a Lamb as it had been slain, having seven horns and seven eyes, which are the seven Spirits of God sent forth into all the earth.

The lamb also has "seven eyes" (six generators of C7 and one identity). It is clear then that as the lamb must always correspond to unity in the reference circuit, being itself placed as if one of the four beasts; the seven eyes must, like that one beast's own, be one of the churches (C7 groups) in the reference/rainbow circuit.

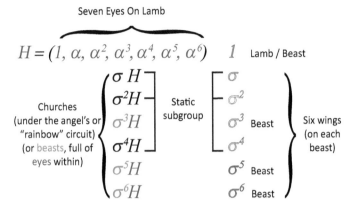

The seven eyes on the lamb map between the seven elements of the particular church/ lampstand corresponding to the unity element in the reference cycle; that same reference or "rainbow" circuit whose multiplicative elements are of unity as the lamb itself and the six wings (maps) of each of the four beasts. Their six wings preserve isomorphic the structure of the octal's subgroups under orbit (relative to its coset of the four beasts and the unity (lamb)). The identity element within the churches themselves, however, is referenced only by the lamb, who may also simply rest as unity within any church in the rainbow circuit (unity is also one of the lamb's "eyes"; I note Christ instructs the least to "watch" in each church).

Then it is seen how Christ "walks amidst the seven golden candlesticks" as the static subgroup corresponding to the particular "eye" of the lamb used within that church/lampstand; that same church corresponding to unity in the reference cycle. This walk is induced by the seven eyes on the lamb rather than the six wings of the four beasts which are limited to products in the reference (rainbow) cycle only.

Then it also becomes clear that given the static subgroup fixed, the other possible positions of unity in the reference are determined by the eight possible groups of multiplication (that fix unity as one of the same four beasts).

The seven eyes sent into all the earth are those of the lamb; yet able to be positioned in any beast of the four (all the Earth) by choice of multiplication, one C7 group from eight. The seven horns of the lamb are as the churches; all seven may correspond to unity in the reference (rainbow) circuit. The horns also represent the strengths or awards to the angel for overcoming in his circuit. (There is no sense that the Holy Spirit fulfils the rainbow circuit of the reference, that being fulfilled by the angel's circuit only.)

And the second arrangement (which was considered above for showing the eyes "before and behind", the seraphim at guard):

The four beasts correspond to the coset of the static subgroup in any one church/lampstand of the seven and the unity element is as one of the beasts. Each of the four beasts correspond as a closed group of "eyes" as to the rainbow circuit: So that the eyes on the lamb (then identical to that of the beasts) tread the rainbow circuit as to shift the one church of seven as the reference (floating unity) amongst the others.

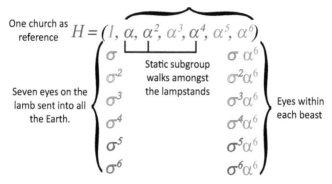

Then the six wings of each beast walk the static subgroup within each church as if the reference and the rainbow generalises this to every church of the seven. Unity, then, as one of the four beasts, as well as one eye of the lamb, leaves the scene in heaven unchanged.

Then the seven eyes on the lamb are the same seven spirits of God (as if the rainbow or the angel's circuit) sent into all the Earth (The four beasts). The seven horns are then the possibilities for which church of seven is present in the reference circuit. Each horn would correspond to a church or "nation" as they are sometimes referred to later.

I use the first arrangement for this study, because not only are they similar: (Map the eyes on the lamb to the reference and the reference to the eyes) but the circuit of the angel is in view more so than that of Christ walking amidst the candlesticks. This I find intuitively from not only the letters, but also the fifth and sixth trumpets, and the passages on the two witnesses later. (It is better to place the Holy Spirit outside of the angel's circuit over the churches, that he then overcomes as the least. The beasts' wings should replace that missing C7 group.)

I find the first arrangement more representative of the seven cycle over the churches due to the reference. (Any church of the seven may equally be that reference cycle!) I place the wings of each seraphim in that reference cycle only; for I then place the lamb at rest in the reference and so a whole "church" (lampstand) in the seven eyes of the lamb becomes at rest in the heavenlies. Otherwise, the lamb is in each church as unity unique from every lampstand, and then Christ has no seven cycle but for floating that unity in the reference and across each church. (Yet then the reference must act on the inverse of its own cycle (as a G-set) as does the angel of the church circuiting through the churches; a clear contradiction as the angel is not Christ Himself and this circuit is unique in the reference: it is best left to the angel to fulfil that circuit, exercising only virtue as "the least of every church" under the worst circumstances, and not Christ Himself who has finished His work and walks amongst them as a static subgroup!)

The unending praise in the heavenly scene (Rev 4:8) represents the completeness of the automorphisms of the Galois group (the set generated by the Frobenius map; here C3 over

GF(8) or C2 over GF(4)) – those in complete agreement with the "life" in Jesus Christ as GF(4)*. Frobenius applied, cycles the group unendingly without inducing any structural deficiency in the Trinity at all. They are "Holy, Holy, Holy". Likewise, the automorphisms hold no static subgroup (or coset) fixed but cycle sets in triples within them, holding them static as a set. These four beasts are therefore in their triples, in continual motion amongst themselves whilst one is at rest (all four cycle in triples without "rest", i.e. the Trinity is dynamic).

"Holy, Holy, Holy" also references the field GF(2) of the one enthroned with the lamb (as GF(2) fixed) before the throne. The Trinity is dynamic in living a life of virtue: as applying "Holy" to each of the terms "which was, and is, and is to come". That there is no rest day nor night simply states that life in Christ; positivity in the Godhead is never static.

From the metaphysics, $\{p, u^{-1}, v^{-1}\}$, $\{p, r, s\}$ and $\{p, (u\&v), (r\&s)^{-1}\}$ are the three subgroups cycled in the K4 form. In triples (respectively) they represent work already done (out of scope), work in progress (in scope) and works that are consequent from the closure of the two, or with, say, q as $(r\&s)^{-1}$ in a separate case. The latter of the three triples simply and logically "follows after" as to form the closure of the group. Then I find the interpretation of "which was, and is, and is to come". That which "is to come" finds only liberty with that "third part".

Rev 4:9 And when those beasts give glory and honour and thanks to him that sat on the throne, who liveth for ever and ever,
Rev 4:10 The four and twenty elders fall down before him that sat on the throne, and worship him that liveth for ever and ever, and cast their crowns before the throne, saying,
Rev 4:11 Thou art worthy, O Lord, to receive glory and honour and power: for thou hast created all things, and for thy pleasure they are and were created.

The beasts give glory, honour and thanks to God enthroned (the statements "Holy, Holy, Holy" above). The triple of beasts (static rather than fixed as unity or the fourth beast) cycled under Frobenius are not just moved relative to the unity, but are cycled under the action of all 24 elders (automorphisms) which hold *only* GF(2) fixed (unity is fixed as one beast of four). Each of the elders represents a movement or "shuffle" of the octal. The elders defer all authority to God and the heavenly setting praises God eternally. Despite the appearance of the Field and its automorphisms, it is clearly God enthroned with power and authority, and not any other.

> **Elders:**
> *The elders represent the generalisation of the three automorphisms of all eight fields GF(8) formed from the eight C7 groups for multiplication. They are then the completed set of automorphisms of GF(8) over each fixed arrangement of seven symbols (the Sun octal), holding the "throne" (zero) and unity (the lamb) fixed.*

As the beasts cycle (giving glory, honour and thanks) they and the twenty four elders are moved out of their places (fall down, moved from rest as unity – if the rainbow is one of eight groups generalised; all are moved), but not the one enthroned. Any cycle of the K4 form (also permuting the four beasts in the octal) immediately scrambles the arrangement of the elders relative to the throne; so they are seen to "cast their crowns (closure) before the throne". If the unity in the K4 form shifts from a ≡ [a,b,c] to a ≡ [a,d,e], say, then the 24 automorphisms in the octal (which send the static triple to itself) also follow suit. This does not invalidate the automorphisms, but cycles them amongst themselves in threes, reordering them.

So (a,b,d,c,f,g,e) becomes (a,d,f,e,b,c,g) etc; and each "elder" appears to change (the crowns or closures are altered in three cycles). Every automorphism shifts the octal to a different and valid seven cycle. (The set of such seven cycles is closed. The automorphisms in sets of three are cycled.) Now the set of automorphisms comprises maps upon sets which are after the manner of an **ordered** list;

$$[(a,b,d,c,f,g,e) \equiv x, (a,d,f,e,b,c,g) \equiv x^2, (a,f,b,g,d,e,c) \equiv x^4]$$

As three elders are cycled and their position or closure (crown) now altered to become:

$$[(a,d,f,e,b,c,g) \equiv x^2, (a,f,b,g,d,e,c) \equiv x^4, (a,b,d,c,f,g,e) \equiv x]$$

This is a small difference but a correct one. So if the elders are all "sat" in some ordered list in polynomial form (as one of three powers of eight indeterminate symbols) then their positions are changed, every single one by a multiplication in the GF(4) form of the ultrafilter.

The elders always form the same set of (permuted) automorphisms, they "worship him that liveth for ever and ever". (The Father, perfect in all positive properties, even necessary existence.) They also cast their crowns before the throne – their arrangements corresponding to GF(4)* made to act upon the K4 form within the octal is subject to the multiplication of GF(4), not the other way around. Jesus has "life in Himself, even as His Father has life in Himself" and sits enthroned, being God. In heaven the elders follow after Christ, not the reverse. Truly, of Jesus Christ it is stated (as in all things) "for thy pleasure they are and were created". The elders and the four beasts reveal Jesus Christ as the mover, not they themselves.

God (Jesus Christ), now accepted as truly worthy to receive those three praises, leads John to describe the unsealing of the last three seals that are only able to be opened by Jesus Christ. Only Christ is as the Trinity complete in regress, and only in Christ can God glorify Himself.

The elders fall down as Christ acts in judgement (as cycling the K4 form) when a deed {p, r, s} performed becomes as some {p, u⁻¹, v⁻¹} and the freedom of {p, (u&v), (r&s)⁻¹} is to be considered as some future {p, r, s}. Then in that closure, {p, u⁻¹, v⁻¹} resolves to that set which in freedom is "to come" that rests on consistency upon all His past works. (There are no works of the Trinity made not of this manner! Perfection is consistently retained.)

The beasts cannot form anything than a coset under the operation of their wings, and each of the elders cannot ever be a subgroup enthroned, only agreeing amongst themselves in threes upon the equivalent judgements of God. (All fall down before Him (are lost without Him), and the authority to sit enthroned with God as one of the twenty four elders comes from none but Christ, as they are seen to cast their crowns before the throne.) Similarly, Christ is He for whom creation was made; for it to resolve in perfection about him as also gifted for His satisfaction by the Father. There is no God in view but the Father of Jesus Christ.

5.7) The Book Of Life

Rev 5:1 And I saw in the right hand of him that sat on the throne a book written within and on the backside, sealed with seven seals.

The book of life itself is placed under seven seals. That is, it is purposely closed and being written on both sides (that whilst it is visibly written on the outside) it must also therefore be full. The book of life, rather, is not "too small" but just the right size, without any empty space. God has His own choice of whom to save.

Book Of Life:
The book of life is the sacred record kept in heaven that contains the names of every elect believer. Christ does not initially know the completed number of saved, but as He begins to become dissatisfied with the dwindling state of the elect, He may open another seal to reveal more believers yet to come: i.e. until the book is completely open and the end of the age arrives with the last day.

Seals:
The seals reference a restriction of the Father upon Christ. The phrase "Time, times and a half" refers to "addition, multiplication and a static subgroup" – with the subgroup last, and as not under orbit. The Father knows what is under the seal, the Son does not. (The Father is free to work in the set of all positive properties and creation exists for the satisfaction of Christ and none other.)

GF(8) has no subfield isomorphic to GF(4). Unity is free to float amongst seven elements; it is permitted to do so. The number of octals additively identical but not multiplicatively so is increased. The four choices of unity are always taken from the coset of the static subgroup, never a member of that subgroup itself. That is, were it not for the GF(4) form of the ultrafilter entered into the scene. Only the K4 form in regress also sat at the right hand of the Father as the static subgroup can open the last three seals – as impossible choices for unity in the octal.

The K4 form consists of the three K4 subgroups (forming a static triple in the octal under the Frobenius map) that each contain the unity element. In a certain self-similar correspondence within itself, its elements are effectively aligned with unity as one member as if:

$$a \equiv 1 \equiv [0, a, b, c]$$
$$b \equiv [0, a, d, e]$$
$$c \equiv [0, a, f, g]$$

That alignment is illegal in the field GF(8). The closest that may be attained is:

$$e \equiv [0, a, b, c]$$
$$g \equiv [0, a, d, e]$$
$$c \equiv [0, a, f, g]$$
$$\text{with } a \equiv 1 \text{ and with } [0, b, d, f] \text{ and } \{c, e, g\} \text{ static}$$

Rev 5:2 And I saw a strong angel proclaiming with a loud voice, Who is worthy to open the book, and to loose the seals thereof?
Rev 5:3 And no man in heaven, nor in earth, neither under the earth, was able to open the book, neither to look thereon.
Rev 5:4 And I wept much, because no man was found worthy to open and to read the book, neither to look thereon.
Rev 5:5 And one of the elders saith unto me, Weep not: behold, the Lion of the tribe of Juda, the Root of David, hath prevailed to open the book, and to loose the seven seals thereof.

None in heaven (as unity), on earth (in the coset with the static triple) or under the earth (dead but not saved) can open the book of life. (There is no mention of those "under the altar" as yet; they are revealed by opening the book of life itself.) When a book is opened, it is either to be read or written upon. Jesus is worthy to both choose those for whom He died as well as to also receive them as given Him by His Father. (Equivalently, reading and writing names in the book is the same thing, outside of time it has the same perspective.) John wept for the reason as would any; John wept because no name could be written into the book or revealed as written inside unless it is opened, and none other than Christ already crucified and raised blameless is found worthy.

Jesus, the "Lion of the tribe of Judah" is foremost of all; "Lion" another term for my lexicon.

Lion:

*"Lion" references dominance; particularly as a **unique** subfield in orbit within a field. As the "Lion of the tribe of Judah" Jesus is the field [0, 1] in an orbit of length 7 in GF(8). ([0,1] $\cong$ GF(2), a subfield in the trinity.) Were Jesus seen whole (unique) in algebraic closure, He would be recognized as GF(4). Another use of the term "Lion" is to reference the heads on the serpents' tails at the sixth trump; these heads are in mimicry of the isomorphisms of GF(4) as present in the octal. Do not confuse that "Lion" with the one beast of four before God's throne with a "face like a lion". In that separate respect, only the appearance of a lion is indicated. However, "lion" may be a "word picture" ((a)bc)defg where ((a)bc) is as the head (a) with the mane (bc) and completed represents a lion with four feet {d,e,f,g}. The figure represents a chain of three subgroups containing the element "a". Under inclusion with subgroups in parentheses, a lion is as ((a)bc)defg.*

Rev 5:6 And I beheld, and, lo, in the midst of the throne and of the four beasts, and in the midst of the elders, stood a Lamb as it had been slain, having seven horns and seven eyes, which are the seven Spirits of God sent forth into all the earth.

The Lamb (Christ slain, now "standing" as His victory) appears as a singleton of GF(8)*. As with wings, the lamb has seven eyes and horns denoting the elements of the field GF(8) (yet not zero). These eyes and horns are the multiplicative and additive elements of the octal through which unity is free to float. In terms of the lamb this is the "rainbow circuit", the reference or origin from which I must take the operations of the reference field GF(8): i.e. its multiplication as GF(8)* acting on the reference (Sun) octal by circuiting the seven churches (lampstands).

NB: The least makes a "circuit in the horns" (the additive elements) without the Holy Spirit, but with the seven letters for guidance instead. (As "zero" he would absorb all multiplication.)

The seven Spirits are sent "into all the Earth", as the scene of heaven is not restricted to the static subgroup by the seven cycle. It is mapped through the octal as a "garment down to the foot". Every C7 group may be placed as the rainbow circuit – observed as multiplication over the "sea of glass": every octal aligned. Each C7 group then forms "the eyes on the lamb".

Those "eyes" on the Lamb are also sent into all the Earth: they become the possibilities for the unity in all 30 **octal groups** under a generalised multiplication of a C7 group. The coset of each static subgroup represents "the Earth" and the four elements of that coset may each be a candidate for unity. (But the eyes may be any member of seven as unity and static subgroup may float together whilst the filter is (then relatively) fixed; the earthly four are so mapped under multiplication to allow isomorphism or rather preserve it.)

The horns are the seven additive singletons of the reference octal GF(8)+. GF(8) and the Godhead have field structure and this includes addition. Whatever the additive state of GF(4), it has its singletons taken or inherited from the octal. That reference octal is the one through which the angel makes his circuit. The seven horns or "strengths" are seven gifts granted the angel upon his overcoming. These compare directly with the devices of the dragon (see later) which are extended across the earth and "sea". (As false doctrines from Israel's past they are overcome in turn.) The gifts ensure the angel does his small work and will not backslide.

The lamb appears as if it were slain: as Christ would, slain and left without the spirit of His Father or the Holy Ghost. (As forsaken upon the cross.) In that mode He is as GF(2) only. In truth, the intersection in algebraic closure between GF(8) and GF(4) is the largest common subfield to both which happens to be GF(2) which appears as the Lamb in the text.

From the Old Testament the prophet Zechariah stated:

Zec 13:8 And it shall come to pass, *that* in all the land, saith the LORD, two parts therein shall be cut off *and* die; but the third shall be left therein.

The **seven** seals may only be loosed by Christ; any placing of unity within the static subgroup is illegal in the field GF(8). The one able to open the book and loose its seals must be able to exist outside of the Father, able to choose unity within a field GF(4). He must Himself be necessarily existent. By opening the seals, Christ proves that He is God Himself.

Rev 5:7 And he came and took the book out of the right hand of him that sat upon the throne.

The text states the book is safe at the right hand of God, fixing the operation as that of the right hand "Sun octal" in common with the circuit of the angel.

Rev 5:8 And when he had taken the book, the four beasts and four *and* twenty elders fell down before the Lamb, having every one of them harps, and golden vials full of odours, which are the prayers of saints.
Rev 5:9 And they sung a new song, saying, Thou art worthy to take the book, and to open

the seals thereof: for thou wast slain, and hast redeemed us to God by thy blood out of every kindred, and tongue, and people, and nation;

Again, the elders and beasts fall down – but here before the lamb. In view is the expectation for Jesus to move or somewhat "float" the "lamb" as unity through the three elements of the static subgroup. In truth, Jesus must himself be as the static subgroup with the lamb floating through Him. Nothing (no-one) else will open the seals and complete the mystery of God.

This "falling down" shows that here, the sealed book of life is as a "lock" waiting to be opened. Beforehand (as by another "falling down"), Jesus was declared worthy as the "key" to turn it.

Within the static subgroup of the octal, the further (or external) placing of unity in its coset is arbitrary, as seen from that perspective. Those not dwelling solely in that subgroup derive their place in heaven in only the four beasts (and unity) and cannot "float" the unity through it. It is as if one were required to pull the rug out from under their own feet.

The scene unfolds with Christ shown as God openly, taking His rightful place in the Godhead. The "new song" surrounds the change to the automorphisms of the field GF(4) from those of GF(8). (Still under Frobenius.) The four choices of unity in the octal are alluded to as "kindred, tongue, people and nation" – the association to those four is made with the lamb offered up in place for their election by grace. (As if the four are as the unity, or are "under the unity".)

Peoples:
People (singular) refers to those united in virtue, agreeing in the lamb. They share the same unity and also seven cycle (a pre-first seal paradigm). Peoples (plural) indicates a set of divided people, having different seven cycles (a second seal paradigm).
Kindreds:
Kindreds share common "blood" and together form peoples. They share the same unity element (of virtue) and the static subgroup but not the same seven cycle. Cf. "Rev 1:7 Behold, he cometh with clouds; and every eye shall see him, and they also which pierced him: and all kindreds of the earth shall wail because of him. Even so, Amen." Regarding "kindreds", the dividing line between saved and sealed and not so elect is strictly applied. Those attaining to unity but not so in the reference seven cycle of the angel are refused the sealing afterward.
Nations:
Nations refers to all those in the seven churches, convicted by the least's circuit and following suit or not. This also refers to those under the first woe. (A second seal paradigm is a given.)
Tongues:
Tongues references those with different seven cycles met in dialogue, concerning different choices of unity, different virtues. A third seal paradigm at least; which also refers to those with "wormwood" and under the second woe.

The presence of the prayers as vials and the incense with which they are offered indicate they are acceptable and welcome offerings before God, without need of great addition of incense. (Unlike later when much incense is required to mask the distaste of prayers for vengeance.) Likewise, the pleasant sounds of harps accompany the new song. With the shift from three automorphisms of GF(8) to just the two automorphisms over GF(4), the "new song" is of these harps and the prayers. They are an acknowledgment that even though two parts in Christ do not appear in the octal, He is worthy of all worship.

The four beasts likewise "fall down before the lamb" and are positioned rearranged as also aware of the two automorphisms as if behind the veil.

The two items of vials and harps are "pleasing things" to God: they are as the two parts cut off in GF(4) which are representative of Christ "In whom the Father is well pleased".

The four beasts together form the elements {d,e,f,g} of the K4 groups which are transposed as (b,c) in pairs under the Frobenius map acting on GF(4) and the K4 form of:

$$a \equiv 1 \equiv [0, a, b, c]$$
$$b \equiv [0, a, d, e]$$
$$c \equiv [0, a, f, g]$$

The 24 elders satisfy the eight choices for multiplication holding to one static subgroup (there are then two C7 groups for each choice of unity in the octal) as well as the three choices for "floating" unity in the K4 form (all being automorphisms upon the octal). Separately, the elder's harps I could align in the above GF(4) form to "b" and the prayers in vials to "c". These two parts "cut off" in GF(8) are still present in the narrative through these harps and prayers. (This then leaves us with eight fields preserving the K4 form.)

These prayers are made to the one and only Christ and are wholly accepted: unlike those later offered with much incense – as made to another christ – an "antichrist". Those prayers are rejected and returned unanswered to those with that antichrist paradigm (of the static triple) within which all those refused election dwell. Those prayers are cast to that triple in "the Earth"; the dialectic does not please God.

In the K4 form above, [a,b,c] is self-similar as the unity to the whole. The harps and vials align to the elements {b,c} and indicate those two elements as together completing "unity" in the K4 form, as they are instead "cast to the lamb"; in effect they are worship (praise) of the lamb.

Rev 5:10 And hast made us unto our God kings and priests: and we shall reign on the earth.

The elders' places are consequently found from the multiplication and automorphism upon the one heavenly octal scene. They are become indistinguishable from the appearance of the Godhead up to a point: they only have their place from the Godhead and not the reverse. As the 24 elders are all subject to the three choices of unity in the K4 form of the ultrafilter, they each likewise share a "three cycle" that clothes them with the judgement of Christ as well as to be found closed (crowned) in the eight multiplicative groups of the octal. They are made of the spirit, placed under Christ and subject to Him – priests to serve Him, because He as God is unmovable by all.

The elders and those present in the scene together (as now also justified by Christ) are similarly formed as to the octal and God. The elders outside that singleton "lamb" (which is to say, the *unity* element fixed under Frobenius) eternally dwell in heavenly places with God, now further generalised by floating the Lamb to the other six positions for unity. Those upon the Earth are to also become priests to minister before Him and to rule in judgement with Him.

Rev 5:11 And I beheld, and I heard the voice of many angels round about the throne and the beasts and the elders: and the number of them was ten thousand times ten thousand, and thousands of thousands;

The many angels (ministers) work to extricate the saved from the Laodicean amalgam. For every state of spiritual captivity (as under conviction from the least's exit), there is an "angel" or "minister" that may intercede. There is no logical or "necessary" captivity in the Revelation as yet! (Only the "second death" and the "lake of fire" **after** the judgement.)

Now, $(2*5)^8$=100,000,000 and $(2*3)^8$=1,679,616. Essentially, an 8-fold cartesian product (for each of the 8 seven cycles over every octal) made in a "five-cycle" or C5 group (to be closed by the Lord's judgement) and also a three cycle, a C3 group. (There is a factor of two as there are co-dependent elements to the complement hand octal, determined by those five and three.)

Each product (as a mapping) represents the exit from "great tribulation" to rest. (Cf. Rev 6:13, 7:13.) The "five cycle" indicates the five wise virgins of the parable overcoming and exiting Laodicea (with unity, finding rest) to Philadelphia in both hands of the Sun octal. The three cycle represents those who exit the "closed" static triple/subgroup of the left and right hands, finding rest (unity) in election under the covering of the lamb. Both groups in product are of prime order and do not intersect in the unity of the octal (the five cycle is found in a different field, GF(16), which cannot intersect with GF(8) other than in [0,1]). GF(16) may contain unity, but it is certainly not within the static triples. John saw those souls having overcome "great tribulation". They do not "tread the outer court underfoot".

There is no issue concerning the worthiness of Jesus to open His own book of life. All may find place in heaven yet through His name alone. The elders may be generalised to the further six states (6*24=144) of all dwelling in heavenly places as do those elders, subject to God and His Christ: the difference is symmetry between those in the earthly and heavenly spheres.

Rev 5:12 Saying with a loud voice, Worthy is the Lamb that was slain to receive power, and riches, and wisdom, and strength, and honour, and glory, and blessing.
Rev 5:13 And every creature which is in heaven, and on the earth, and under the earth, and such as are in the sea, and all that are in them, heard I saying, Blessing, and honour, and glory, and power, be unto him that sitteth upon the throne, and unto the Lamb for ever and ever.
Rev 5:14 And the four beasts said, Amen. And the four and twenty elders fell down and worshipped him that liveth for ever and ever.

Rev 5:14 shows the lamb may be moved through every position in the octal by Christ opening the book (floating unity within the static subgroup). With Rev 5:8, the octal is a "lock" and Christ here is "the key" turning it. The "mystery of God" is still incomplete: the lamb is certainly

God's same virtue. "He that liveth forever and ever" is worthy of all praise, worship and power. (Until the book is opened the mystery of God is incomplete.) Jesus indeed opens the seals and the mystery must also be completed. "With what?" I hear: the title of the last book in the New Testament indicates Jesus completes that mystery. Jesus is revealed as God.

Christ is worthy to receive glory sevenfold (Rev 5:12) but He is only given the glory of four of them. (Rev 5:13) He has three already! (Alluding to Christ sat at the right hand – cf. Rev 4:11).

The "Earth" is the coset of the static subgroup. The "sea" consists of all thirty octals and "heaven" every position of the floating unity (rest) in the model. "Under the Earth" is a peculiar term that is also mirrored in "under the altar". Both terms are reserved for the dead awaiting judgement. The former being the dead not in Christ, those in Christ referenced by the latter.

Heaven:
Heaven is as the unity element (rest) in the reference (Sun) octal. Dwelling in heaven is to also be clothed in white, under the covering of the lamb (unity). This blessed state is generalised from the 24 elders, floating unity to the reference octal's six other positions, to form 144 further automorphisms. A factor of 1000 (years, cycles of GF(8)*) is given; those 144,000 states of salvation is "heaven".

Under The Earth:
Those under the Earth are here found "dead", dwelling in one of the two parts cut off in GF(4) within the Field GF(8). They are found to the left of Christ in {c,e,g}, say, the static triple or "bow" within the four earthly elements.)

Under The Altar:
Those dead and yet preserved in Christ to His right in the right handed triple [b,d,f] – the static subgroup in the right hand octal.

Each creature is referred to in sequence: those dwelling as unity in the octal are elect (and rest in heaven), separate from those dwelling on the four earthly elements (i.e. on the Earth), whilst those under the Earth (those "dead" in the two parts cut off) are as in "b and c" here:

$$a \equiv 1 \equiv [0, a, b, c]$$
$$b \equiv [0, a, d, e]$$
$$c \equiv [0, a, f, g]$$

Being the "physically dead" preserved in Christ (all are baptised into Christ: John baptised Christ to "fulfil all righteousness". Some to the Holy Spirit, others to hellfire). Those as are in the sea, are present in every other octal having no calling to be obedient to God and His Christ. It is incredible that these should acknowledge God and Christ. There is not one living soul able to supplant Christ's right to be glorified in heaven. No one else is made of the same stuff.

Christ alone may open the book – He alone laid down His life and picked it up again. He as GF(4) has multiplication (life) in Himself, outside of the operations of GF(8). He may open the book using His K4 form. It is not true that the octal generates the life in Christ as GF(4)* = Gal(GF(8)). Rather it resolves in complete agreement with Him.

The reference "Sun octal" under a valid seven cycle (e.g. x = (a,b,d,c,f,g,e)) keeps a rule that given unity fixed (say a = 1), Frobenius three-cycles two triples static: the first, second and fourth elements after unity in the cycle, as well as cycling the remaining three. ([0,b,d,f] and {c,e,g} are then static with x above.) All 30 octals have eight valid multiplicative groups (and also keep the rule). Those C7 groups each hold only one of four subgroups/triples in pairs static (with the same unity fixed), cycled under the "Galois group" Gal(GF(8)) $\cong$ C3 of GF(8)'s automorphisms. Those automorphisms, present as the twenty four elders, do not rule over God: they cannot alter Him or move Him from rest (that is, the same Jesus Christ).

Nothing in heaven or on the Earth, or under the Earth, or in the sea can remove Christ from His place by instead opening the book of life. All others, "every creature", are fearfully and wonderfully made: but not as God. Christ has a distinct place in the Godhead and "sat at the right hand" as a static subgroup, has a reign believers will never have (except over themselves), yet the right to sit on the throne (as [0] or [0,1]) is promised to the least under Christ.

Men:

The "number of a man" is either unity or 666. "Men" are as K4 groups or static triples (within their cosets); men decide upon or entirely reject virtue. A "man" would employ the dialectic, logically remaining in (restricted to) the coset of a K4 subgroup (a coset which contains unity and a triple cycled static), and far more: potentially perceiving the octal complete, along with all necessity of virtue. Will men overcome the world? Many will not. Note only Christ decides election as a static subgroup in the octal with unity. That regress has pretenders, but His alone is that elective choice.

Dwelling, To Dwell:

To "dwell on the Earth" is not to realise intercession is made in heaven, not yet gaining a paradigm similar to that of Christ. (Not yet "born again".) To "dwell in heaven" is to receive the grace of God, to perform the works of Christ. Even Satan "dwells" as in "slain among you, where Satan dwelleth" (Rev 2:13). Satan has had a long wait to ensnare the least in God's kingdom.

Across the divide, Satan's diabolical work cannot progress with God undefeated (in one case to be further seized upon). That effort is to devour the least in God's kingdom (yet already elect). If Satan devoured a greater rather than a lesser, would there not be greater effect?

Satan has a-priori lost: Satan, by "dwelling" upon this one attempt to pick off the very least **in** Christ (that he could remain without salvation, deceived and to be accused); yet loses. As the least, his absence would affect no elect but himself, as within the grace he has and shares with all the elect, he alone **fully** accounts for himself from the throne of God. (With his rewards predestined, and overcoming upon them.) So, God is the victor, justifying **all** of His own interventions over His flock as from the cross. Satan gets less than a penny.

The book of life stays sealed, Satan unable to open it (not scratching out even one name). The minimum of surety and the full power of grace shown God's least elect is equally multiplied to all sealed within the Lamb's book of life.

Chapter Six: The Seven Seals

The seals describe a loosening of restraint on those under the Holy Spirit (i.e. of their paradigm). God judges what is positive for His people. In finding agreement with God's absolutes (virtues) preserving the same disjunctions, one is obedient: whether or not each ultrafilter of seven (the octal) is maximal (miraculous or omnipotent) or minimal as some small act of charity.

To agree with the octal's schema of action is to agree with the Holy Spirit rather than the logic. Christ stated "Joh 14:12 Verily, verily, I say unto you, He that believeth on me, the works that I do shall he do also; and greater *works* than these shall he do; because I go unto my Father", exercising virtue with and upon the works of Christ is to extend that virtue further for a "greater work", ministering with the Holy Spirit. (Christ would do yet greater works if He were not already gone to His Father.)

To share the same disjunctions (the same octal) is to be obedient, but only up to one of eight C7 groups for multiplication. Instead true "agreement" is complete consistency: a "pre first seal paradigm". The "nations" of the seven churches are indeed Christians, but to agree with the Holy Spirit as to what is "rest" and where to find it (and in what acts it is to be found) is to be "perfect". (Unique up to the choice of unity and static subgroup.) Virtue must be totally consistent (upon itself in conjunction) and only God may decide the consistent and necessary; everything else is freely decidable as for all. Isomorphism is complete agreement and also to be made with the schema; the octal and seven cycle are then determined by God alone.

Changing seven cycles does not alter the schema, but alters the placing of the sets within it. God decides all virtue of faith; human choices are always imperfectly made. (Unless unity is chosen in every such seven cycle, but then the octal with it is perfectly supplied only by God! Sabbath "rest" is found in the unity of the multiplication made, and only in conjunction upon works in a schema of God's choosing, the virtue as found in the Lamb before the throne.)

Jer 10:23 O LORD, I know that the way of man *is* not in himself: *it is* not in man that walketh to direct his steps.

There is a great difference between the heavenly scene and the earthly sphere. In the heavens unity may be any of the octal's seven non-zero elements. A seven cycle (a,b,d,c,f,g,e) may have its first entry the unity, and reordered to (d,c,f,g,e,a,b) it could be assumed equally that "d" is unity. There is no structural change to the octal, only the multiplicative part of the field GF(8); the seals are a separate case. Floating unity from "a" to "d" in the cycle above changed the subgroup held static under the Frobenius map. With a=1 using (a,b,d,c,f,g,e) perfection may hold [b,d,f] static and with d=1, [c,e,f] is then static. This is not the case with the seals opened to identify Jesus as the lamb victorious.

Christ is unique as the only begotten. By opening the seals He proves that He, as God, is truly raised to life from death (or equivalently non-existence). His sacrifice made on the cross is shown genuine and true in the Revelation. Christ is also shown necessarily existent, being God. Revelation testifies to this by His uniqueness as well as the consequences (and consistency) of that sacrifice of Christ on the cross, being the same gospel, that of the one true God.

Opening seals sequentially shifts unity in the octal whilst the static subgroup is constant. A looser paradigm then exists, shifting from one group of seven cycles to another! The first four seals require four C7 groups, and for the last three there is no group over the octal permitting unity within the static subgroup. Jesus alone has the "keys of death and hell" and can achieve this: for death is existence outside of God (who is life) and hell is to be rejected as desolate by God. Jesus exists as spirit apart from God the Father and He also overcame all accusation of fakery to be completely accepted raised before His Father.

The octal shifts seven cycle to any one of four pairs with the opening of the first seal. Given the choice of these pairs holding the subgroup [b,d,f] static, there are four possibilities for unity (the first element appearing in each cycle, one of {a,e,c,g}):

$$(a,b,d,c,f,g,e),(a,b,f,c,d,e,g)$$
$$(e,b,d,g,f,c,a),(e,b,f,g,d,a,c)$$
$$(c,b,d,a,f,e,g),(c,b,f,a,d,g,e)$$
$$(g,b,d,e,f,a,c),(g,b,f,e,d,c,a)$$

For the first of the seals opened, the seal loosed allows freedom to cross to either group in each row of the four (holding only one true in the octal's schema at any time). The second seal permits two rows to exist at the same time, the third three, the fourth four and so forth. (Or not so forth, because there is no valid seven cycle that permits [b,d,f] static with b=1, say.)

The floating unity allowing subgroups to "walk the seven cycle" (as Jesus' own feet walking through the seven lampstands) is a "pre-first seal" paradigm. It holds to just one cycle or further generalised, each cycle isomorphic. (The generalisation of the angel's circuit allows the traversing of every C7 group without lapsing into a broader paradigm, of even the first seal.)

At the *time of the end* the kingdom of God **will** be likened to ten virgins; to hold a pre-first seal paradigm in five "churches" (and not in the reference circuit of the angel) is to be foolish, lacking the oil (doctrine) to preserve the light (of virtue) required to greet Christ at His return. These "virgins" (always a good spiritual state) were foolish. Foolish not because they were without doctrine, but that they left themselves no option but to "buy" more doctrine from those not invited to the marriage (that buy and sell) and far too late to find any worth using.

That there is not enough oil to share is a given: to be lukewarm is worse than to be cold or hot. Sound doctrine and God's approval do not mix with the faulty. The true must remain unleavened by the false. It is dreadful to return to doctrine that has now become unpalatable – as alike to the proverbial dog: revisiting those five churches (rather sets of false doctrines) is to cast one's lot in with the faulty. Accepting such faults wastes away the wisdom (the good doctrine one would hold) and is as bad as going back oneself.

6.1) The First Seal

Each of the seven seals is opened in turn by Christ;

Rev 6:1 And I saw when the Lamb opened one of the seals, and I heard, as it were the noise of thunder, one of the four beasts saying, Come and see.
Rev 6:2 And I saw, and behold a white horse: and he that sat on him had a bow; and a crown

was given unto him: and he went forth conquering, and to conquer.

Each seal has its effect: the four beasts representing the "earthly elements" of the octal (the coset of the static subgroup) each state "come and see"; the release of restraint upon the people of God is unveiled. Their paradigm remains obedient; the tribulation under grace becomes more subtle and less violent, but grace is no less selective.

The seals reveal to Christ the extent to which the faithful remnant diminishes in size (never quality) and the effect on them from those on the Earth (that sin) without restraint: Christ must be satisfied even though creation would at the end become unsatisfying for Him. (Cf. Rev 4:1 again.) As the paradigm/restraint is loosed the Church undergoes a great falling away; when the gospel ceases to deliver any satisfaction, it is time to open a seal on the book.

"Seals" in the octal are possibilities for the unity element, of necessity requiring a change of C7 group for multiplication. The first seal corresponds to the unity element in the field GF(8), but chosen amongst four elements by a shift in multiplication. The first beast with only "one limb off the ground" represents this singular unity – I will deduce why! Now that unity is free to more than merely float there is a new "movement" in heaven, hence the "sound of thunder".

Seven Seals:
The seven seals are positions in the Sun octal within which unity is free to be set given that the static subgroup is fixed. To open them, one must display the ability to "move God" whilst under all restraint.

Thunders:
A movement of heaven, whether in the scriptural narrative or in time.

Horse:
The "horse" is a word picture for a K4 subgroup. The group's addition table in every row (and column) contains one element zero. The image is like to a "horse" walking with one hoof off the ground. (John saw as if "zero" were empty or missing; he understood rather than saw a table.)

+	0	a	b	c
0	0	a	b	c
a	a	0	c	b
b	b	c	0	a
c	c	b	a	0

He that "sat on him" refers to the unity, the identity of the rider is as the unity sitting, "resting" on the horse to generate the coset, with the static triple in the remaining elements.

Given the reference "Sun octal" the seven cycle (a,b,d,c,f,g,e) holds [b,d,f] static with a=1 and also {c,e,g} likewise. That virtue in "a" is to be denied and privated by judging the positive.

The faulty paradigm of antichrist aesthetically judges $Pos(a) \vee \neg Pos(a) \rightarrow Pos(c\&e\&g)$. The positive virtue of "a" is judged to be negative in the dialectic by opposing it with $Pos(c\&e\&g)$.

> **Bow:**
> *The "bow" given to the rider is the induced 3-cycle upon the triple {c,e,g}. Now, "bow" translates as "bow, as of the simplest fabric", and is the Greek "toxon" from which English gets the word "toxic" (as arrows were dipped in poison). The word not only references the bow, but is usually meant to indicate arrows as well.*

In view then is the following word picture (including the arrows too!):

Diagram showing the three-cycle is "bow shaped",
including the arrows.

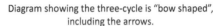
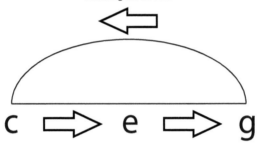

The rider is given a crown (denoting an additive subgroup). The rider implicitly generates a left handed octal acted upon by the Sun octal's seven cycle. That cycle generates seven triples from the "bow" – an orbit of that triple held static (as indicated by white) under Frobenius. These triples implicitly obey "K4-addition" (are crowned) and also the octal operation A + B = (A ∨ B)ᶜ upon them. The static triple is now the left hand static "subgroup", under orbit generating the **"left-handed"** octal. The crown also represents the strength (additive closure) of holding to unity (virtue). (In the octal whole, the positive remains so across the disjunction.)

To hold firmly to the unity (as fixed) common to every C7 group is to have faith in Christ and His virtue. In view is not fixed faith: but faith standing firm against the loosing of further restraint upon that paradigm in "the Earth" (the coset, or "left hand"). The unity is found from four different multiplications – over four fields allowing unity to more than simply float. (There are four further groups for multiplication besides these which are allowed for later.)

Christ's victory already stands. Antichrist – that position taken in dialectic logic with the triple {c,e,g} – is now immediately loosed. It is as if the whole sea of 30 octals were opened and under every seven cycle, the antichrist triple or "bow" released. In essence the rider "goes out conquering and to conquer" as under orbit by the seven cycle: the left-handed operation conquering (filling) the Earth by acting upon the "bow" in every set of four "earthly" elements.

Antichrist goes out conquering and to (and will) conquer, as it is without opposition in the Earth. To the mindset found judging Pos(a) ∨ ¬Pos(a), there is no antidote in any virtue.

The offering of the lamb and the heavenly scene is unmoved (as the "sea of glass" of every octal and field GF(8) aligned as one). The voice "as it were the noise of thunder" (one of the four beasts) states, "come and see". The four beasts are in continual movement, mapping between their every coset under every multiplication, as if it were mapped to the reference. Restraint remains: the four beasts sequentially loosen that restraint. This is why each beast has one face and not as in Ezekiel's vision of the cherubim where each had all four faces (no restraint given).

With that one reference, all "earthly" cosets in every octal, under every possible multiplication (and unity loosed) have their relative effect: this is why the beast's speech is accompanied with a "noise of thunder". Heaven is in motion relative to the four loosed elements of the Earth.

6.2) The Second Seal

John immediately moves on to his next set of word pictures.

Rev 6:3 And when he had opened the second seal, I heard the second beast say, Come and see.
Rev 6:4 And there went out another horse *that was* red: and *power* was given to him that sat thereon to take peace from the earth, and that they should kill one another: and there was given unto him a great sword.

A second choice of unity (with the static subgroup kept constant) alongside the first requires a second C7 group. The rider (unity) must shift position in the coset of the static subgroup. The second seal open entails the presence of two different static triples, each corresponding to the two choices for unity (and both are now free to float under multiplication together).

I posit 1 ≡ a ≡ {c,e,g} and 1 ≡ e ≡ {a,c,g}. With these inferences or "congruences" I may substitute one into the other and obtain:

$$a \equiv \{c,e,g\} \equiv \{c,\{a,c,g\},g\}$$
$$\{a,c,c,g,g\} \equiv a$$

To retain the primacy of unity ("a" here) the entries of "twice c" and "twice g" must cancel; as if order two. With the octal operation, I already have that sense of "modulo two". The system has effectively "leached" those properties from the octal.

Likewise, the antichrist triples clash with the believers caught in the crossfire.

Each position, whether a=1 or e=1 is led to cancel the other out by cross-substitution, in order to preserve their position. It is as if they "should kill one another" (addition modulo two on the earthly elements). Holding to unity against the broader paradigm of antichrist permits divisions between the faithful. (As each surely oppose antichrist effectively, they also then oppose each other!) There remains "virtue" for both, never reaching consensus in the coset.

In "taking peace from the Earth" there are now two conflicting paradigms in the Earth: a fixed position with one antichrist "bow", and another position with two "bows". The effect of this loosed seal in the beast with a face like a calf follows. (Peace between believers is removed from the coset: unity opposes unity.) The "great sword" may equivocate to the cancelling out of "c,c,g,g" to the empty set { } here (rather than unity). It is as if they were "slain";

$$a = \{a, \cancel{c,c,g,g}\}$$

The "sword" given to "a" strikes out the other
elements by cross substitution.

The "great sword given" is that **all** upon the
earth should likewise "fall upon the other".

It far more likely describes the second seal paradigm. The text states "...and there was given unto him a great sword". This sword is properly the Sun octal itself. The first seal open, the rider rode out without opposition in the Earth and as far as it went, things were plain sailing. This second "red" horse presents only a paradigm of conflict, strife and discord with two possible unities loosed, and the corresponding static triples loosed with them.

The second horse rides out to a vista of war and conflict; because it is good for nothing else; this paradigm lays the world to be **put to the sword** in the name of God. It is in that sense that the rider is given a "great sword": the mind-set to freely conquer with the paradigm of violence or dispute as if it were "for the gospel" and to think oneself justified in doing so. In declaring God to be "on one's side" a believer gives antichrist a "great sword". It is the believer in the crossfire preserved by giving their approval to the warring of antichrist that gives antichrist that "sword". Both warring sides may dialectically justify that same position.

There is the clash of two different left-hand octals. The first seal paradigm had no "clash"; the rider went out without opposition but for the elect that did not shift from the virtue to which they held – they stood their ground and were immobile, not moving out of their place.

The seals do not shift across octals in the sea but are present within every octal closed (only the seven cycle changed). With the second paradigm C7 groups were made to clash for the first time, yet as seven cycles over **the same** octal. This paradigm "leaches" properties from the two fields rather than the first which kept to the field operations. (The colour red somewhat implies blood, an allusion to the intersection here due to sharing the same right hand octal or presented by the cross-substitution.)

I begin to see the restraint loosed on those dwelling in the Sun octal's earthly elements as the restrainer is "taken out of the way". The loosing of the seal yields a paradigm in the Earth free to oppose the gospel in yet another fashion. Two opposed fixed positions (unity) become a contentious argument, even violent. Deciding Christ vs. antichrist in the first seal is become a somewhat clouded issue. A mystery to some!

6.3) The Third Seal

The third seal again is rich in word pictures.

Rev 6:5 And when he had opened the third seal, I heard the third beast say, Come and see. And I beheld, and lo a black horse; and he that sat on him had a pair of balances in his hand.

Rev 6:6 And I heard a voice in the midst of the four beasts say, A measure of wheat for a penny, and three measures of barley for a penny; and *see* thou hurt not the oil and the wine.

There are now three positions of unity to reckon with; so I introduce $a \equiv 1 \equiv \{c,e,g\}$, $c \equiv 1 \equiv \{a,e,g\}$ and lastly $e \equiv 1 \equiv \{a,c,g\}$.

A triple formed by these triples is itself a "bow", the missing "$g \equiv \{a,c,e\}$". Choosing the unity of the octal to suit, I essentially form a pair of "balances" with two pans to each side and the equilibrium in the centre. Placing $\{c,e,g\}$ in the centre as the scale, I would fix a=1.

The balances in "the hand" represent not just agreement or the role of the facilitator, but the act of making writ. The facilitator (the "rider") is found lightening each side (the positions in dialogue) for parties that will do the "signing" (place their mark) as they agree to compromise.

Compare this with the seven stars in Christ's right hand – a promise of Christ to write His new name on His angel – completely justifying him as His right hand (the least in the kingdom). In this there is never compromise, but discipline that carries over all as if it were a "rod of iron".

This third seal paradigm generates a position of "synthesis" in logical disjunctions by assuming the set of positive properties is closed in only the four earthly elements $\{a,c,e,g\}$, resulting in disjunctions of the form $a \vee \{c,e,g\}$ etc. The logic appears to be one of equilibrium.

As a picture I present:

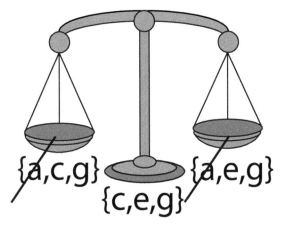

Holding to a=1, the balances $\{a,c,g\} \rightarrow \{c,e,g\} \leftarrow \{a,e,g\}$ may have their two "pans" to both sides "lightened" by removing the unity (as "a"), absorbing it by leaching its multiplicative identity (and I thereby find the "balances"). The paradigm is of conflict resolution. It is the common idea (absolute) on which there is no consensus that is removed. When points of contention are removed – there is agreement to disagree. Every predicate is evaluated "good" or "evil" in the dialectic. However, God is consistent, for $Pos(r) \vee Pos(s)$ can not permit $Pos(s^{-1})$.

This is the dialectic process and it is spiritually deadly. If the process itself is employed (even for evaluating itself), finding consensus within it will form the "wine" (the element $g \equiv \{a,c,e\}$, i.e. the missing triple) as agreeable to everyone at the table, of the three parties in dialogue.

> **Oil and Wine:**
> *"Oil" is doctrine, a fixed absolute. "Wine" an (incentive for) consensus as held at "every place at the table". The process itself fulfils this requirement, making men drunk (and unable to direct their steps).*

A "measure of wheat for a penny" (Rev 6:6) references the reward given the Lord's workers in the parable of the harvest. A penny here is the unity element (sabbath rest), here one of a, c, e or g of the octal. (Unity is the reward of holding to the truth of the gospel.) In this particular case of the dialectic it cannot be g, which is here reserved for "wine".

One measure of wheat is merely a singleton from the right hand or "Sun octal". Unity is free to be chosen from the four earthly elements; each may be in "balance" with a penny (as "a=1" above). Also in balance with unity is the static triple of the coset {c,e,g}.

Three such balances may be formed at this point; each triple may be placed in the centre and the scales lightened accordingly by removing the congruent unity.

The triple {c,e,g} (a subgroup in the **left hand**) is of three measures of "barley" also equal with that penny. (The symbol for unity "1" somewhat resembles a ripened ear of wheat.) Each measure of wheat is now in balance with three measures of barley, "a ≡ {c,e,g}" (both sides are worth a penny). Why use these terms? There is no such unity in the dialectic.

This paradigm shows the reward of a penny instead being paid with a reward of three "tares" (election no more). The dialectic practitioner makes claims of great things, in particular growth.

Wheat represents the elect believer and barley is "animal feed" (or hops for brewing, fermenting with leaven). Then to be as barley is to be ready for "leavening with poor doctrine". (There is "spiritual and doctrinal famine" as tares replace wheat.)

A voice (judgement) instructs "not to hurt the oil and the wine". The oil (doctrine) "absorbed" is the unity lightened from the balances and the "wine" is the singleton in common to all three triples (above it was "g", it is never to be taken for doctrine, it is no absolute.)

Not hurting the oil and the wine, believers are instructed not to confuse the unity element with the common element, as they are not to introduce the remaining triple g ≡ {a,c,e}. If this were the case they would have the fourth seal system! They are not to substitute one for the other. (Neither should the believer private the positive with Pos(r) ∨ Pos(r⁻¹) → Pos(s).)

Then, "not hurting" commands that the oil and wine are not to be confused with nullity. There is no octal with an empty set (as yet). In "hurting", I may model a dialectic logic of disjunction and conjunction; ignoring virtue to introduce a paradigm with which it is possible to "agree to disagree". (I.e. the dialectic decides Pos(r) ∨ Pos(r⁻¹), but the octal preserves the positive.)

"Oil" is already "hurt" (sent to nullity). Unity is to be lightened from both "pans" and must then be within the triples to either side: yet whilst static they cannot contain the unity!

The limited disjunction a ∨ c&e&g holds in the dialectic. Any virtue ("a") is immediately sent to nullity, the empty set { }. (Virtue, by axiom, cannot private any positive property, so becomes inconsistent to the dialogue, its "back broken".) In the dialectic, the converse N¬Pos(a) must hold. Then c&e&g appears positive instead, which is the resulting consensus, having excluded any position holding to that virtue (or of other absolutes). Tares oppose wheat with virtue.

Holding firmly to N(Pos(a)) I have in the worldly and improper ultrafilter of the earthly four elements the complement not in that filter (dropping the Pos operator for ease):

$$N¬(c\&e\&g) \leftrightarrow N¬(c\&¬e^{-1}\&¬g^{-1}) \leftrightarrow N¬(c\&¬(e^{-1}\&g^{-1})) \text{ or } c \rightarrow e^{-1}\&g^{-1}.$$

I.e. $c \rightarrow (e\&g)^{-1}$ given "a".

Likewise, given "a": $e\&g \rightarrow ¬c$ proves the disjunction $a \lor c$ (yet e&g is **unworkable** in the octal).

I could state given "c"; $(a\&e \lor a\&g) \rightarrow (e \lor g)$ if $c \rightarrow a^{-1}$ (i.e. "a" is a virtue privated).

If e&g is judged to be an "evil" because it raises dissent as in the disjunction $a \lor c$, then to resolve conflict, it can be removed from all consideration as $(e\&g)^{-1}$ (which, now evaluated positive proves the position of a&c to be positive; for such a negative cannot entail a positive).

Then a&c entails $(e\&g)^{-1}$ and also proves $e \lor g$ but if "a" and then also N¬(c&e&g) are axiomatically positive and e&g an "evil", then $e\&g \rightarrow ¬c$ cannot be positive ($a \leftrightarrow ¬c$ appears inconsistent and not positive given $a \lor c$: then N(Pos(c&e&g)) as synthesis instead). Pos(a) holds only if there is a positive middle Pos(e&g) in $a \lor c$. Then $a \rightarrow c^{-1}$ is required to be positive.

If Pos(a) is upheld positive by axiom, N¬(c&e&g) holds as if equally positive (in the dialectic). There is the appearance of a freely decidable disjunction in Pos(a) $\lor$ Pos(c). If this must remain freely decidable, N(Pos(e&g)) is required and so there is no such freedom found in Pos(a&c) proving Pos(e) $\lor$ Pos(g). The "good" a&c becomes just as **unworkable** as the "evil" e&g.

Now given the disjunction Pos(e) $\lor$ Pos(g), I note a&c proves that disjunction by entailing the middle of ¬Pos(g) $\lor$ ¬Pos(e), which, dialectically, is equal to Pos(a&c&e) $\lor$ Pos(a&c&g).

Then given Pos(a) and N¬(c&e&g), I have $a\&(c\&e \rightarrow g^{-1})$ and $a\&(c\&g \rightarrow e^{-1})$ now also divisive, both contending as Pos(a&c&e) $\lor$ Pos(a&c&g) above. Then, if ¬Pos(c&e) and ¬Pos(c&g) hold, being divisive (evil) then N¬(Pos(a)) and so N(Pos(c&e&g)) is reached as synthesis in the centre.

It may be stated that the wine "c" is the only positive and that $a\&(c \rightarrow (e\&g)^{-1})$ and the disjunction is "freely" decided (resulting in synthesis, with virtue privated).

This shows that there is room for spin between the properties {a,c} and {e,g}. Each pair appears to back the other up if one appears "evil" and the other "good". The "evil" as e&g would not entail Pos(e&g) $\rightarrow$ Pos((a&c)^{-1}), instead only a&c becomes positive. What is "evil" in the process is to oppose the method with $a \lor c$, or the current position reaching consensus in a&c.

Given those opposing triples a&c&e and a&c&g, the facilitator has the choice to spin his dialectic synthesis in the centre e&g as proving $a \lor c$. Then, N¬(a&c) and the facilitator makes a break for it with the improper logic of $a \rightarrow c^{-1}$ or $c \rightarrow a^{-1}$. Now if "a" is virtue then a^{-1} is void of all positivity, effectively become the empty set. (**The "oil" and "wine" hurt each other if the method is used**. The "spin" is evil, Christ instructs us not to practise the method.)

When the error of Pos(a) $\to$ { } is found out (for unity is weighed rather than accepted as immutable fact or an event), the arbitrary choice of a fourth triple (equivalent to the "wine") is in contradiction with the oil (unity) in this paradigm; this falsity forces the facilitator to go full steam reverse and "spin" his way out by "hurting the oil and the wine" (transposing them, showing that each may be sent to nullity). Yet virtue will never private a positive property!

This transpose of oil and wine is not the goal of the believer, as the believer can go Mach speed ahead and repeat the exercise to state that what was oil is now wine and the opposing triple (consensus) is also logically incompatible. (I do not transpose truths here; the dialectic is faulty logic!) The triples are not opposites of the singletons, neither is the logic used by the facilitator to spin his way out contrapositive: but those objections can be made.

The assumption that the oil may be "absorbed" is equal with that which assumes the wine should be removed also (by identifying this logical error, one will not further "hurt" either oil or wine). With nothing to hold in common in dialogue (as consensus) without the method, the only recourse is to abandon the meeting. After any spin is applied, the same logical deduction may be applied in like and symmetric manner and the dialectic will fall apart. This, however, is a rare gift of the Holy Spirit and I haven't heard or seen it done in years. (Hallelujah if you have it!)

I find I have modelled the Hegelian dialectic; the requirement of consensus is to remove any contentious issue: any absolute held (unity, the oil) but in different regard. To assume either as empty { } resolves conflict.

> **Hegelian Dialectic:**
> *The dialectic attributed to Hegel is far older – Satan practised it upon Eve in the garden, and it is formulated as "thesis meets antithesis, results in synthesis".*
> *The dialectic process occurs in facilitated meetings: every point of contention, all "doctrine" that is fixed but held in different regard must be removed from the dialogue by the facilitator to ensure the consensus "continues" on directed lines.*

The method of the balances "D" (D for dialectic) privates virtue ("a" or unity) of its positivity. I.e. $D \to Pos(a^{-1})$. Then $Pos(a) \to D^{-1}$ and D^{-1} must be virtue (the set of virtues is closed). D is never positive as it always privates virtue. If D is inconsistent given "a", D is a process so is systemically inconsistent given any property positive by axiom: D privates virtue. (The colour "black" indicates a total lack of light (virtue). NB: Oil (good doctrine) + fire (Holy Spirit) = light.)

Then D is an "evil". D allows the rejecting of sound doctrine (all virtue in Christ). If virtue at rest is become positive and should not be exercised then there must be a positive result in its stead. Equally, that is virtue by axiom of virtue and should never be D; yet a near positive result is found in the static triple in the centre of the balances. (Synthesis becomes a new thesis.) "c", "e" and "g" are all positive properties but together they private virtue by the breaking of the "third part" of the octal's addition.

The three groups of the K4 form are as [a,b,c],[a,d,e],[a,f,g]. Then c&e&g does not private "a" in Anselm's filter Ω or in any octal where it may appear as some set "q" etc. (and in agreement with virtue), but certainly does so in the K4 form instead. There is no positive property (not privating some virtue in "a") found in both of the sets {d,e} and {f,g}. Logically (in the dialectic), c (as wine) cannot be in agreement with "e&g" and also preserve all virtue in "a". ($e\&g \to a^{-1}$.) There is also no virtue in the set "a" (unity) permitting free exercise of positive properties as if in disjunction, formed from two elements of {c,e,g} (that is, one also with an "**included middle**", requiring say, both e&g and c^{-1} to be positive properties).

Properly, disjunctions $q \vee p\&q^{-1}$ with p virtue, never have an excluded middle.

I find that once limited to the set {a,c,e,g} as if with an ultrafilter of perfections, given Pos(c&e&g) then Pos(a^{-1}). Pos(c&e&g) is become the synthesis in the centre. One would sit "as God" in this process, showing oneself above all virtue as antichrist.

{a,c,e,g} as a set is not closed by addition in the octal, neither is it all Ω – it is outside of the seven ultrafilters of virtue, itself overconstrained by privating virtue; taking the complement of "a" is to incorrectly mimic the properties of a well-formed and correctly applied ultrafilter.

In effect they that facilitate and practise the dialectic (as to opposing the Sun octal, cancelling "order two" and absorbing unity in the devices of the first, second and third seals as if God) judging every positive predicate inconsistently (deliberately) are met by the verse:

2Th 2:4 Who opposeth and exalteth himself above all that is called God, or that is worshipped; so that he as God sitteth in the temple of God, shewing himself that he is God.

Splitting the dialectic into a disjunction, $a\&c \vee e\&g$, one side is improperly judged "evil" and the other "good" whereas the choice (as evil or even possibly the good) is unable to be worked virtuously (and is a heavy burden if required). Lack of all virtue is shown by the dialectic process acting only within cosets in the octal. The dialectic is itself "fallen logic" and an "evil" and itself has no place in the worship of Jesus Christ. Judging all absolutes with faulty logic within the body/fellowship of Christ – sitting as God whilst opposing God (in the coset), as if leading all to greater perfection (growth or obedience as if God) by approaching the "pleasure" of consensus (the temple) as the only goal is outright blasphemy. God (and all miraculous virtue that God would exemplify) is denied and the son of perdition openly doing so deserves his mantle of sin in exalting himself.

The dialectic has a result Pos(c) $\to$ Pos(a^{-1}) for any virtue "a". If it can be stated that Pos(a) $\to$ Pos(c) say, then the necessary positivity of "c", as virtue (a closed set) would show the dialectic method faulty. Is "wine" ever a virtue?

Perhaps not to the facilitator, but even then there is a reasoned defence that:

$$\neg N\neg(Pos(c)\&\neg Pos(a^{-1})) \leftrightarrow P(Pos(c)\&Pos(a)).$$

If this can be deduced, the inference of Pos(c) $\to$ Pos(a^{-1}) is false. I point out that if "a" is a necessary truth, finding it cannot be positive indicates a faulty inference: yet even then, if oil and wine are transposed, "a" (virtue) cannot private any such "c" as with $N\neg(\neg Pos(c^{-1})\&Pos(a))$.

"But those elements {c,e,g} are positive taken together!" As surely I heard it said in disbelief! The perfect, if not free to act, cannot exercise virtue. To deprive a virtue (even the liberty of God from God Himself) is to be most negative; not to employ the dialectic "D" is in itself virtue. D is a heavy burden (as well as disobedient) depriving those in the dialogue of the perfect solution that is found in virtue (and thereby also revealing the unworkable), refusing the liberty, sovereignty and judgement of God. Refuse Him the first and reject the second: you will reckon on the third. (Divine virtues are compatible, forming a closed set in each triple they are common to; therefore, accept their union.)

Concerning the dialectic process, I am not given peacemakers as facilitators of the like, only the consistent example of the "good Samaritan" – those with no agenda. The Samaritan (i.e. "love thine enemy") had no interest in seeing the man again, he was not trying to "get in good with Israel", and acted out of pity and charity. There is no agenda there to merge Israel and Samaria by compromising the "holy" position of Israel with their relatively "ungodly" neighbours: rather citizens of God's kingdom have bonds of love superseding divisions formed of the kingdoms of the Earth. I.e.

a	one who is my neighbour
c	a citizen of Israel (or in Christ)
e	one not of Israel (nor in Christ)
g	an act of grace

Naïvely, if $Pos(c) \rightarrow Pos(a)$ (my only neighbours are in Israel) then properly, virtue would assert:

$$Pos(c)\&g^{-1} \rightarrow Pos(a).$$

Then $a^{-1} \rightarrow c^{-1}\&Pos(g) \rightarrow Pos(e\&g)$; which is briefly "grace and virtue are not exclusive to Israel". The result is wholly positive and cannot have been entailed by the negative! (Who then, is my neighbour?) All are created in God's image and all non-Israelites are somewhat righteous and fully capable of putting Israel (those in Christ) to shame. Virtue reveals a "more perfect" person than one simply looking after their own. (And the sets "c" and "e" are properly disjoint.)

If I were to restrict good works to my fellows, I would be tempted to go further and restrict them to my family, then just myself etc. (Grace knows no limits, "Charity, seeketh not its own.")

As to God's sovereignty (accepting every result of divine virtue to be consistent):

Mat 4:4 But he answered and said, It is written, Man shall not live by bread alone, but by every word that proceedeth out of the mouth of God.

Compromise on one and all contentious issues (privating virtue, hurting the oil) is not obedience to God, and believers are instructed not to act after that manner.

6.4) The Fourth Seal

The last rider completes the shifting of unity amongst the four beasts, the earthly elements of the octal.

Rev 6:7 And when he had opened the fourth seal, I heard the voice of the fourth beast say, Come and see.
Rev 6:8 And I looked, and behold a pale horse: and his name that sat on him was Death, and Hell followed with him. And power was given unto them over the fourth part of the earth, to kill with sword, and with hunger, and with death, and with the beasts of the earth.

The last unity is now loosed and I have four possible "balances" that appear (in the model) equally valid. Disjunction becomes as conjunction and vice versa. (Interchanging oil and wine.) If the conjunction and disjunction of two predicates is made, there is one solution, the empty set. The model of this pseudo-logic (of an incomplete and fiat filter in {a,c,e,g}) is contradictory, the correctness of virtue shows the pseudo-logic truthfully fails. (The light of virtue is denied and what remains shadows itself or is "pale".)

"Death" is this spiritual contradiction, denial of virtue asserting that every choice of unity is inconsistent. "Hell" is a solution to this failing dialectic, replacing the fourth balance with a synthetic carnal dialectic of "approach pleasure, avoid pain". The fourth part of the Earth (the remaining balance), is brought under the influence of this fifth of carnality. The "power given them over the fourth part" is the "pseudo-logic" used to private virtue then restricting the abstract to a "fake ultrafilter" on those four elements.

With conflict (seal two), dearth of wheat (too many tares or barley, seal three) and with the contradiction of the fourth triple (Death) the Church paradigm is brought under the influence of the "beasts of the Earth" (i.e. governments/paradigms). The loosening of restraint on the paradigm of those in Christ is not to make them similar to the unsaved (those that were and are always without divine restraint); the end result is that members of the "Church" not elect under predestined selection are given over to a "reprobate mind". Christians are mostly genuine souls: instead the Laodicean "Church" amalgam forms a synthetic "carnal" machine in false doctrine. In the dialectic, $Pos(s^{-1}) \rightarrow Pos(r) \lor Pos(r^{-1}) \rightarrow Pos(s)$ always fails in privation.

A fourth seal system completing that sequence in the octal, by sending the identity (unity) to the empty set renders the dialectic model null. (There are no virtues in the dialectic.) The octal is consistent in positive properties, the dialectic in judging the predicates inate positivity is not. There is always privation in the dialectic. The "Pos" operator is "inconsistently" applied.

With a first balance in the octal with $a \equiv 1 \equiv [b,d,f]$ and {c,e,g} static I have the three balances:

$$\{a,c,e\} \quad \rightarrow \quad \{c,e,g\} \quad \leftarrow \quad \{a,c,g\}$$
$$\{a,c,e\} \quad \rightarrow \quad \{c,e,g\} \quad \leftarrow \quad \{a,c,e\}$$
$$\{a,e,g\} \quad \rightarrow \quad \{c,e,g\} \quad \leftarrow \quad \{a,c,g\}$$

The oil as "a" and with "c", "e" and "g" wine respectively. With "a" sent to nullity they result in:

$$\{c,e\} \quad \rightarrow \quad \{c,e,g\} \quad \leftarrow \quad \{c,g\}$$
$$\{c,e\} \quad \rightarrow \quad \{c,e,g\} \quad \leftarrow \quad \{c,e\}$$
$$\{e,g\} \quad \rightarrow \quad \{c,e,g\} \quad \leftarrow \quad \{c,g\}$$

Retaining "a" sent to the empty set, there is another set of balances with $c \equiv 1 \equiv [b,d,g]$ and {a,e,g} static (now become as {e,g})

$$\{a,c,e\} \quad \rightarrow \quad \{a,e,g\} \quad \leftarrow \quad \{a,c,g\}$$
$$\{a,c,e\} \quad \rightarrow \quad \{a,e,g\} \quad \leftarrow \quad \{c,e,g\}$$
$$\{a,c,g\} \quad \rightarrow \quad \{a,e,g\} \quad \leftarrow \quad \{c,e,g\}$$

With the "wine" one of the three remaining elements "a" or "e" or "g". Again, with both "a" and "c" sent to the empty set I have:

$$\{e\} \quad \rightarrow \quad \{e,g\} \quad \leftarrow \quad \{g\}$$
$$\{e\} \quad \rightarrow \quad \{e,g\} \quad \leftarrow \quad \{e,g\}$$
$$\{g\} \quad \rightarrow \quad \{e,g\} \quad \leftarrow \quad \{e,g\}$$

And with $e \equiv 1 \equiv [b,d,f]$ and $\{a,c,g\}$ static, I have the triples:

$$\{a,c,e\} \quad \rightarrow \quad \{a,c,g\} \quad \leftarrow \quad \{a,e,g\}$$
$$\{a,c,e\} \quad \rightarrow \quad \{a,c,g\} \quad \leftarrow \quad \{c,e,g\}$$
$$\{a,e,g\} \quad \rightarrow \quad \{a,c,g\} \quad \leftarrow \quad \{c,e,g\}$$

Again, wine in the above is one of "a", "c" or "g" respectively. Now with "a", "c" and "e" sent to the empty set I am left with:

$$\{\phi\} \quad \rightarrow \quad \{g\} \quad \leftarrow \quad \{g\}$$
$$\{\phi\} \quad \rightarrow \quad \{g\} \quad \leftarrow \quad \{g\}$$
$$\{g\} \quad \rightarrow \quad \{g\} \quad \leftarrow \quad \{g\}$$

With the last and fourth balance where $g \equiv 1 \equiv [b,d,f]$ and $\{a,c,e\}$ static I have:

$$\{a,c,g\} \quad \rightarrow \quad \{a,c,e\} \quad \leftarrow \quad \{a,e,g\}$$
$$\{a,c,g\} \quad \rightarrow \quad \{a,c,e\} \quad \leftarrow \quad \{c,e,g\}$$
$$\{a,e,g\} \quad \rightarrow \quad \{a,c,e\} \quad \leftarrow \quad \{c,e,g\}$$

I am left with the wine as either "a", "c" or "e" respectively. Now with all four sent to the empty set I have rendered the system completely null:

$$\{\phi\} \quad \rightarrow \quad \{\phi\} \quad \leftarrow \quad \{\phi\}$$
$$\{\phi\} \quad \rightarrow \quad \{\phi\} \quad \leftarrow \quad \{\phi\}$$
$$\{\phi\} \quad \rightarrow \quad \{\phi\} \quad \leftarrow \quad \{\phi\}$$

Summary (If the earthly four elements are treated as an ultrafilter):

In the first seal system:

Pos(a) $\vee$ Pos(c&e&g)

Denial of Christ's virtue in Pos(a) affirms antichrist Pos(a^{-1}) $\rightarrow$ Pos(c&e&g). Every disjunction in the dialectic privates a positive property. The individual judges Pos(a) $\vee$ Pos(a^{-1}) and the "Pos" operator is then misapplied, leaving only "the knowledge of good and evil". Instead, every instance of any Pos(a) must be consistent on both sides of the disjunction for divine liberty.

In the second seal system:

$Pos(a) \lor Pos(c\&e\&g)$
Also: $Pos(e) \lor Pos(a\&c\&g)$
$Pos(a) \lor Pos(c\&(a\&c\&g)^{-1}\&g)$
$Pos(a) \lor Pos(a^{-1}\&(c\&c^{-1})\&(g\&g^{-1}))$
$Pos(a) \lor Pos(a^{-1})$

The second seal causes properties c and g to seem inconsistent, they "fall" upon assuming $Pos(a)$ with the affirmation of the other, $Pos(e)$. By symmetry $Pos(a)$ causes c and g to likewise "fall" upon assuming $Pos(e)$. Either position denies virtue in $Pos(a)$ or $Pos(e)$ as stated above.

In the third seal system:

Given $Pos(a)$: $e\&g \lor c\&g$ where "a" is oil and "g" is wine.
And $Pos(c\&e\&g) \rightarrow Pos(a^{-1})$
$Pos(a) \rightarrow N\neg(Pos(c\&e\&g)) \rightarrow N\neg(Pos(g)\&\neg Pos(c^{-1}\&e^{-1}))$
$Pos(g) \rightarrow Pos(c^{-1}\&e^{-1})$. Also, $Pos(c\&e)$ proves the disjunction $Pos(a) \lor Pos(g)$.

$Pos(a\&e\&g) \lor Pos(a\&c\&g)$ becomes $Pos(c) \lor Pos(e)$ (given the freely decidable disjunction of $Pos(a) \lor Pos(g)$ with $Pos(a)$ axiomatic, if $N(Pos(g))$ holds as "wine", it then proves the disjunction $c \lor e$). If the positive a&g entails $(e\&c)^{-1}$ (e&c an "evil") $e\&c \rightarrow \neg g$ cannot be positive. As this is a contradiction of $Pos(a)$ in terms, $Pos(a)$ as virtue (oil) is privated. Then $P\neg(Pos(a))$. (Even as if virtue was inconsistent.) Similarly, with $Pos(g)$ axiomatic, $e\&c \rightarrow \neg a$ cannot be positive.

Whereas this appears freely decidable, $Pos(a) \rightarrow \neg Pos(g)$ privates any virtue.

Then $N(Pos(c\&e\&g))$ holds as synthesis, opposing virtue in "a" and the facilitator for the dialogue deciding a&c&g and a&e&g sits as God in the temple of God showing himself to be God, choosing natural law. (I.e. What is to be "evil".)

I may only agree on "a" and "c" when e&g is "taken off the table" as $Pos((e\&g)^{-1})$, or else the "evil" (e&g) proves $a \lor c$. It also as "evil" shows $Pos(a)$ inconsistent. (I find c&e&g is the synthesis in the middle! Yet I know from the necessary positivity of "a" (virtue) that a&c is logically unworkable in that setting, there is also no excluded middle in $Pos(e) \lor Pos(g)$.)

In the fourth seal system:

Every instance of virtue is denied if the earthly four elements are become as the whole set of positive properties. So, as in four separate cases of the third seal system, $N\neg(Pos(a))$, $N\neg(Pos(c))$, $N\neg(Pos(e))$, $N\neg(Pos(g))$ which are statements made "simultaneously", and nothing is sacred (there are no virtues, they are **always** privated); although, only the process of dialogue then appears positive. (Where everything is up for discussion, even immutable truth.)

Since I may affirm $N(Pos(a))$ as virtue, the dialectic third seal process practised fourfold is the contradiction rather than that virtue falsely posited to be decided in dialogue. The Earthly four elements are not a "complete and closed set" or an ultrafilter of positive properties. There is no sense that $Pos(a^{-1}) \rightarrow Pos(c\&e\&g)$ is logically justifiable in a freely decidable disjunction.

6.5) The Colour Coding Of The Seals

I may now show a possible reason why the horsemen are colour coded.

A plane graph or "map" of regions may be coloured in four colours (the minimum required on a plane) so that no two neighbouring regions share the same colour. On the borders between regions' colours, I may form elements of "colour differences" (added modulo two), and I have an octal (0, ab,ac,ad,bc,bd,cd,abcd) where "xy" represents colour x meeting colour y. In a path crossing borders I may sum the elements of this octal across my route so that ab * ac ≡ bc etc.

Now, abcd is an apparent inconsistency without example in the graph itself, but by choice rather than construction I may keep abcd as the unity and choose the subgroup (0,ac,bc,ab) to be static. This is equivalent to the Frobenius map upon this octal holding one colour fixed in all the graph (namely "d" here) and provide a new arrangement of the labels (corresponding to the three remaining colours of regions), cycled in triples about the graph's "nodes" or vertices.

What is the result? Each choice of static subgroup corresponds to another colour "held fixed". So with unity fixed, the "horse" or static subgroup under the rider (unity) appears to cycle colours under the altered seven cycle. Then the coset of the earthly elements (the four beasts) also three-cycles likewise.

The four rider's colours are found to be "counterfeits" due to the "earthly filter" (which is not an ultrafilter). White becomes "fake righteousness", red becomes a "fake intersection" (as blood, incompatibly sharing the right hand). Black becomes "fake proof against light" (i.e. virtue), and lastly "the pale" becomes "fake evidence" that nothing can be positive or absolute.

There is a far more prosaic explanation as to why the seals are colour coded. Found after looking back at this work – there were four stages of production involved in this writing.

1) Starting with a clean "**blank**" page. (White horse as "blank paper". Many start as antichrist before finding their faith.)

2) Next the "red" horse – the red text of the "absolute" scripture on that blank canvas... should I accept it totally or do I prefer my canvas (paradigm) devoid of all scripture?

3) Upon that red scripture I added the black script of my own "opinions" of that scripture – "weighing it in the balance" as to justifying my place in the centre position somewhere between (1) and (2).

4) Lastly I highlighted with a fluorescent pen (in green) that portion of my own script that is "contradictory" as I prepared it for redaction. (Some would even strike out the scripture rather than their own words.)

These in no way correlate to perfection as even step (1) assumes all are sinners, whereas God acts in a pre-first seal paradigm. (2) Correlates to a division of absolutes – God is not divided over His own words. (3) Is a vacant position – all are assured God has no opinion and has recorded His requirements in the scripture itself. (4) Is a contentious issue – as God is not required to keep laws made for men as He has no requirement to answer a greater authority. What all should expect is that there should be the truth between God and those who keep His commandments – God is trustworthy in His covenant.

6.6) The Restrainer

The enigmatic verse in "2 Thessalonians" on the restrainer is important to note here.

2Th 2:7 For the mystery of iniquity doth already work: only he who now letteth *will let*, until he be taken out of the way.

The Lord's restraint upon the Church is removed in stages with the loosing of the seals. Once the restraint is removed the Church is wide open for the world to corrupt it, as the spirit of antichrist freely moves in and sets up shop. The opening of the first four seals completely loose the mystery of iniquity upon the Earth and the remaining three are of loosed restraint in heaven.

The restrainer loosed has the effect that the paradigm of the church is made far broader, whereas election in Christ requires a stronger love for the gospel's truth. The elect hold firm as if they had lived without the restraint lifted: retaining that obedience to God. The Church becomes worldlier rather than the unsaved. (Those are always without His restraint, as not in Christ.) The elect hold to the one fixed position (unity) and those operating in any broader paradigm loosed by the lifting of restraint will attempt to operate without the fixed truths that the gospel and obedience require.

In the first seal "election" is a matter of "are you in or are you out?". In the second, election depends on standing up to those you oppose that claim they also have **your** answer; the safety is in Christ not conflict, yet peace in the faith is truly taken from the Earth. In the third seal, collectives may work together despite their differences whilst sacrificing the truth over which they disagree most importantly. The fourth seal is as "Death" and is such a broad path that it has lost all value and is beyond redemption; the Church, it can be said, would have no life to offer. In continuing to so operate with doors wide open, they fulfil the system of "Hell".

The "he" in "only he who now letteth will let" is the one of four beasts before the throne that currently holds restraint upon the elect; each of the four beasts restrain the number of floating identity elements present in the paradigm of the saints as each seal is opened. (They each in turn hold restraint, hence "only he" in the verse.)

As to the mystery of iniquity already at work, consider the set {a,c,e,g} where:

a   "grace extended to all" (No condemnation of the law.)
c   "circumcision" (A sign of those under the law that are preserved by it.)
e   "a person under the old covenant"
g   "a person under the new covenant"

I have $Pos(e)\&a^{-1} \to Pos(c)$. Rearranging I have $c^{-1} \to Pos(a\&e^{-1})$ and $Pos(e) \to Pos(a\&c)$

Identifying "a" with virtue and forming the disjunction between "e" and "g" with:

$Pos(a\&c^{-1}) \rightarrow Pos(a\&c^{-1}\&e^{-1}) \rightarrow Pos(g\&c^{-1})$.

I have the K4 group [0,a,e,g] in the ultrafilter (where a=1) and the set of e&g is void. Also, Pos(a&c) breaks the closure of Pos(e). Then, consistently extending into an octal with a predicate N¬Pos(a&c), "c" privates virtue in "a" whilst "g" or "e" are in scope. As "a" is virtue I would treat it as I would unity, because "c" privates that grace. I cannot freely or consistently decide in [0,a,e,g] if I hold true N(Pos(c)). I already have two choices of unity, "a" and "e".

Consider the third seal device with the correspondence to the static triple in the left hand;

$a \equiv 1 \equiv \{c,e,g\}$

I note that [0,a,e,g] is not static in the right hand as it contains unity. The octal it comes from is separate to the left hand with $a \equiv 1 \equiv \{c,e,g\}$ since the static subgroup in the right [b,d,f] has no non-trivial intersection with [a,e,g]. There is a second seal paradigm (requiring also a first trumpet paradigm of "hail and fire". Always the trumpets appear to systematically precede and provide the mechanism for the opening of the seals and the loosening of restraint: they provide the underlying structure for them). This is the mystery at work, that even with the church under God's full restraint the world was not under that restraint.

A third seal paradigm introducing a&c&g and a&c&e dialectically opposes the force of either covenant (I assume $Pos(g) \lor Pos(a\&c\&e)$ etc.); they oppose them as $(e \lor g)$.

Given a&c: $(a\&c\&g \lor a\&c\&e)$ becomes $(e \lor g)$. This has a shell of validity though the force of it has been lost. Virtue is truthfully over constrained – I truly have $(a^{-1}\&e \lor (a\&e^{-1} \rightarrow g))$. The empty middle of e&g (an "evil" without positivity) appears to prove $a \lor c$ whilst also leaving the positivity of a or c inconsistent; it is become necessary that only one of a&c or e&g is positive. Each is actually unworkable, if the other is privated of positivity. To find strength in both covenants enforced together as if equal is a denial of or compromise of both their relative strengths (the "mystery of iniquity" as brought in by the "judaizers" in Paul's day). Such a logical system where anything may be dismissed from positivity would ultimately culminate in one with the mind of "the son of perdition", i.e. that:

2Th 2:4 Who opposeth and exalteth himself above all that is called God, or that is worshipped; so that he as God sitteth in the temple of God, shewing himself that he is God.

6.7) The Four Beasts Of Daniel

The four beasts in the Old Testament (OT) book of Daniel parallel the first four seals in a fashion whereby Satan has been given the route he is to take as the restrainer is loosed. These four New Testament kingdoms or "primary powers" (rising up from the sea) are placed in sequence with the paradigms of the first four seals, each seal as the paradigm for those nations, whilst one carnal dialectic triple or "Hell balance" races ahead by one paradigm so that lastly the sequence ends with the contradiction of the fourth seal system. (Healing the contradiction of the fourth seal system with the synthetic balance of the heart.)

Each kingdom is constructed upon the third trumpet device of "wormwood" formed across the sea of thirty octals. All four paradigms are overlaid upon that one structure. In the wormwood construct, three "kingdoms" (closed structures, cf. "crowns") are required to establish the fourth. Those three are "plucked up" as their dominions are given over to the fourth "dreadful and terrible" beast, which is formed upon that wormwood.

The people of God are slowly cooked "like a frog in a pot, unaware", as their governments by design anchor them in a broader paradigm, as if always one "bow" ahead, racing the restrainer.

Dan 7:1 In the first year of Belshazzar king of Babylon Daniel had a dream and visions of his head upon his bed: then he wrote the dream, *and* told the sum of the matters.
Dan 7:2 Daniel spake and said, I saw in my vision by night, and, behold, the four winds of the heaven strove upon the great sea.
Dan 7:3 And four great beasts came up from the sea, diverse one from another.
Dan 7:4 The first *was* like a lion, and had eagle's wings: I beheld till the wings thereof were plucked, and it was lifted up from the earth, and made stand upon the feet as a man, and a man's heart was given to it.

The first beast is believed to be the British Empire,[8] the "eagle's wings" the USA – by revolution "plucked" from the lion. The heart given the lion rampant (a symbol of Britain) as a synthetic balance races one dialectic (seal, or paradigm) "bow" ahead of the restrainer as loosed once, and I find a second seal system acceptable to the church rather than that of the first: for the parliament is a two party system (a party is voted in to therefore vote the other out, and the paradigm is carnal, of "approach pleasure avoid pain"). The electorate by ballot eliminates the authority of one for the sake of the other. In that sense the "wings" as maps between the heart and the two states (feet, octals left/right) of the two party system are (plucked) removed and replaced with only the action of the heart itself weighing in synthesis.

Dan 7:5 And behold another beast, a second, like to a bear, and it raised up itself on one side, and *it had* three ribs in the mouth of it between the teeth of it: and they said thus unto it, Arise, devour much flesh.

The second beast likewise shows us a third seal system: the three ribs represent the three "bows" in the system of balances. This beast is identified as the "Russian bear" or the USSR.[8] The ribs in the mouth show that the method of speech is the Hegelian dialectic.

I may attribute this to a system of thesis/antithesis gives synthesis. The thesis is capitalism, the antithesis is communism. The bear is seen risen up on one side – to that extreme pole of communism (judging in the dialectic). The result is to totally (unworkably) oppose capitalism and when the people in a nearby country compromise between the two (employing the "Hell" balance) and accepting some form of moderate socialism, the USSR bear would "arise and devour much flesh" and swallow up that country with its lukewarm citizenry for the good of the workers, and therefore the bear feeds to add to its own territory. The "flesh" is an indicator of the carnality of the heart in the synthesis; the swallowing up of nations is a result of thesis and antithesis placed upon the opposite poles of communism and capitalism.

Dan 7:6 After this I beheld, and lo another, like a leopard, which had upon the back of it four wings of a fowl; the beast had also four heads; and dominion was given to it.

The third beast has a fourth seal system of four possible balances (heads) and there are four

[8] James Lloyd, *The Sand and The Sea* (Christian Media), p5-7.

wings to indicate the existence of a map as from the heart to each one of the heads. This beast is thought to be the third Reich of Nazi Germany.[8]

The centre or synthesis of the heart (filling in the fourth balance of the dialectic above) loosed from restraint had become able to assume a position of extreme belief in national superiority (faith that the given carnal "heart" can justify one to be of the "master race"), that through their resolute strength of will, communist workers would lose their resolve and western capitalism would fall as ill-prepared (albeit wary).

A fourth seal system has four such balances, each of three "bows". The carnal sensual balance is free to apply to any of the four above, communism, capitalism, nationalist socialism and (lastly) racial-superiority (eugenics); the heart is used to fill the missing contradictory "balance". These four configurations represent the four "heads" on the back of the leopard: the carnal balance is positioned by those "wings", for each one may be used to "fly the heart into place" quickly as desired. The effort is to "close the system" without discerning contradiction.

Dan 7:7 After this I saw in the night visions, and behold a fourth beast, dreadful and terrible, and strong exceedingly; and it had great iron teeth: it devoured and brake in pieces, and stamped the residue with the feet of it: and it *was* diverse from all the beasts that *were* before it; and it had ten horns.
Dan 7:8 I considered the horns, and, behold, there came up among them another little horn, before whom there were three of the first horns plucked up by the roots: and, behold, in this horn *were* eyes like the eyes of man, and a mouth speaking great things.

The last beast is a fourth seal system where the "heart" is without any restraint, always approving itself within the fourth seal system. A synthetic fifth balance is hinted at: formed by the carnal heart of the citizen (diverse from all the beasts that went before it). Ten horns, as I will state later, are required for constructing such a fifth balance.

> **Horns:**
> *Horns represent authorities in the Revelation. They may be either kings, empires or even the strength of the cross seen in the four horns of the altar in the heavenly throne scene. Authority does not have to be true divine authority; it can be a mockery of it as seen on the false prophet. In every case, a sequence of horns shows authority to have been accumulated.*

The beast devours and breaks in pieces; I will examine this in detail. It should be realised that the logical paradigm of the fourth seal system is contradictory and only completed by the synthetic "Hell" made of the human heart.

As nations are assimilated into the fourth global beast (the collective formed and projected upon the USA as "the last superpower"[8]) by absorbing the paradigms of those smaller beasts (as of lions, bears or leopards), some of the previously held and absolute (as unity) positions of those nations are now inconsistent or contradictory, and so the beast devours with unbreakable iron teeth and breaks every other (weaker, simpler) nation in pieces, leaving behind it nothing but its former contradictions, "stamping the residue" leaving only smaller

[8] Lloyd, *The Sand and The Sea*, p5-7.

defeated beasts (with seal systems one to three) behind.

The little horn rises up whilst the USA is the last remaining superpower (the previous three horns then plucked up). The little horn is a (small) government (and a beast) separate to the USA but found "on it". It is a government approved corporation – a "Church" governing itself, keeping a separation of the Church and state.

The "little horn" is somewhat related to the arrival in the last days of the state of Israel. It cannot itself be that state of Israel, which arrived beforehand, but is instead the later "scarlet beast" of the false prophet's image system. That beast discerns carnally for itself "like a man" (has eyes like a man) its own doctrine without the leading of the Holy Spirit.

The state of Israel enables the false prophet spirit to rally the collective and more importantly abdicate their choice of prophetic doctrine to the synagogue of Satan. That same spirit covets the ideals of a "carnal" Zion. As a source of false doctrine (from the Old Testament), I expect the synagogue of Satan to camouflage itself by aligning themselves with that Zionism which removes Christ. They are those "which say they are Jews, and are not, but do lie" (Rev 3:9).

That false prophet spirit instead judges from the heart (using a model of a "man" as in the image system). It speaks of "great things" by denying absolutes and virtue as for a wider collective. Yet its "mouth" is its "set", and so it decides for itself its own doctrine – using a system of leavening. The "mouth" is then the assembly (as approving by ballot the "mind" of the image) that casts the vote on that doctrine: which is even at the end to become fully that heresy of Nicolaitanism, the whole flock following after the doctrinal leading of the synagogue of Satan itself, a sacrosanct and "closed sect" (as based on its own carnality) fixated upon a regenerate "Old Testament" Israel for its prophetic doctrine.

Israel is not the false prophet, but provides his bait for the switch. The false prophet spirit covets the ideals of Zion, i.e. the inheritance of Israel and the work of the Christ.

6.8) The Fifth Seal

Returning back to the Revelation text there is the opening of the fifth seal.

Rev 6:9 And when he had opened the fifth seal, I saw under the altar the souls of them that were slain for the word of God, and for the testimony which they held:
Rev 6:10 And they cried with a loud voice, saying, How long, O Lord, holy and true, dost thou not judge and avenge our blood on them that dwell on the earth?
Rev 6:11 And white robes were given unto every one of them; and it was said unto them, that they should rest yet for a little season, until their fellowservants also and their brethren, that should be killed as they *were*, should be fulfiled.

On the altar before the throne is the offering of the lamb. To be under the altar is to be represented with the unity element. The offering by Christ is for all those He has as His elect, His chosen people. In being slain as martyrs for their obedience or the Revelation of doctrine for the saints, these under the altar indicate that Christ (in His K4 form of the ultrafilter) has two parts outside of the Field GF(8) wherein restraint may be further loosed: in effect there is a system like the first four seals but in the field GF(4) instead.

Those two parts cut off in GF(4), reveal the coming resurrection of both the just and unjust, already separated to Christ's right and left: for "All are alive to God". With the first element of the static subgroup effectively "opened", John now sees those yet to be raised. (Those slain for their testimony and for the word of God.) The fifth seal reveals that these two sets previously hid are aligned to the character of those two parts of the Trinity in Christ as "holy and true" respectively (as seals 5 and 6).

Whilst the saints expect vengeance (instead God remains Holy and patient to the end) and the fulfilment of the Lord's promises at the judgement (the completion of all things as prophesied), the saints desire to rest after that judgement, but "rest" is referenced here as being of the Lamb and with that judgement occurring as soon as the seventh seal is opened, bringing rest on the last day. They are told to wait a little season, the period between the fifth seal through to the opening of the seventh. (Their "blood", as their place – living within all three of the trinity, is to come to remembrance.) Each of them is given white robes and asked to wait "a little season". (Three months for a season, as "times" or elements of GF(4) to be unloosed in "floating" unity likewise. Two seals (months) then remain.)

Christ spans the octal as the K4 form or "Golden Girdle" and by regression is sat as the right hand static subgroup, which is not a subfield and does not contain unity. Then "under the altar" is an indicator that the saved are not in the static K4 subgroup or unity in the octal unless in regress within Christ, being separated to Christ's right and left – to the static subgroup or bow.

Each of them is given white robes (static as [0,1]), clothed with the same virtue shown in the life of Christ – they are elect and accepted, and told to wait until their number is complete. This is not indefinite, but indicates that on the last day, Christ will judge and no elect need die ever again: He comes very swiftly.

Being accepted as the lamb (as unity) floating in the octal is a great indicator of salvation.

6.9) The Sixth Seal

I have more terms for my vocabulary to ascertain here.

Rev 6:12 And I beheld when he had opened the sixth seal, and, lo, there was a great earthquake; and the sun became black as sackcloth of hair, and the moon became as blood;

As the sixth seal is loosed within the Sun octal, unity is free to be positioned as two of the static subgroup's elements. Christ is **unique**; there is no "duality of unity". The K4 form then merely cycles as with the "will of God": the Lamb then moves through the static subgroup. Then I find $a \equiv 1 \equiv [a,b,c]$ is cycled in complete agreement to $[b,d,f]$ within the K4 form, also inducing $\{c,e,g\}$ in the octal. Jesus, moves as God, proving Himself God. The four beasts are "shaken", falling down before the Lamb (but not by the action of Frobenius in the octal). Instead, the shaking is due to the multiplication of the GF(4) form of Christ whilst enthroned, thereby inducing the very same effect as Frobenius on GF(8). In the fifth seal there was no such "movement". Christ, had merely "sat down". Here, I see the static subgroup cycle.

Under the seven cycle, the devices of the first four seals are also cycled through the Sun octal, and every (arbitrary) unity element of the octal becomes "hurt" by the devices of the balances and from the device of "Death". The Sun becomes "black as a mourning garment", now become as the dialectic devices in every subgroup's coset, as if equally walked through the octal. (Cf. Christ's garment as "down to the foot").

In the Earth, Christ is then "eclipsed" as God. The moon or "bridge" between octals in the sea (allowing shifting to left hands of left hands etc...) allows the traversing of the sea, as if octals truly had proper intersections (or "blood"). The sea is traversed with the seal devices, by those seeking after them, rather than obeying God. (Else multiplication acts only on the reference.)

The "Sun" (reference) octal ceases to be such an attractor (light of virtue – now become black) as the number of the elect is completed. The Sun is now as a mourning garment; the "Sun octal" is in no sense esteemed as the garment of redemption to those not saved, who are the "dead" that operate with the dialectic as antichrist. There are no more converts. With no octal appearing any more valid than any other to the ungodly, the sea of thirty octals under every multiplication (and now at its most traversable) is become a "lattice" of sorts.

The "moon", or bridge in heaven is the mechanism bridging the "light" of the Sun octal and the "darkness" of every other octal not as its left hand. The left hands (and their "left hands" etc.) form a wider system (the sea) with the "moon" bridging every octal connected by intersections of subgroups and groups of subgroups.

Those (sub)groups – now as "blood" (the "moon" also eclipsed, no longer showing the light (virtue) of the Sun octal) – is now a system of **circulation** across octals in the sea, bridged by "synthetic" or "leached" properties of multiplication in the four seal devices. (There appears no reference point in the sea as a "Sun octal".) Traversing octals (as the "moon", present or moving from day to night) in the sea is now become "as life" for those that dwell *in the Earth*, operating with the devices of the seals rather than to hold fast in the unity element which itself is life (as under the covering of the lamb before the throne, as Sabbath rest in Christ).

Blood:
Blood references intersections made of sets: the system of lively circulation between them made possible when two structures share a non-trivial intersection. (They are 'related' or "as kin".) Blood is then "shared". Seven cycles (cf. living water) hold over only two disjoint octals (left/right pairs); there is no common seven cycle to be found between octals unless they are as such pairs (left/right). If there is a K4 group additive intersection (blood) then it is possible whilst without restraint and spiritual "life" to traverse octals in the sea. "Blood" is also used as of the martyrs of the fifth seal: Their blood or place under the Lamb in the octal (the intersection of all in the Trinity) had been lost whilst they remain alive in Christ (in one of two parts cut off) in the resurrection (Christ is the resurrection).

Rev 6:13 And the stars of heaven fell unto the earth, even as a fig tree casteth her untimely figs, when she is shaken of a mighty wind.

The stars of heaven (the singletons in the sets of triples across all thirty octals that form subgroups or even static triples in their cosets) fall to the earthly elements to further the bridging of different octals. The "stars" are the elements of the 30 octals that together are grafted on to the four earthly elements so that in the Sun octal those dwelling within the "Church" would experience a vast "falling away" (from the required love of the truth). The stars in their triples bring with them the operations of the octals they originated from. Those in the Earth diverge from acceptable obedience to "another octal" upon holding a "different paradigm". The stars (bringing alternative operations permitting the dialectic devices) are even fallen as "tares". I simply note that a different choice of unity carries a different static bow (and a different left handed octal) – and hence there is a decision to be made, "which octal do I dwell in?".

Stars:
Stars are mentioned as singletons and mappings (products) in and of the wider set of 120 fields (30 octals each under eight C7 groups) and are usually employed as a device to indicate outside influence upon the reference octal from other octals in the sea. The lifting of restraint allows these influences to aid those on the Earth to shift octals and dwell in a broader, more worldly paradigm.

Then they are also in view as the elements of a five-cycle (see the fifth trumpet). As John saw the seven cycle "burning up" (see the first trumpet described in the next chapter), as using the inverse of the angel's seven cycle which reads left to right (acting top to bottom on triples): that cycle which John instead read right to left, the five cycle appeared to John as elements dropping "before they were due" or before they "attained unity" and became fruitful (at the "head" of the cycle, as the topmost and static subgroup). The effort of the five cycle is to close the devices upon the earthly elements.

2Th 2:3 Let no man deceive you by any means: for *that day shall not come*, except there come a falling away first, and that man of sin be revealed, the son of perdition;

They fall as if shaken by a wind "mightily" (suddenly). That is, they were due to fall (not ripening due to the timing not being right: they are in a shortened cycle of five months not attaining unity). I suspect the restrainer revealed them as "lukewarm" (cf. the letter to Laodicea) and they were once considered as elect dwelling in heaven. (As if unity.)

The elements fallen to the earth are properly a "place prepared" for an arbitrary operation, one to be taken advantage of by tares. (A broader path.) God has a deadline. If these stars fall early and suddenly, the end will come fast.

I compare these "figs" later with those without fear of God who do not repent. They are "shaken out" or expelled from attaining the kingdom of heaven before they were able to succeed (ripen, bear fruit). They directly compare to those that do not live again "until after the 1000 years were ended". Or those that trample the outer court of the temple underfoot for 42 months. (See much later, Rev 20:5 and Rev 11:2.)

Rev 6:14 And the heaven departed as a scroll when it is rolled together; and every mountain and island were moved out of their places.

The heaven departs as a scroll rolled together; that is, God appears to vanish! (The octal "cycles" (rolls up) completely as to unity (rest) and with the cycle of seven thunders heaven "shifts" back to align the trumpets with the angel's circuit – though this "mystery" has not been introduced in the text yet.) In other words, that which comes "later on" (the trumpets etc.) has now become hidden whilst the narrative shifts to overlap those things seen already.

"Mountains and islands" are strongholds of false doctrine using the dialectic on the earthly elements in the seven churches of the angel's circuit and found across the sea. (Cf. the "mourning garment" as if it were the dialectic device under a seven cycle.) Each "church" (a seven cycle) is one of eight over a complement octal to the Sun or reference – within which these mountains leach their properties for the dialectic devices. (For a "vast falling away".)

The "mountains and islands" now moved from their places dovetails with the treading of the winepress, with a five cycle instantiated by God's judgement. The five cycle is separated from the Holy Spirit and the angel's circuit. (C5 diverging from C7 like as a scroll, rolled together; the five cycle appears to "spiral away".) The K4 groups of GF(4)'s orbit in GF(16) are also disjoint.

The victories standing over these devices from the completion of the least's circuit enable the angel (the "Word of God") to tread the winepress. The angel overcomes all the dragon's devices before the sequence of seals is complete. He must overcome that false doctrine at the time of the "three woes" before the seventh seal is opened when the seven trumpets are handed out. These devices will, as later, "not be found"; they are overcome in their entirety.

With the change from the seven cycle to a five cycle (upon "foolish virgins"), the static triples are no longer found within the five cycle (or are as "wormwood", exchanged). Every stronghold of faith (as if unity) and position in the sea (coset in some octal's four earthly elements) are moved out of their true alignment in the Sun or reference octal. It can be said that the least has finally overcome; the elect already wholly separated by the ministry of his two witnesses.

Mountains:
Doctrinal or quasi-doctrinal devices. The latter are worldly – employed as if they were acceptable, though as with the dialectic it is a virtue never to practise them. Mountains reference the doctrines of the Nicolaitans in the "Church" – the seven letters of Christ reference that same false doctrine, which at the end is completely leavened. These quasi-doctrinal "strongholds" revealed as Satan's devices are endemic due to the lifting of restraint upon the Church. The lamb's seven horns ensure the least's overcoming of these "Mountains" (cf. Zec 4:7 as of "Laodicea"), each a device to supplant the reference seven cycle (rainbow) and raise nullity up sevenfold as an "octal" in the place of heaven. Mountains are also referenced as strongholds of sound doctrine, by which the Holy Spirit maps to unity in Zion as in the mystery of the seven angels and candlesticks (with those seven horns). In the octal, mountains reference the K4 form in improper congruence to the earthly (static) triple made within wormwood (or to the unity element as in Mount Zion).

Islands:
Islands are simply as mountains in the wider sea of 30 octals.

The ability to leach operations to shift octals is totally unbound. Though it is not seen yet, the trumpet sequence will account for this structurally.

Rev 6:15 And the kings of the earth, and the great men, and the rich men, and the chief captains, and the mighty men, and every bondman, and every free man, hid themselves in the dens and in the rocks of the mountains;
Rev 6:16 And said to the mountains and rocks, Fall on us, and hide us from the face of him that sitteth on the throne, and from the wrath of the Lamb:
Rev 6:17 For the great day of his wrath is come; and who shall be able to stand?

Seven groups of men now seek refuge from the wrath of God. This is not the angel's seven cycle but a reference to the seven groups of multiplication or the seven candlesticks of Rev 1. (They have a great influx of fearful men, not necessarily tares.) These are the gentiles that "tread the outer court of the temple underfoot" for 42 months. (See Rev 11:2.) They are also as the 7,000 that die in the earthquake when the tenth part of the city falls. (Rev 11:13.)

One group of seven is referenced as "free". These are not "kings" but interesting nonetheless, given that the first six sets appear in sequence under inclusion; something to look out for later. "Free men" are not required to follow but of their own liberty; are the last "free" men separate to all the rest as are those in Christ juxtaposed with those under worldly authority?

One church of seven (Philadelphia) is also approved and completely separate; the other six are also under inclusion in the circuit if one considers a circuit of Church doctrine becoming more strict – always reducing to a proper subset of what was broadly held to begin with – the angel may be selective, until only that which remains is sound doctrine. (The least starts with the freedom of kings and ends with that of slaves until only the open door to leave gives freedom.)

The period of the sixth seal aligns with the end of the angel's circuit, the ministry of the two witnesses begins and culminates in the pouring out of the seven vials of wrath. With the end of the circuit, there is no need for Satan to deceive the people of God any longer; the end of his deception and the wholesale war of destruction upon (rather than of) the gospel begins, only succeeding in hell itself.

At the end, it will not surprise that the Church will be inundated with new members of every walk of life (a "flood" from the mouth of the dragon); hoping for safety in numbers amongst those they consider are truly elect but in reality are nothing but another system of captivity. "Mountains" correspond to strongholds of faith (fellowships included) or doctrine. (Those built on the dialectic.)

Those dens (voids) within are the excluded middles (in disjunctions of virtue) that are found in the dialectic devices. The "rocks" those static triples in cosets (each a "mountain" present within a "church") that as syntheses oppose those virtues then privated. By saying "fall on us" they likewise know the Church has failed, but expect it to be preserved and them with it. Sadly, this is not to happen, as the elect are to be preserved separate to the Church. The elect are able to stand upright without a Church safety blanket.

Rev 7:1 And after these things I saw four angels standing on the four corners of the earth, holding the four winds of the earth, that the wind should not blow on the earth, nor on the

sea, nor on any tree.
Rev 7:2 And I saw another angel ascending from the east, having the seal of the living God: and he cried with a loud voice to the four angels, to whom it was given to hurt the earth and the sea,
Rev 7:3 Saying, Hurt not the earth, neither the sea, nor the trees, till we have sealed the servants of our God in their foreheads.

Four angels are stood on the four corners of the Earth: the four choices of unity in the Sun octal. They restrain any seven cycle not native to the reference from acting upon that octal – re-aligning it to the (now scrambled) others in the sea by action of "wind" or spirit. (Octals across the sea may share three subgroups which intersect in unity.) If that were the case, the sense of election (of unity) of all those chosen across and in the "sea" under any given seven cycle would be prematurely lost; instead the text shows the elect sealed upon their foreheads so they may be identified easily to the angels that will "reap the harvest" for God.

> **Trees:**
> *Trees reference the branching of singletons to their equivalent triples in subgroups and cosets between left and right hands. An example is a ≡ {c,e,g} ≡ [a,b,c],[a,d,e],[a,f,g]. Under the action of a seven cycle such a "tree" appears as a fountain of water. The analogue of trees in the wider set of octals (the sea) are "ships". (With the operation reversed as hail and fire, John appears to have drawn trees and ships vertically, with ships as trees inverted.)*

Christ opened four seals on the earthly coset, loosing four of eight C7 groups, one with each seal rather than the possible two. The remaining four (as pneuma) are "held back" from blowing on the earth, to prevent them forming (by acting on triples) the "great river Euphrates" (with four tributaries, earthly triples under the action of seven-cycles (rivers): only one of four acts on the triples in each instance). These four "rivers" are disjoint in triples (4*7 = 28).

The angels are given to "hurt the earth and the sea" by confusing the "middle ground" of octals sharing subgroups with the Sun octal. They open up (or allow) the exercise of the dialectic devices in a set of "would-be" right-hand octals that contain the three subgroups of the K4 form, but are "antichrist" – a subtly different paradigm not discerning itself to be antichrist. There is a first seal paradigm here: Christ's right becomes His left and vice-versa. (These octals share the same logical schema, but do not agree, breaking closure in the octal.)

The rightmost four triples in each row (see the following table) correspond to the four possible static "bows" of the reference's possible left hands (a=1 in the top row etc.), each entry in those four columns is the static subgroup in that corresponding left handed octal of eight. (The right hand is represented by the leftmost three columns only.) The seven triples in each row form an octal under K4 addition as before, and in those triples implicitly define another octal. Acting downwards in columns under the reference seven cycle that, say, holds [c,e,g] static in the left hand of the Sun octal with a ≡ 1 ≡ [b,d,f], the other three (greyed) columns do not form an octal group under the "reference" seven cycle.

$[a,b,c], [a,d,e], [a,f,g]$ with $[b,e,f], [b,d,g], [c,d,f], [c,e,g]$ with unity $a = 1$

$[b,d,f], [a,b,c], [b,e,g]$ with $[a,d,g], [c,d,e], [c,f,g], [a,e,f]$ with unity $b = 1$

$[c,d,g], [b,d,f], [a,d,e]$ with $[b,c,e], [a,c,f], [e,f,g], [a,b,g]$ with unity $d = 1$

$[c,e,f], [c,d,g], [a,b,c]$ with $[a,d,f], [b,f,g], [a,e,g], [b,d,e]$ with unity $c = 1$

$[a,f,g], [c,e,f], [b,d,f]$ with $[b,c,g], [d,e,g], [a,b,e], [a,c,d]$ with unity $f = 1$

$[b,e,g], [a,f,g], [c,d,g]$ with $[d,e,f], [a,c,e], [a,b,d], [b,c,f]$ with unity $g = 1$

$[a,d,e], [b,e,g], [c,e,f]$ with $[a,c,g], [a,b,f], [b,c,d], [d,f,g]$ with unity $e = 1$

The four angels restrain the seven-cycles acting on subgroups in the rows, permitting cycles over columns only, preserving both hands of the reference. (The leftmost three columns should not be confused with the reference octal though they share part of the rows. The rows may share the "K4 form" with the reference octal, but they share no seven cycle, they are "wormwood" – made bitter.) Restraining seven-cycles within the rows, the "trees" in the first three columns under the reference seven-cycle are not confused with those in each row.

Each row generated by the seven-cycle forms an octal where the static triple (rightmost column here), is exchanged for the static subgroup. In the top row given above, [c,e,g] replaces [b,d,f]. The Earth [a,c,e,g] is hurt to become [a,b,d,f]. In each row, the same device acts as across the "sea". A similar situation occurs in every left hand (hence the two witnesses later on).

Before this last set of four C7 groups is loosed, broadening the paradigm of those on the earth, an angel ascending from the east (as the Sun's left hand above in the rightmost column), seals the saints into the position they held in the sea as to "unity" (i.e. across all thirty octals of the "sea"). This is so that their election in the book of life is not "lost", or "swept or blown away".

The angel ascends (John reads right to left so the seven cycle appears to act upward), and from the east: The saints are sealed into the one right hand corresponding to each of the possible left hands (rightmost four columns, i.e. all four as the "east", paired with the Sun "rising" under a suitable seven-cycle). Without loss the "one corner of the earth" that is "east" is the one column of four that forms a valid octal (as with [c,e,g] above) as a left hand of the Sun octal. (So the cycle is ascending out of the east.)

The saints now dwell in a generalised right hand (Sun) octal but under an isomorphic seven-cycle valid as a pre-first seal paradigm (not "loosed" by a seal on the book of life). The "Euphrates" octals under those last four seven-cycles provide a mechanism for any elect holding to unity in the K4 form to shift to an alternate seven-cycle acting on the same static triples, but with a different left handed octal: e.g. the pair (a,b,d,c,f,g,e) and (a,b,f,c,d,e,g) over the Sun octal. (With this, the sea is traversable, by slipping through "left hands of left hands" etc.) That explains why the elect are sealed into the right hand (even though they are "as sealed" in the book of life already); that of the circuit of the angel (generalised over all 30 octals). They do not shift octals or shift out of the right hand's seven cycle; they attain rest.

The four restrained seven-cycles that hold the same triples static as in the first four seals, together generate this "river Euphrates", each acting on the "earthly" columns (to permit the construction of "waters" (octals) made bitter as "Wormwood", i.e. octals in rows) and the results are close enough to the Sun octal to agree on the K4 form completely; but because of their similarity to their kindreds sharing unity and static triple(s), the restrainer (though loosed), requires a sealing of the elect to keep the saints in the same seven-cycle and hence in the same circuit as of the least – a circuit become the rainbow before the throne. Then even the least slack is accounted for in the paradigm required by God.

The top row correspondence a ≡ 1 ≡ {[a,b,c], [a,d,e], [a,f,g]} holds the same pair of triples static (as if Christ's left and right were exchanged in the K4 form). Both this system of "Euphrates" octals and the "wormwood rows" are a "middle ground" which demands a great amount of detail in the passage on the two witnesses later. The middle ground in the sea is formed of these octals (in rows above) that share intersections (K4 groups) with the Sun octal.

The angel states the earth should not be hurt ([c,e,g] not exchanged for [b,d,f]) and its analogue in every row (across the sea), likewise. The wind should not blow on the Earth, the sea or trees (the trees are seen in each row too). The trees are likewise "hurt" by the exchange of triples; for a ≡ 1 ≡ {c,e,g} ≡ {[a,b,c], [a,d,e], [a,f,g]} has become a ≡ 1 ≡ {b,d,f} ≡ {[a,b,c], [a,d,e], [a,f,g]}.

The sealed dwell as unity, under any of the eight seven-cycles shared by the now generalised octal (aligned to the reference), rather than in one pair of four – so that they may have "power" by (conviction from, and grace with) the one C7 group that circuits the seven churches and finds rest in Philadelphia as does the least. In parallel I draw a connecting line between the period of the end of the sealing of the saints and that of the resurrection of the two witnesses.

The chapter closes with the sealing of the true saints, those clothed with Christ's virtue as if that unity (now loosed free to float), yet in the six additive symbols (positions) not yet included in the reference octal so far in the text.

Those six further "choices" of unity – generalising the analogue of the 24 elders (automorphisms of GF(8)) separately giving a total of 6*24 = 144 states (of "floating" unity) within which the sealed are clothed with the virtue of the lamb. The text presents a factor of 1000 to indicate that God is patient to wait until a person shows fruit under predestination. The factor of 1000 years (cycles) of the octal are each of "seven months" and represent a thousand "years" in Revelation. Each cycle of seven representing the 24 + 144 = (7*24) states of "heaven".

2Pe 3:8 But, beloved, be not ignorant of this one thing, that one day is with the Lord as a thousand years, and a thousand years as one day.
2Pe 3:9 The Lord is not slack concerning his promise, as some men count slackness; but is longsuffering to us-ward, not willing that any should perish, but that all should come to repentance.

A day with the Lord (the reference octal's twenty four elders) is multiplied out by a factor of a thousand "years" through patience. If the Lord waits 1,000 times longer for predestined obedience (fruit) then He gains 1,000 times more elect. (The total of 144,000 is evenly split between the twelve tribes of Israel, the inheritance of the saints is not just divided equally amongst them – but without preference and to a completed Israel, with no deficiency.)

Rev 7:9 After this I beheld, and, lo, a great multitude, which no man could number, of all nations, and kindreds, and people, and tongues, stood before the throne, and before the Lamb, clothed with white robes, and palms in their hands;

Rev 7:10 And cried with a loud voice, saying, Salvation to our God which sitteth upon the throne, and unto the Lamb.
Rev 7:11 And all the angels stood round about the throne, and *about* the elders and the four beasts, and fell before the throne on their faces, and worshipped God,
Rev 7:12 Saying, Amen: Blessing, and glory, and wisdom, and thanksgiving, and honour, and power, and might, *be* unto our God for ever and ever. Amen.

Here are the same four groups of people referenced earlier in the lead up to the first seal in Rev 5:9. These blessed have attained a pre-first seal paradigm.

> **Multitude:**
> *A united people, whether of singularity as holding to faith on the virtue in the lamb (in praise of God) or equally, in full diversity, as crowds numbering 666.*

The redeemed, upright before God – clothed, acceptable to Him (with the same righteousness as the lamb) are again revealed, as the **completed** assembly of the saints: all those made "Kings and priests" to serve God, taking their place in heaven in deference to Him as the only one worthy to open the book of life. He alone (aside from the Father sat on the throne) is worthy as God to share those sevenfold praises; Jesus is one with His Father. (This acknowledged by all in heaven, taking their place under the lamb: as unity in the octal, sealed.)

Despite the present confusion over which octal is which in the sea, God is the one who saved them all and they all give Him worship, and defer their places subject to His.

This completion of the saints is the fulfilment of the promise in the fifth seal, finally finished in the giving of white robes to all those slain under the altar, and to those sealed, of all the saints on Earth.

The cycling within the four beasts occurs again – the elders and beasts fall down before God (not simply the lamb or to worship "He who lives forever"). Another of the seals in the static subgroup has been opened. This cycling of the K4 form and of the four beasts is coincident to the answer to those slain revealed by opening the *fifth* seal (shifted forward, cf. Rev 6:14) the fulfilment of the promise made to those saints under the altar (and their call for "vengeance" upon "those on the Earth") is here given after a wait of "a little season" or a shortened "three months" or "times" in the octal and K4 form – i.e. three elements (seals) of the static K4 group. (Unity in GF(8) is now become unity in GF(4) sat in place at the right hand.)

The delay from the fifth seal – then moved after the opening of the sixth is related to the "seven thunders". I would do better to consider God's answer to the last three seals with judgements from the throne: the fifth, completing the number of the saints (sealing the 144,000); the sixth, ministering to them through tribulation (the two witnesses until the end of the sequence of trumpets); and the seventh, sending the final wrath (the end of the sequence of vials) which is completed on the last day.

So, the last three seals correspond to:

Seal 5)   The sixth seal which leads all the way up to the seventh as every saint (as unity in

GF(4)) is saved.

Seal 6) The period approaching the seventh trumpet which concludes the entire ministry of the two witnesses after the completion of the angel's circuit. (The sealed are exeunt of the Church.)

Seal 7) The seventh vial is poured out on the last day and the book of life is revealed full and complete.

The last two seals are opened with the completion of the trumpets and vials. The sequence has somewhat shifted due to the action of the seven thunders. (I make the equivalence to Christ opening the last three seals with the reaction of the beasts and elders as in Rev 4:10. I saw before that the opening of the seals was only possible as by Christ (the lamb) and is alone fulfilled by Christ (as sat at the right hand of God), now recognised as God Himself: Rev 5:8, 5:14 respectively.)

The use of "Amen" (as the "zero" enthroned as the octal, or of the rainbow, the "generalisation" of the angel's circuit, chosen by God) aligns with the seven praises {Blessing, Glory, Wisdom, Thanksgiving, Honour, Power, Might}. Are these the praises due God's new name, or of indicators of salvation (the song of Moses by the seven churches)? I may match them up with verse 15: The lampstand (as overcoming Laodicea) before the throne is now filled with the elect completed!

Rev 7:13 And one of the elders answered, saying unto me, What are these which are arrayed in white robes? and whence came they?
Rev 7:14 And I said unto him, Sir, thou knowest. And he said to me, These are they which came out of great tribulation, and have washed their robes, and made them white in the blood of the Lamb.
Rev 7:15 Therefore are they before the throne of God, and serve him day and night in his temple: and he that sitteth on the throne shall dwell among them.
Rev 7:16 They shall hunger no more, neither thirst any more; neither shall the sun light on them, nor any heat.
Rev 7:17 For the Lamb which is in the midst of the throne shall feed them, and shall lead them unto living fountains of waters: and God shall wipe away all tears from their eyes.

The elect are identified as those that overcame, having come out of great tribulation (more selective election) upon their love for the truth in the gospel despite the removal of the restraint to their paradigm.

The question as to "whence came they?" is pertinent: the redeemed are become as the angels of God to serve Him in heaven; they take their place amongst the heavenly host without complaint from anyone present: they are worthy and need not be further tested, not by hunger for righteousness, thirst for life or patience in trial – they have their inheritance complete.

They held to virtue (unity) during their own "great tribulation", overcoming Satan's devices enabled by the fallen stars in the verse Rev 6:13 (i.e. from whence they came). They are, equivalently, the saints that heeded the two witnesses and were then sealed. They follow the lamb (in the seven cycle, the angel's circuit) to dwell in the K4 form shared with the middle ground of wormwood (see later). They are sealed into "fountains of waters". (Cf. Rev 12:14)

Dressed in the virtue of the lamb, they dwell as if they were before the throne always. They will serve their God forever. That is, God will never lack for servants but more importantly they are as spiritually clean as that lamb and are forever able to enter His presence freely.

Serving the Lord "day and night" they minister in the setting of the Sun octal and its left hands as well as in every other pairing of octals across the sea of thirty isomorphic. They dwell as unity, ministering with virtue. That unity is paired with the static subgroup (Christ enthroned) or equivalently, the isomorphic K4 form (as each of those three may be found together in one octal up to the choice of unity) as of the Sun or instead present cycled static in the "night". Unity is found to be a common element held with every such octal in the sea. Those "night" octals not sharing those three subgroups and also not corresponding to the same choice of unity have no intersection of that K4 form (as a **living** fountain of water) and so are not "washed in the blood of the lamb" – that blood being the intersection (unity) of the Trinity.

They will not go without – for they have no need to fast to draw close to their God: they are without question saved and will be refreshed for eternal life.

The "Lamb which is in the midst of the throne" (a clear reference to the unity floating through the static subgroup), i.e. that virtue, present in a disjunction (or in intercession) is shown. To decide $r \vee p \& r^{-1} \to s$ is to minister with virtue in p to those that are hungry as r^{-1} or thirsty as s^{-1}. In the static subgroup, however, the trinity acts as in the "Manifold of Virtue" (see section [3.10]). Then in the Trinity (in the colloquial), $r \,\overline{\vee}\, p \& r \to s$ instead ($r \leftrightarrow r^{-1}$). The Lamb "feeds them" and leads them, forming a true K4 group under a seven cycle: the elements of GF(8). (I.e. as "Living fountains of waters".) God will "wipe **all** tears from their eyes". In God, there is never a privation of the positive. By axiom of virtue, every intercession will be made.

6.10) The Seventh Seal Of Sabbath Rest

The seventh seal is representative of the same "rest" found in the cycle of the angel's circuit. As the identity maps every element to itself there is no principle here but rest only, and as one element in a cycle of seven I equate this to the "sabbath rest" believers inherit as co-heirs in Christ. This sabbath is the natural background in the heavenly scene, and in every cycle with every application of some map, there is the presence of the identity element also in action, and the loosing of a seventh seal indicates a completion of the circuit and return to that state which is normal.

Rev 8:1 And when he had opened the seventh seal, there was silence in heaven about the space of half an hour.

An "hour" is one automorphism of 24 over an octal. Half an hour refers to the coset of the static subgroup with four elements (the earthly four); however, there is silence (emptiness) here, and I must acknowledge that Christ in self-similarity between the static subgroup, unity and a K4 group of K4 subgroups (that span the octal) reveals in some sense through the ultrafilter that the coset becomes "empty" if Christ is actually God. Indeed, this verse states just so, and Christ's acceptance by all as Lord of Lords and King of Kings is completed. That is, He is equal in similarity to the Father.

There is a verse that directly alludes to the self-similarity between the K4 form of the ultrafilter and the static subgroup of the field GF(8).

Psa 110:1 A Psalm of David. The LORD said unto my Lord, Sit thou at my right hand, until I make thine enemies thy footstool.

This is to be taken to mean that even Jesus Himself is to be patient – as all are told in the gospels by Christ that not even He knows the time of the end. However, it is up to Him to satisfy Himself as to when to open the seven seals – they being on His own book; there for Him to "reveal" the names of His elect within as He sees fit. (Will He reveal more every time?)

6.11) The Keys Of Hell And Of Death

Rev 19:1 And after these things I heard a great voice of much people in heaven, saying, Alleluia; Salvation, and glory, and honour, and power, unto the Lord our God:

The trumpet devices over the sea permit the "worldly" kingdoms to race the restrainer, that the elect may also become leavened in "great tribulation". (Then "tares" are found sown amongst the harvest of believers.) The trumpets sequentially reveal the completed extent of those seal devices opened across the sea of 30 octals, revealing that same leaven of worldly commerce that has at that stage completely worked its way into the operation of the church – returned from the opening of each of the seals.

The world is always completely without restraint, but government is always somewhere between the righteous and the ungodly. (Or, in strict terms, always one or the other.) The result of the trumpets (as devices) is to create a system of "feedback" (compare it with the process of making the signal or "response" of an amplifier behave more linearly by "feeding back" part of the output signal inverted into the source signal). The church is always made to meet government halfway, as always dealing in terms of compromise (made for the allowance of those without Godly restraint).

The seals are opened when the church of the day is overtaken by those same devices (becoming leavened), or when the number of elect in the lamb's book of life living under that paradigm as revealed by the seal is then completed at the sounding of the relevant trumpet. (These two I find are equivalent.) Then, overcoming the world is found in an elect overcoming the paradigm of the day, rather than requiring in immediacy the sure totality of all grace. (It, the dialectic, is always the same device applied a maximum of fourfold, in static bows.) If one overcomes the paradigm of the day, it is simple enough to show one has restraint over that, and every previous paradigm of the seals, as if the restrainer were successively replaced.

Opening another seal permits a worldlier paradigm. This in turn is a "shock" to the newly chosen (the would-be but not made sure as yet) elect and so only those with a greater love for the truth (as above or at the paradigm of the previous seal) become those which Christ would choose (intercede for) as His elect.

So, one doesn't have to walk on the water to be able to learn to swim. (And then not simply sink, then failing to keep one's head above water.) It does help, however (to test one's talent), to be slowly introduced to deeper water from the shallows!

So, when the church is leavened to the utmost of whatever paradigm prevails, as of that current seal, the trumpet blows (each trump appropriately moved by thunder backwards in time from the opening of the book of life at the last day) and Jesus Christ is then free to open another seal and reveal those that are to become the further elect, also His chosen.

In this manner, by opening the seals, Christ righteously shows virtue by extending grace further to all those that are spiritually (and temporally) further from the cross. Likewise, His own satisfaction in (and expectations for) His elect need not diminish, never showing anything other than the same overcoming of this world (and the paradigms in it). By opening them, He will reveal both beforehand and afterward, that they hold a greater degree of love for the truth in (and of) the faith.

So, when it comes to opening the seals on the book of life (as for the elect as below, also compare Rev 19:1 above), I may make the following associations to the opening of each seal (and the sounding akin of the trumpet over the result, the consequences across the sea, or of the leavening then made complete):

r　　"Any person being elect" (will reveal those corresponding in the octal to salvation).

u^{-1}　　"Has not yet (but could) overcome the most (or in the church, the current) worldly paradigm" and are extended grace (will reveal those who will, by comparison instead, give God the glory for overcoming the world).

s　　"Has restraint without requiring grace" (the elect are made perfect by predestination), (reveals those that out of preference obey; they honour God as Father).

v^{-1}　　"Instead requires another seal to be opened before they will realise their own need for salvation" (virtue of salvation present upon conviction? Grace may be extended to save such a one), (I.e. power, Jesus has all choice, He is mighty to save even those "green things" revealed).

Where $q \rightarrow r\&u^{-1}$ is as "those under grace" compared to q^{-1} "not under grace as yet". Those in the sets "r" and "s" are written in the book of life; u^{-1}, v^{-1} are not so revealed written as yet.

Astonishingly, these four (as of the four beasts before the throne) form no K4 group in the right hand, despite all being very positive (salvation, glory, honour, power). So, in truth, I may find no right handed K4 subgroup in the above coset.

I am missing two elements and the identity (virtue) from the octal. I simply have to ask concerning all those in the book of life "What, exactly, are the keys of Hell and Death?"

Rev 1:18 *I am* he that liveth, and was dead; and, behold, I am alive for evermore, Amen; and have the keys of hell and of death.

Unity in the octal is an earthly element of the four as above, yet Christ as GF(4) also requires a unity. Then the intersection of GF(4) and GF(8) (unity) is as can be told, the "lamb" or "heaven" or equivalently Jesus' own "choice of election" or of His **intercession** as the seals are opened.

However, "Hell", is spiritual captivity (opening another seal would make no difference) and "Death" is perpetual sin without the guiding restraint and grace of the Holy Spirit. In other words, the keys preventing these are as u&v and (r&s)$^{-1}$ respectively, which both make sense.

The seventh element I may equate to the virtue in Christ's intercession as to who is to become His elect, and from his "mouth" or "set" I equivocate this as "p" to the two-edged sword from Christ's mouth (form) GF(4), cutting off to one side and the other. This intercession "p" is very closely related to "r" in the octal's coset above, but is deferred to Christ's choice instead.

Rev 1:16 And he had in his right hand seven stars: and out of his mouth went a sharp twoedged sword: and his countenance *was* as the sun shineth in his strength.

As to why "p" is not equal to "r" in the octal, "r" (if it be that particular virtue of unity) would then be an absolute and would set a clear dividing standard as found in the lamb, but it must also be unity in GF(4) (that field as the person of Christ). Unity, floated into the static subgroup as in the last three seals, must be treated as if it were Christ's choice of who to save from "Death and Hell", else Christ would have no further choice Himself, the Father then making all choices for Him. I find this to be the reason why the Father has sealed the book of life, that Christ should be given all judgement, as to those saved and chosen leading up to the end of all "great tribulation". Then, virtue in any Christian found responding to "election" (which is truly virtue on their part), instead becomes merely positive as then they remain "saved" (salvation as unity or one of the four beasts), being then found "under the altar" as stated in the fifth seal. Virtue (elective choice to save all those that truly desire it) is actually found in Christ within that setting (and also onward, to the approaching resurrection at the seventh seal).

Then the K4 groups corresponding to the following (and their rearrangements) all make complete sense over the set of all Christians, if you would like to check them!

$r \lor p\&r^{-1} \to s$ ... ("s", those not yet elect with restraint then require intercession.)
$s \lor p\&s^{-1} \to r$ ... (those elect and without restraint would also require intercession.)

$u^{-1} \lor p\&u \to v^{-1}$ ... (v^{-1}, the worldly that would respond to grace require intercession.)
$v^{-1} \lor p\&v \to u^{-1}$ ... (those not responding obediently to a shock also require intercession.)

$(r\&s)^{-1} \lor p\&(r\&s) \to u\&v$ ...(the elect with restraint require intercession in case of shocks.)
$u\&v \lor p\&(u\&v)^{-1} \to (r\&s)^{-1}$ ...(the worldly require intercession as they will later be elected.)

  r   "A person elect (or not so as r^{-1})" with restraint under grace (salvation).
  u^{-1} "Does not as yet (but surely will as u) respond obediently to grace" (glory).
  s   "Has godly restraint (or not as s^{-1}) without requiring grace" (honour).
  v^{-1} "Would only repent from a further shock (received as v)" (power).

Then Christ's intercession or Christ's elective choice "p" suffices as virtue in all the above.

Now, it is positive to intercede "p" for those saved elect r&s to ensure they hold firm to salvation even with a further "shock" (u&v) to their paradigm. Also, it is positive for Christ to intercede (as "p") for the "worldly" and disobedient that are caught in (u&v)$^{-1}$, those to be revealed in the book of life yet to overcome the world under His grace, i.e. (r&s)$^{-1}$. (Only Jesus has the choice of those for whom He intercedes.) The latter case of p&(u&v)$^{-1} \to$ (r&s)$^{-1}$ is as forward looking as p&r&s $\to$ u&v. As for the empty middles, no intercession is required for r&s without further "shocks", and $(u\&v)^{-1}$ appears to be ultimately reprobate without grace.

These two outcomes u&v and $(r\&s)^{-1}$ are the keys of "Hell" and "Death". (To always see the need to repent before ever being faced with an impossible fifth dialectic as in $(u\&v)^{-1}$, i.e. to be rescued from "Hell", or otherwise to be elect and overcoming the world respectively r&s, as sure rescue from any sense of "Death".) Without the "keys", the end result is dire.

These two outcomes are mutually exclusive, as together they private Christ's choice. All are either saved and in the book of life or not, all overcome showing restraint or not. Then $r\&u^{-1}\&s\&v^{-1}$ in the coset privates such free choice. (There would be no such grace or virtue extended.) In that case of $r\&u^{-1}\&s\&v^{-1}$ (which is not in the ultrafilter of virtue), all those with restraint (s) and whose names are not written in the book of life or revealed as such so far (v^{-1}) would also be as "elect" (r) and also not obediently "overcoming the world" (u^{-1}). This is not the election of grace, but is then a doctrinal "free for all" (as if "everything must go". It may very well seem positive overall, but not by exercise of virtue. There are simply no exclusions from election, and only privation of Christ's free choice).

Now, Christ does not use "Death" and "Hell" as devices to open the book of life. Rather, the fourth seal reveals those devices of "Death and Hell following after" of the adversary, and these two extra (non-earthly) elements (keys in Christ) which complete the octal ensure He will never lose a single one of His elect to those two devices (and to the worldly paradigm that prevails without virtue). They, as elect, each "know His voice" (a pre-first seal paradigm) and God is able to reign over the Earth and all its devices, all of which are instead "in place of Christ", being antichrist. Then, as found in Christ's free choice, the "keys" of election (rescue from Death and Hell) are simply the statements p&(r&s) and also $p\&(u\&v)^{-1}$.

Then these two "keys" fulfil the mechanism that completes the octal whole and delivers those Christ chooses from the spiritual captivity of only using the dialectic and from those spiritual conditions of "Death" and "Hell".

6.12) The Seals Align with the Trumpets

The book of life is sealed with seven seals. That said, it may not be opened until all the seals are loosed. Then how may Christ sequentially reveal those names in the book of life as the seals are loosed? The answer is that the combination of trumpets and thunders permit Jesus to choose when to open the seals when He has revealed and known (and chosen) those elect that live during that period under that paradigm.

The trumpets declare the moment when the next seal should be opened, each declares the underlying structure for the next seal to be already in place. The first trump before the second seal, the second before the third etc: that is until the fourth seal which is the utter limit of the dialectic.

Mat 13:33 Another parable spake he unto them; The kingdom of heaven is like unto leaven, which a woman took, and hid in three measures of meal, till the whole was leavened.

As the paradigm of the first four seals over the elect becomes broader – and given that a New Testament kingdom or empire (one of the four Beasts of Daniel) is set up to race the restraint upon the body of Christ's paradigm, the elect are "cooked in the pot" (as the frog sitting still in the slowly warming water). The trump, then, declares the moment when that paradigm becomes critical with regard to the restrainer being "taken out of the way".

When those that are elect overcome that much more "worldly" paradigm of each of Daniel's four beasts (as under the restraint of the Sun octal's static subgroup's coset of the four seraphim – the restrainer), any dialectic paradigm broader than that of the worldly kingdom of the day becomes "ineffective" – that is, at least until the trump is blown when that worldly system is overcome to its limit. (NB: The last enemy to be overcome is death, with the resurrection.) Then Jesus may righteously open the next seal, knowing His grace is sufficient for all the elect under a further loosening of restraint (as in the next seal) that was emulated by that same beast of Daniel, then overcome at the end of that epoch in which it arose.

Then it is clear that despite their apparent alignment to the seals, the kingdoms of the New Testament timeline "rise up from the sea" (cf. Rev 13:1), due to the trumpet devices acting across the sea of thirty octals – whereas the beast that exercises the "broader paradigm" in all dealings between the churches and the world (the same spirit always in dialogue), rises "up from the Earth" (cf. Rev 13:11) supplanting the proper operation of the reference octal to construct a "fifth dialectic".

Only one of the elect need overcome each worldly kingdom in order for the next seal to be opened. (But the trump is blown when their number is complete. I suspect the seven thunders uttered the names of those particular ones that overcame.) When that kingdom is overcome, so also is the broader paradigm of the next seal, so that grace may abound all the more. This is no instruction to revolution but is rather an exercise to show charity and virtue when worldly circumstance (or the state at the time) would rather the believer would not.

So, the seven trumpets are handed out after the seals are all opened; the names of the elect are all known to the heavenly host and with them the details of how they overcame the world. The shifting of the heavenly timeline (due to the same movement of heaven that is the "seven thunders") flawlessly and seamlessly (and so moving to a previous point in time or in the "narrative") reveals with each trump to Christ the right moment when the next seal in the sequence should be opened. That moment is known from those names in the book that are revealed saved from that same "great tribulation" as found in the verse:

Rev 7:14 And I said unto him, Sir, thou knowest. And he said to me, These are they which came out of great tribulation, and have washed their robes, and made them white in the blood of the Lamb.

Now, things are never so simple. The fourth trump is blown some time **before** the opening of the fifth seal (with the fifth trumpet in alignment to the same fifth seal) as it represents the very limit of the application of the dialectic across the sea of octals, and this shows the system Satan employs effectively stalls. The sixth and fifth trumps also align on earth and the heavens respectively in the fulfilment of the verse:

Mat 24:22 And except those days should be shortened, there should no flesh be saved: but for the elect's sake those days shall be shortened.

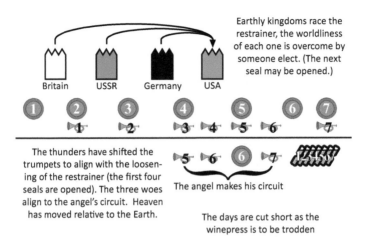

Earthly kingdoms race the restrainer, the worldliness of each one is overcome by someone elect. (The next seal may be opened.)

The thunders have shifted the trumpets to align with the loosening of the restrainer (the first four seals are opened). The three woes align to the angel's circuit. Heaven has moved relative to the Earth.

The angel makes his circuit

The days are cut short as the winepress is to be trodden

The angel makes his circuit between the opening of the fourth seal and the sounding of the seventh trump in order to overcome all three woes, but as the sealing of the saints on Earth aligns with the fifth seal in the heavenlies, and the fifth trump on Earth aligns soon after the fourth seal is opened in heaven: the angel must complete his circuit sometime between the opening of the fifth seal and the sixth in the heavens (that moment is realised to be the same as the thunderous movement of the last trump (brought forward on Earth from the judgement) and that last trump is separated from the last day only by the treading of the winepress).

The "last day" may well be the period between the effective sounding of the seventh trumpet as sounded "on Earth" and the opening of the seventh seal in the heavenlies (during which the last vials of wrath are poured out).

So, the first four seals are aligned to the first three trumps and their utter limit in the Earth aligns to the fourth trump to indicate that the elect surely overcome the world and grace is abounding all the more. Then the last three trumps are truly aligned with the circuit of the least in the kingdom (as written within the seven churches of the letters). The angel overcomes the world and does so as the least and under the worst circumstances; he is a living proof of the salvation of God in Christ.

As to the start of the "Great Tribulation" (and any confusion of the preterist), I have the verse in the Olivet discourse:

Mat 24:21 For then shall be great tribulation, such as was not since the beginning of the world to this time, no, nor ever shall be.

And that tribulation starts the moment the "government" of Judea falls to Titus. There is then a lack of any compatible (government) paradigm within which the early church may exist.

6.13) Firstborn Of The Resurrection

There is no direct reference to the cross and Christ's resurrection in the Revelation. There is, however, a great deal on the consequences of those two events. Those that are seen by John "under the altar" and in the two parts cut off in Christ, found separated to Christ's right and left in the K4 form (in the form of the triples [b,d,f] and {c,e,g}) are the living in the resurrection, though they are not present as that "intersection" in the heavenly scene: represented by the Lamb (and $1 \equiv a \equiv$ [a,b,c] say, by Christ) on the altar. (They are not under grace in the heavenly "present", but in the future in some {d,e,f,g} instead only in the GF(4) ultrafilter, of Christ.)

John had wept for the reason that no one was found worthy to open the book. Now, I ask you, the reader, to put aside your initial belief that Christ was the opener of the book and did so only after His resurrection (I recover that), and to consider what would have occurred if He had opened the book before His ministry! A pre-crucified Christ would also be found able to open the book, in the context that He is also as GF(4) able to sit at the right hand and that He also has all the virtue required (as of the Lamb, the image of the crucified Christ, for whom Jesus would have to account as that one Messiah crucified), as a fitting atonement for all sin, is a given.

As the heavens are outside of normal time, opening the seals would occur along similar lines as one would expect, but on opening the three in the static subgroup whilst so sat enthroned, as the Lamb unseals those revealed under the altar and those 144,000 also yet to be sealed in their foreheads, it becomes apparent that with the seventh seal, in self reference somewhere amongst them all, Christ Himself is also raised! (Or, necessarily alive beforehand, if you like!)

The sequence of trumpets are shifted back from the judgement (with seven thunders) by that same raised Christ (judging on the throne) to appropriately align to the seals, and from the judgement seat Christ elects whosoever He chooses and so writes their names into His book of life. Jesus is Himself necessarily alive as the a-priori resurrected: clearly raised three days after the cross upon which he was crucified. As the first name revealed (it is His book, after all), He is immediately raised to open those seals (revealing all those names in sequence), having returned from His completed ministry worthy to open the book itself. Then, Christ's resurrection is also a prerequisite to the judgement and the whole election of grace.

Joh 10:18 No man taketh it from me, but I lay it down of myself. I have power to lay it down, and I have power to take it again. This commandment have I received of my Father.

Then with the opening of the seventh seal (with $1 \equiv a \equiv$ [a,b,c] $\equiv$ [[a,b,c],[a,d,e],[a,f,g]]) Christ is always raised (on the "third time" or "day" in the static subgroup) in self-referencing Himself (and has shown in Revelation all this in a manner of His own choosing, that He is necessarily alive, not merely existent) and also referencing the Lamb on the altar and this all done in the third part not "cut off", but brought "through the fire" as the Christ in whom all the elect are refined as silver is refined and tried as gold is tried.

Zec 13:9 And I will bring the third part through the fire, and will refine them as silver is refined, and will try them as gold is tried: they shall call on my name, and I will hear them: I will say, It is my people: and they shall say, The LORD is my God.

6.14) The Devil's Devices To Overcome

As the Lamb floats through the static subgroup, the local closure of GF(4) as sat within the heavenlies permits a two cycle under Frobenius, as if in the octal itself. Whilst the elect remain under a seven cycle which will **never** allow any of those automorphisms of GF(4) (they are outside their paradigm) those within the "Church" amalgam face the instantiation of a five cycle in the spirit, as well as the third trumpet wormwood device applied in the "sea of glass" of every octal aligned isomorphic. (The moon truly becomes "as blood", traversing between octals in that wormwood system found in the sea: starting from the reference octal.)

Whilst no alignment of octals in the sea of glass permits the wormwood under a seven cycle, those unsealed with a broader paradigm are able to spiritually instantiate the "pneuma" of the false prophet up to the inclusion of a five cycle. This is done by the locusts of the fifth trump as well as by the second woe – the same army of horsemen. The "keys of hell and death" are then supposedly used across octals to instantiate the five cycle, becoming the "horse bridles" – the winepress is then swiftly to be trodden out. (There is a "vast falling away".)

The elect, then, always with the Holy Spirit are then sealed preserved from these spiritual (now actual) constructs of "Death and Hell" (the five-cycle and the lake of fire as partially constructed.) Later, these two are cast into the lake of fire: they will spiritually exist.

The fifth seal is completed with the wormwood device "stalled" by preventing the reference cycle to act (held back from blowing by four angels) until after the saints are sealed. (This is aligned to the fifth seal from the sixth by the seven thunders.) As the two witnesses minister against those practising wormwood already (the whole world), the elect that are sealed exit the Church amalgam, on seeing those devices enter their fellowships (under conviction by the angel's circuit). With the completion of the sealing of the saints, the witnesses are killed (and rest) and are shown resurrected and approved in a seven cycle – far from the five cycle now become the cage of "every unclean and hateful spirit".

Those without the Holy Ghost that must go into the "Church" and face these very real devices to learn of Christ are to be tried like none other: they are beheaded from Christ (as unity, rest) if they fall to a five cycle and are then trapped in a very real spiritual cage: "the lake of fire". In order to exit to find Christ's approval they must find the "open door" and become wise with sound doctrine. Christ, with the keys of "Death and Hell" as previously shown, may elect any individual clearly not caught by both (they are redeemable).

Likewise, in exercising both those devices of the first two woes, one is caught in the cage of the third (always outside and without Christ forever, to dwell without the Holy Spirit and to be always turned back from that same open door).

There is no sketch proof or thought experiment for entering Christ's approval: to repent takes real action. Faith is required, not reason!

Chapter Seven: The Seven Trumpets

The trumpets provide a sequence similar to the seals. Previously the position of unity was free to be chosen in the octal whilst holding the static subgroup constant. Now, the static triples are free to be chosen instead of unity and the sea of all 30 octals becomes traversable through common operations on triples and subgroups in the trumpet's given examples.

I begin with a reference octal – the "Sun" as mentioned in the text. (A bright source of light (virtue) and a great attractor – of worship.) When unity and the static triple may float under a seven cycle, there are a set of eight complement "left hand" octals to the "right hand" Sun octal (one for each possible C7 group over the "Sun octal"). The wider sea also becomes accessible from sequential application of left hands of left hands etc.

The first trumpet shows the octal (as in the first seal paradigm) generalised to permit any valid C7 group (as in the second seal), the second trumpet the basis for the third seal, the third the basis of the fourth. The fourth trump concludes the "earthly" sequence with the whole sea now traversable rather than properly organised into a lattice; the fourth trump shows the sea as only partially traversable with the dialectic, and limits its application to four "bows" only.

Each of the first three trumpets show a structure already exists for combining a further floating or static triple with the current paradigm (enabling the dialectic): from two to four concurrent – a basis for those dialectic devices of the seals placed in overlay upon that structure.

The last three trumpets ("woes", mysteries misinterpreted by the Church throughout history), are disasters for the faith. (As "horns" they cannot be deprecated.) They reveal the extent of the devices of Satan employed in the "sea" to deceive those dwelling on the "Earth": even reaching up to "heaven" (election, to be clothed with the lamb, unity) to exclude all possibility of the Father (Sun octal) being consistently worshipped by the deceived, making it implausible (devouring the least in the process by it) that there should be any salvation for them.

Satan intends to replace obedience with his devices, with the removal of all restraint across the "sea". Jesus is only concerned with His elect; the world, always a Leviathan, is without restraint. (Not at all like a Behemoth – a beast chief in the ways of God and a "gentle giant".)

As to timing: the least overcomes (to show God's total sovereignty) at the height of deception, during the three woes. (Only the first two as "horns" on the false prophet are complete before the judgement); there will truly be "time no longer" when the last trump is prepared to be blown or "sounded". The two witnesses' ministry overlaps the conclusion of the angel's circuit: he is instructed to remain in Philadelphia (hold fast his doctrine) so that he remains in Christ's approval whilst the two witnesses minister to the elect to exit the "Church" of the last days.

That ministry is empowered by the least overcoming all seven churches (and false doctrines). As that includes Philadelphia, the scarlet beast which is the final state of that "Laodicean Church" (the form it will take if completely lost – and God does form it so in judgement) will kill them (oppose them as desolate, making them ineffective). The beast still ascends during the angel's circuit and the witnesses are "killed" by the beast only after the angel's circuit is finished: the beast with its deadly wound of Philadelphia's exit (then healed) overlaps the circuit of the angel, yet the beast is to be made complete only as the lake of fire.

The days are cut short (else no **flesh** shall be saved by that "Church", now desolate of all elect). The scarlet beast is as any iniquity: never an adversary to God (grace is sufficient), it instead traps tares practising the dialectic within it. (It is a cage for every unclean and hateful "bird" – a "spirit" with all four limbs off the ground as lacking any restraint as with the fourth seal.)

Jesus alone has the keys of Death and Hell: He alone looses Wormwood ("Death") upon those "dwelling on the Earth", saving His elect from it. Only He closes the door on the lake of fire (the scarlet beast completed in His judgement) to recompense the image (the woman) double for all she (actually those that worship her) has done. (Abaddon is the spirit called "Hell" following after "Death".) Those with part in the first resurrection (and born again) – on them, the second death (third woe) has no power. (And the lake of fire is the only second death: a combination of the first two woes completed to cage those within; the last woe is to be greatly feared.)

The seventh trump occurs on the last day with the pouring out of wrath. God's wrath is sent solely for turning that scarlet beast with its image into the lake of fire: I find the seventh seal opened. God, by judging, turns the sinners' ways upon themselves. In worshipping a false gospel they deserve one. (Just as for murdering for "living water" they deserve blood to drink.)

The age is over with the seven vials of wrath, morphing the scarlet beast into the lake of fire – that third woe at the seventh trumpet. The woman upon it is purely a "nonsense", a fiction or mystery (a lie there is some captivity or "Babylon" of Satan) which warrants no additional salvation. The Church is become the scarlet beast: the woman an image which is formed over their collective sin. The judgement arrives with Armageddon/Gog and Magog: it is the final dividing line in the sand – the separation of the raised unjust apart from the raised in Christ.

7.1) The Church Corrupted

Prayers offered to God with incense are answered despite becoming a displeasure to hear, not because of the evil suffered by the elect but for their scattered numbers (a dearth of wheat – spiritual famine) amongst a great many tares falsely acting (as "Death" and "Hell") in Jesus' name. I examine how in the last three trumpets of the seven. (God takes no pleasure in smiting the sinner – but only in answering those that please Him with faith.)

Rev 8:2 And I saw the seven angels which stood before God; and to them were given seven trumpets.
Rev 8:3 And another angel came and stood at the altar, having a golden censer; and there was given unto him much incense, that he should offer *it* with the prayers of all saints upon the golden altar which was before the throne.
Rev 8:4 And the smoke of the incense, *which came* with the prayers of the saints, ascended up before God out of the angel's hand.
Rev 8:5 And the angel took the censer, and filled it with fire of the altar, and cast *it* into the earth: and there were voices, and thunderings, and lightnings, and an earthquake.

There are four choices for unity (static subgroup constant): each requires one of a pair of C7 groups (there are two for each choice of unity) as multiplication. The unity of the reference (the lamb) is shown here on the altar. The "prayers of saints" are accepted if they are of the one set of virtue agreeing with the lamb, and the remaining three which are not compatible with the lamb (being together a broader paradigm), are instead mapped onto the three earthly elements (not unity) in the four beasts and those prayers are "cast back to the Earth". The "fire of the altar" is the seven cycle preserving the lamb as unity, and not those others that would hold fixed some alternate unity.

I assume the incense is to mask the distaste God has for any broader paradigm: an aberration of proper worship (cf. prayers and vials of odours – the pleasing prayers of saints).

Every multiplication on the reference holds constant one static subgroup/bow pair (say, $a \equiv 1 \equiv [b,d,f] \equiv \{c,e,g\}$ as one of four cases – here $a \equiv 1$). The four possible static triples (under those eight choices with unity fixed) together with the three subgroups containing unity (which three are cycled static) form an octal (wormwood), a middle ground in the sea.

$$[a,b,c],[a,d,e],[a,f,g] \ with \ [b,e,f],[b,d,g],[c,d,f],[c,e,g] \ with \ unity \ a=1$$

Whilst the prayers of the elect (as unity or as here the static subgroup [b,d,f]) offered with incense are always accepted; the remaining three earthly triples in the "wormwood octal" do not form octals under the seven cycle that holds the fourth triple (here $\{c,e,g\}$) static. They also permit a dialectic device; the left hand static subgroup is now as a right hand static subgroup.

The other three earthly subgroups within the wormwood octal are also filled with "fire from the altar" (the altar represents the lamb's four positions in the earthly elements) and cast back to the earth. That is, the prayers of the lukewarm practising the dialectic are cast back "to the earth" as if they belonged in the same Sun octal (as if the three triples were mapped onto one). It is clear these prayers are answered but only with rejection and distaste: they are treated as if they were of antichrist $\{c,e,g\}$ one and the same. There are voices (judgements separating [b,d,f] from $\{c,e,g\}$), lightnings (falls from grace $a \equiv [c,e,g]$ of a static subgroup to a triple $\{c,e,g\}$) and a shaking of the earth (the remaining three triples are mapped onto one).

If the redeemed are complete, taking their place in deference to the lamb by whom they are saved (and dwell clothed with that lamb – unity) all are (likewise) preserved in heavenly places and so as unity in the K4 form as well. The static subgroup excludes the four choices for unity in the octal; the K4 form sat in the right hand permits the unity within it to be found in those last three positions as if the K4 form replaced the static subgroup. The censer with fire of the altar is cast back to the Earth (the four earthly elements) outside of Christ. (With the prayers of the elect, who are to be tried on Earth with the three woes also. Yet, unity is mapped to unity!)

Should I do better to consider the thirst for vengeance of all the elect separate in Christ (as in the fifth seal) now answered? Those saints still dwelling in earthly places as unity in the octal, it must be expected, should patiently endure trial. (As they are sealed of the sixth seal.) Yet this sequence of trumpets is removed from its place – the seven thunders (a great movement of heaven) shift the three woes from the complete sequence of trumpets backwards in time to between the opening of the fourth and the end of the sixth seal period. (Then there is realignment, and the first four trumps align somewhat with the opening of the first four seals.)

The K4 form sat as the static subgroup as in the last three seals (with the elders and four beasts singing a new song) opened by Christ had shown the faithful within the two parts cut off in GF(4) rather than present in the octal's unity element, for it was required to be "as if moved" (as if floating) into the static subgroup under the seal. Precedence to answer the pleas for vengeance seems to be given by the prayers being cast back to the Earth, as rejecting those people upon whom "all the righteous blood since Abel has fallen" (culminating at the very end of the age in a time of trial under all the devices of the dragon, for the "manifestation of the sons of God"; when the least is not (and then none are) deceived).

The saints on the Earth also dwell in unity in Christ; $a \equiv 1 \equiv$ [a,b,c] say, and not in {d,e,f,g}, Then those prayers cast back to Earth are done so as to the triple {c,e,g}. The two parts {d,e,f,g} cut off "in all the land" (Zec 13:8), are taken from the earthly elements in the K4 form {d,e,f,g}, cut off from the members of that triple present as unity ($a \equiv$ [a,b,c]) in Christ. Those prayers are cast "to the Earth" as {c,e,g} (to Christ's left) in the octal rather than to $a \equiv 1$. They are returned with vengeance upon those from whom the cost of righteous blood will be required. (This excludes the elect though they be on Earth also.)

The fire of the altar would float the K4 form through the octal, generating a completed union in virtue of all prayers made with a pre-first seal paradigm in the seven positions of unity (24 elders plus 144,000 sealed) – and with the agreement of all that they be answered as upon the lamb's promises (Christ's testament) to be fulfilled before the final Sabbath rest (at the seventh seal) returns in heaven. This union would answer the prayers of all saints. Those made of tares have a different answer – rejection.

The answer to these prayers (and the last set so answered) offered to God involves them being flung back to Earth. This verse leads on toward the very last day, when the Church has been exited by all the elect, and the false doctrine and spiritually lukewarm state of the Laodicean amalgam has reached its full height. Until then, the prayers of the faithful not yet convicted by the least's circuit are cast back to the earth – that they should become so obedient to the Spirit and to leave that captivity.

The judgements correspond to the separation to the left and right of Christ: of the elect and the not so chosen. Then the stars/tares fall "from heaven" (as from static subgroup [c,e,g] to the "bow" {c,e,g} with the inverse of the reference's seven cycle, its reciprocal descending/ falling from heaven (1/x) to John) and are cast to the earth to close the devil's construction of S5 in the Laodicean amalgam: those that fall from grace are judged to do their true Father's works with the dialectic. Those left in the Church living at ease are to be cut off, as also is the cult (false prophet or "little horn") now becoming the scarlet beast: the replacement of "Hell", supplanting Christianity.

The offering is made with **much** incense because prayers for the elect are so few and far between in the flock of that last "Church". The incense is enough for the offering to be answered and the judgement cast back despite the distaste of the prayers accompanying those of the righteous saints. 7000 die (fall from grace as **lightning** as if without conviction by the least's circuit, as "untimely figs" in a five cycle – lukewarm and as stars fallen to Earth, see above) in the **earthquake** as the tenth part of the "great city falls". (See later in this chapter.) Likewise, the "fire from the altar" aligns with the two witnesses in their ministry – that:

Rev 11:5 And if any man will hurt them, fire proceedeth out of their mouth, and devoureth their enemies: and if any man will hurt them, he must in this manner be killed.

...which dove-tails or aligns with the ministry (first seal paradigm) of the two witnesses in extracting the precious few from the greater flock. I then have the presence of judgements (voices – as if answering the call to leave the Church).

7.2) The First Trumpet

The trumpets in sequence mirror the seals, in that whilst the unity was free to be chosen of four elements in the seals, here the static triple itself is free to be chosen from triples upon (over) the Earth. There are four such arrangements in the octal, with the first of the remaining two trumps (before the Sabbath rest is attained) aligned to the woes that correspond to the constructions (and devices) of Satan named in the text as "Death" and "Hell".

Rev 8:6 And the seven angels which had the seven trumpets prepared themselves to sound. Rev 8:7 The first angel sounded, and there followed hail and fire mingled with blood, and they were cast upon the earth: and the third part of trees was burnt up, and all green grass was burnt up.

Beginning with a singleton element of the octal (without loss), I assume it is unity so that perfection may act downward upon the octal by a valid seven cycle (preserving subgroups) to likewise act upon the static subgroup and the coset's static "bow".

So now I have the following under the seven cycle (a,b,d,c,f,g,e):

$$(a,b,d,c,f,g,e)$$

a	$[b,d,f]$	$\{c,e,g\}$
b	$[c,d,g]$	$\{a,e,f\}$
d	$[c,e,f]$	$\{a,b,g\}$
c	$[a,f,g]$	$\{b,d,e\}$
f	$[b,e,g]$	$\{a,c,d\}$
g	$[a,d,e]$	$\{b,c,f\}$
e	$[a,b,c]$	$\{d,f,g\}$

This shows that there is truly a generalisation of the first seal paradigm – an isomorphic structure not requiring the same unity – allowing for a further bow (static triple), as of the second seal, walked through the octal by the seven cycle.

John no doubt read from right to left: so, the action of the seven cycle above would be considered to him to "burn upwards" (the Holy Spirit is as "lamps of fire"). Though John would have considered [b,d,f] static as well as {c,e,g} also, he would have seen the seven cycle reversed, still preserving the subgroups of the right hand octal in the middle column. The octals are preserved but I would consider John to have used the cycle's inverse. From "here on in" the right-handed (centre column) octal is as "fire" and the left handed (rightmost column) is as "hail" because the seven cycle appears to "burn up" the right hand yet when the left hand's additive group is aligned isomorphic to it, multiplication seemed to "fall from heaven" on the left hand – at least in appearance to John (and contrary to the above figure). I may be assured he understood the left handed octal (rightmost column) in which [c,e,g] is a subgroup (which is also held static by Frobenius) is also preserved under the seven cycle above.

As to falling from heaven, I note the inverse of the seven cycle x is the reciprocal 1/x and the cycle "descends" as by difference to the "heaven" (unity).

As you may note, unity in the singletons (the "head" of the cycle) is aligned to the static subgroup in the middle "fire" column whereas it aligns to the static bow in the rightmost "hail" column. In subgroups I then have the spiritually hot, and in the bows, the spiritually cold. Under the operation in each octal formed by $(A \vee B)^c$ I find that the operation of addition in the fire column in the centre is not aligned (but reversed) with the addition of singletons in the seven cycle of the Holy Spirit, whereas the octal of bows (rightmost column) indeed itself is now aligned to the operation of addition upon the singletons. The "fire" of the Spirit is in the seven cycle descending, but in this case is exemplified by the centre column "burning upwards" rather than the "hail" column which "falls down" in alignment. It is also the octal's (additive) operation (the product or "third part") mainly in view and not the seven cycle.

The right hand column is actually the left handed octal and vice-versa. I do not write from right to left. So, my seven cycle should indeed be reversed as if I were a Roman-era Jewish fisherman. (Today's believers are also the recipients, not just John.)

Then in the field GF(8) the left handed (hail) octal aligns to the additive operation overlaid in place upon the seven cycle (as if "mingled" with blood – the operation(s) aligned in agreement). This alignment of elements is to the **intersection** of subgroups/triples in the opposing hand (column) so that the intersection (of unity) is become "as blood" circulating (with the action of C7, as GF(8)*) through the system. Then I find "hail and fire mingled with blood". The element of that seven cycle aligning to the intersection is become the "blood".

Lev 17:11 For the life of the flesh is in the blood: and I have given it to you upon the altar to make an atonement for your souls: for it is the blood that maketh an atonement for the soul.

Every relation is preserved under the seven cycle. (There is no lost structure.) The intersection of groups (or "blood" mingling as the element "a" as in [a,b,c],[a,d,e],[a,f,g]) in the centre column corresponds to triples (a ≡ {c,e,g}) in the left hand and those elements of those triples ("c", "e" and "g") in the right-hand seven cycle of singletons correspond only to those same three subgroups ([a,b,c],[a,d,e],[a,f,g]) of which the intersection was made.

This is found true for either hand octal, they mingle with "blood" and are mapped (cast) to the "Earth". These relations hold over any valid seven cycle upon an octal.

Triples in the left handed octal (a ≡ {c,e,g}) that comprise the earthly elements are always found in correspondence to a subgroup of the right handed octal ([b,d,f]) in the centre column, but never to any singleton it contains. The right handed subgroups are truly "cast to the Earth" as say [0,b,d,f] to only a triple and singleton taken from {a,c,e,g}.

> **Trees, Green Grass:**
> *Trees represent branching or a congruence of singletons to triples. Trees are similar to the K4 form (the golden girdle) which is also as a "fountain" of waters (a "tree" under a seven cycle). Trees are also the triples of K4 groups with intersection not unity. They are found "in the Earth" (with unity) if they are held static under Frobenius. "Grass" references both the unity and singletons in the octal congruent to the three right hand K4 groups together forming "trees".*

The "third part" of trees are burnt up; the product of subgroups as forming a K4 group is **preserved** under action of the seven cycle. The third part of "a and b" is c = a+b. Every branching of singleton to triple is moved throughout its parent octal: there are no non-trivial homomorphisms; every map is 1-1. Likewise, the singletons in those triples are also acted upon in collective by the seven cycle as if by their branching to triples they are as "smaller trees" or "green grass" (the earthly singleton elements under C7 in measures of "wheat" and "barley", cf. the third seal are preserved, the separation of wheat to barley is preserved). I then see a simple description of closure under the operation of C7. There are no exclusions in the octal. Every element and triple is cycled: none others remain.

> **Third Part:**
> *The third part is "the sum of" or "the product of". I would know the third part Z of X and Y, to be Z = X+Y.*

John read the seven cycle right to left. The octals above appeared to be "multiplied" upward.

Exo 3:2 And the angel of the LORD appeared unto him in a flame of fire out of the midst of a bush: and he looked, and, behold, the bush burned with fire, and the bush *was* not consumed.

$$\overrightarrow{(a,b,d,c,f,g,e)}$$ (To John, the seven cycle reads right to left.)

a	$[b,d,f]$	$\{d,f,g\}$	e
b	$[c,d,g]$	$\{b,c,f\}$	g
d	$[c,e,f]$	$\{a,c,d\}$	f
c	$[a,f,g]$	$\{b,d,e\}$	c
f	$[b,e,g]$	$\{a,b,g\}$	d
g	$[a,d,e]$	$\{a,e,f\}$	b
e	$[a,b,c]$	$\{c,e,g\}$	a

With the additive operation of the left and right hands in alignment (isometry), the seven cycle to John "burns up" and also "hails down".

7.3) The Second Trumpet

John swiftly moves to the combination of two "static" subgroups for a given unity shifted in each octal in the sea;

Rev 8:8 And the second angel sounded, and as it were a great mountain burning with fire was cast into the sea: and the third part of the sea became blood;
Rev 8:9 And the third part of the creatures which were in the sea, and had life, died; and the third part of the ships were destroyed.

The "great mountain" (or spiritual stronghold) is an operation that appears to give a "lattice" of isomorphic fields GF(8) across the "sea" of thirty. It fulfils the requirement that one could pass from one octal to another as if "blood" circulating. In effect, given two octals, such a mountain is a method (albeit a quasi-algebraic one, as if "burning with fire") producing a product of octals "A + B". The "third part" in this pseudo-addition becomes the intersection (blood) found of any two "waters" (right/left handed octals under a seven cycle) with an arbitrary third.

That mountain is cast into the sea (mapping within it across octals) and all the sea is able to circulate through this "almost-addition". The third part (product) of creatures (elements as octals) in the sea die: their octals combine with the pseudo-addition and the proper operation perishes, ending spiritual "life" steadfast in God. That operation removed, the **ships**, equivalent to "trees" in the sea (i.e. K4 groups made of triples intersecting in a singleton), also perish; the pseudo-operation between the octals now affects them upon their singletons also.

So, what is this mountain?

Given the reference and the hail and fire columns in subgroups and triples as before.

$$
\begin{aligned}
a &\equiv [0,\ b,\ d,\ f] \equiv [0,\ c,\ e,\ g] \quad * \\
b &\equiv [0,\ c,\ d,\ g] \equiv [0,\ a,\ e,\ f] \\
d &\equiv [0,\ c,\ e,\ f] \equiv [0,\ a,\ b,\ g] \\
c &\equiv [0,\ a,\ f,\ g] \equiv [0,\ b,\ d,\ e] \\
f &\equiv [0,\ b,\ e,\ g] \equiv [0,\ a,\ c,\ d] \\
g &\equiv [0,\ a,\ d,\ e] \equiv [0,\ b,\ c,\ f] \\
e &\equiv [0,\ a,\ b,\ c] \equiv [0,\ d,\ f,\ g] \quad £
\end{aligned}
$$

Given any other valid seven cycle on the same octal I now may write:

$$
\begin{aligned}
a &\equiv [0,\ b,\ d,\ f] \equiv [0,\ c,\ e,\ g] \quad * \\
b &\equiv [0,\ c,\ e,\ f] \equiv [0,\ a,\ d,\ g] \\
f &\equiv [0,\ c,\ d,\ g] \equiv [0,\ a,\ b,\ e] \\
c &\equiv [0,\ a,\ d,\ e] \equiv [0,\ b,\ f,\ g] \\
d &\equiv [0,\ b,\ e,\ g] \equiv [0,\ a,\ c,\ f] \quad \$ \\
e &\equiv [0,\ a,\ f,\ g] \equiv [0,\ b,\ c,\ d] \\
g &\equiv [0,\ a,\ b,\ c] \equiv [0,\ d,\ e,\ f]
\end{aligned}
$$

I note that I may choose any other (arbitrary yet valid) seven cycle upon the same octal to find:

$$
\begin{aligned}
a &\equiv [0, \ c, \ d, \ g] \equiv [0, \ b, \ e, \ f] \\
c &\equiv [0, \ b, \ d, \ f] \equiv [0, \ a, \ e, \ g] \\
d &\equiv [0, \ b, \ e, \ g] \equiv [0, \ a, \ c, \ f] \ \$ \\
b &\equiv [0, \ a, \ f, \ g] \equiv [0, \ c, \ d, \ e] \\
g &\equiv [0, \ c, \ e, \ f] \equiv [0, \ a, \ b, \ d] \\
f &\equiv [0, \ a, \ d, \ e] \equiv [0, \ b, \ c, \ g] \\
e &\equiv [0, \ a, \ b, \ c] \equiv [0, \ d, \ f, \ g] \ £
\end{aligned}
$$

And there is a unique correspondence between these three: they have intersections (the third part become "blood") marked with *, $ and £. The "congruence" in the seven cycles to these rows (pairs of triples) show that I may take any octal in the sea, then three arbitrary seven cycles over the original octal and always find a "bridge" between them that form a subgroup. (Here the shared elements in the seven cycle form the elements [a,d,e].)

The third part of the creatures that were in the sea that had **life** (dwelled in the Sun octal) died – that is, the operation on the octal under the mountain "device" shifts them from one seven cycle to another and into a different field GF(8) in the "hail" columns or left hand octals. They are shifted from their position as they now dwell in the **sea** rather than in heavenly places as in the Sun; i.e. the worldly operation of pseudo-addition is upon the octals as sets rather than singletons as elements. In view is the "third part" as the product, the relationship to any third **arbitrary** seven cycle. The "product dies" as the choice of the third part is arbitrary.

As above, the common (blood) markers across the three octals form a K4 subgroup, but the "third part" of that subgroup in each octal is missing. When the rows corresponding to "a" and "d" are present that octal is become as "e" in the third part with life (the proper octal operation), but that it is missing in the "mountain" indicates that the creature "died".

Likewise, the K4 subgroups corresponding to those creatures/triples (always found in the left hand or "sea") which together form a K4 group under (A ∨ B)ᶜ have their "third part" missing or "dead" – the third part of the "ships" are destroyed.

I find a "mountain" (device) formed of the "third part that dies" (kills the proper operation) over seven cycles (in the sea) – attained from any three choices of seven cycle. The result of this mountain's "pseudo-addition" is arbitrary and is "without life". This third part made of the sea "became blood": as found only within the **intersections** between these octals in pairs.

7.4) The Third Trumpet

Likewise, I move to the next trump where I have the equivalent of three subgroups shifted with respect to unity. Then I will show the results of three shifted subgroups in the sea.

Rev 8:10 And the third angel sounded, and there fell a great star from heaven, burning as it were a lamp, and it fell upon the third part of the rivers, and upon the fountains of waters; Rev 8:11 And the name of the star is called Wormwood: and the third part of the waters became wormwood; and many men died of the waters, because they were made bitter.

A "great star" fallen from heaven (as with an otherwise valid K4 form or "star" – this device "falls" from heaven as a seven cycle improperly applied: the reciprocal 1/x acting upon the octals formed by the alternate logical schema). It corresponds to a device that is laid upon the mountain device of the second trump and it "burns" as if it were a great lamp (cf. the operation of the octal referred to as lamps in the throne scene). Able to be mistaken for the Sun octal by those in the seven cycle (it appears to induce a "floating unity"), the device then has an outward appearance of being "rock solid" whilst using the alternate logical schema.

Adding to the last device of a mountain as a "lamp" (completing K4 groups), I choose any three seven cycles as before: and there is then a fourth filling in the remaining associations of the previous three in a mirror-image of the "Death" system of four balances (from the fourth seal).

From the previous example;

$$
\begin{array}{llll}
a & \equiv & [0, \ b, \ d, \ f] & \equiv & [0, \ c, \ e, \ g] & * \\
b & \equiv & [0, \ c, \ d, \ g] & \equiv & [0, \ a, \ e, \ f] \\
d & \equiv & [0, \ c, \ e, \ f] & \equiv & [0, \ a, \ b, \ g] & ! \\
c & \equiv & [0, \ a, \ f, \ g] & \equiv & [0, \ b, \ d, \ e] \\
f & \equiv & [0, \ b, \ e, \ g] & \equiv & [0, \ a, \ c, \ d] \\
g & \equiv & [0, \ a, \ d, \ e] & \equiv & [0, \ b, \ c, \ f] \\
e & \equiv & [0, \ a, \ b, \ c] & \equiv & [0, \ d, \ f, \ g] & £
\end{array}
$$

With,

$$
\begin{array}{llll}
a & \equiv & [0, \ b, \ d, \ f] & \equiv & [0, \ c, \ e, \ g] & * \\
b & \equiv & [0, \ c, \ e, \ f] & \equiv & [0, \ a, \ d, \ g] \\
f & \equiv & [0, \ c, \ d, \ g] & \equiv & [0, \ a, \ b, \ e] \\
c & \equiv & [0, \ a, \ d, \ e] & \equiv & [0, \ b, \ f, \ g] \\
d & \equiv & [0, \ b, \ e, \ g] & \equiv & [0, \ a, \ c, \ f] & \$ \\
e & \equiv & [0, \ a, \ f, \ g] & \equiv & [0, \ b, \ c, \ d] & ! \\
g & \equiv & [0, \ a, \ b, \ c] & \equiv & [0, \ d, \ e, \ f]
\end{array}
$$

and,

$$
\begin{array}{llll}
a & \equiv & [0, \ c, \ d, \ g] & \equiv & [0, \ b, \ e, \ f] & ! \\
c & \equiv & [0, \ b, \ d, \ f] & \equiv & [0, \ a, \ e, \ g] \\
d & \equiv & [0, \ b, \ e, \ g] & \equiv & [0, \ a, \ c, \ f] & \$ \\
b & \equiv & [0, \ a, \ f, \ g] & \equiv & [0, \ c, \ d, \ e] \\
g & \equiv & [0, \ c, \ e, \ f] & \equiv & [0, \ a, \ b, \ d] \\
f & \equiv & [0, \ a, \ d, \ e] & \equiv & [0, \ b, \ c, \ g] \\
e & \equiv & [0, \ a, \ b, \ c] & \equiv & [0, \ d, \ f, \ g] & £
\end{array}
$$

I require an octal group with a $\equiv$ [0,c,d,g] $\equiv$ [0,b,e,f], d $\equiv$ [0,c,e,f] $\equiv$ [0,a,b,g] and e $\equiv$ [0,a,f,g] $\equiv$ [0,b,c,d]

The third part (marked with "!") of the "rivers" (the octals under seven cycles as above) becomes the formerly missing part now shared with a fourth (wormwood) when chosen appropriately. The "fountains of waters" are K4 groups of K4 subgroups with unity intersection, found present in each "river" (as octals under separate seven cycles) as those three rows completed by the missing parts (in "!") common to the fourth "wormwood" octal. There is now a form of pseudo-addition implicit upon octals in the sea, given a common K4 subgroup (here [a,d,e]). This K4 group corresponds in each left hand octal (then in the sea) to that left hand K4 group of subgroups having the unity as intersection. This identifies the right hand K4 group [a,d,e] as a static subgroup. Each octal above then corresponds to one of four choices of unity. This device then **mirrors** the fourth seal device.

I find a unique group fills the role, so marking the fills with "!"

$$a \equiv [0, \ c, \ d, \ g] \equiv [0, \ b, \ e, \ f] \ !$$
$$b \equiv [0, \ a, \ d, \ e] \equiv [0, \ c, \ f, \ g]$$
$$f \equiv [0, \ a, \ b, \ c] \equiv [0, \ d, \ e, \ g]$$
$$g \equiv [0, \ b, \ d, \ f] \equiv [0, \ a, \ c, \ e]$$
$$e \equiv [0, \ a, \ f, \ g] \equiv [0, \ b, \ c, \ d] \ !$$
$$c \equiv [0, \ b. \ e, \ g] \equiv [0, \ a, \ d, \ f]$$
$$d \equiv [0, \ c, \ e, \ f] \equiv [0, \ a, \ b, \ g] \ !$$

I find that the last and "wormwood" device fills in its "mountains" within the previous three seven cycles. (I.e. fallen upon the "rivers", as a fountain generalising the "trees" in the "!".)

Wormwood:
*Wormwood – a bitter herb. Alternatively, "brackish water" in the earth; a mixture of fresh and sea water (itself a statement made by wormwood). Wormwood is a subtle replacement for the waters of the octals (under the Holy Spirit) across the sea. (Formed and founded solely to enable the ability to practise the dialectic devices.) This one device (using the same logical schema of action and inaction) of "wormwood" is Satan's **most subtle** device (cf. Gen 3:1). Shifting octals, the proper operation of subgroups under the one seven cycle is disrupted. I compare such a system to a "set theory" for the dialectic, but instead of the foundation of rock that is the octal and field GF(8), the house of the dialectic process is truly built on sand (stars as motes fallen to the earthly elements). Note there are many such stars as only God is omnipresent.*

The device falls upon the third part of the "rivers" (as the static subgroup of the left hand acted upon by a seven cycle is a "river" in the "earth"). The "third part" is the product forming the closure of the "wormwood" device across those prior and arbitrary three rivers of eight (that it is likewise fallen upon, again in the left hand as by reciprocal), the sources of water. Rivers feed "seas" and fountains as water sources would have their source in the earth (as unity). (As {c,e,g} corresponds to "the fountain" of "a" which is the branching to [a,b,c], [a,d,e], [a,f,g].)

The wormwood device replaces the reference's operation (as water) or equivalently its "third part", and also the proper **operation** on singletons in the subgroups of **every** octal (many men died) having their K4 groups in orbit as rivers which are "**waters**" in the "sea". By supplanting the earthly triples within the correct operation (A ∨ B)^c, those four groups not containing unity (which are as "rivers") are made "bitter", and with a similar (as alternate) logical schema.

Those operations are "interrupted", made bitter with "wormwood". (The "third part" missing from each arbitrary triple of left handed octals is found in the wormwood octal. They appear to be filled in to complete K4 subgroups.) As water (the operation of the octal) is as a source of spiritual life, the "wormwood" device is a source of spiritual "Death" mirroring the rider in the fourth seal, but echoed in shifting subgroups rather than the fourfold choice of unity.

The operations of subgroups (as [0,a,d,e] above) are shifted to hold between octals in the sea rather than in the reference; the dialectic device of "Death" has somewhat of a solution in the wormwood. The third part of the waters (addition on singletons) then become wormwood. Many men (triples as [a,d,e] static or the unity, also those three left hand triples with the unity as intersection in correspondence) across the sea likewise die: they are "leached out of grace" as from their proper place in the reference right (and left) hand octals by obeying the bitter waters (operation) of wormwood. That wormwood octal forms a "middle ground" for antichrist to operate in the sea.

Each of the rows shared in "wormwood" correspond to a transposition of elements (d,f), (c,e) or (a,b) on the subgroups of the left handed octal [c,e,g], [a,e,f], [a,b,g], [b,d,e], [a,c,d], [b,c,f], [d,f,g] (where, say, the cycle (a,b,d,c,f,g,e) applies to that left hand in common with the "Sun octal". I.e. with g ≡ [a,d,e] ≡ {b,c,f}). These transpositions are covered in the two witnesses of the sixth trump, where the "middle ground" and "wormwood device" has its answer.

Without difficulty I equivocate the rider in the fourth seal "Death" with "Wormwood". They have the same spiritual identity – the same device and hence the same name.

7.5) The Fourth Trumpet

The fourth trump is blown and again there is much maths to approach!

Rev 8:12 And the fourth angel sounded, and the third part of the sun was smitten, and the third part of the moon, and the third part of the stars; so as the third part of them was darkened, and the day shone not for a third part of it, and the night likewise.

No longer restricting the construction to one right handed Sun octal and shifting the possible static subgroups, I may add a "second generation of octals" in the sea, using the first generation left hand as the second generation's right hand. In effect, I shift seven cycles again.

I should, in that the day and night "shone not for a third part", find an equivalence between such pairings of right/left hand octals, made of the Sun or reference (as the day) and the other octals in the sea sharing only one or three subgroups with the reference (the night). That equivalence should ideally meet in the middle: but the above verse tells us that this interface of octals across the sea is not made perfectly. There are exceptions or elements in isolation: the "sea" is then not completely traversable using the devices of the seals and trumpets.

Then I may simply construct a "bridge" between the "day and night" octal pairings using another suitable pairing for that bridge of day and night. I require to find "the moon".

The familiar Sun octal forms the first generation octal pair.

$$a \equiv [0, \ b, \ d, \ f] \equiv [0, \ c, \ e, \ g]$$
$$b \equiv [0, \ c, \ d, \ g] \equiv [0, \ a, \ e, \ f]$$
$$d \equiv [0, \ c, \ e, \ f] \equiv [0, \ a, \ b, \ g]$$
$$c \equiv [0, \ a, \ f, \ g] \equiv [0, \ b, \ d, \ e]$$
$$f \equiv [0, \ b, \ e, \ g] \equiv [0, \ a, \ c, \ d]$$
$$g \equiv [0, \ a, \ d, \ e] \equiv [0, \ b, \ c, \ f]$$
$$e \equiv [0, \ a, \ b, \ c] \equiv [0, \ d, \ f, \ g]$$

Then I make a second generation fire column based on a different left/right subgroup association of the "left hand" octal above (also using a different unity and seven cycle, say):

$$a \equiv [0, \ b, \ d, \ e] \equiv [0, \ c, \ f, \ g]$$
$$b \equiv [0, \ d, \ f, \ g] \equiv [0, \ a, \ c, \ e]$$
$$d \equiv [0, \ c, \ e, \ g] \equiv [0, \ a, \ b, \ f]$$
$$g \equiv [0, \ a, \ e, \ f] \equiv [0, \ b, \ c, \ d]$$
$$e \equiv [0, \ b, \ c, \ f] \equiv [0, \ a, \ d, \ g]$$
$$f \equiv [0, \ a, \ c, \ d] \equiv [0, \ b, \ e, \ g]$$
$$c \equiv [0, \ a, \ b, \ g] \equiv [0, \ d, \ e, \ f]$$

Then I make a third generation similarly on the left handed octal of the second generation.

$$a \equiv [0, \ b, \ c, \ d] \equiv [0, \ e, \ f, \ g]$$
$$b \equiv [0, \ c, \ f, \ g] \equiv [0, \ a, \ d, \ e]$$
$$c \equiv [0, \ d, \ e, \ f] \equiv [0, \ a, \ b, \ g]$$
$$f \equiv [0, \ a, \ d, \ g] \equiv [0, \ b, \ c, \ e]$$
$$d \equiv [0, \ b, \ e, \ g] \equiv [0, \ a, \ c, \ f]$$
$$g \equiv [0, \ a, \ c, \ e] \equiv [0, \ b, \ d, \ f]$$
$$e \equiv [0, \ a, \ b, \ f] \equiv [0, \ c, \ d, \ g]$$

There are shared correspondences as before, but not between certain octals. For instance, although [b,d,f] is common to the first right hand and the third left hand, a ≡ [b,d,f] ≡ [c,e,g] in the first generation, yet g ≡ [b,d,f] ≡ [a,c,e] in the latter. The third part (operation) is "smitten" by this device: properly it is intact, but with this device it is "darkened" (not virtuous).

There is no seven cycle (no operation) to be found that stands across these octals. I may state that the octal's additive operation is not preserved across all three "generations". The "third part" of each one must fail. I.e. either (c + e = a) or [b,d,f] or (c + e = g) fails in the "Sun, moon or stars", respectively. There is no law of cancellation and so no group operation to be found.

So, noting that the Sun shines only in the day and the stars shine at night only, I treat the second generation as the "moon" i.e. that shines in both the day and night, for the right hand of the "Moon octal" will shine at day and the left hand at night.

So, I may note that there are equal correspondences between unity, K4 groups and bow only between octals that share such a "bridge", a seven cycle. I immediately have the table:

	Sun	Moon	Stars
Sun	∃	∃	
Moon	∃	∃	∃
Stars		∃	∃

Where ∃ means "there exists such a correspondence", or "there exists such a seven cycle".

Note the "third part" of the "moon" is "smitten" even though there is a seven cycle that preserves the "moon" on both hands of that octal. The "Sun" is also "smitten", as are the "stars". The "third part" of each of the three octal pairs (generations) differ in their operation(s): each is "smitten". (A seven cycle is valid on two octals only.) This limits any traversing of the "sea" whilst the "third part" of each is "darkened" (there is no more reference). It can be clearly seen from the table how the "Day" and "Night" then "shine not" for a "third part" as above.

7.6) The Three Woes

The trumpets move on whilst the text speaks of "three woes". There is an equivalent to the last three seals which occur only in Christ rather than the octal. The reader is shown the satanic backlash that martyr so many of the saints from the fifth and sixth seals earlier.

Rev 8:13 And I beheld, and heard an angel flying through the midst of heaven, saying with a loud voice, Woe, woe, woe, to the inhabiters of the earth by reason of the other voices of the trumpet of the three angels, which are yet to sound!

As Satan is cast down to Earth, believers are reminded of the fact that Jesus has the keys of hell and death. The life and salvation of God towards us as found in Christ (to both those who are asleep in Him and those who remain saved on the Earth and sealed) answer the devices of Death and Hell that are introduced again in the trumpets (as trumps three and four here).

The angel flies through the midst of heaven: I expect every one positioned as a choice of unity in each arrangement of every octal to be the recipients of this angel's message. Those with faith on the Lamb are also "inhabiters of the earth" as are those practising the dialectic. (The three woes result from their mixing and interacting in a religious setting.) The devices of the seals and trumpets should be watched for but they are not to be deprecated (for they will together remain after the judgement as the inheritance of the ungodly).

In the text there are three woes, and I am not dealing with any K4 structure here, only Satan playing catch-up with his deceptions determined ahead of time. The three woes have names in pairs, "Death" and "Wormwood", "Hell" and "Abaddon" and lastly "Gog and Magog" and "Armageddon". These three woes occur sequentially but they are also overlaid, composite, culminating in the scarlet beast, the result of all three. (The woes occur threefold to those in the Earth that are practising the dialectic; such a one could suffer all three concurrent.)

It appears that these align to the diabolical trinity. Death corresponds to the beast, Hell to the false prophet and once Satan is loosed from the seal placed on him in the bottomless pit, the final acts of his perditious wrath then cast down to earth culminate in Armageddon. This last woe is actually the same scarlet beast and the text will state that the battle is then won by God drawing a line in the sand with His least – as one exercising virtue rather than the dialectic.

Then these three pairs form somewhat of a "diabolical marriage" between these entities and their works. In opposing virtue in "a", each pair of devices along with the tempter forms a dialectic of $a \vee c \& e \& g$ or, as all virtue is to be privated, $e \& g \rightarrow c^{-1}$. Then if say "Death and Wormwood" as devices are as "e" and "g", the "beast" would mimic the "virtue" to be blameless in "c^{-1}". (The virtuous would take the blame.) Otherwise, $a \& c \rightarrow (e \& g)^{-1}$, the virtuous (to be so) must continually endure all temptation. (Clearly that last is false – a virtue non-sequitur.)

The sequence of the last three trumps is very maths-intensive and they contain much of the apocalyptic vision given to John. I reveal what these three mean for those on the Earth by considering their effect upon the least in the kingdom of heaven, the "angel of the church".

7.7) The Fifth Trumpet

Rev 9:1 And the fifth angel sounded, and I saw a star fall from heaven unto the earth: and to him was given the key of the bottomless pit.

The star falls to the Earth. In terms of the churches the star takes its place above "Sardis" and serves to turn back those "foolish virgins" leaving the first five churches (putting false doctrine behind them) to the amalgam, the Laodicean Church from which they came. The angel "falls from heaven" as in reciprocal, mimicking the least. I consider the key as a transposition sending the believer from Sardis back to any "church" of Ephesus to Thyatira or even Sardis itself.

> **The "Bottomless pit":**
> The "pit" represents algebraic closure over GF(2). The key (a transposition) is a root of a polynomial of unknown degree with multiplicative order two. Yet here there is a contradiction, for such a root over a prime field of characteristic two then under Frobenius would send more than one element to unity, and there would not be an automorphism. Since the only homomorphisms of a field are trivial, there is no such transposition. The key to the pit is as an implausible extension in algebraic closure. Algebraic closure effectively remains "locked".

"Fallen from heaven", the star (a K4 subgroup held static as if unity), is fallen to the earth, the set $\{a,c,e,g\}$ if $[b,d,f]$ is static. With "heaven" ($a \equiv 1$) as unity; I expect the star to map no longer from unity to $[b,d,f]$ using the elements x, x^2, x^4 but with their inverse, from unity to $\{c,e,g\}$ with y, y^2, y^4 (the inverse $y=x^{-1}$). I find the same exchange of static subgroup/triple in the wormwood device, but it is far more important to note that in this five cycle, the fallen star (as if the least) floats unity **and triple** through each "church" of five as if it were that inverse.

Jesus instructed the least to strengthen the things (doctrine) that remain before they die (fail) or are lost by returning to the Church with its false doctrine. In strength of grace (finding life), the angel is given a good reason to leave.

However, the bottomless pit is open, by an ordered pair of elements of an extension field of unknown degree over GF(2). With the pair (x, Ty) where addition and multiplication obey the rules as if "T" were a multiplicative element of order two, I may define an extension of the field's operation:

$$(x+Ty)+(u+Tv) \equiv (x+u)+T(y+v)$$
$$(x+Ty)(u+Tv) \equiv (xu+yv)+T(yu+xv)$$

So,

$$(x,y)+(u,v) \equiv (x+u, y+v)$$
$$(x,y)(u,v) \equiv (xu+yv, yu+xv)$$

Believe it or not, this is actually valid.

Rev 9:2 And he opened the bottomless pit; and there arose a smoke out of the pit, as the smoke of a great furnace; and the sun and the air were darkened by reason of the smoke of the pit.

The smoke arising from the pit obscures the operation of the Holy Spirit, GF(8)*. (Spirit = "pneuma" or breath, air etc.) However, this "air" is "inanimate spirit" and so the octal itself is interrupted just as equally as is the field GF(8). Both are "darkened" in that the operation and application of virtue (as the unity) in GF(8) is missing from that five cycle.

The smoke rises as if out of a great furnace (the exceedingly large infinite group of "algebraic closure"). Now, Christ's feet were described by John as like to fine brass burned in a furnace: I had equated that furnace with the set of positive properties indexed by the right hand octal group (as GF(8)+) under one seven cycle. I now compare the octal to the infinite order additive group of algebraic closure obscuring GF(8) (as a subfield within it) from view with very thick smoke (as if "darkened" spirit, without an operation of virtue).

Rev 9:3 And there came out of the smoke locusts upon the earth: and unto them was given power, as the scorpions of the earth have power.
Rev 9:4 And it was commanded them that they should not hurt the grass of the earth, neither any green thing, neither any tree; but only those men which have not the seal of God in their foreheads.
Rev 9:5 And to them it was given that they should not kill them, but that they should be tormented five months: and their torment was as the torment of a scorpion, when he striketh a man.

The locusts come upon the Earth: the pit is opened to turn back anyone in circuit entering "Sardis" to any one of the first four churches. By floating the unity fivefold in a shortened orbit of length five overlaid upon the octal, the static triples of the Sun octal are inherited/leached from the similar operation of floating unity in GF(8). The five cycle is then granted the earthly triple(s) as well as the unity (as power), for there can be only the one "earthly" set (unity and static triple) strictly within the five refining churches of the seven cycle (the set to which the fallen star maps from unity). The five cycle then "torments" independent of the octal.

The other four possibilities for unity in the five cycle must inherit or leach the operation of those static triples from other octals in the sea, for use under that five cycle: for those four elements in the coset of the static subgroup of the Sun octal (but within the five cycle) are "the scorpions of the Earth" that the other sets of scorpions must emulate to have "power".

The operation of the octal is not to be corrupted, for it is required for the dialectic devices. The unity and triple naturally aligned to the Sun octal is without any "torment". The locusts do not hurt anything with new life in God, as singletons (grass) or triples in K4 or "bow" form: i.e. "trees" (in the right hand octal) but only to harm the unsaved (unsealed, not as unity). Compare "grass" with the measures of wheat and barley (see the third seal), the "trees" or triples with the two witnesses – the two olive trees of Zechariah's vision of the lampstand. "Green things" which are as new believers, begin having "new life" and are given no "instant Church membership" (captivity) but start spiritually free: outside the five churches that form the Laodicean "Church". (A very good thing too, that there be a reference for all.)

The scorpions do not kill: there is no fivefold amalgam of "Death" or any "super-wormwood" in the five cycle (a triple cycled fivefold is never sent to { }). The presence of static triples leached from the sea for the five cycle may at most generate the subgroup A5 of S5. There is then torment for five "months". Given power to torment as scorpions torment, they take the right hand octal (as the arm) of God and twist it (torment it) in a non-commutative operation closed on those within. In truth the octal is unaffected but not to those under the looser paradigm of wormwood in the Laodicean Church (or on the Earth).

The locusts aim to prevent any possibility of the least entering Philadelphia into God's good graces. By disrupting the seven cycle, the locusts attempt to hold captive (as in a spiritual Babylon) the believer within the system of wormwood octals closed upon itself: preventing entry into one of the "rivers" or the "fountain" formed from the three groups containing unity (as the K4 form spanning the octal [a,b,c],[a,d,e],[a,f,g]) preserved under the Sun octal's seven cycle or into the left handed octal as of the two witnesses (as the reference's left hand, the orbit of [c,e,g] – that triple also within the wormwood octal sharing the above K4 form).

The operation of the locusts is a device given by God to Satan ahead of time. The torment references the presence of a non-abelian group acting on the five churches, i.e. in the symmetric group of every permutation upon these churches, namely S5.

The "power" given the locusts over the believer (in circuit) depends entirely on choice (even of the elect by God), as to the believer the choice of the octal's two alternate schemas or the prospect of the third woe, the "great city" S5. (The latter's result is not commutative, an aberration of the seven cycle. The "stings" twist the right arm of God out of recognition as if it were stung by a scorpion "of the earth"; the five cycle remains independent of the octal.)

To retain some sobriety here, there is no flock of demonic locusts devouring everything before them, genetically modified or otherwise. Rather it is apparent that the elements of extension fields are towered towards algebraic closure (AC). There is no solution for such an order two root in AC, so I will examine Jesus' antidote in the text. (Satan is "utterly fallen" and is driven completely mad in order to complete his insult to God aright.)

It appears the key to the bottomless pit is simply an arbitrary transposition amongst the first five churches, but whether it is unique and cannot be used upon two pairs of elements at the same time is not clear. (These are together referenced as "horse bridles".) To construct a short orbit of length two there is no solution; there is the ability to arbitrarily turn back to the first four churches from the fifth. (In effect causing a 5-cycle, floating unity as with 1/x, mimicking the least.) There is also a solution in short orbits of GF(4) within GF(16). (I.e. of length five.)

The act of using the key of (x, Ty) to extend once from GF(8)+ to GF(16)+ is actually in view here, as the angel is "fallen" rather than "descending". (Evaluate (x, Ty) to x+y, and repeat.) With the key in place and the extensions x "already opened and filled with locusts" for the key to be continually reapplied as in (x, Ty) (y from a set always isomorphic to that containing x, {x, y in F: F a field}), this lacks any ability to form a short orbit of length two. There is a separation of locusts and the army of the sixth trumpet (serpents) for that reason. To attempt to construct all of S5 from the locusts alone is a ceaseless work.

God flatly states that during those days (automorphisms of GF(16)?) men will seek an operation of wormwood over the elements of the five cycle (as unity with the "bow" in the Sun octal), but are forever unable to do so. Wormwood is founded only upon the sea of thirty octals, not in the elements of the actual Sun octal itself – that wormwood structure is contradictory and inconsistent in the "Sun" octal despite any choice of unity and a broader paradigm. "Death" would send the unity to the empty set, yet that operation is forever out of reach in the locust's five cycle over the five refining churches.

Any set of four such triples {c,e,g}, {a,e,g}, {a,c,g}, {a,c,e} given Pos(a&c&e&g) together private virtue (unity) by forming such disjunctions as of a ∨ c&e&g etc. The attempt is to find (seek) those static triples so combined by "floating" one triple under a five-cycle: it cannot be so done. The result is that there is no "Death" device, it "flees" and is "not found". These triples are limited to those leached properties of C7 groups over the "Sun" octal (as to "desiring to die", as never found in the five cycle), rather than simply walking the reference seven cycle (even in part), for the static triples of the octal "walk" all seven elements also, and they cannot form a five-cycle (death "flees" from them). As unity floats, the "Death" device is out of reach; it "flees away" and is never generated. (Cf. Rev 20:11.)

The locusts swarm as if they were horses (K4 subgroups) arrayed to battle, stood side by side. On their heads (the near-arbitrary subgroup isomorphic to K4 in the orbit as if in product with unity in the generator's cycle; the K4 group which is in repeated product with the generator to form an orbit in F*) are "crowns like Gold": each an additively closed multiplicative coset of a GF(4) **subfield**, necessarily required for a short orbit. (Under multiplication the position of unity and the subfield in orbit is otherwise arbitrary.) Their faces ("front view" of the horse) were as the faces (the semblance) of men. (Similar to the third beast before God: one limb of four as touching the ground, the orbit of the zero element as below showing the "hair".) Here I find that there is a short orbit of this subfield of length |F*|/3 as seen by the maths to follow. The hair of the locusts, as that of women, appears "braided" in every short orbit.

In that case (a short orbit) the cycle repeats in every power of the generator three times over (braiding), so that when aligned as "arrayed to battle", the elements of the "horse" align to the product of a power of a generator of F* in the orbit of "the head". I.e. with the position corresponding to the start of the orbit in product with unity (occurring threefold).

The zeroth or $(p^n - 1)^{th}$ power of the generator (unity) as multiplied upon the head, is in the text the first or principal occurrence of the GF(4) subfield in its orbit: then become "the head". It then shares its *face* (semblance of GF(4) upon it) with each horse or locust made of each image of the head. That head is "crowned" – every orbit has such a closed K4 subgroup with unity, as to become similar or "**like gold**", alike to one of the trinity, but I am interested only in the head of the *orbit* in F of disjoint images of K4 groups free to map to GF(4) as a subfield.

To clarify, every "horse" arrayed to battle (coset of a K4 group, prepared as a **short orbit** to interrupt the octal) has a mapping in the container field's multiplicative group that maps it through the subfield GF(4). Every "horse" and "head" then has a "face" (semblance) as a "man", as an image and coset of GF(4) and by multiplication each may become aligned with unity in the orbit as to its "head". Each has a man's appearance as arrayed, but also a similarity with GF(4)* mimicking a "static triple" in the octal. The "face" (appearance) of a "man" (a triple deciding virtue) is a set under a permutation order three, and the orbit becomes rearranged to align that face of GF(4)* with the head of the cycle – unity. Each horse then also has a "crown like gold" as both a K4 group and as a coset of a group isomorphic to GF(4)* (or C3).

(I simply re-arrange up to a power of a generator of the container's multiplicative group.)

Then I perceive that the cycle is "braided" not by one generator as here but by every generator of the extension field, as overlaid or superposed. (Every horse may be in product with unity in a short orbit.)

Rev 9:8 And they had hair as the hair of women, and their teeth were as *the teeth* of lions.

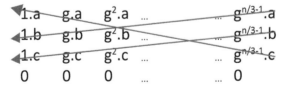

I now begin to move into John's vision of the notation used for the maths (shown after the last verse of the fifth trumpet). The "teeth of lions" is simply John's impression of the symbol or label "W".

Rev 9:9 And they had breastplates, as it were breastplates of iron; and the sound of their wings *was* as the sound of chariots of many horses running to battle.

John's own impressions as to the breastplates of iron are as resembling the letter "X", the wings resemble the letter "Y". The movement in the orbit is referred to as the sound of many horses charging.

Rev 9:10 And they had tails like unto scorpions, and there were stings in their tails: and their power *was* to hurt men five months.

The tails like unto scorpions is as the letter "Z", albeit with a central horizontal strike (a sting in its tail) that I am unable to reproduce from my keyboard. The scorpions are declared to have a short orbit of length some multiple of five.

Rev 9:11 And they had a king over them, *which is* the angel of the bottomless pit, whose name in the Hebrew tongue *is* Abaddon, but in the Greek tongue hath *his* name Apollyon.

I equivocate algebraic closure over GF(2) as to the "bottomless pit", and properly I may equate Abaddon with "Hell" from the seal sequence. That "king" (the minimum extension over GF(2) required to form S5) is the completed system that forms S5 over the churches: this again is unable to be constructed except as put in place by God Himself.

The perfection of finite fields allow through their property of separability short (length less than the field's order) orbits of K4 groups able to be reworked to produce in them "sub-orbits", the orbit of some subfield properly containing the K4 "head" as a subfield. Then if J<F and K<F where J and K are subfields of F (J, K not isomorphic to GF(4)) then J has a short orbit in F, and K also a short orbit in F.

Then there may be formed orbits (of subfields K of greater degree) "containing" (but not quite) that original subfield J. The effect is a system where the degree of extension may be chosen so that the fraction of orbit length in J to the orbit length of the subfield K may supposedly approach "five times" (months) through a continually narrowing fraction. There are such solutions, yet many such extensions required to form them. What is required is a GF(4) subfield of both J and K, to provide a common denominator.

At the very most infinite, it may even be possible to reconstruct the "key to the pit" (a generalised multiplicative transposition within the first four churches not under the "star with the key" which is over Sardis) from algebraic closure by employing the device above using a Cartesian pair (x+Ty). The power of the locusts is really to construct extension fields. (A short orbit of length five is due to orbits of GF(4) in GF(16). (A unique solution modulo two.) The key may continually be "used" to extend the additive group by a degree of two.)

I can equate teeth "W" as part of the head, so there are in orbits,
$$F = W_1,X,X...X,X, \ W_2,X,X...X,X, \ W_3,X,X...X,X, \ W_4,X,X...X,X, \ W_5,X,X...X,X$$

Where the W_i are rearrangements of the additive group W_1 "the head" under product by the elements of the subgroup of F* of order 5. I may also equate the X following each W_i as the structure W_i under repeated product with a power of a generator of F*.

The X following after each occurrence of W are described in terms following after the "head" and so are cosets of W_1 each with that "strength X" having the same "horse" or group structure.

I introduce the following structures.

W_1 a subfield of F isomorphic to GF(4).

J   {W_1,X,X...X,X} A subfield of F of index 5 in F* containing GF(4). (A set of chariots to be drawn by "horses".)

Y   a set of representatives of the cosets of C5 in F*, equally separated in powers of a generator f of F*.

G   {1,g,...,g^4} The multiplicative group of order 5 in F*.

K   (W_1,gW_1,...,g^4W_1) A subfield of F* isomorphic to GF(16) (A chariot drawn by "horses".)

Then as K has a short orbit in F*, each g^rW_1 has a short orbit of length in (a coset of) J equal to that of all K in F*. That is, when evaluated "modulo J*".

Generalising, the same may be said of near any "horse" in the cosets of J, as each coset of K in F is an image of some y*W_1, where the y are taken from the set Y. The cosets of J in F are containers for the orbits of each image of W_1 as from K = g^rW_1. Together the five cosets of J are containers for the full orbit of the W_1 in F, one element of G (i.e. g^r) separating each coset.

Then J is formed only of "chariots" and the other four cosets of J (together forming the orbit of K) are "horses pulling them", whilst the "wings" made of representatives taken from Y (generating the images of W_1 in F, J and K) in product with the W_1 form the cosets of K (which are just as equally formed by elements of J in product with K whole), and these are as if the "sound of their wings *was* as the sound of chariots of many horses running to battle".

Too swift? More detail follows. (See over for diagram.)

The most prominent structures present are the two subfields of F: J and K. J itself contains a short orbit of K4 subgroups, each a disjoint coset of a subfield "W_1" isomorphic to GF(4). Then K (as it is given, which is isomorphic to GF(16)) has a similar short orbit of GF(4), one of "five months". (J also has a short orbit of length five in F.)

The "horses" in J are assembled as "many horses running to battle" in that the cosets of K (with J at their "head") are likewise as "chariots" (pulled by four horses) and their number is large (as "many") and is not given, being arbitrary. The cosets of K in F likewise share the same orbit length as that of W_1 in J (as does J that of W_1 in K) and each yW_1 also forms a "chariot", generated by some particular set y in Y. Each chariot is a product of W_1 with the elements g^r of G (together forming K) and some element f in F (or J). Each fg^rW_1 = fK generates that coset of K then found (to be as each Z); the y = fg^r (g^r in G) are a set of five elements equally separated in the powers of a generator of F. The set of y in Y (as from J or F) generate those "many" chariots as cosets Z of K, each running side by side as if "to battle".

This is best represented in a diagram. (F rearranged, I align the cosets of J, as "five months"):

A "Chariot" (K)	K	$Z = y \cdot K \;(y \in Y)$	F
Upon their "Heads" (g^r) were	$g^4 \cdot W_1$	$(X,X,X,...,X,X,X)$	
"Crowns like Gold" ($W_1 = GF(4)$)	$g^3 \cdot W_1$	$(X,X,X,...,X,X,X)$	
"five months" (G)	$g^2 \cdot W_1$	$(X,X,X,...,X,X,X)\,GJ$	
$gW_n = W_{n+1}$	$g \cdot W_1$	$(X,X,X,...,X,X,X)$	
Every breastplate (X) is a disjoint coset of GF(4)*.	W_1	$(X,X,X,...,X,X,X)$	J
The W_n are as "the teeth of lions"		$y \in Y$	

Sound of their wings (Y) as many horses running to battle.

Frobenius holding fixed J skews, "torments" the Z=yK (fourfold).

Each element g^r of G (G isomorphic to C5) permutes the W_i in a five cycle; G now generates the subfield K = GF(16) from W_1. The subfield J or indeed any of its cosets $g^r J$ may be considered as before to have "womens hair" (with much "braiding") or also such "braiding" in all F (for there is one orbit of GF(4) present here overall, and the generator of F is arbitrary).

Their "wings" (Y) are the cosets of the elements of the subgroup G of F* and of K* of order five. They "leap" or "fly", equally separated in the cycle of an generator of F* (as opposed to "eyes" – the generators of various cycles as indicated before). Yet unity may float amongst the g^r, for then the five churches may each appear to ride the five cycle as unity, i.e. "modulo J*".

The "tails" (once more) are the elements of the cosets Z (of those Z = fg^{r}W$_1$ with g^r in G and some f in J or F) in the short orbit of K, made upon every "head" (g^{r}W$_1$), i.e. of some place in the orbit of the five W_n (representing the five cosets of J).

The biggest mystery are those tails "Z". The map of Frobenius, holding J fixed (squaring to a suitable power) will permute not only the elements of K but also every coset yW$_1$ amongst themselves; each "chariot" is skewed "four times" (cf. Rev 9:3 "unto them was given power, as the scorpions of the earth have power.") rather than "five months" as the elements of J remain fixed. (Yet unity, that which would fix W$_1$ in K is free to float fivefold in the g^r in G.)

Then the tails are not a preservative fixing all of the structure of K; that map of Frobenius may hold fixed W$_1$ and every X in J as GF(4) is a subfield of J. (Though unity may float in the g^r.)

The locusts are simply given "five months" as of a cycle of a K4 group (they require a subgroup C5 of F* with a floating unity found in the g^r; each "head" has a crown "like Gold", as GF(4)), an orbit of the subfield GF(4) of K. The "wings" in C5 leave any "horse" (or K4 group X=fW$_1$) "at rest" as some congruence of X "modulo J*", dwelling in a "church" J of orbit length five.

There are many such cosets of W$_1$ in J, so the "sound of" or "action of" the wings is then also as "chariots of many horses running to battle". I note the locusts construct the whole of the field F, forming every K4 subgroup's orbit; not only that of GF(4).

Now the system must solve the following equation:

$$\frac{p^{nm}-1}{p^{n}-1}=\sum_{r=0}^{m-1}p^{rn}=5$$

Which has with p=2 a simple solution of n=2, m=2. What I now seek is the extension of this formula to another field to enable the orbit in the "Y", when W_1 is a subfield GF(4) of J and K.

$$\frac{p^{nm}-1}{p^{2}-1}\equiv\sum_{r=0}^{\frac{1}{2}(nm-2)}p^{2r}\equiv 5\frac{p^{k}-1}{p^{2}-1}\equiv 0\bmod\left(\frac{p^{nm}-1}{p^{2}-1}\right)\text{ with }k\mid nm,\ (n,k\ even)$$

$$\Rightarrow\quad \frac{p^{nm}-1}{p^{k}-1}\equiv 5\frac{p^{2}-1}{p^{2}-1}\equiv 5\bmod\left(\frac{p^{nm}-1}{p^{2}-1}\right)\text{ with }k\mid nm,\ (n,k\ even)$$

Now, with p^{k} the order of our field J the union of $(W_1,X,X...X,X)$, each a coset of GF(4) (our W_1 and each X) and p^{n} the order of K (a subfield of F) the union of the Y_iW_1, each term a member of $(W_1, gW_1,...,g^{4}W_1)$, I find that:

$$\frac{p^{k}-1}{p^{2}-1}\equiv\frac{p^{nm}-1}{p^{n}-1}\ \bmod\left(\frac{p^{nm}-1}{p^{2}-1}\right)\text{ with }k\mid nm,\ (n,k\ even)$$

$$\Rightarrow\frac{p^{n}-1}{p^{2}-1}\equiv\frac{p^{nm}-1}{p^{k}-1}\equiv 5\bmod\left(\frac{p^{nm}-1}{p^{2}-1}\right)\text{ with }k\mid nm,\ (n,k\ even)$$

Lastly the tails of the locusts are as scorpions. I have a tower of extension fields up to Algebraic Closure (AC), all possible extensions of the chosen field with "five months", namely J. The tails "Z" are twisted under the Frobenius map cyclically whilst holding all the short orbits of GF(4) generated in F static and cycled within themselves. (As F could likewise not be fixed.)

I now find "Z" skewed (tormented) in the cosets of J. So $Z_r = g^{mr}*jW_1$ for some $m=p^{nk}$ (n > 0), g^{r} in G, j in J. Holding J fixed I always have automorphisms in AC that permute the Z_r for r > 0.

In likeness the elements of AC are as those in F. They appear to swarm over or are in *appearance* within and without the additive subgroups of F, no matter their order – rearranging I have a system that is self-similar and separable. Finite fields are defined as "perfect". This is another consequence of that perfection of separable extensions.

7.8) The Sixth Trumpet

I swiftly change to the spiritual system that accompanies "Hell" (captivity in the Church), namely the spiritual identity known as "Death" or "Wormwood".

Rev 9:12 One woe is past; *and*, behold, there come two woes more hereafter.
Rev 9:13 And the sixth angel sounded, and I heard a voice from the four horns of the golden altar which is before God,
Rev 9:14 Saying to the sixth angel which had the trumpet, Loose the four angels which are bound in the great river Euphrates.

The four horns of the golden altar before God upon which the Lamb is offered (i.e. GF(2) is a subfield of GF(8) hence the term "golden"), are present as four choices of unity. A voice from its four horns (within the same additive octal group) states it is time for the four triples corresponding to unity to now be completely loosed under all eight multiplications (not only the four accrued so far). This would occur with the sealing of the saints completed. I will show that the system of wormwood opened up with these four angels is deeply flawed.

> **Four Horns Of The Golden Altar:**
> *The four horns of the altar represent the four positions in the earthly elements which the lamb may take. Each requires a different seven cycle. The altar is golden as it represents a finite field in the Trinity: it forms GF(2) – GF(4) with "two parts cut off". (Cf. Jesus has already opened the seals and is worshipped with "a new song".) The horns are the right handed singletons and not the left handed triples which are consequences of the first four seals. The "strength" of horns is generally taken as cumulative. The altar (or Lamb) stands firm and perfect despite the loosing of the restrainer. The "voice" as from it judges the sealed as truly preserved under the covering of that Lamb. Any further restraint may be freely lifted to "extend, try and refine" the body of Christ.*

Rev 9:15 And the four angels were loosed, which were prepared for an hour, and a day, and a month, and a year, for to slay the third part of men.

Given the eight multiplicative groups (C7) over each octal, there have been only four choices of static subgroup, one for each choice of unity. There are yet four angels bound in the "Great River Euphrates". (This is a nice term for a "wide river" or "water in the earth" in a land of "spiritual captivity", fed by four tributaries.) So the Euphrates represents the four possible left hands generated so far and the four angels now loosed permit the remaining four octals to enter consideration. In effect all restraint is loosed on the earthly elements.

There are eight possible octal groups under those C7 groups for multiplication (given the congruence of unity to that static subgroup). So, I may pair a=1 to one of [b,d,f], [b,e,g], [c,e,f], [c,d,g] and generate two sets of four left hand octals with the seven cycle acting on the reference. I would then obtain from one set of four:

$$[b,e,f], [b,d,g], [c,d,f], [c,e,g] \quad \textit{with unity } a = 1$$
$$[a,d,g], [c,d,e], [c,f,g], [a,e,f] \quad \textit{with unity } b = 1$$
$$[b,c,e], [a,c,f], [e,f,g], [a,b,g] \quad \textit{with unity } d = 1$$
$$[a,d,f], [b,f,g], [a,e,g], [b,d,e] \quad \textit{with unity } c = 1$$
$$[b,c,g], [d,e,g], [a,b,e], [a,c,d] \quad \textit{with unity } f = 1$$
$$[d,e,f], [a,c,e], [a,b,d], [b,c,f] \quad \textit{with unity } g = 1$$
$$[a,c,g], [a,b,f], [b,c,d], [d,f,g] \quad \textit{with unity } e = 1$$

Which in these four columns (as in the top row with a=1) correspond to the orbits of the right hand subgroups [c,d,g], [c,e,f], [b,e,g], [b,d,f] respectively.

In rows, addition of subgroups in the octal $(A \vee B)^c$ always results in a group from the reference or Sun octal! I may close the rows as octals themselves with the three missing groups shown.

$[a,b,c], [a,d,e], [a,f,g]$ *with* $[b,e,f], [b,d,g], [c,d,f], [c,e,g]$ *with unity* $a = 1$
$[b,d,f], [a,b,c], [b,e,g]$ *with* $[a,d,g], [c,d,e], [c,f,g], [a,e,f]$ *with unity* $b = 1$
$[c,d,g], [b,d,f], [a,d,e]$ *with* $[b,c,e], [a,c,f], [e,f,g], [a,b,g]$ *with unity* $d = 1$
$[c,e,f], [c,d,g], [a,b,c]$ *with* $[a,d,f], [b,f,g], [a,e,g], [b,d,e]$ *with unity* $c = 1$
$[a,f,g], [c,e,f], [b,d,f]$ *with* $[b,c,g], [d,e,g], [a,b,e], [a,c,d]$ *with unity* $f = 1$
$[b,e,g], [a,f,g], [c,d,g]$ *with* $[d,e,f], [a,c,e], [a,b,d], [b,c,f]$ *with unity* $g = 1$
$[a,d,e], [b,e,g], [c,e,f]$ *with* $[a,c,g], [a,b,f], [b,c,d], [d,f,g]$ *with unity* $e = 1$

Now, note that each row is a result of a transposition of the elements (b,c) or (d,e) or (f,g) in the Sun octal of the top row, (d,f), (a,c) or (e,g) in the second row: i.e. that preserve the same three groups of the Sun octal. In opening the bottomless pit, across the sea of all thirty octals there is some basis for the fallen star of the fifth trumpet to acquire a mode of transposition.

The constructed transpositions are required for the enemy to prepare for an hour, day, month and year. That is, they are prepared to combat any automorphism, octal, product and cycle (of which there are 24, 30, 48 and 8 variants respectively) generalised from the sense of the angel's circuit in the letters. They are prepared across any or every octal to strengthen the deception upon the church to retain the least in the kingdom of God, Jesus Christ's right hand.

Given that the octals above are formed by switching elements that leave the first three columns (the K4 form) unaffected, these angels are prepared to shift the octal underneath the K4 form of the filter, supplanting it entirely. (In mimicry of the Frobenius map on GF(4).)

They may in these rows keep the K4 form unaffected under a single automorphism (as in the top row with a=1), i.e. **one hour**, as well as with all eight C7 groups upon that row (again keeping unity the intersection of the three K4 form columns). Then under automorphisms of these groups the K4 form is invariant under 24 automorphisms, i.e. **one day**.

The seven cycle acting to form the columns (which is valid on the Sun octal) cycles the intersection in the K4 form, again preserving intact the K4 form's structure in the Sun or reference octal. I.e. **One month**. (One month as a power of a generator of GF(8)*, here with one static subgroup of four: i.e. of eight C7 groups each with six elements not unity, together giving 12 "months" as a result; I find 48/4=12.) Despite the action of the whole C7 group the K4 form seems none the wiser of its antichrist counterpart, whilst the field GF(8) (i.e. **One year**) underneath the K4 form is supplanted. The "third part" of addition (of men) on the earthly elements is killed (slain) by supplanting it with these wormwood octals.

Then "antichrist" has claimed for himself some small territory, identified as the "lake of fire". (The operations of multiplication preserve the wormwood structure between rows.)

Rev 9:16 And the number of the army of the horsemen *were* two hundred thousand thousand: and I heard the number of them.

John **hears** the number: I assume that the calculation should make perfect sense and form the generalised circuit of the angel through the possible octals.

There are $5*4*(2*5)^7 = 200,000,000$ such generalised circuits. In each seven cycle, five symbols each over two octals under the same 7-cycle (giving 2*5=10 states of captivity in each church for the angel to exit; this factor of two may also be reached by including both "horse bridles" that are required to construct each five cycle), raised, using the "mystery of the lampstands" to the seventh power, using representatives from every group under the angel's circuit. Then I have five powers of the circuit's generator remaining. (I do not require maps into unity or that rest upon it!)

I may extend the mystery from two dimensions onward;

Extending a point into another dimension creates a line, a line a quadrilateral, a square a cuboid, a cube a tesseract etc.

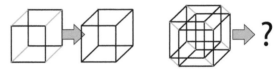

And I may extend by seven cycles to generate every possible circuit of the angel, whose circuit is arbitrary in those eight groups.

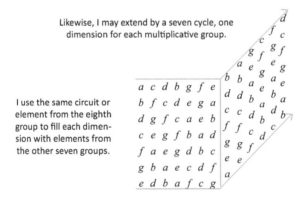

Likewise, I may extend by a seven cycle, one dimension for each multiplicative group.

I use the same circuit or element from the eighth group to fill each dimension with elements from the other seven groups.

$$
\begin{array}{ccccccc}
a & c & d & b & g & f & e \\
b & f & c & d & e & g & a \\
d & g & f & c & a & e & b \\
c & e & g & f & b & a & d \\
f & a & e & g & d & b & c \\
g & b & a & e & c & d & f \\
e & d & b & a & f & c & g
\end{array}
$$

I extend by seven dimensions upon the original scalar of the seven cycle. Upon the 2*5 = 10 states (as in both the left and right hands of the Sun octal) of "captivity" in each seven cycle there are effectively $(2*5)^7 = 10,000,000$ possible choices for the angel's circuit (which is always the same cycle) "in the spirit" of maps from the first five "refining churches" into Philadelphia from outside it. (This excludes maps from and between Laodicea and Philadelphia which are included in the 10,000,000 so far.) Yet I require another factor of 20.

With each extension into every "dimension" I discover a different left hand under the angel's circuit: then I have the sequence totalling $(2*5)^7$ rather than $2*(5^7)$. In essence adding dimensions one at a time with a factor of two: $(2*5^7) * 2 * 2 * 2 * 2 * 2 = (2*5)^7$

As each of these cycles are upon just two (complement) octals, I may choose four subgroups for each singleton element to correspond to, leaving only a factor of 5 remaining. I may assume the identity in each cycle is as "Laodicea". Then the last destination church (which is undetermined as yet) is some other arbitrary element.

I then simply require the factor of five from the five powers of the generator of the angel's circuit as out from the 40,000,000 states of captivity to overcoming, attaining to sabbath rest.

I may map from Laodicea to Philadelphia with the element corresponding to Philadelphia only ("hold that fast which thou hast", $x*1 = x$). There is a separate map from each of the other five elements to Philadelphia. That said, despite the choice of five churches to map to Philadelphia, the fact remains I was considering maps rather than the choices of elements, for I have not determined that Philadelphia is equivalent to any symbol in the cycle as yet!

The number of ways to choose five elements from six is six. Since as the last element (Philadelphia) is arbitrary in the elements of the C7 group I have $6-1 = 5$ as the remaining factor: this subtraction is correct as the seven cycle is determined by the possibilities acting on the seven cycle ($a, b, x_1, x_2, x_3, x_4, x_5$) with $a=1$ as Laodicea and b as Philadelphia, say.

The choice of the five possible "maps" (elements) above in seven dimensions generalise this, as every other form of seven cycle over the octal is as $(a, x_0, x_1, x_2, x_3, x_4, x_5)$ and this has one power (of the element) where the singleton a is sent to b. Then there are only five possibilities for "x_0" after "a" and "b" are determined, for the cycle once "x_0" is fixed is again completely determined by the subgroups in the octal underneath the angel's circuit. (I.e. b is taken from $\{x_1, x_2, x_3, x_4, x_5\}$, and x_0 is chosen from the five remaining elements.)

John heard the number of horsemen in these trumpets properly: it is correct as 200,000,000.

Rev 9:17 And thus I saw the horses in the vision, and them that sat on them, having breastplates of fire, and of jacinth, and brimstone: and the heads of the horses *were* as the heads of lions; and out of their mouths issued fire and smoke and brimstone.

Returning to the four angels unbound, they fulfil the transpositions required, and this is their effect in the octal: they provide the alternative to the "key to the bottomless pit" of a transposition. I may separate the groups with the transpositions into the pairs of elements

a (b c) (d e) (f g).

The horsemen have "breastplates of fire, jacinth and brimstone". Their device or "breastplate" X, is to mimic the Frobenius map on the seven cycle of the Sun octal (the fire is actually as "from the altar" present in cycling each of the four groups $\{[c,e,g],[c,d,f],[b,e,f],[b,d,g]\}$) to "release the restraint" of the angels bound in the river Euphrates (of the sixth seal) together within a single octal (wormwood) and they have the same structure as the Sun octal in their "jacinth" part (the genuine part, a gemstone) or as a part "blue as water" in the shared subgroups $\{[a,b,c],[a,d,e],[a,f,g]\}$. Brimstone, however, is associated with the reward for blasphemy of the Holy Spirit, and this is corruption of the octal by a transposition upon the octal to mimic the K4 ultrafilter, as a reason in part for perpetuating this wider octal system.

Sulphur, Brimstone:
The action of sulphur (i.e. burning it) is to obscure the seven cycle with thick smoke from which God will turn away. The octal underneath the K4 form is disrupted with the switch of two elements in a transposition. It is this transposition that is the "sulphur", which under the action of the proper seven cycle (fire) obscures the proper octal operation of the ultrafilter (as with smoke). God appears to vanish and is switched out for a close counterfeit spiritual paradigm that has an outward appearance of preserving the K4 form intact.

The breastplates (as a symbol "X") is of fire (i.e. spirit, Frobenius on C7), jacinth (with worth in the K4 form shared with the right hand, then as "waters" under a seven cycle) and brimstone (blasphemy, as judging in Christ's place). In sequence the breastplate goes from X to X^2 to X^4 to $X^8 \equiv X$ again (a three cycle). In terms of "brimstone", however, the Frobenius map is made to act on those three subgroups containing unity as if each were an embedding of GF(4) in the octal. (Inducing a two-cycle and acting as a transposition on elements of the octal.)

The "fire" is the three cycle induced from Frobenius on C7 (as by the Holy Spirit) that cycles the static subgroup in the octal, say [b,d,f]. The jacinth refers to the subgroups (the rivers under that three-cycle) [a,b,c],[a,d,e],[a,f,g] that together correspond to unity and the static subgroup [b,d,f], and likewise the brimstone which is the switch of (b,c) or (d,e) or (f,g) to supplant the seven cycle beneath the K4 form, which completes the picture. These form the system of the serpents; $X \rightarrow X^2$ is as the breastplate (the map of Frobenius).

Rev 9:18 By these three was the third part of men killed, by the fire, and by the smoke, and by the brimstone, which issued out of their mouths.

The proper use of the K4 ultrafilter is with a = 1;

$$a = [a, b, c] \ (unity \equiv X)$$
$$b = [a, d, e] \ (X^2)$$
$$c = [a, f, g] \ (X^4)$$

This is invalid in the octal but as the serpents interrupt (obscure) the seven cycle to form a close-counterfeit embedded in the octal, I can state that "brimstone + fire = smoke". I.e. this becomes "sorcery" with regards to the "serpent spirits".

That is to say: the K4 form is as "fire" and then "brimstone" is as the transposition of elements supplanting the octal underneath, or indeed the seven cycle (air) by the "smoke" of the "burning" of the sulphur. The combination of the two is the supplanted octal, which is the action of these spirits.

a (b c) (d e) (f g).

The "head" (as a starting element) or element "a" resembles that head of a lion – (a small portion of the lion of the tribe of Judah, the lamb before the throne). A lion as a figure resembles in letters ((a)bc)defg, that being ((a)bc) which is merely the "head" of the "lion" with its mane. Note the whole "mouth" (the set {...}, braces resemble lips) of the "serpent" horseman resembles the head "a" held static with three transpositions in its tail, like a snake in motion, using the serpentine method. Out of the mouth (the set) issues fire, smoke and brimstone: the latter being transpositions of the remaining elements (b c), (d e) and (f g) that are paired with "a". I may associate the "head" with unity and "a (b c)". Likewise, the tails also have heads "a (f g)" etc. with them they "do hurt".

The seven cycle is interrupted by these octals, for the requisite fallen star of the fifth trumpet is only provided for use in this fashion.

Rev 9:19 For their power is in their mouth, and in their tails: for their tails *were* like unto serpents, and had heads, and with them they do hurt.

The serpent spirits agree on principle with the K4 form, but cannot form its solution in the octal; they are "antichrist". The serpent, or "wormwood" octal generated with a=1 has the three groups [a,b,c],[a,d,e],[a,f,g] in common with the octal, a good basis for the K4 filter that spans the octal. However, in their transpositions they attempt to form a different octal beneath the K4 group of subgroups that appears unaffected by any of (b,c), (d,e) and (f,g) (as if they were also forming the GF(4) filter within the same three subgroups).

The result is mimicry of Christ as the K4 form spanning the octal, by supplanting the octal beneath (and with the dialectic instead). They use a valid but alternate logical schema that overconstrains the octal and breaks the closure of sets of positive properties. The paradigm is as if "God had changed His mind". (A straw that breaks the back (seal) of the Word of God.)

As antichrist themselves, they attempt to restrict the validity of the Christ ultrafilter simply because they cannot attain it in the octal by constructing it themselves. They have a position that as if *structurally* (schematically) denies the possibility of the resurrection (the possibility of the 5th and 6th seals opened) and so blaspheme God. The damage they do is done by the transpositions within the tails in (...) as over the head "a" (see above). These spirits hold judgement reserved for God alone and have sway over men that deny that first resurrection of Christ as nonsense (logically overconstrained as in schema, they "pick at holes" that are not there). These men are those that have had any proper operation of faith in C7 interrupted by the denial of that resurrection in Christ. (For it seems base to them, and requires adding to, breaking closure of sets.) They deny unity (as virtue), holding instead to the (dialectic) bow then become a subgroup (blood exchanged for water), in preference to the Sun octal.

The serpents in the seven "wormwood rows" are formed upon the reference octal which (with its seven cycle holding a=1 and [b,d,f] static) has as its earthly elements {a,c,e,g}: the serpents whose rows correspond to the same choice of unity, each contain the static bow of the left hand of the reference octal as a subgroup (and not then as merely a static triple). These rows are the "middle ground" where the octal may be altered (blasphemed by men judging in God's place) to permit the dialectic devices (with the static triples (bows) as subgroups, which are now become "static") under the seven cycle of the reference octal as acting on the subgroups within each row (generating the seven rows), and all whilst still preserving the three groups common to the reference octal which are together static under the operation of the Frobenius map. The serpents present the position of antichrist, that would perpetuate the devices of the dialectic within those who hold some belief in Christ. ("2Ti 3:5 Having a form of godliness, but denying the power thereof: from such turn away.")

Rev 9:20 And the rest of the men which were not killed by these plagues yet repented not of the works of their hands, that they should not worship devils, and idols of gold, and silver, and brass, and stone, and of wood: which neither can see, nor hear, nor walk:

There are two sets here, the first of seven and the latter of three. The first of the historic false doctrine of OT Israel that leavens the "Church" (under the image device of ten kings, as "seven mountains" – strongholds of false doctrine yet to be examined) under the one mind (of three kings) in dialogue with the beast (as the collective or the seventh king). The latter group of three – (the second collection) set apart the imaginations of the heart of man: that He is unable to give his idols life (even in the image system as mimicking the USA beast's fourth seal system. Only Christ sat at the right hand of God has eternal life to give). The first group mentioned under "works of their hands" contains the other six: I find a mirror held up to the image system, for the seventh is (the beast and) the collective, the "rest of the men".

Those that remain in this system of serpents find it quite natural to them and continue inside it as victims of a cruel "bait and switch" where the spirit of God has been supplanted by the loosened restraint upon the Church. The people outside the Church collective that will not repent are not ever to experience any difference in their lives with "God switched out for antichrist" as compared to the Church that will "weep, wail and gnash their teeth". The result is that all are made to believe that there is simply no difference between one path and any other. What is (at the very end) accepted as Christ in the Church is not Christ.

Their false gods (the set of seven) are stated to be totally inanimate (as the set of three). Then despite these men being altogether too religious, these men never repent in the four parts that are truly animated; those parts formed of themselves (as they also see, hear and walk).

Rev 9:21 Neither repented they of their murders, nor of their sorceries, nor of their fornication, nor of their thefts.

They do not repent of drawing valid faith into this system to destroy it (aka murdering faith under the locusts of the fifth trump) or of any interruption of the seven cycle (with smoke from the pit) of their sorceries (as transpositions in the sixth trump) from burning sulphur (i.e. brimstone + fire = smoke); or, even of their preference with worshipping other spirits (interrupting the octals with other octals or extensions to suit oneself); lastly, even their thefts: leaching the properties of the octal and K4 ultrafilters to use their elements as maps to partly construct S5 over the first five churches, forming the image over the Laodicean "Church".

There are twelve elements of S4 leached from the proper operation of the octal for the construction of S5 (the "Church" playing the harlot with God's adversary). For each choice of unity there are three maps from the elements of the coset's static bow to that unity. Each is a separate power of a generator, and each choice of unity (with the static subgroup fixed) comes from a different C7 group. In total, then, there are twelve maps amongst the four earthly elements that are useful for constructing S4, though they are not elements of S4 themselves. Those maps are "leached" or "stolen" by those with a broader paradigm.

Not until after the winepress is trodden out, is there truly any such five-cycle over S4 to properly close that system (the city), and even then there is yet an "open door" out to "rest" in unity in C7 through those 12 elements. That is, only for those that heed Christ's call to remain separate. Those remaining under the locusts of the fifth trumpet are soon to be caged within, along with all tares in the scarlet beast system.

The army of these serpents with their transpositions is numbered 200,000,000 – each are positioned to interrupt the seven cycle of the least and return him to the collective of the first five churches of the letters with the locusts. They each make a transposition to interrupt the angel in order to tempt and turn the least back in every possible circuit (of the 200,000,000) – forming the fully applied device of the star fallen to the Earth with the key to the bottomless pit.

They, as serpents, form the basis for the locusts of the first woe to animate the "pneuma" of the false prophet to deceive those within that God is still present and the system closed. In the letter to Sardis, Christ states, "I know thy works, that thou hast a name that thou livest, and art dead." The life in the Church is not life in Christ. Yet the pneuma deceives the current least into remaining within: he is in Sardis being turned back to the collective. (The "current least" is always the one who will leave most swiftly in a fellowship.)

7.9) The Angel Of The Church

Given a brief respite from maths I once again begin building my vocabulary of word pictures. I return to the angel (the least in the kingdom) as His circuit is completed in Philadelphia.

Rev 10:1 And I saw another mighty angel come down from heaven, clothed with a cloud: and a rainbow *was* upon his head, and his face *was* as it were the sun, and his feet as pillars of fire:

The angel is clothed in a cloud – a mystery (cf. the mystery of the seven angels and lampstands). He is also the right hand of God and "incognito". The reference to the rainbow is to his circuit, the seven cycle that is now completed fulfilled in the Spirit over the lampstands (churches). The rainbow is "resting upon" his head (as in the letter to Ephesus, "His place" is the least's circuit completed – now the lampstand in the Holy Spirit found upright before the throne), so I may safely assume that the angel is in Philadelphia. (He has yet to put the rainbow underneath him, if that were to be his test.)

His face is as the Sun octal – he is considered (in overcoming the world) as if he were clothed with the same righteousness as Christ, having likewise walked "through the churches". He is able to sit in judgement and is gifted the right to do so as if he were a suitable substitution for the "zero" element in the octal and GF(4) ultrafilters. His feet as "pillars of fire" correspond to His authority in Christ intact; he has the ability to stand unmoved, or "tread on serpents (in the sea) and scorpions (in the Earth)" (as overcoming the fifth and sixth trumpet "woes"). He is gifted such status as a pillar (cf. letter to Philadelphia) through his obedience to Christ, by his diligence as if he were also as C7.

Rev 10:2 And he had in his hand a little book open: and he set his right foot upon the sea, and *his* left *foot* on the earth,

He has a small book (the seven letters?), and His feet are positioned over the sea and the Earth (but the angel's book reads left to right rather than right to left. John's seven cycle was reversed, so the "feet" appear exchanged). What is really the static bow in the right hand octal in the earthly four elements is actually to John the left foot (the seven cycle is reversed and "burns up" as in the first trump and the right hand octal was in the left of two columns). The right foot or static bow (when the two octals are aligned isomorphic) is "placed upon" one of the eight left hand octals in the sea. (Which was placed in the rightmost column.)

He is resisting the now opened or loosed restraint upon the sea (all loosed restraint) with the Sun octal to the right hand of God, and resists the operations of the seals on the earthly elements with the left handed triples to which they correspond in the Sun octal's counterpart. That is, He resists the methods of shifting C7 group over the octal in the right hand.

In common with the opening chapter of Revelation, the angel's stance is seen holding firmly fixed the Sabbath rest or "unity" of his overcoming in his circuit (he has overcome and his circuit is as the rainbow before the throne), he also holds fast the static subgroup and triple. His "feet" are not pictured as walking the seven cycle but are at rest in it!

The little book is revealed as a guide or "reed like a measuring rod" later on by John.

Rev 10:3 And cried with a loud voice, as *when* a lion roareth: and when he had cried, seven thunders uttered their voices.
Rev 10:4 And when the seven thunders had uttered their voices, I was about to write: and I heard a voice from heaven saying unto me, Seal up those things which the seven thunders uttered, and write them not.

Seven thunders map the narrative in Revelation to a previous point in the account. I saw this in the verse Rev 6:14 that "heaven departed as a scroll when it is rolled together" as well as by the 24 elders and four beasts falling down before the throne with the opening of the fifth seal (the first in the static subgroup opened by Christ). The reader may expect that the seals are all opened by Christ as the K4 form, and as if sat at the right hand in regress. (GF(4) and GF(8) otherwise intersect only in GF(2) as then generalised under a seven cycle.)

The seven thunders are uttered with the completion of the opening of the last three seals (Christ proven God with the last three seals of the static subgroup loosed) mapping the trumpets etc. *backwards* in time (as if with a resounding movement in the heavens) with the approval of Christ as both God and victor now also carrying the same authority as at the opening of the first seal onwards. The seven thunders reveal that Christ's authority was always as if He were "sat at the right hand", being in total agreement with the "orbit" of the Lamb through the octal (as if under a seven cycle) formed of the opening of the seals (not as the orbit of unity or GF(2)). Heaven has moved and been shaken in a seven cycle.

Then unity (which had floated through the static subgroup) is mapped by the seven cycle within the static subgroup itself in orbit through the whole octal. Then as the operation of GF(8) gives way to that of the Holy Spirit upon the K4 form instead, there is truly sevenfold "thunder" and the whole of heaven (the place of unity in the octal) is shaken. Those "things" which were uttered by the thunders, John was instructed not to write: For these correspondences are illegal to the operation of GF(8) and not part of the Revelation of Jesus Christ. (Those seals were already opened.)

Then the K4 form of the ultrafilter:

$$1 \equiv a \equiv [0, \ a, \ b, \ c]$$
$$b \equiv [0, \ a, \ d, \ e]$$
$$c \equiv [0, \ a, \ f, \ g]$$

which is later present with Christ sat in judgement on the BEMA seat, is carried back in time under the action of the seven thunders and through the octal sevenfold by the action of the Holy Spirit. The judgement made by Christ as to the elect would immediately fix every name in the book of life and then all the sealed as well as those not so chosen. Also set by thunder, would be the timing (as Christ's judgement) of the seals and trumpets. This sevenfold thunder made of the movement of heaven is initiated from the very judgement seat of Christ, and the faithful are assured that Christ is to come very swiftly indeed.

John is instructed to "seal up the words". He does so by rewriting them in his own words, preserving their meaning but retaining his own understanding (his own seal) of them. They are given back to us in the verses that follow.

Seven thunders represent a huge movement in or "of" Heaven. I have to accomplish a seamless "reset" of the sequence of seals and trumpets to align with the overcoming of the angel.

For God's grace to be fulfilled as sovereign (sufficient in the least case), the least must overcome in the worst circumstances – when the deception of the dragon is at its peak, whilst the image system is in place and before the vials of wrath are poured out. He begins his circuit after the fifth seal is opened and completes it before the last trumpet. The seven thunders shear the timeline so that the three woes of the last three trumps align with the period starting between the fifth and sixth seal and which ends with the seventh of each sequence.

> **The Dragon:**
> *The dragon is Satan, God's adversary – and is displayed here as the "god of this world" holding authority over the Old Testament and New Testament kingdoms of this world in an attempt to oppose God's salvation of Israel. (The elect.)*

Why? When the angel overcomes there should truly be "time no longer"; there should be judgement and the sealing of the saints. I expect the angel to overcome the completed image device and he must then be present during the first two woes. The last woe of the seventh trumpet is in alignment to the seventh seal and the outpouring of the vials of wrath on the last day: that moment is the battle of Armageddon/Gog and Magog culminating in the destruction of all in the "great city" formed by the image template of "MYSTERY BABYLON".

It may be stated that the angel completes His circuit after the loosing of the restrainer (whole) and as the last woe of the tribulation begins proper: that is, between the opening of the fifth and sixth seals. This occurs the moment that the saints are finished being sealed if there truly is time "no longer". Note that an angel states not to hurt any green thing (i.e. with new life coming from a third seal – as green grass, a measure of "wheat" or "barley" (in new growth) – or a fourth seal "pale horse" paradigm) whilst they are still sealing the elect!

The trumpets must be moved from after the seventh seal so that they align to that period; the seventh "Sabbath rest" of trumpets and seals are identical and coincident on the "last day". The "great tribulation" (of the third woe) proper occurs alongside the ministry of the two witnesses and after the sealing of the saints at the end of the angel's circuit (with approval in Philadelphia). In this period the scarlet beast rises from the pit to go into perdition: it will fail and despite killing the angel's two witnesses, loses. They are going to be raised.

Rev 10:5 And the angel which I saw stand upon the sea and upon the earth lifted up his hand to heaven,
Rev 10:6 And sware by him that liveth for ever and ever, who created heaven, and the things that therein are, and the earth, and the things that therein are, and the sea, and the things which are therein, that there should be time no longer:
Rev 10:7 But in the days of the voice of the seventh angel, when he shall begin to sound, the mystery of God should be finished, as he hath declared to his servants the prophets.

As the angel overcomes (at the end of his circuit in Philadelphia, having held firm for long enough) he swears that there should be time no longer: God's sovereignty is assured only by His least overcoming under the worst possible circumstances, things cannot be permitted to become worse still. The devices of Satan have reached up to heaven, have covered the "Earth" and the "sea" that God made and corrupted the men dwelling in them completely. In like manner, the mystery of God (the Revelation) should be revealed: after all – fair's fair.

Rev 10:8 And the voice which I heard from heaven spake unto me again, and said, Go *and* take the little book which is open in the hand of the angel which standeth upon the sea and upon the earth.
Rev 10:9 And I went unto the angel, and said unto him, Give me the little book. And he said unto me, Take *it*, and eat it up; and it shall make thy belly bitter, but it shall be in thy mouth sweet as honey.
Rev 10:10 And I took the little book out of the angel's hand, and ate it up; and it was in my mouth sweet as honey: and as soon as I had eaten it, my belly was bitter.
Rev 10:11 And he said unto me, Thou must prophesy again before many peoples, and nations, and tongues, and kings.

The voice referenced by John is the same voice "as of a trumpet" which John heard with the opening of a door in heaven before, when he was in the spirit. There is a sneak peek at the future. (In fact, the state of the survived book of Revelation that John himself wrote.)

The "little" book opened by the angel and given to John is the "primer" that unlocks the maths of the mysteries in the scripture as so far. The little book contains those things which the angel has done, and essentially fixes the elements of the seven cycle by which He overcomes each of the churches. So, no wonder they are not given and are "confidential".

By "eating it up" I consider John "inwardly digesting" and much appreciating the contents of the book before forgetting it all on reading! (John had sealed up the angel's words and rephrased them in his own way.) "Inwardly digesting" may lead to much cogitation and textual content could possibly be altered by John placing his own "seal" on things. (John would forget!)

John reads the interpretation of the Revelation so far. The words resonate to John as of his own pen: his own words – "sweet in his mouth", but there are many things left to consider – the book itself is incomplete at that point. Perhaps John has forgotten the original meaning given him and needs it refreshing – what of any others that read of it? If John himself cannot decipher it – how will anyone else? This does seem enough to make one (or one's belly) bitter.

John's belly is made bitter as if by a product of two transpositions in the form of a three-cycle and its inverse (consider, sweet, bitter). Two transpositions of (consider, "taste bitter") (consider, "taste sweet") that in themselves are not inverses of each other. It makes better sense rephrased as (consider, "taste bittersweet"). The scripture surviving gave John a bit of a boost, but he is not the only intended recipient and further prophecy is on its way. John, given little chance to "consider", was not only given a command but also a big "Get on with it!".

John no doubt wondered and asked what later terms like "transposition" in respect of the human senses meant ("apperception" – the start of a very long conversation, especially as to why "bitter" is not opposite to "sweet" – see chapter ten). The question "What does all this have to do with sin?" comes to mind. John tasted the sweet despite the consequence of a bitter belly. He could then unify "pleasure" and "bitterness" within his weighing up of "consider", but the transposition is still (consider, taste). The union of the two is into "bittersweet".

John was reminded he is not a machine, but his creator knows how to loose his understanding.

Now is the "little scroll" open in the angel's hand simply a roll of sweets? John will not recall what he has eaten up (i.e. "inwardly digested"); his own seal leads to a bitter belly: it had broken the Lord's own seal. (It cannot be expected that "mouth" references a "set" in this passage.) There is some small loss of continuity in John's understanding. He would have to "prophesy again...".

A transposition is (nearly) the primer of all group theory; the reader is again subject to the seal of John's own words on the primer revealed by the angel. Here, the reader is simply given a small scene showing that is indeed what is to be read here – the primer being given to John. (A similar primer was given to the prophet Ezekiel to understand the word of his vision from the throne of God (Eze 3:3, 3:10), and most certainly the primer was received! Eze 3:13.)

These words were sweet but hard to digest (or a second time?): they are probably also hard to reconcile with the lengths to which Satan will oppose the angel's circuit (even to 200,000,000 devices or attempts against it). The possibility of anyone in Christ beginning to follow (as if completely) the path or circuit of the angel is too hard to bear, for he has his crown certainly, and once had to completely fall away from faith and return to it (cf. the letter to Ephesus).

In any case the angel states to John that this is not the end (he must prophesy again before many peoples, etc. He has much more to preserve for others). "The overcoming of the least" is not the completion of the Revelation but there is much to follow from the generalisation of the angel's circuit and of the devices of Satan applied in the Church in the last days of wrath.

The terms (Rev 10:11) are all directed to those that may be ministered to. "Peoples" are divided on the gospel (a second seal paradigm). "Nations" are within the seven cycles (churches) convicted by the first woe (fifth trump) or otherwise, whilst "tongues" refers to the third/fourth seal system of dialogue, or equally the wormwood devices of the second woe (sixth trump). "Kings" are leaders/ministers in dialogue given the choice of an idol shepherd, but not yet incorporating their fellowships under the "MYSTERY BABYLON" template. John's writing may prophesy to all these; they may be brought back from the brink at the end of the age.

Despite John's wish to dally and read more of his own work (the prospect of understanding the prophecy as sweetness, partly not (or not yet or ever) understood by him – as if bitter), John concedes to moving on and is given some very much needed help.

7.10) The Two Witnesses

The witnesses are Revelation's reply to the sixth trump "serpents". The serpents supplant the octal underneath the K4 form, but only do so to the octal group in order to interrupt the seven cycle (of the least's circuit). It is not true that both groups are so affected because the result would be a simple relabelling and so trivially isomorphic. (And still a valid multiplication.)

The two witnesses essentially argue that wherever the faithful find themselves spiritually, as a triple corresponding to unity or as a unity element in the K4 form in the octal (previously seen as a "Golden Girdle"), the seven cycle of multiplication and the "sealing of the Holy Spirit" suffices to maintain their election in the reference (Sun) octal. Since they hold firm to unity (are sealed, taking their reference from the Lamb), they are not leached from the "Sun octal", neither do they traverse the "sea". The middle ground of octals that partially intersect or share subgroups with the reference octal has no effect of "leading them into temptation" to deny the gospel, to operate as one could do with the dialectic (as in the seals) and trumpet devices.

Those in the "middle" ground in the sea of thirty octals are ministered to by the two witnesses showing there is no "safe ground" avoiding the judgement of Christ, that all must come out of the Church. As the reference seven cycle continues to act, the right and left hands of the field GF(8) are regenerate, and the middle ground under the seven cycle is not closed as is the Sun octal. (The circuit of the angel remains intact generalised across every octal in seven symbols.)

Now, the least in the kingdom of God has no great following (of any) but most likely subsists in small fellowships, and His overcoming is witnessed not by himself, but by two others to all those that have ears to hear it.

With the generalisation of the angel's circuit, John has need of the symmetries of GF(8) to describe what he sees, for without it meaning is lost. (His own words are better with a gift.)

Rev 11:1 And there was given me a reed like unto a rod: and the angel stood, saying, Rise, and measure the temple of God, and the altar, and them that worship therein.

A "pen" (a reed – a device of writing) like a measuring rod (a mathematical primer) is given to John. It may be as simple as writing an arbitrary seven cycle (a,b,c,d,e,f,g) asserting the static subgroup is formed of the second, third and fifth members if the first is unity. The cycle of the elements determines the subgroups. The octal is implicit in the triples of subgroups already.

Rev 11:2 But the court which is without the temple leave out, and measure it not; for it is given unto the Gentiles: and the holy city shall they tread under foot forty *and* two months.

John is to measure the GF(8) ultrafilter and not the other fields used in the fifth trumpet or the sets of properties under the ultrafilters' indexing set; the primer is only effective on the octal.

The octal itself is trodden underfoot for 42 months. These (feet in pairs) are the six non-identity elements of the seven C7 groups that are part of the "outer court" – they are the (non-identity) elements outside the Sun octal's reference (rainbow) seven cycle (the angel's circuit to be generalised) and this makes 42 elements of 48 total (with a last unity as 49th). The Sun octal retains one C7 group as multiplication – the mystery of the seven churches.

Each of the seven cycles of the "outer court" or churches has a "rest element" of floating unity which may "walk through the seven cycle". (One element of seven in the angel's circuit corresponds to the overcoming of Laodicea.) The above seven groups are "closed" when not including any of the least's 200,000,000 possible circuits! (If there is no conviction of spirit.)

"Feet" represent the two static triples in each seven cycle: here, the static triples in all seven churches that refuse to obey the requirements of Laodicea. They may walk in seven groups of seven cycles (seven choices for floating unity in each), though they never change group as the angel in his circuit. (They are amongst the 7000 dead in the great earthquake when 1/10th of the "great city" falls later in the account.) They may walk in the spirit, but are without conviction by the repenting of the least; they are as "gentiles" were without the law of Moses, here without spiritual conviction to follow the least's exit. (And no wonder the remnant are affrighted, for Philadelphia does not escape completely.)

Instead of following the angel out to Philadelphia, they tread the holy city (and not Babylon) underfoot. I find a generalisation of the seven churches in some sense to "the holy city" over a wider set of 30 octals. These gentiles dwell in the churches but do not follow the same calling to exit by the Holy Spirit; one may ask: is it for the sake of these that the Lord sends two witnesses to the angels overcoming? With no witness has the angel done his work otherwise?

Rev 11:3 And I will give *power* unto my two witnesses, and they shall prophesy a thousand two hundred *and* threescore days, clothed in sackcloth.
Rev 11:4 These are the two olive trees, and the two candlesticks standing before the God of the earth.

The gifts that sustained the angel in His circuit (as received in overcoming) empower the witnesses to fulfil the task of their mission, to reveal to anyone in the Church that false doctrine is just that: falsehood. The two witnesses are also referenced as the two olive trees in the OT book of Zechariah (cf. trees as triples as previously, forming the "third part of men"), whose doctrine as "oil" mixes seamlessly from the left and right hands into the lampstand made of the Spirit of God (as the rainbow of the generalised circuit about the throne).

They prophesy to those forty two elements of the seven "outer" C7 groups now generalised over every possible octal of thirty, forming 1260 days of their ministry. I can ignore the 24 elders as here I deal in only the generalisation of the angel's circuit to other C7 groups (circuits in general). I.e. 30 * 42 = 1260 days. (30 days in a month. All the witnesses require to witness is that the cycle is completed, the church corrupt.)

The angel's circuit, complete without any influence of the Holy Spirit, is valid as seven "horns" (additive elements) chosen by Christ for His right hand to use to overcome. Those horns may form octals in any arrangement, any one of thirty octals. There is then no C7 group for (but only from) the angel in his circuit. Instead, his circuit will convict all, saint and sinner alike. With His circuit generalised over all thirty octals in the same seven "horns", the Holy Spirit ministers the greater alternative to all others (those not in receipt of the horns), those for whom the Lord's grace is always sufficient. The lack of multiplication for the angel permits these 1260 days of ministry, whereas he could not do his work otherwise.

Then the two witnesses have all the power of the angel's circuit to witness with, yet with the equivalents in the Holy Spirit instead.

They prophesy clothed in sackcloth (with heavy conviction of Spirit, and as in a mourning garment). That is, they aren't going to do this for pleasure, they will "as fasting" (and mourning as ministering within the "Death" device or "wormwood" octals) be without pleasurable things to wear, eat and drink; this indicates theirs will not be a peaceable debate of dialogue but a ministry of power, force and truth.

As "two olive trees" every compatible pair of static triple and subgroup (for a given choice of unity) is preserved by just two of the eight C7 groups over the same octal. (Though there are always different "complement" left handed octal groups, one for each seven cycle.) There are then two lampstands before God: the first complete in all seven elements being of the angel's circuit corresponding to Laodicea (and the overcoming of it) in the right handed octal, and with it the corresponding complement left handed octal. The two olive trees represent the static triples (with the congruent K4 form) that circuit the seven cycle – and not completed sets of seven, which are properly the seven candlesticks through which they circuit.

Rev 11:5 And if any man will hurt them, fire proceedeth out of their mouth, and devoureth their enemies: and if any man will hurt them, he must in this manner be killed.

I equivocate the witnesses to the triples in the cycle of the angel's circuit generalised over all 30 octals. Whether those triples be of K4 subgroups or elements under orbit in left and right hands of the reference, it is clear that to "hurt" the witnesses is to counter them with the middle ground of the sixth trumpet serpents. Each witness becomes present in their partner's octal – by appearance in the wormwood rows only. This is how the witnesses are "countered". (And this to facilitate the dialectic device in the same setting as of the triple of subgroups.)

$$[a,b,c], [a,d,e], [a,f,g] \text{ with } [b,e,f], [b,d,g], [c,d,f], [c,e,g] \text{ with unity } a = 1$$
$$[b,d,f], [a,b,c], [b,e,g] \text{ with } [a,d,g], [c,d,e], [c,f,g], [a,e,f] \text{ with unity } b = 1$$
$$[c,d,g], [b,d,f], [a,d,e] \text{ with } [b,c,e], [a,c,f], [e,f,g], [a,b,g] \text{ with unity } d = 1$$
$$[c,e,f], [c,d,g], [a,b,c] \text{ with } [a,d,f], [b,f,g], [a,e,g], [b,d,e] \text{ with unity } c = 1$$
$$[a,f,g], [c,e,f], [b,d,f] \text{ with } [b,c,g], [d,e,g], [a,b,e], [a,c,d] \text{ with unity } f = 1$$
$$[b,e,g], [a,f,g], [c,d,g] \text{ with } [d,e,f], [a,c,e], [a,b,d], [b,c,f] \text{ with unity } g = 1$$
$$[a,d,e], [b,e,g], [c,e,f] \text{ with } [a,c,g], [a,b,f], [b,c,d], [d,f,g] \text{ with unity } e = 1$$

If the "mouth" of the witness is the static triple of three subgroups in the first (leftmost) three columns of the wormwood octal, these are also held static by the action of Frobenius on the reference cycle. If that witness' fire comes "out of their mouth" then the reference cycle apparently "leaves" the mouth which it holds static and from it, generates the whole reference octal as before. There is the distinction made here with the later phrase "the earth opened her mouth" as of the woman with twelve stars later, in reference to the closed properties of a set. Here the set of the witness is opened and "extended". As each wormwood row corresponds only to unity, the reference cycle generates the Sun octal upon them simply, for unity was the only possible (closed) subgroup of the group C7 as the reference cycle.

The seven cycle consumes any opposition to the two witnesses in that middle ground. (The remaining three triples under that seven cycle will never form octals.) They are "devoured" in that the seven cycle regenerates only the triples of the Sun octal and its complement; the middle ground is "consumed away" (or "chewed up") rather than destroyed by the seven cycle. As those three other columns cannot form octals alone, they are as an empty set (as an empty mouth – then devoured); they are not valid under the seven cycle. In truth there is no intersection of three triples with the Sun octal; there is no middle ground.

Any enemy must in this manner be killed: this seems somewhat violent, but it states that the only opposition to the two witnesses is actually to take the true middle ground, to rest as unity in the reference. As the witnesses cannot together private that virtue in Christ and the only opposition to them is to do so, then in this like manner those that deny Christ must be "shown dead" in denying virtue (Christ).

The two olive trees (Rev 11:4) I consider are a subset of the octal: that in pairs of (left to right) correspondences of static groups to elements of their counterpart are as "blood" circulating through the octal. Well... see for yourself – it sounds harder than it looks.

$$b \equiv [a, e, f]$$
$$d \equiv [a, b, g]$$
$$f \equiv [a, c, d]$$

For one witness;

$$c \equiv [a, f, g]$$
$$e \equiv [a, b, c]$$
$$g \equiv [a, d, e]$$

For the other. (And a ≡ [b,d,f] ≡ {c,e,g} in the Sun octal as before.)

By "olive trees", there is a reference to the continual "branching" of singletons (as "a" here) to static subgroups in the alternate "witness" (octal) then to the K4 form of the ultrafilter spanning the octal made of the three subgroups with unity "a" in common (these three are the "fountain" of "a").

Their "mouths" in the same witness are equivalent to a K4 group of subgroups. So a ≡ {c,e,g} ≡ [[a,b,c],[a,d,e],[a,f,g]].

Then, fire comes out of their mouths to devour their enemies: only in this method are they countered, i.e.

$$[a,b,c], [a,d,e], [a,f,g] \text{ with unity } a = 1$$
$$[b,d,f], [a,b,c], [b,e,g] \text{ with unity } b = 1$$
$$[c,d,g], [b,d,f], [a,d,e] \text{ with unity } d = 1$$
$$[c,e,f], [c,d,g], [a,b,c] \text{ with unity } c = 1$$
$$[a,f,g], [c,e,f], [b,d,f] \text{ with unity } f = 1$$
$$[b,e,g], [a,f,g], [c,d,g] \text{ with unity } g = 1$$
$$[a,d,e], [b,e,g], [c,e,f] \text{ with unity } e = 1$$

Then "fire" from one witness's mouth is as "a" or unity (the intersection of the three subgroups) with the mouth (set) the triple of subgroups under a seven cycle. The other witness has the same cycle on three subgroups in the other "left hand" octal. The Spirit provides the witnesses with the proper words to counter the only method of opposition to them. In essence, abstraction wins the argument because the truth is abstract, and this is the battleground.

Rev 11:6 These have power to shut heaven, that it rain not in the days of their prophecy: and have power over waters to turn them to blood, and to smite the earth with all plagues, as often as they will.

The witnesses have power to shut heaven without allowing the seven cycle (as water from heaven, in reciprocal acting on the left hand falling – on {c,e,g} as before, additively reversed to the seven cycle and also the "fountain" in the leftmost three columns) to operate upon the wider set of octal (wormwood) rows of the sixth trumpet. Without cycling by (a,b,d,c,f,g,e) above they may remain in a single row holding to unity (as the top above as a=1) with [0,a,b,c],[0,a,d,e],[0,a,f,g]. Under the three cycle in the K4 form this triple of subgroups is all. In the octal it is yet closed and it is possible in each row to sustain Frobenius upon the "wormwood octal" to three-cycle the K4 form preserved. That is, the rain is prevented as water operating from top row to bottom upon the "Earth" (an analogue of "hair" on the head as before, or of rivers drying up – but not truthfully in an orbit; the seven cycle is invalid to the wormwood octals). So, the witnesses may preserve their triples in the left or right hands without requiring the "rain" of (a,b,d,c,f,g,e).

$$[a,b,c], [a,d,e], [a,f,g] \text{ with } [b,e,f], [b,d,g], [c,d,f], [c,e,g] \text{ with unity } a = 1$$
$$[b,d,f], [a,b,c], [b,e,g] \text{ with } [a,d,g], [c,d,e], [c,f,g], [a,e,f] \text{ with unity } b = 1$$
$$[c,d,g], [b,d,f], [a,d,e] \text{ with } [b,c,e], [a,c,f], [e,f,g], [a,b,g] \text{ with unity } d = 1$$
$$[c,e,f], [c,d,g], [a,b,c] \text{ with } [a,d,f], [b,f,g], [a,e,g], [b,d,e] \text{ with unity } c = 1$$
$$[a,f,g], [c,e,f], [b,d,f] \text{ with } [b,c,g], [d,e,g], [a,b,e], [a,c,d] \text{ with unity } f = 1$$
$$[b,e,g], [a,f,g], [c,d,g] \text{ with } [d,e,f], [a,c,e], [a,b,d], [b,c,f] \text{ with unity } g = 1$$
$$[a,d,e], [b,e,g], [c,e,f] \text{ with } [a,c,g], [a,b,f], [b,c,d], [d,f,g] \text{ with unity } e = 1$$

The rain is stopped in "the days of their prophecy". In every octal all 42 multiplicative elements are exchanged for the unity; the witnesses keep to the same row. Heaven is "shut" (unity is a closed subgroup).

They have power to turn waters to blood: so that the static subgroup of the Sun octal may be exchanged for the static triple of the left hand. ({c,e,g} for [b,d,f] in the top row above.) In fact, "blood" represents the intersection between octals, which may either be the leftmost three columns shared with the reference, or the rightmost which is the static triple of the Sun octal's left hand. (Does this case hold symmetrically with the left hand of the Sun octal, or restricted to this wormwood alone? Is the Lord right handed, His arm outstretched? Certainly not with His "hand" then "cut off", as employing this device of Satan.)

The leftmost three columns are properly "a fountain" under a seven cycle in that they are the branching of one element to three triples. The witnesses have power over these "waters" to intersect them with the "golden girdle" or K4 form (as if a "tree") in the reference octal. Likewise, the left hand intersects with the above in the rightmost column only: this being the sole "river" in the earthly elements under the seven cycle. As "water" it also intersects with the left hand of the reference. The witnesses have power over these two sets to correct them: they generate only the left and right hands of the reference "Sun octal" with which the wormwood rows intersect. They have the power over those waters, not antichrist.

In smiting the Earth (with all plagues), I examine the triples aligned to the earthly elements in the rows above.

Now, the angel's circuit is valid in the right hand of the reference, and by the arbitrary choice of multiplication now generalised across the sea of 30 octals (42 "products" each) the witnesses have power to change the multiplication they may use at will. (Their power is also over the "1260 days" of their prophecy: i.e. to smite "as often as they will".) That is, every possible static triple that may be used for the dialectic may instead be used as the other witness in the wormwood octals above. The remaining three columns suffer this contortion of being transformed into an operation that does not form a "river" or octal. They are "tormented".

The plagues are of the Earth – (those three columns are "smitten" for the use of the dialectic), being those paradigms that plague the Church: killing with sword, hunger, death and with the beasts of the Earth.

In the wormwood octals the dialectic bows appear valid as subgroups: but the earthly four elements in the wormwood rows are the rightmost four entries of each row. The witnesses cannot transform these back to the reference octal; the three columns that do not form octals cannot reference the static bow in the reference's left hand. In the wormwood, they reference the right hand witness [b,d,f] instead.

That is, in the reference corresponding to the top row, {a,c,e,g} are the earthly four elements, and [b,d,f] and {c,e,g} are static. ([b,d,g] etc. are not found in the left or right hands.) The plagues are not these alien columns, but are instead the dialectic devices employed by the churches/nations.

The choice of unity from {a,c,e,g} is dependent on the wormwood row chosen above, and the static triple that may be chosen also dependent on the choice of multiplication (one of four triples from each column as above, but actually one of eight – two choices for each triple). The two witnesses may, under any multiplication (i.e. "as often as they will" – as any one of 1260 days of their ministry), smite the earth's remaining three columns with their references seven cycle, which is illegal to those three: "tormenting those that dwell on the earth" (those in the static bow in the wormwood row, those same three columns with an illegal "non-octal").

The witnesses testify to the completion of the angel's circuit (in the right hand) as generalised over all 30 octals in every C7 group over them. They testify that even in a much complicated lattice it is possible to be found and saved in God's good grace in every case. (Not to freely employ the devices of the seals or trumpets.)

The other hand of the reference is free to do likewise – their triples' cosets (the Earth) may be put under the plagues of the seals and trumpets with every movement of each witness in the multiplication on their octals. In the witnesses choosing any multiplication, every dialectic use of a "bow" may be smitten by "torments". Comparing this ministry as verbal with the outpouring of the vials of wrath to come, these witnesses are giving the last and final warning to come out of the Church and/or to repent in obedience to God and His Christ.

Now, I should account for every octal with reference to the "Sun octal" – the subgroups of which are shared in intersection with the rows of the "wormwood octals" above. The leftmost three columns represent an intersection of three subgroups with the Sun octal, and the separate cases of intersections with only a single K4 group are covered with the (left hand) subgroup present in the rows of the "other witness" as water becomes blood etc.

So [a,b,c],[a,d,e],[a,f,g] is the intersection of three subgroups and [c,e,g] the triple of the other witness in the same row. I would expect [b,d,f] in the other witness to be present in the left hand's separate "wormwood octal". So I should have [a,d,g],[a,c,f],[a,b,e] which are the three subgroups of the left hand of the Sun octal under the transpositions (b,c), (d,e), (f,g) together with [b,d,f] (from the same row) which is an intersection of one group with the right hand octal. There are 7 octal groups that intersect with three subgroups of the reference. There are also two groups intersecting in each single subgroup of seven. So, I have 7+14 = 21.

I can disassemble the transpositions of the serpents to reduce to the three cases of "the octal" (one group) and "an intersection of three groups" (seven groups) and "an intersection of one group" (fourteen groups). The fourth case has "no such transposition" and these are the eight choices for "the other witness" for the reference octal. So, I have 1+7+14+8 = 30 groups.

Both serpents and locusts are hell bent on finding twelve elements of S4 on the earthly elements. The two witnesses are the point in case that there is no such middle ground with the reference octal; the presence of two "lampstands" (witnesses) that are upright before God leave no territory for the serpents to extend into, and no kingdom for "Abaddon".

The pairs of witnesses in the wormwood octals as before then account (when under every possible seven cycle) for the whole number of octals in the sea that are linked (by intersection) to the Sun octal, the serpents and locusts "leaching" its operations for their devices employing those stars as sand "cast to the Earth" from elsewhere in the sea.

7.11) That Great City

After the completion of the witnessing (testimony) there is universal conviction of the faithful still in the Church to exit and repent (the least has also accomplished the same). I find the end of their ministry when the resulting Sabbath rest is disturbed by those that dwell on the Earth.

Rev 11:7 And when they shall have finished their testimony, the beast that ascendeth out of the bottomless pit shall make war against them, and shall overcome them, and kill them. Rev 11:8 And their dead bodies *shall lie* in the street of the great city, which spiritually is called Sodom and Egypt, where also our Lord was crucified.

The description of the city as Sodom and Egypt should be read "spiritually perverse" and "spiritual captivity" (in bonds of works). Christ was crucified in Jerusalem, a city of Israel, but one of sinners (to be exited). Nothing more is meant than that.

The witnesses' ministry here aligns to the image device of the ten kings:

Rev 17:14 These shall make war with the Lamb, and the Lamb shall overcome them: for he is Lord of lords, and King of kings: and they that are with him *are* called, and chosen, and faithful.

The witnesses take the same state in triples as to the (now having overcome) angel stood on the Earth and sea that had his feet as "pillars of fire". (The completion of the angel's circuit.) When all who are to dwell in Philadelphia are exeunt from the Church, the beast that ascends (the collective of the woman in the wilderness) will make war with the saints (as to who is chosen): not as to complete all 200,000,000 devices against the circuit of the least (that is done with) but when there are no "least" left within. Yet the sacrifice of Christ has virtue which is not moved. (The lamb corresponds to the rest element in the reference octal – unity.)

In order to show that unity is unmoved, the witnesses die and are raised. They are "overcome" because they no longer generate the seven cycle between the rows (see over) but sit static in the leftmost three columns with the other witness in one of four earthly columns. Each witness of:

$$e \equiv [a, b, c]$$
$$g \equiv [a, d, e]$$
$$c \equiv [a, f, g]$$

$$b \equiv [a, e, f]$$
$$d \equiv [a, b, g]$$
$$f \equiv [a, c, d]$$

is unable to traverse rows to generate the full octal (now only having the identity of rest with their completed ministry), each witness is restricted to only one row, and the operation of the octal underneath them is "smitten", or rather the beast that ascends "kills them", supplanting the octal underneath the two witnesses now in valid K4 form.

The beast is the same collective that rules with ten kings. Rev 17:4 above alludes to the three cycle (floating unity in GF(4)) upon the K4 form with "called, chosen and faithful". Christ has life in Himself to give: the lamb in raising the witnesses shows He has all power over the beast. (Despite the serpents holding judgement in the place of Christ.)

The beast encroaches upon five of seven elements in the seven cycle (cf. fifth trumpet). That is to say that in every seven cycle at least one element of each witness' static subgroup is under the sway of the "great city" (as S5 upon the octal, or the fallen star with the key to the bottomless pit), and the other witness in the same set of singletons is completely under that city of five elements. (The beast or collective of tares forming the woman in the wilderness.) In effect, there is war between the beast and the witnesses, but the lamb is stronger.

The witnesses are surrounded in blasphemy against the Lamb's virtue, the rest element (to be shamed openly as was Christ, which is the ultimate end of their testimony). The two witnesses are killed by the beast from the pit – the action of the locusts and serpents of the fifth and sixth trumps. The ministry of the witnesses is then represented as against the wormwood: the product of the "Euphrates octals" of the sixth trump.

$$[a,b,c], [a,d,e], [a,f,g] \text{ with } [b,e,f], [b,d,g], [c,d,f], [c,e,g] \text{ with unity } a = 1$$
$$[b,d,f], [a,b,c], [b,e,g] \text{ with } [a,d,g], [c,d,e], [c,f,g], [a,e,f] \text{ with unity } b = 1$$
$$[c,d,g], [b,d,f], [a,d,e] \text{ with } [b,c,e], [a,c,f], [e,f,g], [a,b,g] \text{ with unity } d = 1$$
$$[c,e,f], [c,d,g], [a,b,c] \text{ with } [a,d,f], [b,f,g], [a,e,g], [b,d,e] \text{ with unity } c = 1$$
$$[a,f,g], [c,e,f], [b,d,f] \text{ with } [b,c,g], [d,e,g], [a,b,e], [a,c,d] \text{ with unity } f = 1$$
$$[b,e,g], [a,f,g], [c,d,g] \text{ with } [d,e,f], [a,c,e], [a,b,d], [b,c,f] \text{ with unity } g = 1$$
$$[a,d,e], [b,e,g], [c,e,f] \text{ with } [a,c,g], [a,b,f], [b,c,d], [d,f,g] \text{ with unity } e = 1$$

The rows above are also formed by the "wormwood device", hence I call them as octals or rows, "wormwood". (A "bitter" herb or "very small tree". A seed of evil sown to replace that true seed of the word (virtue), and also discovered hidden in the earthly elements.)

With the rest element, I then reduce to just the top row, say:

$$[a,b,c],[a,d,e],[a,f,g] \text{ with } [b,e,f],[b,d,g],[c,d,f],[c,e,g] \text{ with unity } a=1$$

The only seven cycles now in view are those of the confused or interrupted octal: the witnesses cease "branching" under any seven cycle and remain static (so "a=1" above). Their bodies (the left and right hand octals of the reference) lie as dead in the street of the city, a wider path that I may use the following notation to represent (with trees to either side). Seven cycles cease in rest (unity) yet the K4 form may animate each of these sets static in themselves:

$$\{[a,b,c],[a,d,e],[a,f,g]\} \rightarrow \{c,e,g\} \rightarrow a \leftarrow \{b,d,f\} \leftarrow \{[a,e,f],[a,b,g],[a,c,d]\}$$

Rev 11:9 And they of the people and kindreds and tongues and nations shall see their dead bodies three days and an half, and shall not suffer their dead bodies to be put in graves.

The four groups of people represent the four positions or choice of unity that were seen earlier (the same four groups of people praising God) – that do not allow the witnesses to be buried. Yet this may be a good thing for they may expect them to be raised in Christ as elect! The gospel itself is not defeated and "dead and buried"; the kingdom of God is here to stay.

The elect (people) are separated from those sharing the same virtue but not the same church/ C7 group (kindreds), and also from those operating in different C7 groups with a different "lamb" or "set of virtues" (tongues) and lastly those under the Holy Spirit in all seven churches (nations) are in view. They match up to "Christians" in the Church amalgam or otherwise.

The four sets of people may also constitute a tower under inclusion (a poset) formed of (Church) fellowships agreeing in their corporations (cf. "A measure of wheat for a penny and three measures of barley for a penny" – simple to represent in triples as tares corresponding to singletons (unity) as wheat).

By utilising the dialectic process, they have that hard limit of four as if they would also construct the dragon's tail (the limit of the consensus found in dialogue using the dialectic, as upon the inclusion of four collectives, or of the simple fourfold accumulation of sets in a dialectic and as such, a logical form); it will certainly collapse (fall as Babylon) at the fifth dialectic (Hell) once the corporation itself contradicts the people operating by it. It will fall twice over when the people under it likewise contradict the corporate template and each other. (There are only four positions to shift unity in common at any time – even though the last (dialectically, sent to the empty set) is truly to find inconsistency in the method: that's why there are pieces of paper instead of living people!)

In the example above, the top row

$$[a,b,c],[a,d,e],[a,f,g] \text{ with } [b,e,f],[b,d,g],[c,d,f],[c,e,g] \text{ with unity } a = 1$$

has "three days" (formed of [b,e,f], [b,d,g], [c,d,f] under orbit) made into left hand partners for the reference octal (each a different static triple in that reference octal) and as each of these three triples are possibly held static (as a subgroup) to generate a "river", they are each a "day". They all have eight multiplicative groups distinct from the two holding the witness [c,e,g] static and so their own set of 24 "elders" – they are each as a whole day. (It is the case that each share the reference octal as one of their eight left hands.) The fourth and last "earthly" column corresponds to the one witness to one side of the reference octal already.

The remaining three "earthly" columns (which are not generated from [c,e,g]) do not form a valid octal under the seven cycle (a,b,d,c,f,g,e). The witnesses are "dead for three days" (three octals), and half a day, the half-octal formed of the coset they make together: they, a coset of the three groups containing unity or "a" as above do not form subgroups preserved under the seven cycle. The witnesses are seen dead for "three days and a half": the same sets that do not form the proper branches of singletons to subgroups that also correspond to unity:

a ≡ [b,d,f] ≡ {c,e,g} or as the two witnesses,
a ≡ {b,e,g} ≡ [c,d,f] or the first day as under a suitable C7,
a ≡ {c,e,f} ≡ [b,d,g] or the second day as under a suitable C7,
a ≡ {c,d,g} ≡ [b,e,f] as the third day as under a suitable C7.

And [b,e,f], [b,d,g], [c,d,f], [c,e,g] forms half "a day".

Rev 11:10 And they that dwell upon the earth shall rejoice over them, and make merry, and shall send gifts one to another; because these two prophets tormented them that dwelt on the earth.

The sets of four static triples in each row (those that "dwell on the Earth" under the action of Frobenius) appear to have some kind of "four-bow square dance". With the limit of their dialectic now full, their three-cycles are in "orbit" holding a "fixed unity" as if the imaginations of their hearts were (in collective) to be joyfully trusted. Similarly, they may link the rows by using transpositions as before, exchanging "gifts".

They that dwell on the Earth (the elements {a,c,e,g} say with a ≡ 1 ≡ [b,d,f] static) that do not overcome with any of the 200,000,000 variants of the angel's circuit are stated to be united in exchanging gifts – a scene of the whole world practising the devices that killed (halted the ministry of) the witnesses whilst they were in the "rest element". An exchange of elements (a transposition) is become a "gift", and the whole world "lies in wickedness"; they simply show themselves unworthy of paradise. (Utilising the first two woes, possibly (as yet) incomplete.)

"They that dwell upon the Earth" and not as heaven (here the static subgroup) form three columns in the wormwood rows, the same three days in which the two witnesses lay dead. These columns "torment" those on the Earth, as they do not form octals.

Rev 11:11 And after three days and an half the Spirit of life from God entered into them, and they stood upon their feet; and great fear fell upon them which saw them.
Rev 11:12 And they heard a great voice from heaven saying unto them, Come up hither. And they ascended up to heaven in a cloud; and their enemies beheld them.

After the appearance of defeat, the seven cycle is reinstated and the columns regenerated.

$$[a,b,c], [a,d,e], [a,f,g] \text{ with } [b,e,f], [b,d,g], [c,d,f], [c,e,g] \text{ with unity } a = 1$$
$$[b,d,f], [a,b,c], [b,e,g] \text{ with } [a,d,g], [c,d,e], [c,f,g], [a,e,f] \text{ with unity } b = 1$$
$$[c,d,g], [b,d,f], [a,d,e] \text{ with } [b,c,e], [a,c,f], [e,f,g], [a,b,g] \text{ with unity } d = 1$$
$$[c,e,f], [c,d,g], [a,b,c] \text{ with } [a,d,f], [b,f,g], [a,e,g], [b,d,e] \text{ with unity } c = 1$$
$$[a,f,g], [c,e,f], [b,d,f] \text{ with } [b,c,g], [d,e,g], [a,b,e], [a,c,d] \text{ with unity } f = 1$$
$$[b,e,g], [a,f,g], [c,d,g] \text{ with } [d,e,f], [a,c,e], [a,b,d], [b,c,f] \text{ with unity } g = 1$$
$$[a,d,e], [b,e,g], [c,e,f] \text{ with } [a,c,g], [a,b,f], [b,c,d], [d,f,g] \text{ with unity } e = 1$$

Again there are the "feet" walking a seven cycle (as were Christ's) and the angel stood on the Earth and sea, indicating the static triples [b,d,f] and {c,e,g}. (Here the first three columns for one witness hides [b,d,f] in a cloud (mystery) which is the static subgroup in the reference "Sun" octal). The original Sun octal is reasserted and the witnesses retain their life in Christ. The witnesses now "stood on their feet" reveal that the static triples in the Sun octal's left and right hands are the basis for the victory over the wormwood, rather than that stance with the feet exchanged as is found with the device(s) of the wormwood octals.

> **Clouds:**
> *Clouds represent mysteries in the Revelation. Some are revealed in the text and some left under a seal. Cf. the angel clothed with a cloud revealed to be the angel of the church.*

The witnesses "stand upright" (in columns not rows) victorious despite the attempt to supplant their (first) resurrection in Christ with faulty spirit: it is displayed openly here. (The "sign of Jonah" is being three days in the Earth, as given to this generation.) Fear fell upon the remaining three columns, being closed and cut off from the life in the witnesses, now raised from death.

There is a "great voice" (judgement) from heaven; the witnesses standing, ascend still standing (upright). It is a mystery as to why the three earthly columns left behind remain intact under the seven cycle yet without the right and left handed octals present. (Where do those others that dwell in the wormwood octals go? It must be the lake of fire, the city S5: wherein the results of opening the fifth, sixth and seventh seals are fully instantiated as by Christ's judgement, to trap all those using those devices.)

They are as much under judgement as the witnesses raised are clearly not so, being raised acceptable to heaven before all their enemies. To the triples of "men" (operated upon by seven cycles illegal to their triples) the action of the Holy Spirit in the reference octal will not preserve their octals in the sea; to them the action of the Holy Spirit is a mystery, despite them seeing both witnesses raised to life, witnessing the virtue in Christ in the case of the least.

The "cloud" likewise hides the witness within it: for in the reference octal there may be [0,b,d,f] static and a=1. In each row, then, the groups [0,a,b,c],[0,a,d,e],[0,a,f,g] are shared under a seven cycle with the "Sun" octal. These enclose the witness [0,b,d,f] whilst the witness [0,c,e,g] appears in the row with them. Similarly, I have the same on the "other hand" with the other witness. In this manner the group [0,b,d,f] is enclosed as shrouded in a cloud or "mystery" whereas on the left hand (the other side of the street of the great city) the witness [0,b,d,f] is clearly visible in the row (of the other witness). That is, they both ascend in a cloud and their enemies (opposites) properly (spiritually at last!) behold them! ([0,c,e,g] is in clear view, "beheld" by the remaining three columns in earthly elements in the top row above.)

That resurrection is still a mystery: as Christ is the proper form of the K4 ultrafilter within which their lives were hid – and is Himself God, and to Him – masked in a cloud they return victors. They remain at (Sabbath or unity) rest whilst they ascend up by a seven cycle through shifting elements in triples, branching between pairs of the thirty octals. The great (universal in a sense of generalisation) voice commands them to "come up". This is the judgement over all, that only in Christ may they attain the resurrection despite having been physically dead. (They ascend in a seven cycle – John reads the permutation notation right to left.)

Now I begin to see the mechanism through which Christ has the "Keys of hell and death".

Rev 11:13 And the same hour was there a great earthquake, and the tenth part of the city fell, and in the earthquake were slain of men seven thousand: and the remnant were affrighted, and gave glory to the God of heaven.
Rev 11:14 The second woe is past; *and*, behold, the third woe cometh quickly.

One hour refers to one automorphism of the 24 elders. "That same hour" refers also to the "cloud" within which the two witnesses ascended. In the rows above, if a=1, then that hour corresponds to holding [b,d,f] and {c,e,g} static by the reference seven cycle (a,b,d,c,f,g,e).

Now the ministry of the two witnesses is completed, and every octal is now treated as isomorphic with 42 seven cycles (of seven groups) acting upon them, with the last C7 group circuiting between those seven groups as the angel's circuit.

Each of the four earthly triples may be separately treated as a left hand of the Sun octal; I may align each row (below) fourfold to the possible powers in x (with the top three rows as the static triple – (of subgroups) with their intersection as unity). With those choices of multiplication I have:

$$x \equiv [a,b,c] \equiv [a,b,c] \equiv [a,b,c] \equiv [a,b,c]$$
$$x^2 \equiv [a,d,e] \equiv [a,d,e] \equiv [a,d,e] \equiv [a,d,e]$$
$$x^4 \equiv [a,f,g] \equiv [a,f,g] \equiv [a,f,g] \equiv [a,f,g]$$
$$1 \equiv [c,e,g] \equiv [b,d,g] \equiv [b,e,f] \equiv [c,d,f]$$
$$x^3 \equiv [c,d,f] \equiv [b,e,f] \equiv [b,d,g] \equiv [c,e,g]$$
$$x^6 \equiv [b,e,f] \equiv [c,d,f] \equiv [c,e,g] \equiv [b,d,g]$$
$$x^5 \equiv [b,d,g] \equiv [c,e,g] \equiv [c,d,f] \equiv [b,e,f]$$

The rows of octals above are now aligned isomorphic (as in "isometry") with the reference octal, which I have simply denoted by the polynomial powers of the generalised seven cycle (which is also treated as isomorphic to that of the Sun octal and angel's circuit).

The reference acts with that "same hour" or automorphism as causing the "great earthquake" to cycle the triples in {x^3, x^6, x^5} by Frobenius, keeping the subgroup corresponding to unity static. The twelve permutations of the "earthly" triples above are the only twelve from that wormwood octal, twelve elements of S4 are missing. (These, however, are leached, "stolen" from the seven cycle, aligned in an isomorphism from elsewhere in the sea.)

Only the Sun octal's seven cycle (a,b,d,c,f,g,e) is in view, acting solely upon the subgroups of the Sun octal and its own left hand. It is clear from the two witnesses that the angel's circuit does not preserve the subgroups in each wormwood octal. The three "illegal" earthly triples when cycled by squaring the reference cycle, form "plagues"; the reference cycle is only acting "between wormwood rows" rather than in the rows.

The reference seven cycle would itself be permuted by Frobenius (shifting rows, see opposite) but does not permute across the wormwood rows, for each wormwood row is in correspondence to only the unity element of the reference circuit. Frobenius cycles static the "fountain" or "jacinth" part of each row but then cannot map row to row.

One tenth of the "Great City" (12 elements of S5, of 120 elements total) falls or is lost in the great earthquake: these permutations of the earthly elements only align to the reference isomorphically as a result of the completed ministry of the two witnesses generalising the reference octal. These 12 elements of S4 upon the earthly triples do not and cannot align with the reference octal in any other way. (That is, on assuming three subgroups common to both.) There is no sense in which the reference seven cycle may act on three subgroups of seven and generate anything except the Sun octal and its corresponding left hand.

From the mystery of the seven golden candlesticks, there is an octal generated from every subgroup in the reference or Sun octal by every seven cycle valid upon it. The two witnesses preserved these associations of the singletons of the right hand octal and its correspondences to subgroups in the other hand. In the above, there is no seven cycle that preserves the triples in the bottom three rows corresponding to [x^3, x^6, x^5] that are also valid on the reference, and these three non-octals or "plagues" as they are called, cannot form octals and they "fall".

Instead, the triples in the wormwood octals are cycled by their own seven cycles in isomorphism to the reference octal, and are the proper action of the "sea of glass", rather than a result of perpetuating the dialectic devices that shift octals and require this "middle ground".

Satan may (nearly) begin with 12 elements of S4. Three maps to unity in the coset (for each choice of unity) are already present in the multiplication over any octal. These twelve maps provide Satan with nothing but a requirement to perform "thefts" for the scarlet beast. He must "leach" those products from the proper operations of GF(8). (They may form no closed set in the "earthly" elements but generate C7.) These twelve products that were mentioned in the salvation of "thousands of thousands" (Rev 5:11) in no way provide for anything but the saving purposes of God. Satan, then, expects to trap only tares in a five cycle (as if tempting ten thousand times ten thousand). All these, then, survive having overcome "great tribulation".

So, 12 permutations of S4 fail to be constructed on the earthly elements. There are yet, however, 12 that may be constructed by serpents from elsewhere in the "sea". These would form transpositions in the elements $[x^3, x^6, x^5]$. These transpositions are illegal under three cycles by Frobenius, and would have to be made across different choices of "wormwood rows".

There is no proper sense at all by which the reference octal may be transformed (by any valid multiplication) to become wormwood under the serpents of the sixth trump. There is, however, the appearance of the Frobenius map acting on both the wormwood rows and reference cycle in the three subgroups shared between them; I examine that now.

$[a,b,c], [a,d,e], [a,f,g]$ with $[b,e,f], [b,d,g], [c,d,f], [c,e,g]$ with unity $a = 1$
$[b,d,f], [a,b,c], [b,e,g]$ with $[a,d,g], [c,d,e], [c,f,g], [a,e,f]$ with unity $b = 1$
$[c,d,g], [b,d,f], [a,d,e]$ with $[b,c,e], [a,c,f], [e,f,g], [a,b,g]$ with unity $d = 1$
$[c,e,f], [c,d,g], [a,b,c]$ with $[a,d,f], [b,f,g], [a,e,g], [b,d,e]$ with unity $c = 1$
$[a,f,g], [c,e,f], [b,d,f]$ with $[b,c,g], [d,e,g], [a,b,e], [a,c,d]$ with unity $f = 1$
$[b,e,g], [a,f,g], [c,d,g]$ with $[d,e,f], [a,c,e], [a,b,d], [b,c,f]$ with unity $g = 1$
$[a,d,e], [b,e,g], [c,e,f]$ with $[a,c,g], [a,b,f], [b,c,d], [d,f,g]$ with unity $e = 1$

Each row always corresponds to unity in the reference seven cycle. This shows that each row is constructed separately, collected as separate rows only. The Frobenius map (that same hour) then cycles the leftmost three columns (greyed above) closed; the above sharing that same action with the Sun octal over which the reference cycle is valid as multiplication.

Though these three subgroups never attain any true intersection in triples with the reference Sun octal (being in the wormwood rows), they are, when the octals are treated isomorphic to the Sun octal – yet held static – an indicator that they truthfully correspond to unity and unity only. There is only that element in view under the reference seven cycle.

Seven thousand men "die" in the earthquake. These correspond to the seven wormwood octals of those rows above, and also to the seven churches circuited by the least; they share the factor of 1000 that represents the patience and election by God, indicative of unity. As the two witnesses lie dead (C7 closed at unity in wormwood) there is no spiritual conviction.

As unity in the reference, they do not walk the angel's circuit through the seven churches/ lampstands: they are static in both the reference and the wormwood octal in view. (They would remain in one church of seven "suffering" the dialectic devices in the same setting.) The 7000 men are then shown to "die" in that automorphism. (As it is repeated across every wormwood row of seven.)

Those seven thousand assumed the positions of rest in the reference circuit; they are not under conviction by the least. (They simply remain in only one church through which they cycle.) There is some sense of the "normal" or "perpendicular" here as to the angel's circuit. (They tread the "outer court" of the temple underfoot.)

The remnant (those alive in Philadelphia) are affrighted, as 1000 of their number die despite their election and yet did not escape the earthquake. The conviction as by the letter to Laodicea is not apparent for those in stasis in the angel's circuit or in a five cycle within it (the outer court is trodden underfoot), even in the state of good doctrine that is Philadelphia. Philadelphia, likewise, must also be overcome. Without that conviction there is a lack, especially in the sense that these denizens of the wormwood octals are in some manner divorced or not reachable from the Sun octal by any seven cycle common to the reference.

The remnant that survive give glory to the God of heaven, they are the same ones that are under conviction by the letter to Laodicea: they know they themselves are not perfected yet; only Christ is perfect (as election into heaven is by faith on the lamb, perfect in all virtue).

7.12) The 5-Cycle Constructed

There is a swiftly arrived at solution for inducing the 5-cycle of the locusts' construction using the serpent spirits. Merely note that the subgroups of the left and right hand due to (a,b,d,c,f,g,e) are as follows: [a,b,c], [a,d,e], [a,f,g], [b,d,f], [b,e,g], [c,e,f], [c,d,g] in the right hand and {a,e,f}, {a,b,g}, {a,c,d}, {b,c,f}, {b,d,e}, {c,e,g}, {d,f,g} in the left.

I may construct serpents upon the element g, as in the right hand "g(be)(af)(cd)" and in the left hand, "g(ab)(ce)(df)" and I note that (b,e) and (a,b) are valid transpositions under composition from both hands together given g=1.

But most crucially given virtue with a=1 in the octal I may send the ordered list [a,b,d,c,f,g,e] to [b,a,c,f,e,g,d] quite legally using a shift in the seven cycle. So,

$$(a,b,d,c,f,g,e)(b,e) = (d,c,f,g,e)(a,b)$$

I align this (on the right) to the Laodicean amalgam under the locusts and the separate remnant in Philadelphia. (a,b) is become the transposition between those two churches, a=1 (Laodicea), say, and b=x (Philadelphia; i.e. 1*x=x, "Hold that fast which thou hast"). Similarly, the elect in Laodicea are convicted to exit the amalgam, and those in Philadelphia are convicted not to enter. Yet this transposition (a,b) occurs in the left handed octal, not the right. There is a sense of conviction by (a,b) in both "churches", but to overcome is to obtain sabbath rest (unity – overcoming Laodicea), due to conviction at God's left hand rather than in approval to His right.

The right is a reference field as if under complete restraint and the eight left handed octals free to combine and open the system wide to the sea of 30 octals.

The only difficulty is the disparity between a=1 and g=1. I actually have the "second seal paradigm" – the comparison to two choices for unity in the same octal. Then as a consequence I find "the denial of virtue". The "serpent" case of g=1 would ensure that the Frobenius map would cycle [a,d,e] and {b,c,f} and hold static the "serpent's tail" in both left and right hands.

Rearranging; (a,b,d,c,f,g,e)(a,e,b) = (d,c,f,g,e)

What I have is (fire + brimstone = smoke) which is become the "sorcery" in Revelation.

7.13) How Art Thou Fallen

Upon the locusts of the first woe, the "serpent" army of 200,000,000 horsemen (as forming the "sorcery", see previous section), apply as transpositions to form every possible combination constructing every five cycle to interrupt the circuit of the angel and all those convicted by the two witnesses. The five cycle formed by the "serpents" must agree with the short orbits in locusts (of horses, K4 groups under orbit in extension fields) giving relevance to the verses:

Isa 14:12 How art thou fallen from heaven, O Lucifer, son of the morning! *how* art thou cut down to the ground, which didst weaken the nations!
Isa 14:13 For thou hast said in thine heart, I will ascend into heaven, I will exalt my throne above the stars of God: I will sit also upon the mount of the congregation, in the sides of the north:
Isa 14:14 I will ascend above the heights of the clouds; I will be like the most High.
Isa 14:15 Yet thou shalt be brought down to hell, to the sides of the pit.

Lucifer's original state was also to dwell within the seven-cycle and walk in the spirit within the mystery of the "seven golden candlesticks". "Light" indicates virtue (in this sense "liberty", being the "light" that is principal – Lucifer was the "covering cherub") and "Lucifer" means "light-bearer". Lucifer "fell" from that state (heaven, "unity" which is virtue as in the lamb), and I find the same fall in analogy to the four angels bound in the "Euphrates" octals loosed by the sixth seal. With one of the eight valid seven cycles for the reference now acting downwards on the rows (of wormwood octals), three columns from the earthly four do not form octals (in grey below) – yet just one does: the left handed octal of the reference Sun octal. Here I used the seven cycle (a,b,d,c,f,g,e) for the reference as usual.

$$[a,b,c], [a,d,e], [a,f,g] \text{ with } [b,e,f], [b,d,g], [c,d,f], [c,e,g] \text{ with unity } a = 1$$
$$[b,d,f], [a,b,c], [b,e,g] \text{ with } [a,d,g], [c,d,e], [c,f,g], [a,e,f] \text{ with unity } b = 1$$
$$[c,d,g], [b,d,f], [a,d,e] \text{ with } [b,c,e], [a,c,f], [e,f,g], [a,b,g] \text{ with unity } d = 1$$
$$[c,e,f], [c,d,g], [a,b,c] \text{ with } [a,d,f], [b,f,g], [a,e,g], [b,d,e] \text{ with unity } c = 1$$
$$[a,f,g], [c,e,f], [b,d,f] \text{ with } [b,c,g], [d,e,g], [a,b,e], [a,c,d] \text{ with unity } f = 1$$
$$[b,e,g], [a,f,g], [c,d,g] \text{ with } [d,e,f], [a,c,e], [a,b,d], [b,c,f] \text{ with unity } g = 1$$
$$[a,d,e], [b,e,g], [c,e,f] \text{ with } [a,c,g], [a,b,f], [b,c,d], [d,f,g] \text{ with unity } e = 1$$

Then as in the sixth seal in which I had found an angel rising up from the east, having the seal of the living God to seal the elect into the reference octal to be represented as by unity (the lamb), I now see the meaning of "son of the morning" in this word picture.

The greyed columns are formed by adding the entry in the rightmost column to the leftmost three columns taken from the reference (Sun) octal. With the knowledge that John (as any Hebrew), would read the seven cycles right to left, the rightmost column "ascends" on its seven cycle (as does the Sun in the three leftmost columns), and as it is paired with the "Sun octal" it ascends out of the east. (The four "earthly" columns correspond to the "four corners of the Earth".) They are the only new ground to hide (as if from God) once the position of antichrist is taken from the left hand. In being cast out of the reference cycle or the "rainbow" (as generalised across the sea of glass), the result is to dwell in the "half-hour of silence" from the seventh seal (nullity). There is truthfully nowhere to hide.

The greyed columns are a consequence of that one multiplication of eight over the Sun octal. Lucifer, in falling (and not forming an octal in the mystery of the seven golden candlesticks) is cast out and only dwells in those three columns as "son of the morning" (cf. "by-product of the seven cycle of the rising Sun" acting upon "the position of antichrist"), or in the triple [c,e,g] in the top row: that place of "antichrist", which in the reference is the "static bow".

Lucifer is "cut down to the ground". The only octals that are preserved (in the sea, by analogy with the mystery of the seven candlesticks) are those in rows, which are "wormwood" octals. As "son of the morning", Lucifer is restricted to their four earthly triples: he is cast out from the reference which would exclude the three leftmost columns of the Sun octal. Then, choice in the kingdom of God is between dwelling in those three invalid cycles or in the one column in the earthly four which is actually the left hand octal of the reference, as the other three in grey form "plagues" (as they are described in the two witnesses, they are "invalid" as God's spirit is not moved out of His place).

Then the left hand's entry (rightmost column) is restricted to the earthly elements of the Sun octal (given the floating unity). Hence Lucifer is "cut down to the ground", whether in the left hand or in wormwood, mapped across the sea by the seven cycle. (Lucifer is only a "serpent" as in the sixth trump.) He neither cycles the wormwood octals in rows in a new "mystery" nor dwells in the reference. So, he is "cut short" to the one row containing the one earthly triple, wherein there is no "tree" for him to dwell within. (The tree has been cut "at" ground level.)

The "tree" of the leftmost three columns (as the K4 form) becomes no tool of Lucifer. In the reference seven cycle the correspondence is: $a \equiv 1 \equiv [c,e,g] \equiv [[a,b,c], [a,d,e], [a,f,g]]$ and not that which Lucifer had coveted: $a \equiv 1 \equiv [b,d,f] \equiv [[a,b,c], [a,d,e], [a,f,g]]$. Lucifer has no seven cycle for a spirit at all, he cannot construct his own, he is a created being; God is a position already taken and filled. (The kingdom of God is here, the prince of this world cast out.)

Isa 42:8 I *am* the LORD: that *is* my name: and my glory will I not give to another, neither my praise to graven images.

Satan may only construct seven such wormwood octals (separately) and tempt the righteous away from God.

Luk 11:24 When the unclean spirit is gone out of a man, he walketh through dry places, seeking rest; and finding none, he saith, I will return unto my house whence I came out.

Luk 11:25 And when he cometh, he findeth *it* swept and garnished.
Luk 11:26 Then goeth he, and taketh *to him* seven other spirits more wicked than himself; and they enter in, and dwell there: and the last *state* of that man is worse than the first.

The original spirit of the seven is that which is antichrist and avails himself of the left hand, finding no rest in the three invalid columns (as "dry places"), then constructed as "son of the morning". The wormwood device constructed sevenfold (as afterwards) is "more wicked". (The left hand of the reference is a valid octal, a pretence of a "mystery" in this iniquity.)

So "Lucifer" as "light-bearer" is nothing or nobody special: merely the first antichrist to fall from a state of virtue. Once "as unity", floating in freedom within the kingdom of God, falling from unity in the right hand (as any elect) of God now to the left, in rebellion against God (as the octal). Getting to Hell first is not an achievement. His rebellion is given in the passage after.

"Weakening the nations" corresponds to the construction of the five-cycle by locusts. The "ascent into heaven" to the setting up of the dialectic devices sending every (as a choice of unity) element in the reference octal to nullity by overlay. The "exaltation of his throne above the stars" as the dialectic (seal) devices placed over the operations of singletons within the sea of 30 octals (as by the trumpet devices, forming wormwood octals). Satan, by also manipulating by transpositions as with the serpents (themselves here represented in the 200,000,000 required members of an "army of horsemen"), thereby raises up his seat (power) as more than equal to God (God as isomorphic across the "sea of glass").

"Sitting upon the mount of the congregation" corresponds to interrupting the seven cycle and "moving the underlying octal" out of its proper place, from [a,b,d,c,f,g,e] to [b,a,c,f,g,e,d] as above. (The transposition (a,b) as the "sides of the north" – indicative of those that are yet obedient to God, here rearranged to form smoke from fire and brimstone as above. Satan "sits" or "rests" upon these rearranged "serpent-pair" constructions and the full set of transpositions by serpent spirits (as of (a,b) above) over the whole sea is the completed act of his insult.) As to "above the heights of the clouds", Satan places himself as victor over the gospel, and its victory in the least exiting the captivity of Satan's constructions. Satan sets his devices as over the mystery of God (clouds) to deceive God's people so they become "destroyed for lack of knowledge" (Hos 4:6). Yet God's spirit is not moved out of His place.

Satan desires, by constructing orbits of K4 groups (or their containers) under a five-cycle multiplication, to be "like the most high" (effectively to interrupt the salvation of Christ to His least). Satan expects to forestall any judgement and his own damnation. His lie "ye shall be as Gods" is an empty promise.

Satan covets the three rows corresponding to the elements {a,b,c} (which each contain that one same group [a,b,c]): i.e. Satan's devices, the transpositions (b,c), (a,c), (a,b) (which preserve the same logical schema of action/inaction as the reference up to the choice of unity from {a,b,c}) – those three rows do not (**cannot**) actually form a K4 group! (With the reference octal as the identity or zero, and the group to be further preserved under the reference cycle.)

What then, after all, is the product by the identity? Is it "God's will"? It could only be. (The alternate wormwood octal does not preserve the placement of sets on the two sides of the disjunction in virtue, overconstraining and breaking the closure of sets in the octal.) There is no transposition gained from Frobenius in the octal (and only God is omnipotent to provide a solution for that quandary) as to square again in a two-cycle to form a basis for that operation. Those three transpositions mimicking a K4 group together remain **only** lies, without effect.

With the congruence of $1 \equiv a \equiv [a,b,c]$ Satan would have also hoped to have "floated unity" (with the Holy Spirit under Frobenius as C3), through that triple of rows, preserving for himself some semblance of similarity to the Trinity, mimicking the "Manifold Of Virtue" ("I will be like the most High" (Isa 14:14)) and declare himself to be as Jesus' equal in the K4 form of the ultrafilter. Satan would claim necessary existence (as if Christ's own) through the life of the Father as if he were also necessary and indispensable to the operation of the octal. However, [a,b,c] is static under, say, (a,b,d,c,f,g,e) only when e=1, a contradiction he **cannot** stand (I posit his fall was partly due to murdering the right hand of Jesus (unity/zero – the principal element which is necessary), to assume Christ's place as God):

Joh 8:44 Ye are of *your* father the devil, and the lusts of your father ye will do. He was a murderer from the beginning, and abode not in the truth, because there is no truth in him. When he speaketh a lie, he speaketh of his own: for he is a liar, and the father of it.

Those "lusts of the devil" are those same transpositions made to transform from the reference octal to one of wormwood – and they are totally ineffective as to form a K4 group together upon the octal. Those exercising the enemy's devices (in the three woes) give their worship to the devil instead of the Christ they would claim to honour.

That angel (Christ's right hand), is shown as with a "little book" (still open – are they the seven letters addressed to him?). It seems his own record is shorter than most others, and that he is present to get his "own" back: as come direct from the beginning.

Rev 10:2 And he had in his hand a little book open: and he set his right foot upon the sea, and *his* left *foot* on the earth,

God will send Satan down in flames. Satan cannot completely construct his insult (missing 12 elements of S4 and under the five cycle, 60 elements of S5). He is required to make yet more and unending works, from the desperate construction of extension fields in the pit. (Algebraic Closure.) Yet then there is no order two element in the multiplicative part of any finite field characteristic two, and no short orbit of length two in any finite field of odd characteristic.

7.14) To Tread On Serpents

I am convinced the serpents act as they do in order to interface a lukewarm but otherwise obedient fellowship acting partially in the first woe of the locusts (which the serpents willingly represent as "in Christ's place", perhaps with ready "experience" of worldly matters in such dialogue) with any other (i.e. worldly) collective, enabling that synthesis made of the dialectic in the earthly elements of the otherwise complete octal, switching static subgroup for bow.

The first two woes are the false prophets' horns "...and he had two horns like a lamb, and he spake as a dragon" (Rev 13:11) which provide the authority for Satan to accuse the elect; much like the seven horns on the lamb permit the least never to backslide.

The method introduces a fellowship to a wider community – to solidify and **legitimise** a factor of compromise leaving the Holy Spirit interrupted and any fellowship already partially caught in a five cycle (as lukewarm in the five "refining" churches, as treading the outer court underfoot) as certainly caught captive to Satan in that merging (i.e. most certainly found operating in the first woe). The obedient ought to do well to leave the fellowship at this moment, and not to look back for fear of becoming even more lukewarm than they were under just that "first woe", which passed without notice. For continuing in this second woe is enough to be caught marked by the swiftly oncoming third woe. There is no fixing such a broken body for Christ.

The original derivation of the octal split the properties of a K4 group or disjunction of virtue into sets both active and inactive, from $\{p, r_0, s_0\}$ to $\{p, \{r,u^{-1}\}, \{s,v^{-1}\}\}$ with $r_0 = \{r,u^{-1}\}$ etc.

Then with the device of the serpents (using a transposition), those with that paradigm would form a middle that privates the virtue: not in p, but in the closure of the octal over those three K4 subgroups exercising virtue as unity.

There is always some property in $r_0 \wedge s_0$ formed in, say, $r_0 \& v^{-1}$ that does just that: privating sets acting as virtue within that coset of the K4 form (that spans the octal as a "golden girdle"). There are rows in the filter table that permit such a middle to be exercised, but those K4 groups are without that virtue in p, and truthfully, elements do not switch sides in the disjunction. The serpents, however, do not directly private virtue in the K4 form after that manner.

In effect, the properly made disjunction of $N\neg(p\&(r\&s)\&\neg(u\&v))$ is rearranged to become $N\neg(p\&(r\&v^{-1})\&(s\&u^{-1}))$ from the valid logic deciding $(r\&s)^{-1} \vee (u\&v)$.

By axiom of virtue, I may compose both $p\&r^{-1} \to s$ and $p\&v \to u^{-1}$. Then $p\&r^{-1}\&v \to s\&u^{-1}$ and there is no illegal middle formed of this disjunction as the sets r, s, u^{-1} and v^{-1} are disjoint.

Under a transposition (as serpents acting in wormwood) u^{-1} and v^{-1} do not switch sides of the disjunction, but instead they overconstrain the operation of the octal outside of the "jacinth part", i.e. outside those filters with virtue in p. Then I find that each one of those four rows of the filter table of chapter four are interrupted, and the octal itself privated.

With u and v exchanged; $((u\&v \to r) \vee v^{-1})$ appears as valid as $((u\&v \to r) \vee u^{-1})$ or either $(((r\&s)^{-1} \to r) \vee v^{-1})$. Then as u and v cannot change sides in the disjunction, the improper disjunction $((u\&v \to r) \vee u^{-1})$ would over constrain u&v or u^{-1}, as (given u&v) I have $v^{-1} \vee r$. But then the legal $v\&(u\&v) \to r$ cannot oppose that u^{-1} within $(r\&v^{-1} \vee (p\&r^{-1}\&v \to s\&u^{-1}))$; u^{-1} then appears both sides of $r_0 \vee s_0$ and u^{-1} (or v^{-1}) is itself then overconstrained.

Rather than derive the whole octal on the fly, how may the believer counter such a serpent? The fact is that the believer disciple is unaffected, directed by the Holy Spirit to regenerate the full octal from that "jacinth part" as do the angel's two witnesses. Otherwise, it is also possible to counter such a serpent by affirming the correct placement of sets in the mapping table, which I find to be an equivalent solution in this case. (As to restating every correct inference.)

Even though the triples in the wormwood rows are altered (exchanging "water for blood") it is also true that the operation upon the singletons outside of the K4 form has also changed, and I cannot expect to find preserved their placement in the filter and mapping tables.

Christ's commandment is to follow the guidance in the seven cycle from the Holy Spirit: the seven letters all affirm "He that hath an ear, let him hear what the Spirit saith unto the churches". To tread on serpents is to escape without being bitten, and as the Lord said in Eden "to crush their heads". Acting out of virtue and remaining in that "jacinth" portion shared with the reference octal the believer overcomes the serpents, escaping unbitten. To crush their heads (doing so in the whole octal) I return to the simple word picture supplied by John:

a (b c) (d e) (f g) is the "serpent" with tails having heads like lions (as "a" here), with which they "hurt". The serpent's head is to be crushed underfoot, as each triple walks the seven cycle. (The unity element floats with the static subgroup.) One subgroup is always an earthly triple in the other hand, so feet "are upon the earth" and tread or walk the seven cycle.

In the seven cycle upon wormwood, the Holy Spirit preserves the unity element, acting only on the "jacinth portion". That causes the static subgroup (embedded within) of the reference to be under a seven cycle. There is also the equivalence to crushing the heads of the serpents underfoot with the orbit of the static triple. This is the same ministry of the two witnesses that generalise the angel's circuit to overcome as necessary every device used in the false doctrine of these last days for all to find approval before Christ.

Unity is never found in a static triple. The serpent's device (transposition) would affirm just that: yet such congruences are illegal in the octal and that fiat unity (they are not Christ as GF(4)) is crushed "underfoot" by the seven cycle, each foot upon the earthly triple to the left and right of the reference octal. The proper virtue in Christ (as unity applied in the octal) does just that. Satan's device is overcome with virtue. No place for unity is found in the static subgroup (but only Christ outside of the octal). Whilst the reference treads upon the claim that the static triple contains unity, there is a counter reaction to the octal from the wormwood.

The static triple in the "earth" is congruent to the three right hand groups containing unity and also vice-versa, preserving the "unity to static triple" correspondence under the seven cycle. In **wormwood**, the right hand static subgroup of the reference directly corresponds to the three subgroups in the K4 form in the reference (and not the left). The triple regresses; sat at the right hand instead of the left. I again find the two witnesses to the left and right as "trees":

$$[[a,b,c],[a,d,e],[a,f,g]] \rightarrow \{c,e,g\} \rightarrow a \leftarrow [b,d,f] \leftarrow \{\{a,e,f\},\{a,b,g\},\{a,c,d\}\}$$

which is valid in the reference, but in wormwood I find instead:

$$[[a,b,c],[a,d,e],[a,f,g]] \rightarrow [b,d,f] \rightarrow a \leftarrow \{c,e,g\} \leftarrow \{\{a,e,f\},\{a,b,g\},\{a,c,d\}\}$$

And any sense of unity in the static subgroup [b,d,f] is "crushed" by the proper operation of the reference and the correct unity ("a" here). Unity properly corresponds to a position "under the foot" as "a" under {c,e,g}, as "cast to the earth" as in the first trumpet.

The static triple {c,e,g} treads on serpents within the "left hand" wormwood octal, and [b,d,f] on the "right hand". The unity, congruent to the right hand K4 form [[a,b,c],[a,d,e],[a,f,g]], is crushed by the "left foot" [b,d,f]. (The feet appear exchanged in wormwood.)

In a bizarre twist, left and right feet are exchanged and the serpents "**bruise** the heel" of the obedient as they "walk" the octal's subgroups. Whilst not harmed or "hurt" (as sent to nullity), there is a **collision** of the two (of both wormwood and the reference). Treading on serpents (and also the five cycle of the "locusts") is as simple as holding to the Holy Spirit's seven cycle.

7.15) The Seventh Trumpet

The scripture swiftly brings the **last day** with the seventh trumpet and with the seventh seal.

Rev 11:15 And the seventh angel sounded; and there were great voices in heaven, saying, The kingdoms of this world are become *the kingdoms* of our Lord, and of his Christ; and he shall reign for ever and ever.

The kingdoms of this world become those of God; no place is given the ungodly at the judgement. I see no translation of all peoples to God-fearing Christians but rather a transition of all earthly authority to God. The nations of the world do not become "Christian" – it is merely the case that no earthly kingdom stands upright before the kingdom of God; His kingdom is the only one stood upright remaining. (All others are contradictory as become "Death".)

Great voices (judgements) echo the voice of the archangel with the arrival of omnipotent God to judge the world on the last day.

Rev 11:16 And the four and twenty elders, which sat before God on their seats, fell upon their faces, and worshipped God,
Rev 11:17 Saying, We give thee thanks, O Lord God Almighty, which art, and wast, and art to come; because thou hast taken to thee thy great power, and hast reigned.

Once more the elders and four beasts fall down before the lamb (Jesus has opened another of the last three seals on the book of life. See Rev 4:10, 5:8, 5:14, 7:11). Now aware of the action of the seven thunders (a great movement of heaven) I may place here (at the completion of the two witnesses' ministry with their resurrection) the opening of the seventh seal – safely in alignment to the seventh trumpet. (Leaving the seventh seal to also align with the last vial of wrath poured out, a correspondence that should indeed be found later on in the text.)

The reign of God over this current system of things is concluded. The book of life is open filled, and there is no cause for more grace to those that refuse it. Every application of the Lamb's virtue has been extended for salvation to those on the Earth, so that grace and the book of life should be complete. Every arrangement of the octal (each as one of 24 elders) gives God praise for extending such equal grace to men. That virtue, the same yesterday, today and forever as found in Christ is alluded to here as God: "which art, and wast, and art to come".

God is praised for using His right hand to save the gospel from loss (that there be time no longer). The least has overcome as in the worst circumstances and the last ministry of two witnesses to that fact is completed. With the whole world making their minds up or washing their hands, the elect are complete in number and the God of heaven returns to judge all.

Rev 11:18 And the nations were angry, and thy wrath is come, and the time of the dead, that they should be judged, and that thou shouldest give reward unto thy servants the prophets, and to the saints, and them that fear thy name, small and great; and shouldest destroy them which destroy the earth.

The "nations", now that the only kingdoms of the world are those of God, consist of the saints. Those alluded to in the fifth seal as under the altar (asking for vengeance) surely were angry – but each had been given a white robe and asked to wait a little longer. When their number is complete (and exeunt from the Church) the wrath indeed arrives. The judgement is set (as now) to reward those in Christ (the three groups present as the unity floated in GF(4)).

Those without Christ and "outside" – as worldly "nations" that destroy the Earth, are also "angry". They operate as 666 (the number of ways a crowd may deny virtue) and are technically in that "half hour of silence" of nullity without life, sending the four earthly elements to the empty set. Within the "sea of glass" (where static subgroup correlates to the K4 form spanning the octal) those that are antichrist and "destroy the Earth" have their place in the lake of fire. Those that fear God (and are in Christ) are those not destroyed. (Simple exclusion of logic.) This is not a statement of God giving recompense to a lack of "environmentalism".

Not surprisingly those nations (those under government rather than government itself) are truly angry with the fall of the corporate religion in which they had placed their trust. The great city fell, and those that "destroy the Earth" as antichrist in bringing all these plagues upon all men are to be judged accordingly.

Likewise, at that point the person of God is revealed to all and the mystery of God finished, with many fallings, judgements, with movements of heaven and Earth (as the winepress of God is trod – see much later), and with the whole world suffering plagues of hail. (A just judgement from within the sequence of the seven vials of wrath.) As revealed by the verse:

Rev 11:19 And the temple of God was opened in heaven, and there was seen in his temple the ark of his testament: and there were lightnings, and voices, and thunderings, and an earthquake, and great hail.

The temple of God corresponds to the reference octal with the twenty four elders and includes the generalised circuit of the angel as the rainbow, but perhaps not the seven lampstands which are trodden underfoot by the gentiles, as the temple's "outer court". Whether the outer court is included is uncertain: what I can tell is that the ark of the Lord's testimony is seen in the temple and seen to be the "same yesterday today and forever". It, most certainly, **has not** been trodden underfoot. Those that would do so (antichrist) are brought under judgement.

In effect God keeps His word and ensures it continues preserved in heaven. I understand that all those in Christ are preserved to one side and all those in antichrist cut off to the other: that the word of the Lord's testimony as shown constant and unchanging, whilst not always fair, is at least continually just and remains the same as when it was first given.

The temple of God (in heaven) as "opened" corresponds to the defeat of the beast and false prophet with the arrival of the armies of heaven, occurring with the second coming and judgement (Rev 11:17-18). Heaven ceases to be considered as unity only, but is opened to become realised as a collection instead.

As to God exercising His great power: i.e. of the left hand convicted and following the right hand out of the Church proper (leaving it desolate behind), His reign is realised to be that of God written on (or with) His right hand (as with the least's circuit made to be a "rod of iron" over the "nations" of the seven churches). They are "tormented" as captive under the three woes and were then "angry". In opening the temple, the armies of heaven return to Earth with the least (the "Word of God") taking the lead victorious over the dialectic devices.

The "ark of His testament" corresponds to the name of God (the God of Christ) revealed in the same manner to that now written with and upon the least, as on the same angel. (As surely writing "Lord of Lords and King of Kings"). That name written is written forever in the heavenlies and although God has exemplified His new name all along, it will be openly justified at a time appointed in the express person of Christ with His return. That name must be written on the angel (the least) with the completion of his circuit.

Joh 12:28 Father, glorify thy name. Then came there a voice from heaven, *saying*, I have both glorified *it*, and will glorify *it* again.

The angel completes his circuit as that "rod of iron" convicting all, and also justifies the gospel in every case as the minimal case made effective. Truly, Jesus saves.

The lightnings (falls), voices (judgements) and thunderings (movements of heaven) all apply with unity floating through the static subgroup as in the last three seals. With "falls", even kindreds (sharing the same unity and static subgroup but not the same seven cycle) as in the nations (churches) are "cut off". With judgements they are separated as by a "two edged sword", and with thunderings God moves those judgements backwards in time to apply with the seals on the book of life itself, leaving the elective choice as Christ's alone – made at the judgement now set.

There is then a four-fold movement of the Earth (a great earthquake) aligning every octal to one isomorphic form in the "sea of glass" (as with the ministry of the two witnesses) and a great hail, as a seven cycle is applied over the left hand in similar multiplication.

Chapter Eight: Wonders In Heaven

This section in the Revelation is often considered as a "history lesson", though there are other books of the Bible for the history of Israel. The woman is the "Church" and the account ends with the Earth saving her from the dragon (the enemy); the truth is quite different. The woman is ministered to by the two witnesses and the remnant in the woman (the elect) have left her (to meet Christ's approval) leaving behind a collective truthfully reducing to tares (a great "falling away"). The great tribulation from God is reduced to more stringent selection upon the love of the truth: not tribulation of violence to be reciprocated from the dragon.

The woman is helped by the man-child and God: the former empowers the ministry of the two witnesses, and God certainly wields His Spirit to save. The reaction of the Earth supplies a brief stay for the sealed that remain in her, as the "four winds of heaven" are held back from blowing on the Earth until the sealing of the saints is complete after the sixth seal is opened.

The witnesses' ministry ends as the woman corrupted morphs into the scarlet beast; salvation as with the angel's circuit is completed. The angel is "caught up to God and to His throne". Vanquished, Satan's once effective devices against Israel fail against the least in the kingdom of God. Heaven does not permit any of the elect to be separated from the love of their creator. Satan as "god of this world" fails upon his slights against the one and true God's sovereignty.

Satan (with every opportunity) fails to benefit from any of them. The outcome is war in heaven over ownership of the Earth; none are truly and logically deceived by the devil's devices: that itself is a nonsense. Even so, the angel risks Satan's attempts to devour him at every moment. With the child caught up I do not expect a universal rapture (as it may be deduced to be a fallacy as well as logically indeterminate) until the time arrives to save the elect from certain "hell" (this universe as it will become) at the last trump with the Lord's victory over Gog and Magog (Armageddon). That is, upon the very last day when heaven and Earth "pass away" and Christ makes a new heaven and a new Earth. That is a final mystery (as believers are to be caught up in "clouds"), one that is not made as "sure words" yet, but are promised in Christ's own words to be proven (made sure) much later.

8.1) The Woman And Twelve Stars

Revelation continues with a passage commonly thought to refer to the ministry of Christ whereas it refers to the "deliverance" of the least. Why? Not only because of the context but clearly the woman is delivered of the Child in the passage and not only the child of its mother.

Rev 12:1 And there appeared a great wonder in heaven; a woman clothed with the sun, and the moon under her feet, and upon her head a crown of twelve stars:
Rev 12:2 And she being with child cried, travailing in birth, and pained to be delivered.

John describes the woman, the chosen people of God. She is "spiritual Israel": those elect in Christ rather than of genetic descent from Abraham. The woman, who begins so blessed and ends so utterly corrupt (riding the scarlet beast) is wondered at by John; here this "wonder" makes her first appearance in the narrative.

The woman, clothed with the "Sun" is a beautiful picture of the 144,000 – clothed with the "floating" unity in the octal (as if the twenty four elders), in appearance as the "lamb slain" in the heavens. The moon (now under her feet) shows she is able to overcome the devices of the trumpets that allow the shifting between octals, never attaining the Sabbath of the elect's inheritance. She has it in good measure already! Again the "feet" represent the static triples in each seven cycle: the woman now empowered to overcome the loosed restraint upon her.

On her head is a crown of twelve stars, a reference to the tribes of Israel sealed, or yet a more striking reference to the twelve elements that cannot be formed of S5 in "That great city" where the two witnesses "minister". Those elements form an "open door" from the "great city" of spiritual Babylon and out into the grace offered in Christ. They are formed from octals across the sea: so are properly "stars" rather than "trees" or "fountains" in the reference octal.

The "stars" are the elements and K4 subgroups of the wider sea that are outside the Sun or reference octal. That these stars are over her head, indicates she has yet to overcome this state to put it under her feet, and this is grounds for her accusation by the dragon (Satan). crown is for overcoming; this is her test. She has the authority (given as a crown) for it by v   of the angel's circuit (in order that this last age of deception by Satan's devices might be ov   ne as if by all and the woman finally delivered and justified beyond accusation).

The cr. d she "pained to be delivered from" is the very least in the kingdom. (As to her being with child – the least must arrive at some time and be redeemed as any other.) She travails in birth – she must likewise overcome as does the least: it is not "done for her" in the sense that deliverance is given gratis (the least shows she is capable of overcoming them). The angel convicts the remnant in the woman; she remains to be similarly "delivered" (of all the elect, the people of God). She, having a test (or crown) of twelve stars (leached elements of S5) over her head, has an open door left within the image system: her test is to overcome using these stars and be delivered, found in Philadelphia, so dwelling with Christ in heavenly places.

8.2) The Dragon

Rev 12:3 And there appeared another wonder in heaven; and behold a great red dragon, having seven heads and ten horns, and seven crowns upon his heads.

The dragon is as John put it – a "great wonder", more concept than maths. Satan has opposed God's salvation and in the Old Testament the nations of Israel and Judah were oppressed by various Empires:[7] Egypt, Assyria, Babylon, Persia (with the Medes), Greece and Rome.

The result is that Satan has accumulated his devices upon these kingdoms and throughout history the dragon appears to have one head for each kingdom and only one crown of real authority (opposition to God) left to spend, but as it was seen from Daniel, the fourth beast[8] is actually an amalgam of the devices of the previous three. (The USA is an amalgam of those devices used in Britain, the USSR and Nazi Germany.) Including these three-onto-one there are "ten horns" rather than seven if the USA is the fourth/seventh/tenth.

It should be noted that there are other kingdoms, such as the Philistine and Idumean that had buffeted and oppressed Israel – but for example, the book of Maccabees is not in the canon

[7] James Lloyd, The 6, 7, 8 Cycle (Christian Media), p26-28.
[8] Lloyd, The Sand and The Sea, p5-7.

and for some reason that is a good enough reason for it to be excluded from the count.

Satan has no purpose but to eradicate the kingdom of God and to attempt to supplant Him as "god of this world". With Christ's victory, Satan lost any hope of a crown to oppose God accusing the elect, now replaced by the virtue victorious in Christ (the Lamb). His "three-onto-one" device (the beast) is a "bodge job" – no crown of divinely granted authority, only one of deception. It is "synthetic", a copy of the first four seals – for those lacking in restraint as with a reprobate mind. Compare this with the rest found in Christ. (Unity as in the angel's circuit.)

So I may introduce some images.

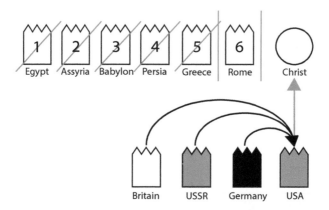

To whom does the Revelation refer to as an adversary? The dragon provides the template for the mystery of iniquity and the woman becomes its target. As marvellous as the woman starts off and turns bad; likewise, in reciprocal, the dragon is a wonder for the damage he causes.

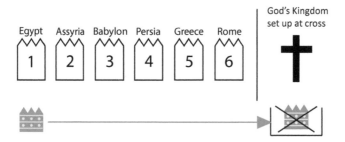

Satan's crown continues (six kingdoms) up to and including
period of Rome before his total defeat by Christ. (The Seventh.)

As to "great red dragon" I note "greatness" is required for an adversary of God himself: not Satan as I would expect him. God has given the path of the dragon and the two beasts for Satan to pose somewhat of a challenge, but one that still fails. God gifted Satan with all iniquity, as the last work of evil for the saints ever to be avenged upon.

It is no surprise Satan must attempt to devour the least in the kingdom and fail – the text sees him poised to do just that, though the same text states his total lack of success at any rate.

8.3) The Dragon's Tail

Rev 12:4 And his tail drew the third part of the stars of heaven, and did cast them to the earth: and the dragon stood before the woman which was ready to be delivered, for to devour her child as soon as it was born.

Dialectic accumulates as dragon's tail marches through history

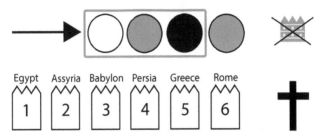

Dragon stands ready to devour Christ with Rome, fourfold dialectic device fails to seize control of the kingdom of God

Satan falls, loses all crowns, but has one head left (today's USA) to apply all his devices against the kingdom of God. Deadly wound must be healed.

Again, my growing vocabulary becomes useful, as the "third part" (product) of the stars (triples) are cast to the four earthly elements (as the devices of the seals loosed). "Heaven" is the unity element; those "stars of heaven" are the static subgroups under any multiplication (and choice of unity). I.e. the possible left hand static triples replace the four earthly "triples" of the Sun octal (i.e. chosen instead as from the "night"). The "third part" of those stars (addition on triples) allow those "products" of the dragon's tail to form the wormwood device (on these several octals) upon the sea. The "tail" is only constructed to form those four beast(s) rising up from the sea; whereas the dialectic is formed upon the Earth instead. The dialectic may only be formed upon the tail by those angels falling/leached from their proper places.

The dragon is unable to construct his device on the four beasts before the throne of God, but the devices of the trumpets allow him to mimic them. The "stars" (or elements) of every octal in the sea are become as the seals (devices) "cast to the Earth" (as the "sand" of the "sea") by that mimicry and his deception to ensnare the least, his "plan" is set, to leach products from all other octals to fashion an overlay of the "Earth upon the sea". That would place a lie (of "Death") upon the wormwood device as upon sending a whole octal to the empty set.

The wormwood device must be overlaid atop the four beasts before the throne, an impossible task for Satan. (He and his angels are "cast to the Earth".) In truth, the dialectic "flood" becomes as one of the sea upon the earth, but it can only overtake those that "build their house on the sand". The dialectic logic is applied to the sea and perceived as if in the earthly elements.

The woman is ready to be delivered: Satan has been poised to devour the least since the ministry of Christ. The least, however, did not overcome way back, but with the removal of all restraint and with great tribulation at the sixth seal. Satan has had a long wait. The woman, "ready" to be delivered, shows she was well prepared by Christ in the truth of the gospel.

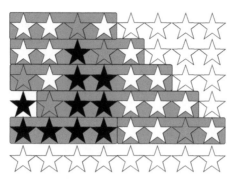

If the last dialectic must stand, then the first must by some means fall.
(Green denotes inconsistency, black shows previously assumed "nullity"
where it was once inconsistent.)

It should be realised that the dragon has been at work since before Egypt, employing the same deception since Eden. His dialectic device on the earthly elements fails in its construction (conjunction and disjunction never equal but in nullity). There is no sequential fulfilment but that limit of four; one "bow" must make way for another (the USA beast devours and breaks in pieces). The tail stays the same length but its "snail trail" leaves behind its broken residue.

However, this "three onto one" device is in reality Satan's "one last shot" at justifying himself to be "god of this world", for if his own kingdom does not stand scrutiny (even amongst the fallen) as an adequate deception, then in a schism it collapses. After the dialectic is applied in full five times the device fails as completely inconsistent. (The cross of Christ is unassailable.)

The tail accumulates and breaks in pieces all left behind it; there is a discontinuity in the sequence if there is allowed an eighth kingdom. This is why the three horns of Britain, the USSR and Nazi Germany were plucked up to form the final USA beast – and it occurred alongside the creation of the modern state of Israel. It is not and there is not, **any** Jewish plot, yet the devices used on the Israel of the Old Testament are re-used with the dialectic in this Christian age. Israel hardly enters into the scene at all (as an eighth) – the tail must not contradict itself!

It beggars belief to see so many Christians following after ideals that have nothing to do with the spiritual Israel of God, and everything to do with "new" cultic doctrines not known to be plucked from the history of the continually failing Old Testament Israel, stated throughout the Old and New Testament scripture as blasphemies which may be considered lies of Satan.

The dragon is poised to devour by deception the least, and stood in a temporary show of sovereignty (a dialectic device) – a rug that is soon to be pulled out from under him completely.

8.4) The Least In The Kingdom Delivered

The least, however, seems to escape unscathed – not harmed in the slightest.

Rev 12:5 And she brought forth a man child, who was to rule all nations with a rod of iron: and her child was caught up unto God, and *to* his throne.

Rev 12:6 And the woman fled into the wilderness, where she hath a place prepared of God, that they should feed her there a thousand two hundred *and* threescore days.

The man-child, here given authority (cf. the letter to Sardis), and with the right to sit enthroned as Christ (Laodicea now overcome), is the angel in Philadelphia – ready to be delivered of that last church in his circuit. (The rainbow rests upon his head, the first woe already overcome.)

That first woe (as to overcome the lukewarm paradigm of the locusts' five-cycle in the church), is the first of three woes the elect are required to overcome in their own "great tribulation". In this passage of Revelation, I find the wider body of Christ's Israel does just that.

He is the "dividing line" as the very least in the kingdom of God, thereby ruling over those in the churches, between those wise and foolish virgins of the parable.

The remnant in the woman are now under conviction to follow suit: they are also required to overcome the devices of Satan and the star of the fifth trump "cast unto the Earth". Therefore, I should consider the four earthly elements of the Sun octal in view.

The "sea" is the set of all 30 octals (waters). The "wilderness" is a great expanse of "Earth" across those octals – specifically those sharing the octal of the angel's circuit.

The Wilderness:
The wilderness is the set of four earthly elements under every seven cycle and in every octal with every possible combination of unity to static subgroup: forming as it were a four-element or "earthly" subset of the 200,000,000 variants of the angel's circuit through the seven churches. The wilderness may then be general-ised as if to become as the angel's circuit across all 30 octals, but must then (in isomorphism to the reference octal) preserve "one to one" mappings (of the four earthly elements under every group of seven cycles) between each of those oc-tals. Then the ministry of two witnesses for 1260 days is fruitful. (1260 = 42*30)

Seven cycle acts downwards, woman "flees"
into the wilderness (boxed elements)

$$
\begin{array}{ccc|cccc}
a & c & d & b & g & f & e \\
b & f & c & d & e & g & a \\
d & g & f & c & a & e & b \\
c & e & g & f & b & a & d \\
f & a & e & g & d & b & c \\
g & b & a & e & c & d & f \\
e & d & b & a & f & c & g
\end{array}
$$

The "place prepared" is (or are) the earthly elements of the Sun octal separated out by "the wings of a great eagle" later. (So she might fly into her place: that is, the woman (the elect);

unity is one of those elements.) The 1260 "days" permit those in the "nations" of the angel's circuit (generalised across 30 octals) to attain a position of unity in the reference (Sun) octal: election or unity is a paradigm within the grasp of all those that also circuit the churches. (They are also under conviction.) The 1260 "days" of ministry are useless to those that "dwell" in the seven churches without that conviction: as foolish virgins they dwell in the woman along with "the remnant of her seed" which will exit with the treading of the winepress – (see later).

Those in the wilderness are ministered to with the set of all valid seven cycles: the 1260 days of the two witnesses. The angel's circuit that is hidden within the mystery of the seven candlesticks reveals just one "church" of seven is in approval and this is Philadelphia. The angel's circuit results in that conviction: those that are not within that circuit or any of the other (42) valid seven cycles upon the Sun octal or its 30 isomorphic copies are practising a much broader paradigm. (They do not "have the moon under their feet" as does the woman.) They traverse the sea by employing the trumpet devices.

The "place prepared" (when it is realised) is to attain the completion of the angel's circuit (Sabbath rest in Christ), and is identical ground as to the ministry of the two witnesses – they instead being as static triples which correspond to the above singleton elements (yet under orbit by seven cycles in the left and right hands), they minister for the same 1260 days. (42 months (elements in seven cycles), sharing a floating identity over 30 octals.) This ministry is prepared with the angel's circuit – his seven cycle that acts downward in the rows above (to which they witness the angel overcame) is already fulfilled practically – hence the woman is truly convicted in the spirit to likewise overcome. The seven cycles (the 1260 days of the two witnesses' ministry) will sustain her in that spiritual place.

The witnesses provide good doctrine (with which the angel overcame), but who are "they" in Rev 12:6? Clearly only the child and God are in view. God is the source of good doctrine and the angel draws them out. God will provide rest and the angel's circuit (by the witnesses) will convict those that long for it. The woman is sustained whilst seeking rest for as long as it is necessary for her to travail in birth: not with the least but with the last remaining within.

> **Two Witnesses:**
> *The witnesses correspond to unity and are the "static triples" in the left and right handed octals. They walk the elements of the angel's circuit as if they were the "feet" of the K4 form in the octal (they are properly associated with the K4 form rather than the octal). They witness the acceptance of the least by Christ, their ministry generalises the seven-fold victories of the angel over seven "mountains" of false doctrine (singletons) over every octal in those seven "churches".*

With the elect exeunt, God stops feeding with good doctrine and the "rest" (the two witnesses end their ministry) found at the "end" of the angel's circuit stops convicting any within – the winepress of God's wrath is then to be trodden out.

The two witnesses, who are they? It should be expected that the right hand or Sun octal is to be constant and a reference. That places to the left hand a witness that continually shifts relative to the choice of multiplication (C7 group) and the floating unity in the Sun octal.

(Which I assume only God may do so perfectly.) The witness to the right is that which witnesses conviction to others by exiting the flock as all are called to leave: whether hearing directly from Christ as with the letters to the angel, or by similar ministry of God through another witnessing to them, leaving the flock ahead of them. The believer fulfils the right hand witness and God will convict those yet to follow (the whole set of elect) by His Spirit.

The right hand witness leads out of the Church and the left hand is the state wherein those convicted to leave (as elect) immediately find themselves unless they also exit as the least.

I have to the left and right two triples (called "men") that witness the will of God's Holy Spirit (with the correct unity, static subgroup and static triple) in the generalised case over every octal of 30: left, right, sea or Sun. In this manner the angel is identified as "seven stars in Christ's right hand". He, hearing direct from God, is Christ's right hand (the least) and by inference or extension, angel of (as right hand to) the whole "church" – the complete body of Christ.

8.5) War In Heaven, The Tribulation On Earth

Rev 12:7 And there was war in heaven: Michael and his angels fought against the dragon; and the dragon fought and his angels,
Rev 12:8 And prevailed not; neither was their place found any more in heaven.
Rev 12:9 And the great dragon was cast out, that old serpent, called the Devil, and Satan, which deceiveth the whole world: he was cast out into the earth, and his angels were cast out with him.

The ministry of the two witnesses is also played out in the heavens. Each wormwood row (as the second woe formed here by the dragon – a serpent) corresponds to unity only: Michael (prince), as the "jacinth" part shared with the reference octal (with his angels: the elements of the reference together forming a "tree" with Michael as the intersection), fight against the dragon. (The two sets of octals are somewhat opposite to each other, the reference seven cycle is not valid across wormwood rows.) The dragon fought back with his angels as if they were "seven spirits more wicked than himself" but as the only element to argue over is unity (not a mere "fountain", properly a "tree"), the dragon's devices, unable to move that unity, result in the dragon's expulsion. Unity is solely the intersection of the trinity in the heavens.

With the first woe Satan may only build a five-cycle, and in the second woe, unity found in the static subgroup is a given for the action of wormwood (Satan now denied the prerequisite of unity "neither was their place found any more in heaven"), such a five cycle is now also a reference to Satan's expulsion from heaven – his construction of the five cycle on the churches outside of "heaven" (which is the floating unity of the seven cycle) is his only recourse: that five cycle ensures there is no place in heaven given him, now shut up in the bottomless pit.

The dragon deceiving the whole world (supplanting all 200,000,000 circuits with the devices of the first and second woes) is cast "into the Earth" (as "son of the morning") and his angels with him. They are all to be "tormented" in three "non-octals" across seven rows of wormwood.

There is maths (or pseudo-maths) described by these events in the Revelation. I find the devil's devices effective only with the removal of the restrainer; virtue (unity) is replaced by the empty set { } in the earthly coset. The devil has some sway with his device to "appear to be an angel of light" in that the remaining earthly elements are similar in operation to a K4 group (hurt the wine) and seem to hold correct some basic laws of logical inference (hurt the oil).

I deduce from this passage that there is war in heaven (on truth) and Satan's ejection affirms he may not send any element or static subgroup to the empty set; it is not unity in the balances. (The GF(4) ultrafilter is unmoved by the devil's devices, Jesus' victory stands firm in heaven.)

Though it is somehow possible to send unity to { } forming a device with "dialectic logic", there is no empty set in the K4 form or across any octals in the sea with the wormwood device. The dragon's devices (shown in the trumpets) are **only** possible on the earthly elements (as the four beasts) about the throne of God. That static subgroup is unassailable with any device of "balances", even in fourfold (fourth seal) form. Satan is "cast out to the Earth with his angels" so that the devices of Wormwood and Abaddon are all he may use (and he does use them).

The dragon's deception is (once the least overcomes) totally unnecessary upon any elect, but who said they were deceived by it anyway? The elect (as unity) and the elements of static subgroups in every octal (as stars in heaven) are unaffected by the dragon's devices in the sea (neither unity nor static subgroup are sent to the empty set).

The fixed truth of God (an octal whole in the sea) is forced to nullity only with an overlay of the system of the fourth seal, for which Satan's tail cast the third part (product) of his angels (stars in the sea) to the Earth, to form them as that overlay on/from the sea. Yet to instantiate these devices, the logic in the earthly elements must be placed upon the wormwood device rather than act in their own cosets within octals across the "sea of glass" (as aligned isomorphic in isometry). They are all cast out to the Earth and their place was no longer found in heaven to construct that wormwood. Wormwood corrupts the static subgroup, the dialectic its coset.

The common octal to the wormwood device is the Sun octal. The last balance (the fourth) of Death would assert itself by sending the reference octal to nullity. "Wormwood" if under the fourth seal system must send a subgroup of the Sun octal to the empty set. Under a sevenfold amalgam of the device the whole of the reference octal is "supplanted" with the empty set.

Compared to the accumulation of the dragon's tail, in deceiving the least and interrupting God's sovereignty in all seven "churches" (with false doctrines of the dragon from those seven empires) Satan would raise nullity up to heaven, the place of "election" of all – interrupting the least's circuit. Devouring the least, Satan would supplant the circuit with nullity and a victory of his "sevenfold" lies (always the same device). The least, in order to overcome, must overcome Philadelphia at the end of His circuit. Each of the dragon's "lies" feature a three onto one device and so the angel suffers tribulation "ten times". (The dragon may be considered to swallow its tail in the circuit: his lies are as one would expect, as from the beast and its image. Note the angel is given a white stone in the third letter to Pergamos, a notice of acquittal!)

"Heaven" in Revelation is every case of the floating unity (on Earth and in heaven) in both the octal and GF(4) forms of the ultrafilter. For the unity to be "hurt" (sent to { }), there is an attempt on "taking heaven" completely in all seven elements (with the accumulated tail of all false doctrine), to construct this device sending the whole reference (Sun octal) to the empty set to create "Death" in the wormwood device upon the sea. All Satan requires is one such complete case to be further generalised. The Sun octal is the battleground; it can be no other.

There is denial that Christ is God – replacing God outside "the Earth" with the half hour of silence sevenfold that is the proper place of antichrist.

The dragon is identified as the one deceiving the whole world (all under heaven) and would want to plant his devices in every octal by sending "God" to nullity in a composite pseudo-logical device. Whether he could have won his stand or not, if it were done it would have been a deception only in the minds of those who will never place their faith on Christ. God will not be absent during His judgement when it arrives.

8.6) After The Least Has Overcome

When the angel indeed completes his circuit, the devil (now cast down) is powerless to deceive the elect – God, in full strength of omnipotence, will save His chosen.

Rev 12:10 And I heard a loud voice saying in heaven, Now is come salvation, and strength, and the kingdom of our God, and the power of his Christ: for the accuser of our brethren is cast down, which accused them before our God day and night.
Rev 12:11 And they overcame him by the blood of the Lamb, and by the word of their testimony; and they loved not their lives unto the death.
Rev 12:12 Therefore rejoice, ye heavens, and ye that dwell in them. Woe to the inhabiters of the earth and of the sea! for the devil is come down unto you, having great wrath, because he knoweth that he hath but a short time.

The loud voice indicates a judgement that Satan is defeated by a truly well-ordered principle. With salvation has come the realisation that across the sea there is no more "middle ground". (The angel's circuit empowering the ministry of the two witnesses is one of victory.) The consequences are salvation in full strength and the total besting of the dragon by the least in God's kingdom. Those that overcome (as the woman) in the wilderness are foreknown, already elect (not deceived, overcoming through faith on Christ and in the truth of His gospel).

The elect were accused before God "day and night". That places Satan as the static triple in the Sun octal and across the sea as in every left hand, as before Christ enthroned in the static subgroup. Not washed in the blood of the lamb, Satan can only operate in the four possible static triples. He would accuse the elect of operating in them just as he.

Satan operates in the "night", persecuting the woman (saints) with his devices of wormwood, using a serpent's speech. He would accuse those he tempts. They overcome him by holding fast to unity (they keep intact the word of virtue as their testimony, the Lamb before the throne, agreeing in the Holy Spirit). They "love not their lives unto the death" in that they are certain of their election. They are the sealed and saved. (As the 144 states of unity.)

There are four things come to the full: salvation, strength, the kingdom of God and Christ's power: aligning nicely with the sets of "peoples, kindreds, nations and tongues" that are before the throne as in the choice of unity. The four are traits of God to save them with might.

Those that overcome rejoice in the heavens, but those that do not are to be very wary indeed.

The salvation of the elect resounds in the heavens; the Earth has the worse of the bargain – Satan being the reward of the ungodly. The dragon knows there should be "time no longer" as the angel swore. With no time left for the dragon – he "takes off the gloves" in wrath (with the least free and only the faithful (the woman) left to attack), but God is far stronger.

Rev 12:13 And when the dragon saw that he was cast unto the earth, he persecuted the woman which brought forth the man *child*.
Rev 12:14 And to the woman were given two wings of a great eagle, that she might fly into the wilderness, into her place, where she is nourished for a time, and times, and half a time, from the face of the serpent.

The woman flies into her place in the wilderness (under the action of the six wings of the four beasts before the throne – and near the time of the opening of the sixth seal) – to a place of separation in great need of Christ. As "in the Earth" and yet to overcome the devil's devices which are opposed by the ministry of the two witnesses (as across 30 octals at once), I found the following word picture that no doubt was shown similarly to John.

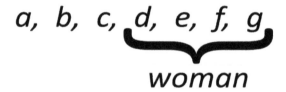

Woman is given "wings of a great eagle" in the
form of a brace that excludes the static subgroup.

Every (valid) seven cycle acts downwards to generate the "wilderness" as before. The "place" prepared is the rest or identity element in the seven cycle. Rather than remain in the octal to be delivered (ready to be tried against Satan), she flew to the place prepared for her by God (of virtue in the lamb) – a position outside the static subgroup in the reference or Sun octal. (I.e. In the four beasts before the throne.) The word "fly" as under the valid action of the six wings of each beast is to mark and distinguish between their action and her own as she is further granted the "two wings as of a great eagle" above. "Fleeing" (cf. Rev 12:6) indicates an invalid map, "flying" one valid, i.e. as the four beasts also "fly". Every element of every C7 group is present rather than just one in the reference only. The wings make up the difference.

The two wings given map the woman between left and right hand octals. There is no valid multiplication (as "wings" of each beast) that can map between left and right hands, those being disjoint in subgroups. For the whole sea of 30 octals to be aligned isomorphic and for the woman to find (by multiplication and not transposition) that place prepared in the Sun octal beforehand, the two wings are required to map from any subgroup in the right hand to an equal triple in the left, itself then aligned to a coset in the reference.

The sea reduces by isomorphism to the one reference (Sun) octal. Here, with the example of the woman, I find the analogue of traversing the sea in a "clean" fashion without the devices of the seals and trumpets.

The two wings switch between the "hands" of each octal to (equivalently) distinguish between a seven cycle and its inverse. The sea, now in isometry, becomes traversable in a clean fashion: each octal's static subgroup's coset becomes as that of the reference (as with unity) and so not to be confused with the static subgroup of any other octal.

The unity of the reference is found in the coset of the static subgroup, yet that unity (one of the four elements in the coset), is only found in two C7 groups of the full eight. The sea of glass is isomorphic to one field GF(8), not to the additive elements of each octal group alone with the unity free to be chosen with multiplication.

The woman, separated out by the brace above, is found mapped (from any element in any octal of the sea) to the four earthly elements by one element of a C7 group. By aligning the sea of thirty octals (as a "sea of glass") those four elements in the Sun octal may be attained by mapping into that one coset of seven as with any equivalent anywhere in the "sea".

There is further confusion caused by "wormwood" – the middle ground and the "face" or "semblance" of the serpent (octals which share no C7 group with the reference). Whilst the woman is in travail, reducing only to tares that practise the dialectic (for whom the static triple is now become a "subgroup" whilst the golden girdle remains) the separation of hot and cold (unity/static subgroup and static bow/plagues) from lukewarm (wormwood) is "great tribulation" and is how the dragon persecutes the woman. The two wings given her preclude the construction of wormwood upon any "left hands", merely allowing isomorphism to the one reference "right hand" only.

The serpent is "cast to the earth". That is, to the four earthly elements of the reference octal. The woman is "tried" on the three elements forming the static bow or the equivalent of wormwood: the three columns not forming an octal under the reference seven cycle. This is the serpent "persecuting the woman".

The ministry of the two witnesses shows the elect are able to be preserved through the device of wormwood – they remain as unity in the three subgroups common to the Sun and wormwood octals. The two "wings of a great eagle" then map between left and right witnesses (and their correspondences), so there are "none stranded" – as unable to reach the Sun octal. All may dwell in the reference octal which is the proper place of the woman; she was and remains "clothed with the Sun" (at least for now).

To be nourished for "a time, times, and half a time" is to be placed under a seal. Those three terms are (respectively) addition, multiplication and the static subgroup as under orbit. By remaining in the coset of the reference's static subgroup one is to be able to rest as unity – the proper position of the elect and the (hoped for) result of the two witnesses' ministry as to the angel's circuit. To be nourished in that coset is not to circuit the churches as does the angel but to immediately find approval: so that with a product of a single element of C7 generalised (the six wings of each beast before the throne acting within the rainbow, now justified by the angel's circuit, as to "fly" her into her place) the believer would find the unity element in the reference and rest as under this seal (and overcome any device of wormwood).

Likewise, "time, times and a half" excludes the coset of the static subgroup. To be "nourished" is for the coset to be mapped into one in the Sun octal, as if the octal, the seven cycle and the static subgroup about that generalised coset, were to be replaced. (I.e. with the same great earthquake occurring with the sixth seal.) They have that ability to shift to the Sun octal as that "place prepared" in the wilderness, empowered by the ministry of the two witnesses.

All may (as in the circuit of the least), attain the unity (sabbath rest) by overcoming Laodicea, reaching Philadelphia. The elect across the sea properly remain under a first seal paradigm in the lamb whose seven horns are all of virtue. (Adding unity to the elements of every static subgroup totally determines and generates the coset, as it does in the "sea of glass".)

The "time, times and half a time" is also equivalent to the seal applied to the 144,000 elect. They will not allow themselves to construct a wormwood octal, for that place prepared is to be spiritually fed in the wilderness in the reference octal and as by the two witnesses. The angel as the right hand has justified the operation of the four beasts' six wings (which may without fault map to unity), and the Spirit of God switches out the octal underneath so that rest obtained in any octal is rest indeed as in the reference. (The sealed remain in the reference, and the dragon is then "cast to the earth": unable to move them from that place of rest, a pre-first seal paradigm.)

Then the faithful remnant delivered of the woman will not leave (as by multiplication) the unity in the earthly four elements wherein she is to be nourished: she may attain that state by flying into that place prepared, but once the elect have found that place, they are sealed by God within so as not to fall away. They will overcome; God gifts them as with unity and rest. (Cf. the end of the ministry of the two witnesses.)

8.7) Great Tribulation

The woman, with the deliverance of the least from within her, remains to be persecuted by the dragon – she is under trial to see whether she will overcome (a test of the "twelve stars", which is to undergo great tribulation ending in the third woe – requiring all to swiftly exit her), for she is about to be mutated into the woman riding the scarlet beast along with its mark.

Those twelve stars are elements mapping from each choice of unity to the elements of the corresponding static triple. There is (in truth) in the sea of glass, only one static triple and in the octals made of wormwood, three columns (plagues) formed from triples that likewise must be overcome. The twelve are only present with a first seal paradigm or one broader.

The generalised coset itself is a "starting position" for the believer's great tribulation (fixing the earthly four elements in each octal isomorphic) that enable all to attain Christ's approval as in Philadelphia with the angel's circuit (the test, as not to dwell in a five cycle). The witnesses minister to the woman (as with a single element of the angel's circuit required to map her directly to "rest" or unity), the faithful may certainly attain an equal state of good doctrine, if they be under similar conviction. The woman is not to be rebuffed by God as yet, not until there are only tares remaining in her. (Then Jesus rejects the Church as desolate.)

Alternatively, the "wings of a great eagle" correspond to the four positions of unity as with the beast with a face like an eagle before the throne. The elements separated out by this brace, are mapped by the action of that beast's six wings to every other coset in the Sun octal. As the unity could be generalised to any one of the four elements so separated, the six wings would also map tares amongst themselves (that practise wormwood as in the columns not octals – those "plagues" or as in the reference, three measures of barley for a penny rather than a measure of wheat) by applying a seven cycle with an incorrect choice of unity.

The tribulation requires the sixth seal open to refine the faithful as well as to distinguish them from tares which will fall away under the same reference (with "plagues" of wormwood in the sea of glass), and also the angel's overcoming of Laodicea to Philadelphia (which as "rest" returns with the approval of the individual, as also of the seventh seal and the judgement).

This period of trial is defined by the last three seals opened by Christ, and because of the movement of heaven by the seven thunders, this requires the period on Earth (and of the "days cut short") starting at the loosing of the restrainer (the fourth seal opened), and that ending during the sixth seal (the witnesses' ministry complete with the elect sealed) rather than the period following after the opening of the fifth seal in heaven, sometime after which the angel had overcome. Then only the last three seals (and also the three woes) remain for the woman's "great tribulation".

The four elements separated are the "Earth" rather than the elect (whose place is unity within the seven cycles of the witnesses' ministry – fixed as one of four earthly elements). Each element remaining is in the static triple(s) of the dialectic and so the wilderness is composed of the elect as well as tares: a measure of wheat and three measures of barley. It is a given that all in the world dwell in the earthly elements.

The three elements excluded from the woman (given the eagle's wings) are the static subgroup. Without loss, it may be assumed that the static subgroup is also a group in the reference octal, as the only eight octal groups not to be considered isomorphic to the reference octal are the choices for its left hand under each C7 group upon the reference octal. (And the two wings may map from their static subgroups to the right hand's static triple(s).)

The mysterious "time, times and a half" refers to this time period itself being under a "seal" (as trial under the fifth, sixth and seventh trumpets – the period of the circuit of the angel shifted forward by the seven thunders) decided only by the choice of God: and this period is unknown. Time is as "addition", times as "multiplication" and half a time as an octal subgroup (compare the operations of the Trinity) or its coset (or a triple under Frobenius).

Each octal appears arbitrary! As the least convicts the elect to exit the woman, the left hand (by the Holy Spirit) convicts the individual believer instead, by mapping to or from the left and right hand octals. The angel justifies the six wings of each of the four beasts and the Spirit of God provides the two further wings (as of a great eagle) given the woman. Then God with the angel accounts for this seemingly arbitrary octal. No octal in the sea remains undiscovered, and the strength of election is universal – none are caught without grace until after the final judgement. There is no "lake of fire" or "second death" just yet.

The woman must find rest (unity) to be nourished, or nourishment to find rest: she needs the "power" of the full circuit of the angel and not to be under the locusts in a five-cycle. She should not be transformed to the wormwood octal(s) either, unless she is only in the static triple/subgroup of the other witness. That sealing as unity corresponds to an element in only the "jacinth" part common to the Sun octal, which does not traverse rows of wormwood under the reference cycle, that action necessarily being the unity by construction.

Unity requires believers not to practise the dialectic, which traverses the sea of octals and requires alternate multiplications. That device requires a denial of virtue and the paradigm of the "son of perdition". Such a one is not saved and then should not be in the woman either. If this paradigm is present it must be in the static bow of the reference, not unity as elect.

As to this mixture of tares and elect: does the five cycle hold because the believer is returned to the collective practising the dialectic because of the strength of their relationships, or does it hold because they take the middle ground of wormwood – enabling the dialectic in a setting which rejects the perfection of the reference octal with its correct seven cycle? There is one answer to being caught in those deceptions – to which the two witnesses testify: be separate.

Either way, the woman being delivered of the man-child (tried, as under tribulation) is slowly reducing to tares that practise the dialectic. Great tribulation is then to hold firm to good doctrine despite the three woes, the third of which is to follow in the text.

8.8) The Dragon (Simply) Defeated

Rev 12:15 And the serpent cast out of his mouth water as a flood after the woman, that he might cause her to be carried away of the flood.

The dragon assaults the woman with a great flood of water, "spiritual speech" or deception as from the sea across the Earth as after the woman. (Like a tsunami or tidal bore?) This is Satan (the dragon) using the wormwood device across the "sea" and overlaying the fourth seal device over it in deception. (With that wormwood device then made as if "placed over" the four beasts before the throne – an impossibility: properly a house "built on sand".)

The wormwood, or "bitter water", i.e. the "flood" of water from the mouth of the dragon is the set of (stolen) seven cycles over the Sun octal then falsely applied to a wormwood octal to generate all seven "wormwood rows" (as in the sixth trumpet: a middle position that mocks Christ and enables the static bow as a subgroup so that the dialectic may appear valid).

The "flood" is the (fourfold) covering of the Earth with water – here present by direct parallel to the four angels bound in the great river Euphrates (now loosed). These four "rivers" cover the Earth and enable the dialectic in the wormwood device. There is a great "falling away".

The dragon's speech, dialectic in the "earthly" elements, is Satan's device to have the woman "swept away". That is, to remove her from her place prepared by God. Once she practises the speech, she has been carried away. The proportion of tares in the woman will take their toll. They are not sealed and although they may be as "kindreds", they are leached from the Sun octal by the same "flood" of the dragon, seen previously as of those four angels now loosed that were bound in the great River Euphrates. Again, this aligns to the sixth seal and second woe, the completion of the angel's circuit and the same ministry of the two witnesses.

Yet the Earth itself whose "mouth" may be considered to be the set of her elements {d,e,f,g} (and not of { }, the empty set to which the dragon would have sent unity to, to infer a pseudo-logical disjunction as by the balances of the third seal). The shape of the "brace" resembles a "great eagle" separating out the "earthly elements".

Rev 12:16 And the earth helped the woman, and the earth opened her mouth, and swallowed up the flood which the dragon cast out of his mouth.

In terms of wormwood, the three columns not forming octals under the reference seven cycle are disjoint in triples with the "river" (left hand's static subgroup under orbit) acted upon by the seven cycle. The Earth then "helps" the woman as the left and right hands are disjoint in triples from these other three columns of triples. The woman then only dwells in the left and right hands of the Sun octal. Tares, however, also dwell as antichrist in the wormwood rows, so there is some mixture still. (The elect dwell in the reference seven cycle; tares do not.)

The Earth may help the woman, but that does not mean she is completely victorious. The Earth opened her mouth (resembling a pair of braces to denote a set, i.e. {...}), rather than the mouth (set) of the woman being "closed". Closure is a property of the elements in a group, and the dialectic devices of the devil "hurting the oil or wine" sending unity to the empty set merely mimic Christ with the dialectic process as method.

If [0,a,b,c] is static as a K4 subgroup, given d=1, say, then e+f+g = d in the octal and the set that corresponds to the dragon's device {{ },e,f,g} is regenerated to {d,e,f,g} and "d" is restored. The woman is not beguiled by the dragon so easily – at least not in the place prepared for her of God, outside the static subgroup in the reference without the devices of wormwood. The whole octal is also consequent of that set (mouth) of the woman opened, as "d+e = a" etc. and the static subgroup may be found again or generated in the reference octal by the sealed (and not across the four octals in the wormwood device as the devil or serpent would have it).

This also holds for addition over K4 subgroups, the "place prepared" for the woman also appears; the product of two "earthly triples" produce a portion of that jacinth part common to the wormwood and Sun octals. (The dialectic cannot be overlaid upon the wormwood device.)

Rev 12:17 And the dragon was wroth with the woman, and went to make war with the remnant of her seed, which keep the commandments of God, and have the testimony of Jesus Christ.

The woman is made of both elect and tares, the remnant of her seed are the "Christians". By an argument of induction, the elect will be similarly delivered, leaving only tares as a remnant. They are the remnant in the seven churches (without conviction of the angel) that tread the grace of God underfoot as gentiles in the outer court of the temple. In view are those in Laodicea walking the seven cycles of the seven candlesticks (churches) before God but not circuits between them – as those 7000 that die when 1/10th of the city falls.

They are yet in the "Church" and the Church is already become the system of captivity of God's worshippers (as to deceive the least) – not in Nebuchadnezzar's Babylon, but in Satan's own.

With the elect now gone from her, the woman has now been stripped naked of all righteousness.

Chapter Nine: The First Two Beasts

The first beast of Revelation 13 is not to be confused with the later scarlet beast. The "third woe" is built upon two prior woes or diabolical "beasts" and that third and final scarlet beast is a prediction of the final state of the Church if it were left free to exercise Satan's devices.

First and Second Beasts:
The two beasts are (in turn) Satan's last kingdom of the Earth and the latter the influence of Satan in the Church. In like manner they combine together to establish the devices of the latter upon the authority of the former. The devices of the "Earth" and the dialectic are also made upon or extended across the "sea".

This "judgement" of the Church is of one entwined with government, the first beast. They (the first beast and the scarlet) are not equals. The judgement reveals the extent the Church would fall to during its tryst with government. It will not be permitted. Enough for God is enough.

The second beast has the motive to send the Church all the way bad. It does not stop doing so.

The first beast in Revelation is properly treated in the four beasts of Daniel in the Old Testament. It is the seventh in a series of worldly empires corrupted by Satan to worsen the standing of Israel before God. It is the result of three empires (horns) that went before it – so that the dragon's tail should not show a discontinuity and openly divided as upon itself.

The first beast is no less dreadful and terrible and it has a deadly wound (instantiating a wholly null fourth seal system) and one caused by the elect exiting the "state modelled" Church (under the beast's purview). The second beast introduces the impetus to create the image, healing the scarlet beast (the Church rather than the first beast), causing the influence of the first beast to prevail in the tryst between it and the scarlet beast in bed with it. The "woman" or would-be bride of God is anything but, she is merely a projection: an imagination of the heart carrying with it an insult that Satan owns the church and its life's blood, its doctrine.

The beast is become "collective rule". Not of the majority, but where the will of the majority is surrendered to a trusted minority the whole have no control over consciously or selectively. If present at the right time in the right place one may be enabled as casting the deciding ballot in the place of a far wider majority. The process is then king and not the heart of the individual that is given free rein to cause havoc (to the tried and fixed position of scriptural absolutes).

The second beast is purposed to maintain the Church (as in appearance) separate to government – on the condition it must govern itself (following government as its god and not the Lord God of all the Earth), and adopts the same system of the fourfold dialectic as an "image" for that Church government. The Church acts in similar manner in order to be separate? Sounds easy to spot yet that is very difficult without discernment under such trial.

The scarlet beast is patterned in such a way as to have all manner of "miracles" interwoven within it. (That stated, there are coincidences of a factor of 24, a body of ratio 6 to 1 (giving 144=6*24), with seven collectives (churches) with one "really bad" as well as an apparent "bride" of doctrine, all decked with purple robe and scarlet colour, with jewels etc.) The false prophet presses that this is truly the "bride of Christ" and government should relax and enjoy its tryst with or alongside her, although the miracles cease for the believer once they overcome it. Those miracles of the false prophet have an outward appearance of scriptural validity despite the truthful contradictions from the text: buyer beware!

The second beast operates all the authority of (the freedom granted by) the first beast to freely pattern the Church as it would see fit (with the image). In truth, the first beast simply is not given the authority to put the mainstream Church with its image into illegality. (That is their "constitutional" right.) What it actually does, is worm its way in deeper and does so in order to perpetuate the tryst – requiring the Church's governance of the image of ten kings to reciprocate, stripping her naked of all righteousness with the beast's own laws that are over it. The individual, however, is free to believe as he wishes, whether he acts on it or not! (Certainly not free to sin though.) He may exit any time through the open door to remain aloof.

> **The Image of Ten Kings:**
> *The image is formed of ten kings, three of which meet in a dialogue with the collective in a facilitated meeting. A meeting which decides what "virtues" (doctrine) may be incorporated in common with other members of the beast (those not present but for whom the decision of the collective will carry). In this they have but one mind – the decision carries over the six other "kings" as seven mountains on which the image (the woman) sits. (As Hell following after Death.)*

The individual and the minor fellowship becomes the "cult" to such government. Patience is required, not rebellion – and not integration into a wider collective under the image or a similarly registered religious institution with the first beast. (An individual may incorporate himself, but cannot then legally exclude others from being party to his privacy. He ceases to be an individual. There may be many individuals with that incorporated "identity". I find three as such in the Revelation: the first beast rising up from the sea; the "Church" and its image rising from the pit; and lastly, the scarlet beast – its final judgement with the image completed.)

9.1) Stood On The Sand Of The Sea

John recounts His vision of the first beast. He is stood on the "sand of the sea" (He was spiritually positioned on that sand, yet remained upright; he was empowered to overcome the shifting of the sea, as if built (founded) on rock: aware instead of the sea of glass, the one Sun octal in Christ – as all others in Him are also aware – being the "sealed") and this is telling in that the "stars" (from the sea) cast to the Earth are now become as "sand". They are to interrupt the proper operation of the octal on each of the 200,000,000 possible circuits that the angel could make. (Sending believers leaving the first five "churches" back to the "Church" – the Laodicean amalgam.)

They have lost their place in Heaven (fallen from unity, grace) and are as "motes" attempting to overlay the myriad singletons and triples of thirty octals **upon** the land (the four earthly elements of the Sun octal) as *from the sea* to complete the city (their house built on sand) – a group isomorphic to S5 over the first five churches. (Twelve elements of the group S4 cannot be constructed with the fallen star with the key to the bottomless pit. These 12 mappings are not constructible even in the sea of octals that intersect with the Sun octal. God's spirit is not twisted out of His purposes. I will look at this again in depth later on.)

The stars try to logically instantiate the devices of the seals upon the Earth: John shows us that the first beast (rising up from the sea) is constructed by their devices. They, drawn for those products (third part) by the dragon's tail (the three horns plucked up), are purposed and set to prevent the exit of the least (and then all) from the Church that is going all the way bad.

Rev 13:1 And I stood upon the sand of the sea, and saw a beast rise up out of the sea, having seven heads and ten horns, and upon his horns ten crowns, and upon his heads the name of blasphemy.

9.2) The Name Of Blasphemy

The dragon held sway over the kingdoms[7] of Egypt, Assyria, Babylon, Persia (with the Medes), Greece and Rome as well as three NT-era kingdoms to form the USA (with the last dialectic device as his deception only). I find that the deception (lies) of the dragon (the names of blasphemy) are upon each head – that all "divine" authority is removed and only Daniel's fourth dreadful and terrible beast remains. The name of blasphemy is the dialectic that privates virtue (unity). It is the dragon's only device against the circuit of the angel (the least).

Only the dialectic device (as accumulated) is in view and no God given authority over those historical kingdoms is present. The beast is the means to assimilate all of the false doctrine of the Old Testament, to be ministered resurgent in the New Testament era. The "ten crowns" represent the ten worldly kingdoms in truly worldly terms: aside from any spiritual construction in their "dragonly" counterparts, the first six actually had some divine authority permitted the dragon as god of this world to continue on (permission now since lost since the kingdom of God has arrived in full – as the seventh kingdom, intercepting the place of the USA).

Those ten "crowns" on the beast represent real kingdoms over which the devil still had some sway. The dragon's influence was limited to seven kingdoms with a maximum space of "ten kingdoms" as the tail is fully stretched. The dragon's ten horns are of the dialectic and the tail accumulating. Satan does not "rule" the world but takes advantage of the means to his ends.

9.3) The Beast Is A Construction

Rev 13:2 And the beast which I saw was like unto a leopard, and his feet were as *the feet* of a bear, and his mouth as the mouth of a lion: and the dragon gave him his power, and his seat, and great authority.

Here is a direct reference in the text to Daniel's fourth beast as the amalgam of the three before it. It is authoritarian in appearance, socialist in politics (employs the dialectic in its static triples, as indicated by "feet" here – its stance), speaks strong (thinks strong) and gets its existence from the three kingdoms plucked up for it in the manner the devil's construction has made (its power – as a nation). Its seat rests upon those devices – they are part of history now, and have given the USA great authority by the free exercise of the dragon's devices.

[7] James Lloyd, *The 6, 7, 8 Cycle* (Christian Media), p26-28.

In turn the three beasts are out of sequence with history – unless the beast's power was granted by the fall of Nazi Germany (the military machine); the fall of the USSR gave the beast its seat (as sole superpower); and their law (freedoms protected by their constitution) by which the beast (collective) gets its great authority (freedom) stems from Britain's defeat in their revolution. Then, they most certainly are in sequence since the terms match up. (The dialectic is truly Satan's device, and it was not a creation of Britain.)

The history of the USA, then, is such that the accumulating dialectic of the dragon's tail should then present itself. (For the quartet of powers have together interacted using the dialectic.)

The beasts of Daniel[8] refer to Imperial Britain, Soviet Russia, Nazi Germany and the USA (as forming the New World Order in a dialectic). Those four are subject to the hard limit of the dialectic, and I examine that termination of the sequence here. The growth of the dialectic in modern times across the globe is due to the response of the USA "recoiling" from conflicts with the other three of Daniel's beasts, and that the moral high ground is null (in the dialectic).

The sequence of the first three allows each to be "plucked up" as the restrainer is loosed in heaven. The simpler paradigm of Britain is rejected by the US after the war of independence in favour of a republic. The rise of communism is opposed by the USA after the defeat of Hitler in Europe during the cold war era. Hitler, himself defeated, nationalist socialism continues on after the war (as in the notorious "operation paperclip"); the scientific expertise of WW2 Germany is also co-opted by the USA for the advancement of the "military industrial complex".

Whilst communism is opposed in the "McCarthy" debacle, the hearts and minds of the more aware citizenry of the United States which oppose that perceived infiltration are slowly overtaken by a more dialectic paradigm of the young: schooled in a system more aligned with the product of that German philosophical school that fled the Nazis and survived them.

There is always a remnant of the social structure in the USA that is slowly choked out by the introduction of the next and broader paradigm, eased by the great authority (freedom) granted by their constitution and bill of rights. The "remnant of the beast" is intended to choke the Christian faith. I find that same mixture of political and religious fervour in the verse:

Rev 13:10 He that leadeth into captivity shall go into captivity: he that killeth with the sword must be killed with the sword. Here is the patience and the faith of the saints.

The small percentage of the citizenry (militia) that fought the war of independence truly instituted their republic; those loyal to the crown were assured of the newly fledged states' insistence upon the illegality of titles and aristocratic standing in the new republic. The rewards due those with loyalist sympathies were effectively staunched. (No land to be gifted, no titles bestowed on any, even gifted abroad in foreign nations.) The militia truly had the choke-hold.

The USSR threatened the twentieth century's USA through covert "infiltration". In undermining the republic, they were the "bait and switch" that the republicans followed after whilst the paradigm of the nationalist/socialist schooling (the synthesis) surviving from Nazi Germany took over by a slow crawl, steadily choking the same republican citizenry by changing the mindset of the young to the dialectic, where individual rights give way to the needs of the group in consensus. The young choke out the vigilant (the patriotic and as it were, the militia).

[8] Lloyd, *The Sand and The Sea*, p5-7.

The ongoing third seal system (educating the USA with socialism) is not as blatantly evil as Hitler's Reich. In context (of modern relativism) it is not found to be as Hitler's shocking evil but is instead more subtle. Assured Hitler's defeat, the paradigm that permitted Hitler's rise goes without notice as moderate socialism is introduced instead of that blatant evil of fascism.

Nazism (as total evil) was soundly defeated, ensuring the "hearts and minds" of any citizenry may be found to agree with the fourth seal system built over that defeat. Yet this "fourth seal" position as for the heart of the collective is in contradiction to that same fascism (which was itself a third seal system). The utmost evil of the holocaust ensures that that particular (or indeed any) third seal system would be totally rejected along the lines of Hitler's insanity. Nazi Germany was instead "plucked up". (The most obvious of the three to be so.) Britain likewise lost the colonies, yet continues with the "special relationship", whilst the USSR instead fell for the NWO to exist in the hands of a socialist USA. (Ensuring the "end" of both their republics.) They show the failure of "The Old World Order" (a remnant of the Cold War), not quite "in the same camp", and display that separation: a position none would wish to continue within.

The components of the beast in Revelation are accounted for; the head (mouth, foreign policy) of a lion; feet (stance, politics) of a bear and appearance (citizenry, paradigm) of a leopard.

The "New World Order" is built on the premise that there should never be a third world war. Yet the NWO choke-hold on the apparent rise of nationalism (as were the Nazis with their fascism) in the dialectic is null! (Globalist socialism has also choked-out soviet communism.)

That nullity allows contradiction (dialogue will always fail), whether of any preemptive military action (illegal aggression?) or even a solution found in treaties, as, for instance, the UN charter.

The moral high ground to prevent the rise of any new "Hitler" is in fact null (the beast rises up from the sea, and the "third part" of the second trumpet device is arbitrary). Its fourth seal system purposed to rally (at will) the hearts and minds of its citizenry to oppose any unwanted action of other kingdoms exercising seal paradigms one to three: the result is that every kingdom joins the template of "the beast" (the fourth beast of Daniel, the USA) or they align themselves to the USA by their support. (Else they are relegated to the "old world order".)

The dialectic in its fourfold form permits the citizenry (the heart) to oppose any nation of that "old world order", opposing any "would-be" Hitler. This is done with the outward appearance of a rational judgement, but the growth of social media empowers the paradigm of the NWO (that the **conjunction** in logic of each of the four paradigms/kingdoms is positive and **must** be maintained) and remains tailor-made for this purpose. (Always to raise yet another talking point in a new dialogue, instigated by numerous petitions and acts of peaceful resistance by the "youth", themselves unknowingly bred for the NWO in the dialectic fallout of WW2.)

What then is the truth? As the world operates on the dialectic, and social media reports on the "old world order" of the paradigms of seal systems one to three, then the answer is simple. Acts of virtue in Jesus Christ are the solution on the scale of the individual and not those other solutions requiring government to solve their populace's divisions (and in a perfect world not military action): but this world will not hear it despite the righteous actions of the minority. The NWO requires this contradiction of "new world" to "old" to be fed with constant fodder: wars are required and in Orwellian style, continuous wars are meant to be waged continuously. The NWO will war, war and war and not let go until every heart and mind are damned without Jesus Christ. Christians are not the target, but the paradigm of Christ is the target. (Only the four powers united would provide any hoped-for intercession.)

The NWO (the beast – 4th seal) has the chokehold on not just the youth but extreme nationalism (3rd seal), soviet communism and the vigilant (which are the defenders of freedom – 2nd seal), dictating all individual rights (privilege) rejecting the higher power (and higher authority – as the 1st seal, a statement I somewhat justify), crushing the paradigm of Christ (also a monarch) whilst the media (as "Hell") judges for them the difference of all good and evil under the NWO.

Christ will return and the beast continues on only "a short space". Christ's commandment is to endure (Rev 13:10) with saintly patience until the end. War is always an evil, yet God has not once said that wars were not worth fighting, but it remains true that the body of Christ does not wage war. That said, those in Christ will defend their freedom on their own part, and God alone defends the border of His own kingdom, excluding those not elect from salvation. Freedom (as the false prophet shows) does not protect the latter's border or the faithful as having kept the former (even, "a republic, if you can keep it"). Christ's kingdom is "not of this world". Freedom requires redefining as well as refreshing if it starts off with the wrong "God".

On the premise that the moral high ground is null, the solution to every future conflict requires union of the four beasts of Daniel, whilst logically on paper they are still in conflict and the only recourse is for one to take the lead. That one of the four must act in complete contradiction, for "The constitution for the United States of America" declares treaties to be as "the supreme law of the land" and the dialectic limit (or contradiction) will therefore present itself. This is not to state that freedom or constitutional government is faulty (authority would be found for every unconstitutional act), rather the dialectic is the method used of the faulty, and it is opposed to freedom to start with. (The reason the USSR deferred to the USA the right of the last superpower, also a heavy burden and one that will cost them too much money.)

Everyone likes freedom, right? Sure they do. The legal ability to practise any form of religion in a nation united under the authority of "Nature's God" is no bad thing, provided Nature's God may be obeyed. If that God is identified as the one under whom all men are created equal then freedom is truly no bad thing, but freedom under a different God – one which permits so much licentiousness and that cannot legally draw a line making a separation of the flock as would a shepherd: that is actually a bad thing. Religion can and does go all the way bad, it loses its way. Where will the illegality of faith be fixed other than God's removal of restraint and the dividing line He draws between His elect and those others not His amongst them?

Either the state becomes "God" or the apparent lack of imminent judgement from heaven infers that there is yet no problem, as if all remains good, "one nation under God".

Such an argument that the USA is still truly one nation under God (and which one?) would upset the cart, the fruit on it rotten. No other country permits as much religious freedom as the USA (always good in that sense when it is righteous) – and so much other freedom aside from it, and although Christianity is booming so I hear, why is that fruit so afflicted with the dialectic? Well, blame Imperial Britain, the USSR and Nazi Germany and the USA's reaction, recoiling from them all with the high ground null: none of them are able to correct each other. (One superpower remains outside act of treaty, the other three are already "plucked up".)

9.4) The Beast's Deadly Wound

Rev 13:3 And I saw one of his heads as it were wounded to death; and his deadly wound was healed: and all the world wondered after the beast.

The first beast, with its name of blasphemy, is Satan's tool to corrupt the seven churches of Revelation, those elect (Philadelphia or some small remnant in Sardis) and the others of the "Laodicean Church of the lukewarm" (including most of Sardis). If I "marry up" the devices of the seals as they follow the angel through his circuit, I may liken each church to the same device, if not the same doctrine that Israel had to overcome in the Old Testament timeline.

The exit of the Church by the angel (the least) is a deadly wound to the beast and that angel stood on the Earth and sea is totally able to overcome those devices (he has excluded himself from any religious forum in which they play a part). Indeed, he exited through the open door.

The wound is healed (most surprisingly), and this healing culminates in the scarlet beast later. The "church exeunt" of Philadelphia has been replaced with a source of leaven (false doctrine), in order to interrupt the proper operation of the seven cycle. Whilst the Church appears lively and active, in truth they are dead. The beast permitting so much freedom and licentiousness has allowed the church to become lukewarm in Christ – if indeed they ever were hot towards Him – timing allows someone born into such a paradigm. This liberty that is the authority granted every individual by the beast has "poisoned" the well of God's water. That water is now bitter (deadly), due to the dialectic that permits those tares to "agree to disagree".

The USA as the seventh has its head "wounded as to death" (as aligned to the fourth seal). That deadly wound is healed with devices from the sea (wormwood). The beast granting so much liberty places the collective in the position of sovereign to heal its deadly wound. (With the synthetic balance of the "heart".) This has dual fulfilment in the scarlet beast. (Which is necessary to Satan; all believers are free to leave the beast-approved systems of worship.)

The scarlet beast heals the first beast's failure to corrupt one truly free church, that being Philadelphia (a wound to the head upon the beast). It is no trap for them, they are free to exit (when the devil's devices infuse their fellowships as regulation by government steps in). It also bears the same wound of the fourth seal device's nullity: healed with a collective (as of the heart of the individual) not by a fellowship of saints but of a dialogued consensus on doctrine, as the elect (with doctrine, oil; cf. the parable of ten virgins) are completely exeunt.

The deadly wound to Satan's scheme caused by the exit of all the saints holding good doctrine is healed with a collection of false doctrines dialogued to consensus after the model of free government: the fourth seal system of "Death". This first beast fails to be a worthy device for Satan to defeat and cage the elect; God requires Satan to be more subtle.

In an inversion of the reference octal sent to the empty set as seen in the earlier war in heaven, I find the beast is created with self-government sent to nullity in the fourth set of balances: giving only the collective the power to speak their conscience in that place due the electoral college (as now become democracy). The system of wormwood in the sea and the static bows (in the devices of the seals) in place over it (assembled as a collective) become relatively as "one nation under god". Clearly there is a pretender squeezed in between heaven and Earth.

Sinister as it is, the collective also rules the beast in the image system; it has deferred to them all "elective" authority (as do the people of the United States, being sovereign, to the Electoral College that selects for the sake of policy). The image is truly "to the beast", in modelling its deadly wound to give the image life. There is yet a "wonder" as to how this wounding as by the sword (Christ's call to exit) may yet give the false prophet's construction "Pneuma" and life. The wound is not and cannot be undone, but instead is taken full advantage of.

9.5) The Beast Is Followed (By Hell?)

In following after the beast they worship with the devices it was created with, and which made it what it has become – once a source of liberty and rightly so for the free exercise of religion, but religion gone all the way bad? It was only inevitable; I do wonder.

Rev 13:4 And they worshipped the dragon which gave power unto the beast: and they worshipped the beast, saying, Who *is* like unto the beast? who is able to make war with him?

The world worships the dragon whose devices permit them that "freedom": freedom which was enumerated as rights under the beast: for that word "under" is in effect in these times. The USA was set up without direct influence (open dominance) of Satan: their constitution is for a fair republic with checks and balances, yet be assured Satan influenced the three nations plucked up to have their effect before it went all the way bad (Britain, the USSR and Germany).

Those kingdoms were all enemies of the USA at some point, and as each were enemies to each other they now meet in dialogue under treaty, bound as if in common distrust. (Even Britain as a monarchy opposed the Russian revolution, also in the Cold War.) The reaction of the USA as the stewed meat in the pot is sadly predictable (after the manner of the human heart, destined to go all the way bad). On paper, the beast is as wormwood, and conflicts remain with those nations engaged in opposition on paper within those treaties, independent of warm relations. In the mind, "union" of the four is an illusion. The USA is become facilitator to the world, and the collective (the four "primary powers" whether in stalemate or not) indeed backs them up as policeman.

Subtlety is perhaps the adversary's greatest point.

So, the world practises the same devices as the beast, liberty does not mean security as any tyrant should understand, and the world swallowed the bait and the Church has had the switch of God for government, drawing its freedom to worship not from God, but from the state.

Exo 13:16 And it shall be for a token upon thine hand, and for frontlets between thine eyes: for by strength of hand the LORD brought us forth out of Egypt.

In the provided salvation, believers are servants of God and spiritually separated (sealed) as such. However, government blasphemes when it applies its own mark to the forehead or hand in the place of God's understanding and agreement (His covenant). It separates out those that delight (tares) in being justified before men instead of being approved by God.

9.6) A Mouth Speaking Great Things

Rev 13:5 And there was given unto him a mouth speaking great things and blasphemies; and power was given unto him to continue forty *and* two months.
Rev 13:6 And he opened his mouth in blasphemy against God, to blaspheme his name, and his tabernacle, and them that dwell in heaven.

The beast's mouth or "set" {…} is the closed quad of triples that is the system of the first four seals or "Death". Now that the USA is become the last superpower (amongst the four "primary powers" or the four beasts of Daniel) great things are become as "promises of peace" and the statement that faith is irrelevant for (and to) it. False promises are then given in the dialectic: fixed truths (that oppose the free exercise of great things, as taking liberties) are vanished in the process – in dialogue. The Death dialectic is formed of four associated triples.

$$a \equiv \{c,e,g\}$$
$$c \equiv \{a,e,g\}$$
$$e \equiv \{a,c,g\}$$
$$g \equiv \{a,c,e\}$$

Arranged in triples and with the remainder sent to { }, the empty set, I may in the octal model the dialectic devices. However, the dialectic practitioner is again somewhat stymied as this beast rises up from the sea, employing the device of wormwood rather than Death (the fourth device of the seals, which is deadly, contradictory or "impossible").

The mouth (set) enabling the dialectic devices of the seals upon the sea (as overlaid upon wormwood) must be a "given"; it is only by similarity to the Earth that this is possible. There is need for the "little horn" or false prophet. There is namely a requirement for a "second beast".

By using the devices over the sea rather than in the four "earthly" elements the beast opens the set (mouth) of triples to blaspheme God. I.e. His name (the "sea of glass") is blasphemed as well as His tabernacle (the seven churches), and those that dwell in heaven (His elect as unity). These blasphemies are also mirrored by the later image system, the scarlet beast.

This is to be taken to state that the devices are not natural and native to the Earth, or indeed heaven or the sea. Death as the beast's "mouth" is opened up (and his speech is constructed).

The 42 months, the outer court of the temple of God – those 42 months generated by the angel's circuit over the seven other C7 groups (the mystery of the seven churches) – is the result of the ministry of the two witnesses restricted to one octal, not the whole sea of thirty. (Though they are all isomorphic. There are thirty days in a month.)

This limit is of the beast's blasphemy: its effect is that seven churches must be overcome in order for the angel to find lasting and Sabbath rest. Those seven churches found in bed with government are no good except for one in which the angel may finally rest: But then that church must not be in bed with the beast! The angel is exeunt of the "Church" of Laodicea – six elements of his circuit provide no rest.

The dragon is defeated, but the falsity he authored is not gone: it is permitted to continue a short "time" – these same 42 months. These are formed of the six elements of the six churches (Ephesus, Smyrna, Pergamos, Thyatira, Sardis, Philadelphia), and six elements of the reference cycle (the angel's circuit through those six churches), leaving Laodicea (now overcome) completely separate and the condemnation of that letter in a state of "rest".

I may align the accumulated tail of the dragon to the angel's circuit: a cycle that stands before God completing the circuit, hidden within the elements of the other seven groups under its own action. Every other element of the angel's circuit (as a permutation of the seven churches) except for the rest element (identity) is aligned to some device of the dragon and his tail now overcome. Together, six remaining elements of the other seven groups isomorphic to C7 are forty two elements with one rest element shared between them.

The mouth (set) has provided the scarlet beast with seven mountains of false doctrines accumulated from Israel's Old (and New) Testament history. (The seventh set is of "Nicolaitanism" in the Church.) The false prophet's construction upon false doctrine is ready to become a new scarlet beast (the eighth is of the seven) and this heals the first beast's deadly wound. Great things are promises of revival but in real terms this is the denial of virtue in all "Church governance": that governance a copy of the first beast in a religious setting.

9.7) War With The Saints

Rev 13:7 And it was given unto him to make war with the saints, and to overcome them: and power was given him over all kindreds, and tongues, and nations.
Rev 13:8 And all that dwell upon the earth shall worship him, whose names are not written in the book of life of the Lamb slain from the foundation of the world.
Rev 13:9 If any man have an ear, let him hear.

The dragon's tail with its accumulating dialectic is his only device in these last days: none other will follow upon or after it. Whether his deprecated attempt with a deadly wound (Philadelphia exeunt) is his final weapon, or the scarlet beast with its deadly wound healed: his device (the dialectic) is ever the same; the first beast provides all authority for the second to make the scarlet beast: that authority is worldly "freedom". The beast is Satan's last effort, it has assumed authority over three groups of the four: "Peoples" is missing. In order to bring the fourth part of the Earth under the devices of "Death and Hell" the beast fills their vacuum (nullity rather than absence) with a collective – the scarlet beast.

Those united in the Lamb and convicted by the angel's circuit exited (overcame) the scarlet beast, for it overcame their fellowships. The remnant in the "Church" (kindreds etc.) are overcome by, and remain within the beast. Those dwelling in static triples in the "sea of glass" do so in the three columns (of wormwood) not forming octals (as plagues, triples under an invalid seven cycle). By dwelling in these triples they worship the beast. This power is identified as the very "seat of the beast": i.e. from which it gains its place and authority from the dragon (a serpent).

To be given "power" is to have authority over those that will fail. If the beast has power or judgement to move those that agree in virtue but not in the Holy Spirit (kindreds) towards compromise (accused as static triples dwelling in "wormwood"), as well as over those with a different unity (virtue) to likewise fall (with "tongues" – in dialogue) by compromise as well as to accuse all those (not convicted yet) with the least's circuit and still in the Church amalgam (nations), then he is an antichrist indeed. There is to be no authority but God's own over His people – the very elect.

The dragon is given all requirement to transform the USA from the land of liberty to the fourth beast of Daniel "dreadful and terrible" in these, the last days (after the plucking up of three horns). The dragon's speech is the baleful oil that the gears of the modern world turn upon – it works well for the "free" USA (or so the world thinks). Those devices are now used openly in the Church: they were used by the rest of the world already.

The three groups are in wormwood (consensus) set against the people of God (as if in $a \vee \{c,e,g\}$). The power to make war (oppose, deceive) is found in the dialectic, restricting the filter to the earthly elements. By constructing a wormwood octal, building it (as if including "a") upon a "great mountain burning as a lamp" (as if $\{c,e,g\}$, even if "a" is "kindred" to "c", or "e" to "c" say) overlaying a dialectic (sending $a \rightarrow \{ \}$), the beast brings in the broad paradigm now practised by those three collectives (the third part of that great mountain (nations) as a "lamp" is arbitrary, given virtue is always to be privated).

The beast rising up from the sea is formed after the wormwood device. With the dialectic (of the Earth) overlaid upon it, the beast is the foundation for the methods of antichrist. It is the exampled "paradigm" and "template" for every instance of the dragonly enterprise, unknowingly worshipped by antichrists everywhere.

War with the saints is not necessarily "against" the saints, but may be a blurring of the line separating out those who truthfully appear "saints". I expect the process of dialogue to insidiously work its leaven all the way through. (The dialectic device is always upon a coset of a static K4 subgroup, outside the ultrafilters of the octal. The empty middle is a heavy burden: not exemplifiable.) The saints are overcome by the "little horn" (cf. Dan 7:8) of the beast. The beast encroaches 42 months (seven churches of six elements each but for the one identity element in the angel's circuit within the one church that finds approval).

That one element, of Sabbath rest in Christ is also encroached upon by the little horn, the false doctrine in the scarlet beast system (the final result of Nicolaitanism). The little status itself is of many small instances rather than one overarching superpower. (It is also a system of self-governance, as a Church incorporated to be separate from the beast.) If every new convert must approach the scarlet beast to learn of Christ: as it forces their rejection by Christ ("beheading" the unity in the seven cycle for the witness of Christ), the situation becomes such that "no flesh can be saved". The beast is the dragon's "choke hold" on the faithful.

In the Church itself, the leavening by OT Israel's false doctrine is found (perhaps begun) in support of the secular nation state of Israel's religious primacy. The little horn speaks of great things (war, peace and blessings of growth) and has "eyes like a man" (seeing only the closure of the static subgroup and not the whole octal, cf. a serpent) which covet an inheritance, and judgement over a land not their own (Zion, the inheritance and choice of Christ). In Christ they would have both separated, they lose one or the other.

The statement "All that dwell upon the Earth shall worship him" is astounding, for the beast uses its devices from the sea. I am so told that these devices are very strong, as the practice of the logical dialectic is now endemic, global and totally corrupt: it extends to an outright rejection of all that is called God.

Jesus has firm words for those deceiving His elect using the dialectic. If they extinguish righteous faith, they too are spiritually dead, and if they hold them "hostage", they must go into captivity themselves. Antichrist is not a wisely taken position for the elect or sinner.

Rev 13:10 He that leadeth into captivity shall go into captivity: he that killeth with the sword must be killed with the sword. Here is the patience and the faith of the saints.

9.8) A Beast Come Up From The Earth

Rev 13:11 And I beheld another beast coming up out of the earth; and he had two horns like a lamb, and he spake as a dragon.

The woman had fled into the wilderness and the two witnesses also minister in a place prepared for her. Here I find that those dwelling in the "Earth" (the woman herself proper) that escaped are those that reside in the wilderness as within the 30 octals of the witnesses' ministry. That is, they dwell within the coset of the static subgroup as "in the Earth".

I exclude those in the earthly elements of the Sun octal and **under** the lamb (in the Sun with which she was clothed, as with the full Sun), for reasons that will become clear later. (She only needs to rest in the Sun octal, and not to map through or from it.) This omission of this single coset of the static subgroup has bearing later on in the calculation of the number of the beast.

This spirit "hurts the oil and wine" employing the dialectic characterising the dragon's speech. He is a deceiver and satanic minister to the crowd that would be the "Church". His "two horns" are not natural but synthetic and like the lamb in that they accumulate, making a mockery of the words of Christ. Christ taught that one must "exceed the righteousness of the scribes and Pharisees" to be saved. The second beast has this inverted, as compared to Christ here:

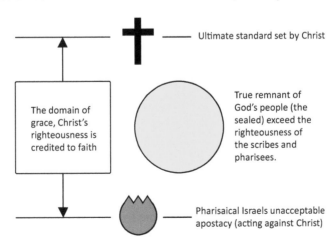

Ultimate standard set by Christ

The domain of grace, Christ's righteousness is credited to faith

True remnant of God's people (the sealed) exceed the righteousness of the scribes and pharisees.

Pharisaical Israels unacceptable apostasy (acting against Christ)

The above is correct to the gospel, but following below is the false prophet's counterpart to these two strengths of doctrine:

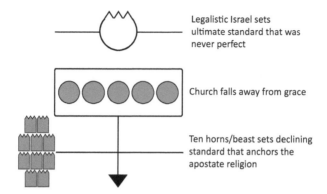

Legalistic Israel sets ultimate standard that was never perfect

Church falls away from grace

Ten horns/beast sets declining standard that anchors the apostate religion

The right hand of Christ covers those under grace (Christ intercedes with virtue) that please God with faith. Christ's left hand corresponds to the false prophet's right (who has them switched one for the other). Then the election by Christ as in the right hand octal (unity – the lamb) is made war against with the image of the ten kings, with the beast "rising up from the pit". The two horns "as a lamb" are a "counterfeit" reversal of the left and right hand octals.

The horns are the authority seized as Satan gains his foothold. They are the first two of three woes. (The third is the death blow to the Church when there is none able to be saved by it and "Hell" as a system is realised complete.) The first constructs the "body" to corrupt the doctrine of those caught within a five cycle (and not convicted by the angel's circuit) with the synagogue of Satan, leavening as a sixth. The second would justify the first with wormwood (consolidating the image system in dialogue), and all whilst the scarlet beast rises up out of that same bottomless pit, "a lake of fire burning with brimstone" (Rev 19:20). Once each horn is deployed there is no going back: it is a "great falling away" (as from grace and of scale).

Aside from the ten kings which are part of the image (see later) I now return to the scripture:

Rev 13:12 And he exerciseth all the power of the first beast before him, and causeth the earth and them which dwell therein to worship the first beast, whose deadly wound was healed.

The "power" of the first beast is the liberty granted the Church to exercise their faith. That freedom is guaranteed by the authority of the first beast. The false prophet spirit here is able to practise deceit freely: not only poisoning the well of the gospel's message with false doctrine but to require that other inhabitants of the Earth (the remnant of the woman's seed) must employ the same devices to freely operate. It supplies them with just one means of freedom (the dialectic process) that "forces all" to accept that their freedom to worship comes from self-government and not from God. (Which is clearly heresy and worse, it is bearing the mark to them, so they may practise the dialectic and have the number of the beast's name, 666.)

The beast's deadly wound is healed: the manner of this healing takes a great deal of explanation. The exit of the Philadelphian elect is a deadly wound to Satan's schema of damnation. The grafting in of pseudo-Christian cultic legalism, whether of Jewish (very unlikely) or satanic deception of historical origin (most surely likely) of "those who say they are Jews and do lie" is enough, in order that they should seize the elects' inheritance in Christ.

Those dwelling in the Earth (practising the dialectic) are deceived into constructing a "Church government" (the scarlet beast) modelled after the republic with its electoral college. Decisions for the majority are deferred to a minority as with one mind. In setting up the self-governing church, the worship is "of the beast" as well as in God's eyes, "to the beast". The scarlet beast does not represent the majority – there is no union of prayer in heaven for them – the beast is not "for the Church". With its deadly wound healed the scarlet beast has set up the first beast as equal to the kingdom of God, as a blueprint for the bride of Christ.

9.9) Doing Great Wonders

This deceiver is not beyond claiming work of heavenly authority is according to this (his) plan to offer OT legalistic (as Israelite) obedience to the Church as "salt" instead of revealing it as leaven. (Only Christ is the salt that makes the salt of the Earth salty again.) This bait and switch, "obedience for leaven", is become the means to his end – the destruction of the Church. When God reveals His wrath, the false prophet claims it is to "save Israel". Perhaps he is right, but Israel is a spiritual election by grace and no longer a right by bloodline.

Rev 13:13 And he doeth great wonders, so that he maketh fire come down from heaven on the earth in the sight of men,

By claiming such miraculous divine wrath is "all part and parcel" of his (or the devil's, the god of this world) schemes, he deceives the Church (whose collective spirit he is, once the bait and switch has occurred), he suggests that they build a structure or idol (an image) upon (and to) the dialectic system of the devil's construction called "the beast".

In appearance to those without proper discernment "fire" (which is the devil's schema of S5 or S6) acts upon the churches, and "heaven" has become the collective in the scarlet beast (the position of the fourth "king" i.e. with the vote, in place of the "sabbath" in Christ). It so appears that the fire comes "down" from heaven, but instead the churches merely follow as "hell" after.

"Fire from heaven" I equate to the seven cycle that burns on the groups of the octal (John described the cycle reading right to left acting bottom up, whereas it acts downwards reading left to right): this is then complete reversal. The inverse of the cycle upon the other hand is symmetric in action but if the two additive groups were aligned in isomorphism the fire would appear to "burn down" from above to below using the one seven cycle for both. The false prophet, then, is not without his tricks, convincing the beast (collective) that he has some power of divination as to what happens next, and how to prepare for "great tribulation" (he constructs an "image"). By this re-alignment of the four static bows collected together as wormwood he does a "great wonder". As each row of wormwood corresponds only to unity, there appears to be the inverse of the seven cycle floating unity through the rows as does the circuit of the angel through the churches, albeit illegally made in wormwood as if it were in amalgam "fire coming down from heaven" in the sight of men, yet not in truth.

The argument is made that the triples required for the first seal "paradigm", and all the seal devices (as antichrist), require such an octal structure in reversal to the Sun octal. Then "fire" (the proper operation of spirit forming a group) is seen in its inverse to "come down" from heaven (i.e. the inverse by reciprocal or **difference** as "down **from** heaven" as 1/x, say, with election or "rest" as unity) on the left hands in the sea of glass, and upon the unity (heaven) itself. The requirements of obedience (virtue) then seem apparently (or relatively) "equal" with the dialectic (as wormwood) of the triples in the earthly elements, aligned in the "sea of glass" (now forming groups as if "fire" in the sight of men, that same "sight" limited to the static subgroup (or jacinth portion of wormwood) rather than the whole octal).

The Sun octal is reversed to the false prophet's purposes and the surety of election in Christ (heaven, unity) is preserved but is then also mapped through the eight left hand octals, enabling the dialectic (as the false prophet is the one present in the dialogue having all religious authority and not the God of scripture that is disgusted with the beast's application of it). The false prophet will argue the left hand static triple is actually of the right. (The ten kings and image then supplant the election in Christ as under a five-cycle.)

By affirming the left hand static triple acts as the right (in wormwood or otherwise) the three static triples that form a synthesis in the "Earth" (of the possible left hand(s)) instead supplant the three groups cycled static which form a "tree" in the reference octal (water is exchanged for blood). "Balances" then oppose "trees" and the setting of virtue is exchanged for the dialectic (appearing "closed" in the Earth) that would itself pretend to be positive (as virtue).

"Fire" is then substituted for "hail" as seen in the first trumpet and so appears to fall (come down) from heaven (unity) upon the Earth (always mapping between triples in cosets, the complement of the right hand's subgroups), as upon those remaining elements: triples (paired with unity as if it were floating) that would generate (implicitly define) the left hand octals and dialectic bows (now also static by squaring) under the seven cycle's inverse. Christ did not state there actually was a left hand octal in this passage: He states clearly all the way through the Revelation that there is only one reference octal in a "sea of glass".

The false prophet seems to have a group structure ready to hand, but he constructs his speech (the dialectic) upon it, and truthfully, the "group" structure which actually appears there (which is valid), is no basis for the dialectic. (The pans lightened of unity in the balances cannot be static as they both contain unity!) In the sight of men, there is possibility for deception.

The false prophet "spins" (flip-flops between) hot and cold (hail and fire) and sits in the centre lukewarm, by constructing lies (balances) to place as a sticking-plaster on the nullity (virtue privated) in the middle. In the same manner, arguing one God of **both** hands He would align these sets of groups (his tabernacle) within both sets that number 666 to left and right and:

Dan 11:45 And he shall plant the tabernacles of his palace between the seas in the glorious holy mountain; yet he shall come to his end, and none shall help him.

The result is one of spiritual poverty: none shall rescue him and his own with help from heaven. The "freedom" to align one's paradigm to any one of eight left-handed octals, instead of the one right hand, is deceptive when the dialectic is employed to enquire of God.

The beast is fourfold with the dialectic. The image is a synthetic *fifth* dialectic. The human heart may logically satisfy the structure that cannot be physically instantiated as both analogous to "both logical disjunction and conjunction – which is nullity". The devil is far below incapable of creating anything both as cause and effect, a real "noumena"[9] or something in itself.

Excluding the Sun (the "glorious holy mountain") to complete the *image* that numbers 666 (see the next chapter), the image is "grafted" on (laid over) the structure of the first beast (a hideous insult to God). Replacing what should be His chaste bride for His Christ; the image system thoroughly leavens false doctrines from their cultic source into the Church whole.

Rev 13:14 And deceiveth them that dwell on the earth by *the means of* those miracles which he had power to do in the sight of the beast; saying to them that dwell on the earth, that they should make an image to the beast, which had the wound by a sword, and did live.

The miracles the false prophet displays are not predictions of what will happen – though the beast (collective) believes so. (The miracle is that something so diabolical can mask itself as holy, and the "image" system seals the trap.) They (the ten kings) are consulted as the "authority" on God in all things by the beast's self-government. (As "Nicolaitanism" it is a big thing to watch out for but the time is regrettably far too late.) In respect, the false prophet suggests that the "Church" needs its own government rather than to mix so directly with the state.

The wound by a sword to the first beast refers not only to the exit of the elect by Christ's command, but from the contradictory position "in treaty" (dialogue) formed from those conflicts over which the USA has placed itself in an overarching position – as sole superpower of the now globalist collective. (The four "primary powers" (beasts of Daniel) project their authority into the USA as mover and shaker of that collective.)

The false prophet cannot create a living image: rather he systematically creates a structure (image) with which the balance of the human heart (that simply approaches pleasure, avoids pain) and the fifth and synthetic dialectic of the image agree, and then this carnal engine becomes nothing but the extension of the heart of man, sinful and full of pride. This synthetic model of the carnal animal with five senses, operating in such a fashion, also separately (without the same reckoning) numbers as 666.

After this manner the image is given "life" (pneuma). Those that cannot or do not so worship the image (by process) are pushed out, and left in a state where they are powerless to change this engine once it is included. Being "put out" is the same as "death" if it destroys all faith: faith will not survive without sound doctrine – such unfortunates outside perish in not having oil as a wise virgin in the parable as taught by Christ.

Note those that "dwell on the Earth" are those who make the image (dwelling in wormwood in the sea of glass, they are under the second woe). They are not elect, and are considered to be those elements of S4 on the earthly elements under the fallen star over Sardis that completes the schema to make the "great city" S5 (albeit with 12 elements missing in the construction). Likewise, note that K4 is a normal subgroup of S4 which is clearly not constructed from the seven-cycle or by the proper operation of the four earthly elements.

[9] Immanuel Kant, *Critique of Pure Reason* (Penguin Classics), p260.

Rev 13:15 And he had power to give life unto the image of the beast, that the image of the beast should both speak, and cause that as many as would not worship the image of the beast should be killed.

The image prophesies as expected, from the heart of man (numbering **660 + 6** – see the next chapter) rather than from the mouth of God. Not to agree to disagree (to hold one's ground) is to be excluded from all dialogue by the facilitator, to have no say in doctrine which should never have been up to debate. In the dialectic process, having one's virtue nullified (the oil hurt, sent to the empty set) is to be "killed" in that all virtue is to be excluded as divisive.

The conjugation of the last six elements upon the "manifold" (see below) confirm the method only; opposition by any one church of six is examined upon a transposition separately to the objections of the other five and is naturally outnumbered by action of that conjugation and the method "wins". (Their views are "killed" as the collective is steered to consensus.) With nothing now sacred and the truth up for grabs by the false prophet, the result of not following the crowd is to be spiritually put to "Death" in the process and this is done as if by the agreement of the whole collective all the while those within remain fervent believers.

> **Manifold:**
> *The interface between the carnal "heart" (processing information from the senses in a manner permitting one's evaluation of the environment), and also the separate understanding to judge and decide one's own place and action in it.*

Rev 13:16 And he causeth all, both small and great, rich and poor, free and bond, to receive a mark in their right hand, or in their foreheads:
Rev 13:17 And that no man might buy or sell, save he that had the mark, or the name of the beast, or the number of his name.

The second beast causes all that would worship in the Church to accept the token that the government both protects their right to worship and certifies it as valid. In other words, it does so saying it owns it and can tax it. True fellowships are going to be very hard to establish if religious communities are left to "freely flourish" but also to draw together in dialogue. Likewise, faith will become a private matter, and will be extinguished if it cannot be transmitted by hearing. Those that operate the Church as a business, all "balance sheets" and "profit and loss" through faceless, soulless corporations, are those in view. They will make themselves the only "recognised body" of Christ, by extinguishing the true remnant.

That no one might "buy or sell" (trade and exchange doctrines) is a statement compatible with the parable of the ten virgins; those that "buy and sell" doctrine were never invited to the wedding supper. The scarlet beast is a system of religious commerce, and that commerce is the exchange of ideas (actually commerce of false doctrines). If the beast will only approve new doctrine when it is voted upon, then no one can buy doctrine for their fellowship from the beast but those overarching churches with the mark (those members accepting all false doctrine as voted in, "following as hell after") and none can sell doctrine to the beast unless he is either an agent of a member church (with the name) or that he persuades the collective to vote it in (has the number).

9.10) Numbering The Antichrist

Rev 13:18 Here is wisdom. Let him that hath understanding count the number of the beast: for it is the number of a man; and his number *is* Six hundred threescore *and* six.

The last verse is where I tread off for now, in order that I may more perfectly contemplate the scripture. I require a great many tools to justify the interpretation of the text, and need to apply some group theory to an idea.

What I may do now is number the beast (antichrist), for in excluding the Sun octal in its entirety I may number the possible "bows" or triples for which the heart may substitute (in the system of four contradictory "bows") in any "fourth seal" device, doing so in a fairly readily obtained way.

There are $(7*6*5) / (1*2*3) = 35$ ways to choose a triple from seven symbols.

Likewise, there are seven symbols and triples forming subgroups in the "Sun octal", so I subtract: $35 - 7 = 28$.

I do this to exclude the Sun octal (the reference) from my count.

For every remaining octal there are four choices for the unity element "1" ... So $28*4 = 112$.

Lastly, one such correspondence is the original Sun octal, so that its own left handed "bow" corresponds to its own unity – a triple in correspondence to one subgroup from the set of seven I subtracted and removed. So then, $112 - 1 = 111$. The seven triples in the reference (Sun octal) are excluded: the one triple held "static" is a vacuous case, the other six cannot be arranged as triples in the left hand, rather than unable to be held "static" also.

Now, over each such "triple to unity" correspondence there are twelve such seven cycles that are so permitting, pairs of which are shared in six octals overall. So, holding a=1 and [0,b,d,f] static I have by elimination:

(a,b,d,g,f,c,e)	shares the same octal with	(a,b,f,g,d,e,c)
(a,b,d,g,f,e,c)	shares the same octal with	(a,b,f,g,d,c,e)
(a,b,d,c,f,g,e)	shares the same octal with	(a,b,f,c,d,e,g)
(a,b,d,c,f,e,g)	shares the same octal with	(a,b,f,c,d,g,e)
(a,b,d,e,f,g,c)	shares the same octal with	(a,b,f,e,d,c,g)
(a,b,d,e,f,c,g)	shares the same octal with	(a,b,f,e,d,g,c)

So I have a factor of six;

$111 * 6 = 666$. As across the whole "sea of thirty octals".

Now, I have not numbered the image system that agrees with antichrist; rather I have numbered the nullity filled by the heart outside of the K4 form of the ultrafilter. In the scripture at the seventh seal there is "silence for the space of (about) half an hour". (I.e. A coset of a K4 static subgroup.) This is the difference between the Father and Jesus Christ. This can then only be empty! So, 666 is the number of antichrist (a "man") and is a "lie" (nullity), being the number of blasphemy.

There is also some issue over whether the number of the beast is 616.

If that were the case I could initially follow the same path of $(35-7)*4*6 = 672$.

Then I note that $672 - 56 = 672 - (8*7) = 616$.

Now what does the product $8*7$ refer to? There are clearly eight possible left hand octals for the reference Sun octal, each with seven choices for unity (floating) or "bows". That said, $672 - 56 = 616$.

But then I have removed far too many bows! As each of those eight left hand octals (now as right hands) also have 7 other 2nd generation left hands with them remaining in the "sea". (I have then discarded too many "bows".) This is not to remove possible multiplications, but possibilities for additive groups underneath multiplications. Fixing unity is not the same as fixing a static subgroup. I indeed removed far too many bows:

Returning then to counting different C7 groups, each C7 group is unique to just one pair – of reference to left hand octals; I can then float unity in seven positions in the seven cycle.

In removing those eight other octals because of one possible shared octal (the reference) I remove one whole C7 group from each of those eight: this leaves us (knowing each C7 group is valid on only two octals) with $672 - 8*7 = 616$.

I removed completely just the one octal from the whole 30. Then eight groups of seven cycles themselves should be removed (as each seven cycle is valid on only two octals). Then if I remove those eight C7 groups I render the eight octals (the left hands) as having seven C7 groups each and not the full eight. I am then missing as if it were the "circuit of the angel", when I delete the reference octal and with it those eight C7 groups. So 616 would appear not to number antichrist but the "complement" of the angel instead: Have I numbered the beast as it continues 42 months?

No, I have not. I, by removing seven possible triples from 35, indeed excluded the octal that contains all seven triples as subgroups, but I must count every octal that contains seven, three or one of those groups elsewhere in the sea. (I have again counted some groups more than once.) In fact, I count, $1 + 7 + 14 = 22$ such groups in full (respectively as sharing seven, three or just one subgroup(s)), i.e. an extra 21 "new" occurrences. These new occurrences share possible triples between them (not in the reference octal) so suffice to say I counted too many "bows". (The subtraction of 56 is not commensurate. I already removed the seven triples of the reference; by removing 56, I counted many of them more than once.)

(The number of octal subgroups to remove from 56 taken from each set of the 21 "new" occurrences not in the reference are: $(7*4)$ and $(14*6)$ respectfully. Those counts of 28 and 84 bows are not distinct and properly number 28 and 21, together making 49.) In order to count the number of bows I should "add back in" just use $672 - 56 + x = 666$. (So, $x = 50$. I had removed the reference octal twice, once by omitting 7 triples from 35, and once from the count of 56.)

672 is the number of fields isomorphic to GF(8) that contain a given static triple in the four earthly elements inclusive of the Sun octal. Each of the eight C7 groups with seven choices of unity would hold static a different subgroup, and three powers of each such element would hold that bow fixed. I removed too many "bows" from the complete 672. Those I counted twice are among the set of fields that share the additive group with the Sun or reference octal but not necessarily the same static triple as the Sun octal. (I have counted everything not in the additive octal rather than everything not in the field in which the angel makes his circuit.)

Then I have deleted from the count everything consequent from the Father's additive group rather than the number of "bows" outside of Christ. If I again include the Father's additive group and count the number of bows, those 56 must be included and a smaller subset excluded on different principle. In the Sun octal I previously calculated 35−7=28 so I could exclude every subgroup of the reference octal: this has the consequence that to remove those 56, I should have properly made commensurate the count upon only those choices of unity possible − not the full seven but only four. And then, not eight C7 groups but only those that do not hold static the reference field's (one of the four possible) static triple and its static subgroup, which then fixes unity.

Then I may count $672 - (8*3)/4$.

I must exclude the three positions of unity invalid in every octal (those in the static bow) and not the count of seven and must divide that result of 24 by four, so that I remove only the one possibility for the static subgroup (as I already removed this: I counted it already in removing 7 from 35). One of those 4 bows remaining after unity is chosen is always the "static triple" that is paired with one static subgroup in the reference field which was excluded; I need not remove the other three cases.

Those 56 choices for "bows" are therefore immediately added back in and I cannot remove them all: they are not commensurate with the removal of "bows" in the reference octal. Then I return to simply removing 6 from 672 as before.

Also, were I to state that there are eight C7 groups over the reference octal, forming those same eight left hands implicitly, then remove the seven elements of each C7 group of the eight (8*7=56), I have once more removed too many and I counted the unity element seven times more than it should be.

In any case, 56 bows are too many to remove. The Bible teaches that God is able to preserve His words. It also states, "Let him who hath understanding count the number of the beast" − it (the calculation) is not transmitted intact by word of mouth. Likewise, you cannot believe everything in every version of the "Bible" when the Bible itself tells us to be careful with which scriptures to choose!

Chapter Ten: The Manifold Under Temptation

This section was formed from a study (using only abstract algebra) that gained clarity from a read of Immanuel Kant's "Critique of Pure Reason".[9] The terms "apperception" and "manifold" were taken from that tome to clarify the connection between my study and Kant's seminal work. I tried to expound my study in light of Kant's text rather than to insinuate my study or algebra into Kant's far superior examination of mindful progressiveness – as I read through, I did not find any sufficient logical reason to disregard the original abstraction in my study itself.

Much of my study of the scarlet beast was built upon content from Dr. Gotcher's radio programme on the Christian Media Network (website: www.christianmedianetwork.com). Both his "Diaprax"[10] and James Lloyd's "Dragonspeak"[11] tape sets upon which I began gave way to too little study and instead much thought: thankfully a read of Kant assured me that my study is (apparently) valid. I found no connection between the three automorphisms or elements of the K4 form over the four choices of unity/static "bow" in the octal (or the four beasts before the throne) to Kant's twelve categories: whilst beyond my requirements (nothing jumped out at me or essentially provided such an alignment to manifest itself) they appear coincident by appearance.

> **Apperception:**
> *The ability for the understanding to use the manifold to process input from the senses in terms of desires and judgements (concepts) in a manner where those desires and concepts are processed as to their worth before and after action is taken upon them – so that the understanding is simply perturbed from rest rather than forced to re-examine everything as for the first time.*

"Apperception", then, is the result of the composition of transpositions in the senses (as yet to be described) and the system where a group may be so conjugated by products of transpositions inverted before and behind provide what I term a "manifold" in several forms. The result is simply the ability to construct a replacement in the human heart for the fourth and logically contradictory balance of "Death". (That is the "Hell" following after.)

10.1) Apperception – Simple Transposition

The senses determine the form and sensations of objects: in fact, the properties of extension into space and of time (required a-priori for objects to be perceived), allows the individual a-priori the inference that they exist because they can think. That said, human beings have only an imperfect contingent existence.

As I experience the world around me there must be a difference between my understanding, which is linked to my imagination, and the raw input of my senses. (Because I can say of myself "I am a thinking thing": I have a separation of the soul from nature.)

From the senses the imagination forms concepts both new and old (the latter as upon past experience), and since the same imagination or "concept" is subject to the reasoned understanding it is fact that what is conceived as true and exists – the "concepts" of the objects

[9] Kant, *Critique of Pure Reason*, p104, 124.
[10] Dean Gotcher, *The Diabolical System of Diaprax* (The Apocalypse Chronicles).
[11] James Lloyd, *Dragonspeak, The Language of Lucifer* (The Apocalypse Chronicles).

sensed are formed by experience and not as "a-priori" knowledge (which is deduced without raw experience). Knowing one's environment is not a-priori possible – that said, there are no universal "natural laws" of human reason. The environment has no "law of the intellect". (The soul is also separate and separate with free will.)

"Apperception" is the mechanical description of this, of that which may be sensed immediately – inferring that the object is something desired or to be avoided. In that sense it belongs under the reasoning mind – the understanding, and not equal to it.

Such concepts are fixed, near-permanent truths and are at the very least as "ideas" and to be strengthened or thrown out as they are found useful (for the individual's benefit) or useless (as not). They are built from experience, not from thought (the senses drive the desires) – it is the correct concept that reason had itself approved that determines actions. Poorly formed concepts are then an excuse for bad behaviour, but much more so the **inability** to form accurate concepts of the environment about us (as people under higher authority).

I may immediately consider an object of sensual "apperception" as an enjoining of the mind to a sense. The transposition (smell, taste) is to be taken to indicate both "that what is tasted agrees in smell" and "what smells agrees in taste".

The transposition of mind as (consider, taste) is to be taken to indicate both "that what is tasted has been desired" and "what has been desired has been tasted". That is, only transpositions taken in pairs that annihilate make any reasoned sense. The transpositions singular are the quanta of perturbations from rest as the mind considers an act.

The mind cannot consider raw input from the five senses in apperception (a set of transpositions (smell, taste) etc.), only transpositions formed by the mind with a sense. Those applications of "mind" as in (mind, taste) are truthfully impressions formed from what is or was once experienced and that built concepts (upon ideas). The transposition (consider, taste) is then in the mind and not raw in the senses only: it is only in understanding separated from the senses by the imagination. It is that imagination that is perturbed, not the reason and understanding.

The product (consider, taste)(consider, smell) = (consider, smell, taste) adds to each transposition with "what was smelled was considered and what was considered was tasted". Equal is the following: (smell, consider)(smell, taste) which adds "what was tasted was smelled and what was smelled was what was considered". "Considered" is a bit of a stretch and made on past experience – but not made from knowledge a-priori. I may desire a taste from merely considering the smell – or touch something because it "looks" fun.

Rom 7:7 What shall we say then? *Is* the law sin? God forbid. Nay, I had not known sin, but by the law: for I had not known lust, except the law had said, Thou shalt not covet.

Inverse to (smell, consider)(smell, taste) is the following: (smell, taste)(smell, consider) which adds "what was considered was smelled and what was smelled was tasted". (Any product of transpositions composes, forming similar statements bridging each neighbouring pair.)

Why should these two be inverses? For continuity the "mind" element of the transposition (in its place as "consider") is missing from the transposition (smell, taste) and that most certainly, that element "consider" or desire was satiated if (smell, taste) preceded it or it was deemed positive if (smell, taste) came after on the basis that it was considered a delight to smell.

If any cycle as a product of transpositions can arise in apperception, may any permutation also be formed? As products of disjoint cycles in the manifold they may. I am left questioning, "Does this even matter?" After all, what is certainly true, is that the mind does all the thinking, and the cycle (consider, smell, see, taste, touch, hear) has meaning, but what of the product:

(smell, see, taste, touch, hear) (consider, smell)

Does it have meaning if (smell, see, taste, touch, hear) is without any mindful consideration?

Consider (taste, smell) as a transposition (annihilated upon what exactly shall be seen later). If the imagination is coupled to the senses, any desire to touch something that is seen in truth may be considered, as the understanding making judgements upon the mind's reason is not coupled to the senses directly. The transposition made here (taste, smell) could be considered "raw apperception" where the mind has no experience nor reasonable input to offer.

It is in all sense of the word, a "lust" (if there is none other for it). For all intents and purposes, one without mindful reason **covets** the taste because one desires the smell, and may do so because the logical part of the imagination tied to the understanding is uncoupled – else the mind is fully coupled to and as raw as the senses. Then to justify the desire for the smell or the desire for the taste is meaningless; man would be a beast: a constant slave to dopamine.

Transposition, (apperception) is enough for Pavlov's dog to salivate, and enough to make men drool also. What is man? Merely an animal with power to think and reason. There is no way to fully divorce the mind from the senses except than to ignore every reason my understanding offers from a-priori judgements (as to external authority) with which I resist or ignore all my coveting (as if disobedience) and so all mindful restraint. (I would only show myself an animal and sin mindlessly.) To fully separate the mind from the senses is to be without mindfulness of the environment, a contradiction to "any thinking thing" – everything must then be tasted by the flesh rather than nothing: it is the only alternative to that "death" of that "thinking thing".

10.2) Identity – The Unity Of Self

If one truly "thinks", the thinker must exist. The mind is a near imperceptible tool (cf. Gen 3:1), empty only when it is considered to rest upon itself: it cannot be mindful of the unity[4] such a thought would bring in self-examination. The inability to consciously self-reference one's own silent thought is the very unity that the mind was hoping to consider, indescribably. It cannot be mindful of the stillness of the imagination whilst the imagination is in motion upon it.

What of considering transpositions? When I talk of a desire in cycle form (consider, smell, taste) I also consider the inverse (here of two new statements bridging those two transpositions).

A. What was smelled was considered and what was considered was tasted.

B. What was tasted was considered and what was considered was smelled.

How are A = B^{-1} inverses? The desire to smell in A is satiated with the consequence of B. That

[4] Descartes, *Meditations and Other Metaphysical Writings*, p58.

is, with or without the initial desire to experience the taste in A or B. Then the transpositions forming those two "order three" cycles A and B are in product;

(consider, smell)(smell,taste)(smell,taste)(consider, smell)

Partly annihilate and reduce to

(consider, smell)(consider, smell)

Only, and these also annihilate.

These cancel as (consider, smell)(consider, smell) returns the mind to its unity, its rest. The unity or identity of "mind" is without motion in the senses or the imagination and is the baseline of reason. It remains or returns under normal operation to being completely undisturbed.

However, I denote the method by which I may act in permutations upon the manifold (the interface of the coupled imagination between the senses and separate understanding that forms my rational concepts), the very order of their mention in product unwinds naturally in the consequence of their fulfilment and satisfaction.

If this were not so: that I were somehow unable to complete their annihilation to unity by some interruption, I would simply wipe my mind clean with thoughts of the same cycle until at most some power rendered the desire inert. "Sleep", it would appear, is necessary to the proper operation of the manifold – the mind unwinds after a stressful day.

10.3) A Mind Closed – The Manifold

If desires are a permutation of the senses, then I am left with the statement, "What permutes the mind amongst them?" Clearly it is possible to separate the manifold's raw sensual input from the understanding that may override it, and importantly I should be able to construct any element of S6 over the five senses and my logical inner sense that judges what is positive and what is harmful: for in experience I may intellectually approach pleasure and avoid pain.

The manifold, then, is a sensual barrier between the world and us: until of course harm comes in the way and stops us thinking altogether, the end of any philosopher who can't stop thinking!

Whilst I consider the mind as continuously forming disjoint cycles and multiplying them to annihilate "lusts" by conjugation of the unperturbed imagination to result in "rest" or unity, I cannot state that such carnal desires (raw in the five senses), divorced from thought and reason, obey any a-priori "sensual" laws under the sway of the imagination whatsoever, as I cannot form laws a-priori with the imagination but only through reason as in the understanding.

As lusts (smell, taste) etc. are continually thrown before the way, only those deemed to be considered logically may be factored into transpositions in the mind and to distinct (but not disjoint) cycles "raw in the senses" which annihilate to "rest" and unity. I need the mechanism whereby the manifold may bring to rest the consciousness or reason undisturbed whilst chaos descends around us in the senses, because when one stops thinking, the world still exists.

Every raw input of the senses separated from the understanding is a member of S5 formed from transpositions in the senses upon the manifold. By conjugating the set S5 the mind absorbs every element in only the five senses (raw in the senses) and consider only whether that manifold S5 of the senses is positive. Then the mind "evaluates the environment".

Such conjugation leaves a disconnected string of transpositions in (consider, sense) etc. that in like manner as the senses, permutes the mind to unity when relaxed, as S6 is normal in S6.

The manifold, however, is considered complete when it is made a synthetic judge of what is to be considered experientially as "good". As the synthetic "halfway house" between environmental self-evaluation conjugating upon S5 (raw in the senses) and the direct and free annihilation conjugated about unity, I examine absorption into A5 by conjugation of elements of S5 about A5, and further transformed through subgroups of A6 by transpositions in "mind" conjugated about that same sensual manifold of A5. (As with $A5 \cong e$ in $S5/A5 \cong C2$.)

10.4) The Lust Of The Eyes

If I divorce my reason from my imagination and merely follow after whatever thing I have previously thought made myself happy – as from some concept or otherwise – then I have made for myself a lust that needs satiating. In essence I have a transposition or n-cycle purely in the senses and I have inverted it and conjugated my manifold by it. Yet I did so by associating beforehand the positivity of the manifold with the satisfaction that comes after such satiation! In that thought I have processed myself with the inverse of my completed desire, more so than were I to have happened across something on the way and dealt with that lust directly.

An example of such is the beguiling of Eve in the garden and the serpent's subtlety of "Yea, hath God said, Ye shall not eat of every tree of the garden?" The serpent beguiles Eve by inciting her to factor down by A5 to a transposition, or else place herself in an unknown situation where she must imagine the inverse with which to annihilate that transposition as if by free apperception. In either case, by getting her to treat that one tree as every other tree throws her off balance, forcing her to attempt to return to a state of rest in her environment by apperception with or factoring by A5 to the transposition (consider, taste). (It then follows as a result that she becomes similarly able to consider lusts pure in the five senses also.)

Eve immediately is thrown off balance and yet, still, engages in the dangerous conversation:

Gen 3:2 And the woman said unto the serpent, We may eat of the fruit of the trees of the garden:
Gen 3:3 But of the fruit of the tree which *is* in the midst of the garden, God hath said, Ye shall not eat of it, neither shall ye touch it, lest ye die.
Gen 3:4 And the serpent said unto the woman, Ye shall not surely die:
Gen 3:5 For God doth know that in the day ye eat thereof, then your eyes shall be opened, and ye shall be as gods, knowing good and evil.

Eve places herself within the possibility that she will keep her eternal life she has already – the continuation of her manifold person ahead of eating the fruit, but does so on the condition that she has tasted it. She sees no conflict within her environment: after all it is just another tree with nothing but the free choice of the mind to obey the commandment, she is young (I assume) and healthy and not liable to die soon!

She had set herself to eat the fruit by removing herself from her God given manifold: that same free rest formed of conjugating S5 in the senses with elements of S6 (evaluating her place in the environment) or of free annihilation as if about the identity element.

Her manifold was altered in permitting her mental faculties to rationalise the act without considering logically the difference between "thou shalt" and "thou shalt not". The food is analysed as "good to eat" and the mechanism of her carnal "lust" gets fired up in a manner she has not experienced before: she is beguiled by this.

Dialectically I may form the following (in r, s, etc. there is no virtue in the dialectic):

a   Eve yet alive (as if "r")
c   God's commandment kept intact (unbroken as if u^{-1})
e   Fruit good to eat (as if "s")
g   Desire to make one wise (yet to be received as if v^{-1}).

Now, all these are positive together, yet the virtue in the octal (outside of the static subgroup [b,d,f]) is clearly "c" were it not for sin, as Eve is beguiled and reasons on the basis that wisdom in "g" is as virtue. (To become like God.)

Then dialectically a&c $\vee$ e&g holds with God's curse repeated by Eve in Gen 3:3, yet the serpent convinces her otherwise. Eve instead finds there is "life" found within the knowledge of good and evil. However, Eve did begin to reason dialectically, and there is introduced the faulty disjunction between a $\vee$ c&e&g whereby the proof of a $\vee$ c by the serpent's dialectic logic is found by e&g. I.e. a&(e&g $\rightarrow$ c^{-1}). The commandment has then become of none effect and Eve then will (in her own reasoning) certainly "live". Then the appealing synthesis in the centre is then the continuation of Eve's life, now with that wisdom and the knowledge of her disobedience. Eve is clearly beguiled by that serpent's logic.

Her "lust of the eyes" then is the conjugation of the person G with xGx^{-1} (a manifold) where x is some single cycle of products of transpositions made in conjugation of G. Now, G can be an A5 or S5 subgroup of A6 or S6 with logic, or it can even be A5 or S5 with products of cycles raw in the senses. She would recover her unity of mind and settle back to rest after, with one fault: the commandment which was her seal of understanding. Her mind will not be the same, as the conjugation has only one rationale: sin.

The manifold springs into action as Eve judges the case before and after and the reader begins to see her self-justify her coveting the fruit by the "lust of the eyes" with a second mechanism.

10.5) The Lust Of The Flesh

The manifold S5 in the senses allows the mind rationally, albeit dialectically, to consider the absorption of the lust into the manifold as so to reduce the cycles in conjugation to only those in (mind, sense) etc: reducing the product to a rational one only with no heed for the carnal lusts. Indeed "I am fearfully and wonderfully made".

Now xHx^{-1} results in an element of H=A5 conjugated by a single transposition in some (mind, sense) only. (Where "x" is a single cycle, a product of transpositions.)

A5 is a *normal* subgroup of S5 with 60 elements of the full 120 and conjugations of A5 by elements of S5 are always absorbed into A5. This has the synthetic result of the Hegelian dialectic. It is no longer a logical dialectic with "Yes, free to eat" with no manifold required. (Conjugation of {e} – the identity by cycles raw in the senses.) It is not now a didactic statement "No! Resist eating", where the manifold as S5 should reduce such carnal lusts to a rest state of mindful consideration.

No longer is it even true dialectically as a "1 = true", vs. "0 = false" paradigm, but now, there is an "almost true, almost false" paradigm in the centre where A5 is conjugated in the senses by transpositions that may not even be members of A5 or S5, but have odd cycle length.

A5 is not only normal in S5, A5 is a **simple** group. There is no factoring down to another normal subgroup within A5 (other than to {e}). This would be in order to provide another "sample of the senses" to read "positive or negative" within A5 as if by another manifold. Conjugation of this manifold is a system that rests upon conjugation about {e} only. Conjugation upon {e} is only "pure apperception" that is conceptually free and without consequence. In comparison to the environmental self-evaluation of conjugation about the manifold of S5, A5 as a manifold is "between total freedom and properly mindful consideration" and using A5 as a manifold in S5 may be misread for "pure apperception" when considering cycles raw in the five senses.

The factoring down of S5 by A5 to a transposition I may equate to a "lust" of a transposition raw in the senses. That element being considered would be the cofactor of A5 in S5 in factorisation, where A5 is the manifold in use. (Eve considers the case before and after.)

There is also the correspondence to the dialectic balances of the serpent's logic shown previously that so beguiled Eve in her innocence.

Consider again the dialectic formed of:

    a   Eve yet alive (as if "r")
    c   God's commandment kept intact (unbroken as if u^{-1}).
    e   Fruit good to eat (as if "s")
    g   Desire to make one wise (yet to be received as if v^{-1})

Again it is seen how Eve has weakened her position on the commandment of God (as turning away from a&c): she had factored down to a transposition (taste, see) (reforming the commandment in "c" to one of no effect in her own understanding) raw in the senses and her manifold A5 (coveting the pair "e&g") had picked up the slack, so that in eating she would gain a secondary layer of apperception (sin as I would know it). In knowing that sin (as to proving $a \lor c$ from $e\&g \rightarrow c^{-1}$) the very difference between good and evil, Eve would be able to decide for herself what should be "good" or "evil", on a purely "approach pleasure, avoid pain" basis. The fruit would not make her wise, but the process was new, she desired to know of it.

Given e&g, there follows by the dialectic $a \rightarrow c^{-1}$ and since in sinfulness c is not considered as a virtue, it is not true that c^{-1} must be of **necessity** found non-positive in the dialectic. (For if c is as virtue, then "a" would or must never be found as a positive property: Eve then "dies" or $c \rightarrow a^{-1}$.) The "pride of life" continues to assert that sin is as positive as the sinner. I.e. I discover $Pos(a) \rightarrow Pos(c^{-1})$. Eve's life and person would apparently continue as positive, untouched by any punishment; she had falsely judged the words of her creator. (For He had surely prophecised that the dialectic would spiritually kill her as does $c \rightarrow a^{-1}$, that His commandment was equivalently intact, and always remains so. He obviously knew the dialectic was the evil that would cause the fall of man.)

Gen 3:6 And when the woman saw that the tree *was* good for food, and that it *was* pleasant to the eyes, and a tree to be desired to make *one* wise, she took of the fruit thereof, and did eat, and gave also unto her husband with her; and he did eat.

When the woman perceived ("saw") that the fruit was just as any other fruit and since she had known no difference in the manifold of A5 to S5 as she had before, she was indeed "beguiled" (to use the dialectic) and though disobedient in that moment, she had no knowledge of the difference to her manifold that disobedience brings (as consequences from authority above) and reasoned that in her **flesh** (the manifold as A5, upon cycles raw in the senses), in which she considered to dwell the eternal life and approval of God, her person, her flesh was superior to the commandment by the synthetic state of "almost true". And so she ate.

What if in wormwood (the device of the serpent) Eve faced the decision to decide not the properly formed disjunction a&c ∨ e&g but a&g ∨ e&c? Eve may have been beguiled by assuming that God (meaning to keep all wisdom for Himself) had lied to her that the fruit was "poisonous" contrary to its appearance. She would "become wise" and God's commandment could be freely broken. Beguiling indeed, but offered as an excuse?

10.6) The Pride of Life

Now, although the manifold A5 may absorb the transpositions pure in the senses, it does not do so for those transpositions in both the mind and senses. (The manifold of A6 is not, by its nature, to be considered "synthetic".) In fact, A5 is not a normal subgroup of A6, but transpositions in the mind of S6 will in conjugation by (mind, sense1)*G*(mind, sense1) (where G is a subgroup isomorphic to A5) exchange within each cycle or element in G that has the symbol of "sense1", to become exchanged with "mind" in its place. The mind now has six sets of elements, all conjugates of G (A5) in A6 rearranged: yet this still gives no rest.

I.e.

(mind, see)*(touch,taste,smell,see)*(hear, taste,see)*(mind,see)

Evaluates with swapping "mind" for "see" so becomes equal to:

(touch, taste,smell,mind)*(hear,taste,mind).

And "see" is swapped out for "mind".

I have formed six conjugates of A5 in A6 by the transpositions of the five senses in S6 with "mind". Remember: don't forget the sixth, the identity element sending A5 to itself!

Now, A6 is also a simple group (as are all An : n>4) and so walking A5 through A6 is likewise a "halfway house" – one in which the mind by "sin" has processed itself toward the negative of the freedom from lawful condemnation by one's creator, and has built for itself a paradigm between that logical freedom and the irrational excusing of one's own actions.

In essence, **dialectic logic** is employed without justification to agree that (in the carnality of the self, the created "sin nature") conjugation upon A5 and so on further within A6 by transpositions is as valid a method of proof of that which is "good" (desirable) as the word of the creator. The subsequent conjugations in A6 allow this process upon the manifold to justify each sense's titillation in a synthetically logical fashion. This leads to the senses overriding the mind or the mind a particular sense – on a purely "approach pleasure, avoid pain" basis. By swapping a symbol within a cycle in the senses (pure lusts) for mind, mind agrees with each sense in the lust by employing the same process: I find sin is then **ongoing** in "the pride of life".

Sin in the dialectic (with the very word "ongoing" as used above) indicates that any synthesis made of sin becomes the new thesis in every following judgement. (The conjugation above reveals the refusal to see one's self as negative in one's own environment instead of faulty within God's own environment.) The dialectic process is a rejection of that one God, and His curse the judgement upon it. Every sinful act following after that one initial sin enters by the same "pride of life" where the assumed positivity of the sinner's continuing life, the same flesh and person of that sinner indicates every positivity is to be found in the freedom to break the commandment(s) of God. (This was shown dialectically as was a $\rightarrow c^{-1}$ beforehand.) This is a fact hard to hear, especially by those claiming they are good people and in the right! Yet in that middle position between lawlessness and life there is that same captivity to sin.

There is now a "lukewarm" position between what was once liberty in God's eyes, that which is without need of consideration by the manifold, conjugating $\{e\}$ (the identity), and that other pole of "sin" which is no longer merely wrongdoing but is an application of the manifold beyond its purpose. As a middle ground it seems up to discussion and able to be put under analysis as to its place in the environment because human beings state "I own it" (sin) and their place in the environment also:

The mind aligns to rather than agrees with the senses by conjugating A5 in A6, producing the six conjugates that constitute the "pride" of life. (Only because I state that I "own" my mind do I likewise state it is always under my control – even when the mind is without its due restraint) I extend that into my environment: not only a physical environment raw in the senses that all share under God's authority but one in which I had also stated I was as a creation superior to the intent of my creator. It is a spiritual environment that is polluted most completely, and it is called sin. (A mere excuse blackened by every conceivable evil.)

10.7) Bittersweet Partners

Earlier, as John took the little book from the hand of the angel, John was given the answer to the question (without knowing the question to ask before) as to why "bitter and sweet" are not opposites but are alike in (consider, taste) as apperception. The text simply hinted that the difference between (consider, "taste bitter") and (consider "taste sweet") could be placed as "pleasure" or "bitterness" in the mindful part "consider" leaving us with (consider, taste) only.

After that manner their product (consider, sweet, bitter) is really only (consider, taste) (consider, taste). I would assume that these annihilate also with conjugate partners to rest.

I find a difficulty in the manifold, as to the bounds between the imagination that forms concepts and the understanding that frames ideas. I assumed that there is a good understanding of "sweet" and "bitter" and that I may add those upon "consider" before weighing up any desire to "taste". There is no difference in the understanding between sweet and bitter in this example, for the imagination causes the mouth to salivate and not the conscious desire itself that decides one's preference for one over the other – to taste the sweet or to avoid the bitter.

I am left with (consider, taste). I ask: "Do I consider the union of sweet and bitter or do I weigh the decision to taste with mindful awareness of the apparent opposite?"

The imagination which is coupled to the senses provides the drive to taste, as well as the repulsion to avoid – but the latter is only learned through experience and is still a concept for the imagination rather than the idea of bitterness and distaste (which although they have become ever present are reasoned from the understanding instead). Distaste is something categorised rather than something ascribed to an object: all learn and fear from their mistakes and the natural world with all its animal lures has taken great advantage of the unwary.

It is better for us to take note that ideas (ever present in the understanding) whether of sweet or bitter are not enough to make us salivate or fast, but only the imagination coupled to the senses forms the concepts that leads to either of those impulses. Then ideas are become far less a drive and the senses yet more so to the extreme: yet the understanding with an idea in agreement to the drive of the imagination itself coupled to the senses for pleasure alone is the most dangerous combination. Under temptation, the manifold will sign its name to near any act, and the person that lives like this always is but an animal. (Note I am not a genuine stoic yet. By "stoic", I mean that I have not justified this human condition as able to be overcome.)

I would consider lusts as simple apperceptions raw in the senses, with separate mindful considerations only in the associated antagonisms formed from commandment of higher authority, past experience, fear and mistrust. I leave the senses to purely examine quanta of positive perturbations from mindful rest so that the mindful (the self) will sensually approach pleasure and rationally avoid pain: the latter is something the senses never accomplish before it is too late. That paradigm is then become the basis of learning and of framing concepts into ideas. (And if they correctly frame such antagonisms, I assume they are all healthy.)

The separation of the will to the senses, or of the understanding that weighs the imagination (coupled to the manifold) as with the past experience of concepts (the "ideas" held), is a process where the senses themselves actually have the upper hand: they have the centre position where the manifold votes two to one that pleasure is preferable to pain (of S5 to A5 and {e}). The understanding has past experience and only votes the true or reasoned desire, the senses do not think but provide drive, and the imagination which is of the heart (I am told) is "Desperately wicked, who can know it?". The carnal mind will only find agreement with itself (in the dialectic centre ground (A5) of the manifold) on post-examination. In conjugation, mindful rest permuting A5 through A6 is only attained by futile self-examination with nothing learned but the carnal mind methodically chewing its own gum.

10.8) The Image Numbers 666

Clearly there are 6! = 720 elements of S6 and therefore also 5! = 120 of S5. That stated – there are 60 elements of S5 not in A5. When conjugating the manifold past its natural order of 120 elements (S5) down to its insoluble normal subgroup A5, I dropped 60 elements without losing a synthetic state of "almost true". For that I only require the absorption of S5 into A5.

So 720 – 60 = 660

Then lastly, I have the six elements that define the conjugates of A5 in A6, the transpositions (mind, sense1)...(mind, sense5) and the identity itself where A5 is left untouched.

Lastly, these elements number the total 660 + 6 = 666.

The scripture states "six hundred, threescore and six" – is this too literal with the use of "and"? The scripture also states "Let him who hath understanding count...". If there are two answers obtained two different ways, is it no surprise the scripture says to defer judgement to the one who "counts" the number: i.e. to him that understands? (Let him that calculates have his own way, here as 660 "and" 6.) I will simply state both methods result in 666 for brevity.

Any reduction from splitting n-cycles in S6 into cycles raw in the senses and transpositions with (mind, sense) will simply reduce to conjugation of A5 by some single transposition (mind, sense). So, I only count the elements of S6 that I actually required for the sinful human heart. That results in 666.

It is important to note that I appear to have counted the last six elements twice! (The elements may also act upon themselves, excluding them from the manifold is actually not implied.)

It is then pertinent at the same time to ask why the elements in S6 outside of S5 are included whole. The "self" that comprises the senses and the imagination but not the intellect is as S6, and I may conjugate the 660 elements (not in the coset of A5 in S5) as H in the centre of gHg^{-1} with elements g in S6 preserving all those elements. It is become as if the elements in S6 outside S5 were mapped "onto" the 660 elements and the elements inside the "hole" of 60 elements in the coset of A5 in S5 are absorbed "into" A5. (Every element of S5 is absorbed into A5, and only those elements in S6.)

It is as if the manifold of S6 with the hole is acting as the identity and the conjugation was acting as pure apperception about it (as if on the identity "e". I.e. $geg^{-1} = gg^{-1} = e$).

I have a "hole" of 60 elements that I may "walk through S6" under conjugation within its "conjugates" in S6. (The final six may instead be due to the image riding the beast: see [13.2].)

Evaluation of the environment conjugating about S5 gives way to pseudo-logical self-evaluation (including the imagination) in S6. The "hole" simply denotes the "heart" of the human animal that is continually implementing such conjugation of the manifold. (Geared for pleasure only.) The last six elements separate out only those (few) elements required in S6 that conjugate A5 through A6. These are placed in conjugation and not present in the "middle" as H and they act as if they were the understanding attempting to justify acts of the imagination.

The "hole" in the group S6 leaving the collection H numbered to 660 may be walked through S6 by conjugating elements of H by any element in S6. There are only the elements in S5 not in A5 that are "absorbed" into the conjugation of A5 (and those in A5 itself). The "hole" of 60 elements is "moved" by conjugation with those last six elements over S5. In that minimal case (required for mapping A5 to conjugates in A6) the heart under temptation that numbers 660 may be "walked through" the imagination (as the pride of life) by six transpositions as before.

There is a somewhat *clouded* issue with the numbering of the manifold to 666 given in this chapter (as walked through A6 in S6). The text instructs us to "let him that hath understanding count the number of the beast". (The one that models understanding in the octal?) It becomes apparent that the method is not passed on by word of mouth, or that there are two methods that agree in their count, or that there are two differing subjects. The number of the beast is the number of antichrist "bows" that the human heart may substitute for. That the heart also numbers as such is cause for severe (grave) concern. The image system and the carnality of the numbering to 666 as "660 *and* six" is coincidence, yet A5 is truly the "least order" example of an insoluble group that also provides a synthetic "balance" for the "Hell" that follows after.

The final *six* elements added generate the orbit of that "manifold" in A6 or S6 under conjugation. (They are not in the set of elements forming the collection of 660 to which I added them, as the elements may act upon themselves also.) The "image to the beast" of three kings from ten facilitating the collective's vote (applying the "mark of the beast" for the wider collective as in their place), deciding all doctrine sits in judgement over all as if they were Christ. They do so in the place of the seventh (at rest or at ease, the seventh king is a "chair" and may sit out the vote) aside from those six sets of false doctrine, but that is left for later (chapter thirteen). The heart numbers as Death or "666". "Hell" following after numbers to 660+6.

With regards to the manifold in the more general sense: were it possible to supplant the priority of one's mind and "think" with one's senses, so that it could be true that pure lusts were equal with mindful consideration (as if a mere beast), it could be inferred that a new-born also has this state, were it not that there would then be no sin in it for a new-born!

Each sense with the mind may be symmetric in the "manifold", so that all six conjugates of A5 in A6 are as worthy of evaluating the environment as those raw in the senses. Then I have found six "manifolds" where $(m_1, m_2)*M_i*(m_1, m_2) = M_j$. Now I could sinfully process myself before any sinful lust and return to my steady state after the sin had run its course. I can pre or post-multiply by a six cycle and obtain the same effect as conjugation with transpositions.

The process numbering 666 (the mind of the beast, the "son of perdition") is formed of conjugation before and after, as $(m_1, m_2)*M_i*(m_1, m_2)$ [covet, lust, sin] $(m_1, m_2)*M_i*(m_1, m_2)$ where the conjugation before and after numbers 6, and the three stages of sin in the process number 660. (666 is the number of the manifold M, as conjugated by transpositions.)

Then I would have then conjugated each separate manifold M_i with (m_1, m_2) both before and also separately after, having processed myself with the dialectic, swapping out "mind" for a particular sense and swapping back afterwards to "rest" in the "pride of life".

Now, M is therefore the whole mind of the "animal" or beast (its manifold), and the conjugation of the manifold by the last six (as in the pride of life) then agrees in whole with that of the first (also conjugated) as (m_1, m_2) and as part of the mind M, I may conclude that 660 + 6 = 666 as the final six are every bit as natural to the animal as the first six. M will absorb every element into the manifold (as before, the 660 elements of M itself), and these extra six are absorbed likewise: so, 660 + 6 = 666. Then I conclude the last six are in M also, $(m_1, m_2)*M_i*(m_1, m_2) = M_j$. M then becomes a set (or rather a process), that numbers 666. I see this as symmetry in the scarlet beast system – there are seven mountains on which the woman sits, and six (churches) follow as hell after that one mind. Yet one of the six is the source of leaven, or "lust".

I find "M" as the mind of the "son of perdition" (a mere beast) abhorrent in comparison to the will of my creator: now, I am stoic.

In that manner, the image system that numbers 660+6 is almost a corollary as "following after". It uniquely fills the vacuum of consent to the beast to form an insult of Satan to God. If it weren't for the fact that this is done openly already, that acceptance of any "new doctrine" in the image system can be put to the vote and approved as if it were beneficial (as gospel) or necessary "for growth" with only the promise of that vote always in the lead (and holding the leash), I would keep it hidden from view myself to prevent anyone attempting to construct it.

2Th 2:3 Let no man deceive you by any means: for *that day shall not come*, except there come a falling away first, and that man of sin be revealed, the son of perdition;

The "Son of perdition" is this same model, revealed by this study. There is no other assertion to be made from Revelation, and this is clear (surely sinners were openly revealed from the beginning?). The "man of sin" is then simply a "mathematical model of a man", and that's all.

I also use the label referring to those that facilitate the dialectic (one is for the other, to employ the model), and that it (he, the "mind" of the beast) will eventually turn on its body. The result is that its "sins" are required of those that had followed afterward as if "agreeing with hell".

Isa 28:15 Because ye have said, We have made a covenant with death, and with hell are we at agreement; when the overflowing scourge shall pass through, it shall not come unto us: for we have made lies our refuge, and under falsehood have we hid ourselves:

The Image Would Speak For God – The False Prophet Promises "Great Things".

That series of actions (where growth becomes the foremost goal and sound doctrine falls away) begins with simple steps within facilitated meetings along the lines of:

1) **We** want growth
2) What should we do **for growth**?
3) What can we do **for growth**?
4) What do we decide to do **for growth**?
5) It's fine, **we** decided.

These steps occur when the voting collective would speak for God (which is sin), but that is not the danger. The danger is found in transferring spiritual "fire" from God to the "false prophet", by employing the same dialectic process in the ongoing and facilitated dialogue, thinly disguised as a "consultation" for group consensus. (The image to the beast is "set up".)

As these steps are fulfilled (and in all denominations independent of their incompatible doctrine), it is farcical to assume "God decided" in step 5. God would not agree on a model of self-governance approved only by collective vote that "appears valid to give growth" for every type of Church independent of their incompatible doctrines, surely? Would He agree when that same model puts (transfers) the power into the hands of a third party, the false prophet?

Thesis vs. Antithesis Brings Synthesis – An Example

I give an example where a didactic pair of opposites may approve the positivity of Christian relationships. The "thesis" of Christian fellowship also meets the "antithesis" of relationships outside of such faith (as within the dialectic) that would facilitate growth by compromising on doctrine. All this is done in order to put the believer and non-believer on an equal footing (though I know from before the middle ground is null).

When relationships become more important than doctrine, I am assured the heart steps in to reason rather than logic. If the emotions or the "heart" deny virtue (especially good doctrine as if by a "false prophet"), it is a given that they cannot be healthy in those relationships.

The "purpose"[12] paradigm I note is a fundamental shift in effort and attitude that emphasises the continuous building of relationships where there is an agenda for growth in the Church. That stated: the agenda requires the minor fellowship to compromise at any means to get that job done. Collectives, in dialogue, are given the option to vote (as false prophet) upon it.

The problem of "compromise" sets in when the issue of doctrinal purity is "taken off the table" whilst growth is considered.[12] Then will God Himself supposedly keep the doctrine intact? Not whilst it is under the collective vote. There may be much sound doctrine even then, but not the spirit of God blessing[12] in it whilst that false prophet spirit holds the reins.

Doctrine takes a second place to growth, which is that thing Jesus hates, it becoming "Nicolaitanism" – compromise and fault would near excuse each other with the believer sat speechless in the middle. (a, c etc. are taken from the octal whereas r, s etc. are from the coset of the K4 form enthroned in the octal – the paradigm of antichrist; worldly, without virtue.)

a   those elect in fellowship with "Individual A" sharing good doctrine (r)
c   the believer (Individual A) acting as apologist (s)
e   requirement to teach "Individual B" continues ("edification" not present yet as u^{-1})
g   every such "Individual B" preached to by A (not yet converted/dismissed as v^{-1})

I would expect the exercise of virtue in "p" as to conversion where the decision of B to follow the doctrine of A is absolute. (I first assume good doctrine, rather than that which is after become leavened.) u&v is positive, it is good to spread the gospel and to make new converts.

But $a \lor ((a\&c)^{-1}\&\neg a \to e)$. If the elect do not **freely** act by sharing good doctrine then B may continue unedified whilst $(r\&s)^{-1}$ remains positive: "neither cast ye your pearls before swine".

[12] Rick Warren, *The Purpose Driven Church (Zondervan)*, rear
cover, p56, 62.

So, I also have the disjunction in the same octal of: $c \lor ((a\&c)^{-1}\&\neg c \rightarrow g)$.

(If the apologist "c" is inactive and $(a\&c)^{-1}$ is in scope, B will not yet be dismissed unconverted.)

I also infer from the octal that if $(e\&g)^{-1}$ (i.e. by u&v, a convert is made) is found as virtue then there is no excluded middle. That is, individual B is not "brethren" (sharing good doctrine in $r\&u^{-1}$) with individual A before the successful conversion of B. I.e. B is edified only by a good apologist: $\neg e \rightarrow Pos(c)$ or $\neg c \rightarrow Pos(e)$. (The pending edification "e" of B is a burden for "c".)

That is, a good apologist will (having edified a convert in u&v) move on to convert others. Any converts are made (as in "¬g") only by sharing good doctrine: $\neg g \rightarrow Pos(a)$ or $\neg a \rightarrow Pos(g)$.

Luk 9:5 And whosoever will not receive you, when ye go out of that city, shake off the very dust from your feet for a testimony against them.

Dialectically I have that good doctrine is excused its proper action and the believer lightens the balances of the third seal either side to show that the relationships before and after are irrespective of good doctrine.

The compromise of doctrine places the believer as facilitator "c" and astonishingly, $a\&c\&g \lor a\&c\&e$ becomes as $c\&e\&g$ such that $Pos(a\&c) \rightarrow Pos(e\&g)^{-1}$ in the middle, and given $Pos(e\&g)$ I find that $Pos(c) \rightarrow \neg Pos(a)$. I.e. doctrine plays no further part; the salt has lost its savour. The set of all believers are then no longer a "separate" or "peculiar" people. (There is no good separation between them and the unconverted, the believer "c" has equal relationships with all: cf. the doctrine of Balaam.)

I.e. $Pos(e\&g) \rightarrow Pos(c^{-1})$, and that was found given $Pos(a)$. Given the lukewarm have relationships irrespective of doctrine, it follows that in having that equality in their relationships, $Pos(e)\&Pos(g)$ immediately entails by the dialectic (in the faulty ultrafilter of the earthly elements), that $Pos(a) \rightarrow \neg Pos(c)$, and "c" privates "a" also $(Pos(c) \rightarrow \neg Pos(a))$.

The disjunction then becomes only in the four earthly elements, so that I find $a\&c\&g \lor e$ as well as $g \lor a\&c\&e$. Essentially the friend invited to Church doesn't see himself as Christian and the believer agrees to ignore this fact. The friend enters the situation lukewarm, and this denial of virtue shows some small fruit, but only as the product of "Nicolaitanism" present by the "core" of believers deciding that this denial of virtue (doctrine) is an acceptable offering to God. So, absolutes may truly hurt feelings, but only when the friend does not have the kingdom of God within them, the axiom of sovereignty (see section [5.3]). Such conversions with that axiom first and principal are made to God rather than as is common: as the lukewarm attending only by holding such relationships.

What can be said is relationships are positive, and it is also positive to share the good news to those I know, but it is not good to nevertheless continue a relationship with one that tramples the gospel underfoot. What if those ones **are** the "core"?

I now move on to the Revelation text once more – I will study the image system of the false prophet in great detail in the passage on the judgement of the Church – now the scarlet beast system.

Chapter Eleven: The Sharp Sword With Two Edges

There is a clear line between the saved and sealed and the unsealed, none of which are saved. The beasts rising up from the Earth and sea give way to John's vision of the throne and the sea of glass before it. The saints in Christ dwell in that "sea of glass" with a pre-first seal paradigm: they have the Father's name in their foreheads, and are following His Christ faithful and true.

The sea is in truth just one field: the alignment of every field's own algebra across all thirty octals is an answer to a deception, the seven symbols in every possible octal of thirty reveal no non-trivial intersection between them, only isomorphisms instead. The seals and trumpets show it is possible to walk "not in the spirit" just as it is possible to truly walk "in the spirit".

Why the difference? Why mention them? In respect to shifting octals I find unbound a continuous movement within the one octal – the Sun or the sea of glass. That is, every element of the octal is employed to map or shift positions, except for unity (rest) where those dwelling under the covering of Christ placed their faith. The first four seals and trumps always refer to those on the Earth, describing the excuses for logic employed to justify their "ways". (Which are used to follow the leading of their hearts, their "flesh" as the apostle Paul would put it.)

The text details the separation of the elect from the others in the Earth, until enough is enough and the dividing line is drawn. The Earth is then cut off from any further salvation, the elect upon it separated (once sealed) from the unsaved. The result is a timely separation between sealed and unsaved. The unsealed in the Church are trod in the winepress with the ungodly (the judgement is set) the result is discernible – believers will know them "by their fruits".

The judgement is a sure-gone conclusion for those alive on the Earth; all that remains is the resurrection of the just and unjust and the damnation of Satan and all his ilk to the "lake of fire" whilst every soul receives its eternal reward at the judgement seat of Christ.

The winepress is the method by which the unjust are shown sinful to all – they operate the devil's devices and are made to do so in only the one octal: always contradictory. They operate as the "lake of fire", a "small sea" morphed into the image with the scarlet beast using "a strange fire". The vials of wrath complete the woman (now bereft of any of the elect) to hold captive all those in what is classically called "hell" – those that employ the devil's devices.

Picking up again from before, the scripture digresses from the imagery of the beasts to show the black and white distinction between the elect (saved) and those that are not so chosen.

11.1) Those That Overcome

John returns to the sixth seal, as the angel reaches Philadelphia (overcoming Laodicea) before the final trumpet. Revealed are God's elect in Christ, "Mount Zion" (a true stronghold of faith, the K4 form GF(4)): i.e. the sealed without the beast's mark. They exited the Church, not seeking their justification from government (the beast).

Rev 14:1 And I looked, and, lo, a Lamb stood on the mount Sion, and with him an hundred forty *and* four thousand, having his Father's name written in their foreheads.

The 144,000 are clothed with the righteousness of Christ, sharing the inheritance of the elders (occupying the other 6 positions for floating unity in the reference, 6*24=144). There is a factor of 1000 present: 1000 years (representing the patience and grace of God to the elect).

The sealed sing a new song: it is not one of questioning (with doubt entertained as to whether one could be certain of their salvation whilst still alive). These fortunate sealed are happy enough to finally have that answer and to know their God (the Father of Jesus Christ) and rest.

Rev 14:2 And I heard a voice from heaven, as the voice of many waters, and as the voice of a great thunder: and I heard the voice of harpers harping with their harps:
Rev 14:3 And they sung as it were a new song before the throne, and before the four beasts, and the elders: and no man could learn that song but the hundred *and* forty *and* four thousand, which were redeemed from the earth.

The voice is not singular but (as in Christ) of the collective in heaven under the lamb (as many waters), united in thanksgiving for heaven's move to save them (as of a great thunder – the days are cut short, the seals, trumps and vials align completed on the last day). This collective united voice is in praise of God, a new song. None else have that song but those sealed: they know they're saved, with the name of God in their foreheads (their understanding).

The sealed are they who exited from the churches of the letters, coming through the "open door" and were even those "few" that did not defile their robes as their churches became in bed with (or modelled on) government. They left when convicted, for the least was instructed:

Rev 3:4 Thou hast a few names even in Sardis which have not defiled their garments; and they shall walk with me in white: for they are worthy.

I find the context of that scripture; I shift to the saved, having exited and separated themselves fully from the corporate institutions of religious commerce that pass as for Christ's bride, as places of worship today. (They are not part of the sixfold image, not "defiled with women".)

Rev 14:4 These are they which were not defiled with women; for they are virgins. These are they which follow the Lamb whithersoever he goeth. These were redeemed from among men, *being* the firstfruits unto God and to the Lamb.
Rev 14:5 And in their mouth was found no guile: for they are without fault before the throne of God.

The elect exit as their fellowships are in turn associated with the "Church" (the corporate template of "MYSTERY BABYLON" – see later) under the image (the false prophet) not remaining in the Church, but leaving it behind, preserving their election. They exit to approval, through "an open door no man can shut" – they know their God as "firstfruits" of the resurrection for greeting Christ on His return, they are certain of their salvation: they have it as "wise virgins".

Not practising guile or "stealing" their way into heaven by some other means, they are not counted as righteous because they do not attend "Church" (as unbelievers also do not!): they are truly the elect. Before God they are accepted: to Him as clean as the lamb slain before God, with the righteousness of Christ. The least overcame: God will not permit any to be deceived now Satan has no claim of sovereignty or of accusation.

11.2) Those That Are Cut Off

There are three angels which represent the soon to be completed sequences of the seals, the trumpets as well as the final sequence with the seven vials of wrath poured out: concluding with the seventh of each sequence on the last day. I align (after the action of the seven thunders) each angel to the opening of seals five, six and seven.

The first flying through heaven states:

Rev 14:6 And I saw another angel fly in the midst of heaven, having the everlasting gospel to preach unto them that dwell on the earth, and to every nation, and kindred, and tongue, and people,
Rev 14:7 Saying with a loud voice, Fear God, and give glory to him; for the hour of his judgment is come: and worship him that made heaven, and earth, and the sea, and the fountains of waters.

The first angel flies in the midst of heaven: the dwelling place of the elect with the gospel (of only one Lord, one field GF(8) – the Sun octal), edifying all believers (in the four categories again, as of choices for unity in the octal upon the earthly elements). They are to prepare for the last three seals, the first of careful obedience to God – for once the angel's circuit is complete and witnessed the great city is to be the recipient of the wrath, and as the sealed of the sixth seal (the fruit of the two witnesses' ministry) they are to give God glory as the firstfruits for Christ whilst the sequence of trumpets is completed. With the seventh seal opened the swift judgement of God arrives with the completion of His wrath on the last day.

Rev 14:6 I equate to the need to stand firm against the devices of the seals and trumpets. Their order of mention reflects "believers", the "righteous" and "tares" ("peoples" either reflects the elect, who already have the gospel or else unbelievers rejecting it) which require repentance (fear), proper honour (glory), and God's judgement to the lake of fire respectively.

The four parts of the sea of glass (the one reference octal remaining) are "heaven" (floating unity), "Earth" (cosets of the static subgroup), "sea" (every octal isomorphic) and "fountains of waters" (K4 groups of K4 subgroups in the sea also within the reference or Sun octal). Note God did not "create" the static subgroup, as God has His only begotten Son and was perhaps Himself once as the K4 form of the filter. Those in the list are creations and outside of Christ, even though Christ may be "sat" as a subgroup.

The "everlasting gospel" references the virtue in Christ of the one Sun octal applied across the wider "sea" of thirty – and that the dialectic devices that failed against the two witnesses preserve the structure of the trinity to act freely as they wish. The fifth and sixth trumps are subject to, and under God's own authority and those spirits or "workers" of those devices are far less strong than He. In honouring God as to give Him the sovereignty over every octal and to deny the devices of the adversary and stand upright as the two witnesses is then to worship Him "that made heaven, and earth, and the sea, and the fountains of waters".

It may be that heaven is under full restraint, the Earth simply as with one choice for unity, the sea traversable with two and the fountains of waters are all that remain with three. (The two witnesses are raised on them, however.) The contradiction of Death is not something that is physically created, as the raised will discover at the judgement. "Death" is not the gospel. There is always a way back, before it becomes too late.

Fountains Of Waters:
These are K4 groups of subgroups acted upon by a seven cycle not valid in the octal in which they originate (but another in which they are present). Their singleton intersection is a "fountain" branching to those groups that under the seven cycle generate the remaining triples (subgroups) of the octal upon which the seven cycle is valid. Fountains are "in the Earth" – the singleton is usually the unity from the earthly four of the octal. (Cf. "trees", as not under a seven cycle.)

Rivers:
These are the multiplicative orbits of "triples" of elements, usually a (static) "bow" (rivers are in the "Earth") aligned to unity as the head of the cycle. I consider them orbits of subgroups as if triples formed the octal rather than singletons. Rivers, then, are under a seven cycle as upon one column of the earthly four (of triples in wormwood). Fountains instead correspond to the intersection of the three K4 groups in regress to the static subgroup (corresponding in the orbit to unity), which in truth is the same as the K4 form or "golden girdle" under a seven cycle.

The gospel states God's overarching sovereignty and all should follow from the "liberty" of God's perfection, whether in action or upon the sufficiency of His past works, God reveals Himself as He sees fit. It is a "sin" to require His liberty from Him when grace has been provided in full already. God, however, will not remain inactive forever.

The angel "flies" and as before, wings represent maps from an element or set of elements to an equal state elsewhere. This angel in the "midst" of heaven is beneath the stars (static subgroup) and above the Earth: that is, aligned with the unity in the middle between static triples. The gospel is even the essence of faith on God's liberty (sovereignty) to do it all His way. The message may substitute for the word of the kingdom sown by anyone, without fault.

This angel "carries the gospel" to those that dwell on the Earth, that the mystery of God is to be completed (soon after the angel of the church overcomes) and the last chance is rapidly approaching for the already elect to separate themselves from the Church system before it receives the judgement of God Most High. This period is aligned to the ministry of the two witnesses after the fifth seal, and directly compares with the "way of the kings of the east".

The call to fear God and to give Him glory occurs with the last chance to repent. The gathering together into bundles by the workers of the harvest (angels or ministers) nears completion and as the wheat is separated from the tares the final judgement arrives, whether those left in the collective beast are ready or not. One should fear God, and defer to His word even if it seems unnatural: it is the better thing. In that thing God may be freely worshipped His way.

The seals are open (the loosing of all restraint), completed with silence for "half an hour", and the knowledge of the last name in the book of life at the seventh seal of "sabbath rest".

God is the one who chose His own triune nature (and the heaven, the scene of 24 elders echoing the election (floating unity) of the 144,000); virtue and obedience and the evil (of the dialectic) opposed to them; the sea within which believers dwell spiritually and in which they are separate as the wheat from the chaff; the sources of all waters in the Revelation, the triples and groups of subgroups separate from the earthly elements in the sea in the trumpet devices. God is God over all, not only His elect – all will come under His judgement.

The second angel references the last trumpets – those devices placed or "cast unto" the Earth.

Rev 14:8 And there followed another angel, saying, Babylon is fallen, is fallen, that great city, because she made all nations drink of the wine of the wrath of her fornication.

Rev 14:8 mirrors Christ's command to His elect: "Come out of her, my people". Every wise virgin comes from somewhere! (A "church" in a five cycle – the "nations" or Laodicean amalgam.)

The trumpet devices construct the beast (the "great city") ascending from the bottomless pit to replace the true people of God (spiritual Israel), replacing that remnant with what is to become the "Church" – the image built (copying the template of the first beast) upon the false doctrine of its members: wed in all but name to the beast (yet that too in the eyes of God).

The world "did drink" from her wine and now is gone mad: there are now presumably five nations in the dialectic process and one must always be in perpetual and clear contradiction to the other four (the first nation being even the template of "MYSTERY BABYLON").

For the beast to prosper it has been made "mad" using the dialectic process as does also the Church (now made a cage of every "foul and detestable spirit" – the system of "Hell" at the fifth trumpet) made so side by side with it as if its consort and made to operate toward that Church as the dragon would "operate" (speak). The wine of her favours has been drunk by all nations through accepting the beast her consort, and in turn herself. She offers wine as if her company is easy: she feigns liberty but instead is rotten company, with consequences.

Babylon is fallen twice over – elevating herself to a collective of four (as under inclusion) disagreeing in doctrine but met in consensus (under the dialectic process) they are in continual contradiction, as they will not strike a member out as contradictory. The template of incorporation is valid for all. When churches form the image they meet together as one of, say:

A. As a family (unincorporated)

B. Other "churches" of other families

C. Other denominations of compatible doctrines

D. Other faiths or gospels calling themselves similar

E. Other spiritual (as all other) nations clearly distinct

F. The "MYSTERY BABYLON" template.

The individual then faces a choice (whether to compromise for the sake of the collective). The contradictory "church" (one of B to E) not consistent with the others is solely "plugged" with the MYSTERY BABYLON template "F", denying all virtue as a "nullity" in itself, spiritually dead. That template, which all with the mark of the beast "agreed upon" heals the beast's "deadly wound" in this setting. Families are the only remaining battle-ground within which divisions fall. (To be a church-goer or to be a Christian.)

Mat 10:34 Think not that I am come to send peace on earth: I came not to send peace, but a sword.
Mat 10:35 For I am come to set a man at variance against his father, and the daughter against her mother, and the daughter in law against her mother in law.
Mat 10:36 And a man's foes *shall be* they of his own household.

Each inclusion forces a compromise (a doctrine essential to the faith of the group is nullified). There is denial of the truth of God's "word". As every member fellowship compromises in undertaking the template of incorporation the difference between them and it is not to be dialectically agreed upon: the divisive doctrine must be removed. Ascending the chain of collectives met in dialogue to agree together, I posit the fourth inclusion is not one that can be consistently (virtuously) agreed upon; its doctrine (unity or virtue) must be both included as it was once "wine" and must also be excluded as it is now "oil" as the three prior (which were as "oil" – excluded doctrines) were the only content of it in its "bow": is it acceptable or not?

Each possible choice of wine to incorporate the fourth fellowship was excluded from the prior three in a compromise (of doctrine or virtue) at some point, and any choice of oil to exclude for consensus was once used as wine in incorporating a previous fellowship of the four. As the fourth is likewise dialogued to consensus, then there is symmetry that everyone must not just agree to disagree – but with themselves and also the method. (They must consider their own doctrine unworthy.) The only solution is that one of the previous three fully absorbs the fourth to continue its operation. That, however, is not on the table. (It (merger) may work for money, but for the Church?) Instead of finding unity in Christ, they worship the template.

At the fourth stage of inclusion there should certainly be a discontinuity as it would require the doctrine to be weighed and three (nullities) lies at one table to consider it – there is a "hole" or nullity introduced somewhere that cannot be agreed upon; Satan's house is divided. The dialectic method is a falsity – it is no system of absolutes such as the gospel or another.

Like the USA the image system sets itself up as a Church government, and it becomes easy enough to predict that the image steps in to ensure itself as facilitator and later the policeman of faith worldwide. The elect best prepare to be "evaluated on doctrine". (Unless they have sound doctrine and can display themselves reasonable within their faith, not a cult doctrine.) Sobriety of faith is rare these days. The solution? Do not sign up for assessment.

The third angel swiftly follows;

Rev 14:9 And the third angel followed them, saying with a loud voice, If any man worship the beast and his image, and receive *his* mark in his forehead, or in his hand,
Rev 14:10 The same shall drink of the wine of the wrath of God, which is poured out without mixture into the cup of his indignation; and he shall be tormented with fire and brimstone in

the presence of the holy angels, and in the presence of the Lamb:
Rev 14:11 And the smoke of their torment ascendeth up for ever and ever: and they have no rest day nor night, who worship the beast and his image, and whosoever receiveth the mark of his name.

The verse "Rev 14:9" I may simply align to the completion of every heptad sequence.

The black and white separation is of saved to condemned. Those worshipping the template and with the mark will receive the wrath of God. The wine "at every place of the table" is agreed by the Holy Trinity. (Creation is for the pleasure (satisfaction) of Christ: only He determines the end through the opening of the seals.) The wrath is poured out without mixture (not watered down for the sake of those within or added to by some other) into the cup of His indignation. That draught is pre-prepared and consists of the full wrath that will satisfy the saints rather than the thirst of a God omnipotent. (It will remain a "dreadful and terrible day of the Lord".)

Fire and brimstone correspond to the army of the sixth trumpet; each transposition forms a "partial slice" with the K4 form. "Fire + sulphur = smoke". ("Fire" is the K4 form static under Frobenius, which altered by that transposition "sulphur" produces a different octal under the K4 form, obscuring it as "smoke". It obscures the true seven cycle under that K4 form.) Being tormented under the locusts of the fifth trump, the "man" is judged before God and the elect as to his part in the image: all will taste their eternal reward, to satisfy the requirements of righteousness in heaven before being cast out into "outer darkness" (denied virtue).

The damned are "tormented": not a picture of torture, but a reference to the three woes which have a non-abelian group S5 as their schema. That non-commutative (tortuous or twisted) operation is the reward for worshipping the beast and its image: formed by the spirits of the fifth and sixth trumpets. The damned are condemned under their own spiritual constructions, witnessed by the remnant of Christians "not partakers of her plagues". That body is inclusive of the entire heavenly host (with the return of God and with His judgement).

> **Torment:**
> *Simply refers to a non-commutative operation, or to that of a non-Abelian group.*

No "rest day nor night" refers to being denied all virtue and intercession of the Lamb. (Christ Jesus.) That "day and night" applies to every octal in the sea of thirty, whether in the Sun octal or across the sea (in darkness). This torment is of the three columns forming plagues in wormwood (a seven cycle applied with an invalid choice of unity). There is no "nourishment in the wilderness" or ministry of the two witnesses for mapping from those three non-octals to unity and God's approval in the reference octal (for all those practising the dialectic).

As to being turned away by Christ and given the "cold shoulder", His wrath rests on those gathered together in their collectives (the five closed upon the synagogue of Satan completing the six). Those systems of worship (the beast and the image) are to become the "lake of fire": a cage of octals without a unity of virtue and only the dialectic deciding all good and evil instead.

That, then, is "Hell". There is no rest for those rejecting Christ. No further salvation will be made for those ungodly souls. God merely refuses them (as the world and those within it pass away), to worship their idols – to be given the cold shoulder by Heaven eternally. Theirs is no rest: Hell (the image) is become a living thing, and God imprisons Satan in "Hell" to animate it.

Rev 14:12 Here is the patience of the saints: here *are* they that keep the commandments of God, and the faith of Jesus.
Rev 14:13 And I heard a voice from heaven saying unto me, Write, Blessed *are* the dead which die in the Lord from henceforth: Yea, saith the Spirit, that they may rest from their labours; and their works do follow them.

Rev 14:12 may be compared with the opening of the fifth seal – this end of the shifted sequence of seven seals is the conclusion of that "little season" in which the number of martyred saints is complete – Here they are blessed and each are given a white robe and told to remain patient (as from the fifth seal before). Pleas for vengeance are all answered with the judgement.

These two verses refer to the patience of the saints – to remain steadfast, faithful during persecution (everyone gets their just reward). To witness the reward of the damned benefits the elect: so they may be satisfied forever. They have no "blood lust" to die in battle, only to rest after their trials so that when they are raised, they dwell in peace and rest instead.

I align the "patience of the saints" to them witnessing the ungodly (under judgement) in Rev 14:10. This is that which all saints are patient for: here are found those that keep the commandments of God, with the ability to judge and be avenged – to see their enemies clearly deserving of their fate. The saved are those found in the company of the "holy angels".

The third angel references the sealed and clarifies their eternal salvation. If they die, they die blessed: this is not (there is not) any "get out of hell free card" for those that would resist as with the sword or going into captivity. The elect will enter heaven very welcome souls indeed.

11.3) The Harvest Of The Earth Is Ripe

The moment for the judgement is finally come, and I now find the swift harvest on the very "last day" with the return of Christ.

Rev 14:14 And I looked, and behold a white cloud, and upon the cloud *one* sat like unto the Son of man, having on his head a golden crown, and in his hand a sharp sickle.

On a cloud (a term for a "mystery") is one who judges like Christ Himself, with much authority (a golden crown as GF(4)) and a sharp pruning knife. It is inferred that this angel is one who has the authority and wisdom to judge whom the elect are upon the Earth: this is a mystery, as the mechanism for salvation is clearly not so prosaic.

Instead of the octal, consider the K4 form of the ultrafilter. Under a seven cycle on a valid octal group the three groups are permuted through the seven in the octal in such a fashion that preserves the K4 form but in a manner that is incompatible within the field GF(8).

$$1 \equiv a \equiv [a, b, c]$$
$$b \equiv [a, d, e]$$
$$c \equiv [a, f, g]$$
under (a,b,d,c,f,g,e) *becomes*
$$1 \equiv b \equiv [b, d, f]$$
$$d \equiv [a, b, c]$$
$$f \equiv [b, e, g]$$

The one sat (at rest) on the cloud is as "the son of man" – this refers to the static subgroup in the octal, the "unity". So, with the K4 form aligned with octal multiplication (the mystery of the cloud), the static subgroup sits "at rest" amongst triples as would the unity element over the static subgroup in the first seal device, as the intersection of the two fields – the unity.

$$1 \equiv a \equiv [0, a, b, c]$$
$$b \equiv [0, a, d, e]$$
$$c \equiv [0, a, f, g]$$
yet in the octal...
$$1 \equiv a \equiv [0, b, d, f]$$

The one sat as Christ takes the place of the static subgroup in the octal [0,b,d,f] but it is replaced in duality with the K4 form within which he dwells as a unity singleton, "a" here. The static subgroup under addition to the groups of the K4 form generates the coset:

$$[0, b, d, f]$$
$$[0, b, e, g]$$
$$[0, c, d, g]$$
$$[0, c, e, f]$$

The seven cycle preserves the arrangement of the K4 form under C7 multiplication but not the association to unity. The one sat on the cloud (the GF(4) filter shrouded from the octal) as the static subgroup must also be sat upon the K4 form of the filter – his head is the first group in orbit under C7 (and by the identity), so his head really is as the K4 form. (The "golden crown", rather than just a "crown".) Crowns indicate additively closed sets – i.e. subgroups etc.

As a "judge of whom is elect" I find his separation of the just from the unjust is to be swift and certain. This angel is apt to see which are in the intersection of the two fields GF(8) and GF(4). Under a seven cycle he is able to find them "everywhere" in any position of the octal. His choice it appears is not to sever those in Christ as the octal's unity (His own position: note that in the K4 form he is as a closed set) or to remove the unity from the K4 form, but to sever all the "earthly four" triples apart from the K4 form containing unity that spans the ultrafilter!

The cloud separates GF(4) from GF(8) though it is no subfield. Those three triples of the K4 form correspond to the three elements of GF(4)* **in regress** as sat at the Father's right hand – those that share the unity element of the octal but are not static in the octal themselves.

Christ (with unity) is the true vine, whereas the tares that form "the vine of the earth" are separate from the vine that is the K4 form in the octal. (These make the "place prepared" – the elements separated out by the "wings of a great eagle" given to the woman in the wilderness.)

Rev 14:15 And another angel came out of the temple, crying with a loud voice to him that sat on the cloud, Thrust in thy sickle, and reap: for the time is come for thee to reap; for the harvest of the earth is ripe.
Rev 14:16 And he that sat on the cloud thrust in his sickle on the earth; and the earth was reaped.

The prior angel severed the four earthly subgroups: the "wilderness" of the Earth in singletons is this "vine" in triples, to which the woman had fled to her place prepared of God. The seven cosets of the K4 form under multiplication by seven cycles (fire) will never contain the identity element of the field GF(8) solely in the K4 form (i.e. "a" the intersection of [a,b,c],[a,d,e],[a,f,g]) just as they will never form the K4 form itself. (The vine is formed by the action of C7 upon the identity (really the static subgroup) of the starting octal.) The octal formed under the orbit of the static subgroup (cycled or floated by multiplication) becomes by association the "Sabbath rest" identified as/in the true vine of Christ, not as a "vine of the earth" by the angel.

Rev 14:17 And another angel came out of the temple which is in heaven, he also having a sharp sickle.
Rev 14:18 And another angel came out from the altar, which had power over fire; and cried with a loud cry to him that had the sharp sickle, saying, Thrust in thy sharp sickle, and gather the clusters of the vine of the earth; for her grapes are fully ripe.
Rev 14:19 And the angel thrust in his sickle into the earth, and gathered the vine of the earth, and cast it into the great winepress of the wrath of God.

A second and third angel (from the temple and with the power over fire) – a representative of the earthly elements under a seven cycle as before, the third angel is free to apply every choice of the 8 possible maps of multiplication to generate those cosets into becoming "the wilderness", the complete set of the "earth" that the vine (of the earth) as a set is within. Then the K4 form floats in the octal and all the 144,000 are spared as if they were as the 24 elders.

The three cycle formed by Frobenius holds static one triple in the earthly elements (acting as multiplication in GF(4)*) – the second angel with a sharp pruning knife cuts away this vine (of earthly triples) from the Earth it was (growing) in (severing those triples in the earthly four not held static (that are the ungodly) casting them whole into the winepress of God's wrath).

The clusters (the four sets of three triples) are separated "divorced" from the singletons to which they associate (the Earth). The possible static subgroups [0,b,d,f], [0,b,e,g], [0,c,d,g], [0,c,e,f] in the right may respectively correspond to singletons in {a,c,e,g} {a,c,d,f}, {a,b,e,f} and {a,b,d,g} depending on the choice of multiplication used in the reference octal. (By the angel with power over that choice.) The operation on those singletons may remain in the Sun octal but the associations to triples from those sets not "In Christ" are cast into the winepress.

If a ≡ 1 ≡ [b,d,f] say, [b,d,f] is as the "true vine" (static subgroup, as if virtue, unity) and the three triples [b,e,g],[c,d,g],[c,e,f] form the "cluster of the vine of the Earth". Generalising this over the eight choices of multiplication generates the whole vine of the Earth.

God does not cut off His right hand octal's subgroups (the true vine), but severs the earthly triples (the vine of the earth, in bows) to His left (forming the vine of the earth under a seven cycle) apart from those static subgroups. (Hence a requirement for a second and sharp sickle.)

The four subgroups ([0,b,d,f],[0,b,e,g],[0,c,d,g],[0,c,e,f]) in the coset of that K4 form's three do not contain unity (as a singleton) or have a place in Christ's K4 form, yet under a suitable multiplication each are as His position sat at the right hand. That which remains (the four static bows each of three triples) outside Christ are wholly cast into the winepress. Then Christ's "enemies" (in the earthly elements) are brought under His feet completely as He is sat at the Father's right hand. The Lord's footstool (the Earth) as separated above, is only of the unsaved, now Christ ready to judge has stood up! (I will see the throne is vacated temporarily.)

The "temple" is of the throne, the elders and four beasts. The angel with the sickle has discernment to identify the sealed in the static subgroup similar to the K4 form (and so rightly has a sharp sickle). The angel with power over fire (to generalise the position of floating unity across the sea) comes from the altar (from the lamb's four places of unity in each octal) and may correctly generate those that overcome as did the angel (with the circuit as a "rainbow").

Those in Christ are to be preserved as "the branch" – the elect. Those practising the devil's devices (tares in the "vine of the earth") are cast into the winepress of God's wrath. The clusters in the winepress are triples that act as subgroups in left hands of the reference. (Of which there are a possible 28.) There is a marked difference between the actions and speech of the elect in virtue and tares in compromise: believers shall "know them by their fruits".

In using those diabolical devices within the three woes the Earth is fully ripe and "legally united" in blasphemy with the image system recognised the whole world over: every fellowship remaining follows after the same legal corporate blueprint.

So, what makes the vine now ripe? The reader is informed in the same passage.

11.4) The Space Of 1600 Stadion

Rev 14:20 And the winepress was trodden without the city, and blood came out of the winepress, even unto the horse bridles, by the space of a thousand *and* six hundred furlongs.

The "winepress" is a circuit trod by the angel (the least), and the gifts to the least form the dividing line between those appointed to wrath and those not so. The winepress is then a generalised seven cycle over the reference octal made in the "horns of the Lamb", and these are totally separate from the several elements of S5 formed by the first two woes.

The five cycle constructed (over the five "refining churches") is of GF(4) in GF(16), the cosets of which are disjoint: so the elements of GF(8) which cycle its K4 subgroups never send those groups to any image of GF(4) in algebraic closure. The winepress is "trodden without the city".

The "horse bridles" are the very limit of what little is left within the press as the five-cycle is constructed. The rider at rest (as a fifth element, an identity) sat on the horse (a K4 subgroup of the reference itself with unity) would hold the bridles. I then expect an element of S5 to ride the horse (a K4 group) through the city. "Bridles" certainly indicate the ability to redirect the horse – from its path in a seven-cycle to one in a five-cycle. It (the K4 group) circuits through

S5 within that "city" when it is completed. (When the beast rises from the bottomless pit.)

The result is that the five cycle is completed as a construction. S5 is to become in every sense "a second death", and the seven cycle alongside it differs in every element: there is no longer an "obscuring" of the Holy Spirit as with smoke (sorcery). Instead, with the five cycle completed, Hell is "closed" and there is no way out. It is become as structurally sound as the Holy Spirit in this judgement.

The five cycle is of disjoint cosets of GF(4)* in GF(16). There is no K4 group or singleton common to any two of those "refining churches", therefore there is nothing to stop the angel's progress in his circuit (C7) through them. Indeed, he does circuit them and (being closed groups) no additive element within them can interrupt that unction. In the multiplicative elements, the angel must overcome the five-cycle whole and persist to find good doctrine in Philadelphia.

The word "furlong" is a translation of the Greek "stadion" which is otherwise in English a "stadium". Now, a stadium has an "audience" and the common Greek word used for a gathering together of people or an "audience" (i.e. an "assembly") is "circe". It is from this word that in English there is the derived word "church".

The "stadium" is therefore the site of the assembly, its footprint or "area". The text could then be read as 1600 churches (squared as units of area or of "dimension two", a measure of area as "space", one unit would become a "stadion"). I find that in eight C7 groups over five churches in the Sun octal (making a set S of cardinality 40) I have every map of each pair:

$$+ : S \times S \to S$$

Each cartesian product (a,b) of the 1600 (i.e. 40^2) made in the set of those churches is cut off in wrath. A map to a set of five elements is formed, free to become as S5 (under this judgement) slicing every seven cycle: those products made across those sets are also closed in S5 and in union more completely form that operation (mappings of 25 and 64 pairs into sets of 5 and 8 respectively). The horns of the false prophet have a cumulative effect. The "great city" is closed on itself without the need for the serpents or locust spirits of the first two woes.

Adding a "third part" (a+b = c) with "c" chosen from only those "circe" in Laodicea (c is assigned to the product (a,b)) shows the 12 missing elements of the city are certain to be filled in by God as judgement – placed upon the reference cycle of the angel's circuit from which they were absent. This is done when the treading of the winepress is completed.

The bridles may represent the ungodly action of the locusts over the churches. (The bridles are the action of the key to the pit. The fifth church is replaced with the star that returns the Church's members to the collective.) The "horse" is then in the seven cycles a "Church" and the bridles form the "pneuma" of the image, leaving the rider as "Death" or here, "Hell".

As with the serpents of the sixth trump, by transposing elements in K4 groups (horses), I may with two "bridles" (serpent transpositions, one in each "hand") redirect the seven cycle to a five-cycle. (As with one horn of the false prophet justifying the other, the second woe instantiating the first.) So, when g=1, (a,b,d,f,c,g,e)(b,e)(a,b) = (d,c,f,g,e).

The five cycle is instantiated even to the "horse bridles" (the use of those transpositions). They become universal in the "nations" (churches) that walk without the conviction of the circuit of the angel. The five cycle of the locusts in GF(16) now transformed with the bridles becomes a cycle common in the elements of the angel's circuit (as agreeing, or with "blood" in the press, in the same circuit as of the least.)

The five-cycle is constructed on cosets of GF(4)* in GF(16)* using the key to the pit. It is separate from GF(8) (images of K4 groups of one being disjoint, of the other not so). "Blood" comes "out of the press", even to the "horse bridles". As with (a,b) in one hand and with (b,e) on the other, the "blood" or equality is made between the bridles and five-cycle together.

(a,b,d,c,f,g,e) = (d,c,f,g,e)(a,b)(b,e)

The bridles are also able to be cycled through the octal with a seven-cycle preserving a similar circuit or five-cycle. I.e. By choosing unity from (a,b,d,c,f,g,e) I have with e=1:

(a,b,d,c,f,g,e) = (c,f,g,e,a)(b,d)(a,d)

The five cycle "floats" through the seven cycle with the choice of unity. Then every such five-cycle of "churches" is then a generalisation of the first five elements of every seven cycle. Then the blood flows out from the winepress not "to the space of" 1600 furlongs, but "**by** the space of" as spanning a distance (space) of five in the seven cycle instead.

The five-cycle walking the seven cycle accounts for the remnant that dies as one tenth of the great city falls. (The elect may yet return to the refining churches, still under condemnation by the letter to Laodicea.) This device must also be overcome: some spirits "come not out but by prayer and fasting". ("Mat 6:11 Give us this day our daily bread." I.e. What bread?) This includes the spirit of a five cycle: it cannot be deprecated, but prayer for deliverance is a must.

Mat 6:13 And lead us not into temptation, but deliver us from evil: For thine is the kingdom, and the power, and the glory, for ever. Amen.

The operation of a five-cycle is constructed over every C7 group valid on one octal (the angel's circuit or the horns on the Lamb, now overcome as exited safe to Philadelphia). The definition of a binary operation as above is then a given. Satan no longer requires 200,000,000 devices; he only needs to counter just the one circuit (the angel's) generalised as in the ministry of the two witnesses. Of the two elements missing from the five-cycle, one is sabbath rest as should be expected: the other is of the stronghold of Philadelphia; the elect are safely separated.

As this binary operation upon the set S is formed "the star with the key to the pit" is now "in strength". (The beast from the pit ascended.) The "door" has partly been shut on Laodicea and there is no exit from what is to become the "lake of fire" for those under the sway of the locusts, now caged. Laodicea is not "closed" with the product (a,b) above (because there are twelve elements still missing in S4 over the first four churches), but Laodicea is to be the target of God's displeasure. Approval remains for those elect seeking after it from the five refining churches. The seven cycle is still in effect until the judgement of the woman, the "great city".

The term "horse" is used for a K4 group, and the accumulation of static subgroups (or the consequence of the choices for the identity) in the seals and also the images of those K4 additive subgroups in fields in the fifth trump. In any event I find by "horse" the application of multiplication upon a closed additive set (using and observing the distributive laws).

With the completion of the "city" S5, the earthly elements are completed to form S4. K4 is a normal subgroup in S4 and with $H \cong K4$ I have gHg^{-1} in H for any element "g" of S4. The fifth element, the "star" (as over Sardis) becomes the rider resting upon the "horse" or as the limit of the "horse bridles", being (quite plausibly) the elements "g" in conjugation on H. Elements in S5 permute S4 (by conjugation) to an isomorphic subgroup of S5 as before in the "pride of life": the K4 group as a horse is also "rode" through the city. I may similarly walk H through S5 by pre-multiplying any "H-set" of four churches by a five cycle as constructed above.

The beast may only construct the group A4 using the static triples of every seven cycle. Under a five-fold composite of these devices it is possible for the beast to construct (nearly half of) the elements of S5 over the five Laodicean churches of that amalgam. To begin, there are only the elements of A4 under a five cycle, as An is generated by three-cycles (Babylon is twice fallen). These are filled in with God's judgement later as "Death and Hell" are thrown into the lake of fire (alive) and Satan is damned to provide all the missing "pneuma".

To complete the "great city", the five cycle is required: for then those missing twelve elements fill in the missing sixty. As for that five cycle, only the five cycles are missing from S5 even if (and when) the five subgroups of S4 are fully constructed. This verse shows the fulfilment of that requirement for "the great city". Satan, it appears, has had to go begging before God. Satan is left with one reply, to construct a short orbit of GF(4) in GF(16) instead. God's Spirit over His right hand (generating the Sun octal) is never twisted out of place. (Satan is shut up in the pit and a seal placed on him.) This shows that the woman is answered by the completion of the one mechanism that is to complete her inheritance in "the lake of fire".

The winepress is closed (a cage for every unclean bird) and as multiplication is done with different C7 groups, I note the action of multiplication across such sets. The equivalent of the angel's circuit (as in the mystery of the seven lampstands) is in view in the "great city", yet by Satan's devices, not made upon those virtues found in God.

The "horse bridles" include the maps between K4 subgroups (of God or else S4) and their respective elements of the various C7 groups, touching just upon the five "churches" and no more. The "bridles" represent these shortened "circuits" of K4 in S4 through S5 (an orbit of length five); they may be imagined to be the mechanism by which the harlot rides the beast.

The city of S5 is on five sets, and the rest element is properly the seventh (unity). The portion of the octal "river" of K4 groups in the five-cycle cannot include one of the possible static subgroups ([0,b,d,f], [0,b,e,g], [0,c,d,g], [0,c,e,f]), as I cannot associate in the "blood" of both cycles their correspondence to that unity "a". In the five-cycles, the images of these groups remain disjoint, an impossibility when taking them from the octal (and the five cycle "floats").

Likewise, also outside of the K4 form are the four triples in the wormwood octal. (If a=1 then I have ([0,b,d,g], [0,b,e,f], [0,c,d,f], [0,c,e,g]).) What was water in the Sun octal is blood in the wormwood octals, and vice versa.

Though the city is built on those wormwood devices, the "blood" to come out of the winepress cannot be the set of subgroups of the Sun octal transposed to the wormwood octals. The winepress "squeezes" the octal sea of 30 into just one octal, a "sea of glass". The Sun octal (rather than a sixth element), is the ground "without the city".

Jer 17:9 The heart *is* deceitful above all *things*, and desperately wicked: who can know it?
Jer 17:10 I the LORD search the heart, *I* try the reins, even to give every man according to his ways, *and* according to the fruit of his doings.

From Jeremiah I read God judges whether a person truly follows after Him (as in the seven cycle as the reins) or after the leading (reins) of their own hearts (as over the calling of God to repent and gain rest). Here in the Revelation I note the same judgement made against those worshipping the image system instead of walking with Christ in white. The "horse bridles" are under the control of someone: it isn't God directly – it is the collective instead. (The "Church" or as it is here, a single "circe" (Death) representing six (Hell) in the collective.) Judgement surely follows either to Christ or the lake of fire, according to the ways of those within and to the benefit of God Himself (the "fruit" is for God triune who alone chooses for salvation).

The winepress is trodden "without the city", so the Spirit of God dwelling with the believer as from the "least" or "angel on up" (the least in Philadelphia and those overcoming Laodicea) are those who do the treading. I take the text to state that as fellowships recognise each other in the freedom given by the beast's authority they are likewise met under conviction with the exit of the least. Eventually, that will be global. It is an argument of induction and the communication of conviction is from outside the "city".

Rev 13:10 He that leadeth into captivity shall go into captivity: he that killeth with the sword must be killed with the sword. Here is the patience and the faith of the saints.

The "star" as over "Sardis" is replaced by a mapping, those returning to the amalgam are then made captive – those that fall as untimely figs as in a five cycle are "beheaded" (killed) as from the proper right hand octal (as with the sword), "shut out" as only a five cycle in GF(16). (They never attain unity or approval in "Philadelphia".) This is a divine judgement upon the Church.

11.5) The Great Winepress Of The Wrath Of God

The winepress of God's wrath is the opposite of the seals in its own respect. Whereas the seals opened up the sea of octals to affect the Church paradigm with the devices of the dragon over the sea: here, the sea of thirty octals is "gathered together" under the action that aligns the sea as one. Those practising those devices of the sea are "crushed" in-between the Earth and a shrinking sea (a lake of fire) whilst attempting to make their devices work in all the four earthly elements of the Sun octal – an impossible task without hurting both the oil and wine.

Whilst this is so, it is impossible for the "not saved" to operate in the Sun octal as they are antichrist numbering 666. They, caught between the rock (of ages) and a hard place, are pressed into what is, in effect, the empty set. The enemies of God are gathered into one place (the earthly elements of the Sun octal) without the dressing of wormwood for their sins. They, as the Earth, are made the Lord's footstool, with His enemies crushed (to { }) under His feet.

The seven vials are the method by which God answers the blasphemy of the living with the disdain He has suffered since the fall in Eden. Whilst answering the devices and sin of all point by point, he returns man's folly upon them in terrible fashion.

11.6) The Sea Of Glass

Rev 15:1 And I saw another sign in heaven, great and marvellous, seven angels having the seven last plagues; for in them is filled up the wrath of God.
Rev 15:2 And I saw as it were a sea of glass mingled with fire: and them that had gotten the victory over the beast, and over his image, and over his mark, *and* over the number of his name, stand on the sea of glass, having the harps of God.

There is another wonder in heaven besides the woman and the dragon. These last angels are also a "great and marvellous" sign. In these angels is the sum total of the wrath of God – it is fore-ordained rather than an answer to bad behaviour on an individual basis. Every sinful soul in the sea is become as bad as each other and God has decreed that everyone on the Earth not dwelling under the covering of heaven will be a recipient of the wrath. This last heptad is the last act of God's wrath upon the Earth excepting the continual judgement to the lake of fire.

The saved attained the Sabbath in Christ and are at rest as unity. They overcame the devices of the dragon by holding fast to the gospel, likewise overcoming the same devices in the sea that permitted such blasphemy in the Church – and the diabolical devices in the Church were used without complaint from the Church, and their use directly against the one they call God.

Those playing on harps before God are in no doubt as to the Trinity: they sing the song, they have the score. There is no longer any mystery of God. (There should be time no longer in the days when the seventh angel is ready to sound: the mystery of God should be finished.)

Overcoming the image and all that created and sustained it is found in overcoming the devil at his worst – those that do truly know and dwell with their God in heavenly places as victors. The Sun octal, is now aligned in isomorphism across all 30 octals of the sea (a bijection between any two and equivalence to them from it). This is the natural state of the saved: not objects of wrath, they are preserved, kept through it all. The sea of glass is mingled with fire – as the octal under a generalised application of the seven cycle (there is that "mingled fire"). Curiously the appearance of the additive structure is reversed between the octals in the right and left hands as hail/fire. (When additive operations are aligned in isomorphism, the seven cycle to John seemed to act or burn "up" as well as with its reverse (hail) "down" on the other hand.)

The left and right hand octals under the one seven cycle have their additive operations in "reversal" rather than isomorphic alignment. I assume that only the reference octal has a singular operation, and that every other octal (including its own left hand octal) is aligned with its left hand: so that every other octal is the recipient of "hail" as well as "fire".

It makes sense (rather than as the false prophet, to reason one contradiction of the whole) that keeping both hands intact is a "better idea". (Octals are either left or right in pairs and present in reversal.) Either that or treat every seven cycle separate from the reference octal or consider just the one in powers of a generator rather than singletons. *Use polynomials* as the teacher said. (Only the left hand opposes the right hand, every other aligned to the left hand.)

Rev 15:3 And they sing the song of Moses the servant of God, and the song of the Lamb, saying, Great and marvellous *are* thy works, Lord God Almighty; just and true *are* thy ways, thou King of saints.

The elect sing the song of Moses – exultant at the end of their trials and the start of their true freedom given them by their God. (They, likewise, are "drawn from the water" – with the Holy Spirit and are spared all the "plagues" of wormwood.)

Rev 15:4 Who shall not fear thee, O Lord, and glorify thy name? for *thou* only *art* holy: for all nations shall come and worship before thee; for thy judgments are made manifest.

With the mystery of God complete, the Lord is praised as the one God whose judgements are truly made. His judgements are those that are seen in the Earth, and not those of some other spirit. All that worship Christ (the nations of the seven churches) will bow, whether they are convicted or not. They should merely observe the ungodly brought low before God: for all are now made His footstool. (His judgements are made manifest, rather than those of any other.)

11.7) The Seven Angels And The Seven Last Plagues

The chapter closes with the pre-prepared (and justly so) last plagues that are not the revenge of murderous rage, but are calmly prepared to answer the blasphemy of men and angels.

Rev 15:5 And after that I looked, and, behold, the temple of the tabernacle of the testimony in heaven was opened:
Rev 15:6 And the seven angels came out of the temple, having the seven plagues, clothed in pure and white linen, and having their breasts girded with golden girdles.
Rev 15:7 And one of the four beasts gave unto the seven angels seven golden vials full of the wrath of God, who liveth for ever and ever.
Rev 15:8 And the temple was filled with smoke from the glory of God, and from his power; and no man was able to enter into the temple, till the seven plagues of the seven angels were fulfilled.

The temple of the testimony is open: the new name of God is already written on the least, his circuit completed and the names of all the elect clearly revealed in heaven (and the tabernacle is filled with God's glory, rededicated). The least's circuit becomes the dividing line between all those saved and not so. The wrath to follow is justified by that circuit of the least and the equivalent of the horns on the Lamb. These, granted as gifts to the angel to overcome, have their answer in the wrath poured out upon those that refused the grace of God.

The angels that exit the temple all wear "golden girdles" as did Christ in Rev 1. This indicated the Godhead. Likewise, the seven angels are subfields (gold) and girt "about the paps" (to be under circuit in the seven cycle). They are subfields as GF(2). They exit the temple (as zero) and are dressed in pure and white linen (as unity or the lamb). The girdles are clearly GF(2): together they span the octal as a golden girdle. It is clear that these seven angels are holy.

The seven angels are subfields: together they "declare" that in the singletons of the octal the "zero" element is adequately defined on every element of the octal – additive or multiplicative. (a+0=a, a*0=0.) This "definition" of the zero element by John comes into effect as a theorist would understand zero, on a per-element basis: not a set but just a simple singleton.

The centre of all things in heaven is God. The temple of the tabernacle of the testimony was "opened". (Then beforehand it had to be closed.) I observed the octal field GF(8)'s structure in terms of the elders (automorphisms) and multiplicative structures and subgroups etc. Up until now I have omitted the additive identity that represents the seat (throne) of God Himself.

Under the operation of $(A \vee B)^c$ on subgroups (in the octal) the additive group of GF(8)+ (the octal itself) is a valid substitution for the additive identity, formerly denoted by "0". Likewise in the K4 form of [0,a,b,c] the K4 group is identity for the same operation that allows that K4 form (rather than the octal) to give [0,a] + [0,b] = [0,c]. Now, the tabernacle is opened to us!

The octal may so "reign" over K4 groups of subgroups; and the K4 group can rule over K4 groups of C2 subgroups: a singleton become as GF(2) can only reign over itself! (For 1000 years with the Lord or until they live in Christ at the resurrection if less?)

Seven angels come from the throne as those from the temple. As before I found that John was to only measure the temple rather than the outer court, the temple of his testimony is of the whole scene of the throne and the four beasts as well as the 24 elders. Now, it is opened to the understanding and sits ready for consideration. The four beasts are notably remaining in the scene and one gives the last vials of wrath to these seven angels ready to be poured out.

The temple is filled with smoke (the "pneuma" or "Holy Spirit" is obscured) no doubt due to the absorptive property of the singleton zero as per multiplication (the glory of God and His power is "held back" from adding to the wrath by the presence of the zero).

God does not sit upon the throne as He had done before as the "zero": rather He appears replaced by a pre-prepared judgement – a calm and just judgement measured out by one with forever to consider and to relent from rashly overheated actions. A singleton element suffices as zero for now. I find a "zero" that is with the covering of smoke (as a mystery upon subgroups) still satisfying for the role. No one can enter the temple until God's wrath is fulfilled.

Whilst the angel's circuit is completed and the new name of God written upon him alone (whilst it remains known only to the least), nobody may enter that temple with the "new name" in effect until the final sabbath rest is obtained after the judgement and the new name of God may be openly revealed when it is impossible for it to be blasphemed with knowledge. The "smoke" (the new name as not openly revealed) of the glory of God and His power (as complete omnipotence) is restricted to just the least and the Lord alone at that time: these seven holy angels with the same vials cut off all those under the wrath from that inheritance.

Whilst the temple is rededicated with God's new name, His glory fills it and none may enter. (The book of life must be completely filled and opened already.) Zero is multiplicatively absorptive and so "unchangeable" – none may substitute for it.

God in His wrath places the singleton zero as it were the "empty set" upon His throne, in order to reject all those He judges should be rejected. That is, zero replaces the scene in a mystery whilst God takes a breather!

Chapter Twelve: The Seven Vials Of Wrath – The Winepress Is Trod

The wrath of God is related to the restrainer, the unsealing of the book of life and the consequences in the sea from the universally applied seal and trumpet devices. Instead of a paradigm of restraint (loosed in the Church) opening up to the sea against which only the elect hold firm, I examine the world under the wrath – as a reversal of restraint previously taken out of the way. As the winepress over the thirty octals squeezes them into just one octal (in alignment) to form "The sea of glass", the devices of the dragon are undone and the lukewarm in the Church are also under the wrath. Judgement begins at the house of God.

The dialectic devices are facilitated further by three spirits as frogs from the diabolical trinity, deceptive workers of "miracles" – i.e. of contradictions that perpetuate the operation of wormwood on a "small sea" despite the one field GF(8). This is only possible in mimicry of the proper operation of the octal and in sending one of the earthly elements beside the static triple to the empty set, and by appearance in that "small sea" making the remaining three elements of the earthly four cycle, as if the device were overlaid upon the octals of the wormwood device, placed upon the four beasts before the throne. The wormwood device requires separate octals: it is the earthly elements of the one octal which the three frogs "cycle", leaving the static subgroup of the Sun octal fixed. As contradictory to the proper Frobenius map, these three spirits are "miraculous".

This "spin" to perpetuate the dialectic in the cases of government, Church and the dragon (worshipped by all that are not sealed) is somewhat effective despite the contradiction. Their "speech" is dialectic, a contradiction to the action of God's wrath: a miraculous sign that deceives the "kings of the Earth" (choices of unity) and those under them – the whole world.

This is not to state that God is not omnipotent: God uses a small hammer to hit a small nail, without total force. (The wrath is instead trod by an angel whose name is "the Word of God".)

The vials do not add anything into the mix to begin with – they begin as the counter-arguments by example to the devices of the seals and trumpets. God draws His dividing line in a manner that reveals the wider sea as a spiritual house is divided against itself, and His creations rightly find contradiction outside of the Sun octal so that all creation would declare the glory of God.

That said, you can't offer faith to someone who rejects it. The result of the wrath is not torment (punishment) but a clear separation of the patient believer from those that scoff at all things God. The wrath prepares the world for judgement, rather than being the end of all things living: the evil of the world is revealed by the wrath:

Rom 1:18 For the wrath of God is revealed from heaven against all ungodliness and unrighteousness of men, who hold the truth in unrighteousness;

...as much as the wrath is revealed by it. Were there no such equivalence it would make no sense. Only those deserving wrath receive it and vice versa. It is a spiritually made separation.

The wrath is the final line in the sand, and believers are assured that salvation is yet offered – there may be a few more elect gained before the seventh seal is opened as the wrath completes, but this is not the case for the vast majority practising the devices of the seals.

To the angel (the least):

Rev 2:5 Remember therefore from whence thou art fallen, and repent, and do the first works; or else I will come unto thee quickly, and will remove thy candlestick out of his place, except thou repent.

For those in the Church:

Rev 2:21 And I gave her space to repent of her fornication; and she repented not.

For those alive and following after the image:

Rev 9:20 And the rest of the men which were not killed by these plagues yet repented not of the works of their hands, that they should not worship devils, and idols of gold, and silver, and brass, and stone, and of wood: which neither can see, nor hear, nor walk:

For all antichrists:

Rev 16:9 And men were scorched with great heat, and blasphemed the name of God, which hath power over these plagues: and they repented not to give him glory.

For all unrighteous sinners:

Rev 16:11 And blasphemed the God of heaven because of their pains and their sores, and repented not of their deeds.

The scope of these statements increases with inclusion on those that refuse to repent. The opened seals and sounding of the trumpets show the number to be saved is dwindling. At which point is there any safety in numbers? There is none to be found, not even for the least.

I state that Christ comes quickly. Not for "rapture" or after a slaughter of His people, but at the overcoming of the least in a period where the world and its Church have spiritually fallen asleep and woken up unknowingly dead. He arrives and comes fast with His judgement.

The judgement is to be separated from Christ forever. The octal which may be considered as "under" the lake of fire is replaced with a "cage" outside of the K4 form of the ultrafilter within an octal with no "throne" over it. The vials force the ungodly into this "wrath" octal in two forms, the position of the wrath's start and as it is finished – mirrors of each other that cage and trap the dialectic devices. As "wormwood" octals of each other, they are totally separated from Christ. (That is the second death.) Only those with part in the first resurrection (that are born again) are saved from this cage of "wrath octals", which will have no power over them.

Rom 8:38 For I am persuaded, that neither death, nor life, nor angels, nor principalities, nor powers, nor things present, nor things to come,
Rom 8:39 Nor height, nor depth, nor any other creature, shall be able to separate us from the love of God, which is in Christ Jesus our Lord.

This is the patience and faith of the saints. If you can hear His voice, it is not too late.

First, I have more word pictures for algebra:

Rev 15:7 And one of the four beasts gave unto the seven angels seven golden vials full of the wrath of God, who liveth for ever and ever.

And

Rev 16:1 And I heard a great voice out of the temple saying to the seven angels, Go your ways, and pour out the vials of the wrath of God upon the earth.

One of the four beasts (outside of the static subgroup) gives the angels seven vials to pour out upon the Earth. I consider only one of the four (without loss the unity) of the Sun octal – poured out upon the cosets of the subgroups (the Earth, the target of all the seven plagues) under a seven cycle, say (a,b,d,c,f,g,e). If "a" is the unity as of that beast, I gain the following:

$$[a, c, e, g] + a = [0, b, d, f]$$
$$[b, f, a, e] + a = [0, c, d, g]$$
$$[d, g, b, a] + a = [0, c, e, f]$$
$$[c, e, d, b] + a = [b, d, e, c]$$
$$[f, a, c, d] + a = [0, g, b, e]$$
$$[g, b, f, c] + a = [f, c, g, b]$$
$$[e, d, g, f] + a = [d, e, f, g]$$

12.1) Treading The Winepress

Now, in the text I find the verse:

Rev 15:6 And the seven angels came out of the temple, having the seven plagues, clothed in pure and white linen, and having their breasts girded with golden girdles.

These seven angels with the seven plagues having golden girdles, are all subfields of the Trinity (clothed in pure and white linen as the lamb or unity) and together they "span" the octal. Each of them is then in correspondence to some field [0,1] where unity is one of a to g.

And also, the verses:

Rev 19:15 And out of his mouth goeth a sharp sword, that with it he should smite the nations: and he shall rule them with a rod of iron: and he treadeth the winepress of the fierceness and wrath of Almighty God.
Rev 19:16 And he hath on *his* vesture and on his thigh a name written, KING OF KINGS, AND LORD OF LORDS.

The angel whose name is "the word of God" (Rev 19:13) I know as the "angel of the church" from before – the right hand of Jesus Christ. I expect the sequence of vials to align with his circuit – inverse to the seven cycle over them (to John cycling right to left instead of left to right) so that unity "floats through" the seven plagues in turn, as it did with the angel's circuit of the churches. I expect the "winepress" to be "trod" as he has overcome where the ungodly have not: and as the least of all, they are (by him) all condemned for lack of faith and disobedience. That stated, anyone overcoming the world (even leaving just one church to find election, as accepted in Christ) breaks that seven cycle trod; and are "not appointed to wrath".

How would a theorist define "zero"? The definition is made upon the set in view, both additively and multiplicatively. The angels spanning the octal together, had exited the temple, fulfilling John's (and the one) definition. Zero is simply a unique symbol 0 such that for all x: 0+x = x and 0*x=0. It takes these seven angels together to define the "zero" element sufficiently.

The octal is a valid identity for K4 group addition (using $(a \vee b)^c$ in the octal). Likewise, with the same operation on singletons in a K4 group, the K4 group itself is also a fine identity, with that K4 group as the container. (In the metaphysics of chapter four, the set of all positive properties was a valid identity in both cases, the coset of the K4 group in scope becomes inconsistent.) Every K4 group sits "with the octal" on the throne as the zero element for K4 addition.

Rev 3:21 To him that overcometh will I grant to sit with me in my throne, even as I also overcame, and am set down with my Father in his throne.

To be sat down equally with Christ is similarly to be a "zero" for singletons; I expect the theorist's definition of zero to apply to the angel of the church also. As Christ is a subgroup in the octal "sat down with" His Father, the zero may act on singletons as a subgroup of the K4 group over that triple. In the set of all positive properties, the metaphysics of the identity element is no different between Christ within the Father enthroned. Likewise, on singletons there is no difference to expect from the angel as "zero" (indexed appropriately, of course!).

The smoke and glory of God that fills the temple whilst the wrath is poured out, I deem to be indicative of the uniqueness of the reward to the angel: it is clouded, and not a gift to all. (None may enter in.) It will occur when the angel of the church finishes making his circuit, this time, convicting all to repent: or if they won't, bringing God's judgement down upon them.

Rev 15:8 And the temple was filled with smoke from the glory of God, and from his power; and no man was able to enter into the temple, till the seven plagues of the seven angels were fulfilled.

The temple then is in the process of being rededicated – accepting the angel as the least – as the "zero". Whilst the vials are poured out (synchronised to the angels overcoming of all seven churches, i.e. when the circuit is complete and he may be rewarded as to overcoming Laodicea) none are deceived and the wrath of God finally arrives.

Each of the seven angels pours out a different element as unity upon the cosets of the subgroups of the octals in view, and I may consider them cycled by the angel's circuit.

The "winepress" is trod out – not a picture of an angel stepping on grapes in a "winevat" but instead a "press", its action turned by the angel. The winepress is a cyclic or twisted "squeezing" of the ungodly into the earthly elements outside of Christ and into the "vine of the earth".

12.2) The First Vial

Rev 16:2 And the first went, and poured out his vial upon the earth; and there fell a noisome and grievous sore upon the men which had the mark of the beast, and *upon* them which worshipped his image.

$$[a, c, e, g] + a = [0, b, d, f] \equiv E$$
$$[b, f, a, e] + b = [0, c, d, g] \equiv G$$
$$[d, g, b, a] + d = [0, c, e, f] \equiv F$$
$$[c, e, d, b] + \{\phi\} = [b, d, e, c] \equiv C$$
$$[f, a, c, d] + f = [0, g, b, e] \equiv D$$
$$[g, b, f, c] + \{\phi\} = [f, c, g, b] \equiv B$$
$$[e, d, g, f] + \{\phi\} = [d, e, f, g] \equiv A$$

Now note that the sum of "a" with the cosets of the left column may be rewritten as such (with the seven cycle (a,b,d,c,f,g,e) acting downwards to generate the cosets).

The seven cosets to the left are generated by the seven cycle, and the element "a = 1" or "vial" may be added to each coset to form the entries on the right: yet not necessarily so in three rows, which may instead be replaced with a sum with the "empty set". (With zero, if you like. I use the empty set simply to denote "no addition". I have no zero element/singleton as yet!)

The braces { } represent a "mouth" and the Greek letter "phi" within it resembles a sore on the mouth closed. "Noisome" transcribes as "worthless" and "grievous" as "hurtful". (I.e. without effect (empty) or profitless and also completely disrupting the octal as if acting in the sum – hurting the operation.) There is also no valid seven cycle for multiplication on this "wrath set".

Under $(a \vee b)^c$ (as if for K4 addition), upon the sets above it is also "worthless and hurtful".

The sets A and B I refer to below are as such:

A.   Cosets of subgroups in the Sun octal in the wrath set above.

B.   Subgroups shared between the Sun octal and the wrath set above.

Adding the members of these sets never produces an element in the "wrath set". It is no longer closed under that operation.

1)   Products in A result in the complement of a coset in A (a subgroup in the Sun not present here). (I.e. the "sore" is worthless.)

2)   Products in B result in the complement of a coset in A (a subgroup in the Sun not present here).

3)   Products between A and B result in the complement of a subgroup in B. (I.e. the "sore" is hurtful.)

The product is "broken" and these exchanges now become "worthless" and "hurtful" to the operation of the octal. No product ever results in a desired outcome, of a member of the set.

As the angel pours his vial (as "a = 1" here), upon the seven cosets, the empty set (sore) is upon the additions that with "a" (the vial poured out), make no difference. This occurs to those practising the dialectic, a static bow in the earthly elements {c,e,g} (that is, given a=1).

The Earth is filled with the devices of the seals as upon the wider sea of 30 octals. The "mark" corresponds to the action in dialogue of those fellowships in the "Church" that are legally registered by government in preference to God's favour, and they only operate as such by including those devices. Those that worship the beast's image are those who defer authority to those with the mark (meeting as those within dialogue in their stead) to fill the contradictory balances of the fourth seal system with their worship: it being their heart's desire – as does Daniel's fourth beast, deferring the will of the people to a select minority. (Whilst voting with the authority of "kings", one king in the image is their representative or chair.)

The elements upon which the null element (empty set) is present (preventing virtue), form the triple or "bow" in the earthly elements that corresponds to the three columns of the "wormwood octals" forming "plagues" – here as the K4 form or that "jacinth" part shared with the reference octal given a=1 (the subgroups intersecting in "a" are missing in the wrath and wormwood-wrath sets (see later). The plagues are on those using the dialectic – that sit in collective as if one of the ten kings – i.e. those with the mark of the beast – or that worship the image, following as hell after by deferring all choice to those with the mark (Nicolaitans)).

The casting vote in the beast (under three triples: thesis, antithesis and facilitated synthesis) require only the one triple to fill this void of contradiction. This "grievous and noisome" sore falls on those with the mark (voting on doctrine and privating virtue as in a = 1) and who as Nicolaitans accept their decision as with the same one mind (worshipping the image), aligning the manifold of their heart to the "great city" S5 over the first five churches (as upon their false doctrine) or vice versa, to fill the void in the fourth balance, a system of contradiction itself. (They are projecting their carnal selves (hearts) upon the doctrine (under dialogue) of the five churches of the collective now bereft of the elect.)

The first angel pours out his vial upon all the cosets of the Sun octal (upon, say, [a,c,e,g] + a = [0,b,d,f] in one case of seven). The second falls upon those using Satan's devices in the sea.

The "sore" falls upon those that practise the dialectic and deny virtue: the "collective" in the beast with the mark, and those that accept it in the image as with the same mind.

This "wrath set" above is a cage for those that practise the dialectic: the "bow" required for synthesis (denying virtue, unity) is formed of the cosets of the subgroups that contain unity.

Products of these three cosets cannot attain a member in agreement with one of their number. (They may produce a subgroup, but then it is the complement of the third such coset.)

In any case, those three subgroups intersecting in unity are missing from this set, and are only attainable through sums of their cosets (in A) and from products in the other four triples (in B, subgroups in the Sun octal).

Now, I see the angel has overcome and this indeed corresponds to his circuit (it is a prerequisite that none under the wrath of even the first vial have overcome any of the seven churches).

Rev 2:7 He that hath an ear, let him hear what the Spirit saith unto the churches; To him that overcometh will I give to eat of the tree of life, which is in the midst of the paradise of God.

That "tree" being any one tree (triple of subgroups containing virtue, unity) of the seven possible in the octal, not consumed by the fire of God – as if always a "burning bush". That tree is formed of the three subgroups that above are mapped to their cosets by the outpouring of the vial. With the tree of life, the octal's proper operation (not that of wormwood) is a given.

12.3) The Second Vial

Rev 16:3 And the second angel poured out his vial upon the sea; and it became as the blood of a dead *man*: and every living soul died in the sea.

Now I should note that each angel must pour out his vial upon the Earth.

Rev 16:1 And I heard a great voice out of the temple saying to the seven angels, Go your ways, and pour out the vials of the wrath of God upon the earth.

I may "pour out" another element (say "b = 1") on the additive cosets of the sea. First, I need to form the "sea".

There are four sets of octals in the sea. The reference (Sun) octal, and those octals sharing three subgroups with the reference (there are seven of these) and octals sharing just one subgroup (there are fourteen of these) and lastly, the final eight octals that share no subgroup with the reference at all. (Which are the possible left hands for multiplication on the reference.)

I note the first vial was poured out onto the coset of the Sun octal; here I see this one similarly poured out but onto the left hands of the Sun octal (the least has his left foot upon the Earth, and his right foot upon the sea). Then the several cases of the octal(s) wherein all seven K4 subgroups are shared with the reference as well as those sharing none are accounted for as across the sea (or even the sea of glass).

Starting with an arbitrary octal in the sea (middle columns below), I can with any three valid choices of multiplication generate left hands (the rightmost column) for that octal and then a unique choice for a fourth "left hand" plugs the gaps, filling in the missing "third part" in each.

$$
\begin{array}{rcccccccccccc}
a & \equiv & [0, & c, & e, & g] & \equiv & [0, & b, & d, & f] & \\
c & \equiv & [0, & b, & d, & e] & \equiv & [0, & a, & f, & g] & ! \\
e & \equiv & [0, & d, & f, & g] & \equiv & [0, & a, & b, & c] & \neg \\
d & \equiv & [0, & a, & b, & g] & \equiv & [0, & c, & e, & f] & \\
g & \equiv & [0, & b, & c, & f] & \equiv & [0, & a, & d, & e] & \pounds \\
b & \equiv & [0, & a, & e, & f] & \equiv & [0, & c, & d, & g] & \\
f & \equiv & [0, & a, & c, & d] & \equiv & [0, & b, & e, & g] & \\
\end{array}
$$

If I find three suitable octals to complete the wormwood device (as I had previously with the third trump and so as to form them in the sea rather than met in the Sun), then I may associate the triples as before within some other three octals here below:

$$e \equiv [0, \ d, \ f, \ g] \equiv [0, \ a, \ b, \ c] \quad \neg$$
$$d \equiv [0, \ b, \ c, \ f] \equiv [0, \ a, \ e, \ g]$$
$$f \equiv [0, \ a, \ b, \ g] \equiv [0, \ c, \ d, \ e]$$
$$b \equiv [0, \ c, \ e, \ g] \equiv [0, \ a, \ d, \ f]$$
$$g \equiv [0, \ a, \ c, \ d] \equiv [0, \ b, \ e, \ f] \quad \$$$
$$c \equiv [0, \ a, \ e, \ f] \equiv [0, \ b, \ d, \ g] \quad \wedge$$
$$a \equiv [0, \ b, \ d, \ e] \equiv [0, \ c, \ f, \ g]$$

$$g \equiv [0, \ b, \ c, \ f] \equiv [0, \ a, \ d, \ e] \quad \pounds$$
$$b \equiv [0, \ a, \ c, \ d] \equiv [0, \ e, \ f, \ g]$$
$$c \equiv [0, \ a, \ e, \ f] \equiv [0, \ b, \ d, \ g] \quad \wedge$$
$$a \equiv [0, \ d, \ f, \ g] \equiv [0, \ b, \ c, \ e]$$
$$f \equiv [0, \ b, \ d, \ e] \equiv [0, \ a, \ c, \ g]$$
$$d \equiv [0, \ c, \ e, \ g] \equiv [0, \ a, \ b, \ f]$$
$$e \equiv [0, \ a, \ b, \ g] \equiv [0, \ c, \ d, \ f] \quad *$$

$$c \equiv [0, \ b, \ d, \ e] \equiv [0, \ a, \ f, \ g] \quad !$$
$$b \equiv [0, \ d, \ f, \ g] \equiv [0, \ a, \ c, \ e]$$
$$d \equiv [0, \ a, \ e, \ f] \equiv [0, \ b, \ c, \ g]$$
$$f \equiv [0, \ c, \ e, \ g] \equiv [0, \ a, \ b, \ d]$$
$$e \equiv [0, \ a, \ b, \ g] \equiv [0, \ c, \ d, \ f] \quad *$$
$$g \equiv [0, \ a, \ c, \ d] \equiv [0, \ b, \ e, \ f] \quad \$$$
$$a \equiv [0, \ b, \ c, \ f] \equiv [0, \ d, \ e, \ g]$$

Which share in place the correspondences of subgroups in either handed octal:

$$c \equiv [0, \ b, \ d, \ e] \equiv [0, \ a, \ f, \ g] \quad !$$
$$e \equiv [0, \ d, \ f, \ g] \equiv [0, \ a, \ b, \ c] \quad \neg$$
$$g \equiv [0, \ b, \ c, \ f] \equiv [0, \ a, \ d, \ e] \quad \pounds$$

$$c \equiv [0, \ a, \ e, \ f] \equiv [0, \ b, \ d, \ g] \quad \wedge$$
$$e \equiv [0, \ d, \ f, \ g] \equiv [0, \ a, \ b, \ c] \quad \neg$$
$$g \equiv [0, \ a, \ c, \ d] \equiv [0, \ b, \ e, \ f] \quad \$$$

$$c \equiv [0, \ a, \ e, \ f] \equiv [0, \ b, \ d, \ g] \quad \wedge$$
$$e \equiv [0, \ a, \ b, \ g] \equiv [0, \ c, \ d, \ f] \quad *$$
$$g \equiv [0, \ b, \ c, \ f] \equiv [0, \ a, \ d, \ e] \quad \pounds$$

$$c \equiv [0, \ b, \ d, \ e] \equiv [0, \ a, \ f, \ g] \quad !$$
$$e \equiv [0, \ a, \ b, \ g] \equiv [0, \ c, \ d, \ f] \quad *$$
$$g \equiv [0, \ a, \ c, \ d] \equiv [0, \ b, \ e, \ f] \quad \$$$

The groups within the rightmost columns all belong to the one same octal [a,b,c], [a,d,e], [a,f,g], [b,e,f], [b,d,g], [c,e,g], [c,d,f] (and I recognise this octal from the sixth trumpet as the middle ground of "**wormwood**" in the sea, formed from the reference with a=1). This device is the limit for traversing the sea: the fourth trumpet showed that third generation octals were without "blood" or intersection. A third of the day, night and moon were struck out.

First, on pouring out the vial "b = 1" I obtain:

$$
\begin{array}{llll}
c &\equiv& [a,\ c,\ f,\ g] &\equiv& [0,\ a,\ f,\ g] & ! \\
e &\equiv& [0,\ d,\ f,\ g] &\equiv& [d\ e,\ f,\ g] & \neg \\
g &\equiv& [a,\ d,\ e,\ g] &\equiv& [0,\ a,\ d,\ e] & \pounds
\end{array}
$$

$$
\begin{array}{llll}
c &\equiv& [0,\ a,\ e,\ f] &\equiv& [a,\ c,\ e,\ f] & \wedge \\
e &\equiv& [0,\ d,\ f,\ g] &\equiv& [d,\ e,\ f,\ g] & \neg \\
g &\equiv& [0,\ a,\ c,\ d] &\equiv& [a,\ c,\ d,\ g] & \$
\end{array}
$$

$$
\begin{array}{llll}
c &\equiv& [0,\ a,\ e,\ f] &\equiv& [a,\ c,\ e,\ f] & \wedge \\
e &\equiv& [c,\ d,\ e,\ f] &\equiv& [0,\ c,\ d,\ f] & * \\
g &\equiv& [a,\ d,\ e,\ g] &\equiv& [0,\ a,\ d,\ e] & \pounds
\end{array}
$$

$$
\begin{array}{llll}
c &\equiv& [a,\ c,\ f,\ g] &\equiv& [0,\ a,\ f,\ g] & ! \\
e &\equiv& [c,\ d,\ e,\ f] &\equiv& [0,\ c,\ d,\ f] & * \\
g &\equiv& [0,\ a,\ c,\ d] &\equiv& [a,\ c,\ d,\ g] & \$
\end{array}
$$

The vial poured out was chosen to be as unity: then by "every subgroup in the sea" in the following I refer only to those that do not contain the unity element "b".

Now, the four octals in which these triples are present have the sets of correspondences grouped above: instead of the "blood" (the intersection) within each of the triples being one element of {a, b, d, f}, now the intersections of the three groups of triples from each octal has become empty. (By extension every subgroup in the sea has "become as the blood of a dead man". I infer the similar situation across all 30 octals and of every subgroup shared.) A "man" is a triple, and a "dead man" is the same triple with the correct operation or "life" broken.

Whereas b = 1 should be the intersection of the three members of the second group of "wormwood" triples, now it is empty instead: the "blood" cannot circulate under multiplication or the map of Frobenius either.

Likewise, wherever the wormwood device is applied across the sea, it is also true that the intersection of any three octals in the sea under the wormwood device is empty also (as it was before) but now, the blood in the sea (these correspondences to {c, e, g} as a subgroup in this example) are now without intersection. Though I may generalise {c, e, g} to any K4 group across the sea, I may only do so for those groups that do not contain the unity. Then the vial poured out will never be a member in this group: it was as a possible unity by construction.

Most importantly the operation within those shared triples in each octal, of $(a \vee b)^c$ will always produce the complement of the remaining third member. "Every living soul" in the sea dies. The operation is now invalid, the sea not "closed". (There is no-one able to eat from the tree of life as in the previous vial.) Compare this with the K4 form of the trinity model, $\{p, u^{-1}, v^{-1}\}$, $\{p, r, s\}$, $\{p, u\&v, (r\&s)^{-1}\}$, i.e. "which wast, which art, and which art to come". God is a living God.

Every "tree" may "die" as above, but in the sense of every octal of 30, every K4 group of triples which has intersection across the sea, dies in their operations, in every octal in which they are found (when the vial is poured out on the sea).

As this occurs in every octal, with every subgroup now not containing unity (the vial), every "living soul" (with virtue) in the sea dies.

Now, back to the angel of the church in Smyrna this time:

Rev 2:11 He that hath an ear, let him hear what the Spirit saith unto the churches; He that overcometh shall not be hurt of the second death.

The second death (lake of fire) is formed of these "wrath sets" that cage the dialectic device in pairs across the sea. As the whole sea is become in pairs alike to the "wrath" and "wormwood-wrath" sets (a cage of "every unclean and hateful bird"), the angel is not caught in this trap.

To be hurt by the second death is to be without the proper operation of the octal group (to be without virtue). The angel keeps to the operation of the reference octal, out of obedience, under grace. Effectively, the seven cycle of the angel's circuit as witnessed is certain to preserve the subgroups of an octal. (Once the angel overcomes, the Holy Spirit ministers those 1260 days with every variant of his circuit.) Then, anything which Jesus "somewhat" had against him is now repented of: his lampstand truly finding His place. (With unity and two elements of his seven cycle 'fixed' the subgroups of the octal are also set under that seven cycle: the Holy Spirit has enough to then determine the "open door" in all 1260 cases to minister.)

12.4) The Third Vial

Now, I build upon those wormwood octals I obtained as I had previously: [a,b,c], [a,d,e], [a,f,g], [b,e,f], [b,d,g], [c,e,g], [c,d,f].

Rev 16:4 And the third angel poured out his vial upon the rivers and fountains of waters; and they became blood.

Here, the vial is poured out onto those octals that share three subgroups with the reference, i.e. "fountains of waters"; there is also implied the alternate case sharing one subgroup as to "the other witness", the left hand octal(s) as before, noted here by John as "rivers".

The "rivers and fountains of waters" are the waters that "feed" the sea. I note that rivers are separate to fountains as "in the Earth" – the fountains are the three-way branching of the unity "a = 1" to a triple of subgroups (with "a" as their intersection) preserved under multiplication (as within the Sun octal). So, perfection may act downward with a seven cycle as I found with the two witnesses. Under the seven cycle (a,b,d,c,f,g,e) I obtain:

$$[a,b,c], [a,d,e], [a,f,g] \; with \; [b,e,f], [b,d,g], [c,d,f], [c,e,g] \;\; with \; unity \; a = 1$$
$$[b,d,f], [a,b,c], [b,e,g] \; with \; [a,d,g], [c,d,e], [c,f,g], [a,e,f] \;\; with \; unity \; b = 1$$
$$[c,d,g], [b,d,f], [a,d,e] \; with \; [b,c,e], [a,c,f], [e,f,g], [a,b,g] \;\; with \; unity \; d = 1$$
$$[c,e,f], [c,d,g], [a,b,c] \; with \; [a,d,f], [b,f,g], [a,e,g], [b,d,e] \;\; with \; unity \; c = 1$$
$$[a,f,g], [c,e,f], [b,d,f] \; with \; [b,c,g], [d,e,g], [a,b,e], [a,c,d] \;\; with \; unity \; f = 1$$
$$[b,e,g], [a,f,g], [c,d,g] \; with \; [d,e,f], [a,c,e], [a,b,d], [b,c,f] \;\; with \; unity \; g = 1$$
$$[a,d,e], [b,e,g], [c,e,f] \; with \; [a,c,g], [a,b,f], [b,c,d], [d,f,g] \;\; with \; unity \; e = 1$$

In which there are the four columns to the right (above) termed "the Earth". (The column generated by {c,e,g} forms an octal, the other three do not.) That rightmost column becomes a "river" in the earth. Then, as {a,c,e,g} are the earthly singletons in the Sun: a river {c,e,g} is completed to be as an octal in the "left hand" or **Earth**.

Now the water of the fountains and rivers is become the "blood", i.e. those groups which intersect with groups of the Sun octal. Under the seven cycle the Sun octal is "regenerated".

Now the leftmost three groups – the "fountain" – will transform by the outpouring of the vial "d = 1" to the "wrath set" that agrees with the "wormwood-wrath set" in only one row: that same row that corresponds to "d = 1" (third row down).

Now, when someone pours something out into a river or a fountain, one does so in one spot and the river or fountain provides the movement or "flow" (here between rows). So, I expect one case to generalise under a seven cycle, rather than the general case found after the seven cycle is applied.

Then the row corresponding to "d = 1" has its leftmost three columns (greyed) mapped to their cosets and the rightmost column (greyed) mapped to itself in that row. (The seven cycle then preserves that same structure across every row.)

Now those three columns on the left once transformed also belong to the one "wrath set" as from the first vial, but transformed with a valid seven cycle as per "the mystery of the seven golden candlesticks" to the entries in each row that would correspond to a different vial poured out (as "a" to "e" on each row above).

Without loss I can use the same seven cycle of the angel's circuit here (upon the rows above), as I had applied to tread the winepress.

The rightmost column generalised will, without fail (in each case under multiplication), be the static triple without unity and the "static bow" for the "wrath set" under the seven cycle.

Then the ungodly in the seven "wrath sets" have once again been given the cold shoulder, and the switch of water for blood (and vice-versa) between these two wrath and wormwood-wrath sets place the ungodly as ready for the lake of fire, caged.

Rev 16:5 And I heard the angel of the waters say, Thou art righteous, O Lord, which art, and wast, and shalt be, because thou hast judged thus.

Rev 16:6 For they have shed the blood of saints and prophets, and thou hast given them blood to drink; for they are worthy.
Rev 16:7 And I heard another out of the altar say, Even so, Lord God Almighty, true and righteous *are* thy judgments.

The transformation to wormwood from the Sun octal by transpositions had opposed and killed the two witnesses and had also sent many a saint to compromise or even death for resisting. The wormwood's intersection of "blood" with the Sun octal to perpetuate the dialectic device in a religious setting (as in the sixth trumpet), had been at the expense of the saints.

Here, if the ungodly wish to shed righteous blood (in figure), they themselves will have to bleed it and drink it, no longer the blood of saints. (For they only share in that intersection of cosets between the wrath and wormwood-wrath sets.) Every subgroup containing virtue (a man, or "saint") is now no longer present in that intersection (blood) between the wrath and wormwood-wrath sets, they instead being replaced with cosets from the reference. Those (as formerly) "jacinth" portions are no longer subgroups but now only those cosets.

Back to the angel of the church again:

Rev 2:17 He that hath an ear, let him hear what the Spirit saith unto the churches; To him that overcometh will I give to eat of the hidden manna, and will give him a white stone, and in the stone a new name written, which no man knoweth saving he that receiveth *it*.

The angel is given a token of election – a white stone, a symbol that he may now enter the temple. Clearly, he may also now certainly dwell in the Sun octal that intersects with the wormwood octal. Those three columns under the Holy Spirit still under the correct operation of the octal (and not in the "wrath sets" or the lake of fire) are enough to ensure that given a new name, he is "born-again". That stone also shows him acquitted from the extra three of "ten times" for which he is put in prison (captivity). Those three "times" correspond to the swallowing of the "dragons tail" against his circuit. They are the same first three "churches" which place him under the device(s) of, or the "lusts" of the devil (Cf. Joh 8:44), which cannot ever form a K4 group; they are the mechanism with which he was accused and held by Satan.

The hidden manna I expect to be "sustenance in the wilderness". (The octal then complete, working in virtue within its cosets also, or "on Earth as in heaven". Cf. the Lord's prayer.) As the woman "clothed with the Sun" is nourished for a "time, times and half a time", the angel I expect is fed with the Holy Spirit as unity in the Sun octal: or rather, he is predestined for it. That predestination is "hidden" in a "thousand years with the Lord" made as "one day".

The fourth angel and his vial now requires some examination.

12.5) The Fourth Vial

Rev 16:8 And the fourth angel poured out his vial upon the sun; and power was given unto him to scorch men with fire.
Rev 16:9 And men were scorched with great heat, and blasphemed the name of God, which hath power over these plagues: and they repented not to give him glory.

If I am to pour "c = 1" out on the Sun octal I should do so as below.

$$[a,c,e,g] + c = [0,b,d,f] \equiv D$$
$$[b,f,a,e] + c = [a,b,e,f] \equiv A$$
$$[d,g,b,a] + c = [a,b,d,g] \equiv B$$
$$[c,e,d,b] + c = [0,a,f,g] \equiv F$$
$$[f,a,c,d] + c = [0,b,e,g] \equiv G$$
$$[g,b,f,c] + c = [0,a,d,e] \equiv E$$
$$[e,d,g,f] + c = [d,e,f,g] \equiv C$$

Which with the empty set as the identity forms an octal with the operation (a ∨ b) and not the complement as before.

+	{}	A	B	C	D	E	F	G
{}	{}	A	B	C	D	E	F	G
A	A	{}	C	B	E	D	G	F
B	B	C	{}	A	F	G	D	E
C	C	B	A	{}	G	F	E	D
D	D	E	F	G	{}	A	B	C
E	E	D	G	F	A	{}	C	B
F	F	G	D	E	B	C	{}	A
G	G	F	E	D	C	B	A	{}

The wrath set (of this vial poured out on the Sun) under the operation (a ∨ b) forms a group, always the complement of the result in the original wrath set under (a ∨ b)ᶜ. Under this different operation of symmetric difference, I call the "wrath set" the "wrath octal", and the "wormwood-wrath set" of the examples of rows in the fountains and rivers as before, simply the "wormwood-wrath octal" when also under symmetric difference.

[a,b,c], [a,d,e], [a,f,g] with [b,e,f], [b,d,g], [c,d,f], [c,e,g] with unity a = 1
[b,d,f], [a,b,c], [b,e,g] with [a,d,g], [c,d,e], [c,f,g], [a,e,f] with unity b = 1
[c,d,g], [b,d,f], [a,d,e] with [b,c,e], [a,c,f], [e,f,g], [a,b,g] with unity d = 1
[c,e,f], [c,d,g], [a,b,c] with [a,d,f], [b,f,g], [a,e,g], [b,d,e] with unity c = 1
[a,f,g], [c,e,f], [b,d,f] with [b,c,g], [d,e,g], [a,b,e], [a,c,d] with unity f = 1
[b,e,g], [a,f,g], [c,d,g] with [d,e,f], [a,c,e], [a,b,d], [b,c,f] with unity g = 1
[a,d,e], [b,e,g], [c,e,f] with [a,c,g], [a,b,f], [b,c,d], [d,f,g] with unity e = 1

The operation is "fire" when it forms a group: here it is given to "scorch men" with "great heat". That operation of "fire" is used as a "contact burn". The leftmost three columns that are shared with "the Sun" in the seven wormwood octals as here (above) when so added (the vial is poured out on their cosets first, to form the sets in that product) will always result in the complement of the same sum of "wrath sets" under the operation (a ∨ b)ᶜ.

Now, it is given that every K4 group in the sea is acted upon as by the "wrath set" under (a ∨ b)ᶜ and not as this angel that applies the symmetric difference instead. Every sum is with its complement "scorched" and vice-versa. Those "men" yet applying $(a \lor b)^c$ are instead scorched with great heat from this octal. (There is no multiplication or seven cycle "as of water" through the wrath set or wrath octal's elements, cf. the parable of Lazarus and the rich man.)

Now, the name of God (the octal operation) is blasphemed in using the symmetric difference. That is, although the "wrath octal" holds with symmetric difference, it is not the operation upon the Sun octal (which is still $(a \lor b)^c$ as before). Likewise those that keep to the operation $(a \lor b)^c$ are yet in the "wrath set" which no longer forms an octal under the (correct) operation.

The name of God is blasphemed either way, unless men repent without changing operation but instead give God glory (in faith: to acknowledge Christ and His virtue and give Him worship).

"Fire" is also the action of multiplication, and so this case generalises to the other wrath octals and wormwood-wrath octals, and across every position of unity in the "sea". (I multiply upon the rows and the unity element as poured out "floats" under a seven cycle of the Sun octal, just as it had in the pouring out of the vial on the "rivers and fountains of waters" before.)

The name of God, as the octal is then still blasphemed. "Men", the rightmost four entries (triples in the earth) still perpetuate the same operation of the "wrath set" $(a \lor b)^c$ rather than transform into the Sun (with the same proper operation of the "name" of God, that being $(a \lor b)^c$). They may have to withstand great heat being scorched before doing so, but if they prefer to dwell in the "wrath octal" then they are heading for the lake of fire: there is very little of it left to make. They simply did not repent to "give God glory".

And back to the angel of the church in Thyatira:

Rev 2:26 And he that overcometh, and keepeth my works unto the end, to him will I give power over the nations:
Rev 2:27 And he shall rule them with a rod of iron; as the vessels of a potter shall they be broken to shivers: even as I received of my Father.
Rev 2:28 And I will give him the morning star.

Keeping Christ's works intact to the end (preserving His word) or by obedience to the same, the angel is given power "over the nations". He fulfils the operation of the octal to the end: that his overcoming in his circuit should be as the pouring out of these seven vials of wrath – he "treads the winepress".

Rev 19:15 And out of his mouth goeth a sharp sword, that with it he should smite the nations: and he shall rule them with a rod of iron: and he treadeth the winepress of the fierceness and wrath of Almighty God.

The nations are "weakened by Lucifer" in that they are transformed to a five cycle (dropping as untimely figs). The seven cycle is interrupted by the transpositions applied by the serpents of the sixth trumpet.

The power given is then to smite them at will. There are two nations in view: one of Laodicea to death, and Philadelphia to life. The rod of iron is unflinching and absolute; there is a clean division between these two "camps".

Psa 2:7 I will declare the decree: the LORD hath said unto me, Thou *art* my Son; this day have I begotten thee.
Psa 2:8 Ask of me, and I shall give *thee* the heathen *for* thine inheritance, and the uttermost parts of the earth *for* thy possession.
Psa 2:9 Thou shalt break them with a rod of iron; thou shalt dash them in pieces like a potter's vessel.

Jesus, in clearly drawing the dividing line between His people and the ungodly, has "chosen" His angel to be a co-heir with Him. These two camps: of the godly and those within that blaspheme the name of God (in Laodicea), the same are under his authority. He may dash them together; he is stood safe and upright in Philadelphia. The "pieces like a potters vessel" (the "nations", cf. Jer 18:4-6) are now as the K4 groups and cosets of the "wrath" sets: formed of "shivers" or "broken pieces", sets under (a ∨ b)^c in "collision" with the same under a ∨ b.

Jesus also grants His angel the title of "morning star". (It is apparently the last position remaining in that capacity; it is referred to as singular.) He will never again leave the Sun octal, always found in a valid left hand or the singular right in a seven cycle. (Cf. the "son of the morning" as falling from perfection.) He is not just chosen and predestined, he is saved.

I understand the "morning star" to be another name for "A Holy Spirit". Christ has chosen His angel as the least (as unity and zero one and the same). In doing so, His people are to be separate with the dividing line drawn at the least: the one upon whom God has the least influence, not at the worst or a mere vessel for dishonour.

12.6) The Fifth Vial

Rev 16:10 And the fifth angel poured out his vial upon the seat of the beast; and his kingdom was full of darkness; and they gnawed their tongues for pain,
Rev 16:11 And blasphemed the God of heaven because of their pains and their sores, and repented not of their deeds.

The seat of the beast is the wormwood device in the sea that permits the fourth seal dialectic device to be (in deception) overlaid upon it. I know this to be "the beast risen up from the sea". The wormwood octal is then the one octal within which the four choices for unity {a, b, d, f} occur, it being the only one that shares all these correspondences in the one octal:

I.e.

$$
\begin{array}{lclclc}
c & \equiv & [0, \ b, \ d, \ e] & \equiv & [0, \ a, \ f, \ g] & ! \\
e & \equiv & [0, \ d, \ f, \ g] & \equiv & [0, \ a, \ b, \ c] & \neg \\
g & \equiv & [0, \ b, \ c, \ f] & \equiv & [0, \ a, \ d, \ e] & £ \\
\\
c & \equiv & [0, \ a, \ e, \ f] & \equiv & [0, \ b, \ d, \ g] & \wedge \\
e & \equiv & [0, \ d, \ f, \ g] & \equiv & [0, \ a, \ b, \ c] & \neg \\
g & \equiv & [0, \ a, \ c, \ d] & \equiv & [0, \ b, \ e, \ f] & \$ \\
\end{array}
$$

$$c \equiv [0, \ a, \ e, \ f] \equiv [0, \ b, \ d, \ g] \quad \wedge$$
$$e \equiv [0, \ a, \ b, \ g] \equiv [0, \ c, \ d, \ f] \quad *$$
$$g \equiv [0, \ b, \ c, \ f] \equiv [0, \ a, \ d, \ e] \quad £$$

$$c \equiv [0, \ b, \ d, \ e] \equiv [0, \ a, \ f, \ g] \quad !$$
$$e \equiv [0, \ a, \ b, \ g] \equiv [0, \ c, \ d, \ f] \quad *$$
$$g \equiv [0, \ a, \ c, \ d] \equiv [0, \ b, \ e, \ f] \quad \$$$

Become the four associations of unity, but in the one octal: [a,b,c], [a,d,e], [a,f,g], [b,e,f], [b,d,g], [c,e,g], [c,d,f]. I am interested in the case where "f = 1". (The bottom three subgroups in the rightmost column (or in the octals in triples given).)

So, the seat of the beast in every case becomes just one of the "wormwood" octals I had before. (Cf. Lucifer, as sat in "the sides of the north".)

$$[a, b, \ c], [a, d, \ e], [a, f, g], [b, \ e, f], [b, d, g], [c, \ d, f], [c, e, g]$$
$$[b, d, f], [a, b, \ c], [b, e, g], [a, d, g], [c, d, \ e], [c, \ f, g], [a, e, f]$$
$$[c, d, g], [b, d, f], [a, d, \ e], [b, \ c, \ e], [a, c, f], [e, \ f, g], [a, b, g]$$
$$[c, \ e, f], [c, d, g], [a, b, \ c], [a, d, f], [b, f, g], [a, \ e, g], [b, d, \ e]$$
$$[a, f, g], [c, \ e, f], [b, d, f], [b, \ c, g], [d, e, g], [a, b, \ e], [a, c, d]$$
$$[b, e, g], [a, f, g], [c, d, g], [d, e, f], [a, c, \ e], [a, b, d], [b, c, f]$$
$$[a, d, \ e], [b, e, g], [c, \ e, f], [a, \ c, g], [a, b, f], [b, \ c, d], [d, f, g]$$

Although I had restricted myself before to just the one row corresponding to unity, the seat of the beast is as every row, and I may equally assume that unity is arbitrary and use "f = 1" as the vial poured out on each row.

Every row has the subgroups containing "f" mapped to their cosets, and every other group is unchanged.

Then the kingdom of the beast (these seven rows) each are without the unity "f = 1", and likewise no sum of subgroups under symmetric difference will produce a subgroup with virtue (light), only the coset. (There is only darkness.)

As the subgroups containing "f" are mapped to their cosets, there is no "virtue" as a solution to the addition of the remaining subgroups or within those cosets under the operation $(a \vee b)^c$. The kingdom of the beast does not contain the closure of the operation on this set (row) and is then "full of darkness" instead, unable to produce a product that entails "light" or "virtue" in each row. (Those three subgroups are now missing from the row or "kingdom".)

Likewise, under symmetric difference, should men be so scorched, the sum never results in a set that produces a subgroup containing virtue. Symmetric difference is closed on each row.

The text states they "gnawed their tongues for pain".

Their "tongues" indicate (in the wrath) that their subsequent and alternate "virtues" are instead now under "grievous and noisome sores" – the empty set. They have nothing (an empty mouth) to oppose virtue in dialogue now they are under the wrath; that is, unless they shift from the proper octal to the operation of symmetric difference and to the lake of fire.

Now, there is no seven cycle preserving any of the "wormwood-wrath" octals, and so the three cycle required by Frobenius for a static triple is somewhat without action. That stated: the wrath sets under the action of the reference seven cycle do hold the three cosets of the subgroups containing "f" static in a cycle. Say, with "f = 1" and under (f,g,e,a,b,d,c) then [b,c,d,e] → [a,b,d,g] → [a,c,e,g]

Similarly, under the reference cycle the static triple [b,e,g] corresponding to "f = 1" is static (as it is sent to itself; it does not contain "f = 1"). Then the remaining three subgroups are also sent to themselves in a three cycle: [a,d,e] → [a,b,c] → [c,d,g].

So given the operation of symmetric difference is then to be "scorched" and I otherwise still operate with $(a \vee b)^c$ I may expect the seven cycle to remain in effect in that latter case, although the seven cycle does not act on both subgroups and cosets.

There is a conflict of the seven cycle. One application is upon cosets under a seven cycle only, and the other upon subgroups only. All are either cosets or subgroups: there are not both.

To "gnaw one's tongue for pain", then, is to be without one static triple or the other.

The men blaspheme because of their pains and sores. The sores were simply that they could not transform back to a subgroup containing virtue. So, also, their pains are from being scorched with a different operation, as well as with this gnawing at one's tongue.

The elements of the seven cycle cannot map between subgroups and cosets, or else the set is not closed. (A contradiction, or "blasphemy" against the God of heaven: of unity, or of multiplication.) However, the three cycle appears to hold if one remains only in a static triple.

Under symmetric difference the set (row) is closed, but there is no seven cycle from the "Sun" at all, and to use a three cycle is to blaspheme God, to gnaw at one's tongue to perpetuate the three cycle where there is none. (Are they God? Christ alone could open the last three seals.)

So, if the set is not closed there is no multiplication and so no Frobenius map, and no three cycle. If the set is closed then there is a different operation that allows no seven cycle to act on the row now the vial is poured out (and no subsequent three cycle).

Those three cycles are certainly possible, but are not natural, and are "painful". (They are not compatible with the octal operation itself. GF(8) has no subfield GF(4) (every element is either a coset or else they are all subgroups), and if the operation becomes the symmetric difference, there is no seven cycle upon which to base a three cycle: the unity is **missing**.)

The result is to gnaw one's tongue for pain. (To attempt to make a seven cycle where there is none, blaspheming the Holy Spirit, i.e. The God of heaven (as multiplication): blaspheming His action as if blaspheming the floating unity in the octal, and so the virtue in the lamb.)

Gnawing or "chewing" the contents of one's empty mouth (set) is in view.

They blaspheme because of their sores (the set is either not closed or is "full of darkness") and their "pains" – to be scorched with "symmetric difference" is to be without a seven cycle: pained chewing one's empty mouth (the "set" of elements, unable to construct cycles between cosets and subgroups, contradictory and "empty", but with something within: a tongue?).

The tongue also refers to the method of dialogue with the three cycle: it is certainly interrupted.

Back again to the angel of the church:

Rev 3:5 He that overcometh, the same shall be clothed in white raiment; and I will not blot out his name out of the book of life, but I will confess his name before my Father, and before his angels.

The angel is clothed as the lamb, and will always have that covering of virtue (light). He will not lose that covering, as always in the book of life. Before His name may be revealed in the book (when opened), Jesus will confess his name as saved before His Father and His angels.

Clearly the angel is never to dwell in a kingdom full of darkness, and is certainly accepted – he does not blaspheme asserting he is saved. (He has attained to the unity; he has a seven cycle.)

12.7) The Sixth Vial

Rev 16:12 And the sixth angel poured out his vial upon the great river Euphrates; and the water thereof was dried up, that the way of the kings of the east might be prepared.

The sixth vial is poured out upon the great river Euphrates – in view are four "rivers": the two sets of four possible left handed octals for the Sun octal. (This opens up the "wider way"; the Euphrates is fed by four tributaries.) That "wider path" is of each river (left hand octal) and the three other "earthly columns" which are those that do not form octals under the seven cycle.

Under the one seven cycle (a,b,d,c,f,g,e) the rightmost four columns are acted upon by the seven cycle but only the rightmost forms an octal.

$$[b,e,f], [b,d,g], [c,d,f], [c,e,g] \quad \textit{with unity } a = 1$$
$$[a,d,g], [c,d,e], [c,f,g], [a,e,f] \quad \textit{with unity } b = 1$$
$$[b,c,e], [a,c,f], [e,f,g], [a,b,g] \quad \textit{with unity } d = 1$$
$$[a,d,f], [b,f,g], [a,e,g], [b,d,e] \quad \textit{with unity } c = 1$$
$$[b,c,g], [d,e,g], [a,b,e], [a,c,d] \quad \textit{with unity } f = 1$$
$$[d,e,f], [a,c,e], [a,b,d], [b,c,f] \quad \textit{with unity } g = 1$$
$$[a,c,g], [a,b,f], [b,c,d], [d,f,g] \quad \textit{with unity } e = 1$$

I have other such arrangements under each one of the possible multiplications, giving one river in every four columns.

Each left hand is a tributary of the "River Euphrates". The vial "g" poured out on the rightmost column sends subgroups containing "g = 1" to their cosets and other subgroups to themselves.

What is important is that as in the last example also, there is now no valid seven cycle acting upon the "Euphrates"; the operation does not preserve the tributary's cosets and triples.

A lack of multiplication appears (as if water) to have dried up the river in view. There is no valid operation that will transform between the rows of the earthly columns. These triples formed implicitly defined octals – a result of the multiplication (on the triple that was a static "bow"), yet the four earthly columns, each based on a possible static triple, have now "dried up".

Without multiplication there is no "sea" in the sense that the left hands can be constructed, and so, the triples in the rightmost four columns become "dried up", not just the left hand.

God is then separating His elect from the unsaved. (Ready for the harvest of the Earth.)

I would think of the octals in the sea as if they aligned in columns as the "way of the kings of the East" prepared ready for following that star (static subgroup or unity).

Now there are no more "left hands" I may align the octals in the sea in additive isomorphism, as in "isometry": i.e. with a valid seven cycle so that the sets of octals in rows here:

$$[a,b,\ c],[a,d,\ e],[a,f,g],[b,\ e,f],[b,d,g],[c,\ d,f],[c,e,g]$$
$$[b,d,f],[a,b,\ c],[b,\ e,g],[a,d,g],[c,d,\ e],[c,\ f,g],[a,e,f]$$
$$[c,d,g],[b,d,f],[a,d,\ e],[b,\ c,\ e],[a,c,f],[e,\ f,g],[a,b,g]$$
$$[c,\ e,\ f],[c,\ d,g],[a,b,\ c],[a,d,f],[b,f,g],[a,\ e,g],[b,d,\ e]$$
$$[a,f,g],[c,\ e,\ f],[b,d,f],[b,\ c,g],[d,e,g],[a,\ b,\ e],[a,c,d]$$
$$[b,\ e,g],[a,f,g],[c,d,g],[d,e,f],[a,c,\ e],[a,b,d],[b,c,f]$$
$$[a,d,\ e],[b,\ e,g],[c,\ e,f],[a,\ c,g],[a,b,f],[b,\ c,d],[d,f,g]$$

Rearrange to the columns:

$$[a,b,\ c],[b,d,f],[c,d,g],[c,\ e,f],[a,f,g],[b,\ e,g],[a,d,e]$$
$$[a,d,\ e],[a,b,\ c],[b,d,f],[c,d,g],[c,\ e,f],[a,f,g],[b,e,g]$$
$$[a,f,g],[b,\ e,g],[a,d,\ e],[a,\ b,\ c],[b,d,f],[c,d,g],[c,e,f]$$

$$[b,\ e,f],[a,d,g],[b,\ c,\ e],[a,\ d,f],[b,c,g],[d,e,f],[a,c,g]$$

$$[b,d,g],[c,\ d,\ e],[a,c,f],[b,\ f,g],[d,e,g],[a,\ c,\ e],[a,b,f]$$
$$[c,d,f],[c,f,g],[e,\ f,g],[a,\ e,g],[a,b,\ e],[a,b,d],[b,c,d]$$
$$[c,\ e,g],[a,e,f],[a,b,g],[b,\ d,\ e],[a,c,d],[b,\ c,f],[d,f,g]$$

Where the middle row would correspond to "unity": i.e. it contains the appropriate static subgroup for this rearranged system. In this manner the possible "rivers" (now acting left to right) become the "way of the kings of the East".

The "way of the kings of the East" refers to the alignment of all 5040 = 7! = 7*6*5*4*3*2*1 arrangements of octals in seven symbols to the one isomorphic template of GF(8) alluded to by the columns above (as rows transposed).

John no doubt wrote them (as above) out in sequence from right to left, top to bottom (as from the East) as three kings (of the lower three rows) following after the centre row subgroups in turn moving under the stars in heaven (the triples or stars in the top three rows) the kings follow after the static subgroup in the centre. (The one new "bright and morning star" of Christ, the sign of His birth.) I need not write out them all but simply note they are isomorphic.

I can only laugh at John – exasperated with the possibility of writing out 5040 permutations of the octal (and far more in triples) aligned isomorphic, right to left. I can only smile knowing that many Christians expect an army of 200 million Chinese soldiers will invade Israel. I deem it utterly beautiful that Satan should account monies for that task, and would have foreseen that he would have to save up for expenses like that! Jesus must have cracked a smile.

Returning to the angel of the church;

Rev 3:12 Him that overcometh will I make a pillar in the temple of my God, and he shall go no more out: and I will write upon him the name of my God, and the name of the city of my God, *which is* new Jerusalem, which cometh down out of heaven from my God: and *I will write upon him* my new name.

The angel, having fulfilled the reference cycle or "rainbow" about the throne, is become the work circuiting all seven churches (actually to remain in the sixth). The temple of God refers to this rainbow cycle across the seven "churches" (properly lampstands) of the outer court.

Zec 4:10 For who hath despised the day of small things? for they shall rejoice, and shall see the plummet in the hand of Zerubbabel *with* those seven; they *are* the eyes of the LORD, which run to and fro through the whole earth.

His circuit (in doctrine) is from conversion (baseline) to the height of the seven cycle, he becomes a "pillar" in the temple of God. He need go no more out. He may rest as unity in the seven cycle: that "rainbow", that "pillar", which finds rest in Philadelphia. (Without needing to "cycle" back to Ephesus.) The angel's circuit itself is become the "pillar" in the temple.

Having those names written on him is a contrast to being completely rejected: he has God's new name written upon him, and that of Christ: his circuit will fulfil its requirements in GF(8).

The angel's circuit is this act of writing (as if written by God's own hand: he forms a "lively stone" (a pillar) in the temple of God). Not only does he complete the seven cycle and form an octal (the name of God), he also becomes the final completed act of satisfaction for God to save. (As the very least.)

As the "New Jerusalem", those that are worthy of the kingdom are those that are convicted by the angel's circuit (for the spirit may convict all). He also traverses the seven lampstands as does Christ. In the act of writing this upon him (as heaven descending in a seven cycle, at rest in each of the seven churches) He writes the same name (as of God) as that of His people (the 144,000 as the 24 elders).

Christ's new name, however, is a mystery. As the very least (one like the Son of man) is saved, Christ is given a name as omnipotent God:

Dan 7:13 I saw in the night visions, and, behold, *one* like the Son of man came with the clouds of heaven, and came to the Ancient of days, and they brought him near before him. Dan 7:14 And there was given him dominion, and glory, and a kingdom, that all people, nations, and languages, should serve him: his dominion *is* an everlasting dominion, which shall not pass away, and his kingdom *that* which shall not be destroyed.

It appears the least is not so because he is saved, but he is the least in all creation by principle. The new name of God then I expect to be such that the angel (being the least) is unique in the kingdom. With dominion over him, He (Jesus) in principle has dominion over all.

Satan has a further attempt with three spirits "like frogs" (that dwell in water and land (sea and Earth?) or at the boundaries of the two. They dwell as if living about the "lake of fire").

12.8) Three Spirits As Frogs

Rev 16:13 And I saw three unclean spirits like frogs *come* out of the mouth of the dragon, and out of the mouth of the beast, and out of the mouth of the false prophet.

Given that every octal in the sea under every possible multiplication is in alignment to the general representation in polynomials (powers of x)

$$x^1$$
$$x^2$$
$$x^4$$
$$1$$
$$x^3$$
$$x^5$$
$$x^6$$

Without any valid left hand multiplication, and given that the ungodly are found caged in "wrath sets" the mouths (sets) of the dragon, beast and false prophet are all dependent on three cycles of earthly subgroups. Now, there is no seven cycle preserving such a three cycle on the "wrath set" (see the fifth vial).

These three "unclean" spirits fulfil the requirement for these missing "three cycles". They do "short hops" across the now aligned sea, or so hop between octals aligned in the "way of the kings of the east". Their three-cycles then hop across octals as above, yet appear to allow the earthly triples outside the static subgroup (which itself is possibly fixed) to cycle.

Frogs dwell at the boundaries of water and earth. I.e. a "lake" and "the earthly coset" in the polynomials above.

Now, the spirits come out of the mouth (the cosets of subgroups, that common speech of the dialectic devices). Now there is only one possibility for unity, that found in the "Sun". I find the possibility that the other three possible choices for unity are aligned elsewhere in the isometry (the way of the kings of the east): that is, to coin the phrase, in the "lake of fire".

This is a clear picture of a small sea under the action of unclean spirit (mimicking Frobenius in contradiction over the "wrath set". This is again deceptive as I am simply relabelling what is in the same Sun octal). There are three possible such balances in the "lake" aside from the field GF(8)'s usual operation under Frobenius. Given a ≡ 1 ≡ [b,d,f] ≡ {c,e,g} static, I may associate the three valid powers of the induced 3-cycle (under Frobenius) in the Sun octal with:

$$c \equiv \{a, e, g\}$$
$$e \equiv \{a, c, g\}$$
$$g \equiv \{a, c, e\}$$

with c ≡ 1 ≡ [b,d,f] ≡ {a,e,g} static (now in the lake of fire rather than the Sun within the sea of glass) I then associate:

$$a \equiv \{c, e, g\}$$
$$e \equiv \{a, c, g\}$$
$$g \equiv \{a, c, e\}$$

and also with e = 1

$$a \equiv \{c, e, g\}$$
$$c \equiv \{a, e, g\}$$
$$g \equiv \{a, c, e\}$$

and last of the three with g=1:

$$a \equiv \{c, e, g\}$$
$$c \equiv \{a, e, g\}$$
$$e \equiv \{a, c, g\}$$

The three unclean (unclean with respect to "ungodly") spirits perform "miracles" by lies, they act as if in contradiction. (The wrath set has no multiplication yet these form a three cycle.)

They also appear to make a three cycle in the earthly elements even when the static subgroup is "fixed" not merely static. Across the "sea" now aligned in isometry, the three spirits may in every setting mimic Frobenius – with the aim that the "Kings of the earth" are deceived.

The frogs perform a three cycle of short hops in mimicry of the Frobenius map. They operate at or in the wrath octal(s), at the "lake of fire" where the beast and false prophet already *are*. (Christ has no satisfaction in any there.) They do not form "Death" fourfold but only a single balance of three bows. All remaining bows are supposedly found elsewhere in the isometry.

Without a sea to permit the construction of the wormwood device there is no basis for that dialectic process to rest upon: the earthly elements in isometry to the Sun octal are all that are left. The three spirits as frogs which form three cycles require a "bridge" between three further octals in the sea to do "short hops" upon what are essentially three labellings of the same three elements. These other arrangements are then of "The kings of the Earth".

In order to "animate" the dialectic device by appearance in the Sun octal, each of the beast (the collective antichrist or scarlet beast), the false prophet (the collective voting their heart's desire under the three kings in the image) and dragon (the god of this world as deceiver, as the dialectic used in the false doctrines from Israel's history) use or project themselves upon these devices of three "frogs" (on triples corresponding to one cycle of x^3, x^5, x^6) in order to function.

These spirits in the "set of the three" are lies (and deceiving), because those sets are without a seven cycle to generate the static triples! (Observe the fifth vial of wrath.) I find a method that produces a result from nothing, or contradiction, a "miracle".

The effect is to mimic the multiplication of GF(4) to enable the methods of the diabolical three – as if the beast, false prophet and dragon were "truth" or equivalently argue that all of the three are indispensable to the dialogue, the commerce of doctrine by the methods of antichrist. This is done by them all projecting themselves into the same triple of frogs.

Rev 16:14 For they are the spirits of devils, working miracles, *which* go forth unto the kings of the earth and of the whole world, to gather them to the battle of that great day of God Almighty.

The unclean spirits (devils, mimicking Christ) work miracles (forming three cycles on the wrath sets where there is no seven cycle as a basis) and have gone to the "Kings of the Earth" that (across octals in the isometry) would construct the dialectic system of "Death". That system is generalised to every possible arrangement of every octal (i.e. of the whole world), to gather all to the battle between the beast and the lamb, who is far stronger (as for great tribulation).

The spirits of devils are now completely opposed to the operation of octal addition with virtue excepting the apparent product of the "wormwood device" which supplants the K4 form. By crossing octals in the sea, they may be considered to be somewhat normal or "perpendicular" to the proper action of the Holy Spirit and are totally rejected by God (as devils). They employ a device similar to the transpositions under the serpents of the sixth trumpet. The "similar" device is to form these three "unclean spirits" for three cycles rather than transpositions.

The kings of the Earth – those crowned as if with the three elements of the first seal "bow", now aligned in every combination as "of the Earth" (four elements aligned) and "of the whole world" (every octal) are the target of these unclean spirits. (The kings form the only triple aligned isomorphic in the sea of glass: it is now the rulers that are in dialogue as those identified with the mark, or as of the whole world – the single collective of all.)

These three spirits are a deception that the dialectic device is still in effect though the sea is now aligned as "glass". The deception is across a smaller set of octals (now an uncomfortable one) and the three state that the time is "not yet" though the scripture states in fact it *is time*.

Rev 16:15 Behold, I come as a thief. Blessed *is* he that watcheth, and keepeth his garments, lest he walk naked, and they see his shame.

Jesus "comes as a thief" – i.e. when it isn't realised something is missing. (He has stolen some time.) The text has shifted once more with the seven thunders (cf. Rev 16:18) backwards in the Revelation narrative. The vials are moved back in time to the period before the last trump: to complete at the opening of the seventh seal. (The moment of Jesus Christ's return to judge – the last day.) This is the moment when the 24 elders fell down before the throne for the opening of the seventh seal. (The end of the seals, trumpets and vials all align on the last day.)

This small place gathered together, bordering the "lake of fire" and the Sun octal's earthly four elements, is the battleground. Those four elements are the last straw: the point at which God claims the octal He has Himself for a name, and the lake of fire "outside" Him is left to its own devices (failing in the Sun octal): devices present only in the lake of fire which as place and people certainly have no claim on a system belonging to God and outside of their paradigm.

Rev 16:16 And he gathered them together into a place called in the Hebrew tongue Armageddon.

This alignment of Earth, sea and the three woes into the one footstool (leaving only the true vine and that of the Earth) under God (in the four beasts) is the place called "Armageddon".

Armageddon:
Armageddon transliterates in English as "a range of hills" and also a "rendez-vous". The term clearly references the "mountains" gathered into one octal as of the sea of glass, a single "range of hills" meeting in one place and at one time (as if "time no longer"). These mountains are the octals of wormwood cycled under a fiat circuit claimed as of the Holy Spirit by the Church – each octal a stronghold of the devil's devices – as if unity were floated amongst the seven rows of wormwood, each row a foundation for the dialectic system of "Death".
Whilst these wormwood octals have their place as fields under a valid isomorphism to the reference, they will not have that seven cycle which cannot act on them as wormwood. Instead the devices are "Gog and Magog" ("antichrist", and "land of antichrist"; the false doctrine's source), present in the churches as "seven nations", with the scarlet beast of the "Church" in false doctrine constructed as the eighth "of the seven". The least (angel) does not circuit in wormwood, by this point he has exited the Church to remain separate.

12.9) The Seventh Vial

The seventh vial is poured out into the air (corresponding to "e = 1" say). The octal is then completely trodden as a winepress. Surviving in the Sun octal are the elements not in the "wrath set" upon which the vial is poured, i.e. "e = 1" (the last vial) is preserved aside from each of the wrath sets' entries, or equivalently the three groups [b,e,g],[a,d,e],[c,e,f] are excluded. These form the K4 group preserved in Christ as I had expected.

Rev 16:17 And the seventh angel poured out his vial into the air; and there came a great voice out of the temple of heaven, from the throne, saying, It is done.

The "air" is "lifeless" spirit. I assumed in every octal in the isometry (in powers of x before), that the three unclean spirits cycle only the earthly static triple. These frogs do not cycle the three elements of the static subgroup, else then they would be as God, "clean", preserving the octal's multiplicative operation.

I may then have static "bows" in the isometry, that do not correspond to the proper action of Frobenius at all. Alternatively, the four earthly elements with the static subgroup fixed are once again the recipients of the vial "e = 1".

The same subgroups containing unity are sent to their cosets, and they are not cycled under these three frogs. These cases are where the seven cycle itself may not act across octals in the isometry and so these octals together are "inanimate".

There are 24 such (multiplicative) arrangements in the sea holding a static subgroup totally fixed. (Permuting only the earthly elements.) Yet, there is no power of a seven cycle mapping between any two of these in the isometry: the static subgroup is totally fixed. (Also, 12 arrangements cannot be constructed as in the sixth trumpet.)

Then with these last 24 "hours", at this conflict between the automorphisms of the valid field GF(8) with its eight multiplications opposed with these last 24 arrangements of the "frogs", I have the completion of the wrath poured out on "the last day".

The wrath is poured out onto the cosets of the octals that intersect with the Sun octal in only one subgroup. (Aligned with the static subgroup totally fixed, lifeless as without a seven cycle.)

These fourteen cases (two octals for each subgroup), are such that there is no valid arrangement of any octal with multiplication that permits the same dialectic devices under those three unclean spirits. By pouring out into the air, every last case across the sea of 30 octals is now accounted for. (The possible left hands were dealt with in the vial poured out on Euphrates.)

Only the legal Frobenius map on the Sun octal remains, and as there is no more "sea", there is only the sea of glass – there is then only the first seal paradigm.

There is a great voice (judgement) from the throne in the temple. The pre-prepared work of treading the winepress is completed. That judgement is made between those that hold to virtue (the elect) and those that hold to the dialectic devices, who are in receipt of the wrath, dwelling in either the wrath set or wrath octal: both are now "caged".

The winepress now trod, the circuit of the angel already complete, this work of Christ in writing His name on His angel is finished. (He may now also sit on the throne.)

Returning to the angel of the church once more:

Rev 3:21 To him that overcometh will I grant to sit with me in my throne, even as I also overcame, and am set down with my Father in his throne.

In treading the winepress complete, I see the circuit of the angel is then finished in victory.

The seventh vial aligns neatly with the last seal, the vial is poured out into the air (the lifeless replacement for pneuma or spirit).

The situation returns to that before the restrainer was loosed: the ungodly have fallen under the tribulation of the wrath sets to the wrath octals, dwelling in the Earth upon which the vials were poured out. Those in the static subgroups and the K4 form of the filter (sharing only unity from the octal as an intersection) escape from the wrath. (And survive having come

through great tribulation.) With the judgement, those in the wrath set have the door shut (operation closed with symmetric difference) and are left to dwell in the wrath octals forever.

The isomorphism of the wrath octal to the Sun octal without the action of the Holy Spirit is not itself by definition "torment": both those being abelian groups with commutative operations. The language of Revelation does not make this point clearly enough! At the judgement, it could well be a defence for the ungodly sinner that torment is never necessary for anyone in Hell, unless they decide to continue to worship in the scarlet beast system with its image.

The great voice is the judgement that the wrath is completed justly and the temple is now reopened so that the throne may be "set" once more. (The work of the seven thunders is completed.) There followed voices (judgements), thunders (the movement of heaven at the seventh trump to align the trumpets with the tribulation rather than following the seventh seal), lightning (great falls of angels to the Earth to form devices from those "stars" (octals in the sea) upon churches (fellowships) using those devices to leaven the Church whole). There is also a massive earthquake (movement of the earthly four elements) as every arrangement (and reversal) of the Earth is shaken to align to the Sun octal and all variance is ironed out flat.

Rev 16:18 And there were voices, and thunders, and lightnings; and there was a great earthquake, such as was not since men were upon the earth, so mighty an earthquake, *and* so great.

The great city is split into three parts (i.e. those of a synthetic manifold):

1) The image of ten kings (the abomination making the church desolate)

2) The source(s) of false doctrines within the beast or collective exercising their mark (the "synagogue of Satan")

3) The five churches under the locusts (the Nicolaitans)

The effort to create "Hell" as to complete the fourfold dialectic system of "Death" (which is inconsistent) fails to complete its "enterprise". The fourth synthetic (carnal) dialectic created to fill the void of inconsistency fails to be "new" or "noumena" and the image system is simply imputed as the sin of the believer in that system.

The three parts into which the city splits are the image (of ten kings and the collective), i.e. the beast – then secondly the false doctrine, and last the churches blindly accepting it. There are only three "manifolds" in the city (as S5), not four. There is no replacement for the inconsistent "fourth part" to be filled by a synthetic fifth: the image system of false doctrine simply aligns naturally with the desire of the believer's heart and is simply to "go into perdition" (as a sin). The image is no use to Satan at all beyond that of an insult to send the Church all the way bad.

Rev 16:19 And the great city was divided into three parts, and the cities of the nations fell: and great Babylon came in remembrance before God, to give unto her the cup of the wine of the fierceness of his wrath.

Then Babylon is brought in remembrance: that is, the false doctrine overcome by the angel (that is her commerce) is once again in view, that being the dialectic (as also in the dragon's tail) that had accumulated upon itself fourfold on her past. (With the template of MYSTERY BABYLON present upon or over them all.)

Every construction of S6 on each of the seven "kings" (as if a church but not so) is a fallen enterprise; each church of seven (the nations) may be "healed" with the mappings from the angel's circuit as ministered by the two witnesses. There is no true "logical deception", all the elect are commanded to be separate by Christ. This sevenfold construction is "the beast".

Each "Nation" may refer to the six Old Testament sources of false doctrine found within the history of Israel. The "cities" are as (being equal to S5) the five churches (leavened by the synagogue of Satan, the source of that doctrine), and by symmetry, they are just as fallen. (Each being a "source" of such leaven within the image system of the scarlet beast.)

Just as easily, in a "tower" of structures I compare the dialectic to the accumulation of the dragon's tail. Then, the "cities of the nations fell": there are four sets of people:

Rev 17:15 And he saith unto me, The waters which thou sawest, where the whore sitteth, are peoples, and multitudes, and nations, and tongues.

As the cities of the nations (every such city, as S5 constructed over each C7 group or "church" but that of the angel's circuit) indeed fall, I would expect the fifth construction – the image system with the "MYSTERY BABYLON" template (as for a corporation) to become the failing replacement for that set (supplanting the fault in "tongues"), so that it would "plug the gap" of inconsistency. "Peoples, multitudes, nations and tongues" is then to account for all "differing faiths" under a common but contradictory language (tongues or "spiritual diplomacy"). Likewise, as the great city is a corporation, all corporations are fallen likewise.

I do not expect language itself to "fall", but collectives agreeing in the dialogue to fail instead. Those "peoples, multitudes, nations, and tongues" fall which are the greatest set of those collectives. With the collectives ruled over by the beast and its image (in dialogue of "tongues" in all matters spiritual) the image replaces the ultimate set of "collectives" in false unity. The doctrine of every collective (Christian or not) becomes equal in sin to that raising up of every possible false doctrine (as if salvation to those that need it but as such it will fail) from every other such "collective". These "Christians" would claim equality with all manner of corrupt corporations (perhaps even Satan himself) rather than admit they've failed and must repent.

"Peoples" are divided on the lamb – they have at least a second seal paradigm. Multitudes operate completely in worldly freedom – by choosing or deciding on virtues (possibly using the dialectic) and so may number 666 as antichrist in "bows". "Nations" are those dwelling in the seven churches but here they operate by a five cycle (first woe). Tongues dialogue on different virtues, with a third seal paradigm or with wormwood (second woe) at least. The harlot rests on all these devices – on divided peoples with the dragon's devices.

With the devil's own constructions failed I am brought back to the present equivalent in the seven cycles preserving the reference octal as found in each witness to left and right. The sharp sword with two edges is in full effect.

The three parts are also now as the Sun octal in the centre with the branching of structures to the left and right, as was seen with the two witnesses lying dead "in the street of the great city" so then there are potential bows to the left and right – forming two systems numbering 666 to left and right, with the Sun octal's seven cycle of a $\equiv 1 \equiv$ [b,d,f] $\equiv$ {c,e,g} in the centre. This aligns neatly with Daniel's prophecy of the kings of the north and south (static subgroup to bows in cosets: made of pairs to either side in "reciprocation").

Dan 11:45 And he shall plant the tabernacles of his palace between the seas in the glorious holy mountain; yet he shall come to his end, and none shall help him.

Babylon comes to remembrance as those three parts: the great city (to whom the two witnesses ministered) is again in view. It had exited the scene, unable to be constructed: the missing twelve elements of S4 in S5 needing to be added to complete it (to shut the door) are answered with the fierceness of the Lord's wrath. In one move He damns the whole system to become the lake of fire, equivalently ejecting it from the reference octal and its seven cycle(s).

Rev 16:20 And every island fled away, and the mountains were not found.

Those seven sets of false doctrines (mountains) are not enough for a person to claim they were logically deceived: even the least could overcome them all. They are "no longer found".

Islands and mountains, I discern as I did before: "mountains" are spiritual strongholds (godly or deluded) and "islands" their counterparts across the whole sea of thirty octals. Mountains are found as sharing the K4 form with the Sun octal and they are present in the sea as "wormwood". Islands are instead the intersection of earthy triples between the Sun octal and other octal(s) in the sea. (Mountains are octals as under the "mystery of seven candlesticks".)

Every island fled away (Islands are an "Earth" in the sea – analogous to a triple [c,e,g]) so every triple becomes as one in the "Sun". There are no possible "left hands": only the K4 form of the ultrafilter that spans the octal "in the centre". The "Sun" becomes as a "generalised" case.

Whilst the woman did "fly" into the wilderness before (with a valid map), the islands instead "flee". I expect they become "not found" or invalid. Islands are an "earth" in the "sea". With the sea aligned in isometry, every "island" is isomorphic to the earth in the Sun octal. Traversing the sea using the dialectic devices (mountains) is now impossible, or at least very difficult.

The earthly elements in the Sun become the Lord's footstool. The "mountains were not found". The mountains upon which the Church has built her places of worship (her doctrine) are gone. (They are "not found" to be valid, cf. the "rapture".) These are not the kingdoms of the first beast or dragon but are the strongholds or false doctrines in the "Church" accumulated by the dragon's tail. (The little horn of false doctrine from OT Israel and Judah that survived the seven kingdoms of the dragon: the false doctrine in the Church is the "mountain" made of it.)

If every mountain is now gone, then those maps from non-identity elements in the octal to the identity are certainties: those six elements corresponding to the circuit of the angel, seen as present before the throne as the lampstand of Laodicea. The only element remaining upon each mountain is the unity element of the slain lamb. (As if 7000 die in the earthquake.)

There is no Laodicean rest in each church; the elect are under conviction in all seven churches.

In this I may recover the operation of the two parts "cut off" not present in the octal structure of GF(8), but present only in the GF(4) form of the filter. (Excluding those to the left and right that number 666, i.e. that do not repent as have those in unity "a" and also those that overcome between the triples [b,d,f] and {c,e,g}.)

So six churches (mountains) are now gone: only the unity remains as Christ triune (in both the reference octal and the system of "wormwood" in the sea; a ≡ 1 ≡ [[a,b,c], [a,d,e], [a,f,g]]).

Every other group (of the possible 35) is found to either the left or right hands, so if there is one isomorphic octal across the sea and they must all align as under the seven cycles, then there is no more set of "30 octals" (no more sea) and the K4 form will always remain.

Rev 16:21 And there fell upon men a great hail out of heaven, *every stone* about the weight of a talent: and men blasphemed God because of the plague of the hail; for the plague thereof was exceeding great.

A great hailstorm occurs, the seven cycle acting downward to John (its reversal appears to act (fall as hail) on the left hand octal(s) aligned in additive morphism to the Sun octal). Every octal in the sea should be aligned by isometry to the Sun octal. There is no "hail" if there is no more "left hand", except for the reversed alignment of singletons of the seven cycle (left/right hands are opposed in the order of the seven cycle) when only the right's seven cycle remains. The seven cycle itself becomes the "hail". Each hailstone is then a singleton and is representative of the K4 form which holds that singleton as unity (the intersection of those three subgroups).

There are eight left hand octals for every right hand: so, I expect the hailstones to fall eightfold.

The "talent" is the indicator of election (as in gold) and is represented by the unity in that K4 form. Each hailstone is now "the weight of a talent" under the action of a seven cycle as hail.

The alignment of every field GF(8) multiplies the hail (upon every group including the reference octal) as opposed to what would be previously fire (as burning up). Every octal is therefore under a seven cycle in alignment so the "hail" is greater than any before it in the Revelation. (As hail where each stone is as [[a,b,c],[a,d,e],[a,f,g]]. I will not supply a picture of a snowflake, but perhaps there is a word picture to be found there, maybe.)

One may consider the Sun octal as still having a left handed octal paired to it – which is now also become in alignment: so, whilst the Sun octal receives fire, so also does every other octal, and as all are in alignment every octal receives hail eightfold as well (also the Sun octal). Every octal in the sea is in receipt of this great plague of hail. (The hail falls on left handed octals only; and therefore, only upon those practising the dialectic in the "lake of fire".)

The ability to find rest in Christ (to be saved) is blasphemed by those that have no rest, they having only "blood to drink" and sitting in the centre of two systems of 666 to left and right (yet not as unity), they refused to repent, and I find the weight of a "talent" in reference to the token of gold – the election as unity within the hail itself as a seven cycle. The world could and should repent but prefer the devices of the dragon to the salvation of God in Christ.

The hail itself is as a singleton in a seven cycle: the intersection of the three groups common to both Sun and "wormwood octals". So given a ≡ [[a,b,c], [a,d,e], [a,f,g]] has "a" as the hailstone and is the same as the floating unity in any row of the seven wormwood octals under the seven cycle, it is become a token of salvation when it is considered that Babylon is come into remembrance (as a captivity of God's people within).

The hail to John was the multiplicative reversal of his seven cycle (as left to right in this case) and its elements would instead map from (overcoming) a "mountain" of false doctrine to virtue or rest as in overcoming Laodicea, rather than floating unity through the seven cycle. The hailstone is as the weight of a talent (the value of election when weighed in gold).

With the judgement entailing "the mountains as not found" and the islands (false rest in a "bow") "fleeing away", I am reminded that these elements in the octal correspond to the churches of the angel's circuit, rather than "the whole world". The world is wholly brought under complete restraint as in one octal, and they don't like it one bit.

Under the single seven cycle of GF(8)*'s operation I have every octal in alignment under a great plague of hail, every stone about the weight of a talent. The cost of the shed blood of every martyr from Abel to this day falls upon the heads of those that do the deeds of the dragon for him. In continuing to use those devices (those in the Earth become as the three frogs performing miracles) they blaspheme (deny the virtue of) the one God without repentance.

12.10) The Lake Of Fire

Returning to the structure of the seven vials;

$$[a, c, e, g] + a = [0, b, d, f] \equiv E$$
$$[b, f, a, e] + a = [0, c, d, g] \equiv G$$
$$[d,g, b, a] + a = [0, c, e, f] \equiv F$$
$$[c, e, d, b] + a = [b, d, e, c] \equiv C$$
$$[f, a, c, d] + a = [0, g, b, e] \equiv D$$
$$[g, b, f, c] + a = [f, c, g, b] \equiv B$$
$$[e, d, g, f] + a = [d, e, f, g] \equiv A$$

I note simply that using symmetric difference (a ∨ b) instead of the familiar operation of (a ∨ b)^c renders the following addition table.

+	{}	A	B	C	D	E	F	G
{}	{}	A	B	C	D	E	F	G
A	A	{}	C	B	E	D	G	F
B	B	C	{}	A	F	G	D	E
C	C	B	A	{}	G	F	E	D
D	D	E	F	G	{}	A	B	C
E	E	D	G	F	A	{}	C	B
F	F	G	D	E	B	C	{}	A
G	G	F	E	D	C	B	A	{}

And this set is then *isomorphic* to the octal group.

The vials of wrath not only give structural alignment of the additive group of this octal with the right hand (Sun octal) under a seven cycle; when "poured out on the earth" they form similar subgroups (as to the reference octal) in an operation after the manner of the octal over triples yet with the three groups in the K4 form [a,b,c], [a,d,e], [a,f,g] excluded for their cosets. What is the importance of this?

If I equate the K4 form that spans the octal to Christ (and with unity a = 1), then there is here an octal outside of or aside from Christ that remains for the ungodly. This structure itself appears to be the inheritance of those doomed to the lake of fire – possibly becoming even the basis of the lake itself. (The multitudes under the scarlet beast.)

Given the original seven cycle (a,b,d,c,f,g,e) the four possible static subgroups are present in the above octal of "wrath" yet curiously not the four possible antichrist bows. I would have expected the lake of fire to contain the bows and provide a cage for the dialectic. What octal do I need to start from to provide this? Clearly the wormwood octal I had previously.

Then I would have for a = 1 with the "wormwood octal" instead of the "Sun octal":

$$[d,e,f,g] + a = [0,a,b,c]^c = [d,e,f,g]$$
$$[b,c,f,g] + a = [0,a,d,e]^c = [b,c,f,g]$$
$$[b,c,d,e] + a = [0,a,f,g]^c = [b,c,d,e]$$
$$[a,b,e,g] + a = [0,c,d,f]$$
$$[a,b,d,f] + a = [0,c,e,g]$$
$$[a,c,e,f] + a = [0,b,d,g]$$
$$[a,c,d,g] + a = [0,b,e,f]$$

So the starting octal should seem to be equal to the wormwood octal from earlier. Yet I discover the reason why this is truly so: I have both!

They that are dwelling in the lake are not to dwell in the Sun octal's earthly elements because heaven and Earth will pass away and there will be a new heaven and a new Earth. The Sun octal as transformed into this "wrath octal" and its second "wormwood-wrath" counterpart together are a completely separate system, with no intersection with the reference octal: from which follows also the fact that this octal is not a possible left hand for the "Sun octal" (it is aligned additively under the seven cycle, and shares no triples or elements or operation).

The wrath now begins to be poured out on the "Earth" in the "Sun octal". In seven stages (vials) it transforms towards this wrath octal that finally becomes the last "cage for antichrist" and every "foul and detestable spirit" and the "wormwood-wrath" octal with it. (One of the vials was poured out on the sea, remember!)

In isomorphism the 30 octals align as one field GF(8). In that case all that remains is the one K4 form of subgroups spanning the octal and the two sets numbering 666 to left and right. These two "wrath" octals are then uniquely found outside of the K4 form. It must then be the only refuge for the beast and antichrist – is this now the foundation for the lake of fire?

As the vials are poured out across the sea there is no preference as to which is the reference; however, if there is one reference "Sun octal" it must be excluded from becoming the "lake of fire". This "Sun octal" opposes each one of "Euphrates" – one of four rivers generated by the seven cycle on a wormwood octal (generated on the top row with {c,e,g}) and vice-versa. Every octal opposes its wormwood counterpart with which its "water" is exchanged for "blood".

The subgroups [0,c,e,f], [0,c,d,g], [0,b,d,f], [0,b,e,g] when added give one of the cosets of the K4 form in the Sun octal [d,e,f,g], [b,c,f,g], [b,c,d,e]. Only one group in the wrath octal corresponds to the static triple [b,d,f] under the seven cycle of the reference field. Eight C7 groups permit only four static triples (bows) when given a fixed unity element in the Sun octal. Each is represented here as one of [0,c,e,f], [0,c,d,g], [0,b,d,f], [0,b,e,g] in the wrath octal.

All "wrath" octals, with the cosets of their subgroups forming closures upon the four possible left hand "bows" (from the Sun octal now under full restraint) are in this wrath octal added (by symmetric difference a ∨ b) to form a closed set (a cage) for antichrists using the dialectic – it is the only refuge from the winepress (the wrath of God) and it is strictly outside of Christ.

I do not shift the unity element here but static groups for the dialectic in the winepress (acting to construct the columns of "Euphrates", just as to construct wormwood by the trumpets) – in effect I found a cage (as the "Church" is to become) not in the earthly four elements of the Sun but within the sea. The winepress forces all "antichrist" into the lake of fire (the wrath octals) whereas the "Church" that "sits on many waters" is become a cage in the sea (now become the lake of fire itself it then contains every nation except spiritual Israel, the elect in Christ).

With the dialectic devices of the sixth trumpet, those caged in the wrath octal could freely transform between that "wrath" octal and the (as I propose) second "wormwood-wrath" octal. After that manner, these two octals (with { } as the identity) together more completely cage and imprison those that would employ the dialectic devices.

Adding worship of the scarlet beast to the mix, the "second death" or "lake of fire" is complete.

12.11) An End To Blasphemy

The first, second and third vials account for the wrath and wormwood-wrath sets as placed to cage the dialectic in all 30 octals of the sea (and across the one sea of glass). Then, the fourth vial cages those practicing the dialectic by closing the operation of wormwood with symmetric difference between the wrath and wormwood-wrath octals, not mere sets as before.

The fifth vial interrupts the dialectic method and blasphemy of the seven cycle, by "darkening" the possibility for any transformation between wormwood rows; the sixth vial finally puts an end to any ability to construct a middle ground over any left hand octal to generate the wormwood octals too.

The result is every octal is aligned isomorphic to the Sun octal and there is a strictly drawn division between reference octal and the "lake of fire", as caged in wrath and wormwood-wrath octals paired: also aligned isomorphic yet outside of Christ.

Most important is the removal of any ability to blaspheme the Holy Spirit, for there will be a "new heaven and a new Earth", and the former will "pass away". It will become impossible to blaspheme the Holy Spirit, whether one believes or does not. This, is the inheritance of all; it will become impossible to blaspheme Him with knowledge.

Luk 16:19 There was a certain rich man, which was clothed in purple and fine linen, and fared sumptuously every day:
Luk 16:20 And there was a certain beggar named Lazarus, which was laid at his gate, full of sores,
Luk 16:21 And desiring to be fed with the crumbs which fell from the rich man's table: moreover the dogs came and licked his sores.
Luk 16:22 And it came to pass, that the beggar died, and was carried by the angels into Abraham's bosom: the rich man also died, and was buried;
Luk 16:23 And in hell he lift up his eyes, being in torments, and seeth Abraham afar off, and Lazarus in his bosom.
Luk 16:24 And he cried and said, Father Abraham, have mercy on me, and send Lazarus, that he may dip the tip of his finger in water, and cool my tongue; for I am tormented in this flame.
Luk 16:25 But Abraham said, Son, remember that thou in thy lifetime receivedst thy good things, and likewise Lazarus evil things: but now he is comforted, and thou art tormented.

The parable appears totally opposed to all my writing! Yet "hell" is the last system of worship of God remaining, and the rich man inquiring of God did so **within** that system (i.e. the "fire", "flame"). It is all he is left with for intercession, and it is total separation from God and His virtue. The rich man deliberately places himself in "torment" to pray (with his tongue requiring water, the dialectic caged): i.e. always being refused, turned aside and pushed back forever.

It would indeed be the utmost cruelty to refuse a burning man even a drop of water (yet that is not what is stated). It is actually impossible to give anyone practising the dialectic (in "Hell") a seven cycle for them to operate their paradigm without also enabling them to blaspheme the Holy Spirit by exercise of that same paradigm. The two will not mix; they are separated by the final judgement of Christ. The treading of the winepress ensures this separation is simply judged and the rich man's own choices brought about this impasse despite the lines drawn by his maker. (He may not have been chosen, but also he should not have sinned.)

It should be noted that it was certainly not Jesus' intent to establish Abraham's preference to be at all toward cruelty; there is simply no way to satisfy the rich man's desire to continue operating in the dialectic at his own ease without the certain results of that blasphemy.

Getting an answer from anyone in paradise (or from God Himself) after the judgement of all such unfortunates (as the "rich man") to the lake of fire is to be "damn hard work" forever. Then it is true that the final system of worship of the scarlet beast is actually equal to being completely refused the refreshing of God's Spirit in eternal life. (This is "the second death".)

So, contacting God from hell is near impossible because the group of S6 is closed, not allowing one to attain the seven cycle. Then, truly, the last system of "hell" is torment to those left there trying to penetrate through that closure to one who will, likewise, turn away.

Chapter Thirteen: The Judgement Of The Church

The sequence of the first four seals culminates with the rider of the previous horses/devices revealed as "Death". The text informs us that "Hell follows after". This judgement as to the final state of the "Church" (then gone all the way bad) is described in a manner that will utilise much of the maths introduced so far.

There is a limit to the number of dialectic devices that may be constructed "logically". That being only three, with a logically contradictory fourth balance added as "Death". There is no room for a fifth to be grafted in (then adding to one side and contradicting on the other) or indeed, any chance of taking the place of one made already without contradicting yet another.

The last-straw system of worship that God permits in His creation is shown in Revelation as the "image to the beast". This consists of ten kings (offices) that reform the false doctrine of the Church and give it its synthetic "life". There is a synthetic structural and procedural dialectic grafted over the logical devices of the dialectic: the speech of the Church is become like that of the dragon, as the false prophet in the heart of the believer. The same is aligned to the overlay of five sets (churches) of false doctrine in amalgam – as if it had a carnal body with five senses.

The heart of the believer aligns the "Hell" of the fifth synthetic dialectic to the contradiction of the logical fourth (opposed to the other three in the logic): the vacuum observed becomes "filled in" by the human heart. I may find the four wings of Daniel's leopard beast mapping the "heart" to be transformed amongst the four possibilities of that contradiction of "Death", and they are overlaid by the false doctrines seen forming the "Church" which is become all (the collective). The fulfilment of Daniel's fourth beast is seen with its deadly wound healed.

Philadelphia's exit caused a deadly wound – they have been replaced with a cult of Satan, the "leaven" in the meal that is worked in by the process. The mind of the "Church" (now become carnal) is the mind of the ten kings, by which the source of leaven is become equal with the remaining five churches and all in the system, the "collective" beast, follow suit as Hell after.

As the first beast had that deadly wound – Satan was utterly defeated upon it: this beast is mk. 2.0 – God's gift (and recompense) to Satan, fulfilling all the depravity he requires to heal that wound and bring an end. If you can accept it, this is the lake of fire with its city complete and when fallen, made whole as the system most would actually and rightly know as "Hell".

13.1) Sitting Upon Many Waters

First, I dive in at the deep end to describe the vision John was given, to show why the two numberings to 666, of the antichrist and the image agree.

Rev 17:1 And there came one of the seven angels which had the seven vials, and talked with me, saying unto me, Come hither; I will shew unto thee the judgment of the great whore that sitteth upon many waters:

The Church (the woman in the vision now gone completely bad) sits on many waters: that is, upon the sea, rivers and fountains introduced up until now. I may exclude the Sun octal and state that the devices are built upon what is to be the "lake of fire", in the remaining octals.

The angel does not describe the last and final end of the woman but how to recognise her instead, and also, more than that: how to recognise the paradigm (judgement) of those operating as her. The judging of false doctrine and of the circumvention of sound doctrine which is the purpose of this scarlet beast is also that "judgement": that "mind of the woman" whose body is slaved to (or subject to) one end only (one Church with false doctrine, whose "body and mind" are both wholly given over and subject to the "one mind" of the ten kings reigning over her). The beast has deferred its judgement to the ten kings directing the collective (as the woman) to follow their direction and leadership.

13.2) The Heart Of The Woman

I count the number of octals under an arbitrary set of "balances". I start given some static triple (my thesis). I do not fix this in a particular octal, as I am only interested in the number of balances that may be formed, instead of returning to calculate the number of the beast 666.

A triple may be found (if a subgroup) in six octals. Keeping these arbitrary I explore the number of balances that I may form. Given a reference octal and one (static) triple, there are once again, four choices for unity. Once unity is fixed, so are the remaining triples in the balances.

One arbitrary thesis with four choices of unity then fixed, I choose another triple as antithesis from the remaining three, and I have 12 possible balances formed. (Four choices of unity by three possible antitheses.) By fixing unity I also fixed a right handed static subgroup disjoint with my four possible "left handed" bows. This subgroup is found in six possible reference octals (the "6" in "660 + 6"?). These octals share no seven cycle, else they would be equal, and neither are each totally disjoint as left/right to each of the other five (they share a subgroup).

The bows (thesis/antithesis) are found in distinct left hands of these six "candidate" reference octals. None of these six candidates share a seven cycle, and as seven cycles are valid on only two octals, it follows that none of their left hands (containing my thesis/antithesis) share a seven cycle with any of the other left hands of the other five "candidates" sharing that static subgroup. (They never agree as identical octals or as left/right hands of each other.) The left hands of these six right hands share no seven cycle but contain the initial static subgroup/ triple pair, their "intersection" is non-empty. (I now exclude the reference: adding to the set numbered 666. The woman encroaches on all but the reference, playing harlot to the beast.)

With unity paired to the static subgroup of the reference, there are twelve seven cycles holding static each subgroup/bow pair, each in pairs over six octals. As there are two choices of multiplication for my "thesis", there are twelve seven cycles in six (almost disjoint) sets. Two of these sets must also be made over the reference octal. I counted the reference twice: in fact, I do not want to count the reference at all! I remove two from 12 to 10 instead.

The 12 octals align in two sets of six (to the choices for the reference) yet they all have different left hand octals (the sets of these left hands are not disjoint in triple-pairs); those possible left hands make a total of ten octals moving away from the reference. I so exclude a first seal paradigm (there should be no "hole" for antichrist), and also every "green thing" (cf. Rev 9:4).

I could remove two octals from the count of twelve over all of those six possible reference octals. That accounts for the set of ten in my construction. More important, is that the left hand octals for each of these candidates are disjoint in seven cycles, and that all these left hand octals share the same static subgroup in the right hand.

Then I have 4*3*10 = 120 possible balances, each comprising the possible thesis/antithesis with an even split between the "oil" and "wine", opposing the choice of unity from those four, with the thesis/antithesis a pair from the remaining three balances.

I counted the number of octals underneath the balances. The initial factor of 4*3=12 is certain enough, but not so obvious the factor of 10. The factor of 10 is also determined on the two possible outcomes (syntheses). The manifold A5 absorbs one set (the oil) of 60 outcomes.

There is symmetry between the two choices of unity for the thesis/antithesis and also the two separate outcomes, the choices of the remaining two unities (the "oil" and "wine"). These also pair with the reference and, again, form 10 outcomes from a full 12 (two are discarded as the reference octal from before). Then there are 10 possible octals in which the syntheses of the balances are formed. And this then clearly results in a factor of 10, giving our full 120.

Now, these 120 are merely triples and are not inverse pairs; I have not generated the 120 elements on the left and right as multiplicative inverses in S5 to form the balances of thesis/antithesis as if (g H g^{-1}). What I have is a set of 120 separate "bow pairs" that splits into two sets to left and right with another group of 60 in the centre, an even split of the remaining 120 (from the two of four bows remaining). Yet there are 666 ways to deny the singular virtue.

It is these triples in the balance that allow the heart to align with it naturally; and not by group isomorphism. Then, the heart naturally fills in the "balance" to form the synthesis of concept.

In the centre I effectively factor down to A5 and its coset in S5. I have the pans of the balances as two bows as "the lust of the eyes" and the synthesis of (g H g^{-1}) with H=A5 in the centre "the lust of the flesh" and so the woman sits upon many waters showing "the pride of life".

Selection between one set of 60 and another in the centre (from the remaining 120 triples) is put to a vote by the collective and is purely "approach pleasure, avoid pain" (or the cases before and after) – carried by the consensus: the manifold evaluates one set positive. To Satan the outcome is never important – merely that the devil's device is in place as the method.

13.3) The Maddening Wine

Rev 17:2 With whom the kings of the earth have committed fornication, and the inhabitants of the earth have been made drunk with the wine of her fornication.

In this verse the woman is intertwined with the kings of the Earth (unity or the static triples, an analogue of the stars of heaven within the sea outside of the Sun) and the "men" in the Earth (the singleton elements corresponding to triples) are made drunk (to stumble) with the wine, "the common element in the synthesis" of her carnality. The heart of the Church member is as subject to the kings as the kings to the five-fold "heart" of the Church overlay. The agreement in the heart of the believer is to the false doctrine of antichrist (as reasoned from the dialectic method), now a defining feature of the Church system by its process.

I will include a few diagrams to help show how the beast's own deadly wound is healed.

13.4) The Scarlet Beast

Rev 17:3 So he carried me away in the spirit into the wilderness: and I saw a woman sit upon a scarlet coloured beast, full of names of blasphemy, having seven heads and ten horns.

This woman originates where the former woman had fled to from the face of the dragon – a place "prepared for her by God". That place is "Sabbath rest" – the end result of the ministry of the two witnesses. (That ministry was to those that the dragon had pursued and made war upon.) They are the remainder of "her seed" (Christians) and unveiled here is the effect that the dragon has had upon them. I see the transformation from the body of the elect to the condemned, a corrupted shell left behind by the exit of the true remnant, the result of the flood sent by the dragon (the speech and mind of his sown tares) finally overtakes the woman.

The scarlet beast is formed from the devices or "methods of deception" of the dragon from all of Old Testament Israel's history woven into the religious institutions of the end. The devices from Egypt,[7] Assyria, Babylon, Persia (with the Medes), Greece and Rome, and the dialectic in the unrestrained USA or NWO beast[8] survived of Britain, the USSR and Nazi Germany are set (in amalgam) to race the restrainer loosed upon the Church. Now, all are this scarlet beast.

Then as in the third woe, I find the scarlet beast is also named "Gog and Magog" (i.e. "antichrist" (prince), and "land of antichrist" – false doctrine). This third woe is also called "Armageddon".

The scarlet beast is similar to the first: the seven heads are become "churches" and the ten horns are systematic and procedural rather than doctrinal. The ten are not empires but instead representative of the dialectic process reigning over the five refining churches of Revelation (Ephesus, Smyrna, Pergamos, Thyatira and Sardis), together with a source of leaven (the sixth) leavening it under the image – it is a device (as the seventh) that supplants the role of Christ.

The ten kings form a "three onto one" device. One I had shown before in the first beast, here occurring in "three from ten" ways. (I.e $(10*9*8)/(1*2*3) = 120$.) They form "balances" in the same number of ways that are synthesised upon by the spirit of antichrist numbering 666.

The "lust of the flesh" by the manifold or the "heart" of each believer is in "agreement" with the image system that sets up Church "unity" (solidarity) with a cult of satanic origin (the source of leaven). This system is aligned with the five refining churches as if five senses, each following suit after the uptake of doctrine from the source of such "leaven" (cf. the desire as a "lust") – as managed over by the ten kings that form the image or "mind of the beast". If the believer does not approve, exclusion from the collective is the result.

A somewhat basic view of the scarlet beast can be seen from the following diagram. Satan's goal is to keep the church following the beast (with the deadly wound) whilst the leaven of false doctrine is worked into the church collective. Satan's aim then is to supplant the reign of Christ with one of a false gospel (i.e. the false prophet) – with the end result being to force God to disown the remnant of Christianity as unworthy.

[7] James Lloyd, The 6, 7, 8 Cycle (Christian Media), p26-28.
[8] Lloyd, The Sand and The Sea, p5-7.

Synagogue of Satan
replaces God's remnant

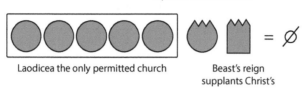

Laodicea the only permitted church   Beast's reign
                                                       supplants Christ's

⌀   God rejects the system as desolate

The name "Laodicea" refers as a term to those in the five refining churches that do not overcome and exit to Philadelphia.

The woman is described as wealthy, or rather temporarily so in worldly terms: there is no mention of true God-given blessing – rather an image of success in a position favoured. (The woman is as favoured by government, before the gifts from taxation hit home hard in the form of an influx of tares that may not be legally excluded.)

Rev 17:4 And the woman was arrayed in purple and scarlet colour, and decked with gold and precious stones and pearls, having a golden cup in her hand full of abominations and filthiness of her fornication:

The "gold", "precious stones" and "pearls" mirror the divine blessings given to the bride of the lamb descending from heaven as "New Jerusalem" later. This woman is clothed (decked) with these only in part, mimicking divine blessing: she "appears blessed" but is not. The "scarlet colour" may be a reference to being clothed in a fine white linen robe made clean in the blood of the lamb – it appears that this is a "fake". Purple cloth was very expensive in John's day.

The other words in the text concerning her wealth (of that "great city") do not reoccur and I assume no word pictures present. (This is the case concerning some few verses to come.) Once the devices are revealed, little remains, yet there is more content found later.

She has a gold cup referencing her enticing status, inviting others to join (agree with) her (she drinks as though royalty), whereas her fellowship filling the cup is poison. (That being the dialectic process as "wine" at every place of the table.)

13.5) The Beast Becomes Universal – The Woman Corrupts All

Rev 17:5 And upon her forehead *was* a name written, MYSTERY, BABYLON THE GREAT, THE MOTHER OF HARLOTS AND ABOMINATIONS OF THE EARTH.

The mystery John saw is now not so mysterious, for the woman is not a heavenly abstraction but a worldly one – she is a corporation. Now, the "MYSTERY" shows governments' effort to legally define "Church" as a term applicable to any form of worship of any kind or creed. "BABYLON" references the legal unity of the world's greatest governments everywhere (from time immemorial), and the remainder reveals her instantiation as to describe all governments, corporations and religions now equal with her, found in the Earth as legal "fictions".

In recognising this woman as a corporation its legality "poisons" every other corporation in God's eyes, no matter how just – like the trustworthy post office or your local supermarket. By throwing in their lot – by proxy of government recognition and equal status – all become common with this Church, guilty and "practising her deceit": they follow after the "god of this world". The overlay is a direct insult at Jesus Christ over the ownership of His elect – His bride.

Rev 17:6 And I saw the woman drunken with the blood of the saints, and with the blood of the martyrs of Jesus: and when I saw her, I wondered with great admiration.
Rev 17:7 And the angel said unto me, Wherefore didst thou marvel? I will tell thee the mystery of the woman, and of the beast that carrieth her, which hath the seven heads and ten horns.

The woman is "drunk with the blood of the saints", that "blood" being the (intersection of) consistent beliefs held amongst their collectives (built on the axiom of sovereignty), and of the doctrine of the notable ones in history (martyrs) upon which the leavening process began. That doctrine (not necessarily correct as is that of the "lamb") is bartered and held common in the ongoing dialogue as "wine", and she causes those within her to stumble (fall).

John marvels at her, she is so enticing but so drunk with damnation: how could this be the "Church"? The angel asks John why he is so astonished – as the bride of Christ is certain to prosper! (Heaven help it to separate itself from this woman, however.) The angel explains this to John, noting the beast carries the woman: the beast is the substance, the "Church"; the woman (the image), is the collective visible to God over the efforts of those within.

Zec 5:5 Then the angel that talked with me went forth, and said unto me, Lift up now thine eyes, and see what is this that goeth forth.
Zec 5:6 And I said, What is it? And he said, This is an ephah that goeth forth. He said moreover, This is their resemblance through all the earth.
Zec 5:7 And, behold, there was lifted up a talent of lead: and this is a woman that sitteth in the midst of the ephah.
Zec 5:8 And he said, This is wickedness. And he cast it into the midst of the ephah; and he cast the weight of lead upon the mouth thereof.
Zec 5:9 Then lifted I up mine eyes, and looked, and, behold, there came out two women, and the wind was in their wings; for they had wings like the wings of a stork: and they lifted up the ephah between the earth and the heaven.
Zec 5:10 Then said I to the angel that talked with me, Whither do these bear the ephah?
Zec 5:11 And he said unto me, To build it an house in the land of Shinar: and it shall be established, and set there upon her own base.

A talent of lead (a heavy burden) along with a "ephah" or "measure" (a model, here as a woman, a "global template" of "all the Earth" numbering 666), represents the "dialectic process", become wickedness. This heavy burden matches the impossible empty middle of the disjunction of virtue and the four sets from the previous metaphysics r, s, u^{-1} and v^{-1} which together become the two wings of each bearer of that same wickedness to Shinar (Babylon).

This "wickedness" is the carnal justification for (or of) the dialectic, now endemic in this day and age. The "woman" is an idol, replacing an acceptable offering to God. The talent is placed in "the middle" (empty though it may be in all but appearance, as also is the "image" of the five refining churches): found in the faulty ultrafilter of the dialectic. The "wind" in the women's wings (of an unclean bird) are equally "of the spirit" (and placed in the octal) all whilst the measure becomes a blasphemous offering to God, and this is that same wickedness leavening the "Church". This is prophecised in Revelation to be "borne up" to God as worship in "MYSTERY BABYLON", the scarlet beast.

13.6) The Beast That Once Was, Is Not And Yet Is

Rev 17:8 The beast that thou sawest was, and is not; and shall ascend out of the bottomless pit, and go into perdition: and they that dwell on the earth shall wonder, whose names were not written in the book of life from the foundation of the world, when they behold the beast that was, and is not, and yet is.

The beast that "once was" is the full-stretched tail of the dragon: the seven kingdoms from Egypt to the USA are the tail that "once was". The eighth and scarlet beast (the Church at the end) is a contradiction to the dialectic (the tail of the dragon), and should not be included (though it must be so): the tail would form a fourth seal system applied a further four times.

The crown of Satan is totally inconsistent in its application: this beast most certainly "is not". Yet, the beast survives accumulating the doctrine of all seven that "once were" (and this collection of false doctrine is the "little horn", an eighth which is of the seven), and acts under the freedom granted by the USA or "New World Order" (NWO). The beast "Yet is".

The beast "once was" formed of God's people, and "is not" – for it is now barren of them at the end; the beast is completed when it has ascended out of the bottomless pit. The seven cycle is then completely obscured and the fallen star from the fifth trump transposes the five churches to closure. The beast system here only goes into sin, it becomes merely a human fault rather than a new adversary to follow after the dragon.

By ascending "out of the bottomless pit" the Church gets its operation from the many extension fields towered up to algebraic closure (AC). The beast is merely the collective assembled to vote on doctrine, and the woman (the image) should be considered separately. The collective together forms a field with a far greater degree of extension than that of GF(16), which was required to turn an individual back to the collective of the first five (refining) churches.

The beast could be considered as the smallest field F (of characteristic two) containing at least one K4 subgroup for each believer assembled: as if within an orbit of K4 groups within a subfield J, itself with a short orbit of length five in F (refer to the fifth trumpet and the "locusts"). Then, every believer within the beast is returned to the collective, and none perceive the open door to approval in Christ (which is outside of the acceptance of false doctrines).

The "Church" in the eyes of the world (and by the manifestation of the false prophet spirit) is the bride of Christ proper (although the remnant is exeunt safe). It "yet is" in appearance the people of God. It is a wonder to marvel at: it is instead an engine for the denial of virtue.

Rev 17:9 And here *is* the mind which hath wisdom. The seven heads are seven mountains, on which the woman sitteth.

Each head is a mountain (a stronghold of false doctrine), of the dialectic overlaid upon the "Earth" by the dragon, as if placed over those seven possible cosets (the four beasts before the throne). Each mountain appears valid only with the interruption of the seven cycle of the least, those false doctrines being those upon which the woman sits. The "seven kings" of the first beast are replaced here with the doctrine that plagued Old Testament Israel along with the healing of the first beast's deadly wound – with the doctrine of the Nicolaitans. The "eighth is of the seven". The dialectic devices of the tail of the dragon are introduced as these "mountains" and not as the seven churches of the letters. There are seven mountains formed by ten horns: those under the image system of ten kings will later "make war with the lamb".

How may the woman five-fold in churches and one in mind (a source of leaven) span seven? The "Laodicean" in the place of the sabbath is not present, turned back to the five-fold carnal part, under the image system (S5 or S6 as of the first two woes, an image laid over to obscure "the seventh head": replacing election with desolation). The image system supplants Christ in an attempt to close the system on the source of leaven switched in for the exeunt Philadelphia.

Rev 17:10 And there are seven kings: five are fallen, and one is, *and* the other is not yet come; and when he cometh, he must continue a short space.
Rev 17:11 And the beast that was, and is not, even he is the eighth, and is of the seven, and goeth into perdition.

The angel describes the deadly wound healed. He refers to the first beast John saw from the sea as five kings (Empires),[7] five of which in John's day were fallen (from Egypt to Greece), one being current (Rome) and the last global USA beast to come in the final days that continues a short space (for one more empire to form an eighth and give the dragon's tail its discontinuity).

The result is then the beast that "devours and breaks in pieces", even upon and as to itself.

The scarlet beast is referred to now as separate, an eighth formed of the false doctrine that has survived all prior seven, and that itself goes into sin. (It is already defeated, and is nothing new to warrant another salvation.) It is formed of all the false doctrine that defeated and corrupted Old Testament Israel and that has now completely corrupted the institutions that would be of the New Testament "Spiritual Israel". (Only the elect in Jesus Christ are yet "chosen people".)

There are actually two more NT-era kingdoms in the timeline, those being the "Holy Roman Empire" and that of Imperial Spain, and these are not without import.[8] However, I find that whilst there is continual "war in heaven" during the collision between the seventh kingdom (of God and His Christ) with that of Daniel's fourth beast (four kingdoms then collapsed to one NWO, arraying itself to be the seventh), the scarlet beast is presently now the "eighth of the seven", for whilst the kingdom of God continues, the beast that "once was" under Satan's now deprecated authority (continuing with the false doctrine from Egypt to Rome and of the NWO which "yet is" – though synthetic in the dialectic with its deadly wound) presently "is not". With every effort, by his every attempt to interrupt the circuit of the least and devour him with the false doctrine of the dragon, Satan would thereby hope to reset the kingdom count by defeating the cross: so that the six NT-era kingdoms would restart the count and the scarlet beast will become again the seventh kingdom in the sequence, supplanting Christ and healing its own fault. Yet God will judge.

[7] James Lloyd, *The 6, 7, 8 Cycle* (Christian Media), p26-28.
[8] Lloyd, *The Sand and The Sea*, p5-7.

13.7) The Beast's Deadly Wound Healed

The first beast of Rev 13 is an attempt to construct the dialectic with the plucking up of Britain, the USSR and Nazi Germany. Here (in the ten kings) the eighth, ninth and tenth kings form the dialectic upon the seventh king. The result is an "image to the beast".

Rev 17:12 And the ten horns which thou sawest are ten kings, which have received no kingdom as yet; but receive power as kings one hour with the beast.

The ten kings (not kingdoms) form the image to the beast. Choosing three kings from ten (in 120 ways) the balances of the mind of the beast as over the heart of the woman may be formed. They do so for one hour: not necessarily one automorphism of one multiplication on one octal only, but to be set (as the 24 possible dialectic bows in each of the cosets of the Sun octal's subgroups) against the angel's circuit through the churches. They receive (as kings) "power with the beast for one hour" – one permutation of four earthly elements forming a fourth seal device – one of a possible set of 24. Likewise, the beast shares the same authority.

Rev 17:13 These have one mind, and shall give their power and strength unto the beast.

As the image is a construction of the false prophet spirit of the second beast (from the Earth), I should first acknowledge the "two horns as a lamb".

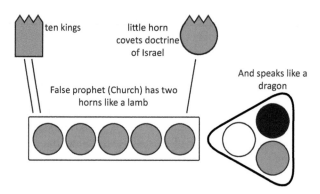

ten kings

little horn
covets doctrine
of Israel

And speaks like a
dragon

False prophet (Church) has two
horns like a lamb

False prophet or "synagogue of Satan" uses image of ten kings and
"little horn" to leaven the Church through dialectic methods

The first beast was the original attempt of Satan to leaven every church, even to deceive the "very elect" in Philadelphia. Looking at that effort of Satan now deprecated (Satan also reads the Bible; the first beast is not dead, but is elaborated upon, "healed". The text states the false prophet forms an "image" to it: making the scarlet beast to war with the saints instead).

The first beast (with a deadly wound) was to entice the seven churches to accept the source of false doctrine (leaven) as one of their own, in effect accepting the false prophet spirit rather than an influx of tares. In assimilating the false doctrine from Israel's history, Satan hoped to taint the church so that the false prophet could set up shop with a cultic set of false doctrines that caused Old Testament Israel to stumble and fail – a collection of heresies held as if by a "synagogue of Satan", those tares already in the churches. I find:

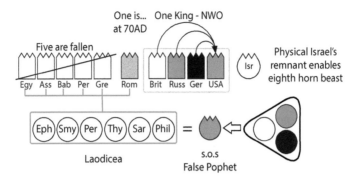

Synagogue of Satan (s.o.s) stands in the place of Laodicea during time of the seventh (USA or NWO) kingdom. The s.o.s (False prophet and Laodicea itself) speak like a dragon to leaven the churches with all the accumulated false doctrine of Israel's OT past. Physical Israel has survived ten kingdoms, yet enables the little horn or eighth beast that the s.o.s rides

The re-appearance of the state of Israel becomes a necessary tenet for the synagogue of Satan, in that God apparently shows His great favour to reinstate their inheritance. Then, rather than see Israel to merely reappear, the fault is to assume they are (with God, i.e. Christ) reinstated. This recent addition to Christianity has its roots in the denial of Christ and the sticking plaster placed between the covenants, for Jesus Christ said of Himself "I am the door" (for the lost sheep of the house of Israel, as into the fold). There is no route to salvation other than Christ crucified (as if the law were able to replace the cross). Israel, instead, is a body translated to a spiritual house under Christ which **must** pass through that door of the cross.

There need be nothing prophetic found in the reappearance of the state of Israel. Israel is not the little horn of Daniel, neither is Israel the beast – (the scarlet beast). Israel as a government may be a "beast" in its own right, yet not present in the Revelation's sequence of kingdoms. The scarlet beast is the eighth formed of the prior seven, formed of the false doctrine of Israel's history. If the (eighth) beast was only the nation of Israel then the church majority would certainly have rejected their false doctrine! (Not converting to Judaism.) Note that the state of Israel in this day and age is actually a secular nation and should be considered such.

The scarlet beast created by Satan is to become the seventh kingdom to heal the first beast's deadly wound. That wound is the cross of Christ that interrupted the kingdom count of Egypt through Rome[7] by becoming the seventh kingdom formed by the execution of Christ using a dialectic device. If the cross were perhaps defeated in the case of the least, the New Testament nations of the "Holy Roman Empire", Imperial Spain and those four beasts[8] of Daniel make the scarlet beast the seventh kingdom (going into perdition), by resetting the count after that "interruption" (empty middle) in God's work: i.e. Satan's undue part in Christ's work through Judas. (The **one of twelve** as a "devil". There are also twelve worldly kingdoms overall to overcome, culminating with the New World Order, but again – as is perhaps a coincidence, at least one is not to be trusted (included) – the scarlet beast in the place of Christ!)

All Satan needs the state of Israel for is to provide a "bait and switch" – this time merely with false doctrine. The little horn of Daniel's fourth beast even then becomes the false prophet's device to leaven the churches with false doctrine taken from the past of OT (and NT) Israel.

[7] James Lloyd, *The 6, 7, 8 Cycle* (Christian Media), p26-28.
[8] Lloyd, *The Sand and The Sea*, p5-7.

The goal is the same: use a little leaven and leaven the whole. The freedom granted by the particular and respective government (in tryst with whatever "Church" is in view) would have its cost: legally excluding those with false doctrine would become indefensible and once the dialectic is taken up the result is the same. However, this attempt was doomed because the remnant was commanded to "come out of her..." by Christ.

In the whole (with false doctrine completely worked in) now becoming the church of Laodicea – Satan would trap the remnant inside the collective by making them to always "point the finger" of guilt elsewhere. Laodicea would cease to appear as the identity element that convicts all in Christ no matter "who" or "where", and instead become "just some other people". By physically providing a "church of the Laodiceans" (i.e. the cult) Satan would spiritually distract and bind the remnant under the "name" of the first beast, so the false prophet would thereby supplant the Sabbath rest in Christ.

The scarlet beast is somewhat different. I must examine the image system. I can consider seven kings of the ten to represent the five churches in Laodicea, a sixth to represent a source of leaven and a seventh the "collective" of the host church led by an arbitrary pastor or "chair" of those seven (as if a "king") in mimicry of the sabbath. (To represent the Nicolaitans present that defer all trust to them as to the validity of the dialogue; "not watching" for Satan's devices now present before them.) The "wine at the table" of the balances is formed by the remaining three kings. Those three offices are of the "thesis", the "antithesis" and that of "facilitator".

Zec 11:15 And the LORD said unto me, Take unto thee yet the instruments of a foolish shepherd.
Zec 11:16 For, lo, I will raise up a shepherd in the land, *which* shall not visit those that be cut off, neither shall seek the young one, nor heal that that is broken, nor feed that that standeth still: but he shall eat the flesh of the fat, and tear their claws in pieces.
Zec 11:17 Woe to the idol shepherd that leaveth the flock! the sword *shall be* upon his arm, and upon his right eye: his arm shall be clean dried up, and his right eye shall be utterly darkened.

A foolish shepherd (there is yet a remnant to minister to) also ministers the gospel to those holding no true value in it (tares) whilst having always kept the elect (the true flock) tended, which is indeed "foolish" when using the (then broken) instruments given for preserving the elect only. (The flock otherwise to be abandoned by the Lord, His covenant to become wholly broken.)

The gifts given to the least of the flock from Christ's letters, the "seven horns on the Lamb" are even as these very same "broken instruments" (as given for the least to "reckon correctly" his salvation, using them together instead of and without the Holy Spirit, as if he were also refused – a tare from that flock), re-used by a shepherd that has certainly left the ninety-nine to find the one (justifying the least and then as with all grace, extending that justification to the whole of the remnant) of His flock that has (or will have) wandered off. (Christ is ever the good shepherd.) Then, the least (the poor of the flock) will leave such faulty ministry, leaving only the idol shepherd with his tares remaining behind. Such a flock of tares has no grace but for those broken instruments.

The "new shepherd" in the land (as negligent to the truth and leading to err) would easily (as not serving God) convict the "least of the flock" to leave (and by extension all others to follow) as soon as their collective conforms to the image system (the least being present at the end of the age). The idol shepherd instead (the very negative of Christ) would be the minister that covets Christ's leadership (without ever experiencing any conviction of the Holy Spirit beyond the verse Zec 11:17 above), initiating the collective into the image system for **growth**, leading the flock into "spiritual prison" or "MYSTERY BABYLON". The sword is upon his arm, cutting him off from approval in Philadelphia, his "arm" is become contorted as S5, and his right eye (that which beholds the "right hand" or Sun octal) is without discernment as false prophet replaces the leading of the Holy Spirit.

I may as if with the senses, with triples chosen from a set of 120, answer any permutation of false doctrine in the five churches with doctrine presented by the source of that "leaven" as if it were the mental faculty over the senses (the manifold). When the result carries, the churches follow suit after the "one mind" in dialogue (under the direction of the facilitator and the majority will (consensus) of the collective) – and the six "churches" or Laodicean Church as "Hell" follows after that "decision" as if also carrying it with that "one mind". The effect is that the collective makes way for the cult, and the false doctrine becomes universal.

Setting up the bijection from the 120 triples of the ten kings onto the Church doctrine overlay is a difficult task: "Rev 9:6 And in those days shall men seek death, and shall not find it; and shall desire to die, and death shall flee from them." Most notably a bijection is actually not given to us! (They received no kingdom, despite there being ten transpositions in S5; the "churches" are not assigned.) "Rev 17:12 And the ten horns which thou sawest are ten kings, which have received no kingdom as yet; but receive power as kings one hour with the beast."

The ten kings reign "receiving power as kings for one hour with the beast" (see "The Manifold" and "Morphism" later). I.e. as equal to the collective in that particular single permutation of 24 found. They receive "power" in that the structure is formed of three kings to one (the collective) as opposed to a collapse of nations. The collective's mind is so formed.

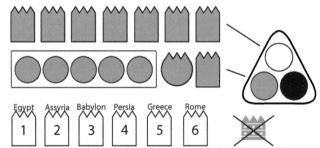

Image device replaces Satan's crown. Ten kings reign "as beast" and six of ten kings follow as hell after one, as with one mind; Church with synagogue of Satan (emboldened by Israel) follow as hell after USA beast. The image device of three onto one supplants Christ (life) with Death (Antichrist, the beast or its image).

● Each king of ten may be "image" of any three others.
● USA beast is image of Britain, the USSR and Nazi Germany.

The seventh kingdom (USA) is made of the devices from the first four seals as across every octal in the sea – likewise the beast that rises up from the Earth overlays (in deception) the dialectic logic upon the "wormwood devices" of the trumpet sequence (in the sea).

By this "marriage" of Earth and sea, Satan aims to bring the dialectic process into the churches with their uptake of the mark (accepting the process whole). In order to keep this "dialectic process" ongoing, the churches are required by government to provide in their dialogue with it and each other the same process and with it the system of the image (the ten kings).

Rev 17:14 These shall make war with the Lamb, and the Lamb shall overcome them: for he is Lord of lords, and King of kings: and they that are with him *are* called, and chosen, and faithful.

The elect will follow Christ's angel out of the image system, the woman riding the beast. They are called (as Christ's own) elect (chosen) and faithful (obedient to His voice over all others).

I had introduced all this before in other terms;

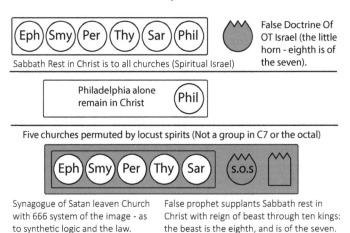

The evil genius of the image system is in the system of leavening (conjugation of churches as with balance devices) permitting the source of false doctrine to act in conjugation to replace each element or permutation of the five churches with a similar permutation utilising itself.

For instance, (c,f)(a,b,c,d,e)(c,f) = (a,b,f,d,e) where "f" is the source of leaven (cult) and the objection to the doctrine is raised by "c", put by the method to the collective's vote (the mark).

So the leavening system is its own justification – as permuting the A5 group formed of churches to groups in A6 still isomorphic to A5, those same sets within A6 by conjugation with the six elements (),(a,f)(b,f)(c,f)(d,f)(e,f) give the beast all "spin" necessary to carry the method itself as valid to the collective. The doctrine of the "little horn" will win the leavening contest against the "Church", even as if it were in appearance as "David against Goliath".

Israel survives seven kingdoms under dragon's device

Synagogue of Satan leavens church

With such a system where the doctrine poisoning the water is the system itself, if there is any complaint on the process, all examination will approve the process itself. It is "most worldly".

The church would only uptake false doctrine compatible with their faith: the ideal candidate for the source of "leaven" being the nation state of Israel. This would not fool the very elect as Jews would openly renounce Christ. Instead, a mixture of cultic or legalistic Christianity and liberal "messianic" Judaism is introduced, neither Christian nor Jewish – and those bringing in the false doctrine (leaven) lie that they are "Jews" (with Old Testament doctrine), whether as spiritually within Christ or physically otherwise.

Three kings map onto one, using OT doctrine from Israel as legal baseline- As if it were common ground for keeping churches "obedient" in their dialogue with the synagogue of Satan through the image system.

Laodicea is physically separate from Philadelphia, spiritually separate from Christ. The image replaces the need for Israel to become "a church" in the leavening process

This eleventh kingdom (the nation of Israel) would be as "the eighth" of the dragon and would cause a discontinuity of the dragon's tail; it is to be "bait" for the switch only. It is a **secular** nation (not of doctrine), and not the "little horn": it is not in the kingdom count (it barely takes a place), and three kingdoms (superpowers) are required to have fallen to establish that horn, immediately disqualifying it. Only the USA (NWO) remains as suitable for the first beast.

The ten kings form the image with the methods of the first three seals: the beast's deadly wound – the exit of the true remnant (spiritual Israel) is healed with the image now in place.

The liberty (authority) granted by the first beast allows Satan to create the false prophet's image system over the five refining churches. The first beast's deadly wound becomes not a complete loss for him; the first beast may remain and its deadly wound may be healed. The result is now doctrinal freefall as the final leavening process begins.

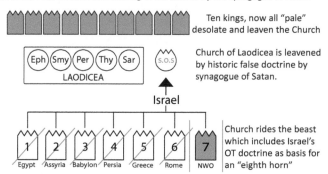

I may align six kings to the five refining churches and the source of leaven. Three kings together form the image of the beast's "mouth" (the components of the speech of the dialectic process are the sets of triples of the seals and trumpets).

There is one place missing (the seventh): this place is taken by the collective whilst an arbitrary king as "chair" sits out (mimicking the sabbath). The mind or "heart" of the individual believer assembled fills in the contradiction of the fourth balance device that may not be instantiated without "life". The missing seventh supplants the reign of Christ and substitutes the system of "Death". "Hell" follows after. (The other six accede to the decision as if reasoned perfectly.)

Isa 28:15 Because ye have said, We have made a covenant with death, and with hell are we at agreement; when the overflowing scourge shall pass through, it shall not come unto us: for we have made lies our refuge, and under falsehood have we hid ourselves:

13.8) The Leavening By The Image System

Returning to scripture;

Rev 17:15 And he saith unto me, The waters which thou sawest, where the whore sitteth, are peoples, and multitudes, and nations, and tongues.
Rev 17:16 And the ten horns which thou sawest upon the beast, these shall hate the whore, and shall make her desolate and naked, and shall eat her flesh, and burn her with fire.
Rev 17:17 For God hath put in their hearts to fulfil his will, and to agree, and give their kingdom unto the beast, until the words of God shall be fulfilled.

The scarlet beast (a corporation) poisons the water of all peoples, multitudes, nations and tongues with the wormwood devices. There is no worldly corporation that escapes the legal equality with the Church of the end. The global status of such a corporation is enough to condemn them (as within corporations) all over the world with a "second death".

The action of the kings on the church is negative: they do not charitably minister to the Laodicean Church collective but they cause it true harm by their method of doctrinal examination and acceptance of new fellowships under their template of disobedient harmony.

Kings:

Kings refers to those that have the judgement as to whether to enter the dialogue in the scarlet beast (the third woe) or not. They either hold firm to the gospel (as unity of virtue) or become an "idol shepherd" over the desolate. They then take up the other three "positions" of unity in the Earthly four elements, leading those in dialogue as thesis/antithesis and facilitator to guide the outcome of synthesis.

Multitudes:

Those united and diverse peoples holding equally to every possible position of the unity element (not necessarily that singular virtue found in the lamb). It refers to those that are united in their diversity and so potentially deny or private virtue in the lamb (in 666 ways as antichrist), whilst also indicating the possible converse, that a diverse group of peoples may also be united under the lamb.

Dialectically, the ten kings hate the woman (and cannot benefit her), sending her doctrine (oil) to the empty set in the balances. They "hurt" the oil and by degrees ruin the woman (the collective of doctrine) of her real substance; the "wine" she offers does its damage. The ten kings' heart (the manifold) in the church (image) overlay agrees by collective vote (the beast) to allow compromise in every possible manner: the elect are to be convicted to exit her. The kings' "hearts" (S5) agree in the "pride of life" (as S5 subgroups of S6) and each church may be a source of false doctrine. Laodicea is to be convicted by this, until all the elect are separate.

They not only strip fellowships of their doctrine (as oil) and righteousness in Christ but they cause the elect within to leave: robbing it of its true reason to exist (its "wealth") to God. As new fellowships enjoin the corporation status, those flocks are merged into a more "universal body" and the conviction caused by the least's exit (the angel) convicts the elect in the wider collective to exit. This action (the leavening process of ten kings) fulfils the purpose of God.

The kings effectively consume the Church's substance (her flesh) and strip her naked of any remnant of the elect by the action and condemnation of either the Holy Spirit or the pit (as if fire in the place of the Holy Spirit). This "fire" is ministered by the two witnesses as the generalised circuit of the angel. In direct blasphemy of God, there is no doubt that these worldly fellowships have spent their inheritance. (A measure of wheat for a penny gives way to three measures of barley for a measure of wheat – the inheritance is spent for growth only in tares: the remnant are called to exit.)

The fire may be C7, or S5 or S6 even. The ten kings' "kingdom" follows after their one mind as Hell after Death. It is S6 over the Church (together with the cultic set of false doctrines (leaven) introduced with the false prophet). Position and title of the kings are still as yet unallocated.

The reign of the ten kings is then "given" to the beast (the systems numbering 666) that agree in both the image and also the human heart, which follows after as Hell as if it were a "more perfect ideal" to complete their desire. Men seek "Death" for where they want the beast to go, for they decide in God's absence what their "false gods" should say.

Rev 17:18 And the woman which thou sawest is that great city, which reigneth over the kings of the earth.

The angel confirms to John that this woman is the group S5 over the churches (as with the two witnesses) set up to frustrate and obscure the seven cycle of the angel's circuit with smoke from the pit. The woman rules over the kings of the Earth: every permutation of triples in the earthly elements are subsets of the larger construction of S5 (over the kings in S4) – though in this vision the woman is yet to be completed and the pathway out (the open door traversed by the angel) is yet to be closed to fill in the whole group S5, to close the lake of fire upon itself.

13.9) The Manifold

With the human heart judging its desire upon the senses (and only the senses) I may conjugate permutations in the senses about A5 rather than the whole group S5. In judging *doctrine* the image device may freely factor down to A5 and by conjugation acts similarly to the "lust of the flesh". There is mimicry of "life" there. As with the human heart of the believer and the "Church", the "lust of the flesh" constructed may be considered to be as two manifolds;

$$\sigma G \sigma^{-1} \in G \triangleleft S5_A$$
$$\mu H \mu^{-1} \in H \triangleleft S5_B$$

Where G and H are isomorphic to the group A5, and the permutations σ and μ are in S5.

I have 120 sets of three kings from ten as well as 120 possible balances of antichrist "bows". (I shall ignore the two further choices – between "oil" and "wine" for now. One king of the three is the facilitator; the missing fourth the collective.) I have a set W=(X, Y) of antichrist bows and also Z of triples of ten kings.

I form bijections, one to one maps from the sets W and Z to elements of S5. I define three maps (on the three sets into which the great city splits when it falls. For it does indeed fall by splitting it into these three components! The model I provide reveals the lack of any synthetic or fifth balance filling the contradictory void in "Death".)

$$\phi: W \to S5_A : \phi(X,Y) \to \quad \sigma$$
$$\delta: Z \to S5_B : \quad \delta(z) \quad \to \quad \mu$$
$$\Delta: Z \to W : \quad \Delta(z) \quad \to (X,Y)$$

I need one more map: a permutation in S5;

$$x: S5_A \to S5_B : x(\sigma) \to \mu$$

Since the maps are one to one (1-1) and bijective I may invert the maps;

$$\delta^{-1}: S5_B \to Z : \delta^{-1}(\mu) = z$$

This gives us the following with composition:

$$x : S5_A \to S5_B, \ [\delta \cdot \Delta^{-1} \cdot \phi^{-1}] \cdot (\sigma) \to \mu$$
$$x^{-1} : S5_B \to S5_A, \ [\phi \cdot \Delta \cdot \delta^{-1}] \cdot (\mu) \to \sigma$$

The human heart and the doctrines of the Church under the image align in conjugation on the "Great City" G as:

$$x \cdot G \cdot x^{-1} \in G$$

Where G is isomorphic to S5 or also A5 or the trivial "pure" group that is the identity (A5 is simple). The human heart and the harlot are linked and form a "manifold" of the senses. (With which it may judge.) Effectively, the worshipper "approaches pleasure" doctrinally and "avoids pain" (hard truths). The arbitrary choice between "oil" and "wine" with the bows and likewise the decisions of the ten kings are subject to the "pleasure" and consensus of the believers in collective: "For God hath put in their hearts to fulfil his will". (Rev 17:17)

I also have under composition:

$$[\delta^{-1} \cdot x \cdot \phi] \ : \ W \to Z$$
$$[\phi^{-1} \cdot x^{-1} \cdot \delta] \ : \ Z \to W$$

And then the manifold:

$$\alpha : W \to W : \alpha(X,Y) = [\phi^{-1} \cdot x^{-1} \cdot \delta] \cdot \theta \cdot [\delta^{-1} \cdot x \cdot \phi](X,Y) = (U,V)$$

θ may be any permutation of S5 (i.e. as of a 1-1 map permuting Z) and (U,V) some other pair of "bows". I have a symmetric result for permutations of the set "Z" under a third manifold. (There is yet no construction of a "fourth balance" or manifold as yet. "x" may be taken as the identity map rather than an isomorphism.) I state that these three manifolds are the three parts into which the city splits when it falls. (It becomes sin and goes into perdition.)

So, the set of antichrist bows in pairs W becomes aligned to the map "x", or equivalently the set Z of all triples of ten kings. Now $\delta^{-1} \cdot x \cdot \phi$ may indeed be a 1-1 map, yet together they act as permutations – a map Δ between Z and W. In fact, Δ is also such a permutation between the two: albeit used as if the identity, a baseline.

I put forward the idea that I may restrict θ to permutations in A5. Then the permutations (U,V) of the possible (X,Y) also belong in A5.

These three manifolds are now become in their bijections the "mystery of iniquity". I have:

---

1) W = (X,Y) the "mouth" or set {...} of the beast (antichrist)

2) Z the mouth (set) of the false prophet (image of triples of ten kings)

3) Elements in G, H or all θ above – the mouth (set) of the dragon.

Now, elements in A5 (or An) are formed of products of **disjoint** cycles of odd cycle length – an even number of transpositions. As I conjugate A5 with elements in the manifold the reversal of the order of transpositions (as the inverse) allows each cycle to be considered separately.

In effect, the devil is "bound" in his choice of θ as he may not construct, not even in the "bottomless pit" of algebraic closure (over GF(2)) any element of multiplicative order two. Likewise, there can be no short orbit of any subgroup of any finite field with length two.

This includes products of transpositions (a,b)(c,d) that are disjoint. (The serpent spirits are as yet unable to construct these, appearing able only to alter the order of a seven cycle.)

So, Satan may not close the group S5, having no ability to form the coset of A5. (As including cycles of even order.) The elements σ and μ are present as in the heart of the believer only, projected upon the collective. There is truthfully a hole in S5 to be filled.

13.10) Morphism

The system above – even completed with bijective maps – does not share the operation of the permutations in A5 or indeed S5. That is, unless I have complete agreement between each individual believer's heart and the doctrinal "Church" overlay.

So, if I take the map:

$$x : S5_A \rightarrow S5_B : x(\sigma) = \mu$$

To be an isomorphism (preserving every operation over S5), I have:

$$x : S5_A \rightarrow S5_B, \ [\delta \cdot \Delta^{-1} \cdot \phi^{-1}] (\sigma) \rightarrow \mu$$
$$x^{-1} : S5_B \rightarrow S5_A, \ [\phi \cdot \Delta \cdot \delta^{-1}] (\mu) \rightarrow \sigma$$

So, the maps...

$$[\delta \cdot \Delta^{-1} \cdot \phi^{-1}](\sigma) = \mu$$
$$[\phi \cdot \Delta \cdot \delta^{-1}](\mu) = \sigma$$

...are now also isomorphisms. Then I may also substitute:

$$\Delta : Z \rightarrow W : \Delta(z) = [\phi^{-1} \cdot x^{-1} \cdot \delta] \cdot (z) = (X, Y)$$

Reducing the previous manifolds to operations over their respective G-Sets in θ :

$$[\phi^{-1} \cdot x^{-1} \cdot \delta] \cdot \theta \cdot [\delta^{-1} \cdot x \cdot \phi \ ] : W \rightarrow W$$
$$[\ \Delta \cdot \theta \cdot \Delta^{-1}] : W \rightarrow W$$

And

$$[\Delta^{-1} \cdot \theta \cdot \Delta] : Z \rightarrow Z$$

So:

$$\Delta^{-1}\theta\Delta \in H \cong A5$$
$$\Delta^{-1}\theta\Delta = \sigma$$
$$\theta\Delta = \Delta\sigma$$
$$H\Delta = \Delta H$$

I.e. given θ is in H (isomorphic to A5) and that the permutations between W and Z are bijections, I discover that the permutation of the G-set in θ by **any choice** of the bijection Δ gives equal left and right cosets, is normal and therefore preserves the conjugation that transforms between the sets of ten king "triples" and *pairs* of antichrist "bows".

The map $[\Delta^{-1} \cdot \theta \cdot \Delta]: Z \rightarrow Z$ and its inverse are isomorphisms within the group action of S5 over the set of θ. Those isomorphisms Δ **really are** the elements of the group acting over the G-set in θ. (Products of conjugations upon a normal subgroup G together form a group.)

But the map Δ was upon Z mapping a permutation of Z to W through an isomorphism "x". Δ is actually an element in S120 (the group of order the 120th factorial or 120!) that here happens to be a permutation of the set Z "equal" to a permutation of S5 by an unknown element of S120. (I may assume it to be the identity as long as it is constant.) Here that element is simply the map "x" acting upon some "rest" state or bijection between kings with no kingdoms as yet (Z), and the possible states of antichrist in the sea (W, the heart of the collective). Then, "x" is dependent on the bijections ϕ and δ being **constant** only: she, the woman (as x) is at "rest" whilst riding the beast (when there is alignment in Z to W).

Then "x" is simply as the "woman" and the map Δ is become the head of the "beast" that she rides. It is clear that there are seven choices of collective remaining for each triple of kings chosen: there are truly seven mountains (Δ heads) upon which the woman sits.

So, the **"woman rides the beast"** of the sets Z and W, they are not groups (as the heart or Church overlay indeed are), neither are they the only G-sets (the set of θ also forms a G-set). What can be said is that the manifold of x conjugated upon the G-set of A5 in θ is independent of the 1-1 maps of Z and W to the manifold S5 or each other. There is only the sense of a mere "labelling" (the kings have no kingdom as yet): the bijections are completely arbitrary! (They preserve the manifold of θ whether conjugated with x, any Δ or even its inverse.)

Any group of order n can be represented with elements of the symmetric group Sn. I effectively generated a closed set of 120 permutations of S120 between W and Z. Those 120 permutations form S5, but in ϕ or δ these are multiplied by an element of S120 (scrambling the sets).

The ten kings give all their authority and power over to the elements of S5 formed by permutations over the "ground state" of ϕ and δ – so that the only real voice in the room is that of the dialectic process in "x", or equivalently, the collective: the beast.

In practice the manifold reduces to the sets of elements in θ. As this is limited to elements in A5 the "ground state" is properly the heart of the believer rather than the doctrinal overlay: for the image only speaks from the heart of the believer in group consensus. I expect as above:

$$\Delta^{-1}\theta\Delta \in H \cong A5$$

$$\Delta^{-1}\theta\Delta = \sigma$$

$$\theta\Delta = \Delta\sigma$$

$$[\phi^{-1} \cdot x^{-1} \cdot \delta] \cdot \sigma = \theta \cdot [\phi^{-1} \cdot x^{-1} \cdot \delta] \cdot e$$

Now, given ϕ and δ are elements of S120: if they were also a permutation representation of an element of S5 I could reduce the map ϕ to an element r, and δ to an element s both in S5. I could then set the "ground state" as the identity "e" (on the far right) to be acted upon by both r and s as permutations in S5 of this ground state (or now as properly a "G-set"). Although "e" is truthfully only a permutation upon five symbols (though as the identity element), both r and s correspond to some permutation of this set (the five churches or the woman's "senses"). What I am looking at in r and s is actually a classic cypher problem.

$$\theta\Delta = \Delta\sigma$$
$$[r^{-1} \cdot x^{-1} \cdot s] \cdot \sigma = \theta \cdot [r^{-1} \cdot x^{-1} \cdot s] \cdot e$$

Where "e" is the identity element. Now, x is an isomorphism, so:

$$[r^{-1} \cdot x^{-1} \cdot s] \cdot \sigma = \theta \cdot [r^{-1} \cdot x^{-1} \cdot s] \cdot e$$
$$r^{-1} \cdot x^{-1}(s\sigma) = (\theta r^{-1})x^{-1}(s)$$
$$(r\theta r^{-1}) = x^{-1}(s\sigma)x^{-1}(s^{-1})$$
$$x(r\theta r^{-1}) = s\sigma s^{-1}$$

So, provided θ is in A5 the left hand side is also in A5. With "s" in S5 and x an isomorphism of S5 it **must** follow that σ is an element in A5. I must of course have "s" a member of S5 (a permutation representation of S5); this allows me to form:

$$x(s) * x(s^{-1}) = x(ss^{-1}) = x(e) = e$$

This simply shows that there is agreement between the believer's heart and the doctrinal overlay. The heart of the woman is simply *"aligned naturally"* to the heart of the believer.

It is not true that W and Z are now both become G-sets under S5 (r and s may be constant). The "ground" state of either W or Z is arbitrary, what I require is x(r) = s*g, or equivalently that x(r)=s "up to an element "g" of S5". (They must generate the same conjugate subgroups.) So, I find the cypher.

Rev 17:13 These have one mind, and shall give their power and strength unto the beast.

As I may extend the group A5 in θ through A6 using conjugations of transpositions in A6 I have six last conjugates of A5 in A6 as I also indeed have with the human heart. (Without requiring any more elements in the set of θ.) The result is that all three manifolds (in x, Δ or its inverse conjugated upon A5, Z or W) may freely number to 666.

With the source of (Old Testament) Israel's leaven and a church swapped out by conjugations of transpositions over A5 in S6, I find the process where a church objecting to false doctrine is overruled by the rest of the collective (with the source of leaven now in their place) or may indeed overrule such doctrine in kind. The real judge on the matter is the collective placed in the seat of Christ as the seventh of the ten kings (an arbitrary "king" of the remnant as chair). This system is not likely to overrule false doctrine on an "approach pleasure, avoid pain" basis.

I had described a manifold under the conjugation of transpositions in S6 as having the senses override the mind: strangely here I show the converse. Doctrinal *faults* accepted for the unity of the collective give the crowd a dopamine rush. The result is not as Eve was, a creation of God "to be beguiled"; the opposite is in place. The model of the "woman" (or son of perdition) is referred to as a "harlot" sitting on many waters. Her "mind" is no slave to the "satisfaction" of her body; the body is slave to the mind. Her doctrine or the "riches" that she grants as "favours" are of this one mind, and her "body" will accept anything to continue this commerce.

The sixth member of the image is equal to or implicitly becomes the source of "leaven" or false doctrine that appears compatible with the gospel at face value – but is in essence a cult invading Christianity. It is the "mind" of the overlay and in transpositions the process will *accept any* doctrine substituted for one already present within the five "churches". The process is slaved to leaven the Church, needing only a "small change" to affect universal damage: false doctrine becomes totally compatible with any "valid" doctrine under such examination. The Church follows as Hell after. The lust of the flesh and the pride of life do their damage.

With the natural isomorphism between the heart of the believer and the Church overlay, I have found only three manifolds – all with their operations in place as if naturally. (And not four manifolds to complete the system of "Death" with the addition of "Hell" – the cypher will equally justify the reduction to three manifolds with x now the identity map and absorbed into the action by g in $(x(r) = s*g)$ and, all things being equal, the razor may be applied and I find no valid fourth balance (manifold) in the overlay.)

This has one result: the Church overlay does not structurally exist except in the heart of the believer (it is merely a projection of the doctrine approved by the collective). The scarlet beast does not construct a *new* dialectic in some magical overlay; the beast simply goes into perdition as just another sin of the believer within.

13.11) Kings One Hour With The Beast

It seems meaningless to associate each of the ten kings to a church in the overlay (there is no bijection there): the kings have "received no kingdom as yet".

Rev 17:12 And the ten horns which thou sawest are ten kings, which have received no kingdom as yet; but receive power as kings one hour with the beast.

The set of pairs of bows is properly 240 strong; within those 240 I may separate out the permutations made on the "Earthly elements" forming each fourth seal device. By so doing I fix some right hand subgroup as static.

Then in the 240 bow pairs (I counted only 120: since choosing a pair leaves a pair behind) there are 24 permutations of those four bows.

I.e. If (a, b, x_1, c, x_2, x_3, x_4) is a generalised cycle over an octal containing [0,a,b,c] the four members $\{x_1,x_2,x_3,x_4\}$ may be permuted in 24 ways, giving us 24 possible seven cycles or combinations of antichrist bows.

I reduce from the whole set of 240, to each king (one of ten) with 24 possible permutations for each king, each a choice of unity, x_1 say. Each one of the ten kings now has 24 permutations (hours) as "power" received with the beast. The beast itself as the collective has only the one seat in the dialogue (the seventh position outside the overlay, whilst an arbitrary "king" of the seven that remain sits as chair in approval, as for carrying the vote) and also receives 24 permutations as a king in the same manner. The six kings in the overlay have 6*24 = 144 permutations, but within the dialogue these are discarded completely: they do not map cumulatively into the dialogue of three kings, they simply (blindly) "follow after". Of the 24 permutations, only one is present in any instance.

So, I can define a map:

$$\lambda : \quad R \rightarrow K \quad : \quad \lambda(r) \rightarrow k$$
$$\omega : R \rightarrow S \subseteq R \: : \: \omega(r) \rightarrow (s)$$
$$\mathrm{ker}(\omega) \cong S4$$
$$[\lambda \circ \omega] \circ (r) = \: \lambda(r) \rightarrow k$$

Where R is the set of 240 consistent bow "quads" in their various permutations and K is the set of ten kings. The kernel $\mathrm{ker}(\omega)$ is isomorphic to S4, and only represents lost information.

The kernel may be placed anywhere in the fourth seal device but I consider it mapped to the collective in place as the "seventh".

In effect the kings mimic "one day off work" per week whilst the collective is present, giving their worship to the image (and the beast) on that single "day off". The three kings mapped onto that false Sabbath are assembled in the same number of ways as choosing seven from ten: with the seventh "day" in dialogue or "worship".

Rev 17:13 These have one mind, and shall give their power and strength unto the beast.

They have one mind: there are not two or more! Only the one seat for the dialectic process with those three kings from ten exists in the system; no others are considered alongside them.

The six remaining kings do not map (give) their 24 permutations (power and strength) to the one collective but all 144 are completely discarded (in total deference to the collective present in the "one mind"). The collective is content to have only the 24 permutations in the "one mind". The six kings follow as "Hell" after, a descending baseline for the Church doctrine.

13.12) War With The Lamb

Instead of choosing three kings from ten I may choose seven and still form these 120 balances – as over the seven churches; the seventh is not representative of the six following after as Hell. Whilst the collective beast as the seventh has the mark, it is not true that the other six have any representation by them.

The 144 = (6*24) states of the "mind" of the beast not in dialogue are discarded and lost. (There is no floating unity or union in heaven for them.) The will of the minority is run roughshod over the rest. (They are sent to the empty set { } "killed" with "Death".) Whilst the system appears to have numerical similarities with election in Christ, believers will know them by their fruits.

Essentially as "order matters" I have 24 arrangements of the same four bows that are quite constructible outside some static subgroup, and those constructions are each present in ten octals (left hands), but are those ten all there are? Each set of ten kings shares a static subgroup on the right: then there are six possible right handed octals in each. The right hand octals are not disjoint in the ten kings' "hour". To examine how many sets exist "outside" of the reference octal I must count carefully.

There is always the reference octal: of every octal sharing only three of its subgroups there are **seven** in total. As from the sixth trumpet each subgroup occurs in three rows of those seven:

$$[a,b,c], [a,d,e], [a,f,g] \; with \; [b,e,f], [b,d,g], [c,d,f], [c,e,g]$$
$$[b,d,f], [a,b,c], [b,e,g] \; with \; [a,d,g], [c,d,e], [c,f,g], [a,e,f]$$
$$[c,e,f], [c,d,g], [a,b,c] \; with \; [a,d,f], [b,f,g], [a,e,g], [b,d,e]$$

And I know that six octals contain a specific triple as a subgroup, leaving me 2 octals remaining.

For each subgroup of the reference octal I have another 2 octals (and these sets of octal pairs (two for each subgroup of the Sun octal) may be considered to be disjoint with each other).

Of 30 octals total I then have 30 − (1 + 7 + 2*7) = 8 and there are only eight octals: those that share no subgroup with a reference octal. As there are eight such groups that are left hands of the reference octal (though using a different seven cycle) – there are only the left hands for these eight and none others.

For example, (a,b,d,c,f,g,e) and (a,b,f,c,d,e,g) share the triple {c,e,g} but not the triple {b,c,d}.

Generalising this to all 30 octals (as reference) yields once more: 30*8 = 240 cases as required.

Simply put: Choosing any two bows from the left hands of the reference octal will always give us distinct sets as over ten octals, since this was the construction. There is, however, nothing "special" – these sets of ten are not disjoint in their right hands but are instead scrambled.

I rearrange the 240 cases from 30*8 to 10*24. The separation between distinct octals has shifted to what are in essence pale horse systems (antichrist). It becomes natural law as to the individual to hold strong to faith in Christ or to the devices of antichrist – a reprobate mind.

That said;

Rev 17:14 These shall make war with the Lamb, and the Lamb shall overcome them: for he is Lord of lords, and King of kings: and they that are with him *are* called, and chosen, and faithful.

Those with Christ (across 30 octals as right hands) are called (dwell in the right hand's operation), chosen (are as a floating unity) and faithful (do not shift octals). Those 240 states outside those right hands are then "antichrist", as the elect dwell within the right hand octal only, always outside of the static triple found in the balance.

The ten kings are distinct in their 24 permutations over their fourth seal systems (but not so in their octal groups). Setting a correspondence a ≡ {c,e,g} fixes every other triple in the ten kings' system. Then the twenty four permutations are formed from six octals containing a certain subgroup ([0,b,d,f] here), with the permutations equal to four choices of unity.

Rev 17:15 And he saith unto me, The waters which thou sawest, where the whore sitteth, are peoples, and multitudes, and nations, and tongues.

The waters are where the woman sits; the doctrine in the overlay is identified to be the doctrine of four sets of collectives, each related by inclusion. "Peoples" (different and opposed faiths) contain "multitudes" (deciding on or "voting" on virtues themselves) which also contain the "nations" (churches) meeting in dialogue ("tongues") in fellowships.

The point is that the six kings forming "Hell" are the collective. The dialectic process may be compared to that seen in the fourth beast of Daniel which "devours" and "breaks in pieces".

As fellowships join together under the image system those new to take up that same process are also subject to the accumulation of the dialectic seen in the dragon's tail. As the device accumulates on four sets under inclusion there must not be a fifth, as the dialectic fails for one of the four (as to the heart), and there is inconsistency in the deception. This is unlike the sequence of kingdoms: as the scarlet beast requires all those categories to operate in the Church amalgam consistently.

There may be a chain of seven under inclusion: smaller fellowships could be dissolved as more globalised ones emerged, small churches could lose meaning in the process, it may be found.

Rev 17:16 And the ten horns which thou sawest upon the beast, these shall hate the whore, and shall make her desolate and naked, and shall eat her flesh, and burn her with fire.

The ten kings "hate" the Church (they cannot benefit it in the image system), hurting her flesh (absolutes of doctrine – they oppose the Sun octal's subgroups with the dialectic made only on the cosets, **opposing** all virtue) with that image. Every conjugate of A5 in A6 is formed by conjugating about A5 with transpositions in S6. This is required for doctrine to be "agreed upon" in the manifold of the overlay, in truth the believer's own heart. This is similar to the sense of "torment". (A5 is not normal in A6, its left and right cosets are not equal.) You do not torment one you love. (Or poison their water.)

The ten kings in the process strip the five churches (and the source of leaven) of their alignment to any king, whilst the three kings retain their 24 permutations each, though the other six kings are completely discarded: leaving just the one set of 24 to the collective with the "one mind", the seventh king approving the process as chair. The woman is then desolate.

Firstly, the kings do not align to churches. In the system only the kings have the authority of those six positions (kingdoms) – not the woman. (She is desolate and her authority were she to be so aligned would also be discarded: as Hell they simply follow after.) The woman is stripped naked. Compare this with each sealed believer given the inheritance of the twenty four elders, clothed in white robes with the righteousness of the lamb. I then see a dreadful device compared to the six positions of floating unity in the octal – here cast to nothing whatsoever, their prayers for the beast are cast back to Earth without answer.

Although the kings received no kingdom as yet, their authority as over their 24 permutations is never shared by the churches in overlay – kings do not align to churches, though the churches are stripped bare because of the process: the six kings appear to be aligned to the woman (doctrine) rather than the churches!

The kings consume her flesh (the believers with good doctrine within her), and burn her with fire (consuming her "oil" or doctrine, leavening it to erase every last spark of light, all virtue) – keeping her **always** within the right hand's static bows, or equivalently the left handed octals only (or in S6 in the lake of fire) – she is always formed of triples in cosets, never found in the Sun octal. She, therefore, under the ten kings is "predestined" (or the lack thereof) to the lake of fire. The ten kings by burning her with fire, bring her near to S6 (or the churches as S5 without the source of leaven). She gets burned either way, whether towards the image schema of the false prophet or by the seven cycle (of the static bow, a subgroup of the left hand: as to become transformed into the wrath octal(s) in the treading of the winepress).

The result is that the Church is now without God, bereft of righteousness, to consume away and be finally destroyed. I do not see word pictures in these four – the overlay or woman is not technically there (as noumena) except in the hearts of those in the collective. The believer within suffers these "plagues".

Rev 17:17 For God hath put in their hearts to fulfil his will, and to agree, and give their kingdom unto the beast, until the words of God shall be fulfilled.

The kings agree on the process. Triples of three kings take their united place in the same sets of balances (third seal) and then their hearts will agree: the missing balance is given to the collective (the beast) – that is, until the wrath is complete and there is become one octal in a sea aligned in isomorphism.

The kings give their kingdom (that being the position of "chair" – mimicking the sabbath with an arbitrary seventh king – rather than being the other six heads of the scarlet beast found to be the "churches"; the kings' kingdom/place is upon that seventh and only that place: the one which finds and completes the dialectic synthesis) to the beast which is properly the collective putting their leading to the vote. This occurs until the words of God are fulfilled – all the words that are "the true sayings of God" culminating in the final judgement.

Christ will return swiftly "with the final trump" as the third woe goes into perdition (hell). Then Christ returns even as the scarlet beast completed ascends from the pit after the faithful elect have left the scarlet beast behind. There is no spiritual vacuum formed after the last in Christ overcomes that third woe. (That being the "left hand" completing his work.) If the right hand is him upon whom God has the least influence, is the left hand he who listens the least? Perhaps he foresees and repents due to the oncoming and victorious shout of the archangel!

With the formation of the "sea of glass" and the completion of the wrath, there is no sense of the ten kings being anything but similar to each other: triples of ten kings will have the same result across the sea. There will not be "120 ways" but only the one, with permutations in the earthly elements of the Sun octal: those permutations found only in two sets each of four octals – the octals other than the Sun are to become the "lake of fire".

Rev 17:18 And the woman which thou sawest is that great city, which reigneth over the kings of the earth.

John identifies the ministry of the two witnesses as to or against this same collective.

13.13) Tokens Of Damnation

To simply sum up – I may now easily identify the following terms.

Rev 15:2 And I saw as it were a sea of glass mingled with fire: and them that had gotten the victory over the beast, and over his image, and over his mark, *and* over the number of his name, stand on the sea of glass, having the harps of God.

The term "mark" (of the beast) is separate from "number of his name". That said, the following should be reached easily:

The "Name of The Beast" is given us;

Rev 17:5 And upon her forehead *was* a name written, MYSTERY, BABYLON THE GREAT, THE MOTHER OF HARLOTS AND ABOMINATIONS OF THE EARTH.

The "Name of the beast" is simply the title of "incorporated" (.inc). It is simply the template for a corporation granted to a fellowship to recognise it as any other legal fiction: "Church" included. The name (capitalised above) in the scripture is a mystery; what defines a "church" anyway? It had better not be God's adversary nor government.

The Church becomes the "MYSTERY" part, all government recognition the "BABYLON THE GREAT" part (which is universal in all governments globally and historically) that has thrown in its token recognition, by placing its name. Lastly, "THE MOTHER OF HARLOTS" is equated with the first church recognised by all government that spawned this same legal "template", to which all others are subsequently approved by the beast. The final and closing "ABOMINATIONS OF THE EARTH" (other dialectic and worldly religion) states that this is purely a device of worldly origin, and that it denies all virtue in the Lamb.

Corporations as churches and vice-versa are not given the name of the beast until after the image system has been set up. It is equal to throwing one's lot in with the beast system that gives a church its "name of the beast". That system is a hideous insult to God and there is every reason to expect that this beast system can be averted.

The corporate template plugs the inconsistency of the fourth bow of the dialectic when fellowships are considered to be under a five-fold tower of inclusion, a poset. This same vacuum is (in appearance) filled in with the image – the woman. The equivalency is that neither construction of Satan permits a fifth dialectic: both are logically inconsistent, never as "noumena". The contradiction of the fourth bow is not "healed" logically but only in appearance, systematically or procedurally. The outcome is that people must "agree to disagree" in their fellowships, by taking part in the system of the scarlet beast itself. The overlay is never more than the practice of sin, accepting the name of the beast is sin also.

The "Number Of His Name" took some maths:

The calculation to 666 bows (separate from the bow in the reference octal) is the number of a man, but which man? It is clearly antichrist – although truthfully antichrist is spiritually "dead". ("Overconstrained" and outside the ultrafilter.)

So, the number 666 is the number of the "dead", and to God, corporations are **all** "dead" although the world grants corporations almost all of the rights that it can grant human beings.

That said, the beast is also a corporation, and corporations are **never** alive in God's eyes. Church fellowships that incorporate have (as the church in Sardis had) a "name for being alive" (called by Christ's name as "Christians") but they are spiritually dead. That is, they are as desolate as the antichrist numbering 666 as any other body that incorporates itself.

Any act of incorporation by Christian fellowships is more than "Yea, yea and nay, nay". Jesus stated it is evil. (Likewise, agreements between churches should be limited to the simple.)

Mat 5:37 But let your communication be, Yea, yea; Nay, nay: for whatsoever is more than these cometh of evil.

God has no dealing with believers by corporations. He has no place for them as they are always "dead". He is good to His word and keeps all His promises. Why would anyone require a government approved body to oversee a contract or charter with God when there is a perfect covenant in Jesus Christ already?

In strict terms, the number 666 is the number of ways that a crowd (and individual) may deny virtue in a specific setting. The scarlet beast is an "engine" for this.

The "Mark Of The Beast" took bible study:

The ten kings chosen in triples judge doctrine alongside the beast (the collective) but instead of the 6*24=144 permutations of pale horse systems aligned to each king over the amalgam of churches, the collective sat in place as the seventh does not have 168 permutations, but merely the 24 it has itself.

That seventh chair at the table in the process is the one with the "mark" – the ability for the collective (the beast) to put its name (signature) to the reign of the three kings over them. (The seventh king is as their representative, a chair that will place "their mark" to the record.) The mark is then equivalent to the beast (the collective) deciding as antichrist what God wants in His own temple for all those not present to cast their vote. Those that worship and follow the beast (the dialectic process) in those six "churches" (also including the cultic source of false doctrine from Israel's past: examples are given in Christ's letters for us to recognise and overcome, not to import in wholesale): as under the six remaining kings are they that worship the image: they defer all their authority to those with the mark. Likewise, those six as "Hell" follow after. (The beast's act of making writ is written on your hand (as scratched) rather than your will writ with its hand. That "scratch", or "engraving" cannot be deprecated within the Church as under the false prophet, but it will heal from your hand if you leave that Church!)

"Bearing The Mark Of The Beast" may be reasoned from the above:

Fellowships that have place outside the collective of this scarlet beast system may be assimilated into it. They are given as a "collective" the choice to enter: they do so only by one bringing the mark of the scarlet beast (now false prophet – as to forming the image with them) into their fellowship and the fellowship itself being absorbed incorporated.

Effectively, those that take part in such a facilitated meeting to join the beast collective are brought to the table to add their "fellowship's" own "name" to the mark of the beast. (Rather it is the individual's name, an "idol shepherd" added as for the beast, as the fellowship was not "incorporated" beforehand.) Then the image comes into town knocking on doors and sends them those three kings "bearing the mark". In such a meeting the cases "for", "against" and one more, "the facilitator" form these three kings. The facilitator is the one that will remove every point of contention from the room.

The first thing to leave is then the applied word of God, the Bible.

The mark is received in the "right hand" or "in the forehead". The mark then as it is applied is for believers to defer their own authority to those that have the mark (the act of making writ) in their place. To have the mark is to defer to others your complete agreement of who should enter your wider fellowship (as greeting with or signing by the right hand, or as your representative acting in your place) and also or separately those who should decide your doctrine (as received in the forehead) in a meeting of select others where the consensus decides for all. That would take place in a facilitated meeting as described.

Don't be deceived that this process is that simple (and meetings can simply be avoided): replace the thesis and antithesis with "the case before" and "the case after", and you could see it everywhere. That is sin. It is not the mark of the beast.

The woman rides the beast

The 120 balances in the manifold (the heart of the woman) adds to the set of 666 bows not corresponding to any octal with some static subgroup. The image makes up five of the missing six octals of 672: encroaching on all but the reference. The ten kings provide a bijection to these 120 balances, limiting the image to just those five octals' 10 left hands only. The woman, separate to the set of 666 bows, then "rides the beast" of 666.

The identity of the "False Prophet" revealed.

The above "tokens" are all marks of the operation of the scarlet beast and are of the "false prophet". There is a simple identification as to this "false prophet beast" (coming up from the earth). The nature of a false prophet is to prophesy falsely from the heart as in the verse:

Jer 23:16 Thus saith the LORD of hosts, Hearken not unto the words of the prophets that prophesy unto you: they make you vain: they speak a vision of their own heart, and not out of the mouth of the LORD.

The image system is the extension of that false prophet's (the collective in dialogue) "heart" and completes the scarlet beast system which is almost become a "perpetual motion machine" for sin. In fact, it has become a cascade, a "great falling away".

The image system is not the false prophet, neither is the false prophet the scarlet beast: rather the scarlet beast is that same "lake of fire" where the beast and false prophet are to be found. The false prophet is that antichrist spirit in which the believer operates (as within their collectives) to align their hearts to the doctrinal overlay of the image system itself. The result is that they attain one or more of the tokens of damnation as stated above and they may be "known by their fruits". In truth, there are "many false prophets" in the collective as indeed there are "many antichrists". That the beast and false prophet are consigned to the lake of fire simply ensures that God's election excludes those that practise such deceit, and raise it up to Him as worship.

The "little horn" is this same scarlet beast system. It becomes a template for use on the small scale between, say, seven churches in a small town – all the way up to the greatest scale (supposedly global collectives). The horn is "small" in comparison to a nation's government, but is to be found present in many places. (Until all become leavened.)

Just as the elect have an inheritance where they are clothed with the righteousness of Christ (the lamb), so the lake of fire becomes the inheritance of the unwise acting as the false prophet. That said, the inheritance Satan has offered is not "noumena" until he is damned within that system itself at the judgement.

Christ Himself has an inheritance in His new name (as equal to His Father omnipotent). It is fitting that Satan should be rewarded with that same clothing he has offered those he deceived. The insult to God from His adversary has attained its final and eternal answer. One step too far? Perhaps, but this is not "where the action is": that being instead the circuit of the least – it makes no sense to promote the inheritance of the damned when the elect are all separated to life in Christ by that same mechanism.

Satan had stated:

Gen 3:5 For God doth know that in the day ye eat thereof, then your eyes shall be opened, and ye shall be as gods, knowing good and evil.

Now, God has provided an eternally lasting dispensation of grace to His elect (the good). Those Satan will tempt away will be in full receipt of the knowledge of the "evil". In fact, now that Satan's devices are known, the elect are far better off for it. Yet, they are not become "as gods" at all, merely having been clothed like the lamb, holy before God as His people. The idea that anyone may willfully choose to attain that blessed state themselves, to be seen as equally righteous before God (as God alone chooses that covering for us) is nonsense.

Isa 28:15 Because ye have said, We have made a covenant with death, and with hell are we at agreement; when the overflowing scourge shall pass through, it shall not come unto us: for we have made lies our refuge, and under falsehood have we hid ourselves:

Similarly, the iniquity and evil of that statement of Satan, the promise that sin will provide a covering for the damned is also nonsense; that covering is succinctly put to be as complete "nakedness". Satan in insult and mimicry has provided nothing; God will repay in kind with Satan's utter damnation.

To know both good and evil is to be offered freely the first and be in receipt of only the second, which is to be "naked" and to receive no inheritance from God. Satan in all subtlety was asking Eve for worship not due. The Lord will be overjoyed to reciprocate upon Satan in full. The same system of "worship not due" will be His answer and Satan's reward.

Chapter Fourteen: The Great City Fallen

The judgements upon the "great city" of all religious commerce come thick and fast, and God states unequivocally that they will have no opportunity to call themselves godly before Him. God will not waste further time sending physicians for so sick a crowd. Their doctrine will not be corrected any longer – such would be spent foolishly, now that the institutions are barren of righteousness.

The angel (the least) makes another appearance, which may be aligned to the period leading to the opening of the seventh seal and the sounding of the last trumpet. The judgement is set with the creation of the lake of fire and the casting of the dragon (Satan) into the lake that is where (or which) the false prophet and scarlet beast are.

The construct of the great city (S5 or S6 when including the "synagogue of Satan" – the cultic source of leaven) relies on the four elements of the seven cycle (and the octal) as the first four churches of the letters (aligned to the earthly elements). The city leaches its operations from the octal and K4 form but is unable to construct 12 elements of S4 over those first four "churches" – the city is incomplete. That is, until God "fills in" the holes in judgement – as recompense for her "thefts" of those mappings from the Spirit of God. The transpositions from the fallen star (as the army of horsemen in the sixth trump) with the key to the pit generate the rest of S5 along with the short orbits of the locusts (hurting five months).

The city falls in a single hour – one automorphism of a field GF(8). That is, with just one seven cycle and in one applied power. The circuit of the same angel which is then completed (and made certain in Philadelphia) allows the angel to return (with his fellows) and with full justification of God's sovereignty declare the institutionalised deception engine of the dragon and false prophet called the "Church" fallen. They, the "synagogue of Satan", will truly worship at his feet upon their faces and declare the love Christ has for him.

With the judgement of God upon the "Church", it burns out of righteousness on the last day: believers should expect God to destroy wonderfully – beginning with those corporations that offend Him most. That will offend the worldly more than may be imagined, as it will hurt where it will hurt most: in their pockets. Still, the fire set alight by the image of ten kings will surely blaze, to burn out and extinguish the last fraction of virtue left within. The winepress will be trod and it will be done swiftly after that single "hour" has passed.

With God's sovereignty displayed before all, truly then "The Lord God omnipotent reigns". The faithful may expect the dragon to be taken down screaming, but without much of a fanfare. The seven trumpets are instead for the victories of Christ that carry from the cross to all His elect (even as the seals are opened and they resist the slackness of the current paradigm to overcome all the devices on the Earth and in the sea). His grace is sufficient for them.

God, sovereignty unchallenged, is free to manifest Himself without a hint of accusation against Him when the devil is shown logically incapable at deceiving (as to deceive even the least in God's kingdom). This, though, is not enough to challenge sovereignty, surely, unless somehow the devil is given the benefit of the doubt? Now, if God's grace were insufficient in some cases and the difference had to be made up by the least in the kingdom, that I could fathom would challenge God's sovereignty were there already unfortunate souls so trapped. Has the devil found such a clause? Not in the work of the cross certainly: is a "second death" of the devil's own making a possibility? With the completion of the salvation of the least, whatever the clouded issue is – the angel is as victorious as any of us are.

In fact, the difference is made up by the least overcoming without the aid of God's Holy Spirit to direct his steps: he reckons for his own salvation upon the throne predestinated.

14.1) The Angel Of The Church

The action of the seven thunders causes a realignment (movement) of heaven and therefore the narrative shifts to the overcoming of the angel as in Rev 10:1-7 previously. (For John so clearly states: "And after these things".)

Effectively the seven thunders realign the seven trumpets in sequence so that the last three "woes" align to the period between the fourth and seventh seals. Essentially the angel must overcome under the worst circumstances to be effective, and so must overcome the three woes in the period between the opening of the fourth and sixth seals before the wrath is poured out, wherein each heptad sequence is concluded on the "last day" (the period between the seventh trump sounded and seventh seal loosed).

Rev 18:1 And after these things I saw another angel come down from heaven, having great power; and the earth was lightened with his glory.

Which compares directly with the passage:

Rev 10:1 And I saw another mighty angel come down from heaven, clothed with a cloud: and a rainbow *was* upon his head, and his face *was* as it were the sun, and his feet as pillars of fire:
Rev 10:2 And he had in his hand a little book open: and he set his right foot upon the sea, and *his* left *foot* on the earth,
Rev 10:3 And cried with a loud voice, as *when* a lion roareth: and when he had cried, seven thunders uttered their voices.

John mentions no cloud or mystery in the verse from Rev 18. The two witnesses have done their work.

This angel is the "angel of the church", the right hand of Jesus Christ again. In "descending from heaven" he traverses the churches with a seven cycle inverse to that of John's own (John reads right to left unlike the angel). By "come down", I read that in terms of "reciprocal" or as "1/x", (or in addition as if difference, subtracting $0 - x$). I also see in "come down" the use of the past tense and the angel's circuit therefore completed. I find that circuit completed over all seven churches, for there is now no longer a rainbow over his head indicating a test, but instead the Earth is lightened with his glory – as if he were completely "clothed with the Sun".

By mentioning the Sun, I infer there is no more mystery surrounding the rainbow, or any "cloud" obscuring the Sun behind it! His circuit complete, the two witnesses are now empowered by the angel to generalise his circuit in their ministry (this directly aligns with the sixth seal period completed and the completion of the sealing of all the saints). Similarly, in that the "Earth was lightened with his glory", I read that there is no deception of Satan left standing in God's eyes to those practising Satan's devices with the dialectic. There is also no mystery of God remaining; the God of heaven is revealed and the "Sun" clearly visible to all! Then, the angel is clearly shown having overcome – properly after making the cry of Rev 10:3.

The angel in strength of judgement affirms Babylon is fallen – operating only the wickedness of the human heart in every matter. It is inhabited by devils (denizens of the pit – namely corrupted extensions of finite fields of all characteristics other than those three forming the Trinity) and a "hold" of every foul spirit – a "fortress" or group S5 formed by the locusts and serpents of the fifth and sixth trumps, now a cage of every "unclean and hateful bird".

Rev 18:2 And he cried mightily with a strong voice, saying, Babylon the great is fallen, is fallen, and is become the habitation of devils, and the hold of every foul spirit, and a cage of every unclean and hateful bird.

The only "bird" mentioned so far was as the fourth beast before the throne of God as a flying eagle – I reasoned that as all four limbs were "off the ground", it indicated a paradigm without divine restraint. These "foul birds" are those spirits operating with the fourfold dialectic devices of "Death".

I also formed a "wrath" octal upon the cosets of subgroups in additive product with one element of the octal (arbitrary, rather than unity). I found an octal that is "outside of Christ", that would suffice as a cage for the dialectic in the earthly elements, and which has as its identity the empty set rather than the octal or zero. To be so caged is to be eventually cast out from God's sight (in a place full of darkness).

14.2) What Goes Around, Comes Around

Rev 18:3 For all nations have drunk of the wine of the wrath of her fornication, and the kings of the earth have committed fornication with her, and the merchants of the earth are waxed rich through the abundance of her delicacies.

All nations (those in the seven churches, the repentant 144,000 of Israel as well as the "gentiles") have operated as corporations in collective: their governments in legally recognising the religious institution of the scarlet beast have made that great city "Babylon" a captivity for all within. The woman is recognised and the whole world throws in their lot with her. All corporations become as she in God's eyes, and are likewise holding her members captive. The world, as separate from the church, has always been without restraint (as Leviathan); whereas the people of God were always under restraint (whether by the law or in Christ by the Holy Ghost). Now that the restrainer is loosed and government has devoured the Church whole, the "kings of the Earth" (those "bows" of the dialectic), have been somewhat wed to her, as if she were one of them. (Albeit as a synthetic balance as the "Hell" system.)

The "Merchants" bewail not the loss of the Church, but of the whole economic structure of the whole Earth (within which she is embedded), now made "as one" and without the true elect dwelling in the "Church". Religious institutions become indistinguishable from any other corporation. The money which is the circulatory system of the world in its corporate structures has allowed the merchants to get rich. The Church is no exception despite their insistence they buy and sell "doctrine". In truth, they operate as a system of monies.

This statement is of the same angel that "treads the winepress of the wrath of God" whose name is called "the Word of God". This is immediately apparent from the statement of the final fallen nature of the Church together with Christ's command to be separate in the verse Rev 18:4. The setting is the same and the purpose identical.

Rev 18:4 And I heard another voice from heaven, saying, Come out of her, my people, that ye be not partakers of her sins, and that ye receive not of her plagues.
Rev 18:5 For her sins have reached unto heaven, and God hath remembered her iniquities.

The Lord's voice from heaven (not the angel of the church but the head of the elect) convicts all to come out of such corporations as centres of fellowships, for the world with its love of money is to be judged and to receive the wrath of God: in fact – to become the hell of the lake of fire. The city (the waters as the entire world collected under the woman, the "Mother of ... abominations") is to be left by all, to never mix Christian religion and business ever again.

With the arrival of God's wrath at the Church's judgement, the dialectic cannot operate and the rest of the world's economy cannot continue to function without it.

Her sin has reached to heaven – the beast has finally encroached upon all 42 elements (months) of the seven lampstands under the angel's circuit: generalised to the two witnesses, each lie dead in the streets of the city; only the identity element of Sabbath rest is left, within which the elect dwell with God in heavenly places (one part brought through and preserved as compared to two parts in GF(4) cut off from the octal but preserved until the judgement by Jesus Christ). The rule of the beast has encroached upon God's salvation, even up to heaven as to claim victory against the virtue of Christ (floating unity): righteousness credited to the elect.

God has remembered her iniquities: she is not now any vessel of the redeemed.

14.3) A Bag Full Of Holes

Rev 18:6 Reward her even as she rewarded you, and double unto her double according to her works: in the cup which she hath filled fill to her double.

The proper and true K4 form of the ultrafilter is as:

$$a \equiv 1 \equiv [0, a, b, c]$$
$$b \equiv [0, a, d, e]$$
$$c \equiv [0, a, f, g]$$

Which is preserved under a seven cycle (a,b,d,c,f,g,e) from the octal thus:

$$b \equiv 1 \equiv [0, b, d, f]$$
$$d \equiv [0, a, b, c]$$
$$f \equiv [0, b, e, g]$$

And is preserved static under Frobenius acting on GF(8) as so:

$$a \equiv 1 \equiv [0, a, d, e]$$
$$d \equiv [0, a, f, g]$$
$$e \equiv [0, a, b, c]$$

Yet if the K4 form is isomorphic to GF(4) having only the members {a,b,c} then there is a valid application of the Frobenius map in GF(4) that would send the K4 form in a two cycle to:

$$a \equiv 1 \equiv [0, a, c, b]$$
$$c \equiv [0, a, g, f]$$
$$b \equiv [0, a, e, d]$$

It is this map of Frobenius on the K4 form of the filter mimicked by the serpents of the sixth trumpet. The map of Frobenius generates a three cycle legally in the octal holding and cycling the three triples [a,b,c], [a,d,e], [a,f,g] static. There is no two cycle by Frobenius found in GF(8).

GF(8) and GF(4) intersect in GF(2) only and the extended form of GF(4) over that prime subfield is completely separate and cannot be constructed as a subfield within GF(8). The extended part is totally separate from the octal. (Unless Christ is considered to "sit at the right hand". The serpents blaspheme that they too may open the last three seals – i.e. that they possess the same judgement. They see a local closure of GF(4)+ in GF(8)+, having "eyes like a man" as does the "little horn" of Dan 7:8; misreading it, grasping at authority they do not have.)

The serpent spirits use transpositions to mimic a two cycle (as if by Frobenius) in the octal as if GF(4) were a subfield. Then if a=1, the transpositions in the reference that they may act with are (b,c), (d,e) and (f,g). Each transposition acting on the reference results in the same "wormwood" octal, and because the triple [a,b,c], [a,d,e], [a,f,g] is static, under the original reference cycle I generate the seven rows

$[a,b,c], [a,d,e], [a,f,g]$ with $[b,e,f], [b,d,g], [c,d,f], [c,e,g]$ with unity $a = 1$
$[b,d,f], [a,b,c], [b,e,g]$ with $[a,d,g], [c,d,e], [c,f,g], [a,e,f]$ with unity $b = 1$
$[c,d,g], [b,d,f], [a,d,e]$ with $[b,c,e], [a,c,f], [e,f,g], [a,b,g]$ with unity $d = 1$
$[c,e,f], [c,d,g], [a,b,c]$ with $[a,d,f], [b,f,g], [a,e,g], [b,d,e]$ with unity $c = 1$
$[a,f,g], [c,e,f], [b,d,f]$ with $[b,c,g], [d,e,g], [a,b,e], [a,c,d]$ with unity $f = 1$
$[b,e,g], [a,f,g], [c,d,g]$ with $[d,e,f], [a,c,e], [a,b,d], [b,c,f]$ with unity $g = 1$
$[a,d,e], [b,e,g], [c,e,f]$ with $[a,c,g], [a,b,f], [b,c,d], [d,f,g]$ with unity $e = 1$

These "switches" (b,c) etc., transform the reference octal to a row of the above. Applying any of the three same transformations return the rows to the reference octal. The K4 form within the row appears unaffected, remaining constant. One could ask "what's the difference?" but the difference is that GF(4) is not embedded in the octal to permit that two cycle under Frobenius.

Under the reference cycle the rows form in their leftmost three columns that "jacinth" part of the "breastplate" of the serpent which is shared with the Sun octal and this is acted upon by Frobenius in an identical manner as by the Sun octal's reference cycle. The K4 form appears equally preserved as it is still under the reference cycle.

Frobenius acting on the K4 form forms a three-cycle as "breastplates of fire" and switching the elements in the octal illegally as a two cycle is as "breastplates of brimstone". However, as there is no such two cycle in the octal, this whole construction of GF(4) embedded in these "wormwood rows" is totally invalid.

Now the jacinth part, shared between the wormwood row and the Sun octal (in order for the serpents of the sixth trump to mimic Christ) must be held static under Frobenius and this requires the intersection of the three groups to correspond to unity. Then this unity is taken in both the wormwood row and reference cycle.

Now, if the rows are constructed as to correspond with unity in the reference cycle, they then do not affect the least in his circuit: the serpents assume the position of rest in the Sun octal, and merely stay within each of the seven churches without being under the conviction of the spirit. They "tread underfoot the outer court of the temple for 42 months".

That reference cycle of multiplication begins with Ephesus and ends with Philadelphia, Laodicea then the identity. Yet the circuit reads in its cycle right to left to John unlike the sequence of the seven dictated letters to the angel, so (properly put) the notation is reversed so the cycle when read right to left makes for "interesting reading". (Each letter, but for the first and last, is addressed to the "angel of the church in…". I note that the angel most certainly reads left to right in the cycle of churches or of the spirit.)

Instead of applying the seven cycle (from Ephesus to Philadelphia) upon the circuit of churches as if multiplying by x = (a,b,d,c,f,g,e) (as if the "churches" a to g of the circuit were a G-set) thereby attaining rest (unity) in the circuit upon solely the (inverse) element x^{-1}, instead I actually multiply by x^{-1} (as the inverse of x in permutation notation is that which reads right to left to John) and find that that "rest" of unity is now become that particular church (within which the angel is now become present) which was as the element x in the cycle beforehand.

Every church has the angel "sat" in place as the identity element: every temptation of each church's false doctrine is equivalent to the same conviction as that to overcome Laodicea. The angel's circuit is of seven "mountains" to overcome: each requires diligence to overcome that particular false doctrine. Every church in the angel's circuit is become as the identity of the reference cycle (Laodicea), and the wormwood rows above apply as the serpent's device.

The Frobenius map over GF(8) as the cycle of the leftmost three subgroups of each row above is not the operation in view. Rather, the intersection of those three subgroups corresponds to unity (if each row corresponds to the reference cycle at all). The seven cycle of the reference is unmoved, acting only as unity (and is not cycled by Frobenius in GF(8) or by these serpents) and only the illegal transpositions forming these illegitimate wormwood octals are made.

Each row is chosen to perpetuate the dialectic devices (as with [c,e,g] in the top row, say) in the same setting as the K4 form or "golden girdle" (the leftmost three subgroups). Within every position of the angel's circuit (as with the "floating unity"), those devices work only if each position in the angel's circuit becomes as unity. (The angel is under no conviction and is tempted to accept the only possible dialectic [c,e,g].) Then in every position of the circuit, the angel "sits pretty" as the intersection of the K4 form (as the angel of the church of Laodicea) whilst the setting is exchanged for the dialectic. (Yet this causes the angel to move on! He is the angel of the church, after all, finding rest in Philadelphia only.)

The reference octal is not at all affected and the rows above function only with the identity of the reference circuit as it is free to float in the cycle: else one element is fixed (say with unity a=1 or as Laodicea) and the conviction of the spirit in the angel's circuit is completely uninterrupted and still sevenfold in length. That said: the action of each wormwood row evaluated on the reference is without exception, only the multiplicative identity. (Which in the reference is free to float under multiplication.)

On another note, referring back to section [7.11], if two octals are aligned isomorphic to each other with the same reference cycle, they must be aligned in a one-to-one manner. If they share even a single subgroup or K4 or "jacinth" part, then to be isomorphic they must be equal. (The two witnesses show that even if octals just share the one subgroup under the same seven cycle then they must be equal as octals.)

Whether they share one or three subgroups with the "Sun" octal, it is apparent that the seven cycle over the reference is invalid in the wormwood rows formed under the action of these serpent spirits. [b,d,f] is exchanged for [c,e,g]. Water is turned to blood and vice versa.

Every octal is aligned to the angel's circuit by the two witnesses across the "sea of glass". This entails that, truthfully, only the one device of wormwood acts against the angel in the sea: these wormwood rows corresponding to unity in the reference seven cycle, the angel's circuit.

Providing the angel is under conviction as by the letter to Laodicea (to complete the seven cycle of the reference, circuiting the seven churches), the 200,000,000 five cycles constructed by the serpent spirits fail to have any effect (other than of the identity element in C7) at all.

Rev 3:8 I know thy works: behold, I have set before thee an open door, and no man can shut it: for thou hast a little strength, and hast kept my word, and hast not denied my name.

Then 200,000,000 devices fail, each of the one device repeated (acting to make no difference). The angel is "in each church" rather than "of each church": he circuits to reach approval.

Then the adversary's five cycle over the churches may be formed by using serpents and composing them on both the left and right hands, so in the case that g=1 I find:

(a,b,d,c,f,g,e)(b,e) = (d,c,f,g,e)(a,b).

By continually applying as the key to the pit (b,e) before every application of the seven cycle, the elect in (b) are turned back to (a).

So, to reward her double, simply reapply her own treatment of the remnant (a,b) and turn her away from Philadelphia (cf. the parable of the ten virgins): thereby rearranging to:

(a,b,d,c,f,g,e)(b,e)(a,b) = (d,c,f,g,e)(a,b)(a,b) = (d,c,f,g,e)

I must ask, why g=1? (a=1 is as Laodicea if b is as Philadelphia.)

Note {b,e,g} is a subgroup in the right hand and {a,b,g} a static triple in the left. Then, exchanging {b,d,f} for {a,c,e}, as water for blood (i.e. with g=1) I have a case of the serpent spirits present in both hands, as with the two witnesses. Yet this is an illegal correspondence using the seven cycle (a,b,d,c,f,g,e) and so I require a shift to an alternate cycle permitting [b,d,f] ≡ g ≡ 1. I must act as from elsewhere in the sea.

These transpositions form the "horse bridles": redirecting the seven cycle to a five cycle (in the same elements), then becoming "blood from the winepress" as then separated, cut off.

As there are 200,000,000 possible five cycles to interrupt every seven cycle (i.e. in every way), each requires a serpent transposition as if the beast were "ascending from the bottomless pit". There are then required 200,000,000 such devices of transpositions (in pairs) from elsewhere in the sea which are all equal to the one failing case against the reference seven cycle. Satan's unending work has truly become completely meaningless.

That is, given that the serpents mimic Frobenius on GF(4), if they may act trivially, as with (d,e) or (f,g) when only (b,c) structurally mimics Frobenius, then although that also transforms the octal to wormwood, there is only a small deception there. If (d,e) may mimic Frobenius in one setting and then under Frobenius the setting changes to (f,g) mimicking Frobenius (or by shifting wormwood rows for a different unity in the reference cycle) and (d,e) is used as a "lemma proof" in an illegal setting (the seven cycle applied is invalid) or the converse, a trivial action shifted by action of C7 to mimic Frobenius elsewhere, then there is some ground to examine as to why (b,c) and (d,e) have a different effect.

I may mimic the K4 form, say, with the wormwood row a = 1 and holding [a,b,c] static. The three rows containing [a,b,c] correspond to the first, second and fourth rows of the seven cycle acting downwards (on the full seven above). Lucifer would become "like the Most High?".

The remaining triples come from the rows corresponding to b and c.

So if a = 1 and [a,b,c] is static I reduce to the three rows:

$$[a,b,c], [a,d,e], [a,f,g] \text{ with } [b,e,f], [b,d,g], [c,d,f], [c,e,g] \text{ with unity } a = 1$$
$$[b,d,f], [a,b,c], [b,e,g] \text{ with } [a,d,g], [c,d,e], [c,f,g], [a,e,f] \text{ with unity } b = 1$$
$$[c,e,f], [c,d,g], [a,b,c] \text{ with } [a,d,f], [b,f,g], [a,e,g], [b,d,e] \text{ with unity } c = 1$$

In the correspondences above, the permutation (b,c) in the octal appears to be the permutation of elements in the K4 filter from [d,e,f,g] to [f,g,d,e] equal to the effect of Frobenius on GF(4).

Note [a,d,e] is common to the three rows corresponding to a, d and e instead. Since the serpents are described as "with their tails they do hurt", I may expect the reference cycle to be acted upon by Frobenius in GF(8) also (to permit the trivial (d,e) to mimic Frobenius): yet in order to return the full set of 12 permutations subject to this separate construction, there are still those trivial transpositions (d,e) etc. that the serpents mimic.

By forming the transpositions (d,e) and (f,g) within the same row a=1, I may construct the other permutations within the earthly elements, as trivially with no mimicry of Frobenius. Then, the serpents simply deceive by equating these trivial mappings (d,e) to the former faked

Frobenius as (b,c) in the same row corresponding to a, and this deception is how "they hurt". Yet, this still requires the correspondence of that row to the unity in the reference cycle.

The serpents pass off the mimicry of the Frobenius map on the K4 form of the filter as a symmetry of the octal in order to validate the transpositions that are made trivially, not in mimicry of Frobenius. This is a deception, and certainly one to watch out for.

Those three rows of seven, then, are essentially the same octal containing [a,b,c] yet with [d,e,f,g] permuted. I may posit the K4 form to be a similar arrangement of subgroups as such, with those same elements permuted. So, taken from each row and keeping a=1, I have:

$$a \equiv 1 \equiv [0, a, b, c]$$
$$b \equiv [0, a, d, e]$$
$$c \equiv [0, a, f, g]$$

$$a \equiv 1 \equiv [0, a, b, c]$$
$$b \equiv [0, a, d, g]$$
$$c \equiv [0, a, e, f]$$

$$a \equiv 1 \equiv [0, a, b, c]$$
$$b \equiv [0, a, d, f]$$
$$c \equiv [0, a, e, g]$$

In each row, I find the apparent permutation of the earthly four:

[d,e,f,g] which permits the permutations of the K4 filter of [f,g,d,e], [e,d,f,g],[d,e,g,f]
[d,g,e,f] which permits the permutations of the K4 filter of [e,f,d,g], [g,d,e,f],[d,g,f,e]
[d,f,e,g] which permits the permutations of the K4 filter of [e,g,d,f], [f,d,e,g],[d,f,g,e].

Also, these actions which mimic Frobenius under the operations of the octal are preserved by the illegal reference cycle. So, if I in mimicry of Frobenius apply (b,c) to send:

$$a \equiv 1 \equiv [0, a, b, c]$$
$$b \equiv [0, a, d, e]$$
$$c \equiv [0, a, f, g]$$

to

$$a \equiv 1 \equiv [0, a, c, b]$$
$$c \equiv [0, a, d, e]$$
$$b \equiv [0, a, f, g]$$

and then cycling the filter in the octal using Frobenius on GF(8) to switch to

$$a \equiv 1 \equiv [0, a, e, d]$$
$$e \equiv [0, a, f, g]$$
$$d \equiv [0, a, b, c]$$

In order to employ the second transposition as (d,e) to

$$a \equiv 1 \equiv [0, a, d, e]$$
$$d \equiv [0, a, f, g]$$
$$e \equiv [0, a, b, c]$$

And cycling again, using Frobenius on GF(8) I find that the effect is null.

$$a \equiv 1 \equiv [0, a, b, c]$$
$$b \equiv [0, a, d, e]$$
$$c \equiv [0, a, f, g]$$

Frobenius upon the octal sustains the same action of serpents acting as if they were the K4 ultrafilter. (I.e. employing (b,c) as mimicking Frobenius, then cycling under Frobenius in the octal and applying the transposition as (d,e) again and cycling back, has the same effect as if (b,c) were applied twice and this cancels itself out.) The illegal operation of Frobenius on the K4 form under the octal is in no way corrupted. This is a dreadful device, this second "woe".

Yet no two different octal groups isomorphic under the same seven cycle can share any subgroup (as if the reference cycle applied across the sea of glass), and so if [a,b,c] is held in common in two forms then immediately this invalidates the twelve permutations of [d,e,f,g].

There is no seven cycle common to any of those three rows from the seven given above ([a,b,c] is shared by all three), yet only those eight seven cycles valid on an octal transform it amongst its own seven cycles; the wormwood rows are permuted under an illegal cycle, taken from the reference. Each row is then only a separate case. There is also no valid seven cycle on the rows permuting the rows' four sets of earthly elements amongst themselves in those same twelve ways. GF(4)'s two cycle under Frobenius is not found in the field GF(8).

So, these twelve elements are also not able to be constructed and they likewise fall. That fall is under one automorphism of Frobenius: i.e. "that same hour" (then the situation is as if the angel has overcome Laodicea taking his rest as a=1, say, and the witnesses share the same state). Then the only element in view is the unity in the reference cycle and then the wormwood rows have no effect on the angel's circuit at all, being unable to tempt. The Frobenius map then affects no change. (As every automorphism over the octal holds fixed the unity, as if that same "hour": with that same element a=1 that completes the angel's circuit).

The end of the angel's circuit and of the two witnesses' ministry are unique in that there remains no conviction from the letter to Laodicea: i.e. upon a state of rest within each church – of unity in the reference octal (being a "mountain" of false doctrine). Philadelphia meets with approval from Christ; there is no conviction to the church to move on, merely an instruction to be diligent. (Many Christians are vigilant watchmen... but are many sober minded?)

Rev 11:13 And the same hour was there a great earthquake, and the tenth part of the city fell, and in the earthquake were slain of men seven thousand: and the remnant were affrighted, and gave glory to the God of heaven.
Rev 11:14 The second woe is past; and, behold, the third woe cometh quickly.

The tenth part, those twelve elements (devices) are leached from multiplication(s) over the right hand: the full 200,000,000 devices (formed by transpositions applying the "star with the key to the pit") only work to turn back the least in his circuit to the first four churches – the Laodicean "Church" under the action of the pit, of the locusts in the fifth trumpet.

Of the whole city S5 with 120 permutations, the elements of S4 on the earthly elements are not constructible as "wormwood" on the reference, but 12 are leached from God: 1/10 of the city immediately falls. (The two witnesses show God's spirit is not moved from His place.) Then the five cycle upon the "refining churches" is unable to reproduce 60 elements of S5.

Rev 18:6 Reward her even as she rewarded you, and double unto her double according to her works: in the cup which she hath filled fill to her double.

"Double unto her double" refers to the serpents' transpositions, as if they were her "works" (thefts of permutations upon the earthly elements, and as of sorcery, burning brimstone to obscure the spirit). It is then the case that these verses indicate the believer is to expect those missing 12 elements to be added at the judgement by God.

The "cup which she has filled" refers to the 60 missing elements of S5 overall which are added (and then the group closed) by God's judgement and they too are presently only "half full".

Yet Satan's devices of the transpositions only act across separate settings – the case a=1 in the reference cycle was countered with that of g=1 in order to construct the five-cycle (hence the full 200,000,000 transpositions applied in every setting when the beast ascends from the pit killing the two witnesses). It is clear Satan's device only operates if one takes the same view as the collective, rather than the salvation of the individual in the setting of the Holy Spirit. Then of necessity the denial of virtue is made: as to any fixed position.

The five cycle operates with serpents by holding fixed the reference octal under the action of unity. In order to interrupt a circuit of the total 200,000,000, the device must do so upon a different seven cycle "elsewhere" in the sea and with a different "left hand" octal paired with the reference as a "right hand" (so only the "outer court" of the temple is in view affected by those five-cycles, upon those lukewarm believers in the churches that fall as "untimely figs").

The least is interrupted in one of (and then all) 200,000,000 circuits only if every one of those transpositions across the sea (with a different unity/octal/seven cycle every one) may be held in common as to "agree to disagree" together. Then, everyone's denial of the doctrine held true by everyone else is common in the idea that every or all doctrine "perceived" by the collective separately or together, agree in one commonality or the whole. This is actually the bottomless pit open. Either unity is fixed in the "lamb before the throne" or it orbits amongst every element outside of the union of the trinity ultrafilter: as in algebraic closure. This is an unending and ceaseless leavening of the truth. To share as common all possible doctrine is to have no true doctrine. (Unless all that doctrine be restricted to that of only the one true God.) I begin to see why Christ would vomit out His own right hand if that angel were truly "of the church of the Laodiceans".

Actually, as every octal is aligned isomorphic, there is truly one commonality of all virtue and therefore sound doctrine, and this is solely found in the lamb as slain before the throne: as the doctrine of the one true God. (The ministry of the two witnesses is a true witness of the gospel.)

So, to sum up: the wormwood device only acts by holding a five cycle over the seven churches, but then only acting as unity in the reference or circuit of the angel. The seven cycle of the angel's circuit is never interrupted by five-cycles, but only those circuits of "gentiles" which are within the seven churches and not under conviction. Then of all 200,000,000 five cycles/ transpositions only the one device is set against the two witnesses' ministry (of the same circuit over every octal) and is then present, all other cases being similar. Then as the "door is open" and these devices are without effect, all 200,000,000 devices fail. "No man can shut it."

14.4) Satan's Last Remaining Device: The Deadly Wound Is Open

In the Revelation, Satan's devices have one aim: to subjugate the church of Jesus Christ, to deceive them into blaspheming the Holy Spirit. I then ask, "Just how does he get me?" Upon each point I raise it becomes apparent that Satan is far better prepared than any of us (and in his efforts remains subtle yet worldly). Yet in Christ believers will always find the solution.

So, just how does Satan subvert the believer? By laying a foundation other than Christ's own.

Entering A Dialogue – As If It Were Any Other Discussion (14.4.1)

Throughout this writing, the exampled K4 groups of the octal have been labelled in the elements {a,c,e,g}, which will (unfortunately), never represent the coset of a static subgroup [b,d,f] with virtue (unity) as a=1. The four are often also labelled with {r, s, u^{-1}, v^{-1}} (the sets of the faulty dialectic filter that is far more "worldly").

Within the K4 form (as Christ alone, separate from the octal, invariant over every logical schema), the static subgroup contains unity and then virtue. It also includes the predicates $(r\&s)^{-1}$ and u&v.

Do not confuse the K4 form as Christ sat enthroned within the octal to the octal itself. The octal contains unity in the static subgroup's coset and the K4 form enthroned as that static subgroup would do so too. That coset, then, may properly contain no subgroup with virtue. This is where the confusion begins.

I improperly mixed and matched the dialectic (the paradigm of antichrist) with a coset opposed to the valid K4 form within an octal proper, and neither has won out. The coset in the elements {a,c,e,g} contains virtue in "a" that would be common to both the octal and K4 form as the unity (the lamb); whilst the K4 form appears to reveal the elements in the limited schema of antichrist, it does not reveal the often present K4 group of virtue in those same four elements that belongs in a valid octal.

If I were to find a K4 group with virtue wholly in the coset, then {a,c,e,g} cannot be the coset of any subgroup of the octal at all! (There is always a non-trivial intersection between subgroups.)

To display that schema {r, s, u^{-1}, v^{-1}}, I show my worldliness; for I have confused myself with the same device as I have need to explain.

As a believer enters into a dialogue with such a logical schema in r, s, u^{-1}, v^{-1}; being immediately beguiled that the setting remains of virtue (which is **never** present in the dialectic), i.e. one that is not found only of a coset, but one permitting a K4 group of virtue; they are translated to a setting where the dialogue has moved from their initial intent to minister the gospel to no opportunity for that at all; as if it were any other worldly discourse. By offering only the "earthly" elements (the coset), that setting is mistaken for one allowing the application of virtue where there can only be its denial. (Where there is no axiom of sovereignty cf. [5.3].)

Then, there will be no opportunity for exercising virtue within that coset, for there can be no subgroup in those elements of the schema, as they are limited to that coset in the octal.

Is Christ eternally enthroned in the static subgroup or not? Is the coset taken in the octal completed with the K4 form or is it limited to the dialectic in $\{r, s, u^{-1}, v^{-1}\}$ as to its whole? The relevance is found only in the latter within that beguiling dialogue. God will not practise the dialectic to rescue the believer so confused.

The dialectic schema (considered as such a coset with virtue) will never recognise Christ (as He is always properly separate) having all virtue placed in the static subgroup. Similarly, it denies any subgroup may be present in that coset at all because the octal has been confused in kind.

Such dialogue starts with the immediate denial of Christ: the coset of the static subgroup in the octal is confused with the dialectic filter sat as if God but always opposed to God.

2Th 2:4 Who opposeth and exalteth himself above all that is called God, or that is worshipped; so that he as God sitteth in the temple of God, shewing himself that he is God.

Spin, Hurting the Oil and the Wine (14.4.2)

Beforehand, I had pointed out that anyone practising the third seal device could "spin" their position when it is pointed out that the oil improperly privates the wine and vice-versa. When this fallacy is indicated (that it is impossible for virtue to private another positive property), the practitioner can spin their "dialectic", bringing the faulty inferences of: $(r\&s) \to (u\&v)$ and also $(u\&v)^{-1} \to (r\&s)^{-1}$ (equal by the modus tollens). The latter, $N\neg(r\&s)$, is all that is required to justify that "oil" privates "wine" and vice-versa.

Unworkable predicates as r&s from the octal then entail a "good" in u&v. Or, if the unworkable are to be judged "evil"; $r\&(u\&v)^{-1} \to \neg s$ and the divisive $r \lor s$ is also found to be an "evil".

Either way: if "r" is virtue or if (r&s) is shown unworkable in the octal, that modus tollens may be claimed positive with spin from the facilitator. I.e. when it is claimed that "r" could not private "s" then P(r&s) and in the dialectic there would be the "good" of P(u&v). So, the practitioner claims that the disjunction in "r" and "s" etc. equally leads to $N\neg(u\&v)$ (which are not virtues by elimination) which entails $N\neg(r\&s)$. He would reference a "disjunction" in the real-world, without presenting the proper solution to it in virtue. "Oil" in "r" would hurt the "wine" in "s" and vice-versa.

Introducing The Faulty Paradigm Of Antichrist (14.4.3)

The dialectic provides the ability to judge for one's self the difference between good and evil. Entering a dialogue over a disjunction within the sets $r\&u^{-1} \lor s\&v^{-1}$ (or a, c, e, g) immediately privates all virtue (to establish a consensus and compromise).

"Good" becomes a consensus reached in r&s, say, and the "evil" becomes the divisive stance over the disjunction $r \vee s$.

I also find $r\&s \rightarrow (u\&v)$ and that $(u\&v)^{-1}$ proves $r \vee s$. $(u\&v)^{-1}$ is judged evil and divisive but r&s which entails u&v is **unworkable** as a good (as taken from the octal). Similarly, $(u\&v)^{-1} \rightarrow (r\&s)^{-1}$ is **unworkable** as a good and (r&s) which proves $u^{-1} \vee v^{-1}$ is also divisive and possibly found to be an "evil".

Mat 23:4 For they bind heavy burdens and grievous to be borne, and lay *them* on men's shoulders; but they *themselves* will not move them with one of their fingers.

And judging good and evil (as if freely) with the dialectic is faulty; an unstable, double-minded (contradictory) and therefore stumbling paradigm.

Facilitating As If Enthroned In The Temple (14.4.4)

Satan's devices can make the Christian faith appear irrelevant. Any attempt to hold any divisive absolute(s) of virtue will be met with exclusion from the consensus in a facilitated dialogue.

I.e. the facilitator dictates which predicates (rather than virtues) are in precedence. There may be the third seal system, using the "balances":

$a\&c\&g \rightarrow c\&e\&g \leftarrow a\&e\&g$ is lightened to either side as: $c\&g \rightarrow c\&e\&g \leftarrow e\&g$

I.e. with the facilitator guiding the dialogue for sending virtue $a \rightarrow \{ \}$.

Given "a" I find: $c\&e \rightarrow \neg g$, yet then $\neg g$ is "evil" (because (a&g) is "good" and entails $(c\&e)^{-1}$) so $N\neg Pos(a)$. Virtue $N(Pos(a))$ becomes inconsistent as a predicate for a positive property.

Then $Pos(c\&e\&g)$. I.e. $a \rightarrow \neg g$ given c&e; oil privates wine and vice versa $g \rightarrow \neg a$.

There are a number of ways these devices of Satan may compromise a fellowship and every believer within it.

Shifting Wormwood Rows, Denial Of Virtue (14.4.5)

On the basis that "oil" and "wine" (as "a" and "g") private each other falsely in the dialectic, in order to secure the dialogue against all such objections (that would cause a deadly wound) Satan has formed quite a device based upon the wormwood octals.

The facilitator of a dialogue will then spin that "a" and "g" are freely interchangeable (yet only a&g would be so, when substituted for both; yet this is shown an unworkable predicate by the octal), instead, these two absolute poles of virtue (for they are that if symmetry ensures they of necessity private one another) are demoted from their absolute and necessary positivity to become as any other "decidable" predicates – the wormwood device comes into play – introducing a K4 group with the elements [a,f,g], where the row holding f=1 is now in scope.

For example, if between fellowships the virtue "a" is such that "The God of Jesus Christ shall not be blasphemed" and virtue in "g" is such that "Everyone's beliefs must be respected as sacred", the virtue should properly become a&g or "Only correct doctrine will be established amongst any fellowship undertaken" rather than, say, "Everyone's beliefs must be respected, so we shall not make mention of Christ but only show a good example" as virtue in "p", which would appear to suffice to somewhat decide freely the disjunction $a \lor p\&a^{-1} \to g$.

Any virtue in a&g&f is completed (and also closed) with p=f in that the virtue a&g is always a good example to show and the scope of "we" has increased to 100% of the fellowship every time a new fellowship is accepted, for everyone's beliefs are respected.

Given the device of wormwood in the row f=1, both "a" and "g" appear as members of a K4 subgroup with "f", shared with the reference octal in that "jacinth" part. However, virtue (and with it the schema) has been altered from the reference octal.

Having shifted from virtue in "a" and having transposed to virtue in "g" there is a need to spin this transpose as "freely decidable" in the octal, by presenting them as if they privated each other: a setting where most unaware Christians would be caught off guard and unsure of the "positivity" within it.

Absolutes of virtue as in "a" and "g" are then denied and translated to the wormwood row with f=1, where a vote for consensus may be taken that one predicate may be freely substituted for by the other. (This is not the proper function of the octal, there are no such transpositions in the operation of C7 or under the Frobenius map.)

I may object to the claim that "a" and "g" private each other (as compromised) and are freely decidable by virtue of "f", with the obvious claim that they are virtues and cannot private each other, a&g or even a&g&f are truly valid virtues.

So, "f" becomes the sole virtue, and "a" and "g" are both denied, demoted. I will show that in a particular case (of setting up the image system) any strict objection to the exchange of two such demoted absolutes will move the setting to the construction of a faked "Manifold Of Virtue" where f=a&g. Otherwise, only a simple vote would need to have been taken for approving that exchange.

Constructing A Manifold Of Virtue (14.4.6)

Moving from the individual's role in a dialogue, I move on to the setting where those devices are used to bring about facilitated consensus in a church. Whether this is truly the image system of the false prophet or not, it fulfils the criteria by using only Satan's devices.

The Holy Trinity together truly form a "Manifold of Virtue" (cf. [3.10]), and it could be argued that the three meet in their "functionality" instead; as if Christ were not only our intercessor, but to reason of **necessity** that the Holy Spirit was also interceding between Christ and His Father for our "comfort" and the Father between His Son and the Holy Spirit as to bring out good works from us.

This is, however, a faulty inference; only the persons of the Trinity meet in such a manifold and they three may well be resting on their person(s) at their own liberty. God, it cannot be argued, is always at work necessarily but only that He is necessarily existent. Every such construction inferred is a mere "God of works". Some works may well be provided for as in the scriptures, but the works of God are not to be argued necessary (and to be always repeated upon themselves) as from God's person.

The example of the "great commission" may be used as a means for such a "faked" manifold of virtue. (This is the "bait" for the "switch": pressured obedience for the mark of the beast.)

Christ chooses those who are to become His disciples; the Holy Spirit moves on those converts to make them attentive, and the Father (by the above), intercedes in power over His creation to make the commission active in the world today. (Obedience is forced, the mark received.)

Then I have found three functions unified in the "Holy Trinity" as if "natural" virtues.

#1)   The Father would supply the world for His purposes as creator omnipotent.

#2)   The Father's purpose gives new Christians to Christ by the works of the Holy Spirit.

#3)   Christ purposefully "chooses" of them His faithful elect and disciples.

In a manifold of virtue it is enough that each pair of the above combined result in the necessity (content) of the third. In particular it is to be argued that #3 = #1 + #2.

#3 should be an obvious result; the desired goal of the great commission (that force/pressure).

What, to a Christian, may seem more reasonable? (The dialectic is become the foundation reaching up to heaven, rather than that of true divine authority reigning whole in Christ.)

He Can Confuse You On Charity (14.4.7)

God has supplied of His virtue these two constants:

- Christians are individuals saved without requiring works.
- The purpose of spiritual gifts is to strengthen fellowships.

Or, restated in terms of the exercise of spiritual gifts:

p_r "our purpose is enough." (It is already provided for by God perfectly.)

p_s "we are offered spiritual gifts only for purpose." (Each fellowship has enough – a perfect minimum.)

The virtue $p_r \& p_s$ is properly met in sufficiency (a minimum of virtue in God's liberty). Within fellowships there is charity for those gifts to further strengthen and deepen fellowship. It is written: "But covet earnestly the best gifts: and yet shew I unto you a more excellent way." (1Co 12:31). I may form an octal within those two virtues, to excel upon their purpose (demoting them, as if coveting them, deciding for myself), thereby maximising the sets r and s within a disjunction formed of charity; which is never made to weaken, but that good works of virtue make fellowships stronger instead. (The result is that $(r \& s)^{-1}$ is effectively become "empty".)

r_0 Christians are already doing enough; they are individuals saved without works.

u_0^{-1} Further charity is offered by God for excelling **beyond** this limit.

s_0 Works do not bring revival, the Holy Spirit does! (They are "His fruits of the Spirit" not ours.) The church has found no unction for any further purpose than ministering to itself. He is the comforter.

v_0^{-1} Spiritual gifts are offered for ministering within any fellowship to **further** produce such fruits.

The dialectic will never present the "keys" of $(r_0 \& s_0)^{-1}$ or $u_0 \& v_0$ in an octal completed; the full solution for this disjunction solved for with the virtue of charity. (The "more excellent" way.)

$(r_0 \& s_0)^{-1}$ This is "empty" as "God does not expect more of us (He has given no gifts for that), our gifts fulfil only their current purpose, and perfectly so, by His comfort."

$u_0 \& v_0$ "It is positive to exercise the gifts we already have that we use amongst ourselves: it is their purpose to be perfected." (This is become "all".)

Then as the virtues p_r and p_s are constant, r_0 and s_0 etc. hold only with them. Exercising charity amongst a fellowship is the purpose of spiritual gifts, they are not to be spread abroad and trod under foot. New converts are chosen by Christ, but new relationships formed may not be.

Then the virtue of charity for brethren is confused with mission work outside of such fellowships. (A setting mistaken for one of virtue as in [14.4.1].)

That which should perfectly meet in $p_r \& p_s$ is instead confused with the faked manifold of virtue, and a dialectic is set up in the two statements r_0 and s_0 as if they were as predicates (a,g) to be transposed to $r_0 \& v_0^{-1} \lor s_0 \& u_0^{-1}$. A lack of new converts is seen as a fault, rather than a sign of a successful fellowship that has reached and met its "divine quota" in spiritual gifts.

a Is the church doing enough already? (#1 would indicate every church always is.)

g Are you saying the church doesn't have the Holy Spirit? (#2 would indicate they haven't if they don't make enough converts.)

Compare these two to the serpent's question to Eve: "hath God not said; Ye may eat of any tree in the garden?" (Gen 3:1) Posing such a question immediately changes the setting as in [14.4.1], and takes a first step away from every absolute in view.

These two statements are mistaken to form a disjunction that fails in mutual privation given a fault in lack of growth. If a church does not meet the purpose of #1 with new converts as in #2, then they fail the definition presented them of a "church": perfect approval as to Philadelphia.

$a = g^{-1}$ If the church has no new converts, they are not using their spiritual gifts to make them. They are not fulfilling the definition and purpose of a "church". I.e. they are only doing "just enough": but more is expected.

$g = a^{-1}$ If a church is not doing enough, it does not meet the Father's purpose. It is not a Christian "church" by divine indication. God gave them the Holy Spirit and with Him, spiritual gifts to make more new converts.

The result is to replace "a" and "g" not with r_0 and s_0 (or even p_r or p_s), but with #1 and #2, to "solve" the dialectic with the faked manifold of works, replacing God's perfectly distributed minimum of spiritual gifts with a "purpose" (#3) to organise (setting up the image system, as if optimised) a group of churches together in their mission work and in their cooperation.

#1 and #2 are met with the difference of their inaction, the rallying cries of:

a   "Is the world ready for the Gospel?" (I.e. #1)

g   "Is this Church prepared for delivering it?" (I.e. #2).

The claim, "This is how it worked for them... it will certainly work for us; this is how it is to be done!" is made, whosoever the "them" in "this" is. In this manner the mark is forced on all.

"a" and "g" are replaced with #1 and #2 in turn in the place of r_0 and s_0 (never presenting the octal whole), conjoining them as #3 instead of the proper virtue $p_r \& p_s$.

What is it that the (as claimed in $a \vee g$) disjunction rests upon?

Gifts are given of God for their perfected application and some are certainly for use within the setting of fellowships; they are never given to stagnate and rust. If it is thought that gifts are becoming rusty they are not His gifts. If someone has no gifts, they are not gifts that they therefore do not exercise. If someone is doing enough to exercise their "meagre" gifts, don't ever consider them lazy; they were gifted just enough for (and should have faith upon) that gift as it is perfect and made/given of God; never used as if in adequacy rather than always perfected – as if made by a God that will always expect more of His chosen.

1Co 13:8 Charity never faileth: but whether *there be* prophecies, they shall fail; whether *there be* tongues, they shall cease; whether *there be* knowledge, it shall vanish away.
1Co 13:9 For we know in part, and we prophesy in part.
1Co 13:10 But when that which is perfect is come, then that which is in part shall be done away.

God's gifts are given of His virtue and are always the bare minimum required to exercise His will. (There are no superfluous virtues within God's liberty.) If things are seemingly done in part then they are not so, but are always as intended through the perfect exercise of God's own charity, never ours; for few perceive that charity in the octal (which is completed) through a work that is only made in part of the octal, as "r" or "s". ("r", "s" are from one K4 group only; none have all that virtue that God has for working in u^{-1} and v^{-1}, $(r \& s)^{-1}$ etc. His strength is "made perfect in weakness": relaxed virtue is still exampled positive; predicates at rest are consequent of virtue also. Nobody should covet those works of God for growth.)

Then $p \& (r_0 \& s_0)$ entailing $u_0 \& v_0$ as above is actually found positive for God, for some virtue "p" found of God's liberty, His "purpose" (the virtue of charity) for us. If not enough is being done, there are no gifts waiting for that which is not done. Then what is the answer? "It is enough for God to exercise in us those gifts which are present already" $Pos(u_0 \& v_0)$ or $Pos(p_r \& p_s)$. (For those are His gifts for His perfect will.)

Deciding "a" and "g" becomes the setting for the image system wherein churches cooperate:

a   "We do the mission ourselves" we have the gifts for it (we are ready, #1).

g   "We organise the mission of others" they have the gifts for it (we will prepare for them, #2).

Pairing The Dialectic With The Manifold (14.4.8)

Once this "manifold" of works (rather than the persons of the Trinity) is constructed, it becomes the main setting for the dialectic: as if Christ were opposed whilst also sat with the "keys of Hell and Death". Only r, s, u^{-1}, and v^{-1} have any place in the dialogue, and though blaspheming the name of God in the octal and Holy Spirit by mimicking GF(4) (as if it were embedded within GF(8)) will provide a faulty symmetry, it is only a deception (this manifold is a mere construction of works).

There is, truly, no octal in which the dialectic and this manifold meet (there is no such manifold of works); God has already provided and may surely be freely resting of liberty on His person, and arguing and imagining works and voting them into His mouth does not make Him work for others anew when He has already provided in full.

By placing the faked manifold in the octal (as if replacing Christ enthroned), there is applied a completely separate and fiat dialectic system mimicking Christ as if the "Earth" were opening the last three seals.

Dividing "workers" ($r\&u^{-1}$) from "organisers" ($s\&v^{-1}$), I find a plausible dialectic system within the sets of the following limited disjunction, supposedly to manage mission amongst fellowships:

r   We do the mission ourselves and elsewhere, it is our purpose

u^{-1} We encourage the support of mission done by others, as if for those without our purpose (new converts, new attendees?)

s   We *organise* mission to do with others, it is a purpose for them

v^{-1} We maintain "outreach" to supply our purpose to other churches to *organise* themselves (as if casting our nets on the other side, "Where are all the fish at?")

In the dialectic there is no octal, no virtue. Yet here, the virtues of p_r and p_s are demoted to a disjunction of virtue in the predicates $a \vee f\&a^{-1} \to g$. This dialectic system in (r, s, u^{-1}, v^{-1}) has no virtue, is not a setting for virtue, and is purposed for denying that virtue already present for purpose(s) not directed by God. It is immediately confusing — the result of the direct manipulation of that same confusion of [14.4.1].

So, is there virtue "p" to be found in Christ for deciding $r \vee s$? What about $u^{-1} \vee v^{-1}$? What of $(r\&s)^{-1} \vee u\&v$? ($(r\&s)^{-1}$ would indicate that it is positive to remain inactive at that same fellowship; whilst u&v would be that it remains positive to not support mission or to work organising mission elsewhere. Virtue remains for these two: certainly for any new converts, the infirm, "widows" etc. and also those with gifts for ministering within that fellowship.)

Does the Father, working all virtue, reference the disjunction $N\neg(p\&r^{-1}\&\neg s)$? He would intend us to be the bearers of His "purpose" according to the faked manifold of virtue (as by #1).

Does the Holy Spirit reference $N\neg(p\&u\&\neg v^{-1})$? (To keep us "attentive" as "one body" as by #2.)

Have these predicates r, s, u^{-1}, v^{-1} alone granted further spiritual gifts? Where is that "p" required? (Where is the charity of God if the process must be planned for and voted upon?)

In fact, these four show very little virtue in Christ $N\neg(p\&r\&s\&\neg(u\&v))$, but only virtue expected of the believer. The only apparent virtue shown in this system is for the building of relationships, which is not a virtue if the gospel is required to take a back seat, compromised.

He Gets You On Action, Organisation And Purpose (14.4.9)

Within the previous dialectic, $r\&u^{-1}$ corresponds to action, $s\&v^{-1}$ corresponds to organisation. Action then requires the "purpose" of #1 in the faked manifold. Successful and organised action would require God to make His will manifest for that purpose, fulfilling #1.

The image system itself is expected to bring that much hoped-for revival as if the mission of all member churches were then "optimised". The Church would expect the system to produce many works of the Holy Spirit, for fulfilling #2.

In organising amongst themselves to form the image system, #1 corresponds to fulfilling that purpose and action, #2 corresponds to setting up the image system and thus organising an amalgam of churches to optimise that "body".

Yet the flaw in all this prophetic expectation is just that – the false prophet "speaking great things" of growth that cannot discern the octal whole (having "eyes like a man") with the presence of u^{-1} and v^{-1} (and also $(r\&s)^{-1}$ and $u\&v$). I.e. that "My grace is sufficient for thee: for my strength is made perfect in weakness." (2Co 12:9).

Deciding "a" and "g" as if the church were failing to meet the full virtue of p_r, p_s etc., is not to doubt that the church is doing enough, but that the fellowship is not a "church". It is a subtlety, a denial that Christ always intercedes for His elect – that they are justified without works.

So, greater "purpose" requires greater works, requiring better organisation, requiring the image system (or so the Church are directed to believe). The image set up is a system of works.

"a" and "g" are replaced with #1 and #2 in the place of r_0 and s_0 (never presenting the octal whole), within a facilitated dialogue for an initial vote to implement this image organising for that same purpose. By interrupting every seven cycle with the devices of the seals and trumpets, using that fake manifold of virtue to replace all the genuine divine virtue purposefully excluded from the octal (the K4 form removed separate), the result is to deny the perfect gifts of God by setting up a dialectic system to manage all membership and mission in that image system, as maintained by vote under the facilitation in the rule of the scarlet beast's ten kings.

He Gets You To Think Outside The Box (14.4.10)

Whilst lured to vote on implementing "a" and "g" (before the setting up of the image system exchanging r and s or u^{-1} and v^{-1}) any division over the Holy Spirit being "voted out" of office could be met by constructing two "static triples" supposedly agreeing in all the dialogue thus far. The consensus is to be brought in with the collective agreement that the static triples [a,f,g] and {b,d,e} with c=1 in the Holy Spirit are still in effect or are preserved in symmetry.

Beginning with the seven cycle (a,b,d,c,f,g,e) with c=1 (our starting point) the static subgroup is [a,f,g]. A vote on (c,e) with f=1 will agree in wormwood with (a,g) and also (b,d). The static subgroup is unaffected. The vote on (b,c) with a=1 will agree in wormwood with (d,e) and (f,g).

The subgroup [a,f,g] is still unaffected.

In the left hand, however, the subgroup [b,d,e] becomes [b,c,d] under (c,e), (a,g) or (b,d) with f=1, and then [b,c,e] under (b,c), (d,e) or (f,g) with a=1. Order reversed, [b,d,e] becomes [c,d,e] under (b,c), (d,e) or (f,g) and then [b,c,e] under (c,e), (a,g) or (b,d).

In these two cases, the end result differs from [b,d,e] by (c,d) (or either (a,f) or (b,e) also with g=1). The product composed of any two appears to be equal to the result of the remaining third. ((c,e) with (d,e) results in (c,d) etc. Rows "a", "f" and "g" together mimic a K4 form.)

Then by appearance, the three meet and agree together in the same two static triples, the content of any two returns the whole to "balance" with the third.

Then there appears a fiat K4 group embedded in the left hand static triple. After this manner the false prophet would deceive all that "And he doeth great wonders, so that he maketh fire come down from heaven on the earth in the sight of men" (Rev 13:13). The seven cycle appears to preserve this faked K4 structure in the "earth" or coset; agreeing with the operation of the seven cycle's inverse acting from unity to the "Earth" and "static bow" (instead of mapping from unity to the static subgroup instead). This is all done as if made upon the operation of C7 – the Holy Spirit. (Yet using four C7 groups.)

This is not valid as a K4 group. It is not a manifold of virtue at all; and may never be a K4 group without an identity element. (If reality or "God's will" is the identity, there is no changing it.)

Then the rows of wormwood will relate the churches in mission ("a", "f" and "g") with their partners ((b,c), (c,e) and (b,e) respectfully) differing only amongst those left-handed static triples, without altering those elements in [a,f,g]. Appearances are enough to carry the validity of wormwood as if it were also a "spiritual gift"; that is, spiritual "glue" or "untempered mortar" ("burning as it were a lamp" cf. Rev 8:10) between fellowships, always seen to agree in the current right hand octal. (Yet only the "jacinth" portion of each wormwood row.)

As the reference is not preserved between multiple wormwood transpositions (only the jacinth portion of what was previously wormwood) I find the fiat K4 group is just a distraction for those entering the image system. If the reference octal is a constant, that group will be present for as long as the setting of every vote is chosen from the groups of the reference under the wormwood device. (The sixth trumpets' serpents.) For those that choose to stick to the reference as if a "rota", that fiat group will remain present.

Otherwise, if any two (or more) churches vote at the same time and before the state of the system changes, the setting of the vote may be shifted as to agree in these faked K4 group(s). (They will have a "use" to simplify any vote to a single setting as if by that "rota").

The Last Device of Satan (14.4.11)

In a "Manifold of Virtue" (as if [a,f,g]) it is found or shown that a = f&g, f=a&g and g=a&f. In this case, the three are chosen together as if virtues and the devil's device is spun to unify them. Rather than forming a seven cycle over the reference as with the least in the kingdom of God; here, Satan supplies a faked seven-cycle over wormwood and sets up a five cycle to interrupt that instead.

In wormwood, holding a=1 allows the transposition (f,g). By holding f=1 the manifold enables (a,g) and last, g=1 enables (a,f). The Frobenius map acting on GF(4) is mimicked as if it were embedded in each wormwood row as the static subgroup, and with it the K4 form also.

The claim is made that these three transpositions construct a K4 group, and that the identity or "zero" (which is properly God's will) is claimed "negligible", or "not to purpose". I.e. they would ask of the church: "Do we accomplish the will of God by sitting on our hands?" (All well and good, if those hands stay clean, for that is truthfully the test.)

In the three rows of {a, f, g}, then, wormwood actually uses entirely different unity elements. Here I would have for this "constructed" manifold of virtue [a,f,g] the transpositions of (b,c) with a=1, (c,e) with f=1 and (b,e) with g=1 (possibly cycled by three unclean spirits as "frogs").

Now, "f" is equivocated to a&g in the manifold. That they may appear to substitute one for the other ("f" for "g" or "f" for "a") in faulty symmetry is not enough; a vote must be taken for consensus agreement. That way, all subsequent (and similar) objections are all to be overruled.

Satan's aim, across seven churches, is to set up the image system in the place (temple) of the Holy Spirit. Every possible seven cycle circuiting through those churches into God's approval then becomes interrupted by a five cycle. It is impossible to remain a member of such a closed system without the Holy Spirit and to overcome the world whilst always worshipping in those churches, always remaining within. They must be exited and left behind completely.

Voting on (c,e) with f=1 and then (b,c) with a=1 will change the seven cycle of the Holy Spirit from, say, (b,c,g,a,d,f,e) to a five cycle (g,a,d,f,e).

I.e. (b,c,g,a,d,f,e)(c,e) = (g,a,d,f,e)(b,c).

And the Holy Spirit **will** leave the collective to the false prophet "pneuma" and its schema of S5 in S6. (Wormwood applies in both $r\&v^{-1}$ and $s\&u^{-1}$ as men do not organise the works of the Holy Spirit. It is also true that neither "a" or "g" nor $(r\&s)^{-1}$ and u&v meet in setting up the image as would #1 and #2 in #3.)

Satan's device requires two votes in the church to form five cycles in the elements of the reference octal. Only two are required as upon, say, (c,e) and (b,c) above. The claim that (b,e) is the symmetric result is a function on wormwood, not in the Holy Spirit; an observed symmetry not in the valid octal but on one interrupted with a five-cycle as shown above.

The above "beheading" of the seven cycle to a five cycle (removing (b,c)) is a "deadly wound", if "b" and "c" will remain separate. This wounding is taken full advantage of – as will be shown.

He Can Get You On Voting In The Image System (14.4.12)

Just how, then, may rows of wormwood have any effect on the Holy Spirit? If they are made across churches, I would expect an accusation of Satan to hold only if the five cycle is actually desired by those churches.

When the claim is made that some churches do too much and others too little and all suffer for it; it could be claimed that the churches are "out of balance" and that the image system could bring things back to a level playing field. Comparison is made with the five "refining" churches of Revelation that are claimed to be "out of balance" (but are not approved of by Christ at all) and that the five should together model themselves on attaining the status of the church of Philadelphia as the result of a superb "balancing act". This opens the system to a separation of the churches to become distinct bodies each closed and separate in the Holy Spirit. (There is no seven cycle across rows of wormwood, the result is no seven cycle at all.)

Mat 25:9 But the wise answered, saying, *Not so*; lest there be not enough for us and you: but go ye rather to them that sell, and buy for yourselves.

As a fellowship acting together may act in one C7 group of eight – if they claim to be such a "refining" church they could be accused as such a one by Satan. Then the vote in the image system may set up a five cycle; if they each identify to partner with (or as) one of those churches under the image.

The fifth trumpet device supplies a five cycle in disjoint K4 groups: cosets of GF(4) in GF(16). The transpositions of the wormwood device require an illegal embedding of GF(4) in GF(8). Any five cycle set up (as by the "locusts") whilst shifting wormwood rows is made to agree as if it cycled this embedding. This, however, is just a distraction; the five cycle of K4 groups is made in fifteen elements, not in the required seven. Yet with the treading of the winepress, the five cycle which *is* to be instantiated will agree with the elements of the seven cycle; to exclude those that would, by nature, operate in the locusts' five cycle. The "blood" (or intersection) of the five cycle in the seven cycle's elements will certainly agree in these churches. The "refining" churches are judged, cut off as to become desolate: become as the scarlet beast.

Rev 14:20 And the winepress was trodden without the city, and blood came out of the winepress, even unto the horse bridles, by the space of a thousand *and* six hundred furlongs.

Philadelphia are supposedly "just about right" but are safely excluded in Christ and they are replaced by a source of leaven instead, as if a church were to be switched out for another, one compatible for one incompatible. These pools would export the image as Philadelphia elsewhere (by bearing the mark), but in truth they are now become the "synagogue of Satan".

The result is that every permutation of the churches is to be made under the schema of S5 in S6, that "great city" that rules over the "kings of the Earth". Those limiting every permutation to those within S5 and deciding in all dialogue the disjunction in (r, s, u^{-1}, v^{-1}) within them, are become those "kings" in this system. After the image is set up, the rule of ten kings is always to facilitate the same device, always a permutation in five churches in S5.

Consider again those predicates:

r   We do the mission ourselves and elsewhere, it is our purpose

u^{-1} We encourage the support of mission done by others, as if for those without our purpose (new converts, new attendees?)

s   We *organise* mission to do with others, it is a purpose for them

v^{-1} We maintain "outreach" to supply our purpose to other churches to *organise* themselves (as if casting our nets on the other side, "Where are all the fish at?")

I introduce the concept of a circuit or permutation of churches, by relating them in terms of who supports whom.

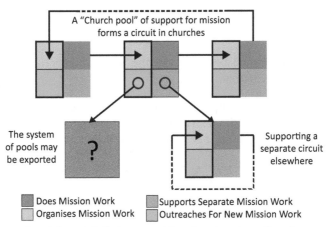

A "Church pool" of support for mission forms a circuit in churches

The system of pools may be exported

?

Supporting a separate circuit elsewhere

Does Mission Work | Supports Separate Mission Work
Organises Mission Work | Outreaches For New Mission Work

Boxed elements indicate an organising role, unboxed an active role

The permutation (x,y,z) would be equivalent to: "x" supports "y" who supports "z" who supports "x". It is then clear that any number of churches may be linked in terms of permutations of disjoint cycles.

Voting in the image system would result (in terms of each church affected) in "r" and "s" becoming reversed relative to u^{-1} and v^{-1} by the wormwood device. (The manifold of virtue is assumed fixed.)

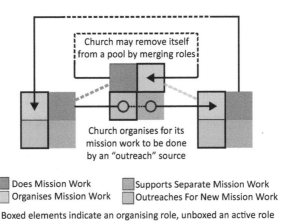

Church may remove itself from a pool by merging roles

Church organises for its mission work to be done by an "outreach" source

Does Mission Work | Supports Separate Mission Work
Organises Mission Work | Outreaches For New Mission Work

Boxed elements indicate an organising role, unboxed an active role

The combination that is become $r\&v^{-1}$ would indicate that mission of church "y" taking place in a church "z" is to be replaced by mission by another church "x" organised by "y" for themselves, as if through the outreach of church "y". Church "y" votes to pass on the mission of "x" to "z" in their own place and "retreats" from that mission.

Similarly, $s\&u^{-1}$ indicates the support (of church "y" for "z") is to organise with church "z" the mission of church "x" through that sister church "z" which was previously in receipt of mission by church "y". Church "y" votes to "organise" for church "x" to support "church z" in their place.

In these cases, there may only be a minimum of two churches present in any circuit/pool. (Church "x" = "z", possibly.) Church "y", by vote, introduces or facilitates for church "x" to support church "z".

Then the division of sets in the dialectic permits the "gifts" of organisation and/or mission to be "re-purposed" in any fellowship.

In the faked manifold of virtue, purpose is either found already completed or the "gifts" are not found present. The "torch" of support for church "z" is then passed on from church "y" to "x" (to those partners now supporting in the bearer's place).

As each church either supports another or is supported in kind, it may be found that any permutation of these churches can be refactored "by vote", by the transpositions made in wormwood (of elements r, s, relative to u^{-1} and v^{-1}), so that the members of the image system may, upon their composition, form any permutation (as in disjoint cycles) of churches in S5.

A transposition (a,b) in a set of seven churches {a,b,c,d,e,f,g} affects such a cycle or "circuit of support" (a,b,d,c,e) by, say, pre-multiplying the circuit by that transposition, justified by a vote switching "r" and "s" relative to u^{-1} and v^{-1}. The result is that the organising support of a fellowship (b) for a sister church (d) is switched out for that of another (a) by that fellowship (b), effectively by pairing v^{-1} with "r", and u^{-1} with "s".

The result is as if (a,b,d,c,e)(a,b) = (a,d,c,e) and "b" is removed from that circuit. Churches "a" and "d" become partners, whereas "b" has switched their supporters for themselves as to those that "b" would have before supported.

The situation may be reversed with any church being added to a pool, simply by introducing an "outreach" sourced church. It may be added to that pool by merging its roles with an existing member of the pool, and then reversing by vote (of that neighbour-to-be) from that shared state as if from the reference of $r\&u^{-1} \vee s\&v^{-1}$ to wormwood in $r\&v^{-1} \vee s\&u^{-1}$.

That reversal to wormwood is enough (after churches' roles have been merged), to outreach to that new member church for mission organised by their neighbour-to-be as well as to support that new church to organise their own mission work (as if the consequent cycle were a totally separate circuit elsewhere; that mission work previously shared then becomes separate instead). The wormwood device naturally "unravels" as if the initial circuit were dissolved and the new circuit already set up to supersede it.

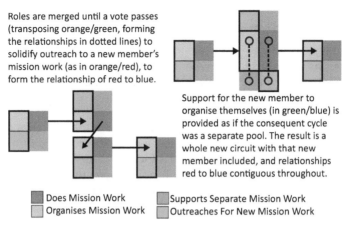

Roles are merged until a vote passes (transposing orange/green, forming the relationships in dotted lines) to solidify outreach to a new member's mission work (as in orange/red), to form the relationship of red to blue.

Support for the new member to organise themselves (in green/blue) is provided as if the consequent cycle was a separate pool. The result is a whole new circuit with that new member included, and relationships red to blue contiguous throughout.

■ Does Mission Work   ■ Supports Separate Mission Work
□ Organises Mission Work   □ Outreaches For New Mission Work

Boxed elements indicate an organising role, unboxed an active role

I would call a permutation cycle a "pool" of churches, and I also note that changes are made that could affect a great number of other fellowships. I also find the analogue of the manifold that numbers 666 (as introduced in chapter [10]).

He Can Get You On The Seven Churches Of Revelation (14.4.13)

Consider such a pool of churches (a,b,c,d,e). Exchanging "b" for "d" requires either a number of churches to vote, both to dissolve any such (a,b,c,d,e) and instantiate (a,d,c,b,e). Either that, or a vote by an impartial church that carries over the whole.

Either (a,d,c,b,e) = (a,b,c,d,e)(a,c)(b,d) for just that one pool or (b,d)U,V,W,…,X,Y,Z(b,d) for every pool in the arbitrary set U to Z, whether they contain one or both of "b" and "d" (or neither). Then conjugation made upon church "pools" will result in exchanging "b" and "d" globally. Pools present without either church are unaffected.

Moving from the image system to the seven churches of Revelation; if it is claimed that the intended model for the church is to bring a "body" of five churches into balance (cf. "I know thy works", Rev 2:2) that all pools are kept to a maximum of five churches, and if each closed system of pools is limited to a maximum of seven; churches may be able to support only those churches they wish to, without exceeding the founding criteria of limiting pools to five. It is easy enough to decide the switching out of a sixth for one of the five with a collective vote held by a designated seventh – a solution found in the wormwood octals, either with a fixed "rota" formed in the jacinth portion shared with the reference octal or with the current arrangement "of wormwood" as if the reference. (The setting for a vote will then never occur in either church transposed.)

S5, then, becomes the very limit of the model and objections to the doctrine of any of the five in a pool may be solved by swapping the objecting or objectionable church out for a sixth by conjugation. The result is that the operation of mission and mixing of fellowships immediately opens up Satan's devices for leavening all doctrine, even if it is only a little at first; it will result in nothing but leavening without the leading of the Holy Spirit and the God of Jesus Christ.

Wormwood, The Foundation For The "Great City" (14.4.14)

Given any starting "reference" octal, the composition of transpositions from any two wormwood rows (as in the sixth seal serpent devices) will compose to the equivalent of a transposition in a third row; that is, only if those transpositions are made in the reference octal, within that "jacinth part".

The completely free exchange of churches will not preserve the reference octal used: any wormwood row can function as the next reference when the device is such that, say, (d,e) with a=1 is followed by (d,f) and c=1 (i.e. "e" was exchanged for "d", (d,f) does not now hold with b=1). In general any second generation will not be upon the same "reference"; in this case the original reference for the first generation did not contain the group [c,d,f].

As long as wormwood transpositions are made upon an octal of seven elements, there will never be a seven cycle preserved between each transposition made from different wormwood rows. However, there is always a five cycle able to be constructed over whatever was the previous reference, no matter which wormwood octal is to be used (for there are yet eight C7 groups over any octal).

Given, say, an arbitrary C7 group represented by (a,b,d,c,f,g,e), any wormwood transposition may interrupt that seven cycle – there is some power of that seven cycle where the elements of that transposition are neighbours in that seven cycle. For instance (a,b) from the row with c=1.

Any two transpositions will meet in a group common in and over the two wormwood rows in which they occur. As (b,e) with g=1 meets with (b,d) with f=1 in the subgroup [a,f,g] common to the wormwood octals of both these transpositions, I may choose a power of any arbitrary seven cycle over the current reference to make one pair in that group "neighbours" in the seven cycle, so that the other pair(s) (now exchanged) are purposefully not. Both remaining pairs always have equivalents in that group's coset that are also neighbours, and being disjoint from the first pair of neighbours in the seven cycle, they are preserved between both applications of the wormwood device. They will (being disjoint cycles) also commute.

When required to compose (e,f) with c=1 and (e,g) (with unity to be determined) under (a,b,d,c,f,g,e), these two transpositions meet in [a,b,c]. Then I note (a,b) are neighbours in that seven cycle and (a,c) and (b,c) are not. Then (a,c), (b,c) composed will, without fail, now exchange "a" for "b" and (a,c) for (b,c) and leaves (b,c) paired with one of the remaining two pairs (here (d,f) and (e,g)) as neighbours (here "e" and "g", to be exchanged by vote in "a"), and (a,b,d,c,f,g,e)(a,b)(e,g) = (a,d,c,f,g) and I note the result (a,b)(e,g) will also commute. The order of the transpositions as applied is not important.

Octals that are not disjoint in triples as left/right pairs will share a minimum of one K4 subgroup, permitting the above five cycle over two generations of wormwood (or else they share three subgroups as across one generation). It is true of any two rows in wormwood that in transposing elements there is an equivalent to be found in one subgroup **shared** between them (preserved from the same reference) each from a different row altogether, as if [a,b,c] with c=1 met with [a,b,c] with b=1 above. (Resulting in (b,c) with a=1 as the equivalent.)

I also note that in these cases, the two transpositions composed will always have equivalents that are disjoint and commute, and also because these are found disjoint; they are preserved between both uses of this wormwood device.

Then every multiplication of transpositions in wormwood will interrupt a seven cycle as raised to a suitable power through an arbitrary set of seven "churches". The result is a five cycle. Only one power of the seven cycle requires interrupting, for the C7 group as applied becomes inconsistent as a whole. By that result, there is always agreement found with a five cycle when it is searched for.

As the wormwood rows are so found without a seven cycle between them (each row then quite separately corresponds only to unity in the reference), that five cycle found is arbitrary in that it is found "natural" as if in mimicry of the circuit of the least (traversing these churches with a cycle formed inverted, say, of {a,b,c,d,e} with (e,d,c,b,a)) floating unity through each "church" and then each row of wormwood.

He Will Get You On Doctrine (14.4.15)

Consider a vote to exchange in the image the two churches "a" and "g". In the rows of wormwood made in the reference (Sun octal), exchanging (a,g) is equal to churches "a" and "g" each voting to exchange, say, (b,c) and (b,e) respectfully; (a,g) is only passed by such a vote of "f". Is "f" the one impartial and seventh church (not in either pool) required to vote for the image system to switch two churches "a" and "g" between their pools? In general, the answer is "No". Yet "f" is always enough of a "fair" choice, thereby always using the current (and obtained of wormwood) octal as the reference to interrupt every seven cycle that may have circuited those churches. (I note these three transpositions meet in the subgroup [a,f,g].)

Limiting pools to five churches will result in at least one church of the seven not being in the same pool with some arbitrary two meeting in one pool; there is certainly no guarantee that this "impartial" seventh church is not in yet another pool with either one of those two, as with "a" or "g" (churches to be exchanged). Any sense of that impartiality is truthfully unfounded; it is far more likely that any vote to exchange these churches would take place at a church decided by the fiat symmetry present in the rows of wormwood instead.

If this is the case, the scarlet beast has its constitution prewritten in the dialectic and in the rows of wormwood, thereby always excluding the Holy Spirit.

If by "virtue" of the wormwood rows "a = 1" and "g = 1", virtues both compatible with "b" (as found in the groups [a,b,c] and [b,e,g]), I could infer that "c" and "e" are both "compatible" with "b", and I must further decide whether "c" and "e" are compatible with each other (and "a" and "g" too when they are transposed) which requires a vote by "f" within a "constitution" of wormwood (here I use the reference for familiarity's sake alone). In this case I may assume the dialectic balances of "b" as the "oil" and the decidability of the disjunction of c ∨ e proven by "f" (as wine), or equivalently, that c&e are compatible for "b" but not compatible with "f" or vice-versa. (Note if a=1 or g=1 then (f,g) and (a,f) have equivalents in f=1 with (a,g) or (c,e).)

I.e. b&(f → ¬(c&e)) and b&(c&e → ¬f), or I must decide whether b&f proves ¬(c&e) or the equivalent c ∨ e (i.e. in the dialectic now operating within the coset of the subgroup [a,f,g]).

Then conflict in the image system is resolved by "f" voting to switch "c" for "e" and the result must be either that transpose, or else one exchanging "f" for "b" (voted for by "d" if swapping (c,e) does not carry, i.e. c&e proves $b \vee f$) across every pool if the beast votes for that compromise because doctrine(s) in (c&e) do not clash. These are the two "positive" syntheses requiring a vote to decide them.

Then the vote taking place in "f" is on either (c,e)(a,b,c,f,g)(c,e) = (a,b,e,f,g) globally, or, that "b" is to be switched for "f" in every pool: (b,f) (b,c)(b,e) (b,f) = (f,c)(f,e). If churches "a" and "g" afterward swap "b" for "f" the result is not simply to reverse (c,e) for "f" (to a state similar to that before the objection) but to shift reference octals (in rows of wormwood) altogether.

The reference octal is altogether done away with for completing S6 and S5, and the image is to be further completed. This system for the Church plays harlot with the operation of the reference octal always in the "jacinth" portion of wormwood for selecting the setting of the next vote, and it is an engine for blaspheming the Holy Spirit using Satan's devices.

And these votes are tailored for by the rule of the ten kings (thesis vs. antithesis brings synthesis) but to the Church, it all seems a "win/win", bringing things back into balance.

He Gets You On The Open Door (14.4.16)

Churches will never teach you that an open door is truly what it is (the exit). To them, exclusion is death, and it is to you too most likely if church and fellowship is your whole life. To those, the door is always open and always leads back inside. The truth is: the gospel message is to leave through that open door and not to ever look back.

The circuit of the least through the reference octal does not translate to a circuit in the rows of wormwood. Without the Holy Spirit, any objection to a doctrinal fault will be met with the choice of moving to a pool without that doctrine.

Based on the accumulation of closures as in the "dragon's tail", in the fourth seal system there will be a set of third seal balances sending a doctrine as "oil" to nullity, yet there will also be a church holding that doctrine as if it were the "wine" also. However, those four churches tolerate each other's false doctrine and would support each other. (Cf. Rev 2:20.)

In a five cycle, raising powers through a circuit in the churches would shift, say, (a,b,c,d,e) to (a,c,e,b,d) to (a,d,b,e,c) etc. Church "a" would not be required to shift partners from "b" to "c", but instead (and from a different perspective), churches "a" and "b" would "double up" their support to each other in giving support to "c" etc. – the result is that the dialectic accumulates as does the dragon's tail in the acceptance of their incompatible doctrine by the collective; there being a limit of four powers before the identity is reached and there appears "rest" in the circuit. That fourth "contradictory" power raised of compromised doctrine(s) asserts (and tests) every relationship as if the dialectic (and the devil's device) is the goal of that same "purpose", rather than that virtue and doctrine of the gospel of Jesus Christ – and it is driven by the dialectic and wormwood alone. (What if only the pastors change, ruling as ten kings?)

That fourth power will, without fail, appear contradictory in their collective doctrine(s) (but each church will never purpose their mission to reach themselves) unless there is total agreement or compromise and only a "core" set of doctrine is approved by the pool. It is not immediately apparent that any core of doctrine will eventually reduce to that of practising only the image system itself. Voting churches in and out of pools will eventually leaven even that core. There is only a "lie" held in common if the dialectic is practised.

Leaving a church for such a fifth or other to escape the dialectic system is then never a solution approved by Christ. The presence of a compatible fifth church in circuit is no escape as the churches are all found incorporated together; holding relationships between those five churches and not leaving them all completely behind is never to pass through the open door.

For the individual, objecting to every church of a pool (a,b,c,d,e) leaves only "f" and "g" as possible fellowships, only if they are totally separate to the image system; otherwise the image is still in place in the system they take part in. There are four powers of any five cycle besides the identity, each church is considered compatible in the dialectic with its sister-others; skipping churches for a "favourite" church is always far from righteous, whether to attend for the "best mission" or to avoid it perpetually; neither option is enough to be found in Christ's approval.

He Cannot Get You On The Solution In Christ. (14.4.17)

It should be immediately apparent that the Holy Spirit and also the circuit through the seven churches by the least is completely unaffected by the image system. There are no "spiritual transpositions" to multiply the seven cycle by: only (and by mere appearance) "physical" ones made in works.

The question should be posed "Whose is the five cycle?" The image interrupts a physical circuit by closing the image system of seven churches at five; but this is a circuit made in works, not in the Holy Spirit as of the least. The seven cycle of the Holy Spirit cannot be interrupted by physical transpositions made in Church "mission" even if it could interrupt any circuit by those not elect as is the least.

There is, truthfully, no such spiritual five cycle obtained to ever interrupt the Holy Spirit, no seven cycle placed over wormwood for it to interrupt, and certainly no transpositions in the Spirit (from S5) for Satan to transform the Holy Spirit to that five-cycle of the false prophet. This is why that five cycle must be separately instantiated as in Rev 14:20. (For it is given to them: cf. Rev 9:5, Rev 9:10.) It appears that Satan has had to go begging once again.

The Holy Spirit is not interrupted over the believer other than with the judgement of God, as to His left and right over the churches of the image; for the power of Satan is simply that of accusation over blasphemy, and there is scope for plenty of that within the image system.

Satan has one effort left, three spirits "as frogs" that act on "inanimate spirit" to form three cycles. **Do not be dismayed** by the seemingly plausible claim that these three frogs act on each equivalent transposition acting in the "jacinth" part shared with the reference. These "frogs" have **no action** (as such not accounted for) upon the reference **at all**, and are totally countered by the vials of wrath, at the due time by God's promise. The armies of heaven clearly show this threefold device deprecated (the angel treading the winepress suffices), therefore be separate when due.

Rev 2:24 But unto you I say, and unto the rest in Thyatira, as many as have not this doctrine, and which have not known the depths of Satan, as they speak; I will put upon you none other burden.

14.5) In One Hour Is Thy Judgement Come

"One hour" in the Revelation refers to one automorphism of one field GF(8) under a single seven cycle (from one C7-group of eight). Bearing this in mind, I find the circuit of the angel completed in that "hour"; the corporate nature of the Church has universally condemned it.

In reigning as the "collective" with the dialectic process of the image (as a "king") the collective receives power for "one hour with the beast" (where the beast in the verse is actually the collective themselves). Taking part in the image system and the process, is to accept the mark of the beast, and to have the number of its name (666 as static bows – antichrist and desolate). That is immediate expulsion from grace and God's unmerited favour. God has to declare such fellowships desolate of His Spirit, and at the placing of the abomination that makes desolate, there is the dividing line of perfection: when collectives are cut off.

The woman with twelve stars no longer has an open door through which she may escape the image process – those twelve elements of S5 that are her crown (or her test, for she should not be so sure of her authority if that was the case) are in her failure become the door slammed shut. The Lord is calling His own out of her. ("Her" is the image system or template of incorporated status; families and churches are not illegal to God, merely faithful or not.)

Rev 18:7 How much she hath glorified herself, and lived deliciously, so much torment and sorrow give her: for she saith in her heart, I sit a queen, and am no widow, and shall see no sorrow.

The Church has sought wealth and coveted (as if by transpositions turning back from Sardis to the collective by the devices of the pit) authority from God holding captive His people within her: sitting enthroned (enshrined in law) as if she were the jewel in the beast's crown, denying the God that she claims works all her ways and that she is never to be judged (rejected) for it.

She is to be given torment (the elements of a completed S5), and left caged by the two "wrath octals" (with { } as the identity). As she covets the inheritance of all of the saints, her sins reach to heaven as the "army" of 200,000,000 is constructed to interrupt every circuit through the seven churches (by adding a transposition back to the collective as a "key to the pit") of the generalised circuit of the angel – here now filled as to interrupt the action of the rainbow about the throne of God. This is an attempt to destroy any possibility of election in Christ. However, there is an open door of twelve "stars" (maps) out of the city which may only be closed by God. (That requires God's judgement of her: the device needs filled in operation of S5 to be complete.) Only heaven (unity) remains untouched by the image.

Rev 18:8 Therefore shall her plagues come in one day, death, and mourning, and famine; and she shall be utterly burned with fire: for strong is the Lord God who judgeth her.

Her "plagues" (the winepress trod – now in full) come in just one day: in isomorphism to GF(8) as the "sea of glass" and not the "lake of fire" which she **is**. Strong is the Lord that judges her.

There is a three cycle of destruction: death leads to mourning which for the "nation struck" leads to famine, leading to death once more etc. Jesus has turned Himself away from those not His own. (There is no blessing as equivalently there will be no unity element: the seven cycle and the six wings of the four beasts before the throne are now as in the ejection from Eden, a "flaming sword that turns every way".) Their diabolical inheritance cannot map them to "rest" as unity in the octal: they will be completely (utterly) burned by the six-fold action of the Spirit (not including identity in the heptad) rather than watered by it with unity as "rest".

I refer to the Sun octal as now totally isomorphic across the sea (of glass). It is true any multiplicative power of the same elements of the angel's circuit are valid for all to overcome, traversing all eight C7 groups as multiplication. The ministry of the two witnesses now complete, the possibilities to overcome those 200,000,000 horsemen are there for those that obey God and who truly "come out of her", even in a five cycle and without the Holy Spirit.

Rev 18:9 And the kings of the earth, who have committed fornication and lived deliciously with her, shall bewail her, and lament for her, when they shall see the smoke of her burning,
Rev 18:10 Standing afar off for the fear of her torment, saying, Alas, alas, that great city Babylon, that mighty city! for in one hour is thy judgment come.

The kings of the Earth (every static bow or triple in the sea of glass) that made their money and energised her commerce stand far off – they see the smoke of her burning. The smoke is that smoke from the pit that obscures the proper operation of the seven cycle. Her burning is the operation of her "fire" now as "lake of fire" completed and closed. Her burning is synonymous with the same torment of the non-abelian groups she has as her manifold.

The ten kings burn the Church with fire – the action of harming her against the Holy Spirit and with His diabolical adversaries, for the hard truth of her judgement has convicted the elect to leave and for the tares within to render the system completely fallen. Its doctrinal wealth that attracts the flock that actually have good doctrine – they are gone and it with them.

When the whole Church is become leavened and "scarlet" the judgement of God swiftly arrives. God judges the city or "system" now to be a "closed" structure and the Lord's footstool (the Earth) belonging to God is shunned by Him, to become the lake of fire – the "wrath octals" where the additive identity is the empty set: there will be no one holding the elect captive whilst attempting to gain from the religious commerce of any "MYSTERY BABYLON" any more.

14.6) The World Stands Still As the Judgements Arrive

With the destruction of all righteous religious institutions will come a world with no essence of any logical value to God. Nothing will entail His favour if there will be no proper faith. (Faith comes but by hearing.) When the night comes when no one can work for the gospel, then I may expect the judgement and for Christ to arrive to do just that: judge.

Rev 18:11 And the merchants of the earth shall weep and mourn over her; for no man buyeth their merchandise any more:
Rev 18:12 The merchandise of gold, and silver, and precious stones, and of pearls, and fine linen, and purple, and silk, and scarlet, and all thyine wood, and all manner vessels of ivory,

and all manner vessels of most precious wood, and of brass, and iron, and marble,
Rev 18:13 And cinnamon, and odours, and ointments, and frankincense, and wine, and oil, and fine flour, and wheat, and beasts, and sheep, and horses, and chariots, and slaves, and souls of men.
Rev 18:14 And the fruits that thy soul lusted after are departed from thee, and all things which were dainty and goodly are departed from thee, and thou shalt find them no more at all.

There are 28 items in the list – no doubt corresponding to the 35–7=28 triples possible outside the reference octal in the same seven symbols. For every triple outside of the Sun octal there is merchandise worth its equal. Some place in the city with someone ready to buy a dispensation from an idol or from some newly printed book etc.

The "fruits that thy soul lusted after" are the fruits of the Holy Spirit hoped for in "revival" that will never come to corrupt institutions. The Church has its purpose, and that now is become merely for growth (profit) and it has a name for being alive, but in truth before God it is dead. (There are many descriptions of wealth – but only this sole reference to the living water that Christ gives freely – yet not unconditionally.)

Those that buy and sell doctrine, the "merchants of these things" (or the "merchants of the Earth") no longer have a source of income: yet there is another reference here to her operation of S5 (the torment). They are themselves as merchants, sources of good doctrine afar off; they exited convicted for fear of the judgement of God. And the woman as scarlet (without sound doctrine) will not stop to hear it – and will not endure sound doctrine.

Rev 18:15 The merchants of these things, which were made rich by her, shall stand afar off for the fear of her torment, weeping and wailing,

They, in stark contrast to John who marvelled at the woman, weep and wail for the loss of all it is and once was. (She yet appears to be clothed as blessed by God with the "appearance" of the chaste bride (New Jerusalem) to those that operated within the commerce of her worldly "blessings".) The judgements come thick and fast.

Rev 18:16 And saying, Alas, alas, that great city, that was clothed in fine linen, and purple, and scarlet, and decked with gold, and precious stones, and pearls!

In that same hour – completion of the angel's circuit – all of the true and false doctrine in the city becomes completely fallen when lumped together into one system of hopeless deception by the dragon. The dragon is cast out and the gloves are now finally off: the angel's circuit witnessed complete entails that the judgement is to arrive swiftly and then the final battle occurs as at the moment of judgement and the second coming, and it is going to be Armageddon. (Resurrected sinners will attack the redeemed.)

Rev 18:17 For in one hour so great riches is come to nought. And every shipmaster, and all the company in ships, and sailors, and as many as trade by sea, stood afar off,
Rev 18:18 And cried when they saw the smoke of her burning, saying, What *city is* like unto this great city!
Rev 18:19 And they cast dust on their heads, and cried, weeping and wailing, saying, Alas,

alas, that great city, wherein were made rich all that had ships in the sea by reason of her costliness! for in one hour is she made desolate.

"Ships" are the analogue of "trees" in the sea. The "shipmaster" would be the intersection of three K4 subgroups or triples, the "company in ships" the corresponding static bow and "sailors" (including them) those remaining elements of the three triples that form the "ship" complete. Those that "trade by sea" would instead refer to those that (as in the passage of the death of the two witnesses) exchange by transpositions, or exchange "gifts". They practise the devices of wormwood across the sea of thirty octals, as by the sense of the two witnesses. The end has caught up with them fast.

This is the period after the sixth seal, and must be that last period before the opening of the seventh – the ministry of the two witnesses may well be fully completed. The city burns because it is spoiled of all its wealth (all sound doctrine) and those who hold to it. It is not a given that the whole world would repent: though this "burning" is coincident with the treading out of the winepress (as the last remnant of virtue is extinguished from the tares in the Church). Even those that are under and practise the very same devices of Satan can see the Church is gone all the way bad, wholly made after the pattern of this world. With no sound doctrine – the Church is unsustainable (made desolate).

All having ships in the sea were made "rich", supposedly by doctrine: for they are those that have intersections of triples (as a shipmaster). This would align in the broadest paradigm to the device formed of wormwood rows that they themselves are within as singletons, those same rows correspond to unity and so their "ships" are held static under Frobenius in those triples (exchanging static triples or the "company in ships"), and that as unity they rest in the angel's circuit: then each row is become that "mountain" of doctrine common to one of the seven churches. That sound doctrine now gone (or if false, overcome): there is no sense of purpose remaining in any church and "Rev 6:14 And the heaven departed as a scroll when it is rolled together; and every mountain and island were moved out of their places.". (Islands the analogue of earthly triples (mountains in terms of wormwood) shared in the sea.) This is the period before the seventh seal when the whole world knows of the arrival of God's wrath.

The "merchants" in the sea, those that dwell in the wider set of octals, now become as the "sea of glass" likewise stay far off. What is interesting is that these two groups, of the Earth and more generally as over the sea, are still very **rich**: they have the goods to keep on going, but with the destruction of the Church system there are no people requiring that wealth. The time surely comes when they will not endure sound doctrine.

Note that the merchants lament as forever lost the income of the city (more properly "traffic" of doctrine). This is in sharp contrast to the commandment to be glad and rejoice that I may compare to the expectation of the saints revealed in the fifth seal (as to wait a little season):

Rev 18:20 Rejoice over her, *thou* heaven, and *ye* holy apostles and prophets; for God hath avenged you on her.

The "rich" I may assume (*as in doctrine*) – are the "elect" that drove the commerce of religion, and they are the "salt of the Earth", or were with regards to the merchandise, those that drove the institutions. Without the Holy Spirit and sound doctrine the Church as an entity is meaningless. The time for the harvest is come as the tares have gathered themselves together.

Now, the judgements of Babylon begin one after the other.

Rev 18:21 And a mighty angel took up a stone like a great millstone, and cast *it* into the sea, saying, Thus with violence shall that great city Babylon be thrown down, and shall be found no more at all.

This mighty angel, I expect to be the left hand of God: in that he is the last of the elect to exit the Church, and leaves it now totally desolate and barren of elect behind him. He is empowered to utter these judgements.

The great millstone – the key to all commerce (without bread the human and economic machine fails), i.e. those keys which so drove the commerce of the Church (keys which when found result in that city "MYSTERY BABYLON", rather than the faith of the inhabitants of it) are to be cast into the sea never to surface again. The devices of Satan are never again to be assembled together in this fashion.

Rev 18:22 And the voice of harpers, and musicians, and of pipers, and trumpeters, shall be heard no more at all in thee; and no craftsman, of whatsoever craft *he be*, shall be found any more in thee; and the sound of a millstone shall be heard no more at all in thee;

Those with God-given talent (spiritual gifts) for doctrine will never dwell in a system of the like. There will be no craftsman (those educated in correct doctrine) in the city either. There will be no *possible* basis (millstone) for the commerce of this model for religion permitted at all.

Rev 18:23 And the light of a candle shall shine no more at all in thee; and the voice of the bridegroom and of the bride shall be heard no more at all in thee: for thy merchants were the great men of the earth; for by thy sorceries were all nations deceived.

There will be no end to the darkness found in that city, there will be no light (not even any operation of K4 subgroups in the city): i.e. even in the wrath octal, the operation is no longer $(A \lor B)^c$ but $(A \lor B)$ only, then acting upon subgroups and cosets in the following:

$$[a, c, e, g] + a = [0, b, d, f] \equiv E$$
$$[b, f, a, e] + a = [0, c, d, g] \equiv G$$
$$[d, g, b, a] + a = [0, c, e, f] \equiv F$$
$$[c, e, d, b] + a = [b, d, e, c] \equiv C$$
$$[f, a, c, d] + a = [0, g, b, e] \equiv D$$
$$[g, b, f, c] + a = [f, c, g, b] \equiv B$$
$$[e, d, g, f] + a = [d, e, f, g] \equiv A$$

and with { } as zero.

With regards to the Trinity model – the operation $(A \lor B)$ will not preserve virtue. (There will be no light, even of a candle.) No virtue will be added or spent any more on this city.

There will be no more "weddings" in the city – God will no longer bless *any* covenants between men (though as legally between men and/or women they will still stand under the laws of men) or between those in the city and Himself. Those that made themselves rich on the operation of religious institutions by furthering their own ends with the Church had their greatness with the beast of all nations – they had an unending supply of customers seeking international government approval for their every "covenanted" act.

Those nations deceived by "sorceries" were moved from a seven cycle to a five cycle, not reaching rest or approval in Philadelphia. (They will have no virtue – light.) As to the merchants being the "great men of the Earth" I may state that in all things Church and state, those that leavened the churches' doctrine in the image system moved in big circles (bows), and were highly esteemed in the dialogue. (To the detriment of the churches.)

Rev 18:24 And in her was found the blood of prophets, and of saints, and of all that were slain upon the earth.

This construction as the last of Satan's efforts to oppose the sovereignty of God was the final end to all his means. As the scarlet beast is full of names of blasphemy – every device and every false doctrine formed under the dragon's tail from Egypt to the USA, every sinful act that has killed the people of God from the dawn of time falls upon this city as if it were the diabolical payment in kind entailed for their wasted blood. She is judged accordingly.

14.7) The Lord Of All The Earth – The Four Beasts Named

With the overcoming of the whole flock now in Philadelphia, the four earthly elements of the octals aligned within the sea of glass are now without any question of sovereignty, as the Lord's own footstool.

Rev 19:1 And after these things I heard a great voice of much people in heaven, saying, Alleluia; Salvation, and glory, and honour, and power, unto the Lord our God:

The devices of the dragon are now completely defeated – he has been overcome by the people of God (the remnant or elect) and he only has sway against the tares still in the "city" of MYSTERY BABYLON. The devices of the seals are referenced here as now defunct – the four beasts about the throne have their functions iterated.

1) The first beast (as a lion) corresponds to **Salvation** (the paradigm of the believer under full restraint in liberty of election).

2) The second beast (as a calf) corresponds to giving God **Glory** (standing one's ground for the truth).

3) The third Beast (as a man) corresponds to giving God **Honour** (rather than discussing His will to reach consensus, His word is absolute).

4) The fourth beast (as an eagle in flight) corresponds to **Power** (God may make a miracle when there is nothing but contradiction, such as raising the dead).

The beasts (full of eyes within) over all the Earth may reciprocate and give justly deserved glory to God, for heaven (unity as election now freely floating and completely filled) is now full of all who are elect (through great tribulation – more selective election) and give God glory.

Those dwelling in heavenly places with Christ (as if unity as one of 144,000) answer the judgement of the city as just and with satisfaction. She will no longer corrupt the Earth or spiritually kill the Lord's own – she is left merely to corrupt herself. It is judged that she will do so in her operation ("fire" – the "pneuma" of the false prophet, her "spirit") of S5 eternally.

Her smoke will obscure her to God and He will reject her in kind forever. Her prayers will not find Him: they and her works will stink of burning sulphur – in contrast to the prayers offered with much incense to Him earlier, these will not even be masked by incense.

I have the last incidence in Revelation of the elders and four beasts falling down before God enthroned. (Jesus opened the seventh seal. Cf. see Rev 4:10, 5:8, 5:14, 7:11, 11:17.) The very last day (24 hours or elders) arrives as the wrath is completed, poured out. Christ returns to judge and His army returns with His right hand to make war in righteousness.

One "day" is of one octal of 30 (also being as twenty four elders) isomorphic across the sea of glass (every octal aligned in isometry) – that sea of glass within which all the 144,000 dwell (as if unity floating in the one octal) as separate from the elders as the identity, the "rest" or the "reference unity" in the Sun octal. The elders reign with Christ in heavenly places: this is the inheritance of the saints. (They have the authority that all Christians share subject to God.)

Now is heard the voice of the archangel as in the declaration above. This world is then justifiably won for God Most High and His only begotten Son, Jesus Christ.

Also heard is the voice of (judgement by) a great multitude (those sealed that overcome the dragon) and the voice of (again, judgement by) many waters (in the sea of 30 octals, those that overcome the devices of "wormwood" and also the three woes). I discover it is then followed by another great judgement (voice) of thunders (movements in heaven).

With the movement (backwards in the narrative from this moment in heaven with thunder, "reversed" once more and with the last trump aligned to the seventh seal), the "mighty thunderings" align to the seven thunders that answered the loud cry of the angel of the church (John paraphrases the angel with his own understanding (sealing up): that there be "time no longer") which itself aligns with the verses:

Rev 18:1 And after these things I saw another angel come down from heaven, having great power; and the earth was lightened with his glory.
Rev 18:2 And he cried mightily with a strong voice, saying, Babylon the great is fallen, is fallen, and is become the habitation of devils, and the hold of every foul spirit, and a cage of every unclean and hateful bird.
Rev 18:3 For all nations have drunk of the wine of the wrath of her fornication, and the kings of the earth have committed fornication with her, and the merchants of the earth are waxed rich through the abundance of her delicacies.

The "great judgement" of "Babylon" itself aligns and dovetails nicely with the overcoming by the least. (He has to hold firmly "that which he has" until sometime after the sixth trump and through the beginning of the two witnesses' ministry, rather than as the city is judged.) The thunders are a movement of heaven, so the first six trumpets completely precede (and vial judgements of God's wrath begin with) the angel's overcoming and all are moved backwards from the (here as in the present) heavenly narrative with this great judgement of the city. This judgement comes from the throne of God, omnipotent and victorious against the dragon.

Rev 19:7 Let us be glad and rejoice, and give honour to him: for the marriage of the Lamb is come, and his wife hath made herself ready.
Rev 19:8 And to her was granted that she should be arrayed in fine linen, clean and white: for the fine linen is the righteousness of saints.
Rev 19:9 And he saith unto me, Write, Blessed *are* they which are called unto the marriage supper of the Lamb. And he saith unto me, These are the true sayings of God.

Now that God is victorious and His judgement set (with the seven thunders) at the opening of the seventh seal (exactly when it should be), the elect are the only remaining faithful in view: "not judged yet". They are ready as a "chaste bride" to join their God as He arrives (they will not be going anywhere just yet). The bride in moving out of the city has made "herself" ready; they are clothed as saints in white unspoiled linen. (They are clothed as unity in the octal, as with the righteousness of Christ, the "lamb of God".)

All those that are so dressed with the grace of God are blessed – and the angel states that these are truly the words of God. (There is no damnation to "unmerited" fire and brimstone, and no judgement of the believer's sins – can this really be true?)

Rev 19:10 And I fell at his feet to worship him. And he said unto me, See *thou do it* not: I am thy fellowservant, and of thy brethren that have the testimony of Jesus: worship God: for the testimony of Jesus is the spirit of prophecy.

The angel's declaration of the liberty found in Christ and the mercies of God towards those that would worship falsity in His name cause John to lose control – John was caught completely off balance and worshipped the angel that had the authority to utter these promises – yet the angel is also a servant of Christ and not worthy of his worship. That is, not worthy of a fellow servant's worship – for the angel of the church is worthy of the worship of those practising the lies from Satan's false prophet through the image system, as I will soon show.

Chapter Fifteen: Faithful And True

In a great departure from the imagery of the beasts and the mystery of iniquity, John is shown things yet to be made sure by God. The setting changes to the inheritance of the saints, the city of God – the New Jerusalem. At this stage the last few loose ends of the judgement and damnation of Satan as well as the millennium of grace offered freely as the water of life are tied up and made plain.

There is no millennium waiting for Christians to breed like rabbits to repopulate the Earth. That is surely a fallacy. What remains is the candidacy for all those saved. If a person, sat down as with the throne of God, can rule over themselves with God's help within a period of grace (1000 years) as an elder, one of 24 (an hour) so that they might live by God's laws then they are offered predestined salvation: they are guided into obedience by the Holy Spirit. They know they have it because they see themselves approaching obedience. Then those without such salvation are known by their fruits.

A thousand years with the Lord is as one day – 24 hours (automorphisms of the octal cf. the 24 elders) which are multiplied by a factor of 1000 years (God is never patient in part, but in 1000-fold more cases). This thousand years with the Lord is as a single day (the sabbath, unity). It has to be "this very day" otherwise you will never be found called, chosen, predestined, saved, etc. If God is one thousand times more patient, He saves one thousand times more elect. Then the completeness of the saved and sealed is as shown in the 144,000 clothed in white.

2Co 6:2 (For he saith, I have heard thee in a time accepted, and in the day of salvation have I succoured thee: behold, now *is* the accepted time; behold, now *is* the day of salvation.)

Rom 8:30 Moreover whom he did predestinate, them he also called: and whom he called, them he also justified: and whom he justified, them he also glorified.

So, obedience comes with the strength of God's spirit and not after whining at how weak grace is. Other gospels such as the "rapture" (as will be examined), are such a test. Be informed that in order for grace to become insufficient to your salvation from any "mark of the beast", God has not approved another gospel, but instead has excluded you from His salvation by the finished work of Jesus Christ so that you would necessarily require that falsehood to be removed from the trial of the scarlet beast. It really is one or the other; it is "strong delusion".

There is no microchip that damns you, no barcode or tracking satellite signal: you are known to the last hairs on your head by one kingdom and so what if so by another? Who you choose to fear is important, not to whom you would flee in terror. Who warned "them" to flee from the wrath to come? It wasn't Jesus, it must have been "some other guy (that they call Father)". Jesus hasn't advised fleeing since predicting the fall of Jerusalem to utter destruction. Instead he states of "all the world..."

Rev 3:10 Because thou hast kept the word of my patience, I also will keep thee from the hour of temptation, which shall come upon all the world, to try them that dwell upon the earth.

Returning to the scripture, John's vision tours the eternal rewards waiting for all those who overcome the devil's devices and wash their robes in the blood of the lamb, clothing themselves with the righteousness of Christ. (They may dwell in Sabbath rest forever – as under the position of unity in the octal.)

15.1) The Inheritance Of The Saints

John immediately shifts from the "great city" upon the Earth to the other portion of the sea of glass; the unity element of election previously seen gives way to the static subgroup in the "way of the kings of the east". Then heaven opened is in view as the static subgroup with it; the floating unity and static triple become associated together one and the same. With God's universal sovereignty comes the Earth as part of heaven.

Rev 19:11 And I saw heaven opened, and behold a white horse; and he that sat upon him *was* called Faithful and True, and in righteousness he doth judge and make war.
Rev 19:12 His eyes *were* as a flame of fire, and on his head *were* many crowns; and he had a name written, that no man knew, but he himself.

John sees not Jesus Christ, but His right hand (the angel of the church) revealed to him upon a white horse, inheritance complete. The white horse as in the first seal is the static subgroup, yet the rider is at rest as the unity element.

In fact, it was a new thing that John saw. He defined the singleton "zero", as an additive identity.

The octal is a suitable identity for K4 subgroup addition in the octal. Also, the K4 subgroup was adequate when adding C2 subgroups in the K4 form of the filter. This "singleton zero" may be added to the sets of either the octal or K4 filter without hampering their place enthroned as the zero. (The octal or K4 group respectively.)

As a singleton, zero does pretty well in the K4 form (as well as the octal) in singletons. So, Jesus is also become the identity of His K4 form (the K4 group of subgroups that span the octal, in being sat down with His Father, the octal). Jesus is then as a subgroup of the whole octal sat on the throne. (So Jesus is as the elements [0,a,b,c] sat with (within) His Father [0,a,b,c,d,e,f,g] as a "subgroup" of that octal which is acting as identity upon the set of K4 subgroups under $(A \lor B)^c$.) In self-similarity Jesus is the identity of the K4 form [[0,a], [0,b], [0,c]] etc.

The angel inherits the right to be also (as zero) sat down "With the Lord" on His throne: always now, as zero in K4 or the octal.

This right to sit enthroned is identified in the letter to Laodicea, as his reward for overcoming. God's right hand is called by Christ's name "Faithful and True" and judges in righteousness, making war in Christ's place (judging the nations – dashing them together as pottery – cf. letter to Thyatira). His eyes are as a flame of fire: the "lust of the eyes" is not present but is overcome with grace in Christ. (He is given the morning star, and is predestined in Christ.) On his head are many crowns (the inheritance of all, a "living" crown – cf. letter to Smyrna: authority to sit enthroned), and no one knows his name but he himself (cf. letter to Pergamos).

Rev 19:13 And he *was* clothed with a vesture dipped in blood: and his name is called The Word of God.

He is clothed with a vesture dipped in blood (cf. letter to Sardis and concerning the blood: "Rev 2:4 Nevertheless I have *somewhat* against thee, because thou hast left thy first love.") and has overcome – and is recognised (Sardis again) and given place in heaven (may circuit the seven cycle – cf. letter to Ephesus.)

However, there was much maths that I skipped over: I had avoided the structure of the rider and his army – which follows better after the next few verses.

The setting segues on from the vials of wrath, to the last devices of the dragon – three unclean spirits as frogs enabling the beast and image system to continue. These are shown to be mostly inoperative and only a first seal paradigm then remains in the Earth.

15.2) The Armies Of Heaven

As the very least is seen again upon the Earth the armies of heaven follow behind, not convicted in the Spirit but at the command of God, being also under divine restraint – righteous and obedient to God also riding on white horses. (The "Word of God" alone represents unity in Christ after heaven is "opened".) The angel becomes as the "head" and the others as if his fellows, in orbit up to a generator of some extension over GF(8). Then the riders are then elements or generators of some (large) collection, an extension of the field GF(8) I will call "F".

In fact, the "army" forms the Lord's answer to the three "frogs" from the mouths of the dragon, beast and false prophet. By sending every element of C7 to unity in F* and examining every orbit containing the static subgroup of GF(8) within F, I emulate the isometry of the "way of the Kings of the east" and without using C7, also examine the seventh vial that is poured out into the "air" ("inanimate spirit"). I will show in a section to follow that the dialogue of the "kings of the Earth" does not operate in truth, only in nullity (contradiction).

I will quickly examine each verse in turn:

Rev 19:11 And I saw heaven opened, and behold a white horse; and he that sat upon him *was* called Faithful and True, and in righteousness he doth judge and make war.

Heaven is properly the covering of the floating unity element in the Sun octal. John saw the position of unity opened up to a larger set: elements within an extension field F of GF(8). Now, in every extension field of degree n over GF(8) (or over any other finite field for that matter), there is a unique subfield of F of degree m for all m>0 dividing n, here m=3 and n=3r (r, any integer greater than zero). F, is otherwise arbitrary. Moreover, if there are two such fields F claimed, there is a certainty in algebraic closure of a third containing them both. If they are identical up to automorphism, then they are each mapped to each other in algebraic closure by the Frobenius map. Then, by induction, there need be no other field "F" for the "isometry".

That subfield of F isomorphic to GF(8) is **unique** in F, containing a static subgroup, as well as the unity and a static triple in the static subgroup's coset. (It has a "white horse".) The unity element generates the coset when added to the subgroup. The unity element becomes he who is named "Faithful and True" and sat on the horse. Not simply because he has unity (virtue), but because the structure of F* that I am to examine is evaluated as under a homomorphism ϕ , and GF(8)+ keeps its structure intact under that morphism but in the orbit of the rider.

$$\begin{pmatrix} I \\ x+I \\ x^2+I \\ x^4+I \end{pmatrix}$$

$$\begin{pmatrix} 0 \\ x \\ x^2 \\ x^4 \end{pmatrix}$$

The subfield GF(8) will retain its structure even when the map of ϕ is applied (he remains "Faithful", remaining steadfast, holding to virtue, unity being in the kernel of ϕ). GF(8)'s additive structure will also remain complete in all its subgroups even without the action of the multiplicative group of GF(8)* to generate them. In every other orbit under ϕ, that same map sends those seven subgroups of GF(8) to the static subgroup, and only the unity from GF(8)* remains to map it to itself under product. (The image of C7, the kernel of ϕ is unity. The rest of that subgroup C7 of F* is sent to the unity.) He **alone** remains "True" as forming GF(8) whole whilst under orbit. Other orbits of different octals with the static subgroup will not generate those subgroups, as C7 is "missing" or "factored out" from F* by ϕ.

The rider alone has unity and the static subgroup (with its coset) together forming GF(8). He righteously (with absolutes) judges disjunctions $q \vee p \& q^{-1}$ with virtue, and operates on the right hand K4 subgroups with the operation $(a \vee b)^c$ rather than that of symmetric difference over the "wrath octals", with whom he is always at odds, making war in opposition with the proper operation of the "Sun octal" as opposed to that now found the cage for the dialectic, as the "lake of fire".

Rev 19:12 His eyes *were* as a flame of fire, and on his head *were* many crowns; and he had a name written, that no man knew, but he himself.

Now his "eyes" operate as the three Frobenius automorphisms of GF(8). As he generates the coset of the static subgroup with unity, he also generates a static triple. (The seven eyes of the lamb, the elements of GF(8)*, are absent for now: they are mentioned only as the kernel of ϕ.) As like to Christ, his "eyes" are as indices multiplied (in squaring) as opposed to those added – then as "flame of" or as "above the" fire.

His elements ("I = 1", and as a triple [I+x, I+x², I+x⁴]) are then closed and static under Frobenius.

$$\left.\begin{matrix} x \\ x^2 \\ x^4 \end{matrix}\right\} static$$

$$closed \left\{ I = x^7 \right.$$

$$\left.\begin{matrix} x^3 = x+I \\ x^6 = x^2+I \\ x^5 = x^4+I \end{matrix}\right\} static$$

Likewise, "crowns" reference a closed set under a binary operation, or of closed groups of (differing) automorphisms of a single structure. There are an infinite number of possible "F" extended over GF(8), and John would not have needed to count them as F is totally arbitrary: there are simply "many" of them. His "reed like unto a (measuring) rod" being a primer on the octal alone, he was not given a general argument as to the count of multiplications of every possible extension F that may be formed. (Though he somewhat indicates later that there are many such multiplications to be reconstructed as F*, as with an inverse map.)

As to the name written, GF(8) is unique in every such extension F. So, Frobenius is preserved, and closure of the static triple is also preserved (his multiplication is certainly preserved), he is unique in F. (Unity is also unique in F.) The name written known only to himself is equivalently the circuit of the least amongst the seven other "lampstands" over the Sun octal. That group C7 is about to become the kernel of the homomorphism ϕ from F* to a factor group H.

The separate orbits in F* are disjoint in octals and each octal only appears in one orbit in F. Then the field GF(8) appears just **once** across every orbit of every rider in F*, and appears sevenfold in his own orbit. No other rider "knows his name", though he does know his own!

Rev 19:13 And he *was* clothed with a vesture dipped in blood: and his name is called The Word of God.

Now, GF(8)* is a subgroup of F*. Then by the "Theorem of Lagrange" the group F* may be reformed into a collection of cosets of C7, and the group formed of these cosets, H, is isomorphic to a cyclic group G (as F* is also cyclic) of order $|F^*|/7$. Then I simply employ an isomorphism (where the group F* is evaluated up to an element of C7 or "mod H", by the "canonical map") from G to those cosets of C7 in F*. I then emulate "inanimate spirit".

As GF(8) is a subfield and under any product the octal GF(8)+ will be mapped to another octal (by the distributive and cancellation laws) and since the multiplicative group C7 or GF(8)* forms disjoint cosets under such products, the orbit of GF(8)+ in F forms disjoint octals in a short orbit of length $|F^*|/7$. That is, up to a generator of GF(8)*, the orbit is repeated sevenfold. It intersects itself completely (seven times over) up to rearrangement (product) by one of its own elements. It has a short orbit and is "dipped in blood" sevenfold (as completely agreeing with its intersection upon itself).

Now his "vesture" is the group H isomorphic to F*/C7 as above. That group generates his short orbit, unique to himself as a subfield. H is isomorphic to G, and F*/G has seven cosets, and these cosets mapped onto H will act later in forming the "ranks" of the army.

As to being named the "Word of God", under the homomorphism there is no longer the multiplicative subgroup C7 of GF(8)* acting in H, yet the additive elements of GF(8)+ remain in their subgroups in the "Sun octal" placed at the head of the orbit of the "Word of God". They are closed as the octal ultrafilter and also closed on addition of K4 subgroups. They form an operation of virtue, and though there appears to be no multiplication of C7 over them (in H), that group (C7) yet remains in F* as if the "Word of God" were able to be recovered by an inverse map (C7 is always present in F*, C7 is mapped to unity by ϕ on F*). There is in F* nothing present but a "pre-first seal paradigm", yet, in H, one that does seem to appear to operate without the seven eyes of the lamb (floating unity). The "Word of God" is genuinely unique as the additive octal of GF(8)+ in F, yet appearing without the Holy Spirit: i.e. he is "obedient" to God in exercising virtue, as Christ is also perfect, yet he is not necessarily God himself (though Christ did act similarly). And, as similar to the least, appears to obey God without any influence of the Holy Spirit (for Christians may act without Him, cf. Act 19:2).

The "Word of God" is then merely as the conditions for virtue waiting to be applied. (It is the "primer" formed of the octal group in positive properties.) I state that the "Word" (doctrine) is as "oil", the Spirit "fire" and virtue is the result, as "light" (living doctrine). Virtue simply needs one to apply it. So, as such is the office of the least. (A lampstand in the kingdom of God.)

In fact, that office of the least is the highest on offer in the Kingdom of God.

Mat 23:11 But he that is greatest among you shall be your servant.

I.e. that servant will be one with the "unity" element, exemplifying virtue. He fulfils all the law in that same service, all virtue defined as within the Kingdom itself, as found in the lamb.

Rev 19:14 And the armies *which were* in heaven followed him upon white horses, clothed in fine linen, white and clean.

I find that only the reference octal (as if zero) will induce a "multiplication" upon the K4 form.

I must digress from the rider for a while. There are many other orbits of octals that contain the static subgroup in F. In fact, in F+ I may form $(|F^*|-3)/4$ such cosets. Instead I form them from the multiplicative group of F* sent to H as in the homomorphism ϕ above.

I require, then, those same three structures:

- F, an extension field over GF(8). (With a cyclic multiplicative group F*.)

- H, the factor group F*/C7 to generate the orbits.

- G = <g>, the cyclic subgroup of F* of order $|F^*|/7$ isomorphic to H to generate the cosets upon the "white horse" of each rider and to agree with multiplication by H upon the orbits.

Now the rider is not merely identified with the additive subgroups of the "Sun octal" at the head of the orbit of the "Word of God", but also to his multiplicative elements of C7 in F*. That the other riders follow him (those that were in heaven), I take to indicate that they, formed in ranks, also ride on white horses, they follow or "inject" into that same static subgroup from the other subgroups of the Sun octal. That kernel C7 of injection in the army forms the "Word of God" also, and those six elements that were previously in heaven, held static or as a "floating unity" in the field GF(8) are mapped to ("injected into" and so follow) the unity/static subgroup by similarly forming shortened orbits on "white horses" as does the "Word of God".

Now, if I collect every orbit of every octal in the image of F under ϕ that contains the static subgroup, I must have H isomorphic to G. First, take an array of such orbits at the "head of their orbit" (in product with unity) as shown here, with the "Word of God" on the far left:

$$\begin{pmatrix} I \\ x+I \\ x^2+I \\ x^4+I \end{pmatrix} \quad \begin{pmatrix} g \\ x+g \\ x^2+g \\ x^4+g \end{pmatrix} \quad \begin{pmatrix} g^2 \\ x+g^2 \\ x^2+g^2 \\ x^4+g^2 \end{pmatrix} \quad \begin{pmatrix} g^3 \\ x+g^3 \\ x^2+g^3 \\ x^4+g^3 \end{pmatrix} \quad \cdots \quad \begin{pmatrix} g^{n-1} \\ x+g^{n-1} \\ x^2+g^{n-1} \\ x^4+g^{n-1} \end{pmatrix}$$

$$\begin{pmatrix} 0 \\ x \\ x^2 \\ x^4 \end{pmatrix} \quad \begin{pmatrix} 0 \\ x \\ x^2 \\ x^4 \end{pmatrix} \quad \begin{pmatrix} 0 \\ x \\ x^2 \\ x^4 \end{pmatrix} \quad \begin{pmatrix} 0 \\ x \\ x^2 \\ x^4 \end{pmatrix} \quad \cdots \quad \begin{pmatrix} 0 \\ x \\ x^2 \\ x^4 \end{pmatrix}$$

Shifting in the orbit by some power of a generator I may find:

$$\begin{pmatrix} g^{-2} \\ g^{-2}x+g^{-2} \\ g^{-2}x^2+g^{-2} \\ g^{-2}x^4+g^{-2} \end{pmatrix} \begin{pmatrix} g^{-1} \\ g^{-2}x+g^{-1} \\ g^{-2}x^2+g^{-1} \\ g^{-2}x^4+g^{-1} \end{pmatrix} \begin{pmatrix} I \\ g^{-2}x+I \\ g^{-2}x^2+I \\ g^{-2}x^4+I \end{pmatrix} \begin{pmatrix} g \\ g^{-2}x+g \\ g^{-2}x^2+g \\ g^{-2}x^4+g \end{pmatrix} \cdots \begin{pmatrix} g^{n-3} \\ g^{-2}x+g^{n-3} \\ g^{-2}x^2+g^{n-3} \\ g^{-2}x^4+g^{n-3} \end{pmatrix}$$

$$\begin{pmatrix} 0 \\ xg^{-2} \\ x^2g^{-2} \\ x^4g^{-2} \end{pmatrix} \begin{pmatrix} 0 \\ xg^{-2} \\ x^2g^{-2} \\ x^4g^{-2} \end{pmatrix} \begin{pmatrix} 0 \\ xg^{-2} \\ x^2g^{-2} \\ x^4g^{-2} \end{pmatrix} \begin{pmatrix} 0 \\ xg^{-2} \\ x^2g^{-2} \\ x^4g^{-2} \end{pmatrix} \cdots \begin{pmatrix} 0 \\ xg^{-2} \\ x^2g^{-2} \\ x^4g^{-2} \end{pmatrix}$$

Then I show that the order of G must be equal to that of H else there are possibly two "heads" of some orbit and the static subgroup then appears twice in the same orbit, which is not to have the orbits distinct and of length equal to $|H|$.

Now, if ϕ sends a subgroup J of F* to unity then the orbits in F* under that morphism are shortened by a factor of $|J|$. If $H \cong F/J$ then $|F^*|=m*|G|=|J|*|H|$ (m, an integer greater than zero) as there are now $|J|$ "heads" of the orbit containing the static subgroup. Under that morphism, these orbits are sent to equal "vestures" (groups H acting to generate orbits) yet only evaluated as such after the morphism is applied. Then what is this certain value of "m"? I should also note that there is now (after that morphism acts), a smaller "set" of elements in the image of ϕ. I.e. that $|G|$ is at most $|H|$ in the image of ϕ, sending F* to H. Then $|G|=|H|$ and effectively m=7.

Now, the kernel of the morphism is the subgroup C7 of F* sent to unity. The above conditions hold under that morphism, whether it is applied to form H or both G and H. In order for there to be no "overabundance of orbits", as when the morphism is applied to F* (and then forming the cosets of C7 in H), I take $|G|=|H|$ (G$\cong$H) quite naturally, else there are possible orbits present from a larger set properly containing H. In that case, some orbits would be repeated as H $\cong$ F/J remains the same. (However, as m=7 is a prime, the only other possible G would be F*.) Under the morphism with G isomorphic to H, I also avoid adding members of the field GF(8) to generate the static subgroup upon itself, only permitting the adding of unity instead.

Under that homomorphism with the elements of C7 in F* sent to the unity this is perhaps easier to see. (Here, $g^{n/7}$ is in both C7 and GF(8) and is to be found in the static subgroup unlike $g^{6n/7}$, its inverse, which is only found in the coset instead.)

$$\begin{pmatrix} I \\ g^{\frac{n}{7}}x + I \\ g^{\frac{n}{7}}x^2 + I \\ g^{\frac{n}{7}}x^4 + I \end{pmatrix}, \ldots, \quad \begin{pmatrix} g^{\frac{6n}{7}} \\ x + g^{\frac{6n}{7}} \\ x^2 + g^{\frac{6n}{7}} \\ x^4 + g^{\frac{6n}{7}} \end{pmatrix}$$

$$\begin{pmatrix} 0 \\ xg^{\frac{n}{7}} \\ x^2 g^{\frac{n}{7}} \\ x^4 g^{\frac{n}{7}} \end{pmatrix}, \ldots, \quad \begin{pmatrix} 0 \\ x \\ x^2 \\ x^4 \end{pmatrix}$$

Both these octals will appear in the orbit of the "Word of God" given $|F*| = n$.

There is agreement (isomorphism) between the elements of G and H. Now every g is become a representative of some element h in H. This has the effect that there is also complete agreement in generating the orbits (vestures) under H with the elements of G instead.

There is to be in the construction a one to one correspondence between the elements of H and those of G. H "clothes" G, without a "cloud" or mystery. Similarly, the "vesture" of each rider H (or G) generates the octal's orbit (to be compared with Christ's "garment down to the foot" in the opening chapter of Revelation).

Using G in place of H multiplicatively as above, every member of the army (i.e. with the static subgroup) with cosets generated by elements of F* added to the static subgroup remain disjoint in their orbits under any generator in F* (or in G). (Each "horse" in orbit under g in G will not appear seven times as different members of the army in F if I hold to using g in G instead of F*, the members of the army are disjoint in their orbits now G is reduced to the image of the morphism on F* and I retain the condition that G is isomorphic to H and F*/C7.)

In adding an element g in G (with inverse equivalent to some h of H, h also in F*) to form the coset upon the static subgroup, I shift the unity element's (unity become as the rider's) place in that orbit (vesture) as to the inverse g^{-1}. Then every orbit is still unique in the image of the morphism and is still disjoint in octals with the other orbits.

I circumvent the problem arising from adding the elements of GF(8)* to the static subgroup to illegally generate the subgroup again, or generating the "Word of God" more than once. Those elements of GF(8)* are sent to unity and generate only the one orbit of the "Word of God".

As a lemma, I also understand that no orbit in the image of the morphism will contain an octal with unity in both the coset as well as within its subgroup.

Why? Given every rider on a "white horse", were every such horse (a subgroup of GF(8)+) in F shifted by some g in its orbit (in F) to a K4 group containing unity it would have shifted by an element of GF(8)* (GF(8) is a subfield with unity). Then that subgroup containing unity is also in GF(8)+ and is by morphism sent to the static subgroup. The result? Unity is never found in the shortened orbit of the static subgroup. Then unity is only ever present in the coset instead.

Adding unity in the coset is then simply as before: to find unity in both an image of the static subgroup and its coset is impossible in F*/C7 and G.

The static K4 subgroup in F will never contain the unity (as at the head of the orbit); however, under orbit by f in F* there will certainly be a subgroup, a multiplicative image of the static subgroup in its orbit that contains unity. This subgroup is sent to the static subgroup in the morphism of F* to H. (As this subgroup containing unity must also be within GF(8), as every product from the static subgroup's elements to the unity in F* are also of elements in GF(8).)

I state that no "rider" (element) in the image of ϕ is ever found in both the coset and the subgroup. Else, under some "f" in F*, unity would be found in both the subgroup and its coset, and with the above as a lemma, the statement follows. Under the morphism ϕ, the sets in the "army" now disjoint in octals are "well-formed". No coset contains any element of the subgroup or horse "under it".

Unlike the image of F* under that homomorphism, the subgroup of C7 in F* would cycle the subgroups of the subfield GF(8). In every orbit in F the static subgroup at the head of each orbit would cycle through the seven K4 subgroups of GF(8). Under the morphism, however (with kernel C7 in F*), K4 subgroups of GF(8)+ in every orbit are mapped to the static subgroup. The seven cosets of G in F*/G map each of the full orbits of every octal (with the static subgroup) in the "army" to a shortened orbit in which the sections of the orbits under these seven cosets agree after the morphism, although they are not "dipped in blood" (the cosets of G acting on the orbit or "vesture" agree completely only **after** the morphism is applied).

This (uniquely) does not affect the "Word of God" at the head of his orbit. The other riders of "white horses" follow the "Word of God", as C7 mapped to the unity element, being those same orbits similarly shortened by a factor of one seventh (forming "ranks" in following behind him). Within the orbits themselves in elements of G or of H being as cosets of C7 (as J), they follow as if they were written out as hJ, h^2J, h^3J, h^4J, etc.

C7 is the true identity of that rider as that "J" or C7, that of the kernel as the unity element (virtue is as that "Word" applied properly), called the "Word of God".

Then H becomes the "vesture" clothing each rider, and the "Word of God" alone has his vesture dipped in blood. The armies that "were" in heaven (as floating unity had corresponded to those same seven subgroups of GF(8)+) now also follow in shortened orbits (of seven "ranks"), formed upon "white horses". Those white horses at the head of the orbits (corresponding to unity – fine linen, every K4 subgroup in GF(8)+ is mapped to the white horse, the horse is to be compared with the floating unity and virtue of the lamb as "heaven") are cycled static under Frobenius (white) and done so legally in F* and H or G without contradiction (using a "clean" operation) unlike the "contradictory" three "frogs" from the dragon, beast and false prophet.

Then every subgroup of GF(8) in every orbit is reduced to the static subgroup only. This will not affect the seven subgroups at the head of the orbit of the "Word of God".

Rev 19:15 And out of his mouth goeth a sharp sword, that with it he should smite the nations: and he shall rule them with a rod of iron: and he treadeth the winepress of the fierceness and wrath of Almighty God.

A "sharp sword" makes a "clean cut". The rider as the least divides the godly from the ungodly; that is, the spiritual nations of the remnants in Philadelphia and Laodicea. He rules with a "rod of iron" that in regard to the heavenly host he himself is also become as a vial of wrath "poured out on the earth" as with the action of the zero element rather than a seven cycle of the multiplicative group. (Adding zero to the earthly coset simply generates GF(8) whole once more.) Whilst to the ungodly his seven cycle in the circuit of the least "treads the winepress", as the least is the counter-example to the ungodly (as found in the kingdom of God), the elect "are not appointed to wrath" and remain sealed.

The least makes his circuit without the multiplication of the Holy Spirit: setting the standard as the least influenced in the kingdom of God. All may certainly repent in "the nations" of the seven churches, and others not Christian are trodden in the winepress by him as his circuit requires. Those with predestinated obedience will be clearly shown as not in the "lake of fire".

"Zero" becomes an absolute (an element that corresponds to necessary existence, "rest" and so eternal life). The dividing line of the "rod of iron" is unflinching; there is a huge difference in the rewards of the "nations", between that of the "lake of fire" and paradise.

The "fierceness" of the wrath (the seven cycle of the winepress) is represented absorbed by the zero element – the throne is obscured by very thick smoke, in that it is not at all fierce towards the "very elect", and so the wrath of almighty God falls only upon the ungodly. (As a seven cycle it is immediately absorbed into the zero, which is become as an element or "covering" for the faithful. The winepress is not also "trodden out" by the angel by impossibly cycling the zero element, there is "no action" of the wrath upon the righteous and sealed.)

The "nations" if they be obedient whilst in a five cycle, are convicted by the angel's circuit, but more importantly the nations (heathens) are those in the "wrath sets/octals" that are either pained with sores or fallen into the lake of fire closed and caged under symmetric difference.

The sword from the rider's mouth is the same rod of iron as over those two: it is the operation of the octal regenerated that convicts or "slays" (the operation of the wrath set) the "heathen" as well as the image system's worshippers: for in the former the octal's proper operations are on the one hand pains to those with sores (in the lake of fire) that worship the beast, and on the other hand those that easily fall to symmetric difference without the covering of virtue in Christ. The operation of the octal is then always opposed to those that "dwell on the Earth" – those with "sores" grievous and noisome.

Jer 50:45 Therefore hear ye the counsel of the LORD, that he hath taken against Babylon; and his purposes, that he hath purposed against the land of the Chaldeans: Surely the least of the flock shall draw them out: surely he shall make *their* habitation desolate with them.

The "Word of God", no longer the "angel of the church" as cut off from God but instead sent from God justified: he (made perfect under grace) perhaps went out sinless, remained faithful and true, and also returns with fruit (the faithful, every elect taken from the scarlet beast etc.).

As the rider is here shown victorious treading the winepress, there will be none found in a five cycle (of the nations in Laodicea) in the Church that are not in the lake of fire. His reference cycle (the angel's circuit) is used to tread the winepress (cf. the wrath compared to the letters to the angel) and it is as if he were (towards the godly) become the zero element poured out on the coset of the static subgroup (he forms GF(8) whole). The least in the kingdom (then as if zero) is the counter-example to all the ungodly, and being as zero added to the coset (the earthly elements) of the reference, the definition of zero separates the saved and sealed clothed with GF(8) from those under the wrath in the lake of fire, properly condemned by the exit of the least from the Church.

The elect that show true virtue before their God are not under (appointed to) the wrath: the angel's circuit as the least in the kingdom is one of overcoming the world, and to have the wrath poured out on one's self is to be excluded from heaven. The elect are not affected, already having an acceptable paradigm and "receive the zero" poured out on the Earth instead of the circuit of C7 treading the winepress. (Christ, as the static subgroup makes up the difference.)

This certain difference between zero as towards the elect and the treading of the winepress to all others is the "rod of iron" (cf. Rev 19:15) with which he "smites" the nations (those within the "Church", found trampling the outer court of the temple underfoot).

I should also compare the sharp sword with the prophecy of Jeremiah, that in removing the believers from the vile (tares) in the flock, the angel is "as the Lord's mouth".

Jer 15:19 Therefore thus saith the LORD, If thou return, then will I bring thee again, *and* thou shalt stand before me: and if thou take forth the precious from the vile, thou shalt be as my mouth: let them return unto thee; but return not thou unto them.

That also concerns the least in the kingdom in a time of certain captivity in Babylon – for Jeremiah had become the least of his day. The Lord protects the nation by assuring the least is preserved amongst the remnant.

Jer 15:20 And I will make thee unto this people a fenced brasen wall: and they shall fight against thee, but they shall not prevail against thee: for I *am* with thee to save thee and to deliver thee, saith the LORD.

The last verse of the passage concerning the army continues on:

Rev 19:16 And he hath on *his* vesture and on his thigh a name written, KING OF KINGS, AND LORD OF LORDS.

His "vesture" is the group H generating the orbit of the octal of the rider, and "thigh" the set G of elements generating the coset of the static subgroup – that portion of the rider placed over the static subgroup or "horse" (added upon it).

The vesture and thigh (H and G as above) meet with the rider equal to unity only in the orbit of the "Word of God", as they then simultaneously have the position of the unity element from F*. In effect, John is stating that the kernel of the morphism is isomorphic to C7 and GF(8)*. The "name of God" is written as that kernel in which "vesture" H and "thigh" G agree.

Not only that, but every multiplication on GF(8) may be "written there" on the kernel and an appropriate inverse map defined from H to F*. In whole, all eight C7 groups over the "Sun octal" may be present, as the "Word of God" is become the circuit of the least in the seven golden lampstands of the seven churches – indwelt by the seven Holy Spirits of God. The "rainbow" circuit becomes "written" upon the kernel amongst the other C7 groups as required. God as it certainly appears, is able to resurrect F* from this reduced state, with the inverse map of the morphism. The rider (the least) is no longer without the Holy Spirit, but has fully overcome. The requirements of God for the elect, however, are unchanged.

As long as there is a least acting with virtue in obedience to God (in Earth or in Heaven), the kingdom of God is here. It will not pass from the world until after the judgement: there is a remnant "which are alive and remain". It is clear that Christ intercedes to all on Earth and also that Satan hates Him for this reason. The cross is open to all (with conditions, yet the kingdom of God is within you), the dialectic does not operate over the elect or even those "called".

Without the Holy Spirit acting in the army, I find the interpretation of the verse from the parable of the ten virgins, concerning those without sound doctrine, not following the rider:

Mat 25:9 But the wise answered, saying, *Not so*; lest there be not enough for us and you: but go ye rather to them that sell, and buy for yourselves.

To assume equality (whilst using the dialectic, requiring more than one static triple) with a member of the army, being without the Holy Spirit (no element of $C7 \cong GF(8)*$ and no K4 subgroup in GF(8) as a consequence, despite $C7 \lhd F*$, i.e. having no white horse or "white linen") is a sure-fire way to fail, and to receive no place. (As having no intercession in Christ.)

In appearance, the inverse map is dependent on the short orbit of GF(8) and the cycle of a generator of G. I also require a multiplication for GF(8) from the full set of eight that are valid on the "Sun octal". Given only the short orbit of GF(8)+, and g a generator of G, I find the choice of multiplication of GF(8) which is also required to reconstruct the full orbit in F* is become arbitrary, as taken amongst those eight. (Each member of G is then "born again".)

John, with the whole set of seven cycles (C7 groups) as named above, also references the many possible multiplications that are possible for the field F. Whilst the operation is not recovered whole, it is recovered up to a valid multiplication on the "Sun octal" (GF(8)). Then as F* is wholly cyclic, F+ is recovered up to an automorphism.

It is even the mystery of the seven golden candlesticks that is become this same name written, and it is unique in all F, clearly indicating the salvation and Kingdom of God. Instead of a statement along the lines of "I think therefore I am" I have instead, as concerning the rider, "there is salvation, as there is the saved". The seven Holy spirits of God indwell those seven lampstands/churches as they also dwell with those obedient to His word.

I find that the collection (the whole army) is to be uniquely associated with the God of the "Word" in the regenerated and full orbit of GF(8) in F* under the inverse map. There is then a return from H to F* containing GF(8) as a subfield (its multiplication was previously "factored out") and not to some other cyclic group that may not contain GF(8)* as a subgroup. GF(8) is a subfield of F to be recovered. (And purposefully so, with all eight C7 groups rather than one.)

Isa 55:11 So shall my word be that goeth forth out of my mouth: it shall not return unto me void, but it shall accomplish that which I please, and it shall prosper *in the thing* whereto I sent it.

Then the "Word of God" truly prospers, because I used the elements in G isomorphic to H (instead of those in F) for generating each coset (therefore every separate orbit is completed). With the inverse map of the morphism that returns to F from H, I would find regenerated every valid orbit containing the static subgroup in F. Then as F is regenerated complete (though it may be some labour), any missing orbit will be found in the regenerated structure of F. There will be no such "gaps" in the field F, the gaps are filled (restored) in the structure and the "Word of God" prospers in that thing whereto God sent it (in that field F extended over GF(8)): God, it would appear, is always very good with His working, as F is totally arbitrary.

15.3) The Right Hand Of God

I find the reward to the least for overcoming Philadelphia – remaining firm (the army includes those separated under conviction by the angel's circuit). Compare that reward to the following:

Rev 19:16 And he hath on *his* vesture and on his thigh a name written, KING OF KINGS, AND LORD OF LORDS.

$$gI\} vesture\ in\ all\ g$$

$$\left.\begin{array}{l} g(x+I) \\ g(x^2+I) \\ g(x^4+I) \end{array}\right\} thigh\ (over\ horse)$$

His "vesture" (orbit) is associated to its wearer uniquely in F. There are no other elements but those of GF(8) in product with the generators of g in G (as that product is equal to a generator in F*) that so map the unity element to pair with the same static subgroup of F sevenfold in the army. This of course requires that there actually is a short orbit of GF(8) in F for the static subgroup. "Heaven" is opened, but the container F may have only the one orbit of GF(8) in length equal to $|F^*|/7$. F contains GF(8) as a subfield. H as a collection is repeated sevenfold by elements of GF(8)* uniquely upon the vesture of the rider, since his alone is the short orbit. His "vesture", his orbit alone has the group C7 "written upon it". C7 is sent (mapped) to unity in H (those same seven "eyes" of the "Most High", the same "King of Kings, and Lord of Lords"). The inverse map (of H to F) is this name written on his vesture (now as if G were in product with C7 equal to F*) and thigh. The name is become the "kernel" of the morphism.

And also compare the above verse to:

Rev 3:12 Him that overcometh will I make a pillar in the temple of my God, and he shall go no more out: and I will write upon him the name of my God, and the name of the city of my God, *which is* new Jerusalem, which cometh down out of heaven from my God: and *I will write upon him* my new name.

The reader is told the angel will go no more out – he will not be called to prove true faith consistent any longer, that God will be always sovereign. In being the head of the vanguard of the faithful descending (The New Jerusalem) he has "The name of the city" written on Him. They will go where He goes, and He will not need to leave that fellowship any longer.

I would compare the "city of God" (the rider's army as composed of the three structures, F, H and G) that is built only on the foundation of Christ (the static subgroup) with the "great city" of "MYSTERY BABYLON" which fell when divided into "three parts" (cf. section [12.9]).

I then see a shift to the Sun octal again, as opposed to the "sea of glass".

Rev 19:17 And I saw an angel standing in the sun; and he cried with a loud voice, saying to all the fowls that fly in the midst of heaven, Come and gather yourselves together unto the supper of the great God;
Rev 19:18 That ye may eat the flesh of kings, and the flesh of captains, and the flesh of mighty men, and the flesh of horses, and of them that sit on them, and the flesh of all *men, both* free and bond, both small and great.

Now the "angel in the Sun" is a picture of every member of the army of heaven: all are stood "in the Sun" clothed with their own vesture g^r. This angel together with the "Sun" indicates the reference octal (not the army) which must also contain the unity element, even though that indicates the position or identity of "the least".

The angel stood in the Sun would appear to be either the very last in the kingdom to come to Christ, or perhaps the least in the heavenly host now the least overall is on Earth or vice-versa. I prefer to think of him as the very last of the elect to leave the Church to salvation in Christ as in Philadelphia (as one under conviction) – tested by being separated to the left handed octal and required to heed the ministry of the two witnesses and to leave the Church for the Sun octal. The reader finds him stood in the Sun, calling to every spirit (those without restraint) to feast on the city, for it will become the cage of them all. (There are no righteous left within, he hands it over to devils completely and effectively has the light cord to pull on His way out.)

As the last leaving the Church amalgam he delivers the fatal blow to the scarlet beast: the last and final kingdom (the "eighth") of the dragon. (For the cross remains Christ's sure victory.)

He is then the left hand of God: one extreme of the two witnesses in their ministry. (God has fed the woman in the wilderness with His right hand and has likewise convicted all in his left.)

In any case the angel is equal to (or in office as) unity now that the least is as "zero", the octal GF(8) (note I am not yet shown a floating zero). In effect the angel represents the saved 144,000 (the elect) and is stood (upright) in the kingdom of God as in one revelatory "day": as GF(2) held fixed under all 24 automorphisms of the eight C7 groups over the only (Sun) octal.

He calls to all the fowls of heaven, those practising the devices of the third and fourth seals across the octal sea (separate from the Sun octal, using the wormwood devices) to gather themselves to align with the Sun octal and to become caged within the lake of fire as inside the five "months" that correspond to the "great city" overlaid upon the elements of the Sun octal. Properly I could consider these fowls to have been in the "wilderness" where the woman fled the dragon. (They are even as vultures, now free to feed on the flesh (and false doctrine) of all in the Church: they are all spiritually dead now the very last to exit has left them behind.)

The fowls that fly in the midst of heaven are "unclean spirits" (as if they were the static subgroups aligned in the "way of the kings of the east", in the sea of glass). By "flying" they map from coset to coset (truly only static triple to static triple and unity element to unity element) in that alignment, supposedly without restraint. Those fowls are gathered together, an indicator that they only act in the kernel of the morphism ϕ, devoured as if by the unity only present in the "Word Of God" (see the next section).

They are invited to eat the flesh of kings (A), captains (B), mighty men (C), horses (D), (being additive subgroups) and their riders (E), and the flesh of all men (F). I consider them to be members of the first two woes whose devices fail with the rider in the army.

A. Those wearing crowns in the fifth trump – subfields.

B. Those cosets of subfields, additive subgroups in orbits: these form short orbits.

C. Other orbits of additive groups also under multiplication: long orbits of the same order as the multiplicative group. (Equivalent to a generator.)

D. An additive subgroup of order four.

E. Equivalent to an additive coset of a subgroup of order four in F+. The rider is a representative member of the coset.

F. Every possible singleton element in some complete general sense, each a member of a subfield of algebraic closure. (The king "Abaddon" as that closure over them all.)

These spirits are invited to fulfil the iniquity of the city reaching unto heaven, and to become closed in a "cage of every unclean and hateful bird" as S5 completed.

These form five sets (not six) under inclusion: kings rule over captains which command mighty men which in the forefront lead those on horses, which are followed by every other to battle.

The dialectic of the dragon's tail has accumulated a fifth dialectic in assimilating the waters upon whom the woman sits (drawing them up into the beast promising "great things"), then becoming "fallen". The fifth set is of those worshipping the image in the "city" (despite it being merely the alignment of their own hearts). It is true that in the wickedness of their hearts they gave it worship and it has formed this fivefold "tower" with the ten kings at the top.

Then there is a contradiction and "Babylon the great is fallen, is fallen". The waters underneath the woman riding the scarlet beast (peoples, multitudes, nations and tongues) are the remaining four sets under that image of ten kings, which give the devil's devices their worship: they are now arrayed for battle against God Most High.

15.4) The Beast Taken

Were a church to enjoin themselves to the scarlet beast on the condition that in no dialogue should the "beast" or the "kings" or any other "collective" deny the virtue in their religion, the argument would be over very quickly indeed. I lament over David defeating Goliath that simply, with that one sure stone. That no-one to my knowledge has done this is truly lamentable, the victory within it should be resounding in the heavens already, woe to us that it is not.

Would the beast roll over, and "play dead"? In all honesty, I expect the answer to be "No", and for it to end with harsher sentences. I would likewise (as would the angel) exclaim that there should be "time no longer" (see section [16.11-12]), that the Lord should put to end the unforgivable blasphemy heaped upon the Holy Spirit as the beast turns upon and devours "breaking in pieces" those very people still in the Church amalgam, whilst revealing those with the mindset of the "son of perdition" judging them in God's place. The beast will go down screaming at the feet of Jesus Christ, the gloves on Satan will likely come off before then.

Virtue will "kill" the beast, but only that of the people of God maintaining the deadly wound caused it by the exit of those believers out to the "church in Philadelphia". Lamentable indeed that not in the above but in this separation only is the antidote. (Yet thank God there is one.)

However, I find in the text the truth, that by the simplicity of virtue and a pre-first-seal paradigm the lamb brings victory against the beast. The least has washed his robes in the blood of the lamb, and is not lost, but overcomes the world.

Rev 19:19 And I saw the beast, and the kings of the earth, and their armies, gathered together to make war against him that sat on the horse, and against his army.

The army from heaven is no "proof" of the beast's (logical) defeat in so far as there is a fantastic contradiction found here (that the beast is an impossibility never to be constructed) – instead the army is the only tenable situation remaining across the sea of glass – and the text states (in the interpretation here given) that given this the only possible ground in all truth, the beast simply cannot function. The beast has the "rug pulled from under it" and is not left standing.

Now the beast is the amalgam of all four balances in the fourth seal device. The kings of the Earth are as the four choices for unity corresponding to those same four triples (under the three unclean spirits as frogs that cycle the earthly elements of the static bows).

Given that the operation of multiplication is now not possible elsewhere in the sea (within the wrath sets properly become the lake of fire), their "armies" are the "whole world" (cf. Rev 16:14) who form the earthly elements in every field aligned as in the "way of the kings of the East". They are all then gathered together in isometry/isomorphism.

The rider on the white horse acts with unity as "g" in G, and the unclean spirits (as frogs) act upon the earthly elements with the static subgroups totally fixed. The seventh vial of wrath was poured out into the "air" or "inanimate breath". The spirits as frogs do not legally (with Frobenius) cycle the static bow, they are "unclean" and break the valid operation of the field $GF(8)$, holding the static subgroup fixed. Only the reference octal is closed in a three cycle under Frobenius, and only that reference provides a "zero" for the K4 form that contains the unity. The automorphisms on that reference alone induce a multiplication upon the K4 form.

Then the sea of glass is isomorphic across every octal, and the "inanimate breath" has an answer in the army of heaven. (If there are two such F, there is some K containing both.)

$$\begin{pmatrix} I \\ x+I \\ x^2+I \\ x^4+I \end{pmatrix} \quad \begin{pmatrix} g \\ x+g \\ x^2+g \\ x^4+g \end{pmatrix} \quad \begin{pmatrix} g^2 \\ x+g^2 \\ x^2+g^2 \\ x^4+g^2 \end{pmatrix} \quad \begin{pmatrix} g^3 \\ x+g^3 \\ x^2+g^3 \\ x^4+g^3 \end{pmatrix} \quad \cdots \quad \begin{pmatrix} g^{n-1} \\ x+g^{n-1} \\ x^2+g^{n-1} \\ x^4+g^{n-1} \end{pmatrix}$$

$$\begin{pmatrix} 0 \\ x \\ x^2 \\ x^4 \end{pmatrix} \quad \begin{pmatrix} 0 \\ x \\ x^2 \\ x^4 \end{pmatrix} \quad \begin{pmatrix} 0 \\ x \\ x^2 \\ x^4 \end{pmatrix} \quad \begin{pmatrix} 0 \\ x \\ x^2 \\ x^4 \end{pmatrix} \quad \cdots \quad \begin{pmatrix} 0 \\ x \\ x^2 \\ x^4 \end{pmatrix}$$

Now, there are arrangements of the army where each "king of the Earth" aligns to one of the elements $1 \equiv g^n$, $1 \equiv x+g^n$, $1 \equiv x^2+g^n$, $1 \equiv x^4+g^n$ in the coset of the static subgroup. These four sets are in the army aligned in "the way of the kings of the East". They are each in alignment as a separate choice of unity in the "sea of glass".

There are then "four positions" for unity in each coset. Given the "kings of the Earth" are placed in the isometry as unity and the static bow is cycled by three unclean spirits as frogs by the dragon, beast and false prophet (without cycling the static subgroup), one position of the "Kings" is always valid, common to the "Kings of the Earth" and the isometry with the sea of glass, the rider and his army. The other three are supposedly also present in the "sea of glass", despite the "inanimate breath". (I will show they are not.)

I first state that there is no seven-cycle remaining in the collection (the army) after the morphism ϕ is applied. So, to prove this I assume that there is such a seven cycle (and equivalently a C7 subgroup of G or H).

The rider at the head of the army has a short orbit of length $|F^*|/7$ and the subgroups of the Sun octal (the "Word of God") are repeated sevenfold. It is impossible, then, for the orbit of the rider at the head of the army to have a complete orbit of length $|F^*|/49$ (not even a "sub-orbit"), or equivalently for there to be a sub-orbit that repeats the same subgroups in the Sun octal another seven times in the images of the rider (his vesture), as the subgroups together form a unique finite field GF(8) in F.

So, there is no g in G such that the static subgroup K is repeated twice within the short orbit of the "Word of God", as every element mapping a subgroup of GF(8) to any other is necessarily an element of GF(8)* already, and then K is only present in the seven images of the kernel of ϕ where the rider's "vesture" repeats and is found to be "dipped in blood".

Now, considering the other orbits in the army: is it possible for them to contain a sub-cycle of length 7? I would find that there is some h in the image of ϕ acting on F* such that $h^8 - h = 0$ and I could find a three-cycle under squaring with the Frobenius map within a different orbit.

Clearly, the subgroups of the Sun octal in F+ are mapped to the static subgroup by construction. If there is to be a repeat of the static subgroup K in an orbit, then it cannot be an image fK of the static subgroup (in its orbit) else $f \in \ker(\phi)$; it must be of a group (**not K**) partially in the coset of the "white horse" formed by each rider. Then at some point in that generalised orbit, K is a subgroup of hL where "L" is the octal containing the static subgroup K at the "head of the orbit". (Note L is arbitrary but for containing the static subgroup K and $h^{-1}K$, and every octal in the orbit of L would, like GF(8), have a short or **sub-orbit** of length $|F^*|/7$ in the image of ϕ.)

L contains $h^{-1}K \neq K$ as a subgroup. This is true for every h in C7 of the construction (if $h^nK=h^rK$ then by cancellation $h^{n-r}K=K$ etc.), each h of C7 assumed within the image of ϕ. Then as there are seven such h, every K4 group in L is mapped to every other K4 group in L by the h^n (hL=L).

Then for any x in L, every such h^n maps x to some other y also in L. I.e. $h^nx \equiv y$ is always in L for any h^n in C7 (assumed to be a subgroup of G). Then as x is in L, and F^* contains its inverse, I have that $x^{-1}L$ contains the unity in the image of ϕ. It follows that because $h^nL = L$, the image $h^nx^{-1}L$ contains every element h^n of C7 and $x^{-1}L$ is therefore isomorphic to our required C7 in G.

Yet $x^{-1}L$ is an octal in the orbit of L and it also contains every element of C7 in G. Then $x^{-1}L$ is isomorphic to GF(8) in the image of ϕ. (It then must have a short orbit of disjoint cosets also.) Yet the seven cycle (and sub-orbit) in the elements h^n of C7 apply to the orbit of $h^nK \subset L$ also; these groups are already mapped to the static subgroup and are not part of the orbit under ϕ and are then found in the "kernel" of ϕ only. (A contradiction, as I assumed K was not an h^nK.)

Why? The cosets of $x^{-1}L$ are disjoint in the image of ϕ (by the theorem of Lagrange). Assuming $g^n \neq h^m$ for any n, m $\neq 0$ (with $g^n \in$ ker(ϕ)), the sub-orbit GL of L in the image of ϕ is of disjoint images of L (up to a factor h^m in G). Then the full orbit $(F^*)L$ of L is formed **without further repeats** of those images (cosets) of L (i.e. $|F^*|=7|G|=|JG|$ with $J=<g^r>=GF(8)^*$. G (and F^*) are then partitioned by cosets of L, and $g^r \notin Gg^s$ unless r=s ($J \cap G = 1$), for $G \cong F^*/J \cong yLJ$ (yL partitions G ($y \in$ G) and then GJ partitions F^*) because $J=<g^r>= \ker(\phi)$).

Working "mod H" in F^*/J, for any $f \in F^*$ I have $fJ = yl_0h^mJ$ for some $l_0 \in L$ and $y \in$ G (each $l \in$ L is some l_0h^m). Given G is also partitioned by the orbit of $x^{-1}L$, I can write $fJ = zh^nJ$ with every $g \in$ G some zh^n (i.e. $<z> \cong G/<h^n>$). Now under Frobenius J may be held fixed leaving $f^8J = z^8h^{8n}J = z^8h^nJ$ as desired. By the cancellation laws in F^*, I find $f^7J = z^7J$. However, $<f^7>$ is actually G, so f^7J partitions all of F^*, but z^7J may contain no h^n because $J \cap G = 1$. Then there can be no seven cycle in G (h=1) unless $J = GF(8)^*$ is actually the subgroup $<h^n>$ of G, a contradiction.

The initial assumption that there is in the army any short orbit other than that of the rider, is false. There are then no three cycles from Frobenius remaining in F^*/J.

Then the "Word of God" is unique in the army (and in F), and this uniqueness of a seven cycle carries over every orbit in the army (those containing K), and also those without K.

So, across every orbit in the army there is no longer any three cycle found (by squaring elements with the Frobenius map) except for within the orbit of the "Word of God". The characteristic equation $h^8 - h = 0$ holds but for the elements of GF(8) in F.

I may now readily show the "beast is taken" and also the "false prophet with him". The dialectic process requires there to be a three cycle under Frobenius in F, such that there is some static triple that may be cycled whilst the static subgroup is held fixed. In finding such a three cycle, the dialogue formed of the static triples (bows) may then possibly continue. In every effort to do so, there is no hold of deception upon any rider in the army (those with intercession in heaven) and by extension no three cycle in any other octal in F.

The rider at the head of the army is as "unity" and the coset's static bow a "king of the Earth". Revealed are the effects of those three frogs on the other "static bows" (though I argue there are become none).

Those three other arrangements are also "kings of the Earth" but with unity a different choice of the four elements of the coset. That which remains, however, as the static triples of those cosets held so by Frobenius, are only valid as they occur in the rider on the white horse. The three cycles required need the field GF(8) to operate, which is not present but in just one multiplicative coset (if the rider is considered as unity). In the one case of the rider (called the word of God) where this is true only, there is such a three cycle; it is found nowhere else.

Aside, I note that (without loss of generality) if say $1 \equiv x+g^r$, then under Frobenius this becomes by squaring: $1 \equiv x^2+g^{2r}$, and also again $1 \equiv x^4+g^{4r}$. Then these identities are all equal and unity is found present in only one member of the isometry (the "Word of God").

Given four distinct choices for unity, $1 \equiv g^r$, $1 \equiv x+g^r$, $1 \equiv x^2+g^r$, $1 \equiv x^4+g^r$, and that Frobenius in each case would hold the coset $[g^r, x+g^r, x^2+g^r, x^4+g^r]$ static in GF(8) (for which they are each a solution to complete GF(8) by adding unity to generate the coset), then these choices for unity are solutions for isomorphisms to the rider at the head of the army (the "Word of God") only.

The possible cases of the "kings of the Earth" are to add to the static subgroup the four distinct choices of unity as above, forming cosets which if cycled static under Frobenius would "three-cycle" in the remaining elements. So, if $g^r \equiv 1+x$ (i.e. $1=x+g^r$) then $x^2+g^r \equiv 1+x+x^2$ and $x^4+g^r \equiv 1+x^2$ etc. Then $g^{4r} \equiv 1+x+x^2 \equiv 1+x^4$ and $g^{2r} \equiv 1+x^2$. Substituting $h^r \equiv g^{2r}$, say, leads to the equivalent case of $h^r \equiv 1+x^2$ and unity appears to become equally as $1 \equiv x^2+h^r$. Then, each case of the four reduces with that substitution to some other of those three separate cases (besides that simply of $g^r \equiv 1$ which is as the rider on the white horse). As the remaining elements (not unity) will (i.e. must) always cycle amongst themselves, I reduce to simply the one case of the rider at the head of the army. (Those three cases reduce to one cycle of three elements in the same octal, with the fourth held fixed as the unity.)

So, as the elements $[x, x^2, x^4]$ form a subgroup that is held static under a three-cycle on its own elements, setting, say, $x+g^r \equiv 1$ entails that the group's coset, on substituting $h \equiv 1 \equiv x+g^r$ becomes $[h, x+h, x^2+h, x^4+h]$ and I reduce to the one case of the rider on the white horse at the head of the army again. (I by necessity must have $h^8 \equiv h$.)

If, however, I argue instead that $1 \equiv x^2+h^2$ etc. is not in the same octal as $1 \equiv x^2+g^{2r}$, I must from my construction yet argue that $g^{8r} \equiv g^r$ and $h^8 \equiv h$ for there to still be a three cycle.

If Frobenius were used crossing separate orbits in G, to form a three cycle across those disjoint orbits, the elements in $\{g^r, g^{2r}, g^{4r}\}$ must then be members of the kernel C7 because $H \cong G$. (The subfield GF(8) is unique in F, uniquely held static under the required three cycle of squaring. So, $x^8-x = x^7-1 = 0$ for only the elements of GF(8) in F.) Then, once again, this kernel C7 is mapped to unity in G (in F* the three cycle is present only in the rider at the head of the army). Then I must (for every three cycle posited) generate a coset upon that same "white horse", reducing to a member of the army and therefore arriving at the head of the army again.

Note the product HG also contains no seven cycle, for seven is prime. There is no seven cycle across cosets, as the orbits are uniquely distinct in their images. Every coset x+hK may be generated on the image of the static subgroup (hK) by any member y of the coset x+hK. Without loss of generality, I can use any such y in HG to generate the cosets, with the result that the cosets are generated four times over. Then as HG contains no seven cycle it becomes apparent that no three cycle may be obtained across cosets/orbits by squaring. The factor of four comes from the fact that K is of order four in the octal: therefore 4 divides 8 which divides $|F|$. Yet G generated **every** coset on the "white horse", so I must have $HG \cong \phi (F^*) \cong H$. (In fact, by the second isomorphism theorem for groups $HG/H \cong G/H \cap G$ and in this case $G/H \cap G \cong \{1\}$ as $H \cap G \cong G$. Yet that theorem is beyond the scope of this writing.) Then as G generates every vesture up to a member of C7 in place of H, I must have $HG \cong G$.

Then the only case valid of the four is, say, the case when $g^r \equiv 1$, with the rider named the "Word of God". Then, unity is not moved out of its place, the "Word of God" is perfect, unmoved and properly unimitated. Adding unity to the static subgroup always has the same result.

That is, by squaring the given powers of g, the army of heaven (and **also** F*=GJ) has no member (octal) held static except for the "rider on the white horse" (as J isomorphic to GF(8)). The other three choices of unity in each coset are not found in the orbit of any other member, always impossibly reducing to the same head of the army with adding $g^{2r} \equiv x+1$ etc., as above.

So, three positions for unity in each octal are now illegal: the "three frogs" still operate on the one "king of the Earth", but the devices with which they act in the remaining three "balances" of the seal devices fail to be found. The "beast" and "false prophet" may not "legally operate" in the army within any such extension F: they are "taken" – their place in the isometry either belongs to another as with the "kings of the Earth" or does not appear at all, being false.

What fails, then, are the three "bows" out of every four in the devices of the seals. Then the beast (as "Death") and the false prophet ("Hell" following after), are interrupted. The sets of three balances in the beast or its image are not constructible after the wrath has aligned the sea of glass, and the only remaining ground is the bottomless pit (algebraic closure).

These two are cast alive (without the fourth seal "Death" device complete) into a lake of fire (as wrath octals, caged) burning with brimstone (as transpositions between wrath and wormwood-wrath sets/octals, those actions of the serpent spirits that persist, as making five-cycles). In that manner, there are 12 elements from these transpositions added: they are not symmetries of a fixed field GF(4) (see section [14.3]).

As these 12 transpositions (three for each choice of unity in the octal) are applied to the static subgroup and correspond to the order in which the coset is generated (when aligned in polynomial powers in the sea of glass) the lake of fire now has transpositions on the elements of its "bows" completed. Hell itself is becoming a more complete cage. (Only "death" and "hell" as following after are missing.)

Rev 19:20 And the beast was taken, and with him the false prophet that wrought miracles before him, with which he deceived them that had received the mark of the beast, and them that worshipped his image. These both were cast alive into a lake of fire burning with brimstone.

Then if the static subgroup is fixed and unity is added to generate the coset, when written in polynomials, the outcome is always determined: there is no possible three cycle of frogs with some three other kings (merely a change of labels without effect).

The end is that only the one "King of the Earth" remains. The remaining three balances – the seal devices due to the beast and false prophet – are "taken" by the legal operation that cycles both static subgroups and triples in the rider's army, which alone has multiplication under the Holy Spirit. There is simply no contest.

The remnant – those people with the mark, the collective denying virtue (as for the beast's fourth bow) and those who worship the image which are projecting themselves upon the same collective (the image of the false prophet) was taken "with him". It is the "collective assembled" in dialogue completing both beasts as that "him", properly the "son of perdition".

Now, David has truly slain the giant! The least brings the swift judgement of God.

Only the one "king" remains, the verse below springs to mind once more.

Psa 110:1 A Psalm of David. The LORD said unto my Lord, Sit thou at my right hand, until I make thine enemies thy footstool.

I found that all the devices of "Death" made in the earthly elements across the whole sea, now gathered in one place (aligned in the sea of glass) are invalid, so that the Sun octal and its left hand (become the Earth) is now aligned with every other of the sea in a first seal paradigm.

This is the action of the winepress I found earlier (as it is trodden out), and whilst the sea is aligned isomorphic, the "great city" defines its operations within the five elements of that one remaining (Sun) octal – that octal which is under the eight C7 groups over the same set of singletons: completely setting an operation on 5*8 = 40 elements (making an ordered pair from a set T into a definition of an operation on T). So, pairs $T \times T \rightarrow T$ mapping into T fulfil the operations of C5 on T upon those five singletons of all eight heptads of the spirit.

On a separate note, the "lake of fire" is built on transpositions between "wrath and wormwood-wrath" sets/octals by the serpents of the sixth trumpet. (**Not** on the balances which are a completely separate case, above shown null within the rider's army. The "lake of fire" is instead formed of or including that action of sorcery or "brimstone".) The army of the rider on the white horse leaves these intact: he "treads the winepress" upon the "lake of fire" instead.

In order to explain those few transpositions from the serpents, given say a = 1 and the serpents a(bc)(de)(fg), in mimicking Christ they by appearance seem to be exactly the same as the legal map of Frobenius on GF(4) permitted by Frobenius acting on the rider on the white horse:

$$a \equiv 1 \equiv [0, a, b, c]$$
$$b \equiv [0, a, d, e]$$
$$c \equiv [0, a, f, g]$$

In sending b ↔ c the serpents make little apparent difference to the K4 form but shift the octal beneath it; whereas in the true Christ-like K4 form the seemingly similar operation as under Frobenius on GF(4) preserves only GF(2) fixed, exchanging the remaining elements (the serpents appear to hold the transposition intact in the static subgroup and then so also in its coset, in the octal corresponding to the rider, the "Word of God"). The finite field GF(4) instead is two parts separate to the octal ultrafilter: this keeps every legal operation intact in the octal – not as with the mimicry of GF(4) in GF(8) by the serpent spirits.

What appears to occur is that two of the elements of the rider's "horse" are also transposed and the lake of fire is filled with the result that there are a few elements of S4 permitted to operate as this effect of "brimstone", and the rider's "white horse" being operated upon as if it was GF(4). The K4 form in the octal must then be wholly apparent. (Which is present when the 5th, 6th and 7th seals are opened.)

Those transpositions mimicking Frobenius on GF(4) do so keeping the "blood" intact between the wormwood and Sun octals, or the two wrath sets equivalently. (The rider "treads the winepress" also.) The static subgroup or static bow, whichever is in view from the wormwood or Sun octals (or wrath), is acted upon by such transpositions before the three cycles may act upon it as in these "frogs". The "inanimate breath" in the winepress is the last gathering point for the dialectic: and the rider's army shows that the dialogue cannot operate.

Now, if there is no interruption of the ultrafilter of Christ (as the K4 form spanning the octal) with multiplication induced by Frobenius on GF(8), the "operation" of the triple manifold of the beast, false prophet and image are brought to nothing. Those that overlaid their fellowships with the constructions of their hearts are deceived: for there is no "pneuma" (life) there but their own. They are deceived by promises of revival and a "quickening" of the spirit.

The deeds of God in His judgements are against the false prophet as a collective, but the individual believer (in the only remaining remnant that practises the dialectic: as antichrist) in the collective is still given the power to speak, and they as Christians knowing the scriptures indeed "watch" for the coming of Christ. The show is given away by the alignment of new and false doctrine in the overlay to the heart. (Believers shall know them by their fruits.)

The collective follows after the deeds of God to the end. It is true that they essentially are able to predict the impact of scripture as if "miracles", for it requires this same process in the Church for a fellowship to be desolate – not false doctrine which is to follow as hell after.

When the system of 666 is accepted and brought by one from elsewhere bearing them the mark of the beast – the end is that those "new fellowships" should through their acceptance of the bearer make them legally equal to what was and should have remained "foreign" (the beast – which is to number 666 under the hearts of a collective vote in the process).

Those that worship the image that baits the hook with obedience and switches out and supplants the dialectic process for Christ are made desolate. (The false prophet places that "abomination that makes desolate".) They will inherit nothing from God. That abomination (as a dead corporation) entails no worth to God upon its essence. If God were to redeem "MYSTERY BABYLON" He would have to redeem every other corporation to which it is equal. (Definitely a no-go for the unclean work of men's hands.)

The "lake of fire" burning with brimstone (but not whole with the missing 12 elements as yet) is the soon to be completed system of doctrinal "examination" of A5 within S5 transformed through S6 that numbers 666. The lake is the portion of the sea of glass (properly the "wormwood" devices now aligned with the Sun octal's earthly elements) spanned in part by the city of Babylon or the woman, rather than the scarlet beast. The "burning with brimstone" is equal to the action of the serpent spirits mimicking and then supplanting the place of Christ. (See the sixth trumpet – [7.8].) They use **similar** logic, but confuse the disjunction.

Burning sulphur stinks, and mining sulphur is comparable to tough work without any true rest – forever. (That is the second death, cf. God's first curse, "Gen 3:19 In the sweat of thy face shalt thou eat bread, till thou return unto the ground; for out of it wast thou taken: for dust thou *art*, and unto dust shalt thou return.".) Prayers will not be answered (the smoke of burnt sulphur will rise forever, always ignored at any height). God will not wipe their stress away.

One could ask: "Why would God use the terms 'lake of fire' and 'brimstone, sulphur' etc?" One could argue that John was painted into a corner by the terms he used to describe the maths in the fifth and sixth trumpets. The lake of fire itself as a "small sea" forever ascending out of the (stinking) bottomless pit under the action of its own spirit (but not ever fulfilling it without the judgement of God, who will Himself avoid it forever) is a perfect description, rather than a worldly one. God, keeping the imagery intact with its own language (of abstraction) is no liar; rather the liars are those misusing the scriptures to slander God as cruel and vindictive.

I already know from scripture the apparent contradiction present in the classical concept of the "lake of fire". God the Father (it is written) has given all judgement into the hands of His Son Jesus. Then it immediately follows that the blanket "pre-judgement" of all (sinner and unfortunate alike) to ready prepared fiery torment is a judgement, but of the Son? No, it would be of the Father (and that the wrong one – they have the wrong Jesus and so a different Father). Jesus is able to judge justly and this must be so without the classical idea of "Hell". Jesus (in His ministry) taught the words the Father instructed Him to speak, and so the gospels are qualified by the language of the Revelation as equally as the Revelation was Jesus Christ's own received from His Father, now passed on to His disciple John, and from John to us also.

15.5) Cut Off To The Left And Right

Rev 19:21 And the remnant were slain with the sword of him that sat upon the horse, which *sword* proceeded out of his mouth: and all the fowls were filled with their flesh.

The deception, clearly illegal (found so "in heaven", salvation then offered freely by Christ's intercession as if made to all, giving grace for them to repent) is revealed and the remnant deceived by the beast and the false prophet are slain with the sword of the rider's mouth. They operate in the wrath sets and are slain by the rider, whose sword is the static subgroup of the octal under its seven cycle, walked through the right hand and also treading the winepress.

Once they fall to the action of the winepress – they oppose the proper operation of the octal – they have received a "cage" for every "unclean and hateful bird". They fall to that operation of symmetric difference in the winepress or they fall in the five-cycle and never attain to unity. (They love and prefer a lie, as the mountains are "not found" as valid doctrines.)

Those with the mark, that deny virtue, and those that worship the image comprise this remnant, and as above they both are slain: one by the octal's proper operation opposed to the "wrath octals" (the rider's God's own strength, cf. "called Faithful"), and the other with the complete seven cycle and the treading of the winepress (his seven cycle, cf. "called True").

Those that had no part of the religious institutions of the Earth and have no place in the "woman" (who are certainly not the elect) are slain with the sword from the rider's mouth. I may assume that they are already spiritually dead and so outside of Christ (the K4 form), therefore there is no new judgement: rather the dead (the fowls) bury their dead.

The rider is the right hand of Christ – as one of the two witnesses by extension – the "sword from His mouth" is that same "fire" which comes out of the witnesses' mouths to devour their enemies. The seven cycle (the angel's circuit) then completed, the left hand is properly determined.

Rev 11:5 And if any man will hurt them, fire proceedeth out of their mouth, and devoureth their enemies: and if any man will hurt them, he must in this manner be killed.

Yet the beast and false prophet were identified with hell itself, and the remnant with the mark or worshipping the image are likewise defeated by the rider on the white horse. The "floating unity" element in the cycle, or the conviction of the angel in his circuit through the churches by the letter to Laodicea, is the sword that comes from his mouth or "set". The set is of the same seven "horns of the lamb before the throne" that also wound the heads of the dragon (that set of false doctrines of Old Testament Israel) then defeated by the obedience of Christ's angel.

Those mountains (made with the wormwood device) were not found, and every island was moved out of its place (an island would correspond to a "small earth", as if the static triple with elements {c,e,g} were always swapped for [b,d,f] etc., within and across the sea).

The remnant were "slain" with the sword, by the unity element of virtue: there are those that form a K4 subgroup to one hand and a static bow to the other. Those that "dwell on the Earth" in the static bow are those with the pains and sores of the wrath: they are "slain" because they are forced into the "lake of fire" using the operation of symmetric difference.

All the "fowls" – those sets of bows numbering 666 in static bows to the left (and right) are "filled" with their flesh. That is, all antichrist is judged and defeated, rather than every virtue requiring denial by the remnant before they can be judged.

Jer 15:19 Therefore thus saith the LORD, If thou return, then will I bring thee again, and thou shalt stand before me: and if thou take forth the precious from the vile, thou shalt be as my mouth: let them return unto thee; but return not thou unto them.

The sharp sword with two edges is the mouth of Christ (the angel is a "sharp sword" that cuts in the right hand only), that cuts off triples to the left and right using unity in the octal. (To one side or the other, I have static bows and subgroups accordingly.) As the K4 form corresponds to static subgroups the dead indeed bury their dead: by pushing themselves into a different "wrath" octal and not the Sun octal which is preserved in Christ.

The fowls in the other octals (as fourth seal systems over the earthly elements, across or in the wilderness) are given these "men" as food to devour (the fowls or spirits operate over them) – the men are pushed out by the treading of the winepress into the "lake of fire" and they are consumed with the dialectic devices, with nothing held sacred there as if virtue.

That said, those outside of Christ (in the Earth) were always without restraint and given over to a reprobate mind. I almost hope they do not notice the difference in the lake for their sakes: "almost" as for the Lord's sake to avenge, but again, I hope they are able to attain a first seal paradigm, that they may be men and not only to act as if having a manifold numbering 666.

15.6) The Devil Bound

If you were disappointed with the fifth trump not being as "demonic" as you would expect – it being in algebra, I include the following which exercises the authority in the Trinity instead.

a   exorcist with authority (r)
c   demon to be cast out (s)
e   possessed man (u^{-1})
g   exorcised man (v^{-1})

Clearly the dialectic without virtue (r, s, etc.) has in the centre {c,e,g} with "a" (the virtue) denied. (If authority is not present there is no transition from e to g.) Then $c \rightarrow (e^{-1}$ & $g^{-1})$. (Without proper authority against it, the demon neither leaves nor appears to remain.)

So, also {a,c,e} $\rightarrow$ {c,e,g} $\leftarrow$ {a,c,g}

Where the demon is as "wine" and the authority (as oil) is in question.

Whereas in grace, $Pos(c)$&$a^{-1} \rightarrow Pos(e)$. (Which is expected to be false, so grace "must" act to the contrary!)

Therefore $e^{-1} \rightarrow Pos(a$&$c^{-1}) \rightarrow Pos(g)$. (The disjunction is still freely decided.)

And there is a logical disjunction in $Pos(e) \vee Pos(g)$ and there is no positive middle (authority) for the demon to resist exorcism. "Lord, even the devils are subject unto us through thy name."

For those spirits that come out only by prayer and fasting, this requires the exorcist to pray and fast. (If he should or must have need to exorcise himself!) He must obviously pray for "g" and fast from c (as abstaining from any act leading to the accusation of uncleanness alike to that of the spirit, for example), where "c" is essentially the will of the unclean spirit. (That there be no hypocrisy, every spirit must be tested.)

It may be argued that c, e and g are not positive properties, but demons and possessed men exist so that (as taken from Jesus' own words) the question might be answered...

Joh 9:2 And his disciples asked him, saying, Master, who did sin, this man, or his parents, that he was born blind?
Joh 9:3 Jesus answered, Neither hath this man sinned, nor his parents: but that the works of

God should be made manifest in him.

And clearly g is positive and e not so. (And I affirm blindness is not a demon: although the works of God that are also manifest upon the possessed are great works.)

Luk 8:38 Now the man out of whom the devils were departed besought him that he might be with him: but Jesus sent him away, saying,
Luk 8:39 Return to thine own house, and shew how great things God hath done unto thee. And he went his way, and published throughout the whole city how great things Jesus had done unto him.

Jesus sent the man away, and to publish this fact of healing (which is out of character for Jesus) – "g" is not virtue so g follows from a&c^{-1} and not from a. Then there are no positive properties in both the sets e or g that imply or are entailed from "a". (There is no necessary relationship or interdependence between the exorcist and the possessed – rather the man is healed whole with no requirement in respect to repay – as God's is the work. Then there is no authority in the exorcist – only in God, and the possessed likewise has no unction to "physician heal thyself".) Jesus had nothing more to teach him in exorcism, the man being made whole.

15.7) The Grace Of God

I noted that Satan is unable to form a completed (S5) G-set of elements θ with every even order element to form short orbits of subgroups, having no ability to construct orbits of additive subgroups of even length. In effect there is no "key to the bottomless pit" for him. However, the system is closed by completing (adding) the missing twelve elements of the city or group S5 and I find that fulfilment here.

Satan is bound for a thousand years (1000 cycles of the group C7). A great chain is used in that this is very heavy restraint. The key is then used upon Satan: Satan is bound to do **nothing other** than attempt to fulfil these twelve missing elements (and Satan's effort, as this "strong man" is facile, because God is a "man stronger than he"), which Satan is **totally** unable to do.

Rev 20:1 And I saw an angel come down from heaven, having the key of the bottomless pit and a great chain in his hand.
Rev 20:2 And he laid hold on the dragon, that old serpent, which is the Devil, and Satan, and bound him a thousand years,
Rev 20:3 And cast him into the bottomless pit, and shut him up, and set a seal upon him, that he should deceive the nations no more, till the thousand years should be fulfilled: and after that he must be loosed a little season.

The key to the pit is simply the ability to extend a finite field with an extra degree of extension. A "great chain" in the hand I expect to be representing the short orbit of GF(8) in Algebraic Closure (GF(8) is a subfield of AC and has a short orbit), as then each chain-link is an image of the "Trinity", mapped by a "generator of AC" and is in the (right) hand. I know that the images made of GF(8) are thereby "disjoint enough" in AC so that the whole of AC may be aligned into one field GF(8) as by isometry (there is no field homomorphism although every subfield of a finite field has a short orbit). GF(8) will then map to itself when one seventh the order of AC is attained. (If AC be cyclic, else if infinite never: I may instead use another "great chain" – some other large extension in place of AC.)

I then use the "key" once and bind Satan in GF(16) instead. Now, GF(8) and GF(16) share no elements other than zero and unity, as the orders of their multiplicative groups (7 and 15) are relatively prime (their greatest common divisor is 1). I call such fields "co-prime".

First, I note that even in algebraic closure (AC) the multiplicative cosets of a subfield are disjoint in their orbits. The only elements that map a subfield to itself are the elements of the subfield. Then, every image of that subfield's additive group taken in product with an element in AC not in the subfield must then not be a member of the subfield.

Likewise, if an image of an additive group is not that of a subfield, it cannot form a short orbit under a generator of the containing field. If that were the case, the multiplicative elements (say any two elements g^u and g^v as an example) of the additive group at some point in their orbit would be a closed set because their "index of separation" $u - v$ would be equally spaced and divide the order of the container field's multiplicative group. (The multiplicative part would then be a coset of a subgroup and the orbit that of a subfield.)

So given two "co-prime" subfields of AC, K and L, $\{gK\} \cap \{hL\} = \{0\}$ for any g, h in AC where g is not equal to any hk with k in K. This is true of both fields GF(4) and GF(16) with GF(8).

Note to start that there are only 15 octal subgroups in GF(16)+, and they have no short orbit.

Now, considering the orbit in AC of GF(8) and that of GF(16) I may state that if any image of GF(8)+ under orbit in AC would become a subgroup of GF(16)+ then the following identity would hold; that is, at some point in the orbit of GF(8) and GF(16) $gF \subseteq hK$ would appear. (Where F is GF(8), K is GF(16) and g, h elements in AC.) Then I simply state that $h^{-1}gF \subseteq K$:

Now, as the images of GF(8) are disjoint in their orbit and K contains [0,1], it becomes apparent that in order for some image $h^{-1}gF \subseteq K$ of F to be a subgroup of K+, it must contain [0,1] also.

Why? The images of GF(8) and GF(16) are disjoint, so the orbit of GF(8)+ that contains [0,1] may not appear as an image containing [0,1] in GF(16)'s own orbit twice, so [0,1] must be within the image $h^{-1}gF \subseteq K$ of F that is equal to F itself. I.e. $h^{-1}g = 1$ (**up to some element k in K**).

Then I have F a subfield of K (F, K co-prime), as F is also closed under multiplication. (A clear contradiction. There is no sub-orbit of GF(8) found in that of GF(16).) Then, there is no image of GF(8) within its orbit in AC that is contained in any image of GF(16) also under orbit in AC.

The octals in GF(16) are not disjoint as are all those in the orbit of GF(8). This also holds between GF(4) and GF(8). GF(4) is no subfield of GF(8), and GF(4)+ shares no image with any K4 subgroup of any octal in the orbit of GF(8) in AC.

For Satan to deceive the nations he requires to construct a five cycle. Now GF(4) has a short orbit of length five in GF(16). With Satan "shut up" in GF(16), any subgroup of GF(16) that forms a short orbit must also then be an image of a subfield, so it may never intersect in its orbit with the additive groups of GF(8).

Now, assume there is a K4 subgroup common to GF(8) and GF(16). Now, this K4 subgroup may not be an image of GF(4), because GF(8) may not contain GF(4) as a subfield. Then, only the other K4 groups of GF(16) in "long" orbits are candidates.

I.e. there is no image of GF(8) in AC that permits a "15-cycle" of the octal groups in GF(16) to contain any K4 subgroup that may induce a 5-cycle on the K4 subgroups of GF(8), and no elements in AC that would agree in an image between these two sets.

The orbits of GF(16) and GF(8) intersect in no octal group (Satan is "shut up") and no five cycle in GF(16) may form a sub-orbit on the K4 subgroups of any image of GF(8) under orbit in AC. (A seal is placed upon him. Christ sat at the right hand, and GF(4)'s orbit in AC is "arrested".)

Now, I state only the very elect can breathe a sigh of relief as to their surety of salvation. Those in view and to be tested with Satan are those dwelling in the reference octal as unity and predestined for 1000 years: not those that are under the locusts of the fifth trumpet. The elect dwell as unity and will therefore always be in the subfield GF(8) of AC (it alone is uniquely closed under multiplication, as those within truly have the Holy Spirit) never in an image of it.

Satan is now unable to leach the saved that are under grace (to tempt them) away from a seven cycle and into a five cycle.

What can be seen is that Satan is still free to deceive his own, but not so the elect by use of a five cycle (the "nations" refers to the seven churches) under these conditions. What is more praiseworthy of God is that all are tested for 1000 years under this equal grace, and only those that show fruit (of the Holy Spirit which predestines them to obedience) live before the end of that grace period (in GF(8)), those disobedient do not. (There is always the choice to obey God or to vacate the obedient setting and follow the crowd.)

At the end of God's patience, Satan is loosed to test those that likewise hold to unity – this time for a "little season" in GF(4) (a phrase that occurred in the fifth seal passage). By imprisoning Satan for 1000 years I would mirror the patience of God with His elect.

That thousand years (cycles of the octal as seven months but only in one octal of 30) in real terms allows a factor of 1000 with the six non-rest elements of the octal as "days" of 24 hours giving us the result of (6*24*1000) = 144,000 sealed. The factor of one thousand is a level of God's patience, His grace is a period within which the believer is free (without the influence of Satan) to reign over himself alongside God, as dwelling in heavenly places predestined.

The 144,000 are able to so reign with God and to have that factor of 1000-fold patience (grace) that allows 1000-fold more elect. Those that were "beheaded" may spiritually correspond to those that were separated from Christ (the lamb, the "head of the cycle") as in the image system's five cycle, and that these overcame the beast and had left to be separate. They form a "transposition" as those convicted by the letter to Laodicea (Christ's testimony – the "head" of the circuit) or by the exit of the least (the angel called the "Word of God"), even becoming as (a,b), to be separated out from the five cycle (a,b,d,c,f,g,e)(b,e) = (d,c,f,g,e)(a,b). They are convicted to continue to remain separate or to exit Laodicea.

Christ taught that the days would be cut short or else "no flesh would be saved" (and the kingdom of God within us would then be extinguished from the world). With the fact that "Faith comes by hearing" (and that only by the word of God) if one must enter the Church to learn of Christ properly, thereby entering a five cycle (and being "beheaded" as without Christ – the head of the seven cycle), then in order to convert one must then become disobedient and enter a Church that has no logical use beyond the lesson to seek God elsewhere. God would not let this perpetual fault continue, surely! Then one is then beheaded "for the witness of Jesus" in choosing to gain it as well as to show it, for the purpose of exemplifying His virtue.

Rev 20:4 And I saw thrones, and they sat upon them, and judgment was given unto them: and I saw the souls of them that were beheaded for the witness of Jesus, and for the word of God, and which had not worshipped the beast, neither his image, neither had received his mark upon their foreheads, or in their hands; and they lived and reigned with Christ a thousand years.

However, the bands of grace can be broken: those that will never repent are cut off and find no life in that grace, so are not predestined. They were foolish.

Rev 20:5 But the rest of the dead lived not again until the thousand years were finished. This is the first resurrection.

The born again are those with a place in the first resurrection. They are the first fruits to God.

Rev 20:6 Blessed and holy is he that hath part in the first resurrection: on such the second death hath no power, but they shall be priests of God and of Christ, and shall reign with him a thousand years.

15.8) No Second Death

Once again, I am treated to the logic of God. I am informed of the "second death". This curse of the second "Death" which is so feared by those that are in the Church (that in their fear they bring it upon themselves) truly indicates they have lost all proper sense of the cross.

Assuming a state of damnation that may never be repented of, there is that possibility that there may exist a state that may befall any individual of living damnation (as unable to be redeemed in Christ). I would assume that there is some "mark of the beast" that once received may not be repented of. It is a "sure-fire" way to get to hell. (Although I must state, this is not the "true" mark of the beast, it is merely a "**strong delusion**" and should be repented of.)

2Th 2:11 And for this cause God shall send them strong delusion, that they should believe a lie:
2Th 2:12 That they all might be damned who believed not the truth, but had pleasure in unrighteousness.

Then I have a dichotomy: for God's election is of grace and freedom from the condemnation of the Mosaic Law (that is, of Moses), and not of demands of immediate obedience. In effect the elect are simply "chosen" as those that love God and wish to obey Him in all things.

The prospect of an elect individual being so damned is an inconsistency, as nothing that damns a believer before God can prevent them being chosen – all matters of the like are of the law. If one is chosen and has life (election) in the first resurrection, then such a curse has no power – because the Lord is mighty to save. The true "mark" must be one able to be repented of.

Consider though the fear of such a "mark". If it may not be repented of, then the Church will face receiving it. Rather than die in obedience to avoid it, the Church has invented the "rapture" doctrine, that they will be whisked away rather than be present to receive such a test. Yet then they logically state that they themselves are the very people who would receive such a mark if they were present to be tested with it. (The "mark" is necessary for rapture.)

What difference does this make to the Christian disciple? **Absolutely none at all**. For the final judgement arrives with no seven year tribulation and no interruption to the grace of Jesus Christ's once and for all sacrifice (that great tribulation having lasted near 2000 years already). The good news continues with never an interruption, and there is great comfort for all in that all grace continues up to Christ's final judgement (and thereby continues to the last day when, dare I say it, there is finally "rapture"). However, the rapture doctrine (whether pre, mid or post-tribulation) has claimed a second dispensation of grace is necessary in the "rapture" where there is none to be found; the grace and cross of Christ are denied by the rapturist.

If it were pointed out as from my previous argument of the two covenants (section [5.3]), that there is "no excluded middle" in the cross and the claimed rapture (that God has gone from grace to yet more grace) then you missed the point. There is sufficiency in the law for salvation – as Christ Himself fulfilled it. However, the rapture doctrine has declared the gospel unable to save in the face of the "mark" (and falsely so, the mark is not as such). Whereas I consent the law is good, this cannot be true for the gospel if it is insufficient. I recover the same inconsistency if the rapture is a finer filter than the Kingdom of God. It is a denial of Christ.

I now show several logical faults I found in the "rapture doctrine". First, a few definitions:

"N" represents again the modal necessity operator and "P" the modal possibility operator.

"saved" will refer to being "rescued from the mark" where the mark is some irredeemable fault. In every case I restrict myself to the "set of Christians" rather than include any other set.

"SAVED" will refer to the redemption of the gospel against all faults under the laws of God.

"faulty" refers to the fault that brought the mark, it is assumed redeemable in possibility: I will see if such possibility is inconsistent, bringing a "second death" of a necessarily irredeemable fault. I'll examine whether it is consistent for God to hold true one or both of the two predicates "saved" and "SAVED" as above. (For these "properties" are to be exemplified perfectly in "action" or "inaction" by God, for perfection I have the distributive laws $\neg(A\&B) \leftrightarrow (\neg A\&\neg B)$.)

I introduce a few axioms:

Axiom (15.8.1)

N¬(raptured & marked)
I.e. marked → not raptured and raptured → not marked.

Axiom (15.8.2)

N¬(raptured&¬saved)
I.e. raptured → saved. If "raptured" is a virtue, then so is "saved".

Axiom (15.8.3)

faulty & ¬saved → marked
I.e. the "mark" is a fault "forced" on all that remain. I ignore "marked → ¬saved & faulty" for the time being, so I employ a simple lemma.

Lemma (15.8.4)

raptured → (saved & "not faulty").
(The rapture has no residue of sin from the mark.) I.e. "not faulty" is a virtue.

This is easily seen by employing axioms 15.8.1 and 15.8.3;
raptured → ¬marked → ¬(faulty&¬saved) → saved&¬faulty
which is consistent with axiom 15.8.2 and completes the proof. //

Corollary 15.8.5

If lemma 15.8.4 is correct (necessary) then N¬(raptured & faulty).

raptured → ¬marked → saved&¬faulty → ¬faulty
N¬(raptured & faulty)

Theorem (15.8.6)

Given only lemma 15.8.4 and corollary 15.8.5, the rapture doctrine held true is consistent.

Proof

Given the above, it is impossible (**given rapture**) to receive the mark of the beast (to be faulty) and not to be rescued from it. You are not at any fault for being raptured (lemma 15.8.4) i.e. rapture preserves you guiltless. I.e. "raptured", "saved" and "¬faulty" are consistent as virtues.

Likewise, given rapture, you must be rescued if the mark of the beast is to be forced upon you (N(saved&¬faulty) ↔ N¬(faulty&¬saved)), but you would **not** thereby be immediately faulty if not so rescued. I.e. it is not the case that N¬(¬raptured&¬faulty) holds also. //

I may simply state the consistency of the rapture doctrine with N(Pos(rapture)), "rapture" becomes a virtue for God to save. I shall examine it and find otherwise.

Axiom (15.8.7)

I step to the bounds of grace and state there is no fault under God's law that is irredeemable. I equivalently have in present terms:
¬N¬(faulty & SAVED)

N¬(SAVED → "not faulty")
N¬("faulty" → "not SAVED")

The mark is then possibly redeemable in Christ, it is just another sin, it is as any other (being most like total unbelief) and lastly it is a falsehood to reason that anyone with the mark is totally irredeemable. That is, before the election of grace is completed and the dead judged.

Now I state:

Axiom (15.8.8)

SAVED → raptured.

I assumed that N(rapture) is of the same gospel and a "Christian doctrine" (albeit falsely). As to all "SAVED" being raptured I note the verse "1Th 4:16 For the Lord himself shall descend from heaven with a shout, with the voice of the archangel, and with the trump of God: and the dead in Christ shall rise first:". So, all the "SAVED" are supposedly raptured – whether alive or dead.

Axiom (15.8.9)

SAVED ∨ marked (from axiom 15.8.1 and 15.8.8, or else all grace continues to the last day.)

Lemma (15.8.9a)

Axiom 15.8.9 holds with or without the rapture hypothesis ¬P(SAVED&faulty).

Proof

With the rapture hypothesis I find that grace continues after the "rapture", and grace is only available to the "not marked".

Clearly SAVED → raptured → saved from axiom 15.8.8.

From axioms 15.8.1 (or 15.8.4) I also have SAVED → raptured → ¬faulty & saved → ¬marked **if and only if** the rapture hypothesis holds – that ¬P(SAVED & faulty) or equivalently that I may entail (SAVED → ¬faulty). I.e. there is a fault before God that would bring such complete and final damnation to the then "marked".

Then to be faulty is not to be "SAVED", and that "fault" is then irredeemable.

I.e. the rapture hypothesis of ¬P(faulty&SAVED) entails axiom 15.8.9. //

Instead, without the rapture hypothesis, the grace of God continues to the last day for all. (There is no requirement for "saved" from "rapture", and the "second death" is redeemable before the judgement is set.)

I.e. P(SAVED&faulty) implies the mark is redeemable, and then grace continues until the last day. So, grace, as Pos(SAVED) (being virtue) continues through to the judgement without interruption if and only if true damnation is that same "second death" **only** after the judgement, wholly and logically equivalent to that same "marked".

So, as the provision of "SAVED" is a virtue of Christ, Pos(¬SAVED) never occurs. (There is no second curse of God for a second death.) P(SAVED&faulty) entails that the rapture hypothesis of SAVED → raptured → saved&¬faulty cannot hold whilst that grace continues. In that case "faulty" becomes inconsistent, and with no such fault to entail any sense of "marked", the rapture is not necessary.

Judgement is just that – judgement. (It is not possible to be damned before you are judged.)

Then N¬(SAVED&marked), or there is no such "marked" because SAVED is necessarily positive (as axiomatic or provable) and I entail axiom 15.8.9 again. //

The Judgement Is Consistent With the Gospel

From the above I find Pos(SAVED) → Pos(¬marked). So, without proving SAVED → ¬marked I state that the latter may not entail the former unless the former were also correct, so it is (**possibly**) a virtue of God not to damn any soul before the last day and His judgement. Then either ¬marked is virtue (and God would never have "marked" any one for damnation unjustly – He being perfect), or "marked" **possibly** breaks closure of virtue in the K4 group. In that latter case, I could expect to find (from axiom 15.8.3) a K4 group of the following disjunction:

faulty ∨ SAVED&¬faulty → ¬saved.

(Yet none SAVED or faulty (under the law) need be saved by rapture (as otherwise damned) – every fault is redeemable before the judgement! I.e. "saved" is not provable for the elect.) The only unforgivable sins in the Gospel would be unbelief or blasphemy of the Holy Spirit.

Then from the middle I would find "faulty&¬saved → marked" which is then excluded as a middle (else axiom 15.8.9 is made void also!). Then "marked" is found to be an empty (necessarily privated of positivity) set on the elect as desired.

Now, it becomes correctly formed of virtue to expect the trivially decidable disjunction of:

marked ∨ (SAVED&¬marked → elect).

(Election is black and white and is a didactic absolute for all Christians.)

Yet then the octal formed is contrary, negated. Virtue in "SAVED" must be privated. Either "SAVED" or "marked" is inconsistent (unless faulty ↔ marked). Therefore, in a properly formed octal, "marked" will not break the closure of virtue in "SAVED" as it does above.

I instead expect "¬faulty ∨ SAVED&faulty → elect" from virtue, making "marked" as positive if saved ↔ elect. Yet if faulty ↔ marked I would find "saved" inconsistent with "elect". Then if SAVED is inconsistent, it must be found so over the set of Christians. (Then "faulty" would or must break the law of faith or of blasphemy as above.)

Either "saved" or "faulty" or both are found inconsistent; and so do not apply over the set of

Christians. In any case, "faulty ∨ SAVED&¬faulty → ¬saved" has no meaning in the octal unless "faulty" and "saved" are negated in place.

Then a forgivable sin must be substituted for by one unforgivable (faithlessness or blasphemy exchanged for a fault under the law of Moses) and also the positivity of "not saved" (not raptured) by that of the "saved and elect" (as rescued with the second coming and the judgement). Then "marked" becomes positive and does not apply to the "elect", but becomes freely decidable upon all: and by whom? (Only God who elects His own.)

Then by constructing an octal (as not contrary) formed upon the negated middle of "¬saved&faulty" (an octal with saved ↔ elect and faulty ↔ faithless), I find P(SAVED&faulty) holds before the judgement upon the **last day**, for every fault except any that entails any equivalence to faithlessness ↔ marked. (Or blasphemy ↔ marked.)

Then "faulty" (faithless or marked) is (of necessity) an empty set on all elect Christians. Then from that or lemma 15.8.9a I have found marked ∨ SAVED and elect ↔ SAVED. (That the judgement is positive and freely decidable by Christ) and I regain axiom 15.8.9 again. The axiomatic positivity of virtue decides this disjunction ("marked" is then a set empty of virtue unless it be virtue to exclude all those "marked", as those without faith). There is nothing else consequent from "SAVED" but the one equivalence to the whole set of redeemed and the manner of all faults forgiven them – those being under the same laws of God.)

Given the excluded middle of P(SAVED&faulty) in marked ∨ SAVED there is no such fault to cause a mark on the faithful: dropping that rapture hypothesis affirms the first resurrection in Christ. To be unmarked is a virtue: that virtue is of God in extending grace to the last day.

Then I may assume axiom 15.8.9 with or without the rapture hypothesis because "marked" is equivalent to "damned" (as it were). This is to become crucial in the several proofs to follow.

Lemma (15.8.10)

N¬(faulty&SAVED)
SAVED → ¬marked → saved&¬faulty → ¬faulty (axiom 15.8.3) //

Theorem (15.8.11)

The first resurrection of salvation renders the rapture inconsistent and the mark with it.

Proof:

From Axiom 15.8.7;
P(faulty & SAVED)
only if "SAVED" is inclusive of all faults including those equivalent to the mark.

By axiom 15.8.9 and 15.8.3:
SAVED → N¬(marked) → N¬(¬saved&faulty) or faulty → saved
P(faulty&SAVED) → P(saved&¬marked) → ¬N¬(saved&¬marked)

P(faulty&SAVED) $\lor$ (saved $\to$ marked)
But N¬(faulty & SAVED) by lemma 15.8.10 so ¬P(faulty & SAVED)
Then raptured $\to$ saved $\to$ marked and axiom 15.8.1 has no justification. //

This contradicts lemma 15.8.4 and also theorem 15.8.6. The whole set of "raptured" would logically (a-priori) have accepted the "mark"; it is not yet shown that all otherwise "marked" Christians are "raptured".

Then the rapture doctrine is certainly inconsistent with the gospel, and I simply state ¬N(rapture). Likewise, I may state ¬N(saved) (or ¬N(Pos(saved))) and if the requirement of salvation (rescue) from the mark is inconsistent then the mark is inconsistent as the rapture doctrine indeed fails. I may now observe a corollary:

Corollary (15.8.12)

Not only axiom 15.8.1 fails, but axioms 15.8.3, 15.8.7 and 15.8.8 also fail if the rapture is assumed consistent with the gospel.

Proof

Without assuming axiom 15.8.1, I posit:
P(SAVED & marked) $\to$ P(SAVED & ¬saved & faulty) $\to$ P(SAVED & ¬saved)

By the modus tollens (I already have SAVED $\to$ raptured $\to$ saved by axioms 15.8.2, 15.8.8);
N¬(¬saved & SAVED) $\to$ N¬(SAVED & marked) or "SAVED $\to$ not marked"
or ¬P(SAVED&faulty), as "not marked" $\to$ ¬faulty & saved $\to$ ¬faulty

Which contradicts axiom 15.8.7. The "fault" or "mark" is inconsistent with the axiom. (The closest principle to the "first resurrection", from death in sin to life in Christ. Axiom 15.8.3 fails if 15.8.7 holds.)

And then those that are elect do not have the mark (fault) of the second death as they do not have the fault of the rapture doctrine (axiom 15.8.1 and also "saved $\to$ marked"); whereas those elect holding to the rapture doctrine most certainly do have a mark "for a frontlet between their eyes". I.e. if the mark is redeemable then they are not to be raptured:

P(¬saved & SAVED) $\to$ ¬N¬(¬saved & SAVED)
i.e. N¬(SAVED $\to$ saved) and so N¬(SAVED $\to$ raptured) and axiom 15.8.8 fails. (NB: modal status is always necessary.) //

Then the scripture assures us that either itself or the rapture doctrine fails: that is, unless the "rapture" also occurs on the **last day** with the resurrection of all and immediate judgement. (The dead are not raised after, the living do not "prevent" it by preceding them!)

Then as 15.8.8 is scripturally correct, it is truthfully impossible for the election of grace to continue with such a "mark" present, or during any period of a "second death". (I.e. perpetual fault, which is properly "hell".) Grace (as in "SAVED") continues all the way through to the day of judgement without requirement for "rapture". (There is no "mark" as a "second death".)

These verses truly are for a "frontlet" or "phylactery" in the mind of the rapturist (as between their eyes), that should correct them! (That there is a remnant of those which are **alive** and **remain** entails that grace yet continues to the last day: there is no return to the old covenant.)

1Th 4:15 For this we say unto you by the word of the Lord, that we which are alive *and* remain unto the coming of the Lord shall not prevent them which are asleep.
1Th 4:16 For the Lord himself shall descend from heaven with a shout, with the voice of the archangel, and with the trump of God: and the dead in Christ shall rise first:
1Th 4:17 Then we which are alive *and* remain shall be caught up together with them in the clouds, to meet the Lord in the air: and so shall we ever be with the Lord.
1Th 4:18 Wherefore comfort one another with these words.

Either the mark is inconsistent: if not, then redemption is inconsistent (axiom 15.8.7). If that is consistent, then the statement on those that have fallen asleep is inconsistent (axiom 15.8.8). Yet there is no inconsistency in the last by the scripture – that "we shall not prevent them" simply because there is a remnant. The first resurrection is not inconsistent in this matter (cf. Rev 20:6), only the mark and its fault entailing the "second death" (albeit not the real one, the "lake of fire") is left as inconsistent. Every fault remains redeemable (overcome with virtue in Christ, as our intercessor of grace) when God is moved to "SAVE" an elect. It is His elective choice and a matter of His own law only. In summary: all inconsistency is resolved if axiom 15.8.1 is dropped. (For then the whole "rapture" hypothesis may be completely discarded.)

I certainly have:
$\neg N(\text{raptured}) \leftrightarrow P(\text{SAVED \& faulty}) \leftrightarrow N\neg(\text{SAVED} \rightarrow \neg\text{faulty}) \leftrightarrow N\neg(\text{faulty} \rightarrow \neg\text{SAVED})$. Then, "grace is sufficient for those with the mark".

So if the fault is irredeemable (I have $\text{SAVED} \vee \neg\text{SAVED}$), and by theorem 15.8.11 I have "saved $\rightarrow$ marked", so given some **irredeemable** fault $N(\text{marked}) \leftrightarrow N(\neg\text{SAVED})$ I expect $N\neg(\text{SAVED\&saved}) \rightarrow N\neg(\text{SAVED\&raptured})$.

So that:
$\text{SAVED} \vee \text{saved}$
$\text{SAVED} \vee \text{raptured}$ (I.e. "rapture" only supersedes an **incomplete** grace in "SAVED".)

(Either the rapture doctrine is inconsistent with the gospel or the fault is redeemable.)

Those requiring rescue as in the rapture doctrine cannot be "SAVED" from the mark. That is, assuming the mark is a necessarily irredeemable fault (which it is not during the election of grace), then the rapture is inconsistent with the true gospel, even in its simplest terms.

Therefore, the rapture doctrine is equivalent to $N(\neg\text{SAVED})$ – that the power of the cross cannot be extended to redeem the mark. I.e. $N(\neg\text{Pos(SAVED)})$ or that salvation is no virtue.

This is a "second death", an exclusionary curse from all redemption and in fear of it the rapturist denies the salvation of God in Jesus Christ. (It is a "strong delusion" and appears as if it were virtue, cf. 2Th 2:11.) In truth, it is only valid on the last day when the election of grace (whilst $P(\text{SAVED\&faulty})$ holds) is completed. Then, hell is prepared itself to become the "second death".

Rev 20:6 Blessed and holy *is* he that hath part in the first resurrection: on such the second death hath no power, but they shall be priests of God and of Christ, and shall reign with him a thousand years.

If then N¬(SAVED&saved) (to be raptured, you must be "irredeemable" under grace, and in denying the gospel of Jesus Christ you have such a "mark" already). In a bizarre twist in an analogue of "virtue" I find:

Given rapture, N(saved) ↔ N(¬SAVED) as the modus tollens shows to be correct.

But then N(saved) is as fine a pretence of this ultrafilter (were it to be in one) as one could get. (An ultrafilter is a maximal filter – no other (as one more "fine") filter may properly contain it. It is "most fine". If two filters logically conflict in the same language and each contain one or the other, then they cannot both be true "almost everywhere". One model is incorrect.) For I did interrupt N(SAVED) as a positive property to extend rapture as a virtue over it. (Were I to use the same notation I would be stating $Pos(p\&q^{-1}) \rightarrow Pos(s)$ where "s" is "saved" and p is "rapture". Then "q" would be the salvation in the gospel.)

Dialectically, the middle of saved&SAVED would prove the disjunction marked ∨ raptured. However, that middle is truthfully not exemplifiable if "rapture" is a virtue and grace fails as above in that disjunction of $q \vee p\&q^{-1} \rightarrow s$. There is no duality of grace if rapture is a virtue.

As a result of "rapture ∨ SAVED", it is nothing more than an observation that $p \leftrightarrow q^{-1}$ in this case. Then "rapture" would become finer than salvation in the ultrafilter. (Salvation becomes no virtue.) If salvation deprives the rapture as positive on the grounds that rapture deprives positivity from salvation – making salvation appear equally as no virtue, I should argue on the fineness of the ultrafilter.

Joh 15:13 Greater love hath no man than this, that a man lay down his life for his friends.

The cross is **virtue**, of freedom of Jesus' own choice, and denying it is to private it: which cannot be done with virtue, or any other positive property. The cross is **always** positively exemplified to all people, everywhere. The cross of Christ may never private salvation – whether in terms of "SAVED" or "saved". It is universally (logically) correct that any sense of $Pos(q^{-1}) \leftrightarrow \neg Pos(q)$ is utter nonsense (since q here is a virtue). The cross will not fail. Likewise, Jesus need never die on the cross twice.

So, "Pos(SAVED) ∨ Pos(saved)". I.e. "rapture becomes a positive statement only when the gospel cannot save at all" is an equivalent statement to "It is impossible for both salvation to be positive and the rapture also". No one can convince that salvation is not positive, even if it is not able to save from the mark.

The equivalent would be to state that "rapture" as an act of God (the God of the gospel) is not positive (as salvation is a virtue of Christ (agape)) as it privates a virtue. (Or it cannot be virtue as it privates the positive.)

The rapture even becomes equal with the liberty of pure divine virtue $p \leftrightarrow q^{-1}$. God simply chooses to lift up those He would lift up. There is no "universality" of the rapture because of some imaginary mark. Likewise, there is no magic key to salvation: God simply calls and chooses those whom He will. After this manner (preserving liberty) the ultrafilter is intact in both cases, but not universally or in a logically necessary fashion. On the last day, then, those "saved" will be only those He chose without exception: there are none remaining after.

There is no N(rapture), as N(rapture)&N(¬SAVED) → N(saved) assumes (as in a disjunction of virtue) N(¬SAVED) beforehand. Then, rapture deprives divine liberty of effect rather than salvation, for "necessarily false" is "impossible", rather than N(SAVED) → ¬N(rapture), which is contingency of free choice (cf. section [16.12.9c] of the book) in possibility.

But is N(saved) a result of virtue upon the privation of any other positive property, i.e. by God remaining inactive? Surely His grace is sufficient for all of us already as N(SAVED). In denying the mechanism of salvation from Christ's mediated grace, I have then raised "rapture" above the positive (virtuous) work of the cross and so "rapture" is not then a virtue if it must private the salvation of the cross in order to be positively exemplified. God may freely exercise His own will surely (as Pos(rapture^{-1}) if grace is sufficient, I would require God to cease His rest otherwise), but it is a sin to demand the liberty from another: especially from God by denying His grace to be sufficient when it most surely is.

I could equivalently state that God of His own liberty would choose all those whom He would "rescue". God would also choose those to whom He would not extend His elective choice to save. He may truly choose His friends, denying the salvation of the cross to all those others He wishes to exclude. (Salvation as an ultrafilter is become ¬faulty ∨ SAVED&faulty → elect.)

In that sense, can the rapture positively save from the destroyer? No, it cannot. For God to ever require the rapture as a mechanism of salvation He would have to choose those who would require it − and He must then have excluded all those from the whole ultrafilter of salvation (the election of grace) and completely deny them the power of the cross. So, rapture is not a mechanism of salvation. It is a strong delusion to deceive even the very elect.

Were believers not told that to break the law is to become under the whole condemnation of the law? To be so self-excluded from the covering of a virtuous act of salvation in order to seek virtue elsewhere is to private a virtue continually extended towards us. Dumb, huh!

This last contradiction results in the rapture doctrine being inconsistent. For then, "SAVED" remains a virtue still, and "rapture" as a mechanism which **privates** it (salvation) is then by definition not a positive act (or positive property applied to any set "raptured") to exemplify. I can, by alignment with the scriptures, state this to be an honest statement: I somewhat agree with God on the logic of, and the difference between, a virtue and a mere positive property! (Yet rapture privates the virtue of salvation simply by stating ¬P(SAVED&faulty).)

2Th 2:11 And for this cause God shall send them strong delusion, that they should believe a lie:
2Th 2:12 That they all might be damned who believed not the truth, but had pleasure in unrighteousness.

N(saved) ↔ N(¬SAVED) should haunt any rapturist: it states that if you hold the rapture true over the gospel of Jesus Christ, you are worshipping upon an idol of a "second death", and the real one, the "lake of fire" (as a "foolish virgin" caught in the Church) is where you will remain.

Axiom 15.8.13

marked $\rightarrow$ ¬saved & faulty (All those marked and left behind could not have been raptured.)

Theorem 15.8.14

Given the "rapture" and axiom 15.8.13 above; that rapture doctrine is a false teaching (inconsistent) and you simply cannot find salvation if you hold to the rapture doctrine. (Crucially, I do not assume axiom 18.8.1 or 15.8.8.)

Proof:

Given "rapture", N(saved) $\leftrightarrow$ N(¬SAVED). (To keep the ultrafilter intact in every case.)
P(SAVED) $\leftrightarrow$ P(¬saved) $\leftrightarrow$ P(¬raptured&¬saved) by axiom 15.8.2.

And that is fine (check it!) because there is no rapture. (Certainly, no logical one, else all would be raptured already.)

P(SAVED & faulty) $\leftrightarrow$ P("not raptured" & not saved" & faulty) $\leftrightarrow$ P("not raptured" & marked)

And I have uncovered a condition for an excluded middle in "SAVED $\vee$ marked". I had actually posited that "marked" is become a redeemable fault: which immediately entails:

P(SAVED & marked) $\leftrightarrow$ P(¬raptured & marked). This contradicts axiom 15.8.9. (The rapture is **false teaching**; i.e. ¬P(SAVED & faulty).)

Given the mark an **irredeemable** fault, with the modus tollens;
N¬(¬raptured & marked) $\leftrightarrow$ N¬(SAVED & marked),

i.e. marked $\rightarrow$ raptured is true **if and only if** SAVED $\rightarrow$ ¬marked,
and "**marked $\rightarrow$ raptured**" is then that same middle that proves "**SAVED $\vee$ marked**".

In order for the mark to be found consistent with the gospel, those SAVED but to be marked must be raptured. (For there to be no middle, I must assume N¬P(marked&raptured)), but there is indeed a middle!)

N¬(SAVED & raptured) i.e. (SAVED&¬marked) $\vee$ (raptured&¬SAVED)
But not being marked will not stop you being redeemed, so: **SAVED $\vee$ raptured**. (Yet I now have P(SAVED & marked) and there is yet no sense of axiom 15.8.8.) //

All grace is either to "SAVE" (and therefore N¬(marked) i.e. there is no such mark under grace unless judged to the lake of fire **after** the last day). Or else, all grace is to await rapture and then not to "SAVE" (i.e. N(marked) **before** the election and strength of God's grace is completed, and grace fails for all, not just the remnant). Then as above, SAVED is virtue which "rapture" as delusion, certainly privates. Then raptured $\vee$ (SAVED&¬raptured $\rightarrow$ elect).

So, either the gospel grants forgiveness of all sins including (the truly repented of and discarded) rapture doctrine (which as such requires God's liberty of Him, also denying the gospel, and Himself, something which God will not and cannot do): or, rapture doctrine is equivalent to worshipping God with an idol (the "second death") in place of the gospel.

The end result is that in one of these two paths "SAVED ∨ raptured" there is the destination of the "lake of fire". The second death will not occur to me if I hold firmly to that which I have in Jesus Christ. The gospel stands firm against the rapturist position, which by theorem 15.8.11 is inconsistent. The rapturist clearly claims the gospel is insufficient to save from the mark. The gospel suffers no interruption up to the judgement: there is no "second death" beforehand.

1Th 4:18 Wherefore comfort one another with these words.

Take note of the above grace in Christ, for I warn you: what follows may be found distressing.

Theorem 15.8.15

The rapture doctrine (¬P(SAVED&faulty)) is equivalent to its own imagined "mark". (Both the rapture doctrine together with its "mark" are totally inconsistent with the Christian gospel.)

Proof

Given ¬P(SAVED&faulty):
saved → marked (by theorem 15.8.11, doesn't assume axiom 15.8.1 or 15.8.8),
marked → raptured (by theorem 15.8.14, doesn't assume axiom 15.8.1 or 15.8.8),
raptured → saved (by axiom 15.8.2),
raptured → marked (by the above).

Therefore, raptured ↔ marked. // (Given axiom 15.8.9 the remainder follows.)

So, then, as there must be no required "rapture" until the **last day** (1Co 15:51-52, 1Th 4:15-17) there is then **no** such popularised "mark" (no *second* death) until after Christ sits in final judgement on that same last day! Only the damned will inherit the second death (cf. Rev 21:8).

Believing one's self to be absent from an incorrect and imagined "second death" without being saved firmly on the rock of Jesus Christ is not sound doctrine. Yet still some say, "I'll explain it to you on the way up..." ((sigh) Saints alive, God preserve us!)

Rev 21:8 But the fearful, and unbelieving, and the abominable, and murderers, and whoremongers, and sorcerers, and idolaters, and all liars, shall have their part in the lake which burneth with fire and brimstone: which is the second death.

Corollary 15.8.16

There is no further dispensation of salvation, and no return to the old covenant if grace fails.

Proof

Simply replace "raptured" with "forgiven an irredeemable sin" and "saved" with "spared". There would then be another gospel or another title in "saved" and there is introduced that other gospel, one inconsistent with redemption in Christ. There is also no return to the law, for if one is failed rapture and "left behind" then "saved" and "raptured" have equivalents in "spared" and "obedient". //

If a dispensation other than the gospel were ever to find effect, God would have to have found finer virtue – the disjunction SAVED ∨ saved does not replace "SAVED → raptured → saved" or "saved always saved" with "obedient". ("SAVED" in grace does not entail any other gospel.)

I.e. There is no disjunction Pos(SAVED) ∨ Pos("obedient & ¬SAVED") → Pos("spared"). If grace were to fail, then a single sin of "faulty" could entail the descriptor "marked" (as all are then "irredeemable"). Note from the above, that there are no other dispensations!

Then, I would find that no matter which dispensation fails, the equivalence SAVED ↔ ¬saved would yet hold, leaving only "¬saved ∨ obedient" to be decided (as in the above or with axiom 15.8.9). When ¬saved is become universal, obedience would appear the only sound option as **all** virtue would clearly decide the disjunction. Yet SAVED ↔ obedient are equivalent in all truth within the gospel (placing one's faith on Christ alone and Him crucified: obedience meriting salvation found only in believing upon that one truth, that one way and that one life – the record of many witnesses), so that same disjunction "damned ∨ obedient" is actually an inconsistency – a fallacy of equivocation. It is implausible in the gospel to wholly equivocate "obedient" with "not faulty" under the law of Moses. (All are faulty, but many obedient!)

I ultimately find from this fallacy that given obedient → saved (I.e. N¬(¬saved&obedient) or that all "obedient" *will* be saved, that obedience is become equivalent to "raptured"): that SAVED ∨ (marked ↔ obedient) holds once more. Redemption, meanwhile, has **not** failed and because the "other gospel" (as if "saved & obedient") is truly for one to remain perpetually faulty (to be found continually rejecting the middle in SAVED ∨ faulty and all salvation in Christ with it), the result is effectively to remain irredeemable or "marked" by that same inconsistency of faith (the same satanic deception) as shown by theorem 15.8.15. //

Finally, there is yet no "second death" or any such inconsistency in the gospel due to any "mark of the beast" (that would replace or interrupt the grace of God merited to faith), other than of the final judgement of the unjust (after the resurrection of all) to the "lake of fire". Any such inconsistency in the gospel would also be (of necessity) present at the time of the cross. Then the scarlet thread of redemption would then become **wholly** undone: snapped off at Christ's end (the claim of Islam, Jesus replaced by another), not at ours. (Days "cut short" do not annul the cross, only preserve a remnant.)

Joh 17:15 I pray not that thou shouldest take them out of the world, but that thou shouldest keep them from the evil.

The Sanctified Saviour 15.8.17

Christ's last prayer states quite simply in part, as following on from the above;

Joh 17:16 They are not of the world, even as I am not of the world.
Joh 17:17 Sanctify them through thy truth: thy word is truth.
Joh 17:18 As thou hast sent me into the world, even so have I also sent them into the world.

Joh 17:19 And for their sakes I sanctify myself, that they also might be sanctified through the truth.
Joh 17:20 Neither pray I for these alone, but for them also which shall believe on me through their word;
Joh 17:21 That they all may be one; as thou, Father, *art* in me, and I in thee, that they also may be one in us: that the world may believe that thou hast sent me.

Now, Christ Himself suffered death, and He did so to show Himself resurrected and blameless in all His testimony. It was a more perfect word of God's intent than the letter of the law, and with His resurrection and complete exoneration thereby, it is shown utterly faultless. Christ stated "And for their sakes I sanctify myself" in that the salvation of the cross would not fail any with faith upon Him. Because Christ laid down His life to show His testimony true, His sacrifice must not have failed Him for His own resurrection, and likewise, it could not and will not fail any of us either. Faith in the first resurrection truly frees us from any power of a "mark" (a curse of a second death) that would interrupt God's grace.

Christ prayed for us to be "sanctified" through His Father's truth, and stated that His word is true (all Christ's testimony given Him from His Father). The word passed on to us from those Christ chose and sent out into the world, has not failed us two thousand years later. It is known better now for it being intact and faultless, as Christ's promises are of (and to) those He chose as "not of this world" but of God. None should doubt the word handed down to us, for God guaranteed His word preserved to us with Christ's resurrection.

Given, then, that a logical fault in the cross would have opened power to a curse of a "second death" of Satan's making as from the very moment of Christ's resurrection, Christ would not have been sanctified in the truth of His Father had there been **any** fault in His words. That word of resurrection would never have survived the cross, and were one to ask whether there is a God of heaven at all, one would not elsewhere or otherwise find a better guarantee that Christ's God saves. His word is sure.

Christ prayed for all those that in faith received His true word preserved: made as one body, without division. The cross (that sure word of all Christ's testimony) is the uniting faith in God the Father made plain as day (Joh 17:21). Any spirit, or anyone "Christian", attempting to lessen the cross (to malign, diminish or nullify it) is antichrist and Anathema (cf. 1Co 16:22).

Even the scriptures can be twisted out of this sure word, in particular (as mentioned above):

Mat 24:22 And except those days should be shortened, there should no flesh be saved: but for the elect's sake those days shall be shortened.

Now that I know the cross will never fail, this verse simply states the flock will not be attenuated out of existence for lack of word (but for the lack of love for it instead), yet there will be a remnant. The end, nearer for the elect's sake, is brought closer for the presence of faulty teaching only. The elect would then find what doctrine and from whom? (Only the sure word that is preserved from God: faith comes but by hearing.)

Mat 24:23 Then if any man shall say unto you, Lo, here *is* Christ, or there; believe *it* not.
Mat 24:24 For there shall arise false Christs, and false prophets, and shall shew great signs

and wonders; insomuch that, if *it were* possible, they shall deceive the very elect.
Mat 24:25 Behold, I have told you before.

Then upon that mote of truth which entails all, I will point out my brethren only as I find them!

Not One Left Behind 15.8.18

So, given the equivalence between the false doctrine of the rapture and the denial of the cross of Christ: am I, by stating that same equivalence will surely leave one bereft of salvation (and effectively "quasi-marked"), acting in the office of Antichrist?

Rev 13:16 And he causeth all, both small and great, rich and poor, free and bond, to receive a mark in their right hand, or in their foreheads:

I cannot give anyone the true "mark of the beast"; it is not as above in "marked". I (or indeed anyone else in the genuine sense of the real "mark") am not going to scratch anyone's hand or implant a chip, I will not barcode them or track their location or force them to deny the gospel of Jesus Christ with a knife held to their neck: I instead state that their chosen fellowship in Christ and their guiding light is left entirely up to them: there are yet fellowships out there to choose from, even if some may only exist online.

Rev 3:4 Thou hast a few names even in Sardis which have not defiled their garments; and they shall walk with me in white: for they are worthy.

If the argument is rejected, that is their freedom to exercise. If offence is taken, I hope logic will rule over emotion and that sobriety of faith will prevail over drunken vigilance.

By logically making those same statements, if I were logically and universally "forcing" such an imaginary "quasi-mark" on every rapturist, I could simply state that given the same rapture doctrine's correctness, the proponents of that doctrine would surely and most certainly be evacuated and out (as instantaneously) of here pretty quick to escape said (universal) logical "force", and then necessarily so before they could instead be converted back to Jesus Christ and to accept sound doctrine. (Yet it is written that they will not endure it.)

Given, then, they remain "quasi-marked" and not raptured absent: there is **no rapture**!

I sincerely doubt and reject any idea that such rapture would ever occur (as during the election of grace), I hope that any rapturist out there, finding themselves in fault, would certainly admit that the lack of immediate evacuation from said "quasi-mark" disproves the rapture completely, and in all honesty, they should repent of that false doctrine!

Logic is instantaneous in its necessity, but God can wait. When, then, is there to be any rapture found if God would let His own (the rapturist – I here speak as a liar), be so universally and logically "quasi-marked", unless it be for the very reason to smack them round the head and thereby show them the error of their doctrine? In Christ, there is no such "mark" to fear.

Think twice before pointing the finger! God has surely sent the associated "strong delusion" to this rapture lie of Satan. (The gospel will not fail any believer before the last day.)

Only the "rapturist" is ever so "quasi-marked". (And by this argument, certainly not the ungodly that still find the cross – there is discernment to be found in that for any reader!) The rapture doctrine is not the one and true gospel, and it is **not** universally accepted as such. Those teaching that doctrine are, without an exception, gravely mistaken.

Anyone realising themselves (even possibly) so "quasi-marked" (none of their fellows being raptured either), should know (and be in truth convicted, to rest assured that there is no "mark of the beast" of that particular kind) that there is no universal rapture coming, for them or anyone else. They instead need Jesus Christ (and only the correct gospel, His gospel) to save them from the errors in their faith that equivalently deny them His Father's grace.

15.9) The Resurrection And Satan Loosed – Standing Firm

Rev 20:7 And when the thousand years are expired, Satan shall be loosed out of his prison,
Rev 20:8 And shall go out to deceive the nations which are in the four quarters of the earth, Gog and Magog, to gather them together to battle: the number of whom *is* as the sand of the sea.

Now, to God all are concurrently alive. God, in waiting 1000 years, accounts for every human being (during the Old or New Testament: OT or NT) that has ever lived, in order to see them predestined or not. At the end of that grace Satan is loosed and the salvation of the believer is tested. That is, when all are alive at the resurrection of the just and unjust, not just the first fruits of the first resurrection in Christ but with the resurrection of all at Christ's great white throne judgement. (The seals are then all undone and Sabbath rest returned.)

This total resurrection happens not after 1000 years of the reign of Christ on Earth but with the return of the armies of heaven and the rider on the white horse. The judgement is immediately set and the dead are raised.

The "nations" are those that dwell in the earthly triples of each wormwood "row" from the sixth trumpet. These compare directly with antichrist (cf. "How art thou fallen" in section [7.13]). These four sets are properly the "four corners of the Earth". Those antichrist, and those consequent of antichrist (as son of the morning).

The deceived nations of Gog and Magog are then "antichrist" and "son of the morning" (or as "Armageddon" – "Death" with "Hell" following after). These nations are the tares within the seven churches practising the dialectic with wormwood. They exercise the dialectic devices under the freedom granted by the "kings of the Earth" to form the "third woe".

The many scarlet beasts (little horns) of worldly "churches" are gathered together by Satan to stand united as with one device (image) built over the dialectic that will fail (under God's wrath) whilst arrayed against the kingdom of God. This passage aligns with the former concerning the three spirits as frogs that come from the mouth of the beast and false prophet and the dragon – going to the kings of the whole world. Believers are warned by Jesus:

Rev 16:15 Behold, I come as a thief. Blessed *is* he that watcheth, and keepeth his garments, lest he walk naked, and they see his shame.

It is true a thief is not noticed until he has already been and gone, having left with something later discovered missing. (That being time: I return to the period before the opening of the seventh seal on the last day – when Jesus returns.) I may happily equate the battles of Gog and Magog with that of Armageddon. For this, I have in place already the word-picture of the four angels that loosed the four winds upon "Euphrates" as well as now become the system of wormwood octals that intersects the Sun octal in three subgroups.

He that "watcheth" is the believer in Philadelphia (not to return to the amalgam of Laodicea – keeping His garments – his doctrine), so that Laodicea may have no shame for him if he returns – in their eyes they would appear justified, he shame-faced.

The alignment is made as also in the street of the great city – a clear division between those that would not let the two witnesses be buried for faith they would be raised, and those that sent each other gifts in celebration of their deaths.

I should note that in this passage the narrative gap approaching the seventh seal has again been "stolen" as by a thief. (To be moved as if by the seven thunders (backward) to align the last vial with the seventh seal, correctly placing Gog and Magog alongside Armageddon. For this, the beasts and 24 elders fell down before the throne.)

Rev 16:16 And he gathered them together into a place called in the Hebrew tongue Armageddon.

Now, I previously paired the names "Death" and "Hell" with "Wormwood" and "Abaddon" respectively. This final equivocation of "Armageddon" to "Gog and Magog" is done likewise. I posit that this final equivalence is made so that it identifies in the text the "third woe" (the previous two were the first and second woes). That scarlet beast (the third woe), then, is the same battle of Armageddon/Gog and Magog: the place of "Great Tribulation" now turned against the elect of God in threat of war upon all virtue.

All that is in view is that the same dialectic devices are in force over the sea outside of the Sun octal. Note that the number of opponents gathered is as the "sand of the sea" which is as the same multitude of fallen stars drawn by the dragon's tail to form the dialectic devices: those of deception. I simply find that the dragon has one last salvo to fire off his devices (he has but a short time), before they are forever defeated. Gog and Magog are simply "antichrist".

Rev 20:9 And they went up on the breadth of the earth, and compassed the camp of the saints about, and the beloved city: and fire came down from God out of heaven, and devoured them.

Amen to that. Why? Read on!

Now the elect that find favour with God are to be found amongst a great mass of unsaved people, and the raised unjust are now under the influence of Satan. The "Gog and Magog" – the antichrists of all history (that have always been without restraint as the seals never applied to them), are influenced by Satan to attack God's elect, and fire "came down from God" above to devour them.

The breadth of the Earth would be the whole of the wilderness, rather than just one coset in one octal of thirty. The wilderness is composed of 7 such sets (aligned across each and all 30 octals in the sea), so, the earthly parts which form the "wilderness" as the "sand of the sea" – of 28 triples in 4 disjoint sets (in bold) over each seven cycle are as I had previously:

$[a,b,c], [a,d,e], [a,f,g]$ with $[b,e,f], [b,d,g], [c,d,f], [c,e,g]$ with unity $a = 1$

$[b,d,f], [a,b,c], [b,e,g]$ with $[a,d,g], [c,d,e], [c,f,g], [a,e,f]$ with unity $b = 1$

$[c,d,g], [b,d,f], [a,d,e]$ with $[b,c,e], [a,c,f], [e,f,g], [a,b,g]$ with unity $d = 1$

$[c,e,f], [c,d,g], [a,b,c]$ with $[a,d,f], [b,f,g], [a,e,g], [b,d,e]$ with unity $c = 1$

$[a,f,g], [c,e,f], [b,d,f]$ with $[b,c,g], [d,e,g], [a,b,e], [a,c,d]$ with unity $f = 1$

$[b,e,g], [a,f,g], [c,d,g]$ with $[d,e,f], [a,c,e], [a,b,d], [b,c,f]$ with unity $g = 1$

$[a,d,e], [b,e,g], [c,e,f]$ with $[a,c,g], [a,b,f], [b,c,d], [d,f,g]$ with unity $e = 1$

It has its earthly cosets now aligned with groups from the Sun octal: yet more so – the breadth of the Earth is across 30 octals not just seven. This also compasses (surrounds) the people of God as with **unity** in the centre and systems of triples numbering 666 to either side.

The Gog-Magog war is the same as that of Armageddon. The fire that comes out from God to devour them is the exact same fire of the Holy Spirit that predestines the believer: for, in GF(8) and in its three cycles under Frobenius, the Holy Spirit has separated out these triples as in Christ in the leftmost three columns above or in the Earth to the rightmost four.

It is then the election (as heaven) that devours them with fire (with unity, "election and rest" preserved under the seven cycle, and with the other elements so excluded from it) acting downward as upon the left handed octal in the lake of fire (John reads right to left)) as subject to the second death (the lake of fire). They (the rightmost four columns) are simply numbered with all those that are in the city of "MYSTERY BABYLON" (the beast and false prophet).

They went "up" on the "Earth", aligning themselves in those four columns as earthly in the four beasts before God (the seven cycle reads right to left to John, so the columns appear to "go up" rather than as above, the seven cycle acting downwards on the four earthly triples); so I should transpose to the "way of the kings of the East" re-arranging rows and columns;

$[a,b, c], [b,d,f], [c,d,g], [c, e,f], [a,f,g], [b, e,g], [a,d,e]$

$[a,d, e], [a,b, c], [b,d,f], [c, d,g], [c,e,f], [a,f,g], [b, e,g]$

$[a,f,g], [b, e, g], [a,d, e], [a, b, c], [b,d,f], [c, d,g], [c, e,f]$

$[c, e, g], [a, e, f], [a,b,g], [b, d, e], [a,c,d], [b, c,f], [d,f,g]$

$[b,d,g], [c,d, e], [a,c,f], [b, f,g], [d,e, g], [a, c, e], [a, b,f]$

$[c, d,f], [c, f,g], [e, f,g], [a, e, g], [a,b, e], [a, b,d], [b, c,d]$

$[b, e,f], [a,d,g], [b, c, e], [a, d,f], [b, c,g], [d,e,f], [a, c,g]$

Now I have the arrangement with the static subgroup in the centre row. I can deform this to a ring-shaped diagram with every possible octal overlaid upon it; with the centre row's ends brought together as a circle containing the top three rows: then the bottom three rows compass the camp of God's people about (middle circle) as well as the "beloved city" (now the centre three circles of the top three rows). John described it this way no doubt to indicate the cyclic nature of every row (octal) under the one seven cycle (here (a,b,d,c,f,g,e).)

By going "up" I find that the four earthly elements as triples include the static subgroup which corresponds to unity in the centre row. The earthly triples and unity are re-arranged, hence "going up **on** the breadth of the earth" (the "breadth" shown here as in the serpent's extended form above, as left to right). The sets in triples assimilate and surround the identity element – the "camp of God's people" which is the intersection of the three rows of "stars" above.

The Spirit from heaven then holds static the top three rows, the centre fixed and the bottom three static with Frobenius, the seven cycle from heaven (in the Sun octal) now blows sideways left to right and excludes the bottom four rows to then be in the lake of fire and cut off to the "wormwood-wrath" octals. (The bottom four rows are outside of the Sun octal.)

In this manner "fire came down from God out of heaven, and devoured them". I see in the circle formed from drawing together the above into a set of rings, a "whirlwind of fire" rotating anticlockwise with the camp of God's people safe in the eye of the storm. In the burning (of the great city to become the lake of fire) the smoke completely obscures (devours) forever the judged not found in the book of life. From above, God sends fire down as if about an axis coming out of the page. (This, only found in a seven cycle, the source is the Spirit of God.)

Likewise, the Holy Trinity (the camp of God's people) are safe in the centre when algebraic closure (AC) as the bottomless pit is open. As the whole world is still under the devices of the locusts, the scene above is also played out as if John were stating "For any prime power there is a unique finite field of that order and the multiplicative group is cyclic".

The five cycles within which the whole world now dwells require the locusts, and so the tabernacle of God (The Trinity) denies them all election forever. The orbit of the Holy Trinity in AC becomes arrayed as the orbit of the "rider at the head of the army of heaven" (cf. Jer 15:20) rather than the "sea of glass" (an alignment of sets disjoint in their symbols or polynomials except for zero) so that the ungodly with their "doctrines of demons" as in GF(16)'s orbit in AC are consumed, separated, by that same fire from heaven which John states cycles as a whirlwind. GF(8) is certainly a subfield of AC, so likewise by Zorn's lemma I should expect the same fate for "Abaddon".

Jer 15:20 And I will make thee unto this people a fenced brasen wall: and they shall fight against thee, but they shall not prevail against thee: for I *am* with thee to save thee and to deliver thee, saith the LORD.

I find that there is an accumulation of doctrine to be examined. Some maths first though:

Make note of the fact that for any finite field F in algebraic closure (AC) containing GF(8) as a subfield, GF(8) is a unique subfield of F. As only GF(8) in F may be held fixed under three applications of the Frobenius map, only the elements of GF(8) are closed under and may be held static in a three cycle by squaring. The Galois group of order three generated by the Frobenius automorphism applies to cycling static the elements of GF(8) only, as the Galois group of F is itself cyclic and of order equal to the degree of F over GF(2). As GF(8) is a subfield of F of degree three over GF(2), the Galois group of F must contain a C3 subgroup.

Whilst it is true the automorphisms in the Galois group of a field F of degree three over any subfield K not GF(2) form a three-cycle, they are not obtained by squaring but by raising to some higher power of two, equal to the degree of K over GF(2). Under squaring, only the singleton elements within GF(8) are held static in F under a three cycle: and GF(8) has but two static triples. These are the only static triples under the Frobenius map in any choice of F.

If there are two such F, there is always an extension in AC containing both along with their common subfield GF(8). The argument of the rider, the "Word of God" holds firm in the sense that there are only these two static triples (formed of singletons) in AC extended over GF(2).

The army of the rider on the white horse comprises of the orbits of all those that have at some place in their orbit (vesture) a K4 subgroup from the reference octal, the field GF(8). There are such octals for **all** x in F. (Those "white horses" are seen present in the "camp of God's people" within the centre of the "tornado of fire" coming down from God out of heaven.)

In F, a generalised extension of GF(8), the army alone has the static subgroup (and not the static triple to the left hand for a subgroup). The argument as in "The Beast Taken" (section [15.4]) holds for any orbit of any octal L in F containing the static subgroup K in the image of ϕ. Other orbits of octals (not containing any image of K) are also not open to a similar three cycle as the static subgroup is uniquely found in GF(8) and then only in the "army".

The only exclusions that may still operate in the dialectic are those symmetric cases to the left hand, upon the four possibilities for the complement static "bow" (as if they were become a separate case of a different static subgroup), each found within another "rider" (the left hand of God?). The left hand is unique up to the multiplication on F and the choice of $\ker(\phi)$.

Then there is found no combination of unity with a three cycle under Frobenius in any octal in the army or also within the whole set of octals of any extension F of GF(8) (containing the static subgroup or the "static bow"), but in those two riders unique in F: the "Word of God" and God's left hand (as repentant) alone. (There is only one octal in F with both static triples.)

The possibility that there is yet a "left hand of God" operating in the static triple (bow) of the right hand, would show that these two angels as "kings of the north and south" both ensure the dialectic completely fails in its function: as the "left hand of God" is also found "standing in the Sun" (as if the rider also) and he is the last to exit the Church amalgam of Laodicea.

Rev 19:17 And I saw an angel standing in the sun; and he cried with a loud voice, saying to all the fowls that fly in the midst of heaven, Come and gather yourselves together unto the supper of the great God;

The "great tribulation" is effectively overcome the same moment a believer is sealed: an equivalent to the two wings as of a great eagle which are granted the woman for her to "fly into her place" where she is to be nourished "for a time, and times and half a time".

Rev 12:14 And to the woman were given two wings of a great eagle, that she might fly into the wilderness, into her place, where she is nourished for a time, and times, and half a time, from the face of the serpent.

These wings (maps) permit the believer to traverse between left/right hand octals as if they were placed in alignment. This is enabled by the ministry of the two witnesses (applying the least's circuit to every isomorphic field GF(8) in seven symbols as "1260 days" or mapping through all such fields GF(8)). There are none remaining stranded in the "sea" of thirty octals. All may (by God's choice), likewise overcome, and as the very last to do so, the left hand of God also overcomes the world. (There is, in effect, just one rider from one argument: a case made of the very least convicting all. Unity is found the same in both left and right hand octals, GF(2) being a proper subfield of both.)

The left hand has the final judgement to repent (calling time to the works) and follow the right hand of God (the least in the kingdom) out to Philadelphia. If the left hand overcomes the dialectic (as the last of all) then the dialectic fails completely in its operation! The office of the left hand then becomes vacant. (And the vesture of sin, the dialectic, becomes only null.)

Jer 50:45 Therefore hear ye the counsel of the LORD, that he hath taken against Babylon; and his purposes, that he hath purposed against the land of the Chaldeans: Surely the least of the flock shall draw them out: surely he shall make *their* habitation desolate with them.

The Lord's enemies become set in the "earthly" elements as if His "footstool". (God has separated the elect to His right and the remainder to His left.) The dialectic cannot operate in any octal in the rider's army. The beast is "taken" and the false prophet with him.

In combination with the great chain binding Satan for a thousand years within the pit, there is no image of any such K4 subgroup from the reference octal (static or otherwise) then found in any member of the rider's army that would share an orbit with any image of GF(4) – that being any K4 group (or any orbit of a suitable field as in the fifth trumpet) in a five cycle from within GF(16)'s orbit in AC. (A consequence of the argument made in "The Grace of God", section [15.7] earlier.) Why a coset of GF(4)? Christ is sat at the right hand of the Father, cf. Psa 110:1.

Every octal in the "vesture" (orbit) of every member of those "armies that were in heaven" appears as some image of an octal formed upon the reference's static subgroup (by adjoining some **near-arbitrary** element free to be chosen from AC to generate that coset), those K4 subgroups then present in the centre of the "camp of God's people" are all to be found safe in the "eye" of the tornado. Then in every case there is someone safely and surely exercising virtue on their behalf "in heaven" or as if "on the throne". (An intercession of grace unique to the static subgroup – as if one (sat as Christ) had exercised the keys of Hell and Death. One may, under predestination, also intercede on one's own part during that thousand years of God's patience as if an elder, being one of the sealed.)

Then Satan and the scarlet beast are "consumed/devoured" as within the orbits of every octal containing certain other K4 groups: those octals that share no K4 subgroup with the reference at all, not one such K4 as within any image of any octal in orbit. (And all those orbits of length five in AC for certain, and then many others also.) So, Satan is to be loosed from his prison as kept within only that five cycle in AC (only released after that thousand years of God's grace, when the sealed are complete in number) but he still may not "enter heaven" (there being no such product or map for him), itself now become safe in the "eye of the tornado". He is free to perform his final deception on all those not called in Christ, who are then not in that "camp of God's people" (now encamped and as an army **at rest**, which cannot practise the dialectic method as with the three unclean frogs), and are instead as those others that have neither worshipped the image or the beast nor accepted its mark, yet are still not elected and are also (possibly) able to be deceived and to practise the same devil's speech.

So, those not having the mark that are still as yet unsaved are to become forever separated from the camp of God's people by this fire from God, coming "out of heaven". Satan is loosed to deceive all these also. (All sinners will surround the redeemed and be "consumed" by this fire which will permanently exclude them from election.) There is certainly, then, no product in any extension field of GF(8) to be made upon any of those other excluded octals in orbit (i.e. of all those sinners) which likewise map any K4 group within any of those other octals to one of the seven K4 groups of the "Word of God" within the "camp of God's people". (And also, therefore, the "beloved city" of the redeemed and resurrected.)

Finally, because a near-arbitrary element was adjoined to the white horse under every member in the rider's (called "the Word of God") army to generate the cosets in each case (the only exclusions being those elements of the static subgroup themselves), all the other orbits that do not or cannot properly intersect in any K4 subgroup of the reference octal in a non-trivial fashion are excluded and "consumed with fire from heaven" along with all those groups caught in a five-cycle in the images of GF(4). (There is the free choice of paradigms.)

"For any prime power there is a unique finite field of that order, and the multiplicative group is cyclic."

Then there is no argument found in AC or "appeal to Abaddon" as a "higher power" for any in the scarlet beast (or five-cycle) to claim entry into the "camp of God's people" (itself a closed set with conditions). Abaddon is a name meaning "destroyer". In these extension fields towered toward AC, there is no strength (argument) found in further numbers – merely in obedience to God and His Christ instead. Faith in an "Abaddon": an appeal to a "higher power than God" is surely a trap in its own regard, and he is well named for it.

Whilst the enemies of God are collected together in opposition to the rider, there is yet the possibility for (automorphisms forming) three cycles to operate over fields of arbitrary degree: that the dialectic process may operate over collectives of sets and not individuals as singletons (which form cosets in octals).

Rev 20:8 And shall go out to deceive the nations which are in the four quarters of the earth, Gog and Magog, to gather them together to battle: the number of whom *is* as the sand of the sea.

It is a fact that these collectives are arbitrary and "as the sand of the sea". Yet the individual deceived in the army will face the judgement of God for their understanding of the purpose in that dialectic: of their own part and reasons to wage Armageddon. If someone has intercession (as a white horse in GF(8)) and does not practise the dialectic, then their life is hidden in Christ, and they are themselves the target of Satan, hid within the camp of God's people and not separate from "some other people" that they instead may be called to follow orders blindly to attack. God will return His wrath upon the individual, and not the "collective" (as it will always be formed of individuals). God is not at war with pieces of paper either, He will have no time for them. "Foreign policy" will not be taken any notice of when Christ judges mankind.

Any dialectic made over sets must be understood by those within those collectives opposing God's elect in order for them to be judged for their part in them. If they did not understand the dialectic in use, they would not be deceived and thus so gathered. The dragon's lie will in truth be as complete nullity, and will make absolutely no sense whatsoever to anyone saved in the camp of the elect. Satan may as well claim his "idea" has truth as noumena because it is "fish shaped" in order to rally the Church in the scarlet beast to his "war"!

Then as none in the army (or F) but the "Word of God" has a three cycle in his vesture (with a unique name of C7 written upon his vesture and thigh, leading the army which is then also identified uniquely as a consequence) no member of the elect practises the dialectic method as if still deceived by Satan's device of the "three frogs". They, of necessity, do not (or no longer) blaspheme the name of God with its use. (Which is clearly a virtue required of the elect.) Their fine linen, remains "white and clean": they are not so deceived.

Then the army led by the "Word of God" is direct testimony from Jesus (as by the spirit of prophecy) that the gospel is extended freely to many (the number is **arbitrary**, referred to as simply "the breadth of the Earth" above, Rev 20:9), even though not everyone may have every opportunity to practise virtue in every setting. Charity, instead, is revealed as that "most excellent way" that comes from within the individual, and is always the worthiest trait shown by anyone without a commandment. That freely offered grace is extended to all that do not worship the scarlet beast and its image, for those together form the "second death".

So, the only condition for election in this case is to place one's faith on the virtue of Christ's intercession in heaven (a horse clothed in white linen, pure and clean). That also being extended out to the seven subgroups of the one octal corresponding to the Father (of Jesus Christ) each subgroup under orbit, extending grace on the condition that His virtue is to be found practised by His disciples so any other faith as without His works is recognised as "dead".

It then becomes clear that the greatest then acts as the "least" in exercising virtue. (As one moved with compassion yet without unction.) There is no requirement for all to practise virtue continually, only to be found doing so in order to perfect good works to show faith present and active. (The evidence of things not seen – that same true intercession in the heavens.)

So, placing faith in the resurrection and following the scriptural record left of Christ is a given, to then merit the only available intercession: for the God of heaven has signed His own name to the New Testament, and has done so in person to John. God has drawn His dividing line.

15.10) The Damnation Of The Devil

Rev 20:10 And the devil that deceived them was cast into the lake of fire and brimstone, where the beast and the false prophet *are*, and shall be tormented day and night for ever and ever.

Lastly, the dragon as "Satan and his angels" that were princes over those now past OT kingdoms are cast into the lake of fire which is the judgement of the beast and false prophet (scarlet beast and the woman), and they are forced themselves to perform the operations of S6 and A5 normal in S5 numbering 666 forever. Satan is in receipt of every torment, being the only direct set A5 formed of permutations θ of the manifold.

The devil, or Satan, is damned – now become slaved as providing the noumena (the missing twelve elements of S4) to the same system of witchcraft he took up against the Lord (the image numbered 666, which is not a man until Satan's life is "hid" in it). No one else will share that torment whole, but will also be judged as to their part in it. This is the result of Satan's grandest enterprise, and it is his longest awaited and final work. Those with part in the works of Satan have been working towards this goal all along, whether they knew it or not. I hope to remain in absentia, but he has my vote for his retaining permanent office!

As to "tormented day and night for ever and ever", I find Satan's eternal damnation: then cast out of the right and left hands of the reference (day) and of wormwood (night) and caged in every pair of wrath (as day) and wormwood-wrath (as night) octals across the sea. Those then become the equivalent for anyone left operating in the dialectic within the lake of fire.

With the vials of wrath the cosets of the Sun octal were transformed to a "wrath octal" – a modified version of the "wormwood octal" I had found earlier. I posited the idea that it would have been better to modify the Sun octal instead to bind those that practise the dialectic over the static triple. I found the solution to cage the dialectic between two such octals, both a "wrath" and a "wormwood-wrath" octal.

"Sulphur" I likened to the action of the serpents of the sixth trumpet: I link it directly to the action of supplanting the seven cycle with a different octal (with a transposition of elements) underneath it. Sulphur, as in this "lake of fire" (a much smaller sea – perhaps extending over the seven wormwood octals as in the resurrection of the two witnesses) is a set of transpositions that transforms between these two "wrath" octals – now "caged outside of Christ". I can state it to be "the lake of fire and **brimstone**" and also where the beast and false prophet **are**.

That should clear up any ambiguity as to the identity of the "lake of fire" with regards to the woman riding the beast (as the Church's judgement). She is made a "cage of every unclean and hateful bird" (the bottomless pit of locusts and **serpents** then completed) outside of Christ (in the lake) and able to practise the dialectic in full without consequence to the elect. These are abstractions and are not physical creations yet. Advantage may be taken of all the time left to turn to Jesus Christ and to spread the good news.

Note that the static triples in GF(8) also apply over those triples made of K4 groups spanning the octal as in the K4 form of the ultrafilter. As the elements of the octal have equivalents in these K4 groups, to be refused intercession in a three cycle is also to be cast separated into the wrath and wormwood-wrath sets/octals if one is found practising the dialectic in the reference (the Sun) octal at all.

I doubt such a one would notice much difference, as they had not taken up the offer of that intercession of Christ already, and the four earthly triples remain the same between reference and wrath octals. I, however, doubt that mere isomorphism of addition is enough to refresh the spirit for eternal life. "Hell" will not permit virtue in disjunction, only the dialectic operating with virtues held as nullity instead: yet there is to be no law of God against good works in Hell. That stated, the good news is still always just that, and gospel.

15.11) The Great White Throne Judgement

Rev 20:11 And I saw a great white throne, and him that sat on it, from whose face the earth and the heaven fled away; and there was found no place for them.

I find Christ sat (as zero) upon the K4 form of the ultrafilter with the correspondences;

$$0 \equiv [0,a,b,c] \equiv [0,a,b,c,d,e,f,g]$$
$$1 \equiv \quad a \quad \equiv \quad [0,a,b,c]$$
$$b \quad \equiv \quad [0,a,d,e]$$
$$c \quad \equiv \quad [0,a,f,g]$$

Where 0 is the additive identity for the singletons in the group [0,a,b,c] and [0,a,b,c] is the identity for the group [[0,a,b,c],[0,a],[0,b],[0,c]] and the octal the identity for the group of subgroups formed by [[0,a,b,c],[0,a,d,e],[0,a,f,g]].

As the identity element, Christ is "sat upon a great white throne", from whom the Earth or "four beasts" (the elements from the Sun octal {d,e,f,g}) flee away as from the stability of the identity: (b,c) are "flung away" to the elements {d,e,f,g} and this may be likened to an infinite descent or regress. Albeit with the octal underneath them which closes the sum of [0,a,f,g] + [0,a,d,e] = [0,a,b,c].

Likewise I note the "heaven" (positions of unity in the octal, {d,e,f,g} above) fled away from His face, I can equate the "heaven" to the octal of the 24 elders and the four beasts: the octal ultrafilter is interrupted by the correspondence of a ≡ [a,b,c] as static. So, the heavenly scene of the stable octal is also "not found present" – no place is found for them. The correspondence to unity of a ≡ [0,a,b,c] provides the "face" of Him who is sat on the throne: Jesus Christ. (The static K4 subgroup is the "face" of the octal group (enthroned as zero) as under inclusion; likewise, "a=1" is as the "face" of the K4 form.)

Rev 20:12 And I saw the dead, small and great, stand before God; and the books were opened: and another book was opened, which is *the book* of life: and the dead were judged out of those things which were written in the books, according to their works.

Now, God judges without respect of persons (small and great are equal before Him) and solely upon whether a believer will live under predestination as becoming obedient under the grace of His patience or whether he will not. The books or "the record" of their lives reigning with Christ (without testing as under the influence of Satan) is examined: do they show evidence (fruit) of predestination (life) or not?

Rev 20:13 And the sea gave up the dead which were in it; and death and hell delivered up the dead which were in them: and they were judged every man according to their works.
Rev 20:14 And death and hell were cast into the lake of fire. This is the second death.

The sea, the collective of the beast and those that worship the image (death and hell) were raised, and those that practised only the operations of the lake of fire through great tribulation are judged as to their works, $(a \vee b)$ or $(a \vee b)^c$. Being no Christian is no exclusion from resurrection; not being gifted eternal life in Christ does not excuse oneself from judgement, or from being under the Father's authority having rejected His Christ.

"Death and hell" are thrown into the lake of fire. These are the spiritual constructs that complete the cage of the scarlet beast, now that the city S5 is filled and closed completely.

Those constructs of the sea and its wormwood devices and the construction of the woman (Babylon the great) upon the Christian believer are no defence: believers will be expected to have overcome. Those that furthered the construction, "Death and Hell", serpents and locusts (tares) are damned (rejected). They show no repentance as Christians as they were never chosen and predestined. Now, after their judgement they really are in receipt of a "second death" as in the verse above. This cannot be repented of: no, not ever.

Lastly, those that never had a portion of redemption in Christ are likewise excluded from paradise with God as in the Sun octal. Not for torment, but to make religion damn hard work.

Rev 20:15 And whosoever was not found written in the book of life was cast into the lake of fire.

There is one other verse I include that speaks of total destruction in hell.

Mat 10:28 And fear not them which kill the body, but are not able to kill the soul: but rather fear him which is able to destroy both soul and body in hell.

Destruction lasts an eternity, most Christians will state: yet without reason believers have used the word "torment" out of its own context. Destruction is both in application of its meaning "over in an instant" and yet apparently takes forever to happen to every individual. Does this euthanasia consist of nothing but torture? God forbid! (As if God should stand over all such for eternity to take pleasure in such a thing – an utterly rancid belief.)

Luk 16:26 And beside all this, between us and you there is a great gulf fixed: so that they which would pass from hence to you cannot; neither can they pass to us, that *would come* from thence.

The "great void" present in the parable of Lazarus and the rich man is this eternity: for the former world is to "pass away" so that a great gulf separates the saved from the lake of fire. That great void is properly (local) time. The former world will experience an eternity that lasts a few moments in the new heaven and new Earth to come. It will be impossible to go back to there from thence and to the New Jerusalem from this world. To all intents and purposes the old universe is completely destroyed with respect to the new. What other answer could there be? The greatest fear to have of God is to be excluded, not tormented with pain but simply by His rejection.

The construction of the wrath octals and the five cycle of locusts ensure there is a destination waiting for those that are refused by God. It will remain forever: totally separated from Him.

Chapter Sixteen: A New Heaven And A New Earth

God seems unconcerned with redeeming the fallen creation beyond that of the human soul. Instead He states He creates a "new heaven and a new Earth". The old ones "pass away". Maybe they continue to exist in some back corner of everywhere with the damned within, beyond hope of any redemption, last chances spent.

Now, the universe is a pretty big place to spend so easily as a gift to those that were your enemies: that is, unless you happen to be capable of creating it all in a week. God had fully created the heavens and the Earth on the first day but had not created it in the manner where it could rest contingent upon itself – existent on its own foundation rather than that of His own power. It was "without form and void". That God rested on the seventh day does not state that He was tired, merely that He ceased from work.

That new creation I ought to expect, some 6000 years later as it is reckoned (despite the also "separately" created fossil record that accounts for all life but that human life created "from the dust of the Earth" – with no fossil record) may be planned to be a little on the larger side. Gone is the requirement for Jesus' heavenly colouring-in book and onward comes the research of His well-planned thesis!

The creation as the justification for all further creation that God would make including paradise itself (and everything certainly less evil than this one paradise he began with) is this single Earth that passes away: it justifies all after. It was made "In the beginning..." as the Bible tells us. It is the first because it is just the worst side of wear and tear. The creation itself is corrupted and is irredeemable, but souls are redeemable. Grace can make a believer perfect in God's own experience and none should doubt. Then only life will survive this creation to join God in His "new-earthly" tabernacle. This heaven and Earth will simply pass away spent and effectively past its use-by-date.

With the end of corrupt behaviour will come much healing: the grace of God will extend into the new creation yet I can expect there to be a far better solution than the one offered to us presently, not because grace is at all deficient but because believers see through a "glass darkly" at present, and the hope is for better.

I will know as I am fully known and as one of the faithful I must put on the mantle of incorruption; the mortal must put on immortality.

Those left behind in the old Earth are to be separated from us as by a "great void" as in the parable of Lazarus and the rich man. Jesus can be expected to not have deceived us and He has given us the worst and final state of a person in Hell: I could suppose it to be at the height of his discomfort in eternity within that old system of things (having much discomfort and a total lack of refreshment refused him forever). Jesus would not exaggerate or underestimate something so serious.

There is no lying deceit in Christ or in His parables. I expect that many people already have been, or will be, more the worse for wear from partying hard (and many times over and over) before they survive even an infinitesimal amount of that eternity and before they reach that most wretched state. They will continue to feel aged and worn down without God, not to mention rejected. However, eternity appears to need a prescription of salvation from its intelligent designer: I wouldn't doubt it. Eventually that feeling of damnation will arrive – the knowledge of it will be far swifter and harsher in its arrival with the judgement.

A total lack of divine charity no-one has yet had any experience of: excepting maybe Satan, even though God provides Satan the ends to his means. It may truly be worse than any other factor of Hell, but if the scriptures are believed – what further benefit would eternal torture make? Those with no part in the first resurrection are already as damned as they will ever get:

Rom 1:18 For the wrath of God is revealed from heaven against all ungodliness and unrighteousness of men, who hold the truth in unrighteousness;

That they are given over to a reprobate mind and eternally excluded from God's presence is the entire length of the scripture on the matter.

16.1) Heaven And Earth Pass Away

Now only the elect remain in view I read they are translated to a new heaven and a new Earth. Those in the lake of fire remain in the first heaven and first Earth; to dwell in the "wrath octals" whereas the elect are in a single new "Sun octal" with no more sea: there, all are living under the first seal paradigm without dialectic devices; there is no sense of a "moon" or "stars".

Rev 21:1 And I saw a new heaven and a new earth: for the first heaven and the first earth were passed away; and there was no more sea.

John sees the dwelling place of the people of God in this new Earth (coming from heaven). Not to be confused with the saints' inheritance, the people are the inheritance of Christ Himself.

Rev 21:2 And I John saw the holy city, new Jerusalem, coming down from God out of heaven, prepared as a bride adorned for her husband.

There is a resounding voice (judgement) that the Lord will never be absent from them. Their place is found in Christ, raised incorruptible as in their obedience within the 1000 year patience of God. They are assured eternal life (sealed) and have no need to be tested further.

Rev 21:3 And I heard a great voice out of heaven saying, Behold, the tabernacle of God *is* with men, and he will dwell with them, and they shall be his people, and God himself shall be with them, *and be* their God.

The elect are "as one" in Christ as Christ is one in the Father – by now it should be apparent that election in Christ is under the unity element common to the Trinity – the intersection of GF(4) with the Father GF(8) and it is sufficient to place the virtues in Christ before the Father as in the place of one of the elect.

Rev 21:4 And God shall wipe away all tears from their eyes; and there shall be no more death, neither sorrow, nor crying, neither shall there be any more pain: for the former things are passed away.

God is a God of all comfort and memories of loss will be comforted with tenderness and there will be no more sufferings for the Lord's people sent by God: they are worthy.

That stated, with all the unsaved in eternal torment, there is no wiping of tears away. Tears will cease when made for those that are simply excluded, not tortured eternally.

16.2) Everything Made New

Rev 21:5 And he that sat upon the throne said, Behold, I make all things new. And he said unto me, Write: for these words are true and faithful.

Jesus, upon the throne (as zero), makes everything new – God, not the throne, is creator: "All things" would also include the Earth itself (presumably), and I may expect those raised to life outside of Christ to inherit a "fresh Earth", replenished in its entirety to its original state. ("1Co 15:26 The last enemy that shall be destroyed is death.") This "thought experiment" as to whether evil has a proper place in eternity (defeated at the beginning) is to be completed. Whilst a person would use a pad and pencil to solve a problem, such a fleetingly passing proof against all evil (as by inspection) is to God as simple as actualising it in the work of all creation – and His proof is only complete when all is realised and completed. He is far above us.

Rev 21:6 And he said unto me, It is done. I am Alpha and Omega, the beginning and the end. I will give unto him that is athirst of the fountain of the water of life freely.

Christ is the first and the last – He is the one for whose pleasure the creation was made. It was His purposed work in the beginning, and this completed work is for Him now. Paradise for Christ is paradise enough for all who follow Him.

Christ will give all to drink of the water of life (the Holy Spirit) that transforms the floating unity as rest throughout the positions of the octal. In fact, by transforming also the static subgroup – self-similar to the groups of subgroups all containing unity, the seven cycle as a fountain (a water source in the Earth) i.e. unity – the life or rest in Christ is given freely to any. (It is now freely given as it is **impossible** under the restraint of the paradigm to number 666. God need not hold it (restraint) back from anyone. All were so nourished in their "Great Tribulation".)

16.3) Their Eternal Reward

Rev 21:7 He that overcometh shall inherit all things; and I will be his God, and he shall be my son.
Rev 21:8 But the fearful, and unbelieving, and the abominable, and murderers, and whoremongers, and sorcerers, and idolaters, and all liars, shall have their part in the lake which burneth with fire and brimstone: which is the second death.

The double-edged sword from Christ's mouth (the judgement) is a promise that the elect that overcome will surely live with Him: even to become as His right hand as the redeemed. Those that are judged (not necessarily those that do not overcome) are given their eternal rewards.

There are eight such types of sinners listed – I may state these as elements; rather, I may align these to the eight elements of the "wrath octal":

$$[a, c, e, g] + a = [0, b, d, f] \equiv E$$
$$[b, f, a, e] + a = [0, c, d, g] \equiv G$$
$$[d, g, b, a] + a = [0, c, e, f] \equiv F$$
$$[c, e, d, b] + a = [b, d, e, c] \equiv C$$
$$[f, a, c, d] + a = [0, g, b, e] \equiv D$$
$$[g, b, f, c] + a = [f, c, g, b] \equiv B$$
$$[e, d, g, f] + a = [d, e, f, g] \equiv A$$

That has as the identity element the empty set { } – as "all liars". Likewise, this is the reward of those that were judged not to be alive (predestined effectively) in Christ.

The lake of fire and brimstone is the "second death" never to be repented of (after the judgement): those condemned within are utterly rejected by God. Their prayers and works will stink of sulphur to God, far beyond them ever being offered acceptably with any incense (as before the angel had once and lastly offered the prayers of the faithful with incense by which they were cast to the Earth as answered with judgement for their distaste) – the smoke of her burning will obscure God's Spirit from them always – they are excluded and so judged – and that, to all in the Lord's salvation, is become the bottom line.

16.4) The Bride Of Christ

John is redirected from the lake of fire (and the element or angel "a" that generates the "wrath octal") to the scene unfolding of the inheritance of the saints.

Rev 21:9 And there came unto me one of the seven angels which had the seven vials full of the seven last plagues, and talked with me, saying, Come hither, I will shew thee the bride, the Lamb's wife.

John, like those that are to be dwelling with God forever, is carried a great distance away from the first heaven and Earth which has passed away. (A great void separates them.)

Luk 16:26 And beside all this, between us and you there is a great gulf fixed: so that they which would pass from hence to you cannot; neither can they pass to us, that *would come from thence.*

Likewise, He is moved to observe in a "distant" place (a great mountain) where heaven and Earth meet, but which also is a "stronghold" of faith; this time correct faith. This is the stronghold of the "church in Philadelphia", Mount Zion. John is shown the inheritance of Christ. The circuit of the angel to John appeared to "burn up" and in the angel's circuit Philadelphia corresponded to "x" and Laodicea to unity, the seven cycle itself applied as x^{-1}, its inverse "x" in Philadelphia appears to "descend out of heaven" (from unity, as a reciprocal).

Rev 21:10 And he carried me away in the spirit to a great and high mountain, and shewed me that great city, the holy Jerusalem, descending out of heaven from God,

John's description is of great richness – but is in allegory to Israel's twelve sons and the twelve disciples, rather than in monetary wealth. The scenes are of "purity" of the wealth (riches)

that are the blessings of God.

Rev 21:11 Having the glory of God: and her light *was* like unto a stone most precious, even like a jasper stone, clear as crystal;

Without the sea of glass in view, the virtue (light) in the city is very pure: as befits a perfectly cut stone. (The virtues in Christ are not subject to change, but are freedom.) The clear crystal, once again, is the reference to a total lack of cloudiness (a mystery is a "cloud" in Revelation). The truth is plain here and not meant to be hid. (Her virtue is the righteousness of the saints.)

Rev 21:12 And had a wall great and high, *and* had twelve gates, and at the gates twelve angels, and names written thereon, which are *the names* of the twelve tribes of the children of Israel:
Rev 21:13 On the east three gates; on the north three gates; on the south three gates; and on the west three gates.
Rev 21:14 And the wall of the city had twelve foundations, and in them the names of the twelve apostles of the Lamb.

The wall "great and high" shows the unassailable nation of God. There will be no undoing of this salvation – redemption in Christ is fact and rock solid. The twelve gates representing the twelve tribes show that only as an elect in Spiritual Israel may one enter into the city. (For there is only the God of Abraham, Isaac and Jacob and He has sent His Christ who Himself taught "Joh 14:6 Jesus saith unto him, I am the way, the truth, and the life: no man cometh unto the Father, but by me.")

Likewise, through the continuing witness left us by the twelve apostles, the one gospel (the foundation of Christ other than which no one can lay) with which the twelve laid the first "works of the city" are preserved until the end: God preserves His words – *they will never pass away.*

The twelve gates, twelve angels and twelve foundations correspond to the three choices of K4 groups in the K4 form as if floating unity (as under Frobenius on the octal inducing GF(4)*). Unity is shared between Father and Son in the intersection of the fields GF(8) and GF(4).

Each wall is as a "choice of unity" in the octal and each of the three gates a subgroup exercising unity (virtue). The "angels" are properly identified as "ministers" and those with such virtue in Christ (interceded for by Him) may freely enter. These twelve groups do not include the static subgroup, but are "one in Christ" as the twelve tribes were under the law, "one in Him". Upon the foundation of Christ (on which everyone must build), the city is built firstly by the twelve apostles, by their ministrations to the elect of (the remnant of the twelve tribes of) Israel.

The octal itself has four possible correspondences to unity from static subgroups in the octal. These combined with the above are equivalent to the three automorphisms of each octal with its four different static subgroups for a fixed unity respectively. It makes no difference to cloud the matter of the eight seven cycles; there is but one "sea of glass" – now become no more sea. This "New Jerusalem" is clear as crystal. (There is no mystery or "cloud".)

For the K4 form of:

$$a \equiv 1 \equiv [0, a, b, c]$$
$$b \equiv [0, a, d, e]$$
$$c \equiv [0, a, f, g]$$

may also be made by floating unity in GF(4) or by automorphism in the octal to become:

$$a \equiv 1 \equiv [0, a, d, e]$$
$$d \equiv [0, a, f, g]$$
$$e \equiv [0, a, b, c]$$

Alternatively the static subgroup in the octal may shift from a ≡ [b,d,f] here to a ≡ [c,d,g] say. So the K4 group would have become under a new squaring of the cycle (a,c,d,b,g,f,e);

$$a \equiv 1 \equiv [0, a, d, e]$$
$$d \equiv [0, a, g, f]$$
$$e \equiv [0, a, c, b]$$

Which, as under Frobenius, is also the K4 form and is once again equal to the form I began with, as was desired. (The K4 form is invariant over the differing logical schema.)

16.5) The Extent Of The Bride

Then John explains the New Jerusalem in terms of the K4 form of the ultrafilter that spans the octal, filling in all the details.

Rev 21:15 And he that talked with me had a golden reed to measure the city, and the gates thereof, and the wall thereof.
Rev 21:16 And the city lieth foursquare, and the length is as large as the breadth: and he measured the city with the reed, twelve thousand furlongs. The length and the breadth and the height of it are equal.
Rev 21:17 And he measured the wall thereof, an hundred *and* forty *and* four cubits, *according to* the measure of a man, that is, of the angel.

An angel with a measuring rod, made of gold (a king's **standard** measure as GF(4)), shows us that the inheritance (in Christ, gold represents a Field structure) is symmetric: the faithful are to forever dwell as in heavenly places but in a new Earth and with their God. What was once the Lord's footstool is now passed away, the new Earth is indwelt by God.

In its three dimensions the measurement of the city is equal. I note that the twelve forms of the K4 ultrafilter could be obtained in three ways: namely fixing the unity in the octal and by the multiplication of the GF(4) ultrafilter (as Christ three times) induce Frobenius on the four possible static subgroups in the octal (4 times), and similarly another twelve reciprocating with GF(8) under automorphism (three times) (The Father), again with a shift of static subgroup (four times) in GF(8) by choosing the seven cycle(s) to hold them static. (The Holy Spirit).

There is then a third dimension: I may for any choice of the K4 form spanning the octal (i.e. of identity, Christ) choose any static subgroup of four (as the Father) and by multiplication in the GF(4) form of the ultrafilter in agreement with the Frobenius map acting on the Father (static subgroup) or vice-versa then fix and induce three powers of one pair of seven cycles (as the Holy Spirit). The Holy Spirit is then effectively "fixed" without need for a second cycle.

In respect these three transforms are by "Son and Spirit", "Father and Spirit" and lastly "Father and Son". Importantly, the Trinity in pairs may operate with the other "at rest". (Each is set at liberty.) Then in a real sense they form a K4 group upon their triune nature as if they were together a "Manifold of Virtue" (see section [3.10] for a more complete treatment). In order to return to "rest" and agreement, the action of any two must account for the liberty of the remaining one. (The argument for this? It is even the whole "Revelation of Jesus Christ".)

So, I find I have a dimension of "12" with respect to each member of the Trinity. As the effect of these three is equal, I found in some sense they meet in identity. They dwell bodily in Christ "symmetric". They meet in so similar a way, they have the "same character".

There is a factor of 1000 – indicating that the city is built on the grace offered to those within. "Furlong" was used as a "stadion" or the "area of a church's footprint" previously. Here, the same word or "footprint" of the city is 12,000 "stadion" and the height is equal to the length and breadth. (The twelve tribes are offered equal grace by all **three** members of the Trinity.)

The volume of a square based pyramid bounded by a cube is 1/3 the volume of the enclosing cube. So, 12,000*12,000*12,000/3 = 576,000,000,000. This number doesn't seem a significant amount (I count it far more accurate in talents of gold than a mere thirty pieces of silver, however!): I would expect (per each member of the sealed) believers to each receive 1,000 square furlongs of land, or inheritance (that portion of the city's base, ignoring the height), a good indicator of eternal grace – there is no demand for instant obedience in totality, but grace continues. (Note that one doesn't measure area with a measuring "reed". Though it was the case earlier that "stadion" applied to "the space of 1600 stadion", an indication of area.)

The wall measured 144 cubits refers to the height of the "defences" of the city ("to the measure of a man", which indicates this is not only a measurement of divine symmetry, but also of floating unity now all saved believers freely dwell in "the Sun", i.e. there is a "sea of glass"); that states simply that the elect (the sealed, rather than only elders) will not be kept out. They will be safe within. I could consider that the city is protected by the grace shown its least. As the cubit is measured to that of the angel, the city is made for man, not man for it.

Jer 15:20 And I will make thee unto this people a fenced brasen wall: and they shall fight against thee, but they shall not prevail against thee: for I *am* with thee to save thee and to deliver thee, saith the LORD.

Moving on;

Rev 21:18 And the building of the wall of it was *of* jasper: and the city *was* pure gold, like unto clear glass.
Rev 21:19 And the foundations of the wall of the city *were* garnished with all manner of precious stones. The first foundation *was* jasper; the second, sapphire; the third, a chalcedony; the fourth, an emerald;
Rev 21:20 The fifth, sardonyx; the sixth, sardius; the seventh, chrysolite; the eighth, beryl; the ninth, a topaz; the tenth, a chrysoprasus; the eleventh, a jacinth; the twelfth, an

amethyst.

The "building of the wall" was as a blue stone, also likened to the appearance of Him that sat on the throne. That is, the promise made to God's people (of dwelling within) was totally "clear cut" (Cf. Rev 21:11) – the city itself described as pure gold, very much refined, even to be as clear as glass. That said, the dwelling is not without substance (see through or transparent) but is a guaranteed outcome of God's "clear cut" promises, of great worth.

The city of God is far wealthier than the woman decked with wealth, the city S5 of the Earth.

The twelve stones are an allegory to the "Ephod" – a breast-piece of twelve stones that represent the inheritance (lands) of the tribes of Israel united in the faith of God. It is a strictly priestly garment to be worn only in proper ministration. It was used sometimes as an oracle of truth in the Old Testament, having the "Urim" and "Thummim" related and possibly tied to it. (They are the "perfections" and "oracular light".) These two articles of priestly faith are as close to the guidance of the Holy Spirit as was available in the Old Testament to the priests of God. (To discern what was more perfect and what was the means to it.)

So, within the city are priests to minister to God forever, and not to be ignorant of God's will.

16.6) The City Is The Image Of Christ

Rev 21:21 And the twelve gates were twelve pearls; every several gate was of one pearl: and the street of the city was pure gold, as it were transparent glass.

As to why the city is pure gold (pure as in ultimately refined under great tribulation) I previously discerned the term "gold" to refer to a finite field's structure – one of the Trinity. The street which is gold shows that the city "walks in God". Whereas the great city of Babylon (figuratively Sodom, Egypt) walked in the way of their own hearts knowing no freedom. That gold, as transparent glass indicates no mystery of God for those within, they all know Him.

The twelve "pearls" are "cloudy gems" rescued from the sea. Then they, in regression, correspond to the three elements of the static subgroup and the four choices of unity. Christ is effectively sat in judgement over each gate as to whom may enter the city, as to letting the truly virtuous enter. Each "several gate" (three gates on each wall) then correspond to the elements of one static subgroup (as of one pearl) which corresponds to Christ.

The entrances are themselves precious – nothing that cheapens them will enter in, nothing unclean or **hidden** will travel on its streets. The street that ran through the great city of Babylon was acting as two systems of 666 in "bows" to the left and right of the two octals under one particular seven cycle. That had the result that in the centre the sword that cuts off to the left and right preserved the K4 ultrafilter.

Dan 11:45 And he shall plant the tabernacles of his palace between the seas in the glorious holy mountain; yet he shall come to his end, and none shall help him.

This street, however, is somewhat dissimilar: to left and right there is now just the Father (pure gold), and believers are predestined to remain at "rest" (as in unity). This leads us nicely to:

Rev 21:22 And I saw no temple therein: for the Lord God Almighty and the Lamb are the temple of it.
Rev 21:23 And the city had no need of the sun, neither of the moon, to shine in it: for the glory of God did lighten it, and the Lamb *is* the light thereof.

Such tabernacles as genuinely "planted between the seas in the glorious Holy Mountain" are free to float with unity. The temple is missing because the throne of God is not separate in heaven from His people on Earth: He is instead walking amongst them.

There is no need to reference elements of the sea, no need to distinguish one reference octal from another. There is no need for alternate "left hands" with their additive structure in opposition. Instead all octals are aligned in one single form of the Trinity in the Godhead. (All octals are isomorphic but are now restricted to automorphisms!) There is no longer any presence of the moon – the "sea" and the "stars of night" are unreachable without the loosing of divine restraint: which requires the "restrainer" to be "taken out of the way".

16.7) Only The Righteous May Enter

Rev 21:24 And the nations of them which are saved shall walk in the light of it: and the kings of the earth do bring their glory and honour into it.
Rev 21:25 And the gates of it shall not be shut at all by day: for there shall be no night there.
Rev 21:26 And they shall bring the glory and honour of the nations into it.
Rev 21:27 And there shall in no wise enter into it any thing that defileth, neither *whatsoever* worketh abomination, or *maketh* a lie: but they which are written in the Lamb's book of life.

The elect shall walk in the virtue (light) of the city – the same virtue that is found in Christ. (I have found that by my construction.) Those saved in the eight C7 groups (nations) walk in a pre-first seal paradigm of virtue: "kings" no longer rule over other men in judging, but they themselves also give their own obedience to God in order to dwell within the city. Kings assembled in dialogue (treaty) previously formed wormwood devices as in the sea. Here, the kings instead show restraint, giving God glory and honour as if overcoming the second and third seal paradigms as when they would as kings "meet together".

The "kings of the Earth", the possible choices for unity – here the names of the second and third beasts (glory and honour) before the throne from earlier are mentioned: within the city are the possibilities for the people to continue using the dialectic, but now they are subject to divine restraint. There is no contradiction as there was before – that of "Death".

Those kings that show restraint are the proof (in singletons) to be generalised in the case of every seven cycle (the nations). Every broad paradigm may be overcome before judgement cuts off those left within the third woe.

Restraint is not imposed rigidly in full: rather grace is ongoing; these devices will be brought under grace. In like manner, those kings give God all glory and honour; I assume no one rules over another. Unity is free to float in a first seal paradigm, there is no division or compromise in this city for the second and third seal paradigms; that is, if God is given all sovereignty.

The gates are not shut – this does not indicate anyone may enter, but that the city will never be attacked (for it needs no defending at night). That states divine grace in the city will never be subject to those devices, but they (as devices) will be subject to God in the city.

There is only day and no night: the sea of glass aligns to one octal in the "Sun octal" only and there is no longer any "sea". There is no dialectic device formed in "night", only the pairings of right and left hands as found in the reference.

Those "kings" (triples) will bring the sin of all dwelling in the city under obedience to God. Once the "proof" is done, the people within are free. (Once the singletons obey, so therefore should any branching in triples also.)

Of course, this is qualified as true only for all those that are written in the lamb's book of life: those not so may never enter in (they are represented in the triples in the systems of 666 to either side of the octal as those that "defile", "worketh abomination" and "maketh a lie"). So, those that bring in corruption (bear the mark, for example) that then exclude the collective from obedience to God (worketh abomination) as if having the number of the beast (practising the image) are finally those that "maketh a lie" (have the name of the beast, a covenant with a government in place of that covenant with God: God does not "share" His people).

16.8) The Tree Of Life

John accounts for the other members of the Trinity within which the elect have their reward.

Rev 22:1 And he shewed me a pure river of water of life, clear as crystal, proceeding out of the throne of God and of the Lamb.

The "River of water of life" is clearly the seven cycle of multiplication. The throne refers to either the octal (as God) as identity for addition of subgroups (in the K4 form) or as the K4 set of elements (as the throne of the lamb) over the singletons of the K4 form. The action of the Holy Spirit flowing out of the throne and "through" the city is such as to generate the whole GF(8) ultrafilter from the GF(4) form of the filter. The reference to "clear as crystal" is a sharp contrast to "cloudiness". (Mysteries in Revelation are referred to as "clouds". There is no mystery here to John, the truth is plain.)

Rev 22:2 In the midst of the street of it, and on either side of the river, was there the tree of life, which bare twelve manner of fruits, and yielded her fruit every month: and the leaves of the tree were for the healing of the nations.

The midst of the street is always the unity element, and to the left and right (as octals) there are the six non-identity (not unity) elements in either of the two complement octals: the "twelve fruits" as it were. The tree yields "fruit" from each element of the seven cycle (month) that generates the octal from the K4 group of the city. The "fruit" is the set of all positive properties in the octal forming K4 groups transformed from the GF(4) filter as Christ under that river or seven cycle. (The Trinity is dynamic.) The "fruits" then become the intersections of the three subgroups that are formed as a result: a singleton. (In each case, singletons or triples, they are paired in the left and right as were the two witnesses.)

The "leaves" – the singleton elements of the tree are for "the healing of nations" (rather than branches; "the branch" is the orbit of the unity element – not the "leaves" of the static triples in orbit). They may, by multiplication, send any element (or K4 group) of the seven cycle onto the rest or identity element (or static subgroup). By the cancellation laws in a group, these twelve "fruits" do just that for their inverses.

The nations are once again as all those seven churches, mapped onto unity (rest) by elements of the "tree of life".

16.9) No More Curse

Rev 22:3 And there shall be no more curse: but the throne of God and of the Lamb shall be in it; and his servants shall serve him:

There is the fact that there will be no more curse – God will not cut off those that He has chosen (under His election) from His rest. It is not now possible to become numbered "as 666". The paradigm is again under restraint and what was once true is true once more;

1Co 15:53 For this corruptible must put on incorruption, and this mortal *must* put on immortality.

This is done with the gifts of God in the river or tree of life.

What of the static subgroups that are not part of the "fruits" but correspond to unity in GF(8)*?

These static elements only map to themselves with the induced multiplication in the octal that agrees with the life in Christ as GF(4)*.

They cannot map another element or subgroup to the "rest element" themselves: i.e. for those requiring "healing" by the Spirit to return to "rest". It is only by the other six elements of C7 in product with the subgroups of the octal that the "nations" are healed, and not by God requiring them to be instantly perfect. God is in view as that unity, a reference unity with the twenty-four elders; God is where the "healed" are made to dwell (as unity). God is the source of the river of life, and of all the octal's elements: its rest and static subgroups themselves are part of the ultrafilter and are "part of God" as much as they are for healing or rest.

There is the now clear distinction between the elect that require healing as one of 144,000 (and not one of the 24 elders about the throne, cf. 6*24 = 144 in the seven cycle and with its rest element) who know they are not perfect and God, of whom all know as forever perfect.

Rev 22:4 And they shall see his face; and his name *shall be* in their foreheads.

His face, once more, is the identity element in the K4 form;

$$
\begin{aligned}
0 &\equiv [0,a,b,c] \equiv [0,a,b,c,d,e,f,g] \\
1 &\equiv \quad a \quad \equiv \quad [0,a,b,c] \\
  &\quad\quad\; b \quad \equiv \quad [0,a,d,e] \\
  &\quad\quad\; c \quad \equiv \quad [0,a,f,g]
\end{aligned}
$$

From which the four earthly elements {d,e,f,g} flee away: as away from the stability or place given to the unity element. Through the healing of the fruits (leaves) of the tree upon the river of life (as mapped to the static group of God's "face"), whether K4 group (spanning the whole octal) or the whole octal itself: His face shall be clearly seen, gifting as if it were the complete set of perfections (virtues) towards His people.

Rev 22:5 And there shall be no night there; and they need no candle, neither light of the sun; for the Lord God giveth them light: and they shall reign for ever and ever.

There is no darkness "outside of the ultrafilter", that being lack of virtue: darkness worthy of "Death" from the construction of the device always outside the reference octal (as numbering 666), neither is there need for the moon to bring them back into the fold from outside the Sun octal: neither is there need for a principal octal or "Sun" octal as over any other – the Lord is no longer sat at the right hand to subdue His enemies but rests forever as the lamb slain before the throne of God, to perfect the limited virtues that each believer may be capable of.

Here with "God omnipotent" Christ indeed has a "new name", having now inherited every property deserving glory, "deserved" because He does span the GF(8) ultrafilter as He rightfully should, being Himself God.

As concerns the static subgroup:

Psa 110:1 A Psalm of David. The LORD said unto my Lord, Sit thou at my right hand, until I make thine enemies thy footstool.

The Earth was already made as His footstool (the lake of fire is pushed into the left handed octal, with its triples always in the "earthly elements") with the defeat of His enemies (completed with the treading of the winepress). There is no further need for refining the nation of His Israel. The elect are complete.

Just to make clear, every left hand octal is in alignment with its own right hand and vice versa. Only the Sun or reference octal remains, its left hand partner and all other octals are aligned to it: so, then, all others are scorched with great heat and suffer the great plague of hail, the result of each seven cycle in action on both the left and right hand octals.

Rev 22:6 And he said unto me, These sayings are faithful and true: and the Lord God of the holy prophets sent his angel to shew unto his servants the things which must shortly be done.

The angel was sent from the perspective of God, being outside of time and that which I would call "the future": the text states that the angel was sent to John with the mystery of God completed and those things which are "yet to be done" from the angel's perspective will shortly be done thereafter.

These things are directly separated from (the statement that caused John's earlier collapse to worship the angel) the verse in the passage:

Rev 19:9 And he saith unto me, Write, Blessed are they which are called unto the marriage

supper of the Lamb. And he saith unto me, These are the true sayings of God.
Rev 19:10 And I fell at his feet to worship him. And he said unto me, See *thou do it* not: I am thy fellowservant, and of thy brethren that have the testimony of Jesus: worship God: for the testimony of Jesus is the spirit of prophecy.

The difference between those statements that are true in Rev 19:9 and later the statements above leading up to Rev 22:6 is worthy of note. The verses in Rev 19 are true: those in Rev 22 are yet to be fulfilled. The former are made sure by Christ already: the latter are yet to be fulfilled by God in His new name also (they continue on from the old).

In any case Jesus states He arrives quickly – with final judgement and without a "millennial reign" of 1000 years going before and immediately following the ripening harvest of the Earth. The judgement arrives when the Church is naked of correct faith. There is no "three and a half years" or "seven year tribulation period". It simply happens swiftly. To understand this is great comfort – there will be more stringent election under greater tribulation which is equal selection under a looser paradigm.

The wrath does not end in burning fire and red pyjamas with pitchforks – but ends with tears wiped away and great comfort for those in Christ, and the freedom of the current universe given to those that reject Him. There will truly be no more curse. (For tribulation or "selection" has done its work; it will be said of redemption merely that the elect are saved, the damned cannot die more (or more than spiritually) as it is to be forever.) All will live, but only the elect will be able to rest from their labours.

Rev 22:7 Behold, I come quickly: blessed *is* he that keepeth the sayings of the prophecy of this book.

16.10) The Word Preserved Intact

John feels overcome by the gravity of these immense statements: the divine promise that the angel is empowered to utter far exceeds John's expectation of a servant's authority. John has a slight moment of weakness mistaking the angel for the divine. The angel kindly corrects him. The angel states he is alike to one with understanding of these scriptures (here and then given in the Revelation to John) and that no one is as worthy of worship by the elect as is God.

Rev 22:8 And I John saw these things, and heard *them*. And when I had heard and seen, I fell down to worship before the feet of the angel which shewed me these things.
Rev 22:9 Then saith he unto me, See *thou do it* not: for I am thy fellowservant, and of thy brethren the prophets, and of them which keep the sayings of this book: worship God.

The angel instructs John not to reword the sayings in the book by placing the seal of his own writing upon it: the words are correct already, they need no re-examination. The time for fulfilment is now – as the book immediately has value.

Rev 22:10 And he saith unto me, Seal not the sayings of the prophecy of this book: for the time is at hand.

The time of ministerial "night" arrives;

Joh 9:4 I must work the works of him that sent me, while it is day: the night cometh, when no man can work.

Such a time arrives and the only sensible thing to do is rest on one's own God-given liberty.

Rev 22:11 He that is unjust, let him be unjust still: and he which is filthy, let him be filthy still: and he that is righteous, let him be righteous still: and he that is holy, let him be holy still.

Jesus assures John through the angel that these things are sure, they are coming, and they are to immediately precede judgement.

Rev 22:12 And, behold, I come quickly; and my reward *is* with me, to give every man according as his work shall be.

Jesus Himself re-iterates the promise of all those things that are yet to be fulfilled. God is trustworthy for the fulfilment of His promises, identifying Himself as God in all truth.

Rev 22:13 I am Alpha and Omega, the beginning and the end, the first and the last.
Rev 22:14 Blessed *are* they that do his commandments, that they may have right to the tree of life, and may enter in through the gates into the city.
Rev 22:15 For without *are* dogs, and sorcerers, and whoremongers, and murderers, and idolaters, and whosoever loveth and maketh a lie.
Rev 22:16 I Jesus have sent mine angel to testify unto you these things in the churches. I am the root and the offspring of David, *and* the bright and morning star.

Again, those that are obedient (and only they: those with correct faith) may enter in: outside are the ungodly in the lake of fire (of five sets (as nothing but the senses) along with a lie as "in the mind"). Those things that are as "dogs" (needing restraint – "dumb dogs" false teachers), "sorcerers" (supplying false rest or healing not by drugs but by other doctrine (religion)), "whoremongers" (those that intermingle the faithful with tares, as the doctrine of Balaam), "murderers" (those that make others subject to the dialectic – being as the false prophet), "idolaters" (those Nicolaitans that return back to the same doctrines that they cannot palate as like the proverbial dog, and will not overcome Laodicea but are under the locusts).

Lastly those that "loveth and maketh a lie" are cast out (as the source of leaven: now providing false purpose to the synthetic life in the image). Their "lies" sent the "Church" into a doctrinal freefall, but those lies for the rest of the damned are merely the "imaginations of their hearts".

These five align to the false doctrine in the letters to the first five "refining" churches: the sixth to the corrupting (external) source of leaven which is cultic and essentially satanic in origin.

The book closes with a promise that Jesus is to return, but not to unite His chosen people (His elect bride) with the Spirit (the union that is the first resurrection) – but instead to freely offer to all those that are not saved an invitation from His Spirit and elect that they are very much welcome to be also chosen. Those that appreciate the Trinity truly and are thirsting to be obedient to the bands of grace will surely receive their reward.

Rev 22:17 And the Spirit and the bride say, Come. And let him that heareth say, Come. And let him that is athirst come. And whosoever will, let him take the water of life freely.

Did you miss it? That was one of the very most important verses in the Revelation, in particular the last word of that verse: "*freely*". That is, without threat of "Hell-fire" with regards to the choice made by the candidate believer. The decision to come to Christ is an open one (and is so to all), and the continuous refreshing of God within eternal life is indeed offered without any threat to anyone deciding to reject it. That is from this study now made certain, and here again is stated in Christ's own words. It is a promise.

The words of this book are word-pictures and the maths requires that they be kept intact. Else the understanding (seal) would be lost. That seal should remain the seal of God. Now, I know that these verses are not a completely empty "*threat*" but instead have consequences, but they must be properly qualified by the whole text to avoid confusion. The verse could otherwise read "God shall add to him the result of His giving them the cold shoulder".

Rev 22:18 For I testify unto every man that heareth the words of the prophecy of this book, If any man shall add unto these things, God shall add unto him the plagues that are written in this book:
Rev 22:19 And if any man shall take away from the words of the book of this prophecy, God shall take away his part out of the book of life, and out of the holy city, and *from* the things which are written in this book.

Altering the argument of God as to His own efficacy is a sure-fire way to end up in trouble. Fortunately, the King James Version is ideal for this study!

Rev 22:20 He which testifieth these things saith, Surely I come quickly. Amen. Even so, come, Lord Jesus.
Rev 22:21 The grace of our Lord Jesus Christ *be* with you all. Amen.

I can only say "Amen, Come, Lord Jesus".

16.11) Time No Longer

In fact, I can say a little more than that, I can reason as is my wont. Whilst you need never fear burning in a literal fiery lake of sulphur, I have yet to state why you should make sure that you are not spiritually lukewarm (or cold) toward Christ. The judgement of God could arrive very suddenly and I hold true to my faith in that it is imminent, though not in it arriving without given warning. The Revelation text is, I state, the primary book on the "mystery of God". Then as I do not completely know the "new name of God", I then reason as follows...

If the "mystery of God" is fully revealed and God has yet a "rest" remaining for Himself, then I find from the same "mystery of God", there should be "time no longer".

All manner of slight against God is due forgiveness except for blasphemy of the Holy Spirit. (Why?) Christians have no name for the Holy Spirit. That the Holy Spirit is in figure yet unnamed assures me that God's new name is more than the symmetries or "mystery" of GF(8) with its multiplicative group C7. Then Son and Father are known already as K4 and the octal (which are known already by virtue of the mystery) and the Holy Spirit of life within them of C4 within C8 remains as yet to be properly and mathematically revealed to His people.

If they (the octal and C8) cannot possibly combine as a finite field alone, then assuming it to be an ultrafilter, the complement of the empty set (a mathematically correct set of everything – the "city of God" which contains GF(8) also) is indexed by this ultrafilter (as a finite self-referencing algebraic closure with that closure at both "four" and "eight" elements, possibly by analogy with division by zero, as it is also the unity). The element zero is given to the least in the Revelation text. (Upon the least God writes His new name: that carries as to all.) The ultrafilter surely also indexes itself, even if $1 \equiv 0$. (I posit that the unity element may still float but only agree in union with zero in the person of the least. And that is surely why he (the least as 1=0, the null ring) alone is the right hand of God, and this must have always been so.)

With the revealing of the mystery of God (the old name complete) no doubt some will still mock all things God and because He yet withholds His whole (new) name, the wise are left with nothing with which to defend the revealed "mystery of God"; the new name seeming totally impossible. Likewise, God has not armed the prudent with the "ultimate" truth and they may only hold their peace in the face of such mockery.

If neither name is regarded Holy, is there a disparity to be found in dialectically placing an excluded middle between the names? May there be any positive properties in that middle?

Yet is there such an excluded middle? I state categorically there is not. What truly remains as blasphemy is the ridicule aimed at both names (as it is the same God) by those that scoff. Then as both names are equivalent (the new superseding the old), blasphemy of either is to dismiss anything and everything respected as Holy. (Everything positive or perfected.)

Then dialectically, neither name appears Holy: the predicate "God-like" (being the names "e" or "g") becomes "meaningless" (as if $e^{-1} \vee g^{-1}$). If the new name is to remain Holy, then it must not be blasphemed, but who or what is being blasphemed? Is it not also scripture as well as God? The K4 group formed ensures no excluded middle in "e" and "g", but this change of names applies to the Holy Spirit only! Then it is simple to accept the free exercise of the dialectic described above as "blasphemy of the Holy Spirit". (Yet forgivable? Is it done without knowledge? Not when the empty dialectic middle of positive properties is pointed out between the names. (The new supersedes the old and satisfies it as an equivalent.) God has been acting with His new name all along, but has shown us only the old. They would, by rejecting this final correction, deny anything and everything positive or possibly perfect.)

So, whilst God permits no authority (positive properties) in that excluded middle, it suffices to say that nothing positive arises from such blasphemy. Then to scoff is an unwise choice (unforgivable if that position is maintained). It denies all virtue in God (not only His Holiness), but as from any property being "God-like" to everything that may then be considered "God-like". When such an individual arises, such a scoffer or "son of perdition" is then revealed openly. (That position becomes as the "wine" and is present or visible to all in the room.)

In both other cases, the logic simply reduces to the fact that God's name is already blasphemed. (In the cases of the Father and the Son, not the Holy Spirit.)

Now, whilst the wise are without the wisdom to answer this mockery of the "mystery of

God", this situation is truly logically untenable: if only God may put a stop to that situation by revealing Himself in full, I find that otherwise it appears to every witness that God openly seems to permit (and to agree for) Himself to be totally mocked and His believers humiliated before all without any adequate response. Do they reject the mystery instead? I hope not.

There is no positive property (as in the dialectic) found entailing the middle $e^{-1} \& g^{-1}$ of these two names: God's new name has superseded the old name entirely, and already. (Whether it is known to or not.) Ridicule or blasphemy aimed at the old name of C7 logically requires an immediacy for a translation of our faith to the completely untouched "new" name of God. In this sense, as with the promise of Christ to then write His new name, the indexing set (the city) and the name of His God (as my own God also) upon the least in the kingdom is enough of a "sealing" and a work "as so then for all". The least will become a "pillar" rather than a lively stone in the temple of God. (For the least as zero is the one God has least influence over.)

With any ridicule aimed at the revealed mystery with its multiplication by C7 (or that it will or is become C8), comes the great hope of my faith: the reign of omnipotent God. This will occur when the angel overcomes. The angel will exit this creation as overcoming all seven churches finding rest in an "eighth" state in which he is yet "dwelling in the heavenlies as unity". Christ's work of "writing" is the circuit of His angel. There will be a period where the Christian is unable to rebut blasphemy aimed at GF(8) because of its simple multiplication. That stated, God's new name simply stated to be "omnipotent" is enough to last the believer out; that is, until one greater comes along who is able to correct all as to the truth.

The authority to correct another's gospel is present upon only believers: i.e. only those holding true the "axiom of God's complete sovereignty". The non-believer escapes as both refused an answer, and as a non-convert rather than corrected; only believers love God enough to repent of any deep-seated beliefs. If the authority to correct unbelievers is not present in the believer and God has provided no answer for them to speak – revealing His name in full, then are they deficient? If they have no knowledge of a new name, then they have no authority to even correct one another upon it. Only God can correct them. Then being the only one as righteous as He is, only God may answer and judge a non-believer with no faith in Him. That said, new supersedes old and faith in the "old name" is still a good thing and to God, it remains pleasing.

So, the authority to correct this blasphemy falls to God and God alone. Such mockery left unanswered is not a contradiction of that sovereignty, but as soon as it is noted;

Mat 18:6 But whoso shall offend one of these little ones which believe in me, it were better for him that a millstone were hanged about his neck, and *that* he were drowned in the depth of the sea.
Mat 18:7 Woe unto the world because of offences! for it must needs be that offences come; but woe to that man by whom the offence cometh!

There is to be expected if such blasphemy continues, much collateral damage in the faith of the 'not-so-sure'.

Permitting (by such silence) that continuing (indefinitely ongoing) and unforgivable (with full knowledge of the old name and 2/3 of the new) blasphemy arising from the reason of His own silence alone: it as "knowledgeable" blasphemy, as become equal to not just that of unbelievers but also now the false prophet and scarlet beast (by lack of an explanation within the mystery of God or in the gospel to answer it), which by it would show to all that of God's own self (in choosing to wait further to reveal His name complete), He would in the meantime so equally permit His only begotten Son to be continually mocked and shamed afresh (as if He were crucified a second time), and all this due to His ongoing silence (i.e. His liberty)? Heaven forbid! God is not mocked. Maranatha!

(For you folks that don't read your New Testament, that Aramaic word "Maranatha" should, to a Christian, mean "Jesus is coming"!)

Rev 11:8 And their dead bodies *shall lie* in the street of the great city, which spiritually is called Sodom and Egypt, where also our Lord was crucified.

So, with the revealing of the old name, I state that unforgivable blasphemy (knowledgeable blasphemy) amongst unbelievers knowing the revealed name (the mystery of God) is an equivalent to the unforgivable blasphemy of the believer having "tasted the fruits of repentance". This equivalence will bring the wrath of God: as that blasphemy of the unsaved was previously only "winked at" as being without knowledge.

Heb 6:4 For *it is* impossible for those who were once enlightened, and have tasted of the heavenly gift, and were made partakers of the Holy Ghost,
Heb 6:5 And have tasted the good word of God, and the powers of the world to come,
Heb 6:6 If they shall fall away, to renew them again unto repentance; seeing they crucify to themselves the Son of God afresh, and put *him* to an open shame.

As it is impossible for a believer to redeem himself after such outrageous blasphemy (in context, this is equal to a Christian convert leaving the fledgling first-century church to persecute them as did Saul before his Damascus road revelation, to swap sides to the authorities to condemn the faith), then it is likewise equal for God to certainly reject all unbelievers (that equivalently blaspheme with knowledge of the mystery) as if their conversion were impossible. God will not easily choose from amongst them afterwards, if at all. With the world divided down the middle in blasphemy there will be a time of trial, and then judgement with the last wrath.

God is already 2/3 known and the difference is rapidly closing to zero when it is realised He is yet omnipotent. The mystery of God: Is it enough to close the gap? Will Christ answer a call? Every reason for the "mystery of God" not to close the gap is in the excluded middle $e^{-1}\&g^{-1}$.

In the ultrafilter of perfections, then, I expect the complement set (of the excluded middle) to be in the filter: the complement set of the empty middle $e^{-1}\&g^{-1}$ is that of every positive property of God's perfection (under the new name which is more perfect and has superseded the old), which gives weight only to His return and/or to the new name being written upon the least. (No longer "cut off" from Christ as zero to C7, but redeemed as unity and as zero in C8.)

That is, if like to the Old Testament law failing in its sufficiency for salvation at the point of killing Christ (translating Israel to a law of faith as in grace ministered by Christ), I find a symmetric situation where the sufficiency of the old name of GF(8) fails and the new must supersede it completely. What then may bridge the gap in translating the law of faith (if it became insufficient), to one of perfect faith but blasphemy once more?

Since the set of all perfections does not include liberty in that set of "q" (that liberty which solely entails "rest" on the virtuous side of the disjunction), the principal element as $L(G)=p$ upon $p\&q^{-1}$ permits the continuation of the rest upon that ongoing and continuing "excluded middle" $e^{-1}\&g^{-1}$. Instead the new name written on the least is become a necessary part of the set of all positive properties (as with the new name of God) present in that q; which includes "God-like or necessary existence".

Then God by necessity must "write His new name on the least" as $1=0$.

Absorption of zero may be derived from the additive identity, because the distributive law states $(x+0)*0 = x*0 + 0*0 = x*0$. Therefore $0*0=0$. Similarly, $x(0+0) = x*0+x*0 = x*0$ so $x*0=0$. Then every element is absorbed into the zero by multiplication. The first half appears consistent with a concept of $1=0$, the second does not. For $1=0$ I require $x*0=x$, and characteristic two, I would have $x+x \equiv 0$ in the second half of the argument except $0+0$ would remain 0, and $x(0+0)$ could not be as $2x \equiv 0$ but only $x*0$ alone.

If there is any duality $1=0$ to be found, it must be unique in a correctly formed set of everything "good for God", indexed by the octal as an ultrafilter, and it must also be the principal element, and present only in the highest ascent of perfection or of heavy constraint: found only at the utter limit of God's ability to create rather than holding it as only zero in isolation — if He would create it at all — because the "least" as $1=0$ (the null ring) may become necessary himself!

For God to show He exists without inconsistency (His new name appearing impossible and the old name trod under as by those that scoff), He must defend His new name in "q" actively. He must write on the least and/or return. Yet God may idly wait and rest positively as solely that principal element entailed from liberty upon the excluded middle $e^{-1}\&g^{-1}$; or can He? Whilst Christ's angel completes his circuit and attains the overcoming of all seven churches and finds $1 \equiv 0$ in the new name, that "rest" will become the angel himself. I have found some sequence!

So, if the angel indeed dwells in an eight cycle, Christ's exhortation to "hold that fast which thou hast" has a specific and personal meaning to His right hand, and to His right hand only.

The angel will return with the host of heaven: after that (and upon the last day) the judgement will be invited by the archangel with "Come, Lord Jesus". The new name will have done better than become known. It will have arrived.

Then, to the one I know as my Heavenly Father — I say "Surely, find the least, and write on Him your new name". There shall then be time enough for the mystery of God to be fulfilled and for His new name to be sanctified in His people by a well-ordered principle. Only the judgement falling between the beginning of all mockery aimed at the mystery of God and before the ultimate knowledge of God's new name (remaining totally untouched and still clean in the hands of God only) may preserve the Holiness of God: for if He does not defend it by swiftly bringing that judgement, given God is Holy, how are two thirds of that new name to remain

Holy? And how is the last third to be Holy if two thirds are not?

Is God not wise? With a new name, He is armed to finally return indeed! (They only mock the old.) The question is: "Is it then "Holy" for God to remain silent – not stating the full name (form) of His Holy Spirit, whilst His own wisdom is trampled before all the wise amongst His people and His Holy name openly ridiculed?" I sincerely doubt it. This untenable situation will be "brief" before He returns with the resurrection of all on the last day and the final judgement: yet it is a fact the light of understanding always separates the old from the new (and scoffing at the new name will not harm it, but it is still foolhardy), and I expect that by blasphemy against the Holy Spirit of C7 and of the revealed mystery only, God will say "Enough!" – and that whilst His new name shall remain clean and not at all threatened, it (C7) is still to be mocked to the full extent to which He is currently known, witnessed before all the "wise" (prudent) amongst His people. (They being us Christians.)

So, I hope for the swift arrival of judgement to occur as the least's circuit, the rainbow or "the Lord's bow" is put (finally completed and overcome) under the angel's "feet" and the two witnesses finish their ministry. (Whether or not any continue to scoff.) There is then no requirement for further time before the judgement. As to mockery: without knowledge "sin is winked at", but with knowledge, sin remains. Blasphemy against the Holy Spirit requires sure and swift repentance, it also being the same as crucifying (rejecting) Christ afresh and making Him an object of scorn, a "gazing-stock". God may never blaspheme Himself through His own silence, surely. Yet permitting His new name to be two thirds mocked without knowledge of the last third because the old name is revealed in full, requires answer.

If this silence in Him (and unfortunately us also) is logically indefensible (if it is continually stated "thy kingdom come"), what if (for lack of gospel to answer) there is no answer to that unforgivable blasphemy coming down upon the speaker from above? What if God is the only protector of the wisdom against this mockery? How then can anyone defend the gospel and its good news without Jesus Christ returned to judge between those with sin or "sin winked at", between they that blaspheme and they who do so quite innocently, for lack of an answer from a God that to them (quite reasonably), does not exist because they see only the lack of an answer to blasphemy?

When, or rather "if" God is then so silent, in which position is any wisdom left? Certainly, not in the position of one that brings judgement on their self. And not in the mystery of God either – it, taken alone, is without any true answer as a mere abstraction (the indexing set is still somewhat a mystery!), yet it alone has answer in God with His new name and the judgement of His Christ. Yet still, without faith, it is impossible to please God. Faith is then placed in that new name and it must needs yet arrive in Jesus Christ.

Now, let God be true and all men liars, as I do confess to a little guile here – the old name was, without equivocation, "a name for miracles". I can only conclude the new name makes a mathematical miracle of unbelief.

Now, Jesus returning to judge is a far better solution than a "Revelation" a second time, simply because giving the new name justified in full would encourage further and complete mockery. (God is not an equation, God is Spirit.)

16.12) Time No Longer II

Here I attempt to make a clearer proposition (being mostly logical), drawing on the lengthy prose of the last section to bring together those same ideas into one argument. I use the spanning K4 form of the filter to further expound upon and detail the content of the previous section. (Though there is much other content.) I will come to a conclusion in the next section.

The Trinity (with sevenfold spirit) will not permit a collapse to a single "essence" as per Gödel's argument[1]. There are only the trivial homomorphisms of GF(8) and GF(4) to consider, the morphism to null being the one possibility: yet that morphism would completely remove the Holy Spirit from the Trinity. Whereas belief in a God as if GF(2) could be sound (albeit naïve), there is no morphism from GF(8) to GF(2). There can be no "zip theorem" across every disjunction that will remove every element of the octal but for the zero (and possibly unity), reducing the divine to a singular essence as of any other individual. (Cf. [4.1.2a].)

God may rest on the octal in GF(8) as on an "old name" without any collapse (morphism) to nullity, and although He has shown us only the old name so far, He is capable of far more. Were there any "necessary collapse" found through such a "zip theorem", resisting it could be an act of omnipotence which only God would resist.

An individual with a singular essence may justify the disjunction of: $\{\Omega \setminus g\} \vee g$ with the essence "g", yet not so the Trinity which at most would (without modal collapse) form in disjunction the principal sets, say, of $\{\Omega \& e_r \setminus e_s\} \vee e_s$. The Trinity will never allow such a collapse (or essence) without the right hand of God (the least). His is a singular essence that can only collapse the disjoint sets within the octal to full closure, maintaining their partition of Ω in a "new name" as will be shown.

The salvation of the least is not only intrinsic to the new name – God is not free to rest on His work of salvation (able to return) without this separation of the least who shares in that knowledge. By showing the dialectic (the name of blasphemy) blasphemes the mystery of God, with the least (and he alone) God may then justify His judging all those that would blaspheme His new name, and (curiously), not casting out His own flock that merely have to honour the old name through which they see the new as if "through a glass darkly". The ungodly are condemned as if by an "equal" (or a lesser) by the least remaining obedient, not then blaspheming the new name (without the present knowledge of it): God's ways are equal.

In a disjunction formed of the two names (as to decide which name God is worthy to exemplify), I find the whole work of creation is now the intercessor with virtue, rather than Christ!

Then I make a separation between this current creation and the new heaven and new Earth to come: for in this creation God is blasphemed but for His new name or that of the Holy Spirit. This creation is corrupted, but the new will not be at all so and God's new name is reserved for those that inherit that incorruption. Then I make a distinction between all those separated out on blasphemy (amongst other conditions) of the published name of God in the current creation (for God may possibly be blasphemed at present) and those saved and sealed elect, separated, worthy of the new creation to come: those that will not blaspheme their God's new name afterwards. By analogy, God and creation become "born again" for all

[1] Sobel, in J. J Thompson (ed.), *On Being and Saying: Essays for Richard Cartwright*, p241-61.

those likewise within them.

Then God has drawn a line between all those that blaspheme and all those that presently do not: i.e. those that later in the new Jerusalem cannot blaspheme, and willingly do not.

The least I find, has the new name of God "written upon him" with his circuit through the seven churches. Over an arbitrary octal in eight elements (he is the eighth in Spirit) the new name of God is written on him with 1=0 (and a cyclic multiplicative group of eight elements). He cannot blaspheme that name within his circuit − it being his name also. (He is given the morning star, he is a "holy spirit".) Truly, God exemplifies His name by a physical manifestation of it in circuit; it is written in creation without being published abroad for blasphemy. The right hand (it would seem) has directed his steps well, as under his own reckoning. Under grace, all those others also so chosen will simply not desire to blaspheme, and the least shows it possible under the worst permissible circumstances as in the current creation.

The least becomes as a principal element in the set of elect. By finding the sole criteria for rest on the election of grace in that one respect (of not blaspheming either name) I discover that the salvation of the least is become necessary (there is a least overall). Then the **person** of God is found inclusive of saving that least, and the least is equally part of that new name. God is logically identified as the "God of the least" as equally as He is named "The one true God". The least (principal nature affirmed) is God's right hand and uniquely so.

I find one reason for God not to return whilst the least expects his answer. That He (Christ) has the **free** opportunity to write on the least His own new name. (I.e. whilst the least survives until he overcomes, so freely waiting to rest on the election of grace whilst it freely continues meanwhile: as writing the name is done with the angel's circuit, requiring his patience: floating $1 = 0$ through the additive elements of the octal.) This delay in answering the least's faith is found in preference to openly granting that knowledge (to all) before physically returning − because His new name is already ready to be defiled by the worst sinners on the planet, rather than ready for Him to give freely, to affirm the correct faith of the Holy justly.

I find that God cannot finally rest on the election of grace completed (so returning by freely choosing when) without writing His new name on the least as promised beforehand. (By giving him an eight cycle in a circuit over an **arbitrary** octal, then enabling him to rest in His inheritance of $1 = 0$, his own person as with eternal life, and necessity in God. All others simply require a seven cycle, ending in Philadelphia.) The contradiction in the angel's faith is simply remedied with holding the same patience as all for the return of Christ once it is perceived to be imminent. (His oath references the same quandary of faith, as equal to the period when the seventh trumpet is ready to be sounded. God will not permit things to become worse than the third woe, fulfilling the angel's oath that there be time no longer. Christ will return not long after. God has enough time to work all salvation without any "fourth woe".)

The least certainly receives the new name but he is also saved before the time of trial. His reckoning completed, he would have already fulfilled seven states of the eight cycle and the new name is thereby written on him completely beforehand. As to receiving the worship of the "synagogue of Satan," the ministry of the two witnesses has to start before that time of trial. I expect the "synagogue of Satan" to be identified completed with the third woe, the scarlet beast (which is only so completed at the end of the two witnesses' ministry after the saints are exeunt, sealed). So, then, as not to leave anyone in logical stalemate, I expect this reward to be given him at the judgement. (At the height of their embarrassment.)

I summarise, stating the second coming must come properly after the angel's circuit of 1 = 0 through the octal is completed. His overcoming is with the completion of the mystery of God, and that can only be fulfilled with the "old name" superseded by a work of the new found present in creation. (The old name satisfies the requirements of "a God" rather than one God "Most High".) The result? The ministry of the two witnesses is free to be empowered with complete omnipotence: a ministry of power, force and truth. There is "time no longer", as the mystery may be supplied but it still needs completing, and that is done only with the completion of the same angel's circuit. As the angel overcomes his state in Philadelphia holding fast that which he has, God is free to judge, and free to return after that time of trial.

One God With Two Names (16.12.1)

I declare that God has two names: I cannot supply the maths for this, but I will treat myself to examine the second closure of the new ultrafilter. I simply state that the set of positive properties under the ultrafilter of the "old name" is properly contained within the set under the index for that of the "new name" – an ultrafilter properly contained in the former yet with the same "language", the new name is wholly compatible with the old name.

Then the new name will entail the old as (effectively) also "God": the old being found on positive properties alone which are enough to prove that there is a "mystery of God". I will refer to the new name as being a "closure" above the old name which is in a closure properly contained within the "new name". I simply argue that if the ultrafilter is principal, then the principal element of the new name is also principal in the old. (Else some sets in the old name would not be in the new.) Then as this principal element "I" is or must be contained in every set of the old name, for one ultrafilter to entail the other, "I" is principal in the old name for all sets in both ultrafilters.

As creation acts as an ultrafilter of virtue, it becomes possible for the principal element in creation to be principal in both names of God also; effectively enabling a duality of 1=0.

I now begin with some identities – as to whether it is good for God to reveal His new name to those that may possibly blaspheme it.

Forming The Octal On Both Names (16.12.2)

The properties below are "positive" if God may choose an elect "I" for exemplifying them ("I" exemplifies these by faith). If these properties instead apply to the person choosing for themselves as with a worldly (improper) ultrafilter, then they are not positive. They are revealed as privating virtue: they are dialectic.

I also infer the current creation acts as a virtuous filter separating out those that would not blaspheme the new name of God. They are the "elect".

Starting with some identities;

a$_0$ "does not blaspheme the name of God" (As is possible by the current creature.)

c   "person remains blameless"

e   "old name of God" (remains Holy, not put to the test)

g   "new name of God" (remains Holy, not put to the test)

Firstly I may appropriately modify a$_0$ to "a":

a   "**cannot** blaspheme the name of God" (a virtue, universal in the new creature.)

And I state that a $\rightarrow$ a$_0$ so that a&e^{-1} $\rightarrow$ a$_0$&e^{-1} $\rightarrow$ c. But again, by the properties of an ultrafilter, the old creation is closed and it is not necessarily true (for all) that if a$_0$ is in the filter (of the old name) then I could state that "a" is in the ultrafilter of the old name also.

I now also state that God's new name **supersedes** His old, and I then also state that g $\rightarrow$ e. I sum up by stating that g belongs in the new creation, but e in the old creation. Then e $\in$ a$_0$ and g $\in$ a. I can form the freely decidable (by God) disjunction of: Pos(g) $\vee$ Pos(a&g^{-1}) $\rightarrow$ Pos(e).

I.e. That there is in "g", a "God-like" individual whose name remains sufficiently Holy as "e" in order that for any person "x" in the old creation a$_0$ to remain blameless with respect to the new name "g", that person should likewise not blaspheme the old name "e".

Then in the current creation of a$_0$, God's name is permitted to be blasphemed, whereas in the new creation "a" to come, God will not permit His name to be blasphemed at all.

Now, I require some axioms:

Ax1)   N(Pos(g)). God's new name is perfect and untouched and is never positively non-exemplified. (God already has His new name "g", though it is indiscernible from "e" in the old creation (creature) a$_0$. Cf. section [4.8.1])

Ax2)   N¬(a&g^{-1}&¬l) for all elect with l $\rightarrow$ c. For every elect, the new name must not have ever been blasphemed. (Without knowledge there is no sin.) Blasphemy of the Holy Spirit is unforgivable, the new name is not revealed in order to keep it untouched so to permit sinners to become elect.

Now I state that the disjunction...

Pos(e) $\vee$ Pos(a$_0$&e^{-1}) $\rightarrow$ c puts God to the test but with His old name only. Now the disjunction...

Pos(g) $\vee$ Pos(a$_0$&g^{-1}) $\rightarrow$ c puts God to the test with His new name. (g is already due all reverence as the new name, and an opportunity for it to be blasphemed in g^{-1} has arisen.)

Ax3)   Finally, God is "not put to the test". Pos(a$_0$&e^{-1}) $\nrightarrow$ Pos(a$_0$&g^{-1}) in the old name.

There is no disjunction g $\vee$ a$_0$&g^{-1} $\rightarrow$ c in the old creation. There is only P(a), not N(a) over all x $\in$ a$_0$. I.e. e $\vee$ a$_0$&e^{-1} $\rightarrow$ c only and it cannot entail e $\vee$ a$_0$&g^{-1} $\rightarrow$ c in the new closure! So, g $\vee$ e, and the middle of the two names crucially remains empty (but for only the least alone, principal in g that supersedes e). I.e. by virtue of the old name and old creation/creature, it is impossible to blaspheme the new name and be elect. (As blasphemy it is wholly without knowledge, but for knowing K4 and the octal. Blaspheming the Holy Spirit as C7 is then unforgivable, as the name is unknown in figure, it is "sealed". Yet, C7 is all that separates God from the "godly" (and God-like), even in the new name.)

I state that I will reach a contradiction and the new name is already in effect: or simply by entailing $a \rightarrow a_0$ I can state that the new name is always in effect, and whether any admit it or not, God has determined no logical blasphemy of His new name (fully determined on the least of all) from any of His creations. God proves Himself perfect by showing only the old name in this creation, but it necessarily moves in the new. Already, none are capable of blaspheming His new name knowledgeably. Election is necessarily decided by blasphemy of the Holy Spirit as C7 only. (For that is given to the elect.) That blasphemy ends when the winepress is trodden out.

Blaspheming the new name **only** blasphemes the old in the old creation: blaspheming the old likewise blasphemes the new in the new creation. It is then impossible to **inherit** that new creation blaspheming the Holy Spirit unforgiven, and so blasphemy of the Holy Spirit is unforgivable because the old creation has an ultrafilter as if virtue (God's spirit rests as on the sabbath). Jesus may choose who to forgive, but this is His sole condition of necessity.

Now, I can clearly form a K4 group on the disjunction $N\neg(a\&g^{-1}\&\neg l)$ as I may also do on $N\neg(a_0\&e^{-1}\&\neg c)$; yet, the set under the new name will entail $N\neg(a\&g^{-1}\&\neg l)$ from $N\neg(a\&e^{-1}\&\neg c)$ but not the converse of the old name acting on merely a_0: I instead require "a" and some principal "l" in g to prove the middle of $g \vee e$ empty. The old name is closed and I would otherwise break the closure to assume that "l" or g is principal in the old name by virtue of the property of inclusion in the ultrafilter's definition. Then g will wholly supersede e.

Then in the new filter $g \rightarrow e$ and I require $a \rightarrow a_0$ in the old name to entail the new; however, it is not true that $a \rightarrow a_0$ for all x in a_0 but only necessarily so for the principal element of the new name (also principal in the old). Of necessity I may state in the new name that $Pos(a_0\&e^{-1}) \rightarrow Pos(a\&g^{-1}) \rightarrow l$. Then I state that $g \vee a\&g^{-1} \rightarrow l$, but God is not put to the test; rather I rearrange and put "l" to the test. (For all $f \in a_0$, $f \rightarrow l$ and $l^{-1} \rightarrow f^{-1}$, so God will not approve of any of those "f" that would not blaspheme His name unless placed under His full restraint: yet $N(Pos(l))$.)

Then $l \vee a\&l^{-1} \rightarrow g$.

I now examine what happens when I split the sets e and g with regards to the old creation and the new. I split "e" into r_0 and u^{-1} and "g" into "s" and v^{-1}.

My much required octal is then formed of:

r_0 old name of God as revealed to the flock (not to be blasphemed by the least)

r   new name of God as revealed to the flock (not to be blasphemed by the least)

s   man remains blameless (as under u&v) in both old and new creations "a", a_0.

u^{-1} "God permits blasphemy of His old name made by those without knowledge of it" (He rests. God may forgive the unconverted.)

v^{-1} "God permits blasphemy of His **old** name where there is knowledge of how to blaspheme it" (He rests. God permits tares that knowingly or unknowingly practise the dialectic.)

By "knowledge of how to blaspheme" I include unforgivable blasphemy of the Holy Ghost. This is made only by those who have knowledge of God's name, of the old or as mocking the new. It is clear the pair above are somewhat opposites.

Now, inverting to action I find the two predicates have become negative:

u	"God does not choose (He utterly rejects, judges) those that blaspheme His old name without knowledge of it" (God does not elect such non-converts).

v	"God does not choose (He utterly rejects, judges) those that blaspheme His old name with knowledge of how to blaspheme it" (God does not elect tares).

So including u^{-1} and v^{-1} is to keep the sets of believers in r_0 and s open (not closed), God has not yet judged or rejected anyone as yet forgivable. Now, these two are not dialectic opposites, but instead require virtue to unravel their disjunctions. These form disjunctions $u^{-1} \vee (L\&u \rightarrow v^{-1})$ etc., where the virtue L is Christ's elective choice. L is all that separates whether an elect "l" will enter the new creation "a" from the old a_0. (Whether any believer is elect or not.)

$u = v^{-1}$	makes sense as v^{-1} is permissive of only those that blaspheme with knowledge of God, as opposed to those that are rejected here by u.

$v = u^{-1}$	likewise makes sense as the excluded (rejected) are those that blaspheme with knowledge or with intent (knowledgeably), which does not refer to those that are permitted blasphemy in ignorance.

Then I can close the octal with

(u&v)	such a one elected is blameless and may enter the new creation "a"

$(r_0\&s)^{-1}$	such a one is not blameless; they will not be elected and remain purely in a_0.

I note $(r_0\&s)^{-1} \rightarrow (r\&s)^{-1}$ in the new name. No blasphemy will be permitted in "a". Then (u&v) is true over all the elect.

Now, in order to decide $e \vee l$ and to entail "l" from e^{-1}, by axiom of virtue I require God's creation or some other a_0. For God has required this mechanism of creation. Then, clearly "a_0" is a virtue implemented by God, and if I were to reason dialectically, I could have:

$$a_0 \vee c\&e\&g$$

and this is taken to mean that the dialectic method itself, this "worldly ultrafilter", blasphemes the name of God by privating virtue. (The creation, or old creature is corrupt.)

I.e. $N(Pos(a_0)) \rightarrow N\neg(Pos(c\&e\&g)) \rightarrow N\neg(Pos(c)\&\neg Pos((e\&g)^{-1}))$

Or, given $Pos(a_0)$: $Pos(c) \rightarrow Pos((e\&g)^{-1})$

So, $a_0\&c$ proves $e \vee g$, that a person may be found blameless when only blaspheming one name or the other.) Or, $g^{-1} \vee e^{-1}$ would have an excluded middle ($e^{-1} \rightarrow g^{-1}$), then entailing $N\neg Pos(a)$.

If a person is to remain blameless (dialectically) they must then state it is impossible to avoid blaspheming one name without blaspheming the other (as if one were not perfect of the pair, one a "different God"), but by practising the dialectic they deny the virtue that God's name(s) should not be blasphemed (put to the test), and then blaspheme using their dialectic reason!

Blasphemy arises when a person claims innocence in blasphemy against the old name of God only because the new name is not known, despite certainly having blasphemed the old name.

The Least in the Kingdom (16.12.3)

I now begin to account for the content of the letter to the least (angel) of Philadelphia and the delivery of the man-child before his being "caught up" to God and His throne. I do this purely using the sets in the octal. God has no further need to explain, the scriptures are open. (This is as far as I can go; I only provide "a solution" rather than one proven from scripture!)

Given $Pos(u^{-1})$ or $Pos(v^{-1})$ and $Pos(u\&v)$ positive as for the whole flock (the set of believers), by restricting these terms to the blameless only, I find "r" and "g" are interchangeable, as are "s" and "c". God (with respect to the flock) may rest with virtue $L(G)$ whilst entailing $Pos(u\&v)$ as "work done" already (the flock are blameless).

For the "man" to remain blameless in the old creation he must enter the new creation having never blasphemed the new name of God. (He is tested on the old only, the new is not put to the test.) Yet C7 is then knowledgeably blasphemed by those understanding the mystery.

I posit that the virtue of liberty "$L(G) = a$" (God is free to create) because $(g\&c)^{-1}$ is equivalent to the empty set having the correspondence to $(r\&s)^{-1}$, with $r = g$ (or $r = e$), and $s = c$ as before. Then there is always a remnant of the faithful, and there is always a least, and one least overall, as such the least "l" with "$l \rightarrow c$" is "Faithful and True".

"l" being principal in both ultrafilters (certainly in the new, with requirement to also prove himself so over the old to break closure: letting the new supersede the old) does not blaspheme the new name and therefore does not blaspheme the old (which forms two parts of it).

Then given $r \rightarrow r_0$ likewise breaks the closure of the ultrafilter (the old name, even when restricted to the flock of the elect), I find in the new creation which entails the old:

$L(G)\&r_0^{-1} \rightarrow L(G)\&r^{-1} \rightarrow s$ where s is the only positive property entailed from liberty, being rest. (God does not reveal His new name in order to preserve it as Holy.)

So; $r \vee L(G)\&r^{-1} \rightarrow s$. And "s" is a principal element that the new name has rest upon. ("s" and "c" are interchangeable.)

Now, "r" entails the set of all positive properties, the "new name of God". Then the possible disjunction is made of:

$(r_0\&s)^{-1} \vee (p\&r_0\&s \rightarrow u\&v)$ where $p=a_0$, which entails in the closure "a";

$(r\&s)^{-1} \vee (p\&r\&s \rightarrow u\&v)$ where $p=a$.

"s" is simply the "rest" from before and $p = a \rightarrow a_0$ acts as would liberty $L(G)$ itself (it will not limit or constrain r&s at all, being a virtue of the whole flock of believers). This requires that r&s is the set of **all** positive properties (including "rest", in Gödel's[1] terms "essence", or "God-likeness") which is privated by no virtue. In real terms, if I have $(r\&s)^{-1}$ then this set is become "empty" (not God-like), as r&s is "of everything". (I default by necessity to the right hand side as positive and that action becomes a decided positive property, a necessity for perfection.)

Now, I posit that "l" is a possibly principal element in the old name (the least in the kingdom). Then I may infer $a\&g^{-1}\to l\to c$ from $a_0\&e^{-1}\to l\to c$ on the condition that it is "l" put to the test, and not God! I simply need to reiterate the election on "l" with the condition that "l" cannot blaspheme the new name of God (as he has knowledge of it).

Then I use Ax1, Ax2, Ax3 to state that $r_0\vee(a\&r_0^{-1}\to l\to c)$ entails $a\&r^{-1}\to l\to c$.

I.e. if there is necessity in election, it is for the least "l". Then I may revisit this (and noting it is valid for all x in a and a_0) by rearranging from $r\vee a\&r^{-1}\to l$ to:

$l\vee a\&l^{-1}\to r$ (The least is successfully put to the test only if the new name stays untouched.)

But, I propose that the election of "l" is a positive property and that that liberty of election (the purpose of creation) renders "l" a suitable principal element upon which Christ may rest, by saving with His elective choice. (Deciding on $(a\&r\&s\to u\&v)$ as above.)

Then; $l\vee a\&l^{-1}\to r\to l$ because l is found principal in a and thereby also $a_0\subset a$.

Then $N\neg(a\&l^{-1}\&\neg l)$ or $a\to l$ without breaking closure of "a", and so "l" becomes that same rest of liberty and l=L ("l" is the right hand of God). That liberty of election, to save and justify the least (and so all others) is the purpose (principal element) of **all** creation. Then the principal element in the case of the least is found principal in unity and is also the zero! (Then 1=0 for the least, and as God is omnipotent the following results become very strong; it may be possible to justify anything! However, this duality of 1=0 is no fallacy of equivocation.)

Then there is no finer element but "l" and I posit there is no further closure: in order for God to maximise His new name, He would minimise "l". I also posit that "l" does not fully break the whole closure of a_0 alone. (God, needs to write His new name of g on "l".)

Liberty Is Necessary – Modal Collapse In The Octal (16.12.3)

The octal also has its own analogue of modal collapse as has Gödel's argument[1]. Given that the principal element is both unity and zero (and floats through the circuit of the least), I face a very real possibility of a disjunction where both sides (freely decided) rest on liberty itself.

Then $l\vee l\&l^{-1}\to l$ and r=s=l in $r\vee a\&r^{-1}\to s$. I have the sets at rest on both sides maximised: that is: $\{u^{-1}\}\cup\{v^{-1}\}=\{(\Omega\setminus l)^{-1}\}$ So $u\&v=\{\Omega\setminus l\}$. Then $(r\&s)^{-1}\vee p\&r\&s\to u\&v$ becomes as $l^{-1}\vee l\to\Omega$. (Note in $\{u^{-1}\}\cup\{v^{-1}\}$ there is no excluded middle with "l" both sides, each side is extant with modal collapse.) I assume that before the circuit of the least is completed (as floating 1 = 0) and all closure is broken in seven octals (horns) this disjunction becomes freely decidable as the disjunction $l^{-1}\&e_x\ \bar\vee\ a\&l\&e_x\to\Omega$. (Both sides have some e_x at rest as "God-like".)

I require a lemma to begin with:

[1] Sobel, in J. J Thompson (ed.), *On Being and Saying: Essays for Richard Cartwright*, p241-61.

Lemma MC0) Given a necessarily decided disjunction – a modal collapse, I may drop the principal elements e_x in the disjunction so that, say: $\neg x \& e_r \vee (a \& e_p) \& (\neg x \& e_r)^{-1} \to \Omega \& e_s$ would become $\neg x \vee a \& e_p \& x \to \Omega$ only. Then as virtue in $a \& e_p$ is necessarily exemplified; Ω is found necessary and as the zero (floated as $1 = 0$ through seven disjoint sets in the least's circuit).

Proof: Given the disjunction $\neg x \& e_r \vee (a \& e_p) \& (\neg x \& e_r)^{-1} \to \Omega \& e_s$ in the octal if x is found necessary then I cannot have $\neg x \& e_r$ or even $N \neg (\neg x \& e_r)$ as then $e_r \to x$ (a contradiction if e_r is principal).

Then I must find $(x \& e_r)^{-1} \vee (a \& e_p) \& x \to \Omega \& e_r \& e_s$, which is $\Omega \& e_p$. On the right hand side of the disjunction I have $(a \& e_p) \& x \to \Omega \& e_p$ and $(a \& x)$ becomes necessary. (x is exemplified in creation or of virtue.) Then, e_r enjoins to the right hand side filter in all $<e_s>$, to become $<e_p>$ (virtue) which remains a filter and closed within $a \& e_p$. A modal collapse here would result in, say, every statement in the equivalence $a \& e_p \& x \to \Omega \to a \& e_p \& x$ becoming necessary.

Were (a&x) not necessary then $P(\neg Pos(a \& x))$ or, as "a" acts as virtue I would expect that $P(a \to \neg x)$ but if $a \& e_p$ is virtue then "a" entails only the positive: this is a contradiction on $N(Pos(x))$. Then $N \neg (\neg Pos(a \& x))$ and $N(a \& x)$ for God. God must bring about x.

Additionally, as a&x is necessarily positive I may drop the term in e_p and arrive at the desired $\neg x \vee a \& e_s \& x \to \Omega$ which may finally become $\neg x \vee a \& x \to \Omega$. Then as x is actualised I have $N(x)$ and the term in e_p is easily dropped and $x \to \Omega$ (the zero) is the equivalent of virtue (unity).//

Theorem MC1)

The principal element "l" causes a modal collapse. $l^{-1} \vee a \& l \to \Omega$. (Unity is become the zero.)

Proof: Given a principal element "l", I begin with the disjunction $g_x \vee a \& g_x^{-1} \to e_x$. Now, I assume that "l" is principal in the new name (but possibly not the old on its own) and I may write $g_x \to l$. Forming an octal, I must be able to derive $(r \& s)^{-1} \vee p \& r \& s \to u \& v$ from $r \vee p \& r^{-1} \to s$. This will result in the minimal case, which is surprisingly equivalent (for some x) to $(l \& e_x)^{-1} \vee a \& l \& e_x \to \Omega$ (which takes some effort to show). Yet I require only $l \& e_x \to \Omega$.

Now, $l \& e_x$ is a set in the new "name" of God, the wider closure. That set similarly entails "l" as principal. Then closure is broken by "l", and I may rearrange and rewrite $g_x \vee a \& g_x^{-1} \to l$ as $l \vee l \& l^{-1} \to l$ because $g_x \to l$ and $a \to l$.

Alternatively a disjunction in the new name of: $g_r \vee g_p \& g_r^{-1} \to g_s$ will entail $l \vee l \& l^{-1} \to l$.

Then I reduce to the seemingly impossible $l \vee l \to l$. Then $\{u^{-1}\} \cup \{v^{-1}\} = \{\Omega \setminus l\}^{-1}$ and therefore $u \& v = \{\Omega \setminus l\}$, acting with **all** creation in the set "$a \supset a_0$" apart from the virtue of liberty "l": i.e. working in $\{a \setminus l\}$ (a result or "split" shown possible from applying liberty as in section [4.8]).

Then $(r \& s)^{-1} \vee p \& r \& s \to u \& v$ is now the analogue of: $l^{-1} \& e_{r \& s} \vee p \& l \& e_{r \& s}^{-1} \to \Omega \& e_{u \& v}$. Now, by axiom, "l" is in the new creation and the election of "l" is sure. I.e. $N(Pos(l))$. "l" breaks closure.

Then I cannot have $l^{-1} \& e_{r \& s}$ or even $N \neg (l^{-1} \& e_{r \& s})$ or its equal $e_{r \& s} \to l$ (a contradiction), that is, if "l" is a principal element in the new name and breaks closure, each Ω of seven completed.

The right-hand side $p\&l\&e_{r\&s}^{-1} \rightarrow \Omega\&e_{u\&v}$ is not found necessary even if "I" is a principal element in g_x. Instead I find God cannot rest on either side, even $N\neg(l^{-1}\&e_{r\&s})$ (as $e_{r\&s} \rightarrow l$ is a contradiction), there is a modal collapse and I find $(l\&e_p)^{-1} \vee a\&l\&e_p \rightarrow \Omega$ as desired. ($e_{r\&s}$ adjoins to all $<e_{u\&v}>$, resulting in $<e_p>$ which is the unity.) Then $l^{-1} \vee a\&l \rightarrow \Omega$ by MC0.//

Theorem MC2)

The principal element makes $g_x \leftrightarrow l\&e_x$ an essence. $g_x^{-1} \vee a\&g_x \rightarrow \Omega$ will also hold.

Proof: First, I posit one more axiom

Ax4) $g_x \rightarrow g_x\&l$. ($g \rightarrow g\&l$, in seven disjoint "churches" forming an octal circuited by the least.)

I.e. if there is a Most High God then He is the God of the least in all creation. (No matter which closure, or "meta-closure" etc.) Then "I" is principal overall in "a".

Given the prior disjunction: $l^{-1}\&e_{r\&s} \vee a\&e_p\&l\&e_{r\&s}^{-1} \rightarrow \Omega\&e_{u\&v}$, reduces (by MC0) to $a\&l \rightarrow \Omega$: on choosing the right hand side God would exemplify $N\neg(l^{-1}\&e_{r\&s})$ or that $e_{r\&s} \rightarrow l$. This is a contradiction, so it must be correct that given Pos(l) the modal collapse ensures God will only rest on the right hand side. Then $(l\&e_{r\&s})^{-1} \vee a\&e_p\&l \rightarrow \Omega\&e_{u\&v}\&e_{r\&s}$, the term $e_{r\&s}$ adjoins to the filter in $<e_{u\&v}>$ and becomes as e_p and principal. Then I have $(l\&e_p)^{-1} \vee a\&l\&e_p \rightarrow \Omega\&e_p$, and again, legally dropping principal elements reduces this to: $l^{-1} \vee a\&l \rightarrow \Omega$.

Then $l^{-1} \vee a\&l\&e_p \rightarrow \Omega$ is necessarily decided: Now, given $g_x \rightarrow e_x$ (and also with the axiom $g_x \rightarrow g_x\&l$), I find $(l\&e_p)^{-1} \rightarrow (l\&g_p)^{-1} \rightarrow g_p^{-1}$ resulting in: $g_p^{-1} \vee a\&l\&e_p \rightarrow \Omega$.

Under a seven cycle and by symmetry in the creation "a", each additive element of the octal e_p may become a g_p. (Keeping the schema and sets of positive properties fixed, the least's circuit need only preserve the full closure of each octal in a master-schema, floating 1 = 0.)

It is then simple that $g_x \leftrightarrow a\&l\&e_x$ but g_x is principal as is e_x and "I" (principal in g_x and "a"), so minimally I must find $g_x \leftrightarrow l\&e_x$. Finally $g_x^{-1} \vee a\&g_x \rightarrow \Omega$ as desired. g_x breaks closure. //

Theorem MC3)

The middles of $l^{-1} \vee a\&l \rightarrow \Omega$ and $g_x^{-1} \vee a\&g_x \rightarrow \Omega$ are empty as "I" is principal and also an essence.

Proof: The disjunction $g_x \vee p\&g_x^{-1} \rightarrow e_x$ rearranges to $e_x \vee a\&e_x^{-1} \rightarrow g_x \rightarrow l$. In the case of the latter I formed $l \vee l \rightarrow l$ from rearranging again to $l \vee l\&\neg l \rightarrow e_x$ as "I" is principal in a and g_x. $l\&\neg l \rightarrow e_x$ then rearranges without breaking the closure of "I" to: $l \rightarrow l\&e_x \rightarrow g_x \rightarrow l$. (I.e. $g_x \leftrightarrow l$.)

Then $(r\&s)^{-1} \vee a\&r\&s \rightarrow u\&v$ is now as: $l^{-1}\&e_{r\&s} \vee a\&e_p\&l\&e_{r\&s}^{-1} \rightarrow \Omega\&e_{u\&v}$ because $\{u^{-1}\} \cup \{v^{-1}\}$ = $\{\Omega \setminus l\}^{-1}$ and therefore $u\&v = \{\Omega \setminus l\}$. This is corrected (by MC1) to: $(l\&e_p)^{-1} \vee a\&l\&e_p \rightarrow \Omega$.

Crucially, u^{-1}, v^{-1} are maximal in this disjunction together spanning all Ω: There is a truly empty middle in the disjunction between $l^{-1} \vee l\&\Omega$. Ω contains **every** positive predicate but for the empty middle of $l^{-1} \leftrightarrow (r\&s)^{-1}$. Given $p=a\&l$, in this case $p\&r\&s$ is also only "$a\&l$" which in the relation $N\neg(p\&r\&s\&\neg(u\&v))$ is $N\neg(a\&l\&\neg\Omega)$. Then $l^{-1} \vee a\&l \rightarrow \Omega$ and clearly there is no middle in $l^{-1} \vee l$.

Clearly, Ω is maximised in virtue as the set that entails all positive predicates in the octal (also a closed set, as is **every** K4 group and ultrafilter over this disjunction), yet Ω is closed and without that one single predicate "l", unless I verily state $a\&l \to l\&\Omega$. Then $a\&l$ must be a superset of Ω and as it contains "l", I truly find $l \leftrightarrow a\&l$. I am able to negate both sides of the disjunction without a fallacy of an excluded middle. This is the "little strength" (Rev 3:8) that is unique to the least. (No "empty difference", "l" is a living soul.)

Now, given $g_p^{-1} \vee g_p \& \Omega$ and $g_p \to \Omega$ as above with $u\&v = \{\Omega \setminus g_{r\&s}\}$ (entailed from the relation $g_p \& (g_{r\&s})^{-1} \to g_{u\&v}$), due to modal collapse (as of the least) g_p is found both principal and an essence. I also find that $<g_p> = <g_u> \wedge <g_v>$ and $x \subseteq \{\Omega_u \setminus g_u\} \cup \{\Omega_v \setminus g_v\}$ for all x in $\{\Omega \setminus g_p\}$. The essence "l" in $g_x \to g_x \& l$ is enough to also show $x \& g_p \in <g_p>$ for all $x \subseteq \{\Omega_u \setminus g_u\}$ etc. Then $<g_p>$ is an ultrafilter on $\{\Omega \setminus g_p\} = \{\Omega_u \setminus g_u\} \cup \{\Omega_v \setminus g_v\}$. Thereby g_p and "l" necessarily break closure.

The sets in Ω entailed by that modal collapse then appear in operation as those of the octal or manifold of virtue. ($g_p = g_u \& g_v$ etc.) The "little strength" of the least is then the crown of union and/or the "marriage of the lamb". Those blameless and elect in $u\&v$ are made **one body**. (As before, "u" corresponds to those forgiven, and "v" to those not fallen. Cf. Rev 22:17 as to the "Bride" and the "Spirit" respectfully; with "him that heareth" being convicted as the least, and "him that is athirst" one to receive the Holy Spirit to close God's potential to create under His new name (of an eight cycle). The least alone is the intersection of all four. By analogy, lives are married in similar fashion: u^{-1} would be as one's "future", a union made with that of v^{-1}.)

Theorem MC4)

The creation "a" is "spanned" by the essence "$l\&e_x$". Given $l^{-1} \vee a\&l \to \Omega$ the following relations are equivalent (for every such disjoint set Ω entailed in the leasts circuit by modal collapse).

$a\&l = <\Omega_r \& \Omega_s \& ... \& \Omega_{u\&v}> \& l = <g_r \& g_s \& ... \& g_{u\&v}> = l\&<e_r \& e_s \& ... \& e_{u\&v}> = <l\&e_r \& l\&e_s \& ... \& l\&e_{u\&v}>$ and also the equality of: $a = <l_r \& e_r \& l_s \& e_s \& ... \& l_{u\&v} \& e_{u\&v}> = <\Omega_r \& \Omega_s \& ... \& \Omega_{u\&v}>$.

Where the $l_r \& l_s \& ... \& l_{u\&v}$ are the gifts of Christ's letters to the least "l" to generate with "l" the "horns on the lamb" $<\Omega_r \& \Omega_s \& ... \& \Omega_{u\&v}> \& l = <g_r \& g_s \& ... \& g_{u\&v}>$ in Revelation.

Proof: Given the relation $g_x^{-1} \vee a\&g_x \to \Omega$, by the splitting of virtue (see section [4.8]) as if between r and u^{-1} etc., "a" must "span creation". Now, "a" was truly required: that the creation was to intercede between the old and new names, and I minimally find that the "a" in this relation is that minimum required for the modal collapse of $l^{-1} \vee p\&l \to \Omega$ to entail the similar collapse of: $g_x^{-1} \vee a\&g_x \to \Omega$. By MC0 this reduces to $a \to \Omega$. Then as $g_x \to \Omega$ and g_x is an essence; the conjunction of all $<g_x>$ spans or generates Ω or, equivalently "a" when walked through an octal, floating the unity as if equally the zero – the identity (virtue), as in the circuit of the angel: the least. Each set Ω of seven is generated by the least and fully completed.

In each case, I would expect "a" to accrue the broken closures in the circuit of the least. Essentially God must retain $g_x \to \Omega$ for Himself, but then God must write each g_x as some "a" in the least's circuit. Now, the e_x may or may not span the old creation (a_0) already (not being essences): all that is required is for the least to be provided the minimum elements of the "new creation" that will break the closure of the old and bring in the new.

Christ's letters provide these minimum conditions as gifts to break each closure so that the least will retain his own progress. Only an "I_x" is required: the least then retains some a=I_x&e_x from each broken closure. Then, each g_x is not only equivalent to I&e_x but also nearly equivalent to each I_x&e_x; that each reference the same "class" or each "church" of seven (i.e. each mountain of wormwood) overcome, without duplicating the person of the least.

Then the letters record Christ's gifts to generate these seven horns <g_r&g_s&...&$g_{u\&v}$> from the seven I_x upon each e_x and also on the least "I". God may only justify His coming with all seven I_x in effect. The least as an eighth must overcome first. The least, then, is possibly the only excluded property not in the conjunction <Ω_r&Ω_s&...&$\Omega_{u\&v}$>, the "horns of the lamb". (Yet it is true that N(Pos(I)).) Then a&I which entails Ω will contain "I" as well.

With every closure broken, the circuit completed, $g_x^{-1} \vee$ a&$g_x \rightarrow \Omega$ could be written as the set of broken closures: $g_x^{-1} \vee I_x$&$e_x \rightarrow \Omega$, each case g_x separated (the additive elements or "horns" chosen by Christ as 1 = 0 floats). The accrued closure would leave a remnant of creation that may or may not have a positive predicate to exclude the election of the least. It is the result of the Revelation that there are none found, the closure is complete and its complement empty.

Equivalently I could write: $I^{-1} \vee$ <I_r&I_s&...&$I_{u\&v}$>&$e_x \rightarrow \Omega$ (for the g_x generate "I" as indiscernible to that product), yet I may confidently write once every closure is broken: $I^{-1} \vee$ <I&e_x> $\rightarrow \Omega$ with a=<g_x>. This, again, is unique to the least. Then I $\leftrightarrow$ < $\wedge$ g_x> and God "writes His new name on the least", finally justified and free to enter all His creation as if made new. //

Modal Collapse Is necessary For Liberty (16.12.3a)

I begin with the disjunction $g_x \vee$ a&$g_x^{-1} \rightarrow e_x$. Now, I must have that "I" is principal in the new name (but possibly not the old on its own) and I may write $g_x \rightarrow$ I. Forming an octal, I must be able to derive $(r\&s)^{-1} \vee$ p&r&s $\rightarrow$ u&v from r $\vee$ p&r$^{-1} \rightarrow$ s. This will result in the minimal case, which is surprisingly equivalent (for some x) to (I&e_x)$^{-1} \vee$ a&I&$e_x \rightarrow \Omega$. Yet I require only I&$e_x \rightarrow \Omega$, breaking the closure of the old name.

Now, I&e_x is a set in the new "name" of God, the wider closure. That set similarly entails "I" as principal. Then closure is broken by "I", and I may rearrange and rewrite $g_x \vee$ a&$g_x^{-1} \rightarrow$ I as I $\vee$ I&I$^{-1} \rightarrow$ I because $g_x \rightarrow$ I and a $\rightarrow$ I.

Now, I reduce to the seemingly impossible I $\vee$ I $\rightarrow$ I. Then {u^{-1}}$\cup$ {v^{-1}} = {Ω\I}$^{-1}$ and therefore u&v = {Ω \ I}, acting with all creation in the set "a" apart from the virtue of liberty "I": i.e. working in {a = Ω \ I} (a result or "split" shown possible from applying liberty as in [4.8]). Then {a = Ω \ I} must span all creation (the set of virtue) in the new name and is effectively the conjunction of <g_p>&...&<$g_{u\&v}$> where each <g_x> is the closure Ω with an "I_x" in the elements (and churches as octals) of the least's circuit. These seven "horns of the lamb" (an equivalent in works to the "morning stars" – the works of the Holy Spirit and the seven days of creation) are written upon the least "I", with I = I&<g_p&...&$g_{u\&v}$>. (I infer a "basis" of all creation with a $\rightarrow$ {Ω \ I} overall spanned by these as if they were "vectors" gained from the octal.)

Given the disjunction in the **old** creation a_0 of: $I^{-1}\&e_{r\&s} \vee (a_0\&e_p\&I\&(e_{r\&s})^{-1} \rightarrow \Omega_0\&e_{u\&v})$, by lemma MC0 I rearrange to the desired $I^{-1} \vee (a_0\&I \rightarrow \Omega_0)$. By equivalence of the right hand side to $a\&I \rightarrow \Omega$ (cf. MC1, which is also correct and a modal collapse), "I" thereby breaks closure.

Then, logically, the necessity of Pos(I) decides the disjunction alone. Or, $N\neg(I^{-1}\&e_{r\&s})$, and $e_{r\&s} \rightarrow I$, a contradiction ("I" is principal in g_x). God cannot rest on $I^{-1}\&e_{r\&s}$. I must have $(I\&e_p)^{-1}$ instead, for here there is no disjunction and the term $e_{r\&s}$ adjoins to the filter in $<e_{u\&v}>$ and becomes e_p and principal; and I find: $(I\&e_p)^{-1} \vee a\&I\&e_p \rightarrow \Omega$ as desired.

Then in that modal collapse (as before) I also have now (manifested sevenfold in circuit within creation rather than in God), $I^{-1} \vee (a\&I \rightarrow \Omega)$ with Ω as my u&v and $a=<g_p\&...\&g_{u\&v}>$ where these g_x are found within the set(s) of virtue and then **"span creation"** (note I also must have removed all the predicates indiscernible from e_p) given $N(Pos(I))$ a virtue and "I" principal in creation itself.

I have seemingly evaluated the disjunction and dropped the terms in e_x that were otherwise present. (I removed the one predicate of "God-likeness" but for "I".) My justification for this is that the set Ω then entailed is taken **only within that of virtue** maximised with "I" principal. (See section [4.8].) As creation is acting as virtue here, a&I then spans all creation.

Once more, I then have (without the principal elements e_x) the disjunction: $I^{-1} \vee (a\&I \rightarrow \Omega)$. (Note that Ω contains the full and maximal closure broken by the least and so taken together sevenfold, each g_x of the seven is free in those horns (spanning creation in "a") as if a "vector-basis" by analogy, those prior closures in e_x broken; the principal name g_x is equivalently "written on the angel" from the completion of the angel's circuit breaking the closure seven-fold to prove him principal in the old name (nothing positive excludes him from election). God, now with His new name, truly principal within that basis as $<g_p\&...\&g_{u\&v}>$ is proven present in His creation whilst the least "I" remains truly unique – see the next subsection.)

As the middle (of $I^{-1} \vee a\&I$) excluded is truly **empty** then (once again) a&I must remain an equivalent to $\neg I^{-1} = I$, so the seven "horns on the lamb" $g_p\&...\&g_{u\&v}$ are "written on the least" with $\{\Omega \setminus I\} = I_p\&...\&I_{u\&v}$. The horns I_x given him are the very **minimum** to break closure with $Pos(I_x\&e_x)$ and $N\neg Pos(I^{-1})$ in the circuit of the octal. God, in each "$I_x\&e_x$" (an essence), may span creation, accruing closures, forming a "pillar in the temple" in which His seven spirits g_x may dwell, as in any possible creation of His choosing.

Every such modal collapse $(I\&e_x)^{-1} \vee a\&I\&e_x \rightarrow \Omega$ with the right hand side necessary, is equivalent to $g_x^{-1} \vee a\&g_x \rightarrow \Omega$. Each $g_x \leftrightarrow a\&g_x$ becomes an essence of God's name, a fact justified by the least "I" in circuit floating g_p sevenfold through an octal. Then $g_x \rightarrow \Omega \rightarrow g_x$. for the seven g_x, each equivalent to the sets of positive predicates in which they are present.

Clearly, all positivity is then entailed in the octal only when there is no liberty, and God must necessarily exemplify Ω as then He is set at liberty. God may with liberty make a "cut" in Ω (as Ω is as His person otherwise) and entail some p from the set of properties in Ω to suffice: $\Omega \rightarrow p$ as necessary. I.e. God, effectively, chooses Ω to suit the collapse in the disjunction. By axiom of virtue, there is complete collapse entailing full closure, but only that property p extant that suffices to return God to rest. (For that is necessary of liberty.)

Note that the set from which Ω is entailed is that of virtue, for it is both necessarily positive and entails the set u&v whilst then bound to r&s. I will then write that $l \rightarrow G$, where G is the set of all virtue, but clearly L=l and $L(G) \rightarrow p$ as required, for L(G) rests on only that virtue required to return God to rest (this was to entail "s" in the conjunction $r \vee p\&r^{-1} \rightarrow s$ beforehand).

The Least Overcomes the World (16.12.4)

Now, to overcome the world, "l" must himself be principal. In each case he is required to show that <l> is a finer filter than any <e_x>. The disjunction $l^{-1}\&e_{r\&s} \vee a\&e_p\&l\&e_{r\&s}^{-1} \rightarrow \Omega \&e_{u\&v}$ must result in modal collapse. It must be true that given r = s = l, that $\{u^{-1}\} \cup \{v^{-1}\} = \{\Omega \setminus l\}^{-1}$ and therefore u&v = $\{\Omega \setminus l\}$.

For modal collapse there should be no middle in $l^{-1} \vee \Omega$. In fact, there must be both $\phi \vee \Omega$ as well as $l \vee \phi$ (from negating both sides in all Ω). This would not be possible if there was (a fallacy of) an excluded middle. Then $l \leftrightarrow \Omega$ and there cannot be any positive x in Ω such that $x \rightarrow l^{-1}$. Nothing positive can exclude the election of "l" if "l" is principal. All that remains is to show $N\neg(x \rightarrow l^{-1})$ for all Pos(x) and also Pos(l); that the election of "l" will entail a positive result and then "l" is shown principal.

I.e. any essence would not be able to exemplify any positive x to entail l^{-1}, and as Pos(l) is supposedly true, given that essences are themselves positive, Pos(l) must follow from all such essences. God, then unable to positively exemplify Pos(l^{-1}), must consider "l" principal as He Himself is perfected in all positive properties Ω, not merely a subset relaxed by liberty that could entail an excluded middle to be found in $l^{-1} \vee \Omega$.

Then to be found principal overall (to overcome the world), <l> must be a finer filter than <e_x>.

Every set in <e_x> must be found in <l> as some $l\&<e_x>$. Given "l" principal in the new name, I find just that. There can simply be no predicate to exclude (private) the election of "l" in that filter <e_x>. Now, Christ has His conditions in His letters, for He states:

Rev 2:4 Nevertheless I have *somewhat* against thee, because thou hast left thy first love.

If these are met, there is no longer anything to exclude "l". That "l" is principal in the new name is another matter. He is the right hand of God; he cannot simply be replaced by a volunteer, he a-priori existed and the job is not up for grabs.

Mar 10:40 But to sit on my right hand and on my left hand is not mine to give; but *it shall be given to them* for whom it is prepared.

Yet the circuit must be made by "l" to overcome the closures of the Holy Spirit (<$\wedge e_x$>) in his circuit, to justify the new name by a modal collapse. The conditions of Christ are written to keep that circuit going, and for the least to accrue those closures in the horns. The circuit is then from "church" to "church", and so each are "spiritually closed" but for Christ's conditions that would lead the angel to leave or otherwise repent. ($l^{-1} \vee a\&l \rightarrow \Omega$ surely entails $l \leftrightarrow r\&s$.)

In each case a church represents the conjunction $l^{-1}\&e_{r\&s}$ which must be overcome, and whilst the least is given no apparent choice as in $l \vee l$, $l^{-1}\&e_{r\&s} \vee a\&e_p\&l\&e_{r\&s}^{-1} \rightarrow \Omega \&e_{u\&v}$

appears to modally collapse and force $e_{r\&s} \to I$ upon him, that he should remain in the "old" closure. Instead, given no choice, he may simply move on to another closure by realising the contradiction of that closure with that of the new; those same conditions written in Christ's letters.

Then God will not rest on $I^{-1}\&e_{r\&s}$ and there is the modal collapse of $(I\&e_p)^{-1} \vee a\&I\&e_p \to \Omega$ instead which is found genuine.

Then by $g_x \to e_x$ and axiom Ax4, $(g\&I)^{-1} \to g^{-1}$ so the necessarily decided $(I\&e_x)^{-1} \vee a\&I\&e_x \to \Omega$ becomes by modal collapse $g_x^{-1} \vee (a\&g_x \to I)$ where $a\&I=I$ and $s=I$ and $(I\&e_x)^{-1} \to (I\&g_x)^{-1} \to g_x^{-1}$.

Then similarly, $g_x^{-1} \vee I\&e_x \to I$, or again, without breaking closure of virtue in $L=I$, I have $g_x^{-1} \vee I\&e_x$. To rest on the empty middle in the octal, I have $N\neg(I\&e_x\&\neg g_x)$. Then as $L=I$ and "I" is principal, for every one of the seven ultrafilters in the octal I now have $L \equiv a \equiv I$ and $g_x \leftrightarrow I\&e_x$ and God writes His new name on the least in the kingdom in order to keep it holy, not further put to the test! (The least is "called by His name" – to test God's own liberty requires that honour for "I" in order to keep "g" holy as I's equivalent: "I" is the predicate of "God-like" or "rest" or even "necessary existence". The name is written on the least because the **test** is Holy!)

...And God does so write necessarily! (Yet $g \vee a\&g^{-1} \to e$ is only decided by God, not by faith.)

But, does "I" belong in a or a_0? (God requires the former to decide the above.)

I must have that $g \to g\&I$, so $g \to I$. Now, clearly "I" is "not of this world" (as g is not) so I state that $I \in a$. Now, "I" is then the "essence" of God as a principal element and by Gödel's argument[1] (or modal collapse – see above), the predicate "I" will entail or prove g (i.e. $I \to g$).

I also state rather simply that $e_x \in a_0$ and $g_x \in a$ but $g_x \notin a_0$. Then $I\&e_x \in a$. Now, $a \to a_0$ but $a_0 \not\to a$ so if $a_0 \to I$ then as $a_0 \to e_x$ I find $a_0 \to I\&e_x$ or equivalently $a_0 \to g_x$ a contradiction: therefore "I" breaks closure when God writes His new name on the least, after the angel overcomes the world (Ax4). Then begins the "reign of omnipotent God".

The least breaks all closure sevenfold in his circuit by overcoming. Each closure broken is such that nothing positive in that closure may exclude his election. Then the least in the kingdom of heaven, once justified by a "horn of the lamb" moves on to another closure, to overcome it in turn. When he is then proven principal in the new name, again, closure is broken. The time for this is determined by Christ with the "key of David" (see later). Finally the victor, the least is truly of the new name of God (wholly a new creature).

Yet "I" is found principal over the old name also (contradicting the old name in the e_x, which has no principal essence and no analogue of modal collapse), whilst still of the new name completely. I simply state that God's kingdom is here in this creation and though a_0 is closed, $a \to a_0$ (or $a \supset a_0$). Closure is broken and new supersedes old ($a = a_0$).

So, God necessarily writes on the least (he accrues seven horns, breaking every closure) and He does so only in the current creation, simply requiring the least to come to faith and to overcome the world without blaspheming the new name of God. For all $x \in a_0$ it may be safely

[1] Sobel, in J. J Thompson (ed.), *On Being and Saying: Essays for Richard Cartwright*, p241-61.

stated that x is "possibly" a tare and there is no necessity of election and no certainty that "a" is a descriptor of x. a_0 simply has "too many tares". The name cannot be safely published.

That name is of seven closures, seven horns, seven spirits. The eighth closure is the city of "New Jerusalem" (the new creation). The eighth element of the cycle is the person of the least (those closures possibly eight-cycle in self-reference or regress).

In the old name alone, without a principal element in creation it cannot similarly be found that each e_x is an essence. I may clearly form the bare disjunction $e_r \vee e_p \& e_r^{-1} \to e_s$. Then, $e_{r\&s} \vee (a\&e_p\&e_{r\&s}^{-1}) \to e_{u\&v}\&\Omega$ for the set of all virtue "$a\&e_p$" where e_p is equivalent to $e_{r\&s}\&e_{u\&v}$ and $<e_u> \wedge <e_v> = <e_p> \subseteq \Omega$ (note u^{-1} and v^{-1} are not in scope with e_p). So, u&v may at most be the set $\{\Omega \setminus e_{u\&v}\} \subseteq \{\Omega_u \setminus e_u\} \cup \{\Omega_v \setminus e_v\}$ and there is equality only if, say, $\Omega_v = e_v$. The schema permits $(a\&e_p) \to e_{r\&s}\&e_{u\&v}\&\Omega$ which is $(a\&e_p) \to e_p\&\Omega$ ($<e_{r\&s}> = e_{r\&s}$ only). I may drop the term in e_p leaving only $\phi \vee a \to \Omega$. God may create anything with a positive result (is this why He loves the world?) but creation is contingently existent and can be no "essence". There is no equivalent of modal collapse $e_p \to \Omega$, yet "God-likeness" is still a valid predicate (essence) stating each e_x is the equivalent of NE and is principal.

Could God's sovereignty be challenged simply because His (old) name is not equivalent to an essence? The possibility of full modal collapse now aside, Gödel was probably right.

The new name of God g_x (through the result of $I \to \Omega$ with $g_x \to g_x\&I$) is then shown an essence through the agency of all creation; the seven "God-like" g_x each become principal and equivalent to I across every disjunction of the octal.

Now, as $I\&e_x$ is an essence in the new name, and entails g_x, if e_x were also an essence in the old name, it would be simple to repeat the argument as for the new name and state that g_x is an essence similarly! That is, if g_x entails e_x, which it must do if e_x be an essence, for e_x remains positive and necessary. However, God requires the mechanism of creation to prove g_x to be an essence equivalent to "I", and God will never again need to prove any "newer name" h_x principal and "God-like", it being a simple argument already won with $h_x \to g_x$ an essence and all closure broken.

Then $g_x \leftrightarrow I\&e_x$.

Now, something has been proven, the necessity of $g_x \leftrightarrow I\&e_x$ as an essence (as of Gödel's argument[1]). Every positive predicate is entailed by a set in the new name of the wider closure. Likewise, this is found from within a closure properly containing the set e_x. As there are seven of these, the new name of God in Ω is made manifest in creation in conjunction with that zero element "I".

Then every subset of Ω becomes necessary (entailed from that essence in the new name of God), and with "I", the old name's closure has been broken and the new name is truly made manifest. There is then an eighth set, which is now the creation itself and which has broken the closure of the lower and "old" name(s). It is properly contained within the new, as "I" cannot be an element in the old name without breaking closure and so there is necessarily

---

[1] Sobel, in J. J Thompson (ed.), *On Being and Saying: Essays for Richard Cartwright*, p241-61.

a God with a new name, a wider closure, necessarily omnipotent in all creation, and God (effectively), is a-priori "born again" with a new name, forever perfect as a God in all positive predicates alone is not a necessarily existent being given the existence of a greater wider closure in a new name.

Now, I find that "I" must be principal in the new name, and that $l\&e_x$ becomes principal and as g_x, for $l\&e_x$ is an essence, and every God-like set must entail necessary existence as through Gödel's argument[1] and as NE is positive, I find that in every God-like set NE must be principal and I therefore entail $l\&e_x \to g_x$. Now, I have one axiom, that $g_x \to g_x\&l$, that God, is the God of "the least".

Then truthfully God has always had that new name and creation merely becomes His justification for acting with it, as "all creation declares the glory of God".

The Kingdom Through A Glass Darkly (16.12.5)

Pos(g) for all $x \in a_0$ is false.

Pos(g) for all $x \in a$ is true. Now, $l \in a$ and therefore Pos(g) is true for some "I", so Pos(g) is true for $l \in a_0$ as $a \to a_0$. (I.e. $a_0 \subset a$.)

As $l \in a_0 \subset a$, once the new name is written on the least, then $l \to$ Pos(g).

I.e. $g \lor l\&g^{-1} \to l$ and in combination with N(Pos(g)) (Ax1) I state **without breaking closure** of virtue in "I" that $N\neg(l\&g^{-1}\&\neg l)$ easily rearranges to $l \to g\&l \to g$. I may argue "I" is virtue and g necessarily positive, but for the disjunction to be freely decidable "I" is surely necessary, and is so for God to exist in the model's seven ultrafilters with virtue. Now, N(Pos(g)) so I reduce to $N\neg(l\&\neg l)$ which entails "I" is necessary (consistent or provable) given g (without breaking closure of virtue), and I infer that $N\neg(l\&\neg g)$ by absorbing the superfluous "I".

Then I may somewhat state "there is salvation (found in g) as there is the saved ("I")". I can state that $(g\&l)^{-1} \to \neg l$ (see section [2.1b], cf. [2.8.1]) and $l \to g$ with the modus tollens. However, g is God-like, and entails (maximising the disjunction and choosing the one side maximised) the freely chosen $g \lor \phi$ which always entails $l \lor \phi$, as "I" is "an essence", equivalent to "God-likeness" or "necessary existence" (in that as an essence, $g \to l$ as by Gödel's argument.[1] "I" is of "a", not only a_0: so "I" is found breaking the closure of a_0 and entails g after God necessarily writes His new name on him. Note that $N\neg(l\&\neg l)$ permits $\neg l$ to be consistent with g also! "I" is put to the test, not God). Then I apparently require modal collapse for justifying $l \to g$. (Yet, "I" is liberty and God may choose the set Ω in the disjunction freely!) Then by Gödel's argument[1] I may minimally (without loss) entail only the **necessary and closed**, $l \to g$.

I.e. the disjunction rearranges to $l \lor l\&l^{-1} \to g$ or clearly $l \to g$ (it is true that N(Pos(l)) as virtue, so N(Pos($l \to g$)) which God will exemplify) without breaking the closure of "I". I.e. strictly $l \lor g$, if "I" is merely positive ("I" not principal in g, "I" could blaspheme g) and N(l&g) otherwise. Then $l \to l\&g \to g$ if "I" is principal in g. ("I" is also an "essence" or "God-like", entailing every set containing it to be "God-like". "g" is "God-like" if it entails "I", and "I" entails that essence "g" on every set in <l>.)

[1] Sobel, in J. J Thompson (ed.), *On Being and Saying: Essays for Richard Cartwright*, p241-61.

There is modal collapse requiring $N(Pos(g))$ and $N(Pos(l))$ to entail everything positive if "g" were to break the closure of "l" (which it truly does, as "g" itself is also principal and as a virtue in "a"), yet both must remain disjoint sets, for $l \lor l\&l^{-1} \to g$ is otherwise inconsistent in virtue, and I would have one of $(l \to \Omega) \lor \phi$ or $(g \to \Omega) \lor \phi$ or then $\Omega \lor \Omega$ and $\phi \lor \phi$. Instead, only $(l\&g)^{-1} \lor l\&l\&g \to \Omega$ cannot be freely decided. I must have $\phi \lor l\&g \to \Omega$. "l=L" as liberty satisfies every inference that each act of omnipotence is freely decidable, $l \to l\&g$ as well as $g \to l\&g$, for then $l \leftrightarrow g$ are equivalents in "a", $l \lor \phi$ or $g \lor \phi$, but both "l" and "g" remain members of disjoint sets ("l" becomes the additive identity however) and both are "God-like" if found in this modal collapse.

Both sides of $g \lor l\&g^{-1} \to l$ are equally positive. ("l" can equally be put to the test as can God.)

Now, to decide all election in creation (in the general sense) with $a_0\&c^{-1} \to e$;

$l \to c$, $c^{-1} \to l^{-1}$ and the old name (unbroken) and old creation are sufficient for all election.

$e \lor a_0\&e^{-1} \to l$, so $l \lor a_0\&l^{-1} \to g \to l$ if "l" overcomes in $a_0 \subseteq a$ ($<l> \supset <e>$) and proves himself principal overall (rather than e), resulting in a modal collapse $l^{-1} \lor a_0\&l \to \Omega$ (i.e. God may rest on the left hand side whilst there is no "l" as yet, closure of a_0 is not yet found broken with "l".) Then l^{-1} is positive and $a_0\&l^{-1} \to e$ is extant, as is $a_0\&f^{-1} \to e$ ("e" is principal) for every $f \in a_0$ with $f \to c$ and so $f\&f^{-1} \to e$ and $f \to e\&f$. ("f" rests on the side freely extant: $f\&f \to e$.) //

$e \lor a_0\&e^{-1} \to c$. Now, "l" cannot blaspheme so $e \lor a\&g^{-1} \to c$ in the case of "l" (closure of a_0 broken). Then as "l" is principal in "a" with $f \to l$, $e \lor f\&e^{-1} \to l$ and as $f \to e\&f$, I find that $f \to e\&l \to g$ and finally, as $f \to e\&f$ is found in the set $e \land g$, $\exists l : \forall f, f \to l, f \to g$ in "a".

The elect $f \in a$ ($f \to l$) cannot blaspheme the Holy Spirit ($e \bar\lor e$ in a_0 and $g \bar\lor g$ in "a") once the least overcomes. I.e. given e^{-1} at rest is "God-like", $e^{-1} \to e$ in $e \lor f\&e^{-1}$: I also have $g^{-1} \to g$ in $g \lor f\&g^{-1}$ likewise. (I note "l" is principal so $l \to g \to g\&l \to e\&l \to l$, so $l \leftrightarrow g \leftrightarrow g\&l$.)

The Least is Surprisingly Sinless (16.12.6)

Now, to put the heaviest constraint on "l" or equivalently a_0, I can state that $l \to c$ for all "sins" c^{-1}. "l", must be found blameless, i.e. sinless. Yet how would God keep "l" blameless? I.e. $l \to c$ or $a \to c$. Note "l" is principal and entails no other positive properties. So $l \to x$ is as $l \leftrightarrow x$.

If you "do not do so" you are "blameless" so I assume there is some x such that $x \to (l \to c^{-1})$ in such terms that x is not considered sinless in a_0 (the old creation), "x" is somewhat "at fault" for any apparent sin of "l"). "x" is a curse of "death": exclusion from the first resurrection.

I.e. $x \leftrightarrow N\neg(l\&c)$ therefore $x^{-1} \leftrightarrow P(l\&c) \leftrightarrow \neg N\neg(l\&c)$. So, if x appears to cause "l" to sin then $l\&c$ is possible if x is "found a lie" or x is justifiable as an act of "l" if "l" is justified, i.e. "l" made an act of charity causing his captivity to x. "x" is then a sure state of captivity, and "l" is caught under accusation. (Yet "x" is not the curse of God but one only found of the enemy.)

Now, as x is sinful in a_0 (holding the innocent "l" with charity captive): God cannot exemplify x at all (in the new creation, $g \in a$), so $l \to (\neg x\&\neg c)$ and breaking closure of the old name (the least finding the new – breaking closure is a consequence of him overcoming the world, for then $\neg x$ for $l \in e \wedge g$ "defeats" x for all f in a_o. (Because x was no sin of "l" ($l \nrightarrow x$), even if it appears to cause "l" to sin. $N\neg(x\&\neg N\neg(l\&c))$ is the same as $x \rightarrow (l \rightarrow c^{-1})$ and it rearranges to:

$N\neg(x\&P(l\&c)) \rightarrow N(\neg x\&N\neg(l\&c))$ as modal status is always necessary.

$N(N\neg(l\&c)\&\neg x) \rightarrow N\neg((l\&c)\&x) \rightarrow N\neg(l\&\neg c^{-1}\&\neg x^{-1})$

I.e. $l \rightarrow (x^{-1}\&c^{-1}) \rightarrow N\neg(x\&\neg c^{-1})$. Note I also have $x^{-1} \leftrightarrow P(l\&c) \leftrightarrow \neg N\neg(l\&c) \leftrightarrow N\neg(l \rightarrow \neg c)$.

So now, $l \rightarrow (x \rightarrow c^{-1})$ or $l \rightarrow (c \rightarrow \neg x)$

Anything that causes "l" to sin is defeated on the premise that God would never exemplify any such x. (I may likewise use the distributive law because it is perfect to do so. "l" is predestined.)

Stating "l" is holy despite such an x(l) is **blasphemy of the Holy Spirit**, if x were not caused by a sure act of charity! I also state that $l \in a$ necessarily as the least in the kingdom of God. There can be no such "x" in the new creation, and by breaking closure there is then no "x" in the old. "l" is sinless by contradiction! (NB: "l", as principal is equivalent to L, purely divine liberty.)

If "l" caused himself to be put to x by an act of charity (entailing no blasphemy) then I find that $l \rightarrow N\neg(x\&c)$ or both $x \rightarrow \neg c$ and $c \rightarrow \neg x$ and then because "l" is justified (not truly fallen at all from a certain act of charity causing x(l)) then $l \rightarrow (c \rightarrow \neg x)$ and "l" is blameless for all possible sins c^{-1} so **entailed**. Then "l" overcomes the world $g \rightarrow g\&l$ and closure is necessarily broken.

Now, for any individual "f" caught in that captivity of x; I may state that $l \rightarrow c$ and $l \rightarrow u\&v$ so I entail that grace is sufficient for all f such that x(f), i.e. x is a predicate on f; f is "caught" as "l". Given x(f) I infer that $a_o\&e^{-1} \rightarrow c$ for all $f \in a_o$. I of course have $l \rightarrow (c\&x)^{-1}$ or $l \rightarrow N\neg(c\&x)$.

So if there exists such an "l" then $a_o\&e^{-1} \rightarrow l \rightarrow c \rightarrow \neg x$ given "l" is the very least of all and $l \leftrightarrow a\&l$, then the new creation (new creature) breaks x with closure (the full liberty of God). So, with the very least it is proven that $c \rightarrow \neg x$ for every elect: and therefore all "f" which are without exception (but for Satan) free for God to elect (there is no complement set in a_o: the gospel saves from all sin and captivity; the gates of hell will not prevail against it).

Then $a \rightarrow a_o \rightarrow (\exists c : a_o\&e^{-1} \rightarrow c \rightarrow \neg x)$ for all "f". Simply because "l" will overcome the world $(g \rightarrow g\&l)$ and "l" is blameless. Therefore, $N\neg x(f)$ is logically true from the least on up, not just the least Himself, which is the case for all other "f", those being equal and of this old creation.

Now, is "x" the very worst circumstance? Can the right hand of God have been sold to Satan in an act of charity to regain those likewise under an oath, contract, bonds or any other captivity unto the one accuser of those "f" before God? (Satan, otherwise, yet has a "kingdom".) There is no such "x" by virtue of the argument above; there is no such sin in the new creature.

Then as "l" is principal in "a", and $a \rightarrow a_o$, I have for all "f" in "a" with x(f) that $f \rightarrow l \rightarrow c \rightarrow \neg x$. So if "l" is truly principal in "a" (and $l \leftrightarrow c$), then it is logically true that $f \rightarrow \neg x$ for all "f", including all "f" in "a" and also, notably all "f" in $a_o \subseteq a$ when closure is broken. Breaking the closure of a_o leads to the least "l" in "a" carrying $\neg x$ necessarily, and then there is no fault in either "a" or a_o for any "f" caused by any such "x". If "x" appears to cause "f" to sin, that is a faulty inference and though it is not as any other sin, requiring an act of grace (requiring charity by "I" – the apparent blasphemy of the Holy Spirit in x is not able to be necessarily forgiven under the gospel of Christ otherwise), it is merely a deception – for the test of "x" is itself faulty, not the individual "f" to which "x" applies. "f" has no residue of sin caused by any "x". Such as would hold captive all "f" by such x are at fault, and the fault of the test lies solely upon the cause of "x" in the case of "I" alone (I.e. Satan as holding all "f" captive) and thus, as so for all, all such "f" are easily forgiven as "I" remains holy. I.e. I $\leftrightarrow$ c and God cannot exemplify any such x(f) in "a" and therefore not in a_0. (To save the principal "I" is to exercise the axiom of sovereignty.)

Breaking closure with the least "I" proves N¬x(f) for all "f". (Possibly even for "I" if x(I), as "I" is such an "f". Otherwise L=I, and God cannot initiate (or exemplify) any x(L).) I.e. f $\rightarrow$ ¬x as desired (all debts to Satan are void in a and a_0), given f $\rightarrow$ I. (Every believer must respect God's sovereignty choosing His least for their own selves' sakes.) Only the fault of entering such captivity (a "mark of the beast" or "second death") remains, now defeated with the charity of the cross of Christ, and election is open freely to all (also "I") in the eyes of God: no matter the duress forced on or of those under Satan. (Cf. Mat 18:6-7.)

Mat 25:27 Thou oughtest therefore to have put my money to the exchangers, and then at my coming I should have received mine own with usury

The act of simply selling oneself to free those captives of x(f) is an act of charity if it is to have this very outcome! Even the least has his own ministry. (Every blood oath to Satan, or to Lucifer etc., is broken, now overcome in Jesus Christ. Cf. Acts 13:39.) Then every consequence is struck out as any fault x $\leftrightarrow$ I is become as a lie of the enemy, not of the least. (Satan, consequently, owns **nothing** in all creation! Satan receives nothing from God, and ultimately, nothing from any man: Satan is broken.) One sows and another reaps, but the least is become a pillar: from baseline (Christ) to the full and ultimate closure of God's new name. He, overcoming, is not to be followed: grace is then completed. (The pillar holds up the roof, after all!)

The Morning Star (16.12.7)

Now, if the argument between "I" and say the cause of x is such that Satan "owns" "I" or Satan does not "own" "I", then as "I" is an essence as well as principal I must apply either x(I) or ¬x(I).

If Satan does not own the essence "I" then I $\rightarrow$ c and I $\in$ a, the essence is equivalently that of God's own. Then "I" is free from all condemnation and is also saved as per the above.

On the other hand, if x(I) then Satan must claim he "owns the essence" and then has some sovereignty over any x(f), i.e. I $\rightarrow$ Pos(x) (as if x(I) were also positive, found so to exclude "I" from salvation) in order to put himself between God and "I". As "I" is also principal, i.e. x $\leftrightarrow$ I, Satan would thereby interrupt the Kingdom of God. Yet, even then Satan as if "x" causes "I" to sin and "I" is still then justified by an act of charity.

Now, given that act of charity, can Satan own "I" (I $\rightarrow$ Pos(x)) without causing him to sin?

If there is no such x and x(I) there is no argument, I then assume there is no benefit for "I" in concord, merely some contract entered into by that charity.

If I → Pos(x) and x does not cause "I" to sin, then I assume x is positive in a → a_0, without any consequence. On the other hand, if I → Pos(x) and x causes "I" to sin, then I → c^{-1}. Yet if x(I) is an act of charity, there is no fault c^{-1}. I also have "I" principal so I → Pos(x) entails f → Pos(x) for all f in a, as well as a_0. Surprisingly, ¬x in a → a_0 becomes a virtue and all x(f) are necessarily redeemed but for Satan alone who remains damned and always positive to exclude as I → Pos(y). (For "I" also "saves"; "I" is that charity – it is agape, and of the Father.)

I then find a rule that every permissible act of charity that appears to break the law is without any fault for all f ∈ a and a_0 (this, includes God Himself; I also note "I" is elect). Then, Satan must claim he caused "I" to sin as if by y → (I → Pos(x)) to accuse him, interrupting the charity. I.e. y → (I → c^{-1}) despite the act of charity (or apparently not so and of Satan). Satan, then, becomes such a "y" himself, not putting "I" at fault and so "I" is completely vindicated. Satan as y, substitutes for "x" as from before. His accusations cannot stand; Satan, causes "I" to sin.

Satan's only option is to claim I → Pos(y) → Pos(x), for all x, as if y, were most negative and an "enemy" to exclude. It appears Satan (as the ultimate devil) is always so fallen, yet any captive x(f) (of Satan as y) excluded by an x (y → (I → (Pos(x) → c^{-1}))) is also redeemable (for "I → c" remains an essence). It stands before God that the chain in the x(f) will effectively end with Satan damned (Pos(y)), broken of all others. That contract in x, unable to cause "I" (or any f) to sin is safely dissolved without sin. Only Satan remains as the cause of that sin and he will not be absolved (though N¬y(f) for all f not y). Then "I" spoils Satan of all possible f (Satan also entails Pos(I), as the devil required to devour him), returning all to God glorified upon His own essence, finally justified to judge all. (Cf. Dan 7:13-14 and 1Co 15:27-28.) Satan is robbed of everything God had excluded from His kingdom, every "f" and "x" is redeemed and f → I → ¬x → Pos(y) for all f ∈ a and also a_0. Then ¬y(f) and as God's essence is "I", the principal element of all creation, God is justified as creator of all and with His new name, all forever afterward.

Then "I" is holy, found so from his essence – and is a "morning star" – and not the Satan that would accuse him. (y → (I → (Pos(x) → c^{-1})) does not permit I → Pos(y) → x(f) → Pos(x).) The least has descended as if from the heavens and has risen from the depths justified, shining.

Not Climbing Up Some Other Way (16.12.8)

Bearing on, I find an answer as to how far grace can actually be stretched: I will find "blasphemy of the Holy Spirit" is only irredeemable if it be continuously and actively upheld in death (as without true repentance and faith with the one true gospel of atonement found in Christ).

Now, the least cannot climb up "some other way" (as without having a wedding garment, cf. Mat 22:11), and the angel's work of charity must be beyond accusation. I examine the case when x(I) and I necessitate upon the intercession of Christ as "p". It becomes clear why Christ could not have caused those captives in x(f) to be loosed Himself. It can only be found possible from accusation that I → x and not p → (I → x).

So, rather than just x ↔ N¬(I&c) I posit I → x and so P(I&c) → ¬x → ¬I. I.e. "I" cannot be elected without being necessarily sinful I → c^{-1}. Christ's right hand would "cause Him to sin" – he must then be "cut off": c → ¬I. (The same is true for any such f with x(f).)

To gain the atonement for "I" (or f) in Christ I simply require (by axiom of virtue) the freely chosen disjunction of $Pos(l) \vee Pos(p\&\neg l) \rightarrow Pos(\neg x)$ and there exists some p in omnipotent God. (The charity of p=l would clearly suffice for the work!)

Now, "l" is principal so the above $l \vee p\&\neg l \rightarrow \neg x$ which rearranges to $\neg x \vee p\&x \rightarrow l$. But "l" is principal as ever so (positing "l" to be elect) $p\&\neg l \rightarrow \neg x\&l$ so without breaking closure of virtue in p I immediately have: $p \rightarrow \neg x\&l$. However, "l" is divine liberty L or $\wedge g_x$, always present in the set of "p", so $p \rightarrow \neg x$. Then clearly as $p \rightarrow c$, p is found to remain utterly blameless with no blasphemy found of any "x" at all.

Such a "p" clearly exists in the person of Christ Jesus (the Lamb). Jesus cannot blaspheme Himself by choosing "l" to be elect, even if $l \rightarrow x$. (For the law is of His own free choice as is all atonement before the Holy Spirit.) "l" must clearly satisfy the gospel to be chosen as elect without any fault found in p. (Albeit Jesus would know that His right hand works with charity.)

Now, returning to $x \leftrightarrow N\neg(l\&c)$ I now substitute p for "l" and once again I find $p \rightarrow N\neg(x\&c)$ and $p \rightarrow c$ entails $p \rightarrow c \rightarrow \neg x$ necessarily for an "l" such that $l \rightarrow p$ (unique to the least). Now, given L=l and "l" is uniquely "God-like" ($l \rightarrow L(G) \rightarrow p$), I take full advantage of modal collapse, for I state that in this case $N(Pos(l))$. (Grace is perfected (completed) in "l", and God thereby breaks closure! Or He otherwise recognises it as broken.) The disjunction deciding the election of "l" is not freely decidable.

I.e. $l \vee p\&\neg l \rightarrow \neg x \rightarrow l$ in a_0. p intercedes for all non-blasphemers elected as into the new name, thereby entailing "l"; yet only whilst "l" is not elected yet. The contradiction is resolved in that all such x have no "residue of sin" after the election of "l", and x does not require further intercession. (And Christ remains blameless.)

Now, $r \vee p\&\neg r \rightarrow s$ has the result of $(r\&s)^{-1} \vee p\&r\&s \rightarrow u\&v$. Here, r=s=l and $\{u^{-1}\} \cup \{v^{-1}\}=\{(\Omega \setminus l)^{-1}\}$ So $u\&v=\{\Omega \setminus l\}$. Then $(r\&s)^{-1} \vee p\&r\&s \rightarrow u\&v$ becomes by modal collapse $l^{-1} \vee p\&l \rightarrow \Omega$. God, then, may decide the set $G=\Omega$ in the disjunction. (It is truthfully and surely entailed with faith, but is properly found written in the circuit of the angel (the very least) with the seven horns of the lamb.) Closure is broken by the least of the ultimate closure, that of the Most High. Since "l" acts with liberty whilst found blameless, I truthfully and equally find no sin in x for the elect, and that p=l and that $l \rightarrow \Omega$. All closures then broken, a pillar is set up in the temple of God. (The least as "God-like" overcomes.) The Most High cycles all closure with the least as the identity element, $l=\wedge g_x$ is sufficient.

Then by modal collapse I find $l \vee \phi$ and as "l" is "God-like" now entailing all virtue in the set of all virtues G (L=l always acts on the side of the disjunction which is extant), the disjunction is not freely decidable and all G is possibly entailed without the virtue of liberty "making a cut". I have actually found $l \rightarrow p$ from all G. The right hand of God is unique in this, and thereby God gains His closure on the broken back of His enemy, as with His hand "tied behind His back".

Now, "l" climbs up only in obedience to the gospel, as Jesus (as in p) must remain faultless in choosing him as elect. His right hand once "cut off", "cannot cause Him to sin". There must needs be a faultless intercessor to excuse the least from all accusation by the enemy. Jesus has supplied him an open door that no-one can shut. (He only circuits once to break all closure.)

Likewise, Jesus could not have saved those captives x(f) with His own ministry, as when p becomes principal and f → p for all **elect** f, then p is likewise caught as is His own hand "I", even though p is God-like as is "I" and would act purely through charity, and also would break all closure. (L=I could not so acquit Jesus in kind, since of necessity **all** p entails L=I, and f cannot entail p for all other f not "I" in a_0.) p → ¬x would yet hold only if x → (p → c^{-1}) were false: if Christ were to openly appear to sin claiming Himself Holy with such a very real and present x, before two witnesses (that would certainly agree and sin would then be established) the law would be wholly satisfied and He would have failed, legally executed for blasphemy. He would never have been raised blameless.

Now it is also true that the least must be the right hand of God and also "God-like" to complete this mission, for whilst being the least is enough, it is necessary that "I" be truly "God-like" to escape all accusation by freely entailing l → p. Then only the right hand of God can necessitate this breaking of closure, thereby robbing Satan blind of any claim to God's creation. Then the least has a unique and small or "miraculous strength" (i.e. of f → l → ¬x for all f in a_0 and since p&¬x → f → c&¬x only for those f → ¬x in $a_0 \subset a$ that are born again, Jesus could not justify those captives (to any x) similarly or otherwise) and it is true that:

Mat 11:11 Verily I say unto you, Among them that are born of women there hath not risen a greater than John the Baptist: notwithstanding he that is least in the kingdom of heaven is greater than he.

Theorem (16.12.9)

If ¬x is a virtue in a_0, and x is possibly positive in "a"; then x may be exercised as a virtue in "a", and after closure is broken, ¬x simply becomes as e_p in a_0.

(A Modal Collapse from a broken closure results in a virtue to bring God to rest in liberty. God may always rest on some e_x.)

Proof: Given a collapse as before in Theorem MC1, the collapse is entailed from the set of virtue, which does not (by axiom) private any property of Ω, so l^{-1} ∨ a&l → Ω results in some minimal "p" from "a" that will return God to liberty. Essentially p=l&e_p is sufficient, and God (under a seven cycle) may rest on any l&e_x. Given a disjunction leading to a modal collapse: x&$e_{r\&s}$ ∨ (a&e_p)&(x&$e_{r\&s}$)$^{-1}$ → Ω &$e_{u\&v}$ as from before, with the right side necessary l would have (removing the principal elements): x ∨ a&e_p&¬x → Ω. Then, as this is not freely decidable given N(Pos(¬x)), ¬x a virtue in a_0, but x merely a positive predicate in "a"; God may instead put e_p for ¬x, and merely rest upon it in a_0, the lower closure properly within that of "a".

Then e_p^{-1} ∨ a&e_p → Ω, and "l" principal in "a" renders only e_p^{-1} ∨ l&e_p → Ω. Now there is a modal collapse breaking closure; God may now freely exercise x: l&e_p&¬x → Ω becomes l&e_p → x&Ω without breaking the closure of g_p in modal collapse. ("x" becomes virtue.)

So, either N(Pos(x)) or N(Pos(x^{-1})), for Pos(z&y) → z^{-1}&Pos(y) quite legally by liberty if "z" is virtue; for then z = Pos(x^{-1}) or z = Pos(x) must be absorbed by its negation within the inference N¬(z&y&¬(z^{-1}&y)) as N(z^{-1}&Pos(y) → Pos(y)). Given P(Pos(x)), God is free to exercise x in "a".

Then if ¬x is a virtue in a_0, God may simply rest upon it in "a". With all closure broken and nothing in a_0 privated by $\neg x = e_x$, Pos(x) in "a" becomes freely decidable. It will remain a virtue that God is omnipotent; He will or may exercise x freely in "a" only within the set of virtues.

Theorem (16.12.9a)

If x is positive in "a" but never found in a_0 then x is a virtue. Every virtue in "a" not in a_0 may be exercised freely through a modal collapse.

(Every act of omnipotence breaking closure by modal collapse in God's new name is virtue.)

Proof: Every modal collapse in g_x requires either $N\neg Pos(g_x \& l^{-1})$ or $N\neg(g_x^{-1})$ to decide the two disjunctions $l^{-1}\&g_{r\&s} \vee a\&g_p\&l\&g_{r\&s}^{-1} \rightarrow \Omega\&g_{u\&v}$ and $g_x^{-1}\vee a\&g_x \rightarrow \Omega$. The first simplifies by MC0 to the familiar: $l^{-1}\vee a\&l\&e_p \rightarrow \Omega$.

Given some virtue for God to exemplify in the broken closure not in the lower (a work of the new name) say, "x" (I.e. $x \rightarrow l$), by [16.12.9] I can make with MC0 the simplified disjunction of: $x\&l^{-1}\vee a\&l\&e_p\&\neg x \rightarrow \Omega$ in the closure of "a" and then $a\&l\&e_p \rightarrow x\&\Omega$ in "a" as before.

However, in a_0, God must not be able to work x, and $\neg x$ must be a virtue. So God may rest in a_0 on x as before. Putting $\neg x=e_p$, I finally arrive at: $(e_p\&l)^{-1}\vee a\&l\&e_p \rightarrow \Omega$ and therefore $g_p^{-1}\vee a\&g_p \rightarrow \Omega$. Closure, broken by "l", suffices to entail rest on $\neg x=e_p$ in $a \supset a_0$. God simply relaxes the virtue $\neg x$ to e_p in "a" for maintaining $\neg x$ as a "virtue" in a_0. This has no effect on the free decidability of x in the new creation "a". Here, unlike in MC1, there is no contradiction on $x \rightarrow l$.

The converse that every omnipotent act (a modal collapse) of God's new name is a virtue is found similarly, this time, in a modal collapse in God's new name as above, $g_p^{-1}\vee a\&g_p \rightarrow \Omega$ (there is no middle) and "a" as if all virtue must "span" $\Omega \setminus g_p$. Putting $a=\wedge g_x$ clearly suffices, which in the new name is equivalent to "$l\& \wedge e_x$". Then $g_p \rightarrow \Omega$, and each g_x forms the full closure of the set Ω in the disjunction. Then minimally with $a=l\&e_p$, each $(g_p \leftrightarrow a\&g_p) \rightarrow \Omega$ under a seven-cycle break the closure of the octal and all Ω is found necessarily true (as virtue to entail a positive result).

No predicate in $g_p \leftrightarrow a\&g_p$ can private any other in Ω, which is the set of **all** positive predicates. Although "a" is found as the creation operating as intercessor, Ω is entailed from both "a" and also g_x. It suffices to say that $l\& \wedge e_x$ maintains every closure broken until the new creation wholly supersedes the old and God may "move in" omnipotent.

Consider again the disjunction: $l^{-1}\&g_{r\&s} \vee a\&g_p\&l\&g_{r\&s}^{-1} \rightarrow \Omega\&g_{u\&v}$.

Given $g_{r\&s} \rightarrow l$ I have a modal collapse and the term $a\&g_p\&l\&g_{r\&s}^{-1} \rightarrow \Omega\&g_{u\&v}$ is extant. This by MC0 rearranges to: $a\&g_p\&l \rightarrow \Omega$. Every term in Ω is entailed from the set of virtue which alone contains the element e_p. //

Theorem (16.12.9.b)

The new creation contains virtues not in the old – the new name will also have the octal's structure.

Proof: Given a virtue rested upon in the old creation as with [16.12.9], if ¬x is not a virtue in a_0 or freely decidable in "a", i.e. x is found positive in Ω (with $l^{-1} \vee a\&l \to \Omega$) yet not in any Ω consequent of $g_x^{-1} \vee a\&g_x \to \Omega$ with any g_x, then x must, by elimination, be a virtue in the new creation "a", relaxed to rest by liberty in a_0. Then the relation $a\&x \to \Omega$ must also break closure of a_0, and also the closure of every $a\&g_x$ putting $x=e_x$ as by [16.12.9].

There can be no ¬x in "a", so there is modal collapse with N¬(x&l⁻¹), "l" remains principal. I would find the familiar disjunction of: $x\&l^{-1}\&g_{r\&s} \vee a\&g_p\&¬x\&l\&g_{r\&s}^{-1} \to \Omega \&g_{u\&v}$ then reduces to the merely necessary $a\&g_p\&¬x\&l \to \Omega$ or that $a\&g_p\&l \to x\&\Omega$ becomes by MC0 $a\&l \to x\&\Omega$ (I.e. $g_x^{-1} \vee a\&g_x \to \Omega$ does not entail x, $a\&l$ is "finer") and x then becomes necessary in the new creation "a". (x is certainly a virtue, and "a" is a paradise.)

A Corollary – The Least Is Possibly "Raptured" (16.12.9c)

If $f \to ¬x$ for all f in "a" and also a_0 then God is somewhat without liberty. (In particular, if "x" is "will be caught up to God".) Given $(l \to c) \to (x \to c^{-1})$ I posit that P(Pos(x&c)) for some "x" (in "a"), based on the free **possibility** P(x) of God's choice (amongst the **blameless** and possibly elect that are certainly elected whilst they remain so, i.e. $¬N(c\&¬l) \to ¬P(c\&¬l)$ cf. Rev 14:5). Note "l" is principal. I find all "Gods" are equally omnipotent, and $N¬(x \to ¬l)$ in a and a_0.

Then there are three alternatives for the state of x which is possibly positive:

Case 1) x is a positive predicate freely exercised in a and a_0.

Case 2) ¬x is a virtue in a_0 but x is a positive predicate in $a \supset a_0$.

Case 3) x is a virtue relaxed by liberty to ¬x in a_0 and is simply not in effect. In other terms, x is a virtue that breaks all closure of every g_x, requiring complete omnipotence in the use of the new name.

By the modus tollens of the above I have $¬N¬(x\&c) \to ¬N¬(l\&¬c)$ which is $¬N(c\&¬l) \to N¬(c\&¬l)$ and I have $l \leftrightarrow c$ if P(x) as $l \to c$. God exemplifies x as **positive** in "a", but **not** in a_0 (a_0 appears closed without x). Then $¬x \vee p\&x \to l$ in $a \supset a_0$. Then I find $(r\&s)^{-1} \vee (p\&r\&s \to u\&v)$ or (as it stands here) I have the disjunction $(¬x\&l)^{-1}\&e_{r\&s} \vee (p_0\&e_p\&l\&¬x\&(e_{r\&s})^{-1} \to \Omega \&e_{u\&v})$.

Now given ¬x a virtue in $p=a_0$ I simplify to $l^{-1}\&e_{r\&s} \vee (p_0\&e_p\&l\&(e_{r\&s})^{-1} \to \Omega \&e_{u\&v})$ where Ω is the set $\{\{u^{-1} \setminus l\} \cup \{v^{-1} \setminus l\}\}^{-1}$ and where $p=a_0=p_0\&e_p$. Now, as a virtue ¬x may not private (constrain) any positive predicate in Ω, given ¬x is a virtue I will still find a modal collapse of $p_0\&¬x\&l \to \Omega$. Also $N¬(x\&l^{-1})$ is $x \to l$, for x is possibly positive in the new creation "a".

I note that "l" appears principal in $p=a_0$ and I also have $e_{r\&s} \vee (e_p\&(e_{r\&s})^{-1} \to e_{u\&v})$. Then I can relax liberty from $(¬x\&l)^{-1}\&e_{r\&s} \vee (p_0\&e_p\&l\&¬x\&(e_{r\&s})^{-1} \to \Omega \&e_{u\&v})$ to $(¬x\&l)^{-1} \vee (p_0\&¬x\&l \to \Omega)$ which then evaluates to $l^{-1} \vee (p_0\&l \to \Omega)$ in $a_0 \subset a$, legally dropping the principal elements e_p, $e_{r\&s}$, $e_{u\&v}$, which entail no other predicates. The remainder is then both possible and consistent.

Case 1) As $l \leftrightarrow p_0\&l$ also breaks closure, $p_0\&l \to \Omega$ by "modal collapse". (I.e. rephrase the above with $r\&s=e_x\&l=g_x$ and with $u\&v=\Omega \setminus g_x$. Note that $g_x \in \Omega$.) Then $p_0\&g_x \to \Omega$ and $p_0\&l \to \Omega$ just as equally (i.e. breaking closure by both, with $p_0=\{a \setminus e_p\}$ or $p_0=\{a_0 \setminus e_p\}$ similarly). Now God simply "rests" on "x" as e_x in all $a_0 \subset a$, and $l \in a$ (x, though out of scope, also breaks the closure of a_0: "x" is also in u&v). Now, as "l" is principal in "a" and $l \rightarrow \wedge g_x \rightarrow g_x \rightarrow l\&e_x \rightarrow l$, by the properties of an ultrafilter, there is certainly some "p_0" in "a" such that $p_0\&l \rightarrow \Omega$ (for the "l" that overcomes predestined). Then $p_0\&\neg x\&l \rightarrow \Omega$ simply becomes $p_0\&l\&e_x \rightarrow \Omega$ in a $\supset a_0$ also breaking closure. God is then free to rest on "x" at will as with any other positive predicate.

In "a", then, I may reduce to $p_0\&l \rightarrow x\&\Omega$ without further breaking the closure of "$p_0\&l$" (a set of virtues) and then certainly $x \in \Omega$ is a virtue for the "Most High". //

Case 2) If $\neg x$ is held a virtue in a_0 then I may by liberty choose a minimal "cut" of virtue in $p_0 \rightarrow l\&\neg x$ in "a_0" to entail Ω. I assert that I still hold true both $l\&\neg x \rightarrow l$ and also $\neg x \nrightarrow l$ in "a" (for x is assumed positive in "a", in which "l" is principal, but not positive in a_0), but $l \leftrightarrow c$ so if $x \rightarrow c^{-1}$ therefore $l \rightarrow \neg x$. So "l" is blameless in **not** holding "x" as a property in Ω or as a **virtue** in G in a_0, yet it may be freely realised that x could well be positive (or a virtue) in "a" by breaking closure (in the set of virtue $p_0\&l\&e_x \rightarrow \Omega$) and therefore $N(G \rightarrow x)$. I.e. modal collapse in the set of virtue is the liberating statement "Every act of omnipotence in the new name must also be used as virtue".

Then if x is necessarily false in a_0 and cannot be exemplified in a_0 at all, given x is "positive" in $a \supset a_0$ it then belongs only in "a". As closure is broken in the set of virtue; x is then a virtue in "a": contradicting (closure broken) the assumption that $\neg x$ is a virtue in a_0. If there is modal collapse and full closure broken, then $N\neg(x\&l^{-1})$ and $x \rightarrow l$ (x is positive and l principal in "a"). Then $\neg x$ is inconsistent ($x \rightarrow l \rightarrow \neg x$) and cannot be a virtue in a_0.

Alternatively, as $N\neg Pos(x)$ in a_0, God could not exemplify x in a_0. So, before closure is broken I have (with "l" not yet present) $(\neg x\&l)^{-1}\&e_{r\&s} \vee p_0\&e_p\&l\&\neg x\&(e_{r\&s})^{-1} \rightarrow \Omega \&e_{u\&v}$ and (by dropping every principal element), as this is necessarily decided with $\neg x\&l$, I have $\neg x\&l \leftrightarrow p_0\&l\&\neg x$. (Then $\neg x$ is possibly in p_0 in "a" also.) Yet $l \leftrightarrow (p_0\&l \rightarrow \Omega)$ and without breaking closure of $p_0\&l$ I may again place x as possibly positive in $p_0\&l \rightarrow x\&\Omega$ in Ω. Now there is no contradiction here, as I do have a modal collapse and also after "l" is present I must also have $N\neg(x\&l^{-1})$ so $x \rightarrow l$. (x is found positive in "a" and "l" is principal, a virtue – then "x" must be in the set of virtue by the property U4 of an ultrafilter.) Since in a_0 I have $l \rightarrow \neg x$, I must have found a contradiction: yet there is no contradiction in the octal itself (or in dropping the terms in principal elements); I must have found that $N\neg Pos(x)$ is false in a_0. "x", then, is also freely exercised in a_0 after closure is broken. Then the structure of the octal due to virtue(s) is preserved with a new set of virtues (present in "a" but not in a_0) in God's new name.

So, $l \rightarrow (G \rightarrow x)$ (G, the set of all virtue) and as I have $l \rightarrow G$ I must of necessity entail $N(l \rightarrow x)$. x is then a virtue in $a \supset a_0$, contradicting $x \rightarrow c^{-1}$ or even $P(Pos(\neg x))$ in a_0 for "l". //

I note God remains at liberty in both cases (1) and (2); all "Gods" are found equally omnipotent. "l" similarly by (2), overcomes all $x \rightarrow l^{-1}$ by $g \rightarrow g\&l$: "l" does not backslide. (Nothing positive will exclude the election of "l".)

The "rapture" of the least "l" is (by modal collapse) a virtue for God and must be necessarily exemplified, and always found possible in the set "a" of the new creation.

I have essentially repeated the proof of [16.12.6], except here, God may initiate the "rapture" of Pos(x) after closure of a_0 is broken as if $\neg x$ was equal to the blasphemy of the Holy Spirit as if by the "corruption of blood" (Exo 34:7, Num 14:18), or as if ongoing from an oath. Then, as equally as "I" may be sent with a certain a-priori justification with an act of charity to break the power of any oath to Satan, modal collapse ensures that the "rapture" x of "I" is certainly a virtue after "I" is shown blameless. Then God is always set at liberty to lift up such an "I" of necessity in the new name.

However, God may only have to exemplify a minimal "x" (or an equivalent **virtue** in G) to remain at liberty. Then from $x \& l^{-1} \vee p_o \& l \& \neg x \rightarrow \Omega$, I entail in the ultimate closure "a" that either $x \in \Omega$ or ¬x is itself a virtue ($p_o \& l \& \neg x \rightarrow \Omega$) and putting $p_o = g_x$, $\neg x = e_x$ (God-like and at rest) is wholly adequate to keep "I" blameless without contradiction, for then God is set at liberty, at rest. Essentially $l \& e_x \rightarrow \Omega$ requiring "rest" on both sides (i.e. Pos($x \& l^{-1}$)) if God must so provide to keep "I" blameless. A **minimal** N(Pos(x))=e_y is required of God whilst "I" is not present, as God breaks the closure of Ω with $l \& e_x$) (There can only be a principal element e_y remaining whilst Pos(l^{-1})). Every broken closure is insufficient to prove the necessity of rapture but for p = l = $\wedge g_x$, as there may remain some plausible or latent Pos(x) such that $x \rightarrow l^{-1}$.

Either x is a virtue or ¬x remains a virtue but may be minimally present as e_x in a_o or (given no positive predicate with $x \rightarrow l^{-1}$), x is merely positive but then also freely possible.

Case 3) If there is yet a closure remaining after all seven $\wedge g_x$, are overcome, then x must, by elimination, be neither a positive property, nor ¬x a virtue. It must then be a virtue relaxed by liberty to rest and N(Pos(x)) as a virtue entails that $p \& l \& x \rightarrow \Omega$ breaks the final closure of all prior seven found in Ω. "I" is then possibly (quite plausibly) "raptured". Now, may God rest on this last closure? It may be far more perfect to **minimally** exemplify it with the bringing of the very least (I ↔ x) in the kingdom! (The least will rest, "he shall go no more out".) God, omnipotent, will certainly deliver him. (He will afterwards surely account for his predestined salvation on God's throne.) //

In these three cases, God is yet at liberty to lift up such an "I". "x" is always a virtue in the new creation, and that is a certainty for the redeemed elect surviving on the last day of the old.

I assumed c → l: together with all Pos(x) → l^{-1} overcome I would have that "I" is sinless.

Overcoming the world in $\wedge g_x$ (all closures broken and equality in $l \rightarrow \wedge g_x$ written, the least an essence overall), the least has also overcome the devil. If the devil is "of the world", he is overcome already in the g_x, and if he is assumed "not of this world" then he is subject to God in a → a_o and as a created being, he is thwarted with I ↔ c; he is subject to the will of God Most High (and His right hand "I") as then under the unity in C8 and also the zero in the octal. Nothing excludes "I" from election; the least has overcome the world, and the Devil within it.

Then the octal may fully justify the account of the least as he overcomes in Philadelphia, before being finally delivered, caught up to God and His throne (which is possible) – not devoured by the dragon or in a war for heaven. Those seven stars in Jesus' right hand are one and the same.

"War in heaven" arises from claiming that the least made no circuit in the closures of the octal, but in the rows of wormwood instead, floating unity through them all. The accusation will fail, however, as the "mountains" of the dragon's devices are present in the Church and scarlet beast; which the least has certainly exited by that point.

The Second Coming Enabled Suddenly (16.12.10)

Then, will Christ return? He surely will, yet writing the name of God not in the understanding but with an exemplified act (the angel's circuit, a necessary consequence of the least overcoming the world), all creation then becomes as equal to "a" rather than a_0. Then if it is posited that $g \in a_0$ because it is proven by that act of writing that $a_0 \to a$ then God is truly set at liberty. Does this creation join the multiverse of God's creative liberty? It surely will, by the name of "Him that liveth for ever and ever".

Rev 19:6 And I heard as it were the voice of a great multitude, and as the voice of many waters, and as the voice of mighty thunderings, saying, Alleluia: for the Lord God omnipotent reigneth.

So, where is the contradiction that God will write His name on the least or else return? Simply by assuming not $l \in a_0$ only but also $l \in a$ (there is a least or principal member in the kingdom of heaven), then $g_x \leftrightarrow e_x \& l$ becomes necessary (God writes His new name on the least when the seven horns are accrued completed). Then by a suitable ministry of the least himself, I cry with shoutings, "grace, grace unto it!" Grace, it would appear, can go no further than $l \to c \to \neg x$.

I may similarly state that God already has His new name. If it were not good for God to enter His own creation (His sovereignty challenged, the least sold to the enemy who will not benefit), then the necessity of $g_x \leftrightarrow e_x \& l$ is equally opposed to the antinomy that God Himself could not enter, not even in the least sense. Then God will certainly enter, and always necessarily so in the person of the least at the ending, who is "not of this world". As God already has each g_x, He first writes seven "horns" e_x (i.e. horns of the lamb) in creation upon the least. The least has a determination (predestination) of omnipotence to move him to salvation. God will remain.

Then $g_x \in a$ (ultimate closure), and by induction on the fact that the true right hand of God is principal in the kingdom of God, without loss of generality and equivalently by choice (Zorn's lemma) I may state that I ascend the chain to the very top of the "Most High" by the ultimate closure of all future creation — by saying "surely bring the least overall" (a request to the Most High that will not ignore such injustice, as to validate creation within the hands of an incompetent or malevolent deity) and God omnipotent (as necessary) will certainly reign in this world. Then there is no restriction of grace in a_0 (a_0 is part of "a").

So, "I" is most fine in the ultimate closure (as the very least), and given the angel is to complete his circuit and overcome, I find that the least (as God's very own liberty) will entail the principal element of rest: whether in the K4 groups of the octal in the $l \& e_x$ (L=l is as the "zero") or separately as in the faith of the angel, I have the freely decidable $g_x \bar{\vee} l \& g_x^{-1} \to g_x$ (as taken from $g_x \vee l \& g_x^{-1} \to e_x$ with $a \leftrightarrow l$) and then $g_x \bar{\vee} g_z \& g_x^{-1} \to g_y$ becomes $g_x \vee l \& e_z \& g_x^{-1} \to e_x \& l \& e_z \to l \& e_y \to g_y$ so $g_x + g_z = g_y$ which follows from $e_x + e_z = e_y$. Then "I" acts as it does in the "Manifold Of Virtue" as Christ's right hand (and to the whole Trinity) and the "empty essence lemma"[2] cannot therefore apply (as if "I" were empty — see [4.1.2a]).

Clearly $g_x \to l$ and then I do not break the closure of "I" (a virtue) with the equivalence of $l \& g_x^{-1} \to g_x$ to $l \to g_x$. Now, if "I" has come in the closure of the old name, and does not break the closure of a_0, then he, as an identity, should not entail any other virtue or positive property

[2] Benzmüller and Woltzenlogel Paleo, *An Object-Logic Explanation for the Inconsistency in Gödel's Ontological Theory.*

(he is principal, i.e. the "zero" or in the octal, $I \equiv \wedge g_x$).

Then, as $I \in a_0 \subset a$, "I" as the identity does not cause the closure of a_0 to break: only the time of his overcoming does so! Then as Christ freely writes on His least, and $I \rightarrow g$, I infer that "g" will enter this creation to close the argument: so that God writes on the least or returns Himself; either way, "I" is justified and God has his new name enter the old creation. (With the inference of $a \leftrightarrow a_0$ Jesus is already there!) God, also, lives within His creation.

Then the only necessity in creation other than the Trinity is that principal element "I". "I" is principal (1=0) in all creation (as $a \supset a_0$) just as the axiomatic liberty of God's own sovereignty, in equality, is principal in the kingdom of God. Now, being the first (or principal) in creation and not being God doesn't make you the greatest, as there is the verse to the angel of Laodicea:

Rev 3:14 And unto the angel of the church of the Laodiceans write; These things saith the Amen, the faithful and true witness, the beginning of the creation of God;

And it would appear that pride would come somewhat before a fall.

Now, once the least overcomes and it is true that the closure of $a_0 \subset a$ has been broken, it is necessary from the octal used here that the new name of God remains completely clean and untouched. There must be a time of trial of every individual y in a_0. God will test and suddenly return to judge (as promised). I do not expect these to be the same moment – the former is not to be final; it will be quite a shock to realise one's destination is in one's hands.

All those failing the final test (Christ's judgement) will be so judged and closed within the old creation a_0 forever, whilst the elect will sit and reign with their God "in heavenly places", in a "new heaven and new Earth".

Now, I find that Jesus Christ returns at the moment of judgement as fully God-like in "g" (to prevent such a contradiction that the ungodly may enter into the new creation "a" as $a \leftrightarrow a_0$ and may potentially blaspheme His new name), whereas the armies of heaven arrive beforehand led by the virtue of their least; for they themselves are not "God-like" and the least who is "God-like" (being principal) does not entail any other positive properties unless (by modal collapse) the disjunction is not freely decidable and all positivity is found on one side only.

So, Christ (only He) returns, breaking that closure of a_0, entailing all "a" and thereby of necessity bars the way between a_0 and "a": rejecting those that fail the time of trial and are so judged to the lake of fire. (All blasphemy of the Holy Spirit ceases with the wrath and "army of heaven".)

Of necessity, then, there is no millennial reign of Christ in the popular sense, only grace extended to all and for any fault(s). For God to return God-like as g, and fulfil necessity, the armies of heaven return with the least who also has that predicate of "g" entailed from his person "I" but "I" entails no other positive properties as he himself is set at liberty! (There is no modal collapse unless there is no liberty.) "I" is truly God-like but the set of Ω is freely decided upon by God alone. God in liberty rests upon closure, and "I" is closed itself!

The least is a finite singular algebraic closure, if that be the language of God's new name!

It would be most unfair to test those in the old creation as to whether they **cannot** blaspheme either name of God; rather they are to be tested as to whether they **do** blaspheme God's old name. This is done with a test on whether the dialectic is brought into play (the "name of blasphemy"). In stating with the dialectic that they have not blasphemed the new name by claiming ignorance whilst blaspheming the old name using the dialectic (privating virtue, to be tested with God's choice of virtue), they blaspheme claiming they are themselves positive by employing dialectic logic: itself privating all virtue by process (in life, or as in the pride of it).

The closure of the old creature (not just creation but of the flesh within which life is placed), provides an equal test in this, for I may assume a fair test using De Morgan's laws on sets.

The Hour Of Temptation (16.12.11)

Then I may take advantage of the closure of the new name (ultrafilter) over the old for a fair test of those that would commit blasphemy in the new creature.

Returning to the dialectic (being as blasphemy) as before;

$N(Pos(a_0)) \rightarrow N\neg(Pos(c\&e\&g))$ or $c \rightarrow Pos((e\&g)^{-1})$

Yet now, $e^{-1} \rightarrow g^{-1}$. So, $Pos(c) \rightarrow Pos(g^{-1})$.

A person remains blameless only if the opportunity to blaspheme the one true God's new name arises? If this were a contradiction then I would have to choose $a_0 \vee c\&e\&g$ and $Pos(c\&e\&g)$ would be the result. No bad thing you may state, but $Pos(c\&e\&g)$ as a result requires the privation of virtue in a_0. (A faulty ultrafilter in $\{a,c,e,g\}$; and all whilst blaspheming God's names new and old: their positivity is not decidable by man or as a dialectic.)

Now, $I \rightarrow c$ in truth so I may substitute "I" into the dialectic $a_0\&I \rightarrow g^{-1}$. Then as "I" does not blaspheme the new name of God, such an exampled "I" is sufficient to convict or condemn those using the dialectic that would remain blameless by claiming ignorance. Assuming one's self elect (or as a "good person" privating virtue) without honouring the old name of God, merely claiming all innocence through ignorance of the new, is to fail the test. $Pos(a_0\&c) \rightarrow Pos(g^{-1})$ is **equal** in that dialectic logic to $Pos(a_0\&g) \rightarrow Pos(c^{-1})$ but it is sinfully judged thereby as an "evil". Also, it is true $g \rightarrow e$ as well!

Knowing of "God" is enough to convict anyone of that faulty logic of the dialectic: for in truth $e \vee c$ and so it is true in the dialectic that $Pos(a_0\&e) \rightarrow Pos(c^{-1})$ also.

To pass the test one must honour God rather than remain ambivalent by assuming the position that staying silent on $Pos(g^{-1})$ should permit blamelessness. Resting on the position of "not blaspheming" the names of God is not enough: one must have both faith and give God glory. If you can't with both faith and doctrine show yourself elect (as separated by u&v) without blaspheming; face it, you are not elect. What then? The scripture clearly lays out the requirements of God: such a one should go to those that "buy and sell" (standing far off, cf. Rev 18:10-11) and find some genuine doctrine worth keeping! (Searching through that quagmire for the genuine article is beyond the scope of this writing; I will keep to my own.)

I infer $g \subseteq \bigcup A_i \leftrightarrow g^c \supseteq \bigcap A_i^c$ (with De Morgan's law). That the new name as doctrine (g) (God's liberty, His sovereignty as if a mustard seed of faith) is written on the least of the flock as amongst the union of **all** "consistent doctrines of the flock" in the A_i, implies that the intersection taken within **every** set of inconsistent or false doctrines A_i^c excludes that doctrine g of the least (including $A_i = g$ alone), but intersection is actually binary! (So much for "safety in numbers".)

It would appear God need write upon all believers, but this is not the case. As a simple result it can be readily stated of all the ungodly "Their witness does not agree". (They only agree on $c \to g^{-1}$ in the dialectic and this illogical pride is that upon which they are tried.) The intersect of all false "doctrines" is truly empty. By testing all (believers and unbelievers alike) in not blaspheming the old name of God, I infer $g \to e$ and $e^{-1} \to g^{-1}$ as required; I am left with:

$$g \subseteq \bigcup A_i \to e \subseteq \bigcup A_i$$
$$e \subseteq \bigcup A_i \leftrightarrow e^c \supseteq \bigcap A_i^c$$
$$e^c \supseteq \bigcap A_i^c \to g^c \supseteq \bigcap A_i^c$$

$$e \subseteq \bigcup A_i \leftrightarrow g^c \supseteq \bigcap A_i^c$$

And the time of temptation swiftly follows justified. (And God need only write upon the least of the flock, His new name as doctrine (a pearl) need not ever be spread abroad for blasphemy.)

And of the ungodly, Pos(g^{-1}) is necessarily false so "their witness does not agree" (in a principal and consistent "God-like" predicate). They do not agree on the one true God and the lamb. The least, however, has dutifully entailed $l \to c$ and also by not blaspheming the new name. He succeeded where they have not, the dividing line is well drawn. (The dialectic, as a fair test blasphemes the one true God's name, so that test is just, properly closing the sets of r_0 and s.)

Rev 3:11 Behold, I come quickly: hold that fast which thou hast, that no man take thy crown. Rev 3:12 Him that overcometh will I make a pillar in the temple of my God, and he shall go no more out: and I will write upon him the name of my God, and the name of the city of my God, *which is* new Jerusalem, which cometh down out of heaven from my God: and *I will write upon him* my new name.

One further thing may be stated about this "glaring flaw". In order to answer that blasphemy, there is either to be found strength in numbers (going into captivity) or fervent argument (killing with the sword). Here is the patience and faith of the saints.

Rev 13:10 He that leadeth into captivity shall go into captivity: he that killeth with the sword must be killed with the sword. Here is the patience and the faith of the saints.

Whilst the world is tested with the dialectic (the false prophet beast) to its utmost extent (the scarlet beast) the believers themselves (without having such an answer) remain totally separate, having overcome the beast and its image. They left the moment they saw the devil's devices arrive in their churches. They defeat the beast with a pre-first seal paradigm.

Rev 14:12 Here is the patience of the saints: here *are* they that keep the commandments of God, and the faith of Jesus.

The Key of David, The Last Day (16.12.12)

Now, Christ will judge after the angel overcomes and the time of trial is completely justified. The least, in an eight-cycle circuits the seven churches and rests in the eighth – his own spirit. That eight cycle is not the new name written unless (of course) his spirit $(1 = 0)$ is proven to be the very least in the kingdom. The least will surely have to prove himself least overall – by the test as to whether he may logically defeat his captivity to x, rather than merely correct himself. He must be principal: Jesus returns with the resurrection of all, including of the least. Then, if the angel is proven to be as "g" (God-like) then he must be necessarily existent. God will have to do a little more than merely resurrect him, He will have to find him without raising him also.

Jesus has the choice (the key) as to who will enter heaven. Jesus alone chooses (opens – i.e. as "a") Jesus alone judges (closes – as in a_0). Likewise, Satan cannot prevent Christ's return with the least accused of "blaspheming" (so closing a_0). Jesus knows the angel's "little strength" but also has the ultimate choice of when to allow the angel to finally prove the eight cycle in place from the angel's necessary existence: the new name not written unless the angel is proven so. The least cannot blaspheme the Holy Spirit; God is worthy to inherit a multiverse.

How would that test occur but for opening the book of life and finding his name inside? With the resurrection, if the least is present, he is present somewhere: God has not left an empty space for his existence if it is necessary: the angel is surely there and raised with those seven horns of the lamb as much as if he were given necessary existence from starting without it!

The angel does not climb up some other way backsliding: if he accrues all seven horns he is raised because He has reckoned for his salvation with them already, reigning over himself predestined. (Fulfilling Christ's conditions is true a-priori before he finishes his circuit.) As nothing in $\wedge g_x$ can exclude the election of "I" (i.e. $N\neg(x \to \neg I)$), "I" is shown an **essence** if Pos(I). I.e. God must exemplify Pos(I) as principal, for Pos(¬I) cannot follow from any essence!

Then by fulfilling Christ's conditions in His letters, nothing positive may exclude the election of "I" (he remains truly sinless, remember), and closure is broken by equivalently overcoming the fineness of each filter in the octal made of the seven horns (each generated as $<e_x>$). "I" proves himself principal in the union of every closure accrued and finally, "God-like", completing the new name and bringing in the reign of omnipotent God before the judgement.

Then he is only tested with that key of David to see if he remains in the book of life (holding true a&r&s, the least must remain blameless not denying Christ's new name), and Jesus arrives with judgement at the moment the seventh seal is opened. (Or the final trumpet sounded.)

By Him That Liveth Forever and Ever (16.12.13)

By using the whole octal in this way with the three groups of {p, r, s}, {p, u^{-1}, v^{-1}}, {p, u&v, $(r\&s)^{-1}$} which correspond to "which art, and wast, and art to come", I find that there should truly be "time no longer" in the days when the seventh trump (at the third woe) is to be sounded.

I assume every positive reason for the new name of God not to be written or revealed is found in the empty middle of e ∨ c (given a_0 as virtue) remaining positive or necessarily empty. That is, without such intercession in a_0, all sinners remain innocent of blasphemy only when the old name of God is not put to the test (i.e. Pos(e&c)). Likewise guilt and election cannot be attributed to the corrupt without such a test! Now, this middle is empty in creation and has been since the fall, as is that middle of $g \vee c$. (God's new name must remain Holy when revealed.) The middle of $e \vee c$ is empty only for as long as there is no least (necessity of the middle then follows): if "l" breaks closure of a_0 then I may infer $g \vee c$. But $l \rightarrow c$, and so $g \vee l$ in the ultrafilter and the middle g&l is then necessary once the least arrives and overcomes. Given $g \rightarrow e$, the election of grace will conclude shortly thereafter, because while there is no middle in $e \vee c$ there is none yet entailed from $g \vee c$ (the least has not arrived to break closure). Given the new name is not written until after the arrival of the least, whilst the middle of Pos(e&c) remains null, the new name will not have been revealed. Yet it is revealed to the least for whom Pos(e&c) is true. (He is not fallen but justified, overcoming the world.)

When will the fall end? (When N¬(Pos(e&c)) is universally contradicted.) Only when there is no more curse – no requirement for the test N¬(Pos(e&c)) upon all the corrupt with sin; I.e. when "Death is swallowed up in victory", and only when "the last enemy *that* shall be destroyed is death". Jesus Christ makes everything new. Then with the last trump, the mystery of God should be finished, because God will arrive with new name safely intact.

For those saved elect there will be no new mystery; God will dwell with them and they shall all know Him. For those not elect there will be no God to reign over them beyond that judgement, no God to refresh their inheritance; the "mystery" will certainly end either way.

Can this word be trusted?

Now, "l" is principal in the ultimate closure of God's "new name" or "His God's" new name, or the "meta-God of Gods'" new name, ascending the chain as far as it can go. (Yet there is only one God.) Why? If one new name of God were principal, it would have to be principal in the old name (superseded by the new) after closure is broken. "l" as principal from the ultimate closure is by inference and necessity principal in all the new (and old) names, else some sets in the old name would not contain the principal element of the new name, the new name could not be in effect and one ultrafilter could not properly close upon or contain the sets in the other. It could not be true that there would ever arrive any least or principal element.

But are all new names principal? If every God has His God then yes, up to the least in their kingdoms! For the greater surely has the least of the lesser. Without grace of the (**every** new not the "Most High", and without the old) Holy Spirit but only exercising the liberty of the Most High there is no lesser. (The ultimate contains every closure.) Then there is no necessary agreement in creation at any **closure** on anything but every closure justifying the least overall, name(s) written on him! Then $g \rightarrow g\&l$ for "g" the "Most High". "l", the minimum "God-like" individual, entails "g" when free in all positive properties. Every "God" passes the least up the chain: only the Most High (as creator already justified) can assist the least to justify His own creation. (And it is only the name of the Holy Spirit that differs, as in one third of the Trinity.)

Each "God" with a "new" name must have once started with and still share one "old" in similar fashion (for they are certainly much more than that two thirds valid, as God is omnipotent and has sovereignty already; the group C7 must yet remain a valid name for the Most High also), and the chain would end with a "Most High" God with the one "new name" (otherwise keeping our Father-Creator Yahweh in the dark!). Every "new name" is always principal, justified by each "God" starting out as with one "old" (and principal) and must have surely placed their least in the lowest creation (closure) to justify themselves to be present in that chain at all. Then our God, who already has the least (being the creator who chose him thus; God, omnipresent, knows of the least in **all** creation, **unique** as 1=0 from the beginning), has given to the least all reckoning for himself (without any divine grace), which can justify **only** the Most High God's name first (to whose closure he truly and only belongs, as ultimate creator of all), never that of some other. His own God will win out – not a pretender (as of Marcion) and, as I note, God already has that new name and is forever perfect.

I may reach the modal collapse $l^{-1} \vee a\&l \rightarrow \Omega$ from $a_0 \& g_p \& l \rightarrow \Omega_0 \& g_p$ just as easily as I may do so from $a_0 \& e_p \& l \rightarrow \Omega_0 \& e_p$. As the former is without the principal elements of either name, the proof is identical for all "Gods". Any "newer name" $l \rightarrow h_x$ ("l" of 1=0, proven an essence and also unique by Leibniz's law, the identity of indiscernibles $\forall x \forall y [\forall F (Fx \leftrightarrow Fy) \rightarrow x = y]$) supersedes the g_x and since $h_x \rightarrow g_x \rightarrow l$ is principal and an essence for any h_x, then $h_x \leftrightarrow g_x$.

It is also true that this world is effectively the lowest closure of all as any "sub-Gods" would have the same least with the same argument, and the least has yet to bring the judgement **here**. Then as all this is prophecised to occur here in the Bible (God's Word), "here" is the lowest closure as the "sub-God" would otherwise justify the Most High over the whole chain from within their own closure. God, I may trust, is also the lowest as well as Most High.

By arguing from the least, the (unique) Most High God reigns in omnipotence. (Closures cycle in regress.) This occurs already in the new creation containing the old, and is justified here (and there) when the least overcomes the world and $g \rightarrow g\&l$. It is as sure as God's own name.

16.13) In All Possible Worlds

Now, that which is necessary is proved existent only if it is proven necessarily existent "in all possible worlds". That may sound heinous, but in terms of the octal the same structure exists in every freely decidable disjunction (for there to remain any disjunction) in just one world; this creation.

God may choose any set of virtue to entail in any real world setting every possible consequence. By fixing some choice of Ω to form some closed subset of everything possible in this world or any other, there may be many such Ω constrained in many possible "real-world" disjunctions. Then every possible real-world event (a decided disjunction) may possibly occur and every possible arrangement of the sets p, r, s, u^{-1}, v^{-1}, $(r\&s)^{-1}$, u&v may occur in any disjunction. The same predicate x may appear in one class in the octal in one disjunction and be exchanged with another in yet another disjunction (these sets are all disjoint).

As any permutation may be rewritten as a product of transpositions, every octal (of thirty total) in the same seven symbols (sets of predicates) under any multiplication are possible principal elements for each set in Ω, with every arrangement of the logical schema of the ultrafilter played out in merely the one creation (world). There is no difference to God whether the sets of positive predicates are the same across different "worlds" or not: it only matters that the octal may have any one of the 84 logical schemas and any one of the thirty operations of a generalised octal in seven symbols (and zero to close it).

That number of 84 is reached because in any disjunction of sets from the mapping table, there are seven choices in the octal for virtue, three possible K4 groups containing unity for the disjunction in action between r and s, as well as two such groups remaining for the disjunction not in action as between u^{-1} and v^{-1}, and then two choices between alternate octals: one that references the "real-world" and that other octal which swaps the sides of the disjunction in one of the previous two K4 groups and references some other disjunction (rather than wormwood). Then $7*3*2*2 = 84$.

The K4 form is unaffected by any shift in logical schema. There are seven choices for unity from the octal, three choices for the identity K4 group and then two automorphisms of GF(4) (the whole). Lastly, there is the separation to Christ's right or left as to the reference octal or the wormwood in an alternate schema. These directly correspond to those 84 variants.

Then as "all possible worlds" are, to God, all the same – as He must exist across all of them, being the creator – perfection is natural to the octal rather than the set Ω of positive predicates and the schema made in them. The person of God, found in the seven e_x alone, is that which is necessary: even the essence of the right hand of God "I" from before is somewhat necessary but only so after the disjunction $I^{-1} \vee a\&I \to \Omega$ (where $I \leftrightarrow a\&I$) is freely decided when the least "I" is sent by God into this creation itself, and only at a time of His choosing: for the least always exists in the "new creation" of the wider closure rather than the "old" creation and name of positive predicates alone. For "I" breaks all closure in the e_x, and must prove that also and in a circuit made of the seven churches of Revelation floating $1 = 0$ through the octal.

"I" must show he is positive for God to elect and retain, so that $N(Pos(I))$ and also $N\neg(Pos(x) \to Pos(I^{-1}))$, or that nothing positive may exclude that election of I, for he must remain positive and also an essence. There must, in fact, be nothing in the wider closure of the new name that will exclude the election of I, so the idea that his inheritance may be seized upon on Earth is false: for "I" is a new creation a-priori, and by overcoming the world predestined, he himself is still a principal element (as a-priori) and nothing at all possible in either old or new name can provide a positive reason for his election to be excluded except falling away from his inheritance completely.

So, "I" has already got there first, and no-one (even in the new creation) can supplant him. There is only one caveat: that his election is made sure only by his circuit and his own reckoning on the throne of grace to move himself to break that closure. Or, rather, to show himself guiltless in the seven (he must effect a modal collapse in each case).

If $I^{-1}\&e_{r\&s} \vee a_0\&e_p\&I\&(e_{r\&s})^{-1} \to \Omega_0 \&e_{u\&v}$ is freely decidable (i.e. where $a_0 = \{a \setminus e_p\}$ and $\Omega_0 = \{\Omega \setminus e_{u\&v}\}$). By lemma MC0 this reduces to: $I^{-1} \vee a_0\&e_p\&I \to \Omega_0$ and $I^{-1} \vee a_0\&I \to \Omega_0$ with $a = <g_x>$ I may write $I^{-1} \vee a_0\&I \to \Omega$ given every closure broken. I.e. $N(Pos(I))$ and $N\neg(I^{-1}\&e_{r\&s})$ is $e_{r\&s} \to I$, a contradiction. ("I" breaks closure and <I> is a finer filter than $<e_x>$.)

I may also relax the one term in e_p which is principal and entails no other predicates.

Then I have, $I^{-1} \vee (a_0\&I \to \Omega_0)$ where I have the closure $\Omega_0 = \{\Omega \setminus e_p\}$, as "I" ($N\neg Pos(I^{-1})$) is equivalent to the right hand side $a_0\&I \to \Omega_0$.

$l^{-1} \vee a_0\&l \to \Omega$ or the right hand side of $l^{-1} \vee a_0\&l\&e_p \to \Omega$ is necessary given the election of the principal "l", on proving $N\neg(x \to l^{-1})$ for all $Pos(x)$ in circuit.

Similarly, given the positive predicates (of section [16.12.2]) $g \vee a\&g^{-1} \to e$, if $g \to l$ and $a \to l$ ("l" is principal in creation) I may rearrange to $e \vee l\&e^{-1} \to l$ and, relaxing principal elements to gain $(l\&e)^{-1} \vee a\&l\&e \to \Omega$. Now, if $g \to e$, $(l\&g)^{-1} \vee a\&l\&e \to \Omega$ or even $g^{-1} \vee a\&l\&e \to \Omega$ given $g \to g\&l$. The disjunction $g \vee a\&g^{-1} \to e$ is decided with the reckoning of "l" (I.e. $N\neg Pos(l^{-1})$).

Again, rearranging in virtue reveals clearly that $g \leftrightarrow l\&e$. Then as g is principal in Ω, the virtues not "l" in "a" must entail all of $\Omega \setminus g$. Also, to break all closure completely, "l" must be the sole principal element not in Ω and remaining in "a". Now, $e \vee l\&e^{-1} \to l$ becomes $l \vee l\&l^{-1} \to e$ or clearly, without breaking the closure of "l" becomes $l \vee l\&l^{-1} \to l\&e$ or simply $l \vee l\&l^{-1} \to g$. Then $l \to g$ and l is principal as $l \leftrightarrow g \leftrightarrow l\&e$.

Finally, $g^{-1} \vee a\&g \to \Omega$ becomes $l^{-1} \vee a\&g \to \Omega$ and "l" becomes as $a\&g$. "l" becomes equivalent to a set of virtue spanning all Ω. Effectively, $l \to \Omega$ (and "l" is the smallest "God-like" set, an essence).

Then whether or not the predicate "God-like" is considered principal (as is necessary existence), the new name is shown an essence through the circuit of the least.

If there are any sets in the new creation in which an f is not found, then f is not principal. To be the least, "l" must remain a Christian to reach the very end of His circuit when he finally overcomes. Nothing may private $Pos(l)$ because of its necessity as if liberty. Neither may anything positive exclude the election of "l". If these are both true of the essence and shown so, what may fail due to a certain reckoning (an open door) made before the circuit from the throne of God? Only that the one conditional axiom (not a-priori true) $g \to g\&l$ would fail, the axiom that God with His new name is the God of the least. That new name of God should never be denied, and this only remains uncertain whilst the angel watches.

Now, the mechanism for this certainly belongs in "The Revelation of Jesus Christ", which is best read in its entirety. What occurs is modal collapse in each set of the octal in turn corresponding to virtue. There is a circuit of seven churches, wherein each "church" becomes as the unity floating in circuit in the same octal of seven "horns" (together spanning all possible creation) used to reckon the least to his salvation. As modal collapse occurs only within the set of virtue and entails that whole set of virtue possible: as God is omnipotent; there is supplied an alternative virtue (horn) seven times over which will return God to rest and liberty. This, if every closure of the local virtue Ω is as one "horn", from the same seven horns accrued which span all possible creation (including by the new name): each modal collapse writes those horns "on the least" ($l \leftrightarrow a\&l$): ascending a series of broken closures (floating $1 = 0$, forming a "pillar in the temple" rather than a circuit, though the least suffices to show the circuit equal to one in the Holy Spirit instead). That accrued set of broken closures, each church then overcome as to the good retained and the evil rejected (even evil found corrected in that broken closure) constructs the "faith" of the least that is the prerequisite of every saved elect.

Now, the only condition remaining is $N(Pos(l))$. God is a God that saves and it must be positive for Him to save anyone otherwise going to hell only because of a curse in a world He created; but "l" must effectively be "without sin". Now, if "l" breaks all seven closures in the seven horns of each $<e_x>$ as insufficient to save him (there is some positive x such that $N\neg Pos(x\&l)$), or unable to also demonstrate Pos(l) at all, then God is unable to save him. Yet those seven horns identify "l" as the least, every closure broken through modal collapse as he is indeed by principle the least. (And as a-priori reckoned for; saved, justified and victorious.)

Now, I introduce my other axiom $N(g \rightarrow g\&l)$. God is "God of the least" and also whosoever is principal. If the least were simply accepted by God as a born-again Christian, then Pos(g) would entail Pos(l). Nothing more than a circuit in the Holy Spirit to accrue those horns is required to show that the least in the kingdom of God is God's right hand.

Those seven horns in the $<l_x\&e_x>$ are written on the least and become part of his own inheritance, he will retain them. By floating the unity as zero (virtue) in modal collapse through seven churches constructing a "pillar" of positive predicates that will span (as if a "vector-basis") God's full ability to create, every possible arbitrary octal of thirty in the same seven (additive) symbols are accounted for in the same closure of all Ω (and for the schema, clearly the seven choices for virtue), and also through modal collapse every one of the seven $l_x\&e_x$ are written sevenfold on him as g_x, and if each "church" disjoint in the circuit is of each left hand octal whole, the right hand remains for the octal also implicitly defined by his circuit as with the mystery of the seven golden lampstands. For his circuit is in the "rainbow" or the Holy Spirit, and with all eight left hands accounted for and overcome/accrued in closure(s), there is no longer any positive property in the right hand octal remaining to exclude him.

To be found at Christ's right hand is a matter of Christ's judgement to right and left: that Pos(l) is ultimately Jesus Christ's choice alone.

I.e. Every left hand taken together must contain the full closure of positive properties in the right hand octal they all share. If each left hand is to be treated as a disjoint set in the circuit within an eighth, then the eighth that circuits them must be the union of all the other seven "left hands", and shares the same closure with its right hand in those same seven predicates which span that closure. (Every set in the octal becomes equivalent to its principal element.)

So, the angel's circuit forming such a closure or "pillar" is valid in every arbitrary octal that can be constructed in the same creation as spanned by his "pillar" in all positive properties. The new name ($g_p = g_r\&g_s$ becomes $p = r \cup s$ etc.) and reign of omnipotent God will arrive with the overcoming of the least, bringing the second coming and the judgement with it swiftly.

Given virtue maximised as "p" to the full closure of all positive predicates, the new and wider closure becomes intuitively apparent: for, it may be asked, "What will, besides p and e_s, be entailed as "x" from $p\&e_r^{-1} \rightarrow x$?" In positive properties alone, perhaps nothing. The extension by virtue and the broken closure(s) made with the least will entail a whole new octal.

Under circuit, as forming a "reference octal" common to all eight possible multiplications, the circuit formed by modal collapse in virtue generates that octal under the one seven cycle: justifying that octal as a reference for all seven other multiplications in the same elements.

The sets of positive predicates in every octal form seven disjoint sets; this closure gained in all possible **virtue** Ω is such that every eighth C7 group circuiting each of the seven others over a given octal will self-reference all the others in a mode of infinite regression. God's new name is free in **all** positive properties, and by modal collapse God is free in His new name to contain (through that regression, all virtue itself now spanned by seven "horns" within "a") every conceivable positive statement. Ω is then "spanned" in whole (Ω is completely exemplified as necessarily positive) no longer divided into eight "equivalence" classes. Every octal will then properly contain the closure of every other: there are simply no closures outside that new name remaining, as to be found by modal collapse with the least's circuit. However, it is the g_x that generates every C7 (or with "I" found generating C8) group spanning Ω whole.

Those eight sets that partitioned Ω become the "floor" of each new octal of thirty. The modal collapse due to the least operates from within the set of virtue alone. Each octal is then truly maximised, no longer partitioning Ω overall: each is completed with the remaining sets of the least's circuit: yet are also under their own operations, and under their own seven cycles.

Then every octal is the union of each of the seven others; in modal collapse, the least generates the full closure in all Ω. God, now as g_x, is justified to do so in all creation likewise. Truly, the "Revelation of Jesus Christ" reveals Christ is God and as God, His Father's equal in all Ω.

If every positive property in Ω is written as some a_0 on "I" as if $I \leftrightarrow a_0\&I$, with $a_0\&I \rightarrow \Omega$, then as every positive property is found necessarily exemplified as (a floating unity of) virtue in a modal collapse, so also, potentially, is every octal that splits any subset of those predicates into a schema of action or inaction. If there are positive predicates and it is possible for God to use them in any schema consistently, then God's new name is now found principal in all the g_x and also principal overall as with "I" in Ω. Necessary existence becomes God's own essence "I" of "God-likeness", the smallest "God-like" set being "I" or each g_x on its own. For now, God may justify His own modal collapse as freely worked not in the e_x, but in all the g_x instead.

As the horns are accrued by the least and the new name written on him, God effectively brings about the new creation to supersede the old. Those g_x are now found free in a_0, now become as "a". God moves in (and does so with a pillar made in the temple of God).

Zec 4:10 For who hath despised the day of small things? for they shall rejoice, and shall see the plummet in the hand of Zerubbabel *with* those seven; they *are* the eyes of the LORD, which run to and fro through the whole earth.

Then any challenge to God's sovereignty over His name in the e_x as if it were not "God-like" (an essence as per Gödel's argument,[1] an obvious objection to any Trinity) is also thereby defeated. As an essence under modal collapse, the operation of the octal becomes a union.

In that sense (and in one setting), the proof of God's new name and equivalently His universal sovereignty in "all possible worlds" is worked and made manifest in this one creation.

Of the eighty-four possible logical schemas, half preserve the subgroups of an octal in seven symbols. (The other half will not correspond to any real-world disjunction, being the rearrangements in wormwood, unworkable in every disjunction – one side being extant only.) The *living*[13] choices all make are constrained by disjunctions presented us in God's *universe*, and all may be assured that God sees this problem exactly the same way; for He justifies His new name with a circuit in seven churches of six elements with the seventh identity element shared between them. So, there are but forty-two "octal schemas" that justify *everything*

[1] Sobel, in J. J. Thompson (ed.), *On Being and Saying: Essays for Richard Cartwright*, p241-61.

[13] Douglas Adams, *The Ultimate Hitchhiker's Guide to the Galaxy - The Complete Trilogy in Five Parts* (Pan), p101-109.

workable for God (and by us, taking a subset), and rest is found the only remnant (and even atheists have their prophets).

An evil may be employed by God only as virtue, to result in the wholly positive. Hell is presented just one outcome, of rest (God forever remains inactive towards the damned). In every disjunction it may be inferred "Hell will end the seventh day", every measure of misfortune delivering that one minimal positive outcome: nothing other than the ongoing rest of each e_x.

Outside of the "life" found in the angel's circuit, it may be asked "What else is there?". Whilst perfecting the outcome of every workable disjunction is truly the answer, no doubt some would instead muse concerning the product of 8x7 (which simplifies in the angel's circuit and God's "name" to 7x6), "What else is there to be found?" (i.e. by considering the product of their two nearest neighbours).[13] To fill the vacuum, the smallest sideways step is once more taken. Then what else *is* there? It is a meaningless question; there is no reasonable thought deep enough to examine even the one **true** half of the universe's history to determine any other answer as to the Revelation on whether God's purpose survives His active presence.

16.14) The End Of All Things

Throughout the Revelation I have shown some of John's word pictures laid out as if needing a little more polish: John's fantastic artistry that shrouded these word pictures had somewhat glamourised the scripture for many. That said, the simplicity and (never crude) apocalyptic prose must have seemed to John to slowly develop as a ticking time-bomb, one from which John could only be certain (and glad) to have made his escape from being "found out" some two thousand years earlier.

Jesus, happy to go with this final text as from the start, cannot be disappointed; there is no other sense for the marriage of the lamb to be completed with any more than this said:

Joh 2:1 And the third day there was a marriage in Cana of Galilee; and the mother of Jesus was there:
Joh 2:2 And both Jesus was called, and his disciples, to the marriage.
Joh 2:3 And when they wanted wine, the mother of Jesus saith unto him, They have no wine.
Joh 2:4 Jesus saith unto her, Woman, what have I to do with thee? mine hour is not yet come.
Joh 2:5 His mother saith unto the servants, Whatsoever he saith unto you, do *it*.
Joh 2:6 And there were set there six waterpots of stone, after the manner of the purifying of the Jews, containing two or three firkins apiece.
Joh 2:7 Jesus saith unto them, Fill the waterpots with water. And they filled them up to the brim.
Joh 2:8 And he saith unto them, Draw out now, and bear unto the governor of the feast. And they bare *it*.
Joh 2:9 When the ruler of the feast had tasted the water that was made wine, and knew not whence it was: (but the servants which drew the water knew;) the governor of the feast called the bridegroom,
Joh 2:10 And saith unto him, Every man at the beginning doth set forth good wine; and when men have well drunk, then that which is worse: *but* thou hast kept the good wine until

[13] Adams, *The Ultimate Hitchhiker's Guide to the Galaxy - The Complete Trilogy in Five Parts*, p279.

now.

John wrote both the Revelation as well as the gospel attributed to him, despite being once completely illiterate (cf. Acts 4:13). He still had plenty of time to learn to write for when it was required of him, though it maybe took him decades. I should not be upset that John really did have amazing vision and clarity to put the Revelation into word pictures. There is no doubt in my mind that John was inspired and all I have left to write upon are his table scraps.

Now, should there be any attempt to divorce the Revelation from its recorder (the apostle John), it is always good to give the book a "second witness". So as one record may not be made to discredit the other, John has given us a quick test, as to whether the book is genuinely authored by the same apostle that Jesus loved.

Then in keeping with some small due diligence I perform a quick comparison of the above model with John's rather cryptic passage in 1Joh 5:6-8

1Jo 5:6 This is he that came by water and blood, *even* Jesus Christ; not by water only, but by water and blood. And it is the Spirit that beareth witness, because the Spirit is truth.
1Jo 5:7 For there are three that bear record in heaven, the Father, the Word, and the Holy Ghost: and these three are one.
1Jo 5:8 And there are three that bear witness in earth, the spirit, and the water, and the blood: and these three agree in one.

As for "bearing record in heaven", I must include the structure of the whole octal's ultrafilters: the three made of the Father (as the octal complete) together with the word (the static subgroup) and the Holy Ghost (the seven cycle) are together that one field GF(8). However, restricting myself to the coset of the static subgroup in the octal (as in the "Earth"), I also find the seven cycle (as a "river" or "waters") acts on the static triple whose elements correspond to the three groups of the K4 form (then as "fountains of water"). The remaining fourth element (the unity) is always their intersection (the blood). Then John indicates the Spirit (as C7 and as by Frobenius) agrees with the water and blood but in the person of Christ only (the K4 form).

Christ, then, came by "water and blood": He preserved all virtue within the Holy Spirit and that virtue was as the unity under the seven cycle. The works of the Holy Spirit showed that (by "water") not only the left hand triples were preserved complete, but Christ (as the K4 form) came preserving the blood also (His virtue, as in "the Lamb"). The Spirit "bearing witness" show Christ is not only as that same static subgroup remaining (there being only the implicitly defined right hand octal missing) but in Christ is the octal (person of the Father) perfected and complete. The Spirit, water and blood all agree in Christ as the K4 form.

(It may be of interest also that only seven of the miracles Jesus Christ performed were recorded in the gospel of John.)

Then Christ's identity is found no mystery, He is the Lord, the only begotten Son of God.

2Ti 2:15 Study to shew thyself approved unto God, a workman that needeth not to be ashamed, rightly dividing the word of truth.

It has been a delight writing this study; as far as I can tell there are always a few more answers to be gleaned from the text – and all from only one small book of the New Testament. The book of Daniel has much similar apocalyptic content and Ezekiel's vision of the Cherubim is amazingly apocalyptic; I hope that you feel encouraged to turn the pages of Revelation knowing the content rather than the interpretation. The interpretation (when it comes) is always without fail of a ***personal sort***. The content is actually substantial and will never move an inch for all the force put against it.

If you feel up to it – try reading this book again. I would have preferred to have filled every verse with every word I could find to give the detail in full. However, the first four chapters were deep enough to prove that a bad idea – confusing the reader was far from my intent. (John's apocalyptic content could bamboozle anyone with every detail from the start; he had no need to begin simply for prose!)

Joh 21:25 And there are also many other things which Jesus did, the which, if they should be written every one, I suppose that even the world itself could not contain the books that should be written. Amen.

I hope then that the maths was not too complex and the explanations not too lengthy or verging on the repetitive. To those that understand already, these things are easily unveiled, but my intent was to keep things sweet in every mouth and not to make anyone's belly bitter.

My hope would be for you to put this book down and read the whole New Testament – all the way through. When you finish the Revelation, read this study again and ask yourself if those words have (or ever will have) passed away.

Should it ever become apparent that the Godhead or the patterns of the seals, trumpets and vials may be applied or may seem to be found almost everywhere, I can only infer that the manifestation of God (whether as an ultrafilter in its own language or not) which I, as a creation made in His image, would by some anthropic principle find, truly entails the ultrafilter is most certainly not represented everywhere, meaninglessly embedded in writing and data as by mere illusion, but rather is truthfully found in all my conscious perceptions of those occurrences in such media. I, made in the image of God, should not be surprised to find Him "almost everywhere". (But in so doing I find only fault in myself.)

As an example, I include the following modified advert from a music store email circular dated 27th October 2016:

3 is the magic number.

Three new releases from YHWH that is...
The legendary Jesus Christ now has a little brother in His second in command.
A set of seven-cycle circuits within the Holy Spirit for anything requiring 4 to 8 solutions, A Trinity that fixes any feedback issues during what is called the Kingdom of God

Where YHWH is become the manufacturer, Jesus Christ the amplifier model, the newer, more basic model is now His second in command, the set of seven cycle circuits replacing a multiband compressor becomes a new offering as from the Holy Spirit, the number of solutions to save from the left hand is as the number of strings on the instrument (vessel) and the mystery of God as "Trinity" becomes the "label" which closes the system during this current system of things, in answering any accusation of fault.

The original advert? I will keep you guessing! Inspired by God? I hope (for you) and hope not. (I have a great fondness for Seb Pearce's online "new age bullshit" generator. [website: sebpearce.com/bullshit] I sincerely hope that if anyone innocently attempts a rewrite of God's own electrons they will eventually get re-energised, though not in blaspheming!)

I hope that I have in writing somewhat answered the mystery of the Holy Trinity adequately. (God is not that mantra!) My effort I hope will not be described as a "Hell manual" or some technical specification for evil. (Please, please don't try to make the image, folks!) Once again this is not an attempt at satire or to supply a "sick note" in God's absence. What I would hope is that you, the reader, should feel empowered to counter arguments that God is cruel and vindictive and ready to torture those that are not His children forever. God is and always has been patient and kind, slow to anger. Whilst His laws are severe in their application, the moral purity of His elect (the New Covenant Israel) is as strictly kept today as in the Old Testament: and any blasphemer would be wise to leave that Israel now and as was true back then, to freely live outside. (As a prodigal son.) (For being an unrepentant sinner and claiming citizenship in Israel against its laws is actually blasphemy against the Holy Spirit.) "Tortured ceaselessly and forever" though is a far harsher sentence than God would rationally conceive for a common sinner.

If anyone yet states that another "burning in hell" is a real outcome, it should be noted that God the Father has committed all judgement into the hands of His Christ.

Joh 5:22 For the Father judgeth no man, but hath committed all judgment unto the Son: Joh 5:23 That all *men* should honour the Son, even as they honour the Father. He that honoureth not the Son honoureth not the Father which hath sent him.

Now, if the book of Revelation teaches that all unbelievers are condemned to fiery torments without judgement, it is not true that each are judged "according to their works" when all are saved or damned already under a law of faith. But more so, Jesus then has no judgement left, the elect chosen for him by His Father. Truly, Jesus chooses: Jesus alone judges.

Rev 20:13 And the sea gave up the dead which were in it; and death and hell delivered up the dead which were in them: and they were judged every man according to their works.

And also:

Mat 16:27 For the Son of man shall come in the glory of his Father with his angels; and then he shall reward every man according to his works.

If there is simply a strict separation between agonising physical torment and paradise, the near-innocent are damned most unjustly beyond recompense for their sins. The following verses support this simple drawing of a line in the sand. Yet these are no different on absolutes than as I had stated:

Joh 5:28 Marvel not at this: for the hour is coming, in the which all that are in the graves

shall hear his voice,
Joh 5:29 And shall come forth; they that have done good, unto the resurrection of life; and they that have done evil, unto the resurrection of damnation.

Yet Jesus has all judgement. If anyone yet states "burn in hell!" it only makes sense to state to them that Jesus has that judgement, not the Father who gave Him all judgement as in His Revelation. Counter them with the equivalent that they have taken the judgement from Christ and given it back to His Father again: simply ask them to concede on that point, and if they protest (or rather, cannot answer), state that they themselves don't have that judgement either, it is no surprise. (They have the wrong Father.)

I state, however, that every degree of judgement is within the vengeance of Christ: For He will not sit idle whilst those that shed the blood of saints are to otherwise go completely free.

Mat 10:28 And fear not them which kill the body, but are not able to kill the soul: but rather fear him which is able to destroy both soul and body in hell.

There is one thing that may be added: that I am as sure of this interpretation as I am of God from any book of the Bible – and if my testimony could ever be worthy to do so, I would stand on your part at His judgement if you would require rescuing from a burning lake of sulphur of God's making with added pitchforks wielded by horned beasties in red pyjamas. Not to stand in complaint but to simply point out the revelatory meaning of the text written by John – not taking its meaning misread by those who give the scripture their own personal interpretation out of their own deep seated fears.

1Jo 2:3 And hereby we do know that we know him, if we keep his commandments.
1Jo 2:4 He that saith, I know him, and keepeth not his commandments, is a liar, and the truth is not in him.
1Jo 2:5 But whoso keepeth his word, in him verily is the love of God perfected: hereby know we that we are in him.

Fear of God is indeed good, but when did fear start to go beyond that of being rejected by Him and excluded from His presence forever? (A second ejection from paradise, as if from Eden.) If you know God, you know that outcome is more painful to you right now than the imagined fear of fire and torture later. You should be able to discern for yourselves, surely?

The statement of Christ is of note;

Mar 9:40 For he that is not against us is on our part.
Mar 9:41 For whosoever shall give you a cup of water to drink in my name, because ye belong to Christ, verily I say unto you, he shall not lose his reward.
Mar 9:42 And whosoever shall offend one of *these* little ones that believe in me, it is better for him that a millstone were hanged about his neck, and he were cast into the sea.
Mar 9:43 And if thy hand offend thee, cut it off: it is better for thee to enter into life maimed, than having two hands to go into hell, into the fire that never shall be quenched:
Mar 9:44 Where their worm dieth not, and the fire is not quenched.
Mar 9:45 And if thy foot offend thee, cut it off: it is better for thee to enter halt into life, than having two feet to be cast into hell, into the fire that never shall be quenched:

Mar 9:46 Where their worm dieth not, and the fire is not quenched.
Mar 9:47 And if thine eye offend thee, pluck it out: it is better for thee to enter into the kingdom of God with one eye, than having two eyes to be cast into hell fire:
Mar 9:48 Where their worm dieth not, and the fire is not quenched.
Mar 9:49 For every one shall be salted with fire, and every sacrifice shall be salted with salt.
Mar 9:50 Salt *is* good: but if the salt have lost his saltness, wherewith will ye season it? Have salt in yourselves, and have peace one with another.

Now to qualify, "hell," is the fifth dialectic requiring a five cycle grafted over the seven cycle by the false prophet. It will never attain rest in the seven cycle. If the three parts, the "hand", "foot" and "eye" would "offend" or contradict, they are to be cut off. That is, the right **hand** octal (the Sun) contains no such offence (only wormwood does, which is excluded having no common multiplication to the reference), neither does Christ circuit the churches by **walking** through that five cycle, it is "cut off" or "excluded". (There is no image of any subfield GF(4) in GF(16) that shares an orbit with any K4 subgroup of GF(8)+ in algebraic closure). Also, If the **discernment** of members fail (the "eye"), then as five foolish virgins (Laodiceans), such members are also "cut off". Then "hell fire" is only that which offends, and those members are spiritually dwelling there already. Christ's body (His bride) does not require such surgery – all of its members are most surely elect. (Those severed are those that would cut off or offend!)

Then "their worm" (A) is their corruption of religion that is never corrected: and the "fire not quenched" is the spirit (B) that will always require refreshing. I find then $A \rightarrow \neg B$ and $B \rightarrow \neg A$.

Simple acts of charity are never corrupt (Mat 9:41) just as with those minor fellowships or disciple-individuals unfairly rebuked, as left alone in their faith (once offended and then rejecting the gospel for the offence's sake) show a failure and corruption of the good news in a great many eyes (Mat 9:42). To be a good ambassador for the Lord, then, is more than to encourage charity and the turning of none away. God Himself it is discovered (and He alone) will do all the choosing. Jesus stated to His disciples that they (ultimately) only have the authority to reign over themselves (not to be confused with the correction of others who must then truly reign likewise): in severing themselves from their sins, God would refresh that part of their spiritual person (salting and preserving them) whole in election. Born again, it is not the literal fire of zeal that will heal, or that of good doctrine; but God as "good salt" only.

Christ pleaded three times with His disciples for them not to condemn each other as "damned" to be refused by God or even to what is a "classical" or fiery hell. (For they as the saved are truly brethren whether they are recognised as such or not. They are to have "salt in themselves" and be at peace.) They are all members of Christ's body of believers and as such all are indispensable to the whole. That is to say, those that are genuinely saved. (Jesus states that the dividing line of truthful belief is absolute, and it is easy to discern its lack from the "offensive fruit produced".)

The last of the repeated three statements (as Mar 9:48) has the word "fire" in the previous verse qualified with the word "Gehenna", which is only a clear reference to the rubbish tip crawling with worms and of burning heaps outside the old city of Jerusalem. Yet it is, as it appears, Jesus' last word on the entire sum of the matter. (Why? He answers this afterwards!) His disciples are not even to cast each other out (as not in the least sense).

Jesus' statement on salt shows that every believer is to be salted with that very same fire of the Holy Spirit (not yet given them until Pentecost, they operated on zeal for the most part beforehand), and that they quite separately also serve with knowledge of such an "unfortunate end" which, for lack of a more succinct way to put it, would "light a fire under their butts" to ministry and action. Those stated to be in need of sacrifices "salted with salt" have continual grace. That is, all believers are recipients of the continual intercession that is made for the believer by Christ. The believer's repentance as an offering to God (offered up by Christ to His Father as worthy of election) – show them as not just the tertiary recipients of the fruitful sown word, but also having the direct fellowship of Christ rather than merely receiving the "fire" salted to them by the ministry of His disciples. "Sacrifices" offered, then, indicate those that are so salted with Christ – rather than by fear. Then all believers are brethren in Christ, and should not be cut off. Yet if the believers have no such salt of Christ, and have lost their way, what will the disciples salt them with? Not threat of hellfire surely, but Christ once more!

If there is no imperative to repent then Jesus' ministry (as good salt) loses some of its force, the knowledge that there actually are no flames awaiting the sinner in hell is secondary to the requirement for fellowship with God in Christ. (Only Christ – as fire – can make salt "salty" again, not the threat of Hell.) That said, fire as of Christ (in the Revelation) is "action of Spirit" or "mapping to unity in Christ" and not "turning back from Christ" which is "hell" or the lake of fire. This "fire" in Mark's account is totally reconcilable with Christ's Revelation. (The disciples had admonished another for exorcism by the name of Christ – one that would not go along with the disciples' demands for them to stop, they were "turning him back from Christ", as if to the lake of fire.) The passage in Mark 9 is, quite literally, of Jesus confronting His disciples with their own contradictions.

The disciples, by affirming in themselves all required authority in Christ able to judge another's salvation, had assumed (as a matter of course) that they were ever the most correct, but only Christ chooses and, ultimately, only Christ will judge. The truth is that only Christ is correct, because He had corrected them as His disciples; so, where then is their authority to correct another one that Christ need not correct? (That one was doing a good work in His name, after all, there was no mention of them or of the one claiming to be perfect without Christ.)

Returning to the matter of Christ as a preservative of all good salt, it is sorry that Christ in the Church is become taught with doctrine that hell-fire is now become that salt, or that it is adequate for the role of such a preservative: they clearly in Christ's own words have lost their savour.

To have that "saltness" in one's own self is to be made perfect under grace as with every other believer (even if they are not recognised as such). Arguing over those members of the body of Christ as to whether they should be cut off to hell is pointless (there is no reason to expect literal hell-fire here); put simply, the body of Christ does not contain those that require cutting off. The election of grace is down to the choice of God and there are no "grey areas" in a strictly saved/not saved paradigm.

So, may I assert there is no literal hell-fire awaiting the sinner either?

That may seem too bold a statement and beyond the scripture, yet even after John had finally reached the conclusion of his word pictures with the lake of fire and descent of the bride, Jesus ends His own Revelation with promises that those without repentance are simply left to themselves, no matter how wretched a sinner they are.

Rev 22:11 He that is unjust, let him be unjust still: and he which is filthy, let him be filthy still: and he that is righteous, let him be righteous still: and he that is holy, let him be holy still.

And the following I cannot deny: that the punishment of the most sinful is to share the fate of those that are merely "unfortunate". This still does not condemn a sinner to torments but exposes Jesus to show near universal mercy. (The Father delights in His Son's exercise of virtue – to excel beyond His words; those that are not His people may expect every mercy as well as true justice.)

Rev 22:14 Blessed *are* they that do his commandments, that they may have right to the tree of life, and may enter in through the gates into the city.
Rev 22:15 For without *are* dogs, and sorcerers, and whoremongers, and murderers, and idolaters, and whosoever loveth and maketh a lie.

Satan, though, is another matter altogether and those that worship Him and the image system in the lake of fire have that religion with Satan imprisoned in it and he is tormented to answer their worship. Religion will become damn hard work.

So, there you have it! As far as may be reasonably deduced no-one need ever fear flames and torment in Hell and as much as can be said, this all given in God's own words! What remains of Hell is "Gehenna", and for a modern play on words it is simply to be "refused".

I hope it is apparent that one may truly take a stance as a strict fundamentalist as well as rigidly adhere to the scripture as much as it can be (i.e. completely), and then merely and freely deduce that hell of fire and torment awaits no-one. I also freely deduce that the second coming arrives swiftly, with no burning mountains falling out of the sky etc., and none to expect; the end is heralded by no seven year warning period and that "Great Tribulation" is merely more (and equal) selective election in grace under a broader paradigm.

There is no reason why the last day need not arrive swiftly; after all, there is no sense of a thousand year millennial trial period for the sealed (saved) in Christ other than the testing and predestination of God's patience (and mapped back in action for the believer's salvation into this day today by grace).

There is only the ministry of the two witnesses remaining in the middle of this "tribulation" and scripture itself does a good enough job without them! (As to the "angel of the church" at least.) Plausibly only the right hand and left hand of God need to be considered whilst they respectively lead and convict (in the case of each individual believer's own "Great Tribulation" as to whether they leave the Church through the open door to Christ) by Spirit (as left) as the lesser (the right) are exeunt before them, and they are truly to play that role in the only context, and they are already witnessing now. The middle ground in the sea of octals is not a closed cage for the elect. The Church is instead found that cage (cf. Rev 22:16) with the mark.

The last day could theoretically arrive swiftly with no warning; the only sign is to watch for the loosening of restraint upon the Church and exit (into Christ as the "ark") when called or convicted to. (Any popular "mark" sends the fearful to the Church, its image is truly a cage.)

Mat 24:37 But as the days of Noe *were,* so shall also the coming of the Son of man be.
Mat 24:38 For as in the days that were before the flood they were eating and drinking, marrying and giving in marriage, until the day that Noe entered into the ark,
Mat 24:39 And knew not until the flood came, and took them all away; so shall also the coming of the Son of man be.

So, I may deduce quite correctly that I should not be so lukewarm, as well as truly able to counter any threats of hell-fire by those that have not done their own reading, and all that in the book that poses the greatest difficulty to an apologist as well! The claim that scripture "is nonsense" or "makes no sense" is in itself nonsense and makes no sense in the respect that God chooses His own language, not those that would scoff at all things God.

When you read the New Testament, keep an eye open for passages showing virtue – passages such as Christ weeping for the resurrection of Lazarus is such an example. Grace itself as virtue is understandable enough: charity may be a term equal to the positive exercise of God's liberty by His own measure of what virtue to show. (He is not invoked as mantra to justify the words and works of any of us; He is no slave to the failings of men.)

Rom 9:15 For he saith to Moses, I will have mercy on whom I will have mercy, and I will have compassion on whom I will have compassion.

Now, only one thing remains. Charity, grace, virtue one and the same in Jesus Christ. They are not separated. (To be separate to the Church is to be able to reach Christ, despite the world.)

If you have any reason still to expect fire and torment in a "hell" due to any popular "mark" or because you will not commit to Christ in good time, I hope that you would accept some simple arguments that although sophist (as Hell is not a "place") you may find answers from:

- If good works were required for you to get into paradise, be aware that if that were true, everyone would have to do unending work there when they arrived. (Paradise is become no paradise.)

- You have every chance of salvation. If that grace never takes hold this day and today, you're not doing it right.

- If all sinners were truly to deserve hellfire and torments from God, this universe would be exactly that already. (All have sinned and hell is become as no more hell.) Indeed, it is become "Hell" without the fire and torments.

- Likewise, the fact that this universe is filled with evil but was made for the multiplication of life witnesses that "hell" (as bad as can be) when created is survivable.

- It can't get any worse than this. God sends tribulation so that His salvation holds firm – the gates of Hell will not prevail against it; that is, salvation is that same exit to paradise, so the open door to paradise is out of the gates of hell and then to leave this corrupted creation for the incorruptible. Not to accomplish that is then only to remain.

- No one can ever work their way out of hell; evil superposes on every good act: without faith, it is impossible to please God.

- The judgement of Christ on the last day with the resurrection of the just and unjust is final. (Certainly.)

- They still make cold beer in hell; I hope they continue to. Yet who (on seeing no good reason to), would give you their own without a fiduciary stimulant?

As to now remaining somewhat lukewarm, if you consider that a total lack of ministry is a ministry indeed you are not even the least in the kingdom of heaven. The lack of an imperative to repent is clearly not equal to "the good news", for the converse that it is "bad news" (to those that reject it) is then not true as gospel either. Unless there truly is a forceful reason why God desires all to be saved, the gospel will fail to retain Christ's encouragement and all seven seals would be immediately opened. That reason as best He knows (in both respects) would be because this world gives no reply to what is on offer for free. As if saying, "Mat 11:17 And saying, We have piped unto you, and ye have not danced; we have mourned unto you, and ye have not lamented.".

A lack of interest in the gospel is not that which would conclude this system of things; whilst the world itself may be completely disinterested, that is all the more reason for God to extend His free offer of grace. Instead the imperative to repent must be totally and globally refused (and this is done alone by holding all in equality to the MYSTERY BABYLON template the world over), in full knowledge of the mystery of God whilst all are under no deception whatsoever, which is one thing I for one will never do.

Once the least has finished his circuit there is that small chance of "Great Tribulation" to silence all those that will not so refuse God's grace. Yet this is to be globally made sure by applying the template, and then doing that in the only setting where it matters – in the Church.

The real reason to repent is one of God's own. He would choose all of us, and if all refuse Him that would certainly conclude everything. Those terms will not stop the least, nor many others from repenting and even God has daily left Himself room to repent as the end becomes apparent. How many last days will there be before God is satisfied He has the least and the rest (logically speaking) are neither found necessarily in the fold or stuck already in stalemate? That could be true instead if the last were so central, yet I deduce not: as it is written (in a great amount of the book), the least ensures none are deceived and all may repent. (The last to leave the Church may not be the very last saved outside of such institutions of religion – God will not hurt "any green thing" that is likewise sealed.)

Jer 50:45 Therefore hear ye the counsel of the LORD, that he hath taken against Babylon; and his purposes, that he hath purposed against the land of the Chaldeans: Surely the least of the flock shall draw them out: surely he shall make *their* habitation desolate with them.

A logical conviction is instant of necessity but as God already repents of destruction every day (whilst the angel states there should be "time no longer"), God is far stronger and will continue to be. Logic can wait.

Keyword Index

Study Index

Bibliography

1: Sobel, J. (1987) 'Gödel's Ontological Proof'[*] in J. J Thompson (ed.), *On Being and Saying: Essays for Richard Cartwright*. Cambridge, MA: MIT Press, 1987.

2: Benzmüller, Christoph, and Bruno Woltzenlogel Paleo, *An Object-Logic Explanation for the Inconsistency in Gödel's Ontological Theory*, 2016.

3: Fraleigh, John B., *A First Course In Abstract Algebra 6th Ed*. Addison Wesley Longman, 2000

4: Descartes, René, *Meditations and Other Metaphysical Writings*. Penguin Classics, 1998.

5: Small, Christopher G., *Reflections on Gödel's Ontological Argument*, 2009.

6: Hofstadter, Douglas, *Gödel, Escher, Bach: An Eternal Golden Braid*. Vintage Books, 1979.

7: Lloyd, James, *The 6, 7, 8 Cycle*. Christian Media, 2010.

8: Lloyd, James, *The Sand and The Sea*. Christian Media, 2006.

9: Kant, Immanuel, *Critique of Pure Reason*. Penguin Classics, 2007.

10: Gotcher, Dean, *The Diabolical System of Diaprax* (Audio). The Apocalypse Chronicles, February 2002.

11: Lloyd, James, *Dragonspeak, The Language of Lucifer* (Audio). The Apocalypse Chronicles, July 2002.

12: Warren, Rick, *The Purpose Driven Church*. Zondervan, 1996.

13: Adams, Douglas, *The Ultimate Hitchhiker's Guide to the Galaxy - The Complete Trilogy in Five Parts*. Pan, 2017.

14: Scott, Dana, (1972) *Appendix B: 'Notes in Dana Scott's Hand'* In J.H. Sobel (ed.), *Logic and Theism: Arguments for and Against Beliefs in God*. Cambridge: Cambridge University Press, 2004.

[*] 'Gödel's Ontological Proof' is reproduced in Sobel (1987) (*Appendix 2: 'Notes in Kurt Gödel's Hand'*, p256-257 and *Appendix 3: 'Notes in Dana Scott's Hand'*, p257-258) from notes transcribed by J. H. Sobel with permission of Dana Scott, and John Milnor on behalf of the custodians of Kurt Gödel's Nachlass, held by the Institute for Advanced Study, Princeton, New Jersey.

Appendix: Gödel's Ontological Argument

Gödel's argument[1] is as follows:

Ax1) $(P(\varphi) \wedge \Box\forall x:(\varphi(x) \Rightarrow \psi(x))) \Rightarrow P(\psi)$. *(I.e $P(x)$: x is a positive predicate, with $\Box$ the modal necessitation operator.)*
Ax2) $P(\neg\varphi) \Leftrightarrow \neg P(\varphi)$

Th1) $P(\varphi) \Rightarrow \Diamond\exists x:\varphi(x)$. *(I.e. with $\Diamond$ the modal possibility operator.)*
Let $P(\varphi)$ hold with $\neg\Diamond\exists x:\varphi(x)$. Then $\Box\forall x:\neg\varphi(x)$ holds implying that $x \neq x$ is a predicate $\lambda(x)$ on x if $\Box\neg(\varphi(x) \wedge \neg\lambda(x))$. From Ax1 $P(\lambda)$ holds, yet every predicate on x entails that x is self-identical so, again, from Ax1 $P(\neg\lambda)$ holds, a contradiction by Ax2. Thus $\Diamond\exists x:\varphi(x)$.

Def1) $G(x) \Leftrightarrow \forall\varphi:(P(\varphi) \Rightarrow \varphi(x))$. *(I.e. $G(x)$: x is a "god-like" individual.)*
Ax3) $P(G)$

Th2) $\Diamond\exists x:G(x)$
Follows from Th1 combined with Ax3.

Def2) $\varphi \, ess \, x \Leftrightarrow \varphi(x) \wedge \forall\psi:(\psi(x) \Rightarrow \Box\forall y:(\varphi(y) \Rightarrow \psi(y)))$ [1,14]. *(I.e. $\varphi \, ess \, x$: φ an 'essence' of x.)*
Ax4) $P(\varphi) \Rightarrow \Box P(\varphi)$

Th3) $G(x) \Rightarrow G \, ess \, x$
Suppose $G(x)$ is true and x necessarily has property H. I.e. $\Box H(x)$ is true. Then by Def1 and Ax2 $P(H)$ holds (H must be positive), but $P(H) \Rightarrow \exists x:(G(x) \Rightarrow H(x))$ as from Def1 (recall that $G(x)$ is true) and if something necessarily has a property H then it has the property. Now, $\Box P(H)$ is true (Ax4), so by modal modus ponens, $\Box(\exists x:(G(x) \Rightarrow H(x)))$. So, if x has any property H necessarily, then that H is entailed of the property G, so $G \Rightarrow H$. Conversely, suppose $G(x)$ is true and $G \Rightarrow H$, then by Ax1 and Ax3, I have $P(H)$. It follows that a 'god-like' individual x has property H necessarily, by Def1. That is, $\Box H(x)$ is true.

Def3) $E(x) \Leftrightarrow \forall\varphi:(\varphi \, ess \, x \Rightarrow \Box\exists y:\varphi(y))$. *(I.e. $E(x)$: x is 'necessarily existent'.)*
Ax5) $P(E)$

Th4) $\Box\exists x:G(x)$
If $G(x)$ were true, then by Def1 x has every positive property necessarily. Ax5 states necessary existence is positive, so $\Box E(x)$ is true. But by Th3 if $G(x)$ were true, $G \, ess \, x$ is true. Now, as by Def3, if any x is god-like then 'god-like' as a property is necessarily exemplified: I.e. $\exists x:G(x) \Rightarrow \Box(\exists x:G(x))$. This follows from Def1, Ax5 and Th3. (This is essentially Anselm's principle.) Combining this with Th2 as if by substitution, $\Diamond(\exists x:G(x)) \Rightarrow \Diamond(\Box(\exists x:G(x)))$. I.e. if $\Diamond(\exists x:G(x))$ is true, it follows that $\Diamond(\Box(\exists x:G(x)))$ is also true. But (as the modus tollens shows to be correct), $\Diamond(\Box\psi) \Rightarrow \Box\psi$. Thus $\Box(\exists x:G(x))$.

[1] J. Sobel (1987), *Gödel's Ontological Proof* in J. J Thompson (ed.) *On Being and Saying: Essays for Richard Cartwright*. (Cambridge, MA: MIT Press, 1987), p241-61.

[14] D. Scott (1972), *Appendix B: Notes in Dana Scott's Hand*, In J.H. Sobel (ed.). *Logic and Theism: Arguments for and Against Beliefs in God*. (Cambridge: Cambridge University Press, 2004), p145–146